T0306122

Twort's Water Supply

Twort's Water Supply

Sixth Edition

Don D. Ratnayaka
BSc, DIC, MSc, FIChemE, FCIWEM

Malcolm J. Brandt
BSc, FICE, FCIWEM

K. Michael Johnson
BSc, FICE, FCIWEM

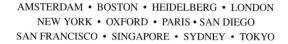
AMSTERDAM • BOSTON • HEIDELBERG • LONDON
NEW YORK • OXFORD • PARIS • SAN DIEGO
SAN FRANCISCO • SINGAPORE • SYDNEY • TOKYO

ELSEVIER

Butterworth-Heinemann is an imprint of Elsevier

Butterworth-Heinemann is an imprint of Elsevier
30 Corporate Drive, Suite 400, Burlington, MA 01803, USA
Linacre House, Jordan Hill, Oxford OX2 8DP, UK

Notice
No responsibility is assumed by the publisher or the authors for any injury and/or damage to persons or
property as a matter of products liability, negligence or otherwise, or from any use or operation of any
information, opinions, methods or ideas contained in the material herein. The publishing of this work should
not be interpreted as an attempt to render engineering or other professional services. If such services are
needed the assistance of an appropriate professional should be obtained.

Library of Congress Cataloging-in-Publication Data
Application submitted

British Library Cataloguing-in-Publication Data
A catalogue record for this book is available from the British Library.

ISBN: 978-0-7506-6843-9

For information on all Butterworth–Heinemann publications
visit our Web site at www.elsevierdirect.com

Printed and bound in Great Britain
09 10 11 12 13 10 9 8 7 6 5 4 3 2 1

Typeset by: diacriTech, Chennai, India

Working together to grow
libraries in developing countries

www.elsevier.com | www.bookaid.org | www.sabre.org

ELSEVIER   BOOK AID International   Sabre Foundation

To the Memory of
Alan Charles Twort

Author of the First Five Editions of This Work

Who Left the World a Better Place

Contents

Foreword

Since the first edition of this book was published in 1963 the procurement and treatment of water for public consumption has become increasingly difficult. There are real worries now that freshwater supplies in many parts of the world are not sufficient to meet the demands of growing populations and the need for increased food production. Water resources in some countries may even be entering a period of decline due to global warming.

In the same period, the production of potable water from raw water supplies has become more complicated with increasing awareness of the pollutants being discharged into the hydrological environment and the harm they may be causing.

These adverse changes pose difficult problems for all water engineers and scientists. Means have to be found of reducing wastage of water by consumers and losses from pipe distribution systems. The former involves encouraging consumers to use only the water they need; the latter involves the costly and difficult task of repairing or renewing many old water mains in densely populated urban areas.

As a consequence of the need to cope with these diverse challenges the 6th Edition of this book conveys the advice and experience of three authors plus thirty contributing specialists and advisers, several of whom have high global standing in their specialisation. The authors and the majority of specialists have experience of working for the consulting engineering firm of Binnie & Partners—which practiced for over 100 years, for Paterson Candy Limited—which designed and built water treatment plant worldwide and for Black & Veatch Corporation of Kansas City—who has engineered water treatment and supply facilities across the USA since the early 20th century.

The importance of using the best specialist knowledge has to be emphasised because of the ever increasing technical and social challenges as well as the relentless pressure to find the most cost-effective and environmentally sustainable solutions. The in-depth knowledge of the expert, when given time to study the task, can help meet these challenges and achieve the results needed in tomorrow's world.

Alan C. Twort
April 2008

Preface

The first edition of *Water Supply* provided the practicing water engineer with a practical treatise on all aspects of the water supply in one book. Since then, the book's readership has broadened to include students of engineering and it is now used as a standard text in many universities. Recognising the contribution of Alan Twort, the lead author and editor of the first five editions, this edition is titled *Twort's Water Supply*. It benefits from the extensive experience of UK and US specialists from across the water industry. The result is an extended and rebalanced coverage, with many new references and illustrations, which brings the work up-to-date. Practice in the US, UK, Europe and worldwide is included.

Chapters are arranged in order of: demand, statutory, financing and economic aspects of supply, water sources and quality, treatment processes, hydraulics and system design, distribution, pipelines, valves, pumps and other MEICA plant and treated water storage.

The institutional aspects covered now include international water rights, the EU Water Framework Directive and experience with private sector participation. Text on hydrological yield has been rewritten to reflect present methods and flood coverage is revised to reflect issues that have arisen with the *Flood Estimation Handbook*.

The text gives up-dated standards for drinking water and describes in detail the significance of the large number of chemicals and organisms, which now present, in raw waters, a potential hazard to human health. Conventional and specialised treatment processes, waste treatment and disinfection are covered in five separate chapters. Treatment chemicals are described in detail. Developing technologies such as membrane filtration and advanced treatment methods for micropollutants are also dealt with. Desalination by reverse osmosis and thermal processes is covered.

Chapters on system hydraulic design and practice are rearranged to make access easier with coverage of new methods. Pipe structural design and valves and meters are presented with new material in separate chapters. The chapter on pumps is broadened to cover other electrical, control and instrumentation aspects of water works and the chapter on service reservoirs is extended to embrace treated water storage of most types likely to be encountered.

The authors are grateful to the many contributors and reviewers who have aided the production of this sixth edition and to Black & Veatch for permitting use of their work.

Don D. Ratnayaka
Malcolm J. Brandt
K. Michael Johnson

Abbreviations for Organisations

ADB	Asian Development Bank, Manila, Philippines
ASCE	American Society of Civil Engineers, Reston, VA, USA
ASTM	American Society for Testing Materials (now ASTM International), PA, USA
AWWA	American Waterworks Association, Denver, USA
AwwaRF	American Waterworks Association Research Foundation, Denver
BGS	British Geological Survey, Nottingham, UK
BHRA	British Hydromechanics Group, Cranfield, UK
BSI	British Standards Institution, London
BW	British Water, London
CIRIA	Construction Industry Research & Information Association, London
CIWEM	Chartered Institution of Water & Environmental Management, London
CRC	Chemical Rubber Company, FL, USA
Defra	Department for Environment, Food and Rural Affairs, London
DETR	Department of Environment, Transport and Regions, London
DFID	Department for International Development, London
DHSS	Department of Health and Social Services, London
DoE	Department of the Environment, London
DWI	Drinking Water Inspectorate, London
EA	The Environment Agency, Bristol, UK
EEA	European Environment Agency, Copenhagen
FWR	Foundation for Water Research, Marlow, UK
HMSO (OPSI)	Her Majesty's Stationery Office, London (Office of Public Sector Information)
HSE	Health and Safety Executive, London
IAHS	International Association of Hydrological Sciences, Wallingford, UK
IChemE	Institution of Chemical Engineers, Rugby, UK
ICOLD	International Committee of Large Dams, Paris
IET	Institute of Engineering Technology, London
IMechE	Institute of Mechanical Engineers, London
IoH (CEH)	Institute of Hydrology, Wallingford, UK (now Centre for Ecology & Hydrology)
ISO	International Organisation for Standardisation, Geneva
IUVA	International Ultraviolet Association, Ontario
IWA/IWSA	International Water Association, The Hague
IWEM/IWES	Institution of Water Engineers (& Scientists), London
NERC	Natural Environmental Research Council, Swindon, UK
NEWA	New England Waterworks Association, Massachusetts, USA
NFPA	National Fire Protection Agency, Massachusetts, US
NFPA	National Fire Protection Association, Ascot, UK
Ofwat/WSRA	Water Services Regulation Authority, Birmingham, UK (Term Ofwat still used)
PWC	Portsmouth Water Company, Portsmouth, UK
STW	Severn Trent Water, Coventry, UK

SWTE	Society for Water Treatment and Examination, London
TWU	Thames Water Utilities, Reading, UK
UKWIR	UK Water Industry Research Ltd., London
UNESCO	UN Educational, Scientific and Cultural Organisation, Paris
UNICEF	UN Children's Fund, New York
USACE	US Army Corps of Engineers, Washington
US EPA	US Environmental Protection Agency, USA
USGS	United States Geological Survey, Washington, USA
WEDC	Water Education Development Centre, Loughborough, UK
WHO	World Health Organization, Geneva
WMO	World Meteorological Organization, Geneva
WRc	Water Research Centre, Medmenham, UK
WSA	Water Services Association, London
WSRT - Aqua	Water Supply: Research Technology – Aqua, IWA, The Hague

Contributing Authors, Reviewers and Advisors

CONTRIBUTING AUTHORS

Black & Veatch

Peter B. Clark MA, MSc, FICE	Technical Director (Hydraulics)
Anthony N. Coe MA, FIMechE, FCIWEM	Consultant Mechanical Engineer (Pumping Plant)
Scott Freeman BS, BA	Membrane Technology Leader (Membrane processes)
Keith R. Gardner BSc, MSc, FIStrutE, MICE	Consultant Structural Engineer (Service Reservoir Structures)
John Hall BSc, MCIWEM	Chief Hydrologist (Surface Supplies and Floods)
Robert Hulsey BS, MS	Director Water Technology (Ozone and UV)
Mike J. Little MA, MICE, MCIWEM	Technical Director (Pipelines)
John C. Maunder BSc, MIET	Consultant, Control & Instrumentation (Control and Automation Systems)
John L. Petrie BSc, MSc, FGS, CGeol.	Chief Hydrogeologist (Groundwater Supplies)
Alvin J. Smith BSc, MSc	Consultant Biologist (Water Biology and Storage)

Others

Peter Mason BSc, MPhil, PhD, FICE	Consultant (Dams)
Claire R. Stacey FRSC, MIWO	Formerly Deputy Chief Inspector, DWI (water Quality Issues and Standards)

Reviewers and Advisors

John Ackers BSc, FICE	Black & Veatch
Roger Brown BSc, MSc, DIC, FICE, FCIWEM	Consultant
Nicholas Burns BS, MS	Black & Veatch
Simon Cole BSc, Dphil, MCIWEM	Wessex Water
Neville A. Cowton BSc, MSc, MICE, MCIWEM	Black & Veatch

Sarah Dyer BA, MA	Black & Veatch
Andrew Elphinston BSc, MIChemE, MCIWEM	Black & Veatch
Terry Heard BSc, MRSC, MCIWEM	Black & Veatch
Brian Hoyle MRSC	Consultant
Heather Landis BS, MS	Black & Veatch
Frank Law BSc, MICE, FCIWEM	Institute of Hydrology
David E. MacDonald BSc, MSc, MCIWEM	Black & Veatch
James Ostrowski BSc, MIChemE, MCIWEM	Black & Veatch
Bill Querns BSc, FCIWEM, FCIM	Consultant
Trevor Peploe B.Tech, MRSC	Consultant
Bryan Townsend BS, MS	Black & Veatch
H. Jane Walbancke BSc, MSc, PhD, FGS, MICE	Black & Veatch
Paul A. Yiannouzis BSc, MIEE	Consultant

Attribution of Illustrations

The names of the organizations acknowledged in connection with illustrations are those current at the time the illustrations were produced. Where an organization has since changed its name or has been acquired by another, the new name is given below:

Sir Alexander Binnie, Son & Deacon	Black & Veatch
Binnie & Partners	Black & Veatch
Binnie Black & Veatch	Black & Veatch
Paterson Candy Ltd	Black & Veatch

The Demand for Public Water Supplies

1.1 CATEGORIES OF CONSUMPTION

The demand for public water is made up of authorised consumption by domestic and non-domestic consumers and water losses.

Domestic consumers use water within the household: for drinking, personal hygiene cooking and cleaning, and outside the dwelling: for cleaning patios, irrigating gardens, filling ponds and swimming pools and washing cars. Domestic consumers include households that are not connected to the distribution mains but rely on collecting their supply from standpipes and public taps located in the street.

In England and Wales in 2007 about 30% of domestic supplies were metered with individual meter penetration ranging between 8% and 66% for the 22 water companies (Ofwat, 2007a). In Scotland and Northern Ireland domestic consumers are not currently metered. Overseas metering domestic supplies is widespread, although not universal. A survey by the ADB in 1996 (ADB, 1997) showed that, of 27 Asian cities serving over 1 million people, only 15 were fully metered and six metered less than 7% of their connections (Calcutta 0%, Karachi 1%). UNESCO report that in 2000 nearly a third of all urban dwellers worldwide, more than 900 million people, lived in slums (UNESCO, 2006). Standpipes providing water to urban slums and rural communities are not metered and the supplies are usually given free.

Non-domestic consumption comprises industrial, commercial, institutional and agricultural demand legitimately drawn from the distribution mains. This category also includes legitimate public use for irrigating public parks and green areas, street cleaning, flushing water mains and sewers and for fire-fighting.

Commercial and industrial supplies are usually metered because they represent a major source of income to a water utility. In the UK small shops and offices occupied only in the daytime used not to be metered but now generally are, even though their consumption is small. In many countries large quantities of water are used for watering public parks and green areas and supplying government offices, military establishments and other institutional buildings. These supplies are often not metered nor paid for if the government (state or city) supplies the water.

Water losses comprise the leakage and wastage from the distribution network; these and other components of non-legitimate use are categorised as:

Apparent losses: source and supply meter errors, unauthorised or unrecorded consumption, and

Real losses: leakage from transmission and distribution mains and service pipes upstream of consumers' meters, from valves, hydrants and washouts and leakage and overflows from the water utility's storage facilities.

1.2 LEVELS OF TOTAL CONSUMPTION

The usual measure of total consumption is the amount supplied from sources per head of population. However, in many cases the population served is not known accurately. In large cities there may be thousands of commuters coming in daily from outside; in holiday areas the population may double for part of the year. Other factors having a major influence on consumption figures are:

- whether the available supplies and pressure are sufficient to meet the demand, 24-hour or intermittent;
- the population using standpipes;
- the extent to which waterborne sanitation is available;
- the utility's efficiency in metering and billing and in controlling leakage and wastage;
- how much of the supply goes to relatively few large industrial consumers;
- the climate.

Many cities, more typically in Asia and Africa, either do not have a 24 hour supply, or the supply pressure is so low that many consumers receive an intermittent supply. The ADB survey of 1996 (ADB, 1997) showed that 40% of 50 Asian cities surveyed did not have a 24-hour supply and that about two-thirds had street standpipe supplies. Hence comparing average total consumption between utilities is not informative. High consumption can be caused by large industrial demand and low consumption by a shortage of resources. However, the general range of total supplies per capita is:

- from 500 to 800 lcd (litres per capita per day) in the big industrial cities of USA;
- from 200 to 500 lcd for many major cities and urban areas throughout the world;
- from 90 to 150 lcd in areas where supplies are restricted, where there are many street stand-pipes or where much of the population has private wells.

In 2006/07 in England and Wales the average total supply was 278 lcd (Ofwat, 2007a). In Scotland it was about 494 lcd in 2003/04 (Scottish Water, 2005) and in Northern Ireland about 440 lcd.

1.3 DOMESTIC DEMAND

Domestic consumption reported by various countries is not necessarily comparable mainly because there is no assurance that the figures quoted are produced on the same basis. In the USA the typical in-house consumption (excluding cooling) is 180–230 lcd but this can be expected to reduce gradually with increased installation of low flush toilets and reduced rates of consumption by other fittings. Average domestic plus small trade consumption reported by European countries for 2002 (IWA, 2004) centred about 150–160 lcd (range 100–330 lcd); France, Italy, Norway and Switzerland were the only ones above 200 lcd. By comparison the equivalent consumption for 21 countries in Asia was about 275 lcd (range 165–465 lcd), the highest consumption being in Australia and South Korea, but it is suspected that some of these figures include considerable non-domestic consumption.

In-house domestic consumption is influenced by many factors including, the class of dwelling, number of people in the household, changes in household income, ablution habits, culture, religion, differences in climate including seasonal variations and the number and capacity of water fittings installed. The influence of dwelling class can be seen in the figures of average domestic consumption in England and Wales given in Table 1.1. The higher domestic consumption reported by the water-only companies reflects the fact that 74% of the population they serve resides in the more affluent southern parts of England where housing standards are generally higher than in the north. The lower figures for metered consumption are not representative because metering is optional for most householders in the UK and householders choosing to have a metered supply tend to be those expecting to pay less for their supply because of their low consumption, for example single occupants or elderly retired people. This is reflected in the household occupancy figures, the 'occupancy ratio'.

When estimating consumption for a whole area of a distribution system, it is the average occupancy which is important. In England and Wales average occupancy according to census figures (Census, 2005) showed a decline from 2.70 in 1981 to 2.36 in 2001 and is expected to continue to fall slowly. The 2001 figures varied from 2.10–2.20 in retirement areas to 2.30–2.60 in the more

Table 1.1 Domestic consumption in England & Wales, 2007/08		
	10 regional water and sewerage companies	**12 water-only companies**
Population in households	44.01 million	10.01 million
Reported average consumption excluding supply pipe leakage (lcd) and range		
Un-metered households	151 (141–163)	165 (128–176)
Metered households	131 (115–142)	141 (111–153)
(Percent households metered)	(32.7%)	(32.7%)
(Percent population metered)	(30.5%)	(30.1%)
(Occupancy ratio un-metered households)	(2.5)	(2.6)
(Occupancy ratio metered households)	(2.1)	(2.1)
Weighted mean per capita consumption	145.2	157.7
All companies mean per capita consumption	147.6	
All companies average occupancy ratio	2.4	

Source of information: Ofwat, 2008

Table 1.2 Domestic consumption per capita (pcc) by household size (lcd)

Number in household		1	2	3	4	5	6
Portsmouth Water Company[1]	lcd	252	183	140	130	108	101
% of company average pcc	%	156	113	87	80	67	63
Thames Water Utilities[2]	lcd	215	184	159	141	135	123
% of company average pcc	%	131	112	97	86	82	75
Severn Trent Water[3]	lcd	189	153	122	117	109 (5+)	
% of company average pcc	%	139	113	90	86	80	

Sources: [1]PWC, 2005; [2]TWU, 2007; [3]STW, 2007

dense populated urban areas. In the USA mean household occupancy reported by the US Census Bureau (US Census, 2005) declined form 2.63 people in 1990 to 2.57 in 2004. Individual US water suppliers report average occupancies varying from 2.4 in multi-family complexes to 3.1 in single family residences, the usual range being 2.5–2.8 people per household. In many countries in Africa, the Indian sub-continent and in South East Asia, etc. the average occupancy is five or six people per household. Table 1.2 illustrates the influence of the number of people in a household on household demand for three UK water companies and shows that smaller households have proportionately larger per capita usage.

Water usage varies both in quantity and timing during the week, the pattern for working days being relatively consistent with morning and evening peaks coinciding with leaving for and returning from work and school. However, at weekends demand tends to increase and diurnal peaks are often later than during the week and are of greater magnitude. Cultural and religious characteristics of a supply area and religious days and public holidays can also influence weekly and seasonal demand patterns. For example during Ramadan, the Muslim holy month of fasting, domestic per capita demand increases and the diurnal pattern switches from day to night with peaks coinciding with sunrise and sunset.

Components of Domestic In-House Consumption

In-house water usage is for drinking, personal hygiene, WC flushing, showers, baths and hand basins, cooking, cleaning and laundry, including washing machines and dishwashers. Water consumption in developed countries has increased as a consequence of trends to install dish washers and modernise bathrooms including fitting high capacity power showers. Conversely during the last decade there has been increased awareness in some sectors of society for the need to conserve water, resulting in a worldwide trend for water utilities and environmental agencies to promote water conservation and encourage the householder to install lower flow, smaller capacity and more efficient fittings.

Table 1.3(A) presents reported domestic consumption in litres per person per day for a selection of countries. The figures are not strictly comparable because they are compiled using different methodologies and may include small business usage and some components of leakage. However, they are indicative of regional, social and economic factors.

Table 1.3(B) includes in-house domestic consumption in litres per person per day from detailed studies. Again the figures are not strictly comparable because they are obtained by different monitoring methods and small sample sets. In the 'diary' method for monitoring household consumption, residents record each day the number of times each fitting is used over a period. The frequencies of use for each type of fitting are multiplied by the average consumption, determined from separate measurements, for the fitting to derive the household consumption. In the 'data logging' method, the flow records from pulsed output consumer meters recording at frequent time intervals are analysed by computer; the different uses being identified by the flow pattern. In a few instances, utilities have installed meters at every point of use in a property. The table shows that, with the exception of a few anomalous total in-house figures, per capita consumption is relatively consistent across the reported figures, the higher figures tending to represent older housing stock before conservation measures had been implemented.

Table 1.3 also shows that toilet flushing represents the single largest use of water. In the UK, water bylaws introduced in 1994 resulted in cistern capacity being reduced from 9 litres (13 litres in Scotland) to 7.5 litres. The capacity was further reduced for new installations in 2001 to 6 litres. More recently dual flushing toilets with 6/3 litre capacities have become available. The water utilities are also promoting displacement devices, for example the 'Hippo the Water Saver', a stiff plastic bag immersed in the cistern that further reduces the flush capacity. The full impact of introducing reduced capacity cisterns on flushing demands will not be immediate but will only be realised over about 20 years, the average life of a cistern, as older ones are replaced with the newer specification tanks. There have been similar trends in reducing flow rates and capacities of other in-house appliances including flow restrictors, efficient spray devices on taps and showers and smaller capacity washing machines. Reduced capacity fittings are also being installed in systems and locations where water resources are limited, including Australia, Singapore and in many cities in North America under the US Energy Policy Act 1992. Table 1.4 compares the capacity of fittings in different countries. However, water efficient fittings are only effective where they are maintained. If an appliance malfunctions or a dripping or leaking fitting is not maintained, as can be common with flap valve cisterns, the water losses from the defective fittings can represent a significant proportion of the household demand.

Water used in air conditioners and humidifiers is not included in the US figures in Table 1.3. In some parts of western USA where the climate is hot and arid, evaporative or 'desert' coolers are used; a fan draws air through a vertical porous pad of cellulose fibre, down which water is trickled. Recirculating units use about 12–15 l/h. Coolers that bleed off part of the surplus water to reduce deposits on the porous pad can use up to 40 l/h. The impact of cooling units on domestic consumption depends on the percentage of dwellings equipped with them and the duration of the hot season. Estimates of the additional quantity above the annual average daily domestic consumption attributable to cooling units vary between about 90 lcd, in areas where the summer climate is exceptionally hot and dry, for example Arizona where the average temperature in July is 40 °C with 29% daytime humidity, and about 5 lcd for coastal areas. In hot and wet climates, as in the tropics with a high humidity, electrical air conditioners are used which do not consume water.

Outdoor Domestic Use for Garden Irrigation and Bathing Pools

Outdoor water uses can represent a significant proportion of domestic consumption. In hotter climates in the US up to 60% of the average annual day demand can be used for irrigation, the demand being sustained throughout most of the year. In Australia 35% of the average annual day demand is

Table 1.3 Domestic consumption per capita (lcd)

(A) Per capita consumption (lcd) (countries and cities with reference year)

Argentina, Buenos Aires, 2003[1]	261	Morocco, Rabat 2003[1]	100
Australia, Perth 2005[2]	294	Nepal, Kathmandu 2001[3]	68[a]
Australia, Sydney 2005[2]	214	Netherlands, Amsterdam 2003[1]	158
Australia, Yarra Valley 2005[2]	202	Netherlands, The Hague 2003[1]	102
Bangladesh, Dhaka, 2001[3]	115[a]	Norway, Oslo, 2003[1]	200
Cambodia, Phnom Penh 2001[3]	104	Pakistan, Karachi, 2001[3]	197[a]
China, Hong Kong, 2004[7]	203	Philippines, Manila 2001[3]	127[b]
China, Chengdu, 2001[3]	138	Poland, Plock 2003[1]	97
China, Shanghai, 2001[3]	251	Poland, Sopot 2003[1]	144
Denmark, Copenhagen, 2003[3]	126	Singapore 2007[4]	158
Finland, Helsinki, 2003[1]	167	South Korea, Seoul, 2001[3]	205
France, Lille, 2003[1]	127	Spain, Bilbao 2003[1]	101
France, Paris, 2003[1]	276	Spain, Madrid 2003[1]	159
Germany, Munich, 2004[7]	130	Sri Lanka, Colombo, 2001[3]	119[b]
India, Delhi, 2001[3]	110[a]	Switzerland, Geneva 2003[1]	228
Japan, Nagoya 2003[1]	223	UK, England & Wales 2005[2]	151
Japan, Sapporo 2003[1]	202	UK, Scotland, 2005[2]	177
Japan, Tokyo 2004[7]	268	USA, South California 2004[2]	550
Laos, Vientiane, 2001[3]	110[b]	USA, Elizabethtown 2004[2]	441
Lithuania, Siauliai 2003[1]	56	USA, New Jersey WC 2004[2]	284
Lithuania, Vilnius 2003[1]	89	USA, Illinois 2004[2]	188
Mongolia, Ulaanbaatar 2001[3]	278[b]	Uzbekistan, Tashkent 2001[3]	328
Morocco, Casablanca 2003[1]	72	Vietnam, Ho Chi Min, 2001[3]	167[b]

(*continued*)

Table 1.3 (*continued*)

(B) In-house household consumption by usage excluding supply pipe leakage (lcd)

Location	Internal total[c]	WC flush	Personal hygiene	Kitchen uses[d]	Laundry[e]
Australia, Perth—single residence, 2003[5]	155	33	51	29	42
Australia, Perth—multiple residence, 2003[5]	166	28	55	40	43
Australia, Sydney[6]	184	45	66	24	49
Singapore, 2004[7]	162	26	64	36	31
Thailand, Bangkok[6]	169	27	68	29	45
UK, Thames, metered h/hold, 2006[8]	140	39	71	10	21
UK, Thames, unmetered h/hold, 2006[8]	153	46	77	11	21
UK, Severn Trent, metered h/hold, 2007[9]	121	41	40	21	19
UK, Severn Trent, unmetered h/hold, 2007[9]	132	45	47	23	18
USA, Seattle, no conservation, 2000[10]	240	71	48	65	56
USA, Seattle, high efficiency, 2000[10]	151	30	43	43	35
USA, Study, no conservation, 1997[11]	220	69	51	43	56
USA, Study, with conservation, 1997[11]	164	39	42	42	40
USA, Westminster, pre 77 houses, 1995[11]	235	70	53	56	56
USA, Westminster, post 88 houses, 1995[11]	179	53	53	23	49

Sources: [1]IWA, 2004; [2]Ofwat, 2007b; [3]ADB, 2001; [4]PUB, 2007; [5]Water Corp, 2003; [6]White, 2004; [7]PUB, 2005; [8]TWU, 2007; [9]STW, 2007; [10]Aquacraft, 2000; [11]AwwaRF, 1999
Notes: [a]Supply not 24-hour; [b]24-hour supply only to some areas of city; [c]Adjusted for net of premise leakage and excludes rounding of component figures; [d]Includes consumption and dishwasher use; [e]Includes washing machine use

Table 1.4 Typical water appliance usages

	USA	UK	Singapore	Australia[5,6]
WC toilets—flush size	6 litres[1]	6 litres, but cisterns installed before 1 July 1999 can be replaced by same volume.[2] Flushing or pressure flushing cistern.[2]	LCFC (Low capacity flush cistern) 3.5–4.5 litres[4,a]	Efficient = average 3.6 l per 5 flush cycles (6/3 capacity). Old = 12 l/flush.
—flush type	Flap-valve[1]	7.5 l/unit/hr + fitted with demand control.[3]		On demand flush control.
—urinals		1.5 l/flush with pressure flushing valve.[3]		Most efficient = 1.5 l/flush
Showerhead flow	11–30 l/min[1] (Low flow = 7 l/min)	Power shower = 12 + l/min.[3] "Water Saver" heads = 4–9 l/min.[3] Ultra-low flow = 1.5 l/min (non-domestic).[3]	7 l/min[4,b]	A = 12–9 l/min AA = 9–6.8 l/min AAA = <6.8 l/min
Taps (faucets): Bath	10–26 l/min[1]	12–16 l/min (19 mm)[2]		For all taps: A = aerator & positive seal
Washbasin		6–9 l/min (13 mm)[2]	2 l/min[4,b]	AA = as A with flow control
Kitchen		6–12 l/min (13 mm)[2]	6 l/min[4,b]	AAA = as AA with self closing
All taps			Self-closing delayed action[4,c]	
Constant flow regulators		Advice[3]	Yes[4,c]	Advice
Washing machines: (5 kg load)		<50 l/cycle[3]	Full load only[4,d]	Efficient = 45 l/cycle Inefficient = 180 l/cycle
Dishwasher: (12 place setting load)		16 l/cycle[3]	Full load only[4,d]	Efficient = 12 l/cycle Inefficient = 36 l/cycle

Sources: [1] USA Energy Policy Act 1992
[2] UK Water Supply (Water Fittings) Regulations 1999
[3] UK Environment Agency, Water Resources, CWB Fact cards, 2006 website
[4] PUB, Singapore Water for All: Conservation, Value, Enjoy, 2006 website
[5] Water Services Association, Buyers Guide to Saving Water, 2006 website
[6] Water Smart Guidelines, Master Plumbers and Mechanical Services Association of Australia, 2006 website (using the Water-Conservation Rating system A to AAAA)

Notes: [a] Mandatory (new premises including non-domestic premises)
[b] Best water conservation flow rates for existing taps
[c] Mandatory (non-domestic premises)
[d] Water saving advice

for outdoor uses including irrigation and yard cleaning. In Europe it represents about 2% of the total demand (Ofwat, 2007b).

In UK garden watering can increase daily consumption by up to 50% during a prolonged dry period but the total amount used in a year depends on whether the summer is a 'dry' or 'wet' year. In the north of UK prolonged dry periods are rare. In the drier south-eastern part of England, garden watering has been estimated to account for about 5 to 10 lcd (3 to 6%) of the total household demand during recent years of prolonged low summer rainfall. However, dry summer periods in UK are often of relatively short duration and the time-lag between the start of a dry period and the build-up of garden watering demand means that the peak of the latter is short lived. The demand for garden irrigation is more significant for the seasonal peak than the average annual per capita consumption.

Water used for swimming pools depends on the incidence of such pools in an area. Coupled with car washing and miscellaneous other outdoor uses, estimated outdoor consumption excluding garden irrigation represents about 3 to 13% of the total average residential consumption, or 5–20 lcd.

In response to the 2007–8 prolonged drought in Australia, some municipalities imposed restrictions on the use of water for filling swimming pools, washing down outside surfaces and cars and for irrigating gardens and lawns. Restrictions include banning the use of hosepipes, sprinklers and irrigation devices and limiting garden watering to alternate days using a hand bucket. Less stringent restrictions on outdoor water use were imposed by a number of water companies in south east England during the 2006 drought. Although specific restrictions many be lifted at the end of a drought, it is likely that governments and utilities will increasingly use restrictions on outdoor water use to support initiatives to reduce domestic demand. Indeed, with the increasing awareness of the scarcity of fresh water and need for water conservation, schemes for rainwater harvesting and recycling household grey water are being promoted by government agencies and water utilities. If the campaigns are successful, they will have a significant impact on reducing the demand for mains water for outside use and irrigation demand.

1.4 STANDPIPE DEMAND

Various researchers have suggested that a minimum quantity of 50 lcd should be available to consumers using standpipe supplies. However, there is no international minimum standard beyond:

- minimum criteria for water supply should be 20 lcd (WEDC, 1998);
- consumers should have access to 'at least 20 lcd from a source within one kilometre of the users dwelling' (WHO, 2000);
- a key indicator in meeting a minimum standard for disaster relief: 15 lcd (Sphere, 2002).

Standpipe consumption is influenced by the distance over which consumers must fetch water, the usages permitted from the standpipe, the degree of control exercised over use at the standpipe, and the daily hours of supply. The range of uses can be as follows:

1. Water taken away for drinking and cooking only.
2. Additionally for household cleaning, clothes washing, etc.
3. Bathing and laundering at the standpipe.
4. Additionally for watering animals at or near the standpipe.
5. In all cases: spillage, wastage and cleansing vessels at the standpipe.

The quantity that an individual can carry and use is directly related to the travel time and carrying distance. Where the supply is for drinking and cooking only, the basic minimum requirement for direct consumption is about 8 lcd (WHO, 2003) but spillage and wastage at the tap cause the minimum consumption to be 10 lcd. Consumption for items (1) and (2) combined is about 15–20 lcd but wastage and spillage raise this to 25 lcd, a suggested minimum design figure where the water is to be carried over more than a few hundred metres. Where there is little control exercised over consumers' usage of water at a standpipe and where bathing and clothes washing takes place at or near the standpipe, then at least 50 lcd needs to be provided. If purpose-built bathing and laundering facilities are provided consumption rises to 65 lcd. In India 50 lcd is the usual standpipe design allowance, in Indonesia the quantity is 15–20 lcd where the water is sold from standpipes.

In rural Egypt where uncontrolled all-purpose usage tends to occur, including watering of animals, surveys indicated consumptions of between 45 and 70 lcd (Binnie, 1979).

In low income communities in the lesser developed countries it is often the practice that one householder has an external 'yard tap' which he permits his neighbours to use, usually charging them for such use. These are, in effect, 'private standpipes' and, because carrying distances are short, the consumption can be up to about 90 lcd based on numbers reliant on the yard tap water.

A public standpipe should supply at a rate sufficient to fill consumers' receptacles in a reasonably short time, otherwise consumers may damage the standpipe in an attempt to get a better flow. The design of the standpipe should permit a typical vessel to be stood on the ground below the tap so that water is not wasted; the tap, typically 19 mm (3/4 in) size, should provide a good flow under low pressure conditions so that it is not vandalized and should be constructed of readily available materials to reduce the temptation for theft. The hours of supply need to be adequate morning and evening for the number of people using the facility and proper drainage should be provided at the standpipe to take away spillage.

1.5 SUGGESTED DOMESTIC DESIGN ALLOWANCES

Domestic in-house consumption for average middle class properties having a kitchen, a bath facility and waterborne sanitation falls into a fairly narrow range of 120–160 lcd, irrespective of climate or country. A system that uses a class of dwelling unit type (flat or house, etc.) and value (size, age, etc.), such as the ACORN geo-demographic classification system (www.caci.co.uk), is the most practicable basis to use for estimating domestic consumption where more accurate data are not available. From visual identification of the predominant housing stock in an area and knowledge of the average occupancy, therein the demand for the area can be estimated with reasonable accuracy.

Suggested design allowances of domestic per capita consumptions by dwelling type are given in Table 1.5.

1.6 NON DOMESTIC DEMAND

Non domestic demands comprise:

Industrial: Factories, industries, power stations, docks, etc.

Commercial: Shops, offices, restaurants, hotels, railway stations, airports, small trades, workshops, etc.

Institutional: Hospitals, schools, universities, government offices, military establishments, etc.

Table 1.5 In-house domestic water and standpipe demand—suggested design allowances (lcd)

Type of property and income group[a]	UK and Europe[b]	Hotter climates
1. Villas, large detached houses, luxury apartments, wealthy villages and suburbs; four plus bedrooms, two or three WCs, bath and shower, kitchen and utility room. (High income groups—25%)	190 *180*	230–250
2. Detached and suburban houses, large flats, rented properties (professionals and students sharing); three or four bedrooms, one or two WCs, bath and shower and kitchen. (Upper middle income groups—10%)	165 *150*	200–230
3. Flats, semi-detached and terrace houses, retirement flats; two or three bedrooms, one or two WCs, bath or shower and kitchen. (Average middle income groups—25%)	150 *130*	180–200
4. Semi-detached and terrace houses and flats, two or three bedrooms, one WC, bathroom and kitchen. (Lower middle income groups—15%)	140 *120*	
Developing countries: Tenement blocks: • Block centrally metered • Individual household metered		160+ 130
5. Small houses, cottages and flats, sustainable housing with separate kitchen and bathroom, high occupancy. (Low income groups—25%)	90–110 *75–90*	
Developing countries: Tenement blocks, high density occupation with one shower, one Asian toilet, one or two taps: • Building centrally metered or free • Individually metered		130+ 90
Developing countries: Lowest income groups: Low grade tenement blocks with one or two room dwellings and high density occupation: • Communal washrooms (unmetered) • One tap and one Asian toilet per household: block metered • One tap dwellings with shared toilet or none; dwellings with intermittent supplies		110 90 50–55
Developing countries: Standpipe supplies • Urban areas with no control • Rural areas under village control • Rural with washing and laundering facilities at the standpipe • Absolute minimum for drinking, cooking and spillage allowance		70+ 45 65 25

Notes: [a] Income category and percentage of UK population are indicative only. Within an area of similar housing stock there will be a range of household sizes and incomes
[b] Figures in italics are for metered consumption
[c] Figures exclude lawn and garden irrigation, bathing pool use and use of evaporative coolers; but include an allowance for unavoidable consumer wastage

Agricultural: Use for crops, livestock, horticulture, greenhouses, dairies and farmsteads. Industrial demand for water can be divided into four categories:

1. *Cooling water demand*—usually abstracted direct from rivers or estuaries and returned to the same with little loss. It is not normally taken from the public supply except for some supplies to water cooled air conditioning systems for commercial and office buildings.
2. *Major industrial demand*—consumption greater than 1000 m³/day for example for paper making, chemical manufacturing, production of iron and steel and oil refining. Large capacity water supplies tend to be obtained either from private sources or a 'raw water' supply provided by the water utility. The raw water is distributed through a public non-potable network or a dedicated pipeline to the industry and may receive disinfection treatment to reduce the health risk to people who could come into contact with it. The user would normally treat the water to the quality required for his processes including additional treatment and 'polishing' where the supply is derived from a potable supply. Non-potable supplies are always reported separately from the 'public' water supply in statistics.
3. *Large industrial demand*—factories using 100–500 m³/day for uses such as food processing, vegetable washing, drinks bottling and chemical products. These demands are often met from the public supply. Generally the supply receives additional treatment on site to meet process requirements.
4. *Medium to small industrial demand*—factories and all kinds of small manufacturers using less than 50 m³/day, the great majority taking their water from the public supply.

All industrial premises provide a potable supply for their staff for hygiene and catering. Generally this 'domestic use' supply is obtained from the public system, but occasionally it may be supplied from the treated water used in the industrial processes.

Estimating industrial demand can be complex. The same industry in a different environment can use significantly different quantities of water per processed unit. For example the specific water use of industrial production of raw steel from delivered ore in 8 European countries is reported to range from 0.6 to 600 litres/kg (average of 90 litres/kg); for paper produced from dry pulp, the range is 15 to 500 litres/kg of product; (average of 140 litres/kg) (EEA, 1999). These broad ranges demonstrate how variations in production process, water use, water efficiency, water recycling and possibly tariff structure can all influence the specific usage by an industry. There may also be differences in how each industry or country reports the statistics.

Existing industrial demand can be best estimated by measuring the daily and weekly demand for the specific consumer at the point of supply using the consumption survey approach outlined in Section 1.14. Ideally diurnal and seasonal variations should also be measured to ensure that the range of demand on the system is understood especially where the industry takes water as required and has seasonal production variations, for example seasonal food processing. Industries without on-site storage can impose onerous operational performance characteristics on a network such as surge and short term high flows; these may adversely affect the system and other consumers. Often it is found that about 90% of the total demand in a large industrial area is accounted for by only 10 to 15% of the industrial consumers. A consumption survey can therefore be selectively targeted to monitor the major users. However, it is important to check the accuracy of the meters used to ensure that the demand is being accurately measured and that usage lost through meter error is not being incorrectly reported as transmission mains losses (Section 1.8).

For new industries, the preferred approach is to adopt the forecast demand for water proposed by the developer, by industrial area or individual development site. Where no other information is available, industrial demand can be derived using typical usage by plot area and by industry category, for example car production, chemical industries, electronics, food production, general industrial and service industries. Only a few service industries for example drinks bottling, laundries, ice and concrete block manufacturing, use large quantities of process water. Many light industries, such as those involved in printing, timber products and garment making use water only for staff hygiene and catering. Typical ranges of industrial usage are given in Table 1.6; however, as discussed above, they should be used with caution.

Commercial and institutional consumption comprises the demand from shops, offices, schools, restaurants, hotels, hospitals, small workshops and similar activities common in urban areas. In England the overall average commercial and institutional demand is equivalent to about 25 lcd over the whole population served. This includes domestic use by people living in non-domestic premises or in attached living quarters. In the USA commercial and institutional demand can be significantly higher because of the high consumption for water cooled air conditioning systems and for outdoor irrigation in the hot climate. For example in office buildings the indoor use for employees for personal hygiene and catering can represent 35 to 40% of the total building demand. Similarly for schools the internal use may represent only 20% of the total (AwwaRF, 2000). Typical allowances

Table 1.6 Commercial and institutional water consumption allowances

Usage	Consumption allowance
Industrial areas	50–100 m³/ha. Product specific demand should be assessed where practical.
Light industry that includes food and drinks processing	1–1.5 m³/day per employee
Light industry that excludes large water consumption	0.25–0.5 m³/day per employee
Light industry and warehouse; cleaning and sanitation use only	50 l/day per employee
For small trades, small lock-up shops and offices in urban areas	In the UK: 3–15 lcd. Elsewhere: up to 25 lcd (applied as a per capita allowance to the whole urban population)
Offices*	50–75 l/day per employee
Department stores*	75–135 l/day per employee
Restaurants, bars	75–120 l/day per customer/table seating
Hospitals	350–500 l/day per bed
Hotels	250–400 l/day per bed; up to 750 l/day per bed for luxury hotels in hot climates
Schools*	25 l/day per pupil and staff for small schools; rising to 75 l/day per pupil and staff in large schools

Note: * Applies to the days when those establishments are open.

Table 1.7 Agricultural water demands—suggested allowances

Livestock	Dairy: Farming	40–100 l/day per animal in milk[a]
	Dairy: process cleaning	20–50 l/day per animal[a]
	Beef cattle	25–80 l/day per animal for drinking[a]
	Beef abattoir use	Average 1500 litre per animal
	Horses	30–55 l/day per animal[a]
	Sheep	5–15 l/day per animal for drinking[a]
	Pigs	10–20 l/day per animal for drinking[a]
	Poultry (eggs)	20–40 l/day per 100 birds[a]
	Poultry (meat)	15–25 l/day per 100 birds[a]
	Poultry meat processing	15–20 l/bird
Crop irrigation		130 m³/ha per week during growing season[b]
Glass house crop production		20–30 m³/day per hectare (or more) in growing season 10–15 m³/day per hectare in winter

Notes: [a] Livestock consumption depends on season, age of animal and production stage. Consumption can increase by 20 to 30% in extreme heat conditions
[b] Demand depends on crop and rainfall during growing season

made for demand in certain types of commercial and institutional premises in UK are also given in Table 1.6.

Most water for agriculture, including crop irrigation, horticulture and greenhouses, is taken direct from rivers or boreholes because it does not need to be treated. The principal use of the public supply is for the watering of animals via cattle troughs, for cleaning down premises and for milk bottling. Table 1.7 gives estimates of such consumption.

1.7 PUBLIC AND MISCELLANEOUS USE OF WATER

The quantities used to water and maintain parks, green areas, ornamental ponds, fountains and gardens attached to public buildings have to be assessed for each particular case in relation to the area to be watered and the demand from the type of cover planted, for example area of grass or flower beds, types of plants and shrubs, etc. The estimate should include potentially high seasonal variation especially in hot dry climates. Often the quantity of water used for public watering is only limited by the available supply. However, in some cities, raw water or 'grey water' is used for these purposes. Other miscellaneous uses include supplies to government owned properties, street cleaning, flushing water mains and sewers and for fire-fighting. Where supplies to public buildings such as government offices, museums, universities and military establishments, are not paid for, the demand can be substantial compared with the usage in equivalent private sector buildings. Water used for street cleaning, flushing, fire fighting and system maintenance activities can be assessed from the records of the time, duration and equipment used.

In the UK supplies to public parks and buildings and to government and local authority offices would be metered and should thus form part of the metered consumption. Unbilled and unmeasured legitimate water usage would be for fire fighting and for routine maintenance activities such as testing fire hydrants, sewer cleansing and flushing dead ends of mains. Temporary connections for building sites, which used not to be recorded, are now metered and the consumption is billed. The total of the miscellaneous unbilled and unmeasured demand is estimated by utilities in England and Wales to be about 1.0% of the total input into distribution, equivalent to about 3 lcd for the total population.

1.8 WATER LOSSES

In order to standardize the different interpretations of terms such as 'unaccounted-for water', 'non-revenue water', 'legitimate usage' and 'losses', the IWA proposed an internationally consistent set of terms and definitions for the components of water losses within the water balance. Figure 1.1 illustrates their proposed terminology. The terms are being used increasingly worldwide. Reducing water losses, leakage and wastage is a water utility's high priority when managing the water supply and demand balance. Water losses, a component of *'non revenue water'* (NRW), are made up of *apparent* and *real* losses and *'unbilled authorized consumption'*. Unbilled authorized consumption, essentially unbilled metered and unmetered consumption, can be managed effectively either by installing permanent meters to measure consumption or by monitoring the supplies regularly to assess demand and identify changes in demand patterns.

Apparent losses represent unauthorized consumption that is not measured or billed to the consumer, for example illegal connections, meter tampering or bypassing and meter inaccuracies.

System input volume	Authorized consumption	Billed authorized consumption	Billed metered consumption (including water exported)	Revenue water
			Billed unmetered consumption	
		Unbilled authorized consumption	Unbilled metered consumption	Non-revenue water (NRW)
			Unbilled unmetered consumption (eg. fire demand)	
	Water losses	Apparent losses	Unauthorized consumption (eg. illegal connections, meter tampering or bypassing)	
			Customer metering inaccuracies	
		Real losses	Leakage from transmission and distribution mains	
			Leakage and overflows at utility's storage facilities	
			Leakage from service connections up to the customer's meter	

Source of information: IWA Publishing, 2000

FIGURE 1.1

IWA standard international water balance and terminology.

Apparent losses also include consumer meter inaccuracy and errors in the meter reading and billing processes that are key to identifying and eliminating unauthorized consumption. Metering inaccuracies, the largest proportion of apparent losses, can be minimized by maintaining the meters (inspection, recalibration and replacement) and by managing the billing procedures to minimize data entry and thereby billing errors. Managing unauthorized consumption is complicated because of the difficulties in quantifying the illegal usage and locating the connections. Consequently unauthorized consumption tends to be included in the legitimate per capita consumption figures.

Real losses generally represent the majority of non-revenue water. Real losses comprise leakage, overflow and wastage from trunk mains, distribution pipework, storage facilities and service connections between the distribution pipework and the consumers' premises. Leaks occur from pipes, pipe joints and fittings; valves, hydrants and washouts; and from service pipes upstream of consumers' meters or boundary stopcocks. The ferrule connections of service pipes to mains are often a major cause of distribution leakage. Hence distribution losses are influenced both by the length of mains serving consumers and the number of service pipe connections per kilometre.

Water losses can not be measured directly but have to be estimated by measuring the total input into a system and deducting the amount supplied for legitimate consumption, including an estimated allowance for leakage from supply pipes and plumbing systems where the consumer's supply is not metered. Water utilities worldwide report a range of figures for non-revenue water and losses because the figures are influenced by a variety of factors such as the age and condition of the pipes, supply pressures, efficiency of leak and waste prevention measures, how the unmetered demand is estimated and the methodology used for compiling the statistics. Reported leakage and non-revenue water figures from utilities worldwide can range from 5 to 10% of the distribution input (the quantity of potable water supplied) for well managed systems, up to 40% and 60% or more for systems in poor condition where there is a history of long term under investment in network maintenance and rehabilitation. Systems with intermittent supplies also exhibit high leakage rates. Table 1.8 lists levels of losses and the circumstances commonly found to give rise to them.

Table 1.8 Typical figures of non-revenue water

Percentage of total supply	Typical circumstances applying
5–15%	Small systems with little leakage; residential parts of large systems with little leakage
16–20%	Usual lowest reported for whole cities, often associated with active leak control strategy
20–25%	Achievable in large systems with active leakage and waste control methods and good system monitoring and network data
25–35%	Reported for large systems comprising old mains and service pipes in moderate to poor condition, lower meter coverage and poor data
35–55% and greater	Systems with many old mains and service pipes in poor condition; inefficient metering, lack of attention to leaks and consumer wastage and limited financial resources

The wide range of figures reflects the variety of methods used to estimate water losses as well as the range of actual losses themselves. Apart from metering errors there is always some unmeasured consumption that has to be estimated. High figures in excess of 30% may be partly due to leakage from pipes and partly due to lack of valid consumption data. Losses in a new or extensively renewed system should be low, say 5 to 10%, but low reported figures could also result from liberal estimates of un-metered consumption, or by excluding trunk mains losses or meter inaccuracy losses from the calculation.

Expressing water losses as a percentage of the distribution input may be appropriate within a utility because the data used and methodology of the calculation is understood and can be applied consistently. However, it is not suitable for comparing different organizations with different physical and operational characteristics and different per capita and non-domestic demands. This can be demonstrated by the following calculation. Company A reports a total per capita supply (total authorized consumption and water losses) of 300 lcd and water losses at 20% of the total demand. The losses represent 60 litres per consumer per day. The equivalent losses for Company B with an overall per capita supply of 500 lcd and reported losses of 15% would be 75 lcd. The calculation illustrates that using percentages as a performance indicator between utilities understates the real losses of utilities with high usage and misrepresents the losses for utilities with lower unit usage, but higher reported percentage losses. Quoting percentage losses can also disguise other system differences such as utilities that deliver a large proportion of their supply through a few connections to industry compared with utilities with no large industrial supplies. Furthermore, percentage losses can decline as consumption rises and not because the losses have actually been reduced.

In recognizing the need for consistent reporting, the IWA Task Force on Water Losses developed the Infrastructure Leakage Index, *ILI* for reporting and comparing real water losses:

$$ILI = \frac{CARL}{UARL}$$

where *CARL* is the current annual real loss derived from the annual volume of real losses expressed either as litres/day, litres per connection per day or litres per kilometre of main per day for the hours in the day when the system is pressurized. *UARL* is the system specific unavoidable annual real losses, the technically achievable lowest real water loss based on pipe burst frequency, duration and flow rates, and system pressures for well run systems in good condition. *UARL* is made up of: Background (Unavoidable) Losses + Reported Bursts + Unreported Bursts, or:

$$UARL\left(\frac{litres}{day}\right) = \left(18L_m + 0.8N_c + 25L_p\right)P$$

where L_m = Total length of mains in km
$\quad N_c$ = number of service connections
$\quad L_p$ = Total length of underground supply pipe in km
$\quad P$ = Average zone operating pressure in metres

The equation can be reconfigured to calculate UARL in different units such as litres per km of main per day per metre pressure or in gallons and miles. The UARL coefficients given in

Table 1.9 were derived from international data for minimum background loss rates and typical burst flow rates and frequencies (Lambert, 1999). The calculation assumes that there is a linear relationship between leakage and pressure and it can be modified to take account of intermittent periods of supply.

ILI is being used increasingly for international performance comparisons. For systems with 24 hour supplies and supply pressure above about 20 metres, it is suggested that ILIs up to 2 represent networks where water losses are being managed efficiently and further reductions would need to be assessed carefully in relation to the cost of achieving additional savings. ILIs between 2 and 8 represent networks where water losses could be reduced, the higher the index the greater the potential for savings. ILIs over 8 represent systems with unacceptably high leakage and where leakage reduction programmes should be implemented as high priority. Equivalent breakpoints for developing countries, networks with intermittent supplies and low supply pressures are less than 4, 4–16, and over 16 (Liemberger, 2005).

Table 1.10 gives guidance on unavoidable real losses for a range of average operating pressures and connection densities for 24 hour pressurized systems.

Table 1.9 UARL (Unavoidable Annual Real Loss) coefficients

	Per day/metre of pressure	Background losses	Reported bursts	Unreported bursts	UARL total
Distribution mains	l/km	9.6	5.8	2.6	18.0
Service pipe to property boundary	l/connection	0.60	0.04	0.16	0.80
Service pipe–property boundary to meter	l/km	16.0	1.9[a]	7.1[a]	25.0

Note: [a] Assumes 15 m average length of underground service pipe within property boundary
Source of information: Lambert, 1999

Table 1.10 Unavoidable real losses (meter at boundary) (l/service connection/day) and suggested performance indicator for developed and developing countries

Connection density	Average operating pressure in metres				
Number per km of main	20 m	40 m	60 m	80 m	100 m
20	34	68	112	146	170
40	25	50	75	100	125
60	22	44	66	88	110
80	21	41	62	82	103
100	20	39	59	78	98

Source of information: IWA, 2000

In the UK the WSRA, also known as **Ofwat**, uses two performance indicators, '*litres/property/ day*' and '*m³/kilometer/day*' to assess and compare total leakage. However, these measures also need to be viewed with caution when making international performance comparisons. Issues that need recognizing include: a service connection may supply a single property or multiple dwelling units but the reported leakage is only on the service pipe; utilities supplying rural areas have long mains serving few connections per km; and utilities supplying urban areas can have a high number of connections per km of main. Table 1.11 presents the 2006/07 leakage statistics for England and Wales and also comparative figures for the UK and selected international countries for 2003/04.

Some of the difference between the leakage figures of the UK regional (water and sewerage) and water-only companies may stem from differences of approach in estimating losses, or because the larger regional companies have a larger scale of problems to deal with. However, physical factors also contribute to the difference. The regional companies supply the largest urban areas in the country, which tend to have older systems than the water only companies do; some include coal-mining areas where ground settlement has disturbed mains and several have to supply hilly areas requiring high distribution pressures.

The water from leaks is not actually 'wasted'. Much of it percolates underground and recharges aquifers. Hydrologists often take account of leakage when assessing groundwater flows (Section 4.5).

Table 1.11 International comparison of leakage performance measures

	l/prop/d		m³/km/d	
	Average	Range	Average	Range
Statistics for 2007/08[1]				
England & Wales (22 utilities)	135.2		9.7	
• 10 Regional companies		80–200		5.6–22.7
• 12 Water only companies		70–129		5.3–12.1
Statistics for 2004/05[2]				
England & Wales (23 utilities)	150	70–248	11	6–29
Australia (8 utilities)	83	44–115	5	3–9
Netherlands (5 utilities)	17	4–35	1	1–2
Portugal	126	62–268	8	1–31
Scandinavia (4 utilities)	80	33–114	15	5–21
Scotland	460	—	24	—
USA (9 utilities)	314	114–657	14	6–26

Sources: [1]Ofwat, 2008; [2]Ofwat, 2007(b)

1.9 REAL LOSSES (LEAKAGE) FROM 24-HOUR SUPPLY SYSTEMS

The level of leakage from a system depends on the success of the water utility's loss reduction programme, the age and condition of the system and the system operating pressure. Policies for loss reduction depend on the financial and manpower resources a water utility can allocate to leakage control, both for 'one-off' exercises to reduce current levels of leakage to a satisfactory level and for continued application of leakage control to maintain that level.

The age of a distribution system is a major factor influencing real losses. High losses quoted by several UK water utilities are primarily due to the advanced age of many of their mains and service pipes. Thames Water, with the highest leakage rate in England and Wales, reports the average age of its mains in London to be over 100 years old, with a third being over 150 years old compared with 60 to 70 years for other UK companies. Many European cities report average ages of between 40–50 years.

Table 1.12 summarizes international burst rates by material type. Table 1.13 presents some typical 'background' leakage levels, which are estimated to occur on UK and international water distribution networks together with indicative leak flow rates.

Table 1.12 International comparison of burst frequencies

Pipe material	Cast iron	Ductile iron	Steel	PE	PVC	AC	All mains
Burst rate per 100 km of main per year							
UK—all companies	21.1	4.2	11.1	3.1	8.7	9.4	27.2
UK—11 companies—range	12–30	3–7	5–23	1–11	4–19	6–31	12–31
Australia—3 cities	22.3	1.6	9.8	—	—	8.5	34.3
Australia—range	13–25	—	—	—	—	7–54	2–72
Canada	39.0	9.7	—	—	1.2	7.3	36.7
West Germany	19.0	2.0	—	10.3	6.0	6.0	—
East Germany	41.0	—	—	74.0	14.0	34.0	—
% material of total length of mains							
UK—all companies	64	10	4	8	1	13	
UK—11 companies—range	43–83	2–16	0–6	2–17	4–25	0–20	
Australia—3 cities	68	10	1	0	1	20	
Australia—range	31–78	0–18	0–2	0	0–2	1–69	
Canada	51	23	0	0	11	15	

Source of information: UKWIR, 2001

Table 1.13 Background leakage and UK industry average leak flow rates

Estimated leakage level at 50 m pressure				
Infrastructure element		Low	Average	High
Trunk mains	l/km/hr	100	200	400
Distribution mains[1]	l/km/hr	20.0	40.0	60.0
Service pipe to meter at property boundary[1]	l/conn/hr	1.5	3.0	4.5
Service pipe to internal meter[2]	l/conn/hr	1.75	3.50	5.25
In-house plumbing losses[2]	l/property/hr	0.25	0.50	0.75
Average flow rate by type of leak and burst frequency[3]				
		Flow (m³/hr)	No./1000 properties/year	
All leaks			16.4	
Mains leak		3.0	3.6	
Detected mains leak		3.0	8.0	
Mains fittings (valves, hydrants)		0.15		
Service pipe—communication pipe		0.4	All 2.5[2]	
Service pipe—supply pipe		0.4	Reported 1.5[2]	
Communication and supply pipe fittings		0.1	Unreported 0.5[2]	

Sources: [1]WRc, 1994; [2]Lambert, 1998; [3]UKWIR, 2006b

Leaks on distribution systems break out continuously so that the total leakage from a system for a period is the aggregate sum of each leak-rate multiplied by the time it runs before repair. Hence the frequency with which all parts of a system can be tested for leaks influences the level of leakage experienced. Obviously there is a practical limitation to that frequency; therefore, some level of leakage is unavoidable. There is also a need to determine the economic level of resources that should be put into leak detection and repair. This is discussed in Section 14.12.

Leakage from service reservoirs can be found by direct static testing. Acceptance figures for a new concrete service reservoir would normally be a drop in water level of 1/500th of its depth (up to 5 m deep) over a seven day test period, equivalent to about 0.03% per day of the average daily supply with diurnal turnover, which is a negligible amount. Overflow discharge pipelines should be so designed that any overflow can be seen and therefore stopped.

The data used to assess non-revenue water and water losses and hence whether action is necessary to reduce leakage needs to be validated. Meter error resulting from the under-recording of consumers' supply meters can represent a significant percentage of non-revenue water. Low flow consumer demands, particularly at night, are frequently below a supply meter's stalling speed and hence go unrecorded. Over a period of time they can, in total, represent an appreciable demand on the system. The methods described in Section 1.15 can be used to assess the accuracy of consumers' supply meters and the magnitude of their likely under-recording. It is essential to monitor and assess all unbilled legitimate uses regularly.

1.10 SUPPLY PIPE LEAKAGE AND CONSUMER WASTAGE

The service pipe connects the main in the street to the consumer's premises. It comprises two parts; the '*communication pipe*', maintained by the utility, which goes up to the boundary stopcock and meter where fitted, and the '*supply pipe*', maintained by the customer, which completes the connection onto the property plumbing network. Leaks on communication pipes are accounted for in the utilities' reported distribution losses.

In the UK many of the service pipes are over 50 years old and made of galvanized iron or lead. In the installation of 50 000 meters on household supplies in the Isle of Wight during the National Metering Trials, 1989–92, it was reported that 8000 service pipes were either repaired or replaced in part or in total, most defects being found on the customer's supply pipe downstream of the boundary stopcock. This represents 1 in 6 of such pipes being found faulty (Smith, 1992). For 2006–07, utilities in England and Wales reported that the leakage estimated from unmetered consumer supply pipes averaged about 43 l/day per property (range 19–76 l/day), but on properties where meters were installed at the property boundary, supply pipe leakage was reported to average 20 l/day per property (range 1–46 l/day) (Ofwat, 2007a).

Within the property, there can also be leakage and wastage, termed 'plumbing losses'. In England and Wales losses from internal plumbing systems are not high because for many years the quality and design of water fittings and plumbing has been controlled by byelaws, exercised by every utility. These byelaws require, among other things, that all WC and storage cisterns are float valve controlled and the cistern overflow pipe must discharge outside the premises. Hence overflows can be easily noticed or heard by waste inspectors and leakage technicians checking premises at night; and the nuisance created by the overspill at the premises may (sometimes) motivate the occupier to take remedial action. Some wastage caused by dripping taps is, however, unavoidable. Where supplies are plentiful or water is cheap, or where waste conservation measures are slack, consumer wastage can be 50 lcd or more. Block metering of multiple occupancy buildings tends to result in high consumer wastage where one meter measures the supply to all the households and a single landlord pays the water charges (recovering the cost through the rents charged) in such a way that individual householders do not pay for their own wastage. Table 1.5 shows that 25–40% extra domestic consumption needs to be allowed for in block metered premises.

1.11 MINIMUM NIGHT FLOW AS INDICATOR OF LEAKAGE AND WASTAGE

The minimum night flow (MNF) to a section of the distribution system can act as an indicator of distribution leakage and consumer wastage. MNF is the measured flow into a controlled area of the network, for example a District Meter Area (DMA) during the period of minimum demand. The MNF typically occurs during the night between 01.00 and 04.00 hours but the characteristics of the area can have an significant impact on when the MNF occurs, for example DMAs containing a high density of bars and nightclubs in tourist areas. There is always some

Table 1.14 Figures for minimum night flow (MNF) Per Connection

MNF per connection	Interpretation
1.7 l/h	Average minimum night use per household (WRc, 1994) (or 0.6 l/h × number of people in household)
5 l/h	About the lowest found in practice in parts of systems in good condition
7 l/h	A frequent 'target level' for distribution districts, indicating good control over leakage and wastage
9 l/h	Experienced on large systems where there is a fair amount of nocturnal demand and/or some distribution leakage and consumer wastage
11 l/h	Indicative of substantial night demand and/or considerable distribution leakage and/or consumer wastage

legitimate demand for water at night, which has to be deducted from the minimum recorded flow. Although MNF tests do not measure quantities of water lost, they are a good indicator of the condition of a system; Table 1.14 shows an interpretation of results obtained. However, the test is impracticable if the supply is intermittent or houses have large storages that fill at night. In the UK domestic house storage tanks are relatively small and therefore are usually full before an MNF test takes place. It is important therefore to understand the characteristics of consumer demand and plumbing as well as how the distribution system performs when analysing night flow records.

The lowest night flow in small residential areas comprises distribution system and supply pipe leakage, legitimate night domestic demand plus unavoidable consumer wastage from dripping taps. However, in larger test areas legitimate use is likely to include non-domestic night-time consumption at premises such as hospitals, nursing homes, police and fire stations, railway stations, airports, clubs, as well as industrial demand from units working night shifts. This non-domestic consumption needs to be monitored or measured during the period of the flow test. Large users may be continuously monitored. The consumption of metered consumers can be assessed by reading their meters at the start and end of a night test or, if not practical because of the number of meters to be read or manpower resource constraints, instead their 'typical' rates of night consumption can be measured before the test takes place and deducted from the area night flow measurement.

1.12 VARIATIONS IN DEMAND (PEAKING FACTORS)

Consumers draw water as and when they require it and consequently the demand for water varies both diurnally and seasonally. Demand over a 24-hour period varies from the minimum night flow to a peak hour demand, albeit that the rate of draw may be reduced where the supply passes through in-house storage. Demand also varies: by the day of the week, weekend demand patterns are different to weekday usage; and seasonally, depending on temperature variations; and for regional reasons, for example city demands can be observed to fall during holiday seasons when at the same time demand in tourist areas rises significantly to cater for the influx of visitors.

The variations in system diurnal and seasonal demand are usually derived as ratios (factors) or percentages of the Average Annual Day Demand (AADD). Peaking factors are influenced by the size of the area being monitored. The demand from an individual consumer can have a significant impact on the flow profile of a small supply area such as a rural community or an area with a single large industrial unit, for example a seasonal food processing factory. However, the overall peaking factor reduces with the number and mix of consumers and the impact of an abnormal user can be less apparent.

Maximum and Minimum Hourly rate of consumption

The diurnal demand depends on the size of the population and the type of commercial and industrial usage in the area served. In the UK the domestic peak period for residential areas is typically between 07:00 and 09:00 hours reflecting households getting up and preparing for work and school; peak hour factors vary between about 2.25 and 1.75 depending on the size of the area. The peak period can start earlier and be shorter in communities with high proportions of workers who commute and in rural areas; peaking factors being up to 2.5 or even 3.0 for small areas. For mixed residential and industrial areas, the peak is typically at mid day with factors typically between about 1.5 and 1.75. Similar peak factor ranges are observed worldwide but cultural, religious and lifestyle differences can greatly influence the amplitude, duration and time of the peak.

Diurnal factors also vary between weekday and weekends. Traditionally domestic peak hour factors were influenced by weekday household water consumption. Increasingly flow profiles are demonstrating changing water use patterns to higher domestic consumption at weekends, the peak hour occurring later in the day, between 09:00 and 11:00 in the morning or in the evening and the higher demand period lasting for longer in the day. There are also indications that the peak day demand can vary between a Saturday and Sunday in areas that otherwise appear to have similar characteristics.

Garden watering demand can create an evening peaking factor of 3.0 or more. In the USA with its hot summers where peak hourly factors for in-house demand are reported to be between 3.0 in eastern states and up to 5.0 for western states, sprinkler demand can increase the factor to 6.0.

Seasonal Variations

The Maximum or Peak Day Demand is the maximum daily demand reported in a year. The quantity is significant for understanding the potential absolute demand on a system and the centres of the high demands. However, it is not as important as the demand during a maximum week in the year, expressed as the Average Day Peak Week (ADPW) demand. The ADPW is typically only a few percent below the peak day demand but it is important because the daily overdraw above the average for seven consecutive days cannot usually be met from the amount of service reservoir storage provided. Therefore, utilities typically manage their resources and storage to maintain an overall weekly resource balance. This means that the maximum output of the source works must be at least equal to the average daily demand for the 7 days of the peak week. Typical system ratios of the ADPW to the annual Average Day Demand (ADD) are given in Table 1.15. In temperate climates a design figure of 140% is often adopted.

Table 1.15 Peak system demand ratios

Location, etc.	Ratio: ADPW/ADD 1
UK. Seaside and holiday resorts	130–150%
UK. Residential towns, rural areas	120–130%
UK. Industrial towns	115–125%
UK. Peak due to garden watering in prolonged hot dry weather	150–170%
USA. Typical peak domestic demands—in-house only	
—western states	180–190%
—eastern states	130–140%
USA. Typical peak domestic demands due to lawn sprinkling:	
—western state	220–340%
—eastern states	200–300%
Worldwide. Cities with hot dry summers	135–145%
Worldwide. Cities in equable climates	125–135%
Worldwide. Cities with substantial industrial demand	110–125%

1.13 GROWTH TRENDS OF CONSUMPTION AND FORECASTING FUTURE DEMAND

The increase in public water supply consumption in England and Wales since 1970 is shown in Figure 1.2. Over the period 1970–2007 total per capita consumption rose from 277 lcd to a peak of 331 lcd in 1995–96 and thereafter has reduced over the last 11 years to about 278 lcd. The figures are the aggregate of the individual water company estimates. The reduction is primarily due to increased work on leakage reduction but the decline in industrial demand and measures to restrain domestic consumption have contributed. Variations in the summer climate, which affects household and garden water use, may also have had an impact.

Long-term forecasting of water demand presents problems. Figure 1.2 shows how a period of sharp economic decline in 1992 reduced domestic consumption immediately and trade consumption the following year. Such incidents interrupt previous trends of increase that, in the case of domestic consumption in developed countries, tend to be asymptotic to some future maximum demand per capita. The ultimate maximum unconstrained level of domestic consumption depends on household wealth and housing and standards of water fittings installed, coupled with the policy of the water supplier. However, where resources are scarce, additional supplies are difficult or expensive to procure; or where construction of a new water supply scheme would meet with strong environmental opposition, measures to restrain the rise of domestic demand may need to be adopted.

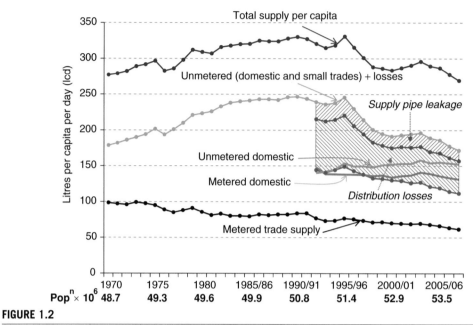

FIGURE 1.2

Change in per capita consumption, England and Wales, 1970–2008.

1.14 WATER CONSERVATION AND DEMAND MANAGEMENT

Water scarcity and the need to conserve resources are recognized worldwide as challenges for the near future. Water conservation and environmentally sustainable use of water will be increasingly implemented to manage the water balance, particularly where water resources are limited and the areas are subjected to droughts. Much emphasis is being placed in Australia, the USA and in the drier parts of south-east England on the adoption of demand constraint measures. In England emphasis is being placed on extending the metering of domestic supplies and on increasing measures to reduce leakage and wastage. In the USA, as required by the Safe Drinking Water Act 1996, US EPA published draft guidelines in 1998 to water suppliers for 'conservation planning', i.e. measures to induce economy in the use of water. The guidelines propose three sequential levels of approach. The first comprises universal metering, loss control (i.e. leakage and wastage reduction) and public education. The second and third levels include such measures as water audits, pressure management, re-use and recycling, and integrated resource management. However, emphasis on demand constraint does not necessarily apply in all countries, where some utilities may be reluctant to curb the demand from metered industrial consumers and from metered households occupied by higher income groups; the payments made by them comprise a major part of the utility's income and are needed to cross-fund supplies given free through standpipes or at below cost to low income groups. However, the conditions and position adopted by each utility vary. Where more plentiful supplies exist, such as in Canada and Scotland, there may be less incentive for restraint.

In many countries restricted hours of supply have to be adopted in order to prevent consumption and losses exceeding available supplies. Metering is adopted for the same purpose but it must be efficient to be effective. Intermittent supplies bring many problems. Consumers store water when the supply is on, but throw away the unused balance when the supply next comes on, believing the new supply is 'fresher'. Consumers may leave taps open so as not to miss when the supply comes on again, allowing storage vessels to overflow. Intermittent supplies make leak detection and prevention of consumer wastage very difficult. Typically the hours of supply have to be reduced to at most 4 hours in the morning and 4 hours in the evening, and frequently less, to gain control of consumption. To some extent intermittent supplies are self-defeating; more consumer wastage and more distribution leakage occur because of the difficulty of maintaining the system in a good state and if mains become emptied, contaminated groundwater may enter the pipes and endanger the health of consumers. The situation is often exacerbated by loss of income due to difficulties with metering and income collection. Nevertheless, many utilities world-wide have to adopt intermittent supplies.

Water conservation measures that can be effective on 24-hour supplies include:

- imposing temporary bans on the use of water for washing vehicles, refilling swimming pools and ponds, and on the use of hosepipes and sprinkler equipment for watering gardens during drought or shortage of supplies. During the 2006 hosepipe bans in southern England, seasonal reductions of up to 15 percent were reported;
- good publicity can achieve a temporary and short term reduction in demand, perhaps as much as 10 percent;
- metering domestic supplies can curb excessive consumption, especially water used for lawns and gardens, provided the tariff structure imposes a financial penalty when consumption exceeds a reasonable amount;
- promoting the use of low water use fittings (Table 1.4) can make significant reductions as illustrated in Table 1.3;
- using pressure management control to reduce leakage and to increase the life of the pipes, improves the reliability of network control valves, which can operate within their designed range, thereby reducing wastage through the valve malfunctioning;
- keeping operational pressures at the minimum necessary to maintain levels of service reduces water taken unnecessarily;
- flow limiters and throttles on service pipes can curb consumption but are not always effective. If set too low, consumers leave taps open to fill containers which overspill. They can also be by-passed in an attempt to get a better supply.

Commercial and institutional demand can be constrained by metering all consumers: both large and small shops, offices and other business premises. Wastage through plumbing fittings from these non-domestic consumers is frequently high because no one working in the premises is responsible for paying the water charges and the premises are unoccupied outside working hours. Many cases have been reported of night and weekend flows to unoccupied premises being nearly as high as daytime flows when staff are present, especially in government offices in some countries. Manufacturers are also often unaware of the potential financial savings they can achieve by adopting water conservation measures, the lower usage reducing both their water purchase and effluent discharge costs.

Table 1.16 summarizes the opportunities and constraints of water saving technologies, metering and tariffs and leakage reduction, the three primary measures for managing consumption.

Estimating future demands in developed countries can be approached by combining the forecast trends in population and per capita consumption growth with implementing water conservation measures within the forecast period. The trends provide an upper limit to the projections. The conservation measures will produce a range of achievable forecasts provided there is the political and social will to conserve water and pay a realistic tariff for water and provided the utility manages, and is seen to be managing, system water losses and leakage.

Table 1.16 Water conservation measures

Component	Opportunities	Constraints
Water saving technologies	• Low and ultra-low water use efficient appliances • Standards and regulations will achieve greatest savings	• Policing installation and retrofit • Does not stop excessive use • Time to promote and develop • Unlikely to achieve 100% acceptance
Metering and tariffs	• Reduces demand, short and long term • Consumers responsible for their impact on environment • Charge real cost of usage	• To be effective, need both • Ability to pay and need to maintain supplies to low income and vulnerable consumers • Cost of meter installation and long-term maintenance
Leakage control	• Proactive control can reduce losses down to economic level of leakage (ELL) • Short term local solution	• Reduction to ELL is one-off win, thereafter marginal impact on resources • Savings depend on maintaining ELL/leakage reduction • Is ELL optimum level?

1.15 THE QUESTION OF METERING DOMESTIC SUPPLIES IN THE UK

Reducing demand for fresh water is an environmental benefit to all. The aim of metering is to restrain rises in domestic consumption, especially in areas where additional sources of water are difficult or expensive to procure. With rising demand for water, increasing difficulty in developing new supplies, recent periods of low rainfall, and with the supply area of one water company in Southeast England, Folkestone and Dover Water Company, being designated a water scarcity area, the question of whether universal metering of the 70% unmetered domestic supplies in England and Wales should be adopted has been debated repeatedly by a cross section of stakeholders. The questions centre mainly on cost, fairness to the consumer and what reduction of consumption could be achieved.

The benchmark cost of installing a new meter in the UK is £113 inside a property and £191 when fitted outside together with a boundary box (Ofwat, 2005). The National Metering Trials, 1989–92, found that only about 30% of properties could be metered inside and for 5% of properties a water meter could not be installed due to plumbing difficulties or high cost. In many older housing areas up to 20 properties can be fed by one common supply pipe: one utility was said to have 650 000 properties fed through joint service pipes (Roberts, 1986). Common supply pipes are often laid at the rear of terraced properties and occasionally the supply pipe is laid through the roof space of a terrace of properties. Multi-occupancy buildings, such as blocks of flats, are also mostly fed by one metered riser supply pipe; inserting meters on short off-takes to individual dwelling units could cause unacceptable disruption of kitchen or bathroom fitments.

The savings in consumption achieved by metering are difficult to assess. The National Metering trials gave erratic results, the best information coming from the Isle of Wight where 50 000 meters installed resulted in 5–9% reduction (WSA, 1993). More recently a report for UKWIR gives potential savings of between 10 and 15% based on trials and compulsory programmes (UKWIR, 2006a). It is probably safe to assume that metering can achieve at least 7.5% reduction, equivalent to about 12 lcd on current domestic consumption in England and Wales of 150–160 lcd. Greater benefit may be achieved through the use of seasonal tariffs to curb both seasonal demand and the peak day and hour ratios, both of which cause utilities supply problems. A reduction of about 30 l/property/day in supply pipe leakage could also eventually be expected where meters are installed outside (Section 1.10). This is a useful benefit, but whether universal domestic metering is justified by the cost is a complex matter. Water companies in England and Wales install domestic meters free when requested by the householder, on change of property ownership and to all new newly built houses.

The average water standing charge in England and Wales in 2007/08 for household customers taking metered water supplies is about £25 per year, which covers meter provision, meter reading and extra billing costs (Ofwat, 2007c). This represents a substantial proportion of a metered consumer's bill (Section 2.27). Any offset saving of cost due to the reduced production of water achieved by metering is unlikely to be significant (less than £5 per annum), unless the reduction in consumption is sufficient to avoid the need to develop a new source of supply. However, while companies continue to have the right to use the rateable value of houses as a basis for unmeasured charges for water and sewerage, voluntary domestic metering is likely to increase only in those areas where rateable values are high and household occupancy is low. Hence substantial extension of domestic metering is likely to occur only in conjunction with a process of designating water scarce areas.

The question of fairness of metering to householders also causes debate. The large family will obviously pay more than the small family; but if a tariff is adopted which provides a large family with a sufficient basic water allowance at low cost, then a single occupant can take several times his basic need at the same low cost. The average occupancy ratio of metered households at 2.1 (range 1.6–2.3) is significantly lower than for unmetered dwellings at 2.5 (range 2.4–2.9) (Ofwat, 2007a). At present charging for unmetered supplies according to property value seems reasonably fair, since there is a strong relationship between property value and consumption per capita as Figure 1.3

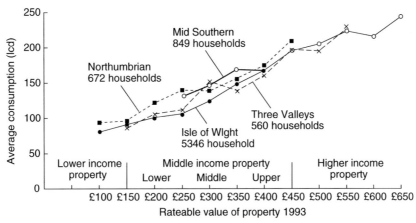

Source of information: National Metering Trials 1989–1993, Final Report 1993 and Halifax Building Society: www.hbosplc.com/economy/HistoricalDataSpreadsheet.asp

FIGURE 1.3

Relationship between in-house domestic consumption and property class, England 1993.

shows and low income families generally reside in low value property. However, the metering of domestic supplies is so widely practised in many countries that on the face of it, it is difficult to maintain that domestic metering creates an unacceptable injustice. When taking into account tariff subsidies, the argument becomes less straightforward and utility specific.

1.16 EFFECT OF PRICE ON WATER DEMAND

In their Environmental assessment report No. 1 titled *Sustainable water use in Europe* (EA, 1999), the EA concluded that although water pricing is difficult to use as a demand management measure, increasing water tariffs is a useful tool to make users more responsible for their water use when applied in conjunction with other water conservation initiatives. They cite a case in Hungary where demand fell 50% over an 11 year period during which the price of water increased from HUF 2 to HUF 120 per m³. The report concluded that the dramatic price rise contributed to the reduction in demand.

It is generally true that the demand for a product or service reduces with increasing price to the customer and vice versa and there are formulae, guidelines and rules of thumb for predicting the impact of a price change on demand. Where the product or service is subsidized, the impact of cost changes becomes more difficult to predict because it is dependent on whether the demand is suppressed or not, the level of subsidy in place and the affluence of the community. However, there is limited documented evidence of case studies demonstrating the relationship between demand and water tariffs.

Price elasticity of demand is the term used to measure the percentage change in demand resulting from a percentage change in price. The relationship is represented by:

$$\text{Price elasticity of demand } e = \frac{\Delta Q/Q}{\Delta P/P}$$

where Q is the demand at price P per unit of consumption. Since price increases tend to cause a reduction in demand, e is negative. When $e = -1.0$ increases in P cause proportional decreases in Q. A value of e between 0 and -1 indicates a high degree of price inelasticity, e.g. at $e = -0.2$ a 25% price increase of P would cause Q to decrease by 5%. Where e is greater than -1 price is considered elastic and a given change in price will result in a greater change in demand.

Many of the documented studies of demand analyses conclude that water tariff changes fall within the price inelastic range: that their impact on demand had been proportionately less than the price increases. However, the conclusions from a study of eight systems, six systems in Europe and two towns in Australia, also concluded that there was more elasticity in increasing block tariffs for residential usage (Metaxas, 2005).

Values of e are difficult to determine accurately because environmental or economic conditions before and after some price rise are often not the same. In addition, the elasticity value must be influenced by the size of the price rise, a large one-off price rise having a greater influence than smaller rises annually. A time lag also occurs between a rise in price and any observable effect on demand. Consequently e values quoted show a wide variation for apparently similar situations. In the Hungarian example discussed above the price elasticity of demand in each year remained in the inelastic range averaging about -0.2 (range -0.01 to -0.5). The generally reported range for price elasticity of demand is between -0.1 and -0.3 for average annual in-house domestic demand.

Despite the difficulty of getting an accurate measure of e, the elasticity of demand is important to a utility. Industrial and trade consumers and the higher income householders are often a major source of income to many water utilities overseas. Hence the elasticity of their demand is an important factor, because a price rise may cause them to reduce their take, thereby not producing the anticipated proportionate increase of income.

US EPA in their *Water Conservation Plan Guidelines* (US EPA, 1998) suggests the following benchmark figures for the impact of price rises on demand:

- 10% increase in residential prices reduces domestic demand by 2 to 4 per cent ($e = -0.2$ to -0.4);
- 10% increase in non-domestic prices reduces non-domestic demand by 5 to 8 per cent ($e = -0.5$ to -0.8);
- Increasing block tariff rates reduces demand by 5 per cent.

1.17 ASSESSING FUTURE DEMAND IN DEVELOPING COUNTRIES

In developing countries consumption is often limited to the amount of water available, with the result that there can be much unsatisfied demand. Estimating the long-term potential demand therefore involves a different approach. Population growth forecasts may be available, but they can be unreliable because of unexpected changes in birth rates. A UN manual (UN, 1956) gives mathematical methods of forecasting population growth based on fertility and mortality rates, sex ratio and age distribution of population, etc. but such data may not be available in a given case or necessarily relevant. A major proportion of the population growth in many cities in under-developed countries is often caused by migration from rural to urban areas. Hence estimating future water demand of such a city involves the following steps:

1. Plot the population trend for the past 10–20 years and assess the likely proportion due to migration and that due to natural increase of the existing population.
2. Divide the supply area into different classes of housing and assess in which classes of housing the main rises of population have occurred and are likely to occur in the future.
3. Assess typical rates of domestic per capita consumption by class of housing.
4. Seek likely figures for future migration and natural increase and allocate these to appropriate classes of housing.
5. Investigate future proposals for new housing development (master plans, development plans and proposals).

This approach can be used to build up forecasts of future demand. However, the estimates may need to be adjusted to ensure that they represent a realistic continuation of historic consumption trends. Generally speaking, consumption forecasts for more than 10 years ahead tend to be unreliable.

Growth of commercial and institutional demands is often estimated as a per capital allowance on the population growth because these activities tend to relate to the size of the population served. Increases in manufacturing and industrial demand are generally dominated by the needs of a relatively few major industries whose developments plans should be ascertained. However, the sum of their estimated individual future demands should be written down to allow for the probability that not all the developments are likely to go ahead in the period of forecast.

1.18 CONSUMPTION SURVEYS

Although the total supply available for an area can seem to be adequate when expressed as the amount available per head of population, this statistic may conceal the fact that, due to excessive wastage and leakage in some areas and by some consumers, there is unsatisfied demand in other parts of the system. Symptoms of such problems include: apparent high demand or water losses; consumers in some areas cannot receive an adequate supply or pressure; metering and billing practices appear to be inefficient. All these situations occur worldwide in utilities constrained by lack of money and technical resources, resulting in poor system performance and service to the consumer. The water balance equation for a suppressed demand is:

Available supply = legitimate potential demand + water losses − unsatisfied demand

In order to understand the demand in an area, it may be necessary to carry out a consumption survey. This provides information for analysis of the performance of the system in terms of delivering an adequate supply to meet the demand for water. The survey addresses how much, when and for what purposes water is being used. The steps involved in a consumption survey are:

1. Using consumer complaints and system monitoring data identify areas in the distribution network where pressures are below a (service) level sufficient to deliver an adequate supply to consumers both at peak demand times and during the whole of the daytime.
2. Check the accuracy of source meters supplying the area. This may involve temporarily installing a check meter adjacent to the meter under investigation or monitoring the source meter while temporarily discharging a measured quantity into a tank or through a controlled fire hydrant or washout point where the flow rate is monitored.
3. Check consumer supply meter readings against observed check-readings over a period and compare with the billing records:
   - Quantify the typical number of supply meters found stopped or faulty at any one time and investigate what billings are made when meters are found stopped.
   - Determine the average age of supply meters and how frequently they are tested, recalibrated and repaired; establish the performance characteristics of meters from different manufacturers installed in the network.
   - Test some meters of typical size and age for accuracy. Where more than 15% of meters are found stopped at any one time or where many meters are over 10 years old and not regularly tested and repaired, a substantial amount of under-recording must be suspected.
4. Assess typical domestic consumption per capita by classifying dwellings into five or six classes and test metering 30–35 dwellings typical of each class (Section 1.19).
5. Test-meter a few typical standpipes and estimate the population reliant on each. Estimate the typical standpipe consumption per capita.
6. Examine the supply meters on all large trade and industrial supplies. Compare observed readings against billing system data and investigate inconsistencies. Check the accuracy of those found in poor condition.
   - Identify the largest potential trade consumers and check their billing records to see whether usage appears reasonable in relation to the size of the supply pipe, the hours of take, and the amount of water likely to be used for their activities.
   - Check for major consumers missed from the billing system; those with two feeds to the premises only one of which is metered; or for which the billing is estimated from a previous lower usage activity (common where there has been a change of owner or activity).

7. Monitor and estimate or meter the amount of water supplied to legitimate un-metered consumers such as government or municipal offices, public parks and gardens.

The information obtained during the survey then needs to be compiled to estimate the probable total potential demand on the system. On a map of the distribution system identify supply districts and mark areas of the different classes of housing. Using appropriate population densities per hectare (or census data) and measuring the area of each class of housing within a district, estimate the total domestic demand per district by using the appropriate consumption per capita (Step 4).

For each district any standpipe consumption, commercial, industrial and other non-domestic demands should be added, apportioned according to the characteristics of the district. Major trade consumers should be individually assigned. An allowance for a reasonable quantity of unavoidable distribution leakage should be added to the total domestic and non-domestic demand. This gives the total average daily demand on the whole system, broken down into sufficiently small supply districts for the demand in each district to be distributed to 'nodal points' of the mains layout, i.e. key junctions of mains. These 'nodal demands' can be used for the basis of a hydraulic analysis of flows in the distribution system, as described in Section 13.15. The whole exercise will determine the adequacy of the distribution system to meet the demands. All meters measuring the output of sources have also to be checked for accuracy; unless regularly serviced they will almost certainly be in error. Consumer meter inaccuracy is categorized as an apparent water loss in the water balance (Fig. 1.1).

1.19 TEST-METERING IN-HOUSE DOMESTIC CONSUMPTION

Sample sets of domestic properties are monitored when individual supplies are not metered or where records of metered domestic consumption are not sufficiently reliable to assess an average domestic consumption per person. The sample set should be representative of the consumers in the area and should include about 35 properties from each of a maximum of five or six household categories, individually or collectively metered. A sample size of at least 30 properties is required to provide a representative estimate of the mean per capita consumption for a given household category. By monitoring 35 properties per category, about 30 valid sets of results should be obtained allowing for equipment failures and data lost from individual properties during the survey. Larger sample sets are difficult to monitor because of the need to ensure that, during the period of the survey, all meters continue to work accurately and the service pipes remain leak free, and to keep a check on the number of people in each household. The consumer categories are usually related to housing type and their number is restricted because of the difficulties of distinguishing between ranges of consumers with any certainty. Furthermore, since the intention is to monitor small sets of specifically identified properties and household categories, the sample set should not be chosen randomly but specifically identified in order to avoid introducing a bias of the sample mean towards the higher or lower end of the range within the class. The test period should be 2 to 4 weeks, avoiding holiday times and, if possible, extremes of weather. Meter readings and occupancy rates should be recorded weekly.

Such tests generally show a wide scatter of results. The mean per capita consumption can be substantially influenced by a few households where the consumption seems extraordinarily high. However, provided high consumption readings are not due to meter reading error, the figures should be included because domestic consumption is so highly variable. The limitation of the exercise is that the sample size is too small to evaluate the incidence of such high consumption which may be, for example, one in 20, so that a sample size of 30 may contain no such high consumers, or one or two. However, if five classes of housing are adopted there will be at least five separate samples from which to judge, roughly, the frequency of such exceptional consumption. The mean domestic per capita consumption is derived from the total consumption in the 30 or so households tested in a given category, divided by the total occupancy during the test period.

Where individual properties can not be metered, the flow through a main supplying a number of properties of the same class can be monitored instead. This approach will not reveal individual high or low household consumption and will include leakage from the supply main and service pipes downstream of the metering point. However, it is a useful supplementary method of assessing mean consumption if reasonably leak-free conditions can be assured.

Statistical analysis techniques can be used to define the size of the sample sets. However, because the per capita consumption varies so greatly the statistical standard deviation of samples is seldom below 30 lcd and often higher. Samples sets would need to be very large to get a useful degree of accuracy in the mean, for example, with a standard deviation of 30 lcd, a sample size of 865 is needed for 95% probability that the population mean lies within ±2 lcd of the sample mean. To test-meter such a large sample is impracticable. If, alternatively, existing billing records are used for the analysis, the mean will include metering and billing errors, consumer wastage and leakage, and it will be difficult to know accurately the population in residence during the billing period. Further-more relying on analysis of existing billing records is inadvisable because an important purpose of test-metering is to check the validity of billings.

In the UK, some utilities have been running long term monitoring surveys usually with the agreement of the householder, often employees of the organisation, in order to establish long term trends in consumption. However, the difficulty for the utilities is maintaining a long term stable sample set, monitoring for changes in household characteristics and continuing the survey when properties change ownership.

REFERENCES

ADB (1997). *Second Water Utilities Data Book: Asian and Pacific Region.* (McIntosh, A. C. and Yniguez, C. E. Eds). ADB.

ADB (2001). *City Profile Statistics.* ADB.

Aquacraft (2000). *Seattle Home Water Conservation Study.* Aquacraft.

AwwaRF (1999). *Resident End Uses of Water.* AwwaRF.

AwwaRF (2000). *Commercial and Institutional End Use of Water.* AwwaRF.

Binnie & Partners (1979). *Report on Egyptian Provincial Water Supplies.* Binnie-Taylor.

Census UK (2005). *Census, Key Statistics for Local Authorities, 2001.* Office of National Statistics.

EEA (1999). *Environmental assessment report No 1 "Sustainable water use in Europe".* EEA.

IWA (2000). *The Blue Pages, Losses from Water Supply Systems and Recommended Performance Measures.* IWA Publishing.

IWA (2004). *International Statistics for Water Services.* IWA Publishing.

Lambert, A. O., Myers, S. and Trow, S. (1998). *Managing Water Leakage*, Financial Times Energy.

Lambert, A. O., Brown, T. G., Takizawa, M. and Weimer, D. (1999). A Review of Performance Indicators for real losses from water supply systems. *J. Aqua,* **48**. IWSA (now IWA).

Liemberger, R. and Mckenzie, R. (2005). Accuracy Limitations of the ILI—Is it an appropriate Indicator for Developing Countries. *Proc. IWA Leakage Conference*, IWA, September 2005, Halifax, Nova Scotia, Canada.

Metaxas, S. and Charalambous, E. (2005). *Residential price elasticity of demand for water*, Water Science & Technology: Water Supply, **5**(6) IWA.

Ofwat (2005). *Water and sewerage service unit costs and relative efficiency, 2003–04 report*. Ofwat.

Ofwat (2007a). *Security of supply, 2006–07 report*. Ofwat.

Ofwat (2007b). *International comparison of water and sewerage services, 2007 report covering period 2004–05*. Ofwat.

Ofwat (2007c). *Water and sewerage charges 2007–08 report, Sample a bills for businesses and households supplementary tables*. Ofwat.

Ofwat (2008). *Service and Delivery, performance of the water companies in England and Wales 2007–08 report, Supporting information*. Ofwat.

PWC (2005). *Annual Report 2003/04*. PWC.

PUB (2005). *Towards Environmental Sustainability*, Chapter 2. Public Utilities Board, Singapore.

PUB (2007). *Key Environmental Statistics*. Public Utilities Board, Singapore.

Roberts, K. F. (1986). Requirements for Trial Programme. *Proc. WRc Seminar on Water Metering*, April. WRc.

Scottish Water (2005). *Public Water Supplies in Scotland, 2003–2004, Annual Water Resources Survey*. Scottish Water.

Smith, A. (1992). The Reduction of Leakage Following Water Metering on the Isle of Wight. *J IWEM*, p. 516.

Sphere Project (2002). *Humanitarian Charter and Minimum Standards in Disaster Response*. OXFAM Publishing.

STW (2007). Personal communication. Severn Trent Water.

TWU (2007). Personal communication. Thames Water Utilities Limited.

UKWIR (2001). *Understanding Burst Rate Patterns of Water Pipes*. UKWIR.

UKWIR (2006a). *Critical Review of Relevant Research Concerning the Effects of Charging and Collection Methods on Water Demand, Different Customer Groups and Debt*. UKWIR.

UKWIR (2006b). *Natural Rate of Rise in Leakage*. UKWIR.

UN (1956). *Methods of Population Projections by Sex and Age*. UN.

UNESCO (2006). *Water a Shared Responsibility*. UNESCO.

US Census Bureau (2005). *Current Population Reports*. US Census Bureau.

US EPA (1998). *Water Conservation Plan Guidelines*. US EPA.

Water Corp. (2003). *Domestic Water Use Study in Perth Western Australia, 1998–2001*. Water Corporation, Australia.

WEDC (1998). *Guidance manual of water supply and sanitation programmes*. WEDC for DFID.

White, S., Milne, G. and Reidy, C. (2004). End use analysis, issues and lessons. *Water Science & Technology: Water Supply*, **4**, (3).

WHO (2000). *Global Water Supply and Sanitation Assessment Report*. WHO/UNICEF.

WHO (2003). *Domestic Water Quality, Service Level and Health*. WHO.

WRc (1994). *Managing Leakage*. WRc.

WSA (1993). *Water Metering Trials: Final Report*, p. 67 Water Services Association.

Water Supply Regulation, Protection, Organisation and Financing

DEVELOPMENT, REGULATION AND PROTECTION

2.1 CONTROL OF PUBLIC WATER SUPPLIES

All governments must exercise some control over public water supply to ensure that everyone, irrespective of income, receives an adequate supply of water that is fit for human consumption. Regulations will usually include service level provisions that aim to ensure that access to water for basic needs is available to the whole population. The quantities of water to be provided are discussed in Chapter 1; service levels are discussed in Section 14.2. The requirement for water to be fit for human consumption embraces physical, chemical, bacteriological and aesthetic qualities. Aesthetic criteria are used so that other unsafe supplies are not used for drinking and cooking.

Controls affecting water supplies are usually set out in legal enactments passed by national or state governments or imposed by law making organisations representing groups of countries, such as the European Union. Controls in developed regions may be very sophisticated and wide ranging but those in poor regions may need to be restricted to the essential elements for minimum levels of service and a *'safe'* supply.

To ensure a safe supply many governments require their water supply authorities to comply with the water quality standards recommended by the WHO as a minimum. Some countries set extensive national water quality standards, which are legally enforceable, and systems of inspection may be used to ensure they are achieved. However, compliance with such standards is expensive; consequently countries with low per capita incomes may need to limit national standards to those that require the water to be free of bacteriological contamination and substances obviously injurious to health.

Since all persons need wholesome water, it may have to be supplied to low income households at a charge below the cost of production, or free of charge via standpipes set up in the street.

2.2 CONTROL OF ABSTRACTIONS

Governments also have to control abstractions of water from natural sources and may need to prioritize use for different purposes such as domestic, commercial, industry and agriculture. In a world of limited resources abstractions must make best use of the available water and do as little harm to the environment as possible. Far sighted governments will want to ensure that abstractions are sustainable and may need to limit 'mining' of fossil water.

Irrigation and impounding for hydropower can have a major impact on stream flows, sediment transport and water quality. Where catchments cross frontiers, numerous international (usually bi-lateral) agreements on impounding and abstraction have been made. Examples are: the 1892, 1956 and 1970 treaties affecting the regulation of the river Rhine; the 1978 Great Lakes Water Quality Agreement; the 1944 Treaty on the Utilization of the River Colorado and the 1926 USSR–Persia Frontier Water Convention (Teclaf, 1981).

It is clear that water resources have long been a potential source of dispute between neighbouring countries. With ever-increasing demand due to population growth, pressure on shared resources is bound to increase. In 1997 the UN adopted the Watercourse Convention, which resulted from nearly 25 years of work by the UN International Law Commission (Eckstein, 2005). The purpose of the convention is to provide a framework for international agreements. It is aimed principally at surface water courses but also covers linked aquifers that discharge to a terminus in common with the associated watercourse. In this respect the Convention recognizes the 'drainage basin' concept enshrined in the 'Helsinki Rules' (ILA, 1966). However, certain types of aquifer are not covered. Examples are the Nubian Sandstone aquifer beneath Chad, Egypt, Libya and Sudan and the 'Mountain Aquifer', recharged in largely Palestinian territories and becoming 'confined' as it flows under Israel.

2.3 PUBLIC WATER SUPPLIES IN THE USA

Public water supplies in the United States are characterized by numerous small supply systems as shown in Table 2.1. The majority of small systems are privately owned; the larger systems

Table 2.1 Number of water utilities in USA in mid 1993

US EPA designation	Population served	No. of water utilities		Total served (million)	
Very small	25–500	36 515	(62%)	5.569	(2%)
Small	501–3300	14 516	(25%)	20.053	(8%)
Medium	3301–10 000	4251	(7%)	24.729	(10%)
Large	10 001–100 000	3062	(5%)	85.035	(35%)
Very large	over 100 000	326	(1%)	109.797	(45%)
Total		58 670		245.183	

Source of information: AWWA, 1997

are predominantly municipal and publicly owned. The share of public ownership of all systems increased from 20% in 1830 to 50% in 1880 and 70% in 1924 (Melosi, 2000). The physical isolation of many small communities in a large country is partly a cause of this fragmentation, but an additional factor (Okun, 1995; MacDonald, 1997) is that developers have often preferred to site new residential communities outside the limits of major urban areas, where land is cheaper and property taxes lower; individual small water supply systems have thus been established instead of connecting the new communities to the nearest existing system.

2.4 PUBLIC WATER SUPPLIES IN MAINLAND EUROPE

The manner of development of public water supplies across Europe was influenced by the interplay between differing administrative and legal systems (Newman, 1996) and by differing cultures. Centralising influences include monarchies, Napoleon and communism. Decentralising influences include federal governments and the strength of municipalities who were instrumental in water supply development in most European countries. In France this interplay led to the operation, by a few large companies, of a large number of supplies owned by municipalities. This contrasts with a process, more widespread elsewhere, of transfer of private commercial water supply systems to public ownership in order to better control quality and meet the needs of all citizens. Private participation in public water supplies in Europe has a long history from early commercial enterprises, to long term concessions in several large cities such as Barcelona and to privatisation in UK and Eastern Europe.

With a few exceptions, public water supplies to large cities in Europe commenced between 1850 and 1900, in some cases using private companies. Initial drivers included fire fighting, industrial demand and public health. By about 1950 most remaining supplies in private ownership had been taken over by municipalities. In Eastern Europe water supplies were generally nationalized between 1945 and 1950 but were re-privatized or transferred to municipal ownership around 1995. Privatisation of some city water supplies occurred generally between 1990 and 2005 but some cases of transfer back to municipal ownership have occurred, for example Grenoble in 2000 (Juuti, 2005).

Members of the EU, 27 states in 2008, have to comply with the same European legislation (EC directives) applying to UK and described elsewhere in this chapter. The institutions responsible for different aspects of water supply and environmental legislation vary widely.

2.5 PUBLIC WATER SUPPLIES IN SCOTLAND AND NORTHERN IRELAND

A single Scottish Water Authority is now responsible for some 200 water supply and 230 sewerage undertakings in Scotland. It has the duty of promoting the conservation and effective use of water resources, a duty previously held by the Secretary of State for Scotland (Anderson, 1997). Unlike the previous publicly owned water undertakings in England and Wales before privatization, the publicly owned Scottish Water Authority has been given powers to seek private (i.e. commercial) finance for capital investment projects. This accords with the UK Government's Private Finance Initiative which began to be widely used in the education and health sectors in 2000.

A Scottish Environment Protection Agency (SEPA) was set up to take over the work of the previous ten River Purification Authorities (RPAs), in controlling abstractions, authorising effluent discharges and monitoring their quality. From 1 April 2008 Scottish water supplies are to be split between supplier and retailers (Section 2.16).

In Northern Ireland, water and sewerage services were transferred in 1996 to the Northern Ireland Water Service, which acts as a separate government Agency under the Department of the Environment in Northern Ireland (DoE (NI)). The Agency's headquarters in Belfast carries out overall control and planning etc., whilst four Divisions perform the day-to-day operational functions. The intention was that the four Divisions would be privatized, but the change was not made. Control of pollution and effluent discharges comes under the Environment & Heritage Service of the DoE (NI), as set out in the Water (Northern Ireland) Act 1972. DoE (NI) consent has to be obtained for all effluent discharges to inland and coastal waters.

2.6 WATER SUPPLY ABSTRACTION AND REGULATION IN THE ENGLAND AND WALES

In the UK up to about 50 years ago the building of water abstraction works for public water supply had be to sanctioned by a Private Act passed by Parliament, so called since it gave powers to a particular (i.e. 'private') body. This procedure enabled interested parties to lodge objections. A parliamentary committee would hear the arguments and approve or modify the proposed works. This procedure was later modified to enable the government to obtain an 'Order' authorising the works after achieving agreement between all the parties involved.

This unplanned approach and the demand for more waterworks led to the setting up of 26 River Authorities in England & Wales under the 1963 Water Resources Act. The River Authorities had control over use of water by licensing all water abstractions in each river catchment (except those for domestic use in a single dwelling). A National Water Resources Board allowed development of inter-basin transfer schemes such as those implemented in the North-west and East of England. In 1973 ten new 'multipurpose' regional *Water Authorities* were set up to cover England and Wales. Each Water Authority took over all the water supply undertakings in its area (except specially licensed water companies), plus all the sewerage works belonging to the local authorities, and all the duties of the previous River Authorities. The authorities had powers to licence all water abstractions and sewage and wastewater discharges in their basins and therefore, were able to tackle growing river pollution. However, with this decentralisation the ability to plan strategic use of water for national benefit was reduced.

2.7 THE 'PRIVATISATION' OF WATER IN ENGLAND AND WALES

The success of the Water Authorities created in 1973 was limited by Government restrictions on borrowing by the public sector, imposed to contain inflation. This made it difficult to continue to improve many of the sewerage systems they had taken over and to meet more rigorous public water supply quality requirements set by a 1980 EC directive. The solution adopted was to 'privatize' the Water Authorities by changing them to private companies. This would move their large outstanding debts and future borrowings out of the 'public sector'. The debts would become the debts of private companies.

The privatisation process overcame three difficulties. A new body, the *National Rivers Authority,* was set up in 1989 to take over public duties that posed a potential conflict of interest—licensing of water abstractions and wastewater discharges, the control of pollution, inland navigation, fishing and amenity protection, flood protection, land drainage and coastal defences. A *Director General of Water Services,* with a department known as the Office of Water Services (Ofwat), was appointed to control charges to consumers and to set service levels. On the other hand, investment in the industry, where organic growth prospects were limited (99% of householders already had a water supply and 96% were connected to the sewerage system (WCA, 1992)), was made more attractive by allowing formation of *'Holding Companies'* who would own the Water Service Companies and also be able to undertake other sorts of commercial activity. The result was a successful stock market flotation of shares in the new Holding Companies.

After privatisation under the UK Water Act 1989 four 'Consolidation Acts' were passed to tidy up the legislation and avoid the need to refer to previous Acts. The four Acts were as follows:

- The Water Industry Act 1991.
- The Water Resources Act 1991.
- The Statutory Water Companies Act 1991.
- The Land Drainage Act 1991.

The Director General (DG) had powers to control charges to consumers and could require the improvement of water and sewerage services given to consumers. The charges were to be set for 5 year periods and charges from one period to the next could not be increased by more than 'the current rate of inflation $\pm$ K', known as 'the K-factor'. It was expected that rising 'real' costs would be partially or wholly offset by efficiency gains, which were to be assessed by comparison between companies, using information provided by the companies in periodic reviews.

2.8 EXPERIENCE WITH WATER PRIVATISATION IN UK

Released from government constraints on borrowing, the new Water Service Companies were able to raise finance and progress with many needed improvement works. However, the Holding Companies had mixed success with other commercial ventures (Isack, 1995; CN, 1994; NCE, 1994). After a period of experimentation they settled with their core activities, plus a few related activities such as solid waste disposal or acting as consultant to overseas water undertakings.

In a drive for profit in a restricted regime, some Holding Companies transferred their design staffs and laboratories to separate subsidiary companies, using these and equipment supply companies they had purchased to supply goods and services they needed. The DG of Water Service decided such supplies had to be put to open competitive bidding from others and that a subsidiary could not be favoured in the bidding process.

With the 'K' factor derived from information provided by the companies, it was inevitable that such information was coloured to produce results favourable to the companies. The DG had to audit these submissions using a team of inspectors, so as to control the excessive profit that could result. In some cases of serious distortion, Plcs agreed to give rebates to customers; in other cases the DG imposed penalties such as requiring additional expenditure to reduce leakage.

Overall, privatisation has worked, but it has been difficult for the regulator to respond to a continual stream of developments, including take-over of some Holding Companies by large foreign firms whose other subsidiaries provide goods and services commonly used in the water industry. Under commercial, regulatory and political pressures the industry continues to change. Dŵr Cymru/Welsh

Water converted to non-profit making status in 2001 but a similar application by Yorkshire Water was turned down by the DG of Water Services.

Under the Water Act 2003 (Section 2.10) the office of the Director General of Water (Ofwat) was renamed as the *Water Services Regulation Authority (WSRA)*. However, in the UK and overseas it continues to be referred to and known as Ofwat.

2.9 FUNCTIONS OF THE US ENVIRONMENTAL PROTECTION AGENCY

US EPA sets drinking water quality standards under the Safe Drinking Water Act (SDWA) 1996. It can issue Regulations stipulating:

- an MCLG (maximum contaminant level goal) which is not mandatory;
- an MCL (maximum contaminant level) which is mandatory;
- a 'Treatment Rule' which is mandatory and which sets out the type(s) of water treatment to be adopted where an MCL is not appropriate or sufficient (Section 6.65) and for minimising the incidence of disinfection by-products.

An 'MCLG' is defined under the 1996 Act as—"a level at which no known or anticipated adverse effect on human health occurs and that allows for an adequate margin of safety". An 'MCL' is defined as—"a level as close to the MCLG as feasible". 'Feasible' is defined as—"practicable according to current treatment technology, provided no adverse effect is caused on other treatment processes used to meet other water quality standards" (Pontius, 1997). MCL values under the 1996 Act are shown in Table 6.1.

Under the previous Safe Drinking Water Acts of 1974 and 1986, US EPA had issued Regulations covering some 85 organic and inorganic substances, and 16 others covering radionuclides and microbial levels. US EPA had not been required to assess either the cost of implementing a regulation or the health benefits it would achieve. This gave rise to widespread criticism that some costs had been imposed on public water suppliers that could not be justified by the incidence of such contaminants or the numbers of persons at risk.

Under the 1996 Act US EPA is now required to support any proposed new Regulation by publishing a report on:

- the risk the contaminant presents to human health;
- the estimated occurrence of the contaminant in public supplies;
- the population groups and numbers of persons estimated to be affected by the contaminant;
- the benefits of reduced risks to health the proposed Regulation should achieve;
- the estimated costs to water utilities of implementing the Regulation;
- estimated changes to costs and benefits of incremental changes to the MCL value proposed;
- the range of uncertainties applying to the above evaluations.

US EPA has to publish and keep updated every five years a list of contaminants likely to require regulation, choosing at least five of them to Regulate every five years and giving priority to those posing the greatest health risk first. A period of 18 to 27 months for public comment and US EPA's consultation with certain authorities must be allowed before a proposed Regulation containing an MCL is promulgated i.e. formally put into operation.

To follow these new procedures US EPA had first to collect country-wide data revealing the incidence of various types of contaminants in public water supplies and the effectiveness of current treatment processes in reducing such contaminants to required levels. All systems supplying a population of 100 000 or more were required to submit detailed reports on the results of 18 to 24 month comprehensive sampling programmes, with lesser sampling programmes being required for smaller systems serving upwards of 10 000 population. Also to help small water systems serving 10 000 population or less, the 1996 Act required US EPA to provide a list of treatment techniques to help them achieve the required quality standards. US EPA must also provide small systems with 'variances' which permit the adoption of treatment processes that will achieve nearest compliance with an MCL, taking into consideration a system's resources and the quality of its source water. Government funds made available to States are to be used to assist in training waterworks operators to standards set by US EPA. The funds are also intended to help these small systems obtain technical assistance with water quality compliance problems and source protection.

The radical changes made by the 1996 SDWA should achieve a more realistic approach to setting of water quality standards, and make it more practicable for utilities to comply with them. However, the performance of the many small utilities may continue to present a weakness, unless they can be provided with sufficient day-to-day technical and laboratory assistance to ensure they achieve compliance with the many sophisticated drinking water standards in force. The evaluation procedures US EPA must undertake to support any proposed new regulation are complex and may raise questions concerning the methods of approach to be adopted, which may need clarification and agreement.

2.10 THE ENVIRONMENT AGENCY (UK)

In the early 1990s it became increasingly evident there was a need to protect the whole environment from all forms of pollution. Therefore, the Environment Act 1995 set up the EA for England and Wales to take over the duties of the National Rivers Authority but with much wider powers. The Agency's three main functions in relation to the use of water resources were:

- licensing all water abstractions and discharges of wastewater;
- preventing, controlling and reducing the pollution of all waters;
- undertaking flood protection and coastal defence measures.

Other duties were promoting the environmental and recreational benefits of water, fisheries protection, land drainage, controlling inland navigation and complying with special protection measures required by the Government in any area designated as a 'water protection zone'. These provisions applied to all water in rivers, lakes and underground and also estuarial and coastal waters three miles out to sea.

To cater for a wider diversity of abstraction licence situations than had been foreseen in the 1995 Act, the Water Act 2003 was passed. Part 1 of this extended the water abstraction licensing powers of the EA. The most important of the changes affecting water and sewerage services were the following (the figures in brackets showing the Sections of the 2003 Act referred to):

- The EA has to issue separate licences for abstraction and for impounding (3)(12), and can require alteration of impounding works if they cause environmental damage (4).
- It can vary a licence for impounding in a way that requires the impounding works to be modified (22).

- It can transfer an abstraction licence from one water utility to another, the latter compensating the former (26).
- After consultation with the WSRA, the EA can propose that a bulk supply agreement between water utilities be entered into where necessary for the proper use of water resources (31).
- Water undertakers have a duty to conserve water resources (82) and must produce water resource management plans (62) and drought plans (63). They must fluoridate water supplied if the appropriate health authority requests it (58).
- The WSRA must protect the interests of water and sewerage customers by promoting effective competition for supply of services required by the Water Service Companies (39), and statutory water utilities must disclose if they link the remuneration of the directors to standards of performance (50).

A further addition to the Agency's responsibilities was made in 2003 by The Water Environment (Water Framework Directive) Regulations (SI 2003/3242). Under these, the EA was appointed the 'competent authority' to carry out the requirements of the EC Water Framework Directive 2000/60.

2.11 THE EUROPEAN WATER FRAMEWORK DIRECTIVE

The Water Framework Directive (2000/60/EC) requires Member States of the European Union (EU) to control, protect and improve the water resources of the State. Member States had to define appropriate river basin districts within their country boundaries, and set up 'a competent authority' for each. For river basins that cross national boundaries, international river basin districts must be set up. In England and Wales, Regulations SI 2003/3242, which came into force on 2 January 2004, require the EA and the National Assembly for Wales to undertake the duties imposed by the EC Directive. In Scotland implementation of the Directive is the responsibility of the Scottish Environment Protection Agency (SEPA), and in Northern Ireland the responsibility of the Northern Irish government.

For each river basin district the EC Directive required base information to be identified and reported to the EC Commission. All forms of water had to be assessed—rivers, lakes, 'artificial surface waters' such as reservoirs, underground water, together with basin-related estuarial waters (termed 'transient waters') and coastal waters one nautical mile out to sea.

A programme of measures (River Basin Management Plan—RBMP) had to be drawn up for each river basin district in order to achieve the Directive's environmental objectives by 22 December 2015. The RBMPs have to be operational by 2012. The objectives are to achieve 'Good status' for surface waters and groundwater within each basin, as defined in Tables set out in Annex V to the Directive. For surface waters the biological quality, hydro-morphological quality and physico-chemical quality (inclusive of synthetic and non-synthetic pollutants) are stipulated. For groundwater, 'Good status' comprises achieving a balance between abstraction and recharge and preventing or limiting the ingress of pollutants.

The provisions of the Directive directly affecting water and sewerage suppliers are:

- adoption of water-pricing policies which provide incentives for users to use water resources efficiently;
- basing water charges on "the principle of recovery of the costs of water including environmental and resource costs";

- setting charges for wastewater disposal "in accordance with the polluter pays principle";
- conducting an economic analysis to decide the most effective combination of water abstraction and wastewater disposal measures that should be adopted.

Compliance with the requirements of the Directive may pose substantial administrative cost on member states, especially those that have not hitherto managed water resources by river basin, or do not already have a national organisation responsible for the water environment. In the UK the use of water resources has been planned on a river basin basis since 1973; consequently considerable information is already available to meet the Directive requirements, though perhaps not to the extensive detail required by the Directive. Draft RNBPs published in December 2008 for 10 basins in England and Wales show £26.5bn of cost and £2.8bn of benefit although much of the cost would have arisen in any case.

2.12 OTHER POLLUTION CONTROL MEASURES IN THE UK

The *Drinking Water Inspectorate* (DWI), formed in 1990 after the privatisation of the English and Welsh water industries (Section 2.7), plays a key role in England (under Defra) and Wales ensuring that the consumer receives a wholesome water supply that complies strictly and continuously with the many quality requirements described in Chapter 6. A similar body regulates water quality in Northern Ireland. The role of the DWI is to monitor water suppliers' performance, to give guidance on best practice in the use of reliable quality monitoring procedures and to advise where new or improved techniques need to be adopted.

HM Inspectorate of Pollution (HMIP) was first set up under the Control of Pollution Act 1974 and its scope was widened under the Environmental Protection Act 1990. It now forms part of the EA and deals more especially with controlling the discharge of industrial wastes (solid, liquid and gaseous), including trade wastes discharged to public sewers. Several EC directives set limits for the discharge of certain dangerous substances to the aquatic environment. In response, the UK Defra maintains a 'Red-List' of substances needing priority control with the objective of reducing discharge loads to the absolute minimum. Except for the metals mercury and cadmium, all the Red-List substances are organic compounds, mostly pesticides and herbicides; although some are particular wastes from certain types of industry. Two different quality standards have to be applied to the discharge of wastes to surface waters—the more stringent governs:

- an 'environmental quality standard' (EQS), or
- 'best available technology not entailing excessive cost' (BATNEEC)—sometimes alternatively denoted as a 'uniform emission standard' (UES).

An EQS is set by estimating the effect the discharge of a substance has on the environment; this involves research and monitoring of such discharges by the EA, followed by the setting of an appropriate EQS by Defra. For the BATNEEC or UES standards, HMIP approaches an industrial polluter who is producing or discharging a Red-List substance to assess what process should be applied to render it suitable for discharge and to set limits for the amount discharged. Neither the relative volumes of the discharge and the receiving water nor the existence of other discharges of similar substances are taken into account, since receiving waters have no acceptable assimilative capacity for Red-list substances. HMIP has to deal also with radioactive wastes under the Radioactive Substances Act 1993.

Agricultural pollution forms another area of pollution control. 'Point source' pollution can occur from slurries, silage effluents, yard washings and vegetable processing wastes. These all have high BOD (biological oxygen demand), silage effluents exceptionally so, and the slurries and yard washings have high ammonia content as well. Guidance to farmers and grants for measures taken are provided by Defra. Its predecessor, DoE, published a *Code of Good Agricultural Practice for the Protection of Water* in 1991. The legal requirements on farmers are set out in the *Control of Pollution (Silage, Slurry and Agricultural Fuel Oil) Regulations 1991* (Statutory Instrument No. 324, 1991). Many treatment methods have been tried on farm wastes but the simplest of them is to store slurries and sludge drainage for nine to twelve months in open ponds, after which they can be sprinkled evenly over grassland in the right weather conditions, avoiding any direct runoff to a watercourse (Barker, 1991).

'Diffuse pollution' (not arising at a point) of the environment occurs from the use of pesticides (including herbicides) and therefore control has to be applied to the usage. Not all such pollution comes from farming. Increasing detection of the presence of the herbicides atrazine and simazine in groundwater was attributed mainly to their use by road and rail authorities. Dealing with the many types of pesticides and herbicides used is a complex matter, as illustrated by the fact that Defra lists about 400 approved compounds. Some compounds have been banned, including DDT, aldrin, dieldrin and chlordane. An added complication is that farmers rotate the use of different pesticides from year to year to avoid build-up of resistance; this increases the difficulty of monitoring for such contaminants.

Defra designates 'nitrate sensitive areas' within which farmers are assisted to adopt practices designed to limit, and possibly reduce, the amount of nitrates found in water. EC Directive 91/676 made a similar policy mandatory on member states, stipulating the need to identify 'vulnerable zones' where underground or surface waters used for drinking would contain more than 50 mg/litre nitrate if no protective action were taken. Among the practices required by the EC Directive are measures to limit the application of nitrogenous fertilizers and manures and to prohibit the application of nitrogenous fertilizers during certain periods of the year.

Solid waste disposal sites are, in the first place, the responsibility of the County Council Waste Regulation Authorities. However, before licensing a site a County Authority must obtain agreement of the EA and of any water undertaking likely to be affected. Compliance with EC Directive Protection of groundwater against pollution caused by certain dangerous substances (EC Directive 80/68) is required. This prohibits certain substances from waste disposal landfill sites entering groundwater and limits the amount allowable for others; it has led to use of membrane seals under fill.

The substances prohibited are: cadmium, cyanides, mercury; mineral oils and hydrocarbons; organohalogen, organophosphorous and organotin compounds; and those possessing carcinogenic, mutagenic or teratogenic properties.

Substances that must be limited (although no limit criteria are stated) include: the inorganic substances normally limited in drinking water supplies e.g. arsenic, chromium etc. (Table 6.1(A)) plus biocides, taste or odour producing substances and toxic compounds of silica.

Contaminated land has to be identified and registered by the local authorities who must inform the owner thereof and the EA. The local authority can require the owner to carry out remediation of the land or can undertake remediation itself, charging the cost to the owner. If, however, the EA decides the land is a 'special site', it takes over responsibility for enforcing or undertaking the necessary remediation measures. A 'special site' as defined by Defra is one which would or might cause serious harm, or serious pollution of 'controlled waters', as defined in the Water Resources Act 1991, Section 104.

ORGANISATION

2.13 ORGANIZATION OF A WATER UTILITY

The organisational structures of water utilities are increasingly diverse, depending on:

- the size of the utility;
- its spread of other responsibilities which could include wastewater disposal, highway and storm drainage, flood defence and regulation of discharges by others;
- how it procures its services; and
- how 'vertically integrated' it is.

It is now rare for water utilities to be responsible for flood defence from rivers and the sea and for them to regulate discharges direct to water courses or the sea by others; generally these duties remain a part of central government. The description below is limited to those water utilities that are responsible for water supply to, and possibly wastewater disposal from, both domestic and non-domestic customers.

There is wide diversity of water utilities in Europe, particularly in the UK, driven by European Union legislation on competition. In the UK most companies remain vertically integrated: that is they are responsible for the supply of all water related services from water resource development to customer billings. However, under the Water Services Scotland Act 2005, water supply in Scotland was split (on 1 April 2008) into the supplier of water (the wholesaler) and organisations responsible for selling and billing the customer (the retailer). The wholesaler remains a regulated monopoly while several retailers compete for the business of providing the customer interface, including billing. In England and Wales competition is developing in a different way. Here water distribution remains the responsibility of the regulated monopoly, but new entrants can develop resources and find customers, agreeing a transportation fee with the distributor. If the two parties cannot agree the transfer fee then the matter is adjudicated by the regulator. In England the development of competition has been proceeding slowly with competition currently being limited to larger non-domestic supplies. However, this may change as the regulator is currently suggesting that within the near future all supplies to non-domestic customers should be open to competition.

In the past water undertakers carried out most duties in-house. However, increasingly, many duties are outsourced. It is now common to see engineering design and construction together with scientific services such as water quality sampling being outsourced. In one extreme case in the United Kingdom all basic duties have been outsourced with the water utility retaining only a small core of senior managers and regulatory personnel in house.

However, the water utility wishes to procure its services, the same functions are needed and are likely to cover the following main areas:

The Board is responsible for policy and corporate governance. The Board will be assisted by an audit committee that undertakes internal audits to ensure that correct corporate governance is being followed.

The Chief Executive and Executive Directors are responsible for implementing the Board's policy. They are therefore responsible for the day to day running of the utility.

Operations are responsible for the technical operations of the undertaking. Water supply operations are likely to be split into a number of departments such as water resources, water treatment, water distribution, leakage control, emergency planning and response and asset maintenance. Where relevant there may be equivalent wastewater responsibilities.

Customer Services generally comprise two parts: customer billing and customer contacts. As indicated by its title the former will prepare and issue bills, deal with customer queries and manage debt recovery. The latter responds to all non-billing related customer contacts. These are generally of a technical nature such as interruptions to supply, water quality and flooding from sewers. Customer Services therefore work closely with operations to rectify technical deficiencies that impact the customer. A major part of customer services is the call centre, which receives telephone calls from customers. A well run call centre is a vital part of any water utility as it is where it interfaces most directly with the customer. Where water supply connections are metered meter reading is the responsibility either of customer services or of the billing section.

Strategic, asset management and capital investment planning. The water industry is capital intensive with large programmes of capital investment. The water utilities' assets have to be managed through the asset life cycle of:

> **Appraise → Design → Procure → Construct/install/replace → Operate and Maintain → Appraise**

Apart from the operational life of the asset, all other parts of the life cycle are the responsibility of the capital investment department. Many of the detailed aspects of the department can be outsourced but key functions, such as overall programme management, strategic planning and client project management, are likely to be retained in house.

Scientific Services comprise water and wastewater sampling and laboratory testing. This is a key function since the maintenance of water and wastewater quality standards are central to the responsibilities of any water undertaking.

These main functional areas will be assisted by a number of support departments, such as: Finance and Accounting; Human Resources; Payroll; Regulation; Health and Safety.

2.14 STAFFING LEVELS

Water utility staffing levels expressed as a number per property served depend on many of the considerations covered in Section 2.13. Staffing levels are influenced by many factors including the size and characteristics of the company supply area, the level of service provided to the consumer, staff salaries, work culture, organisational structure and the extent of control and optimization. Typical staffing levels for water utilities in the United Kingdom range between 0.7 and 1.6 employees per 1000 connections. These levels represent the situation where the majority of the capital programme is outsourced to external designers and contractors but all other functions remain in-house. International staffing levels range between 1.5 and about 3.0 employees per 1000 connections for utilities with similar characteristics to UK companies. International utilities that utilize less sophisticated technology, deliver a lower level of service or intermittent supplies, or where labour costs are low, tend to operate with higher staffing levels, typically between 2 and 20 employees per 1000 connections.

PROJECT APPRAISAL AND FINANCING OF CAPITAL WORKS

2.15 APPRAISAL REQUIREMENTS

For a water supply project involving capital works to proceed, it has to be demonstrated that:

1. There is a need for the project; this would be demonstrated by study of the driving factors, for example demand, quality and levels of service, public health and efficiency gains.
2. Where government regulations or financing institution rules require, the project is justifiable on economic and environmental grounds at the appropriate level—national, regional or sector—and that the selected option is the most favourable.
3. That any loans required for financing the work can be repaid.

The extent of study required depends on the owner's status (public or private), sources of finance proposed and the legislative framework applying. Aspects of some common driving factors are discussed elsewhere, for example: demand (Chapter 1); water quality (Chapter 6); service levels (Chapter 13); energy efficiency (Chapter 17).

Economic and financial appraisals are discussed in the following sections. The reader will find useful background in *An introduction to engineering economics* (ICE, 1969) and is referred to literature published by agencies such as the UN and World Bank. The ADB *Guidelines for the Economic Analysis of Water Supply Projects* (ADB, 1998) and the *Handbook* on the same subject (ADB, 1999) cover the principles and methods involved, as well as specific ADB requirements. Some of the key terms used in economic and financial appraisal are explained in Table 2.2.

Table 2.2 Project appraisal terms

Term	Meaning	Explanation/use	Section
	Least cost analysis	Selection of option with lowest IEC. Can be determined using NPV, IRR, EDR or payback period methods.	
(A)IEC	(Average) Incremental Economic Cost	Method of comparing options as alternative to NPV or IRR. It provides information on long-run marginal costs—useful in tariff determination.	2.21
NPV[a]	Net Present Value	The value today of a future cash flow. Determined by discounting future cash inflows (revenue) and outflows (construction and operation and maintenance costs). A simple method but it provides no information about unit cost of water.	2.16
IRR[b]	Internal Rate of Return	The discount rate that results in zero NPV. It should be greater than the minimum return required on capital (OCC). Requires more calculation than NPV.	2.19
EDR	Equalising Discount Rate	For comparison of two options by finding the discount rate at which preference for one changes to the other. Requires more calculation than NPV.	
	Payback period	Length of time from initial capital outlay (for construction) to when revenue equals that outlay.	

(continued)

Table 2.2 (*continued*)

Term	Meaning	Explanation/use	Section
OC(C)	Opportunity Cost (of Capital)	Value of commodity (such as capital or water) if used for other purposes than the project.	
	Shadow price	Price that reflects scarcity or real value of a resource.	2.16
r	Discount rate	The return which capital could obtain on open market investment.	
STPR	Social Time Preference Rate	Rate for discounting given in Green Book (HM Treasury, 2003)	2.16
	Cost—benefit analysis:	Comparison of costs and benefits using constant:	
	• Financial	• financial (actual) prices	
	• Economic	• economic (true or real) prices after (i) removing distorting effects such as subsidy, shortage, artificial exchange rates and (ii) allowing for opportunity costs.	
NEB	Net Economic Benefit	Economic benefit for each year (for use in discounting).	
NFB	Net Financial Benefit	Financial benefit (revenues less costs) for each year (for use in discounting).	

ENPV = Economic NPV;
[b]EIRR = Economic IRR; FIRR = Financial IRR (shows profitability).

In a world of finite resources development of water supplies may lead to long-term depletion of a resource such as an aquifer. Such an activity is not sustainable by definition but may be justified as a short term expedient. Economic appraisal of a project that produces resource depletion should take account of the impact it would have. One way to do this is to apply a 'depletion premium', which would allow for the cost of replacing the resource once it is depleted (ADB, 1999).

Many of the variables used in project appraisal cannot be accurately foreseen. For this reason studies usually include analyses of sensitivity of the conclusions to possible differences (for example in price) and of risks (for example to construction period or customers' ability to pay). The sensitivity analyses should be taken into account in any decision making.

2.16 ECONOMIC COMPARISON OF PROPOSED CAPITAL PROJECTS

Discounting to compare present values of projects. If two or more different projects are possible to meet an expected demand for more water, they will almost certainly differ in cost and probably also in the timing of the capital outlay required. They may also differ in the balance of capital to operating cost. One way of comparing them is by 'discounting' each scheme's costs to the present, to obtain the 'total present value' cost of each project. This allows the timing of expenditure to be taken into account.

The present value P of £X due to be paid in n years' time is taken as $P = £X/(1+r)^n$, where r is the discount rate expressed as a decimal. This is the inverse of compound interest calculation, i.e. P invested at $r\%$ compound interest accumulates to £X in n years' time. To compare projects (or other options), the compared work must meet the same objectives to the same extent. An example is meeting demand increases over the same period—usually 20 or 25 years. The capital expenditure and works renewal costs, plus the running costs, for each year are estimated and discounted to give their equivalent present value cost and the total is summed for the chosen period.

The UK Government Green Book (HM Treasury, 2003) adopts the concept of the Social Time Preference Rate (STPR) and recommends a rate of 3.5%. This is a 'real' rate applicable to analyses based on current costs without allowance for inflation. The STPR is made up of two components—time preference and an allowance for real future growth in per capita consumption. Both are related to society's perception of the excess of the value of present consumption over future consumption. Sensitivity checks using values either side of the recommended 3.5% may need to be carried out. For very long periods (over 30 years) the Green Book suggests that discount rates reducing with time are appropriate. Outside UK the advice of the relevant government or funding agency should be taken on discount rates to be used. The present discount rates contrast with those adopted in the 1980s at a time of high interest rates (and inflation).

The higher the discount rate, the smaller is the 'present value' of a future cost. Hence high discount rates tend to favour schemes which can be built in stages, or which have a cheaper initial capital cost despite having higher running costs.

Inflation of prices. The treatment of inflation should be consistent. Present prices can be used with a 'real' discount rate (excluding inflation) or costs that are escalated for inflation may be taken with a discount rate that includes inflation. Nevertheless, inflation is not usually taken into account when discounting, on the assumption that it affects all prices proportionately. History, however, shows otherwise. As standards of living have increased, labour costs have inflated more than material prices due to increased machine production. Fuel oil prices doubled in 1973 and then fell back, then rose again later. Hence, possible differences of inflation should be allowed for in sensitivity analyses (Section 2.17).

Shadow pricing of costs is sometimes adopted when market prices do not represent 'true' costs. Thus, if unemployment is high, the 'shadow price' for labour is the wage paid (including on-costs) less the cost to society of that person when unemployed (e.g. the unemployment pay etc.). Taxes are excluded from shadow prices because they represent only a transfer of money from one section of society to another. Where the unofficial exchange rate implies that offshore goods and services cost more in local currency than apparent from the official exchange rate, the shadow price for purchase of offshore goods may need to be based on the unofficial rate because this represents the real cost of offshore purchases. Such shadow pricing has the benefit of favouring a scheme that uses more local labour or inshore goods than another does, an advantage to the country in which the project is to be built. However, carried to extremes, shadow pricing is complex, because each item price has to be broken down into its component parts e.g. labour, fuel, plant, materials etc. and also broken down into offshore and onshore elements, with taxes taken off. Consequently shadow pricing tends to be adopted only when a funding agency requires it, possibly for labour only.

2.17 COMMENTS ON THE USE OF DISCOUNTING

From the end of the 20th Century project evaluation has extended into a wide ranging assessment covering a number of social and environmental aspects as well as economic comparisons. The weight given to economic studies has reduced. One reason is the difficulty in accurately forecasting future costs and the balance between future costs of different types. It is possible to carry out sensitivity checks to test the impact of a number of possible variables such as discount rate, energy costs and labour costs. However, the resulting economic analysis remains only part of the evaluation process. It can help decision making but should not be the basis of it.

Economic comparisons of long term developments have been used in the past to plan sequences of investment to meet demands projected over a long period. However, the result depends on timing of the analysis. Due to the effect of discounting future costs, the earlier projects in the most economic sequence will tend to be those requiring less capital expenditure and more operating costs. If medium term expenditure plans are based on periodic updates of the economic analyses, high capital cost projects with lower operating cost are likely never to get built. This is not necessarily compatible with a desire to conserve resources, particularly energy.

2.18 SHORT AND LONG TERM MARGINAL COSTING

When a utility has spare capacity at its source works, it can meet additional demand at little extra cost because over 90 per cent of a water company's annual expenditure is fixed. Debt repayments are fixed and the operational costs involved in the day-to-day running of a water system are virtually fixed because less than 10 per cent of the cost varies with the amount of water delivered. However, if new major works have to be built, the cost of the supply rises sharply because the loan repayments on the new works have to be met immediately, whereas the initial rise in consumption is usually small and the increase is gradual. To avoid this sudden rise in charges, 'long term marginal costing' can be used, under which account is taken of the cost of the next stage (or stages) of capital expenditure required. This means that, before the new source works are actually required, consumers pay more than current costs; therefore money is set aside to fund partially the cost of the new works; price rises are thus smoothed out. The early increase in charges may induce metered consumers to curb their demand. This may postpone the date when the new works are required.

How much current consumers should be asked to pay for works that benefit future consumers is debatable; but often current consumers benefit from works still in use but long since paid for by previous consumers. Arguments about the monetary value to be put on paid-off assets still in use are only relevant in the relatively rare case when a water utility is put up for sale. However, more importantly, a utility should not become so heavily burdened by long-term loans, that it finds itself in financial difficulty if unexpected heavy financial commitments should arise. This situation occurred in England, where many new schemes had been built in the 1960s and 1970s to meet swiftly rising demand for water. As a consequence, the water authorities were already burdened with long-term debt at a time when they were required to meet new standards of water quality as well as for effluent discharges. Hence, when the UK Government decided to privatize the water authorities (Section 2.7), it wrote off £5.2 billion of their long-term debt (Kinnersley, 1994) in order to make them attractive financially.

2.19 OTHER ASPECTS OF PROJECT ASSESSMENT

The *internal rate of return (IRR)* is another way of comparing projects (Table 2.2). In this case the 'value' of 'the outputs' has to be assessed. For water supply projects it is difficult to establish an appropriate value for the supply. Tariff charges can be used but it may be more appropriate to attach a value to the service that takes account of secondary benefits such as health, environment, comfort, recreation, trade, industry and food production. One option is to relate the 'value' of domestic supply to a proportion of average family household income. The value of water used for trade or industry generally might be judged from the elasticity of the demand for it, if that can be found out (Section 1.15). Often, however, the funding or loan agency will give guidance on what value to take.

Using the 'value' (in 'real' money terms) of, say, a cubic metre of water sold, one can calculate the present value of all the water supplied over the planning horizon. Then by trial and error one can find what discount rate makes the present value in real terms of the output equal to the present value of all the expenditure needed to construct and maintain the project. That rate is 'the return on capital' or the IRR. The usefulness of the procedure to a funding agency or government is that it puts projects on a comparable basis indicating which projects give 'the best value for money' from a funding point of view.

A full *social cost-benefit analysis* tries to evaluate all the benefits and disbenefits which accrue from a project over a period, in order to compare options. When applied to a major water project, such as an impounding reservoir, it is difficult to put any clear 'value' to many of the disbenefits, such as the loss in perpetuity of the valley land covered by the reservoir. An 'environmental impact assessment', now required for most major water projects, is a more informative approach because it sets out all the results of building a project whether such results can be evaluated or not. Whatever the method of comparing the social and environmental impacts of different courses of action, it is important that both benefits and disbenefits are taken into account.

2.20 SUSTAINABILITY AND ENGINEERING CHOICES

The definition of sustainable development adopted by the UK water industry, in common with UK government, is 'development which satisfies people's 'basic' needs and provides a better quality of life without compromising the quality of life of future generations'.

Implementation of the Water Framework Directive in Europe will influence the way water is made use of in the environment and therefore will directly affect water supplies. However, the way water supply developments are carried out must include efficient use of non-renewable resources (including fossil fuels), waste minimisation and recycling. Water companies in UK have embraced the sustainability indicators established by their trade association, Water UK, which publishes an annual report on progress (Water UK, 2005) but a coherent and comprehensive approach is still lacking.

The choices made in implementing water supply developments will increasingly be influenced by sustainability criteria. However, balancing the effects on use of different kinds of resource against different forms of consequence and against cost is a matter for widely different interpretation. Making such choices in a way all can agree will require consensus on the approach and on the weight to be applied to each factor. This is an area where central government can assist.

2.21 FINANCING OF CAPITAL WORKS

A water utility usually borrows money to finance construction of major new capital works because the cost of such works is too large to meet from current income; however, it can reasonably be borne by future consumers who will benefit from the works. Before agreeing to make a loan a financial institution will need to be confident that it will be repaid. The most straightforward way to demonstrate capacity to repay a loan is by use of cash flow analysis. This will form the basis of a financial plan and will show annual expenditures, including construction and operation and maintenance costs, and revenues. It will also include inflows from loan payments and outflows on repayment. If funds are to be raised by sale of equity (share capital) receipts from the sale and future dividends would also need to be included.

Loan repayments are typically 10 to 15 years for plant and machinery; 20 to 30 years for buildings; and 50 to 60 years for dams and land. Routine capital works expenditure on extension of pipelines can normally be met from current income. Three methods of loan repayment are possible: capital plus interest, annuity and sinking fund. If a loan of £X is repayable over n years at r% interest rate (where r is expressed as a decimal) the annual amounts payable under the three different methods are as follows.

Method	Annual Payment
Capital plus interest	$X/n + r$ times the loan balance outstanding at the start of the year
Annuity	$Xr(1 + r)^n/[(1 + r)^n - 1]$
Sinking fund	$Xr/[(1 + r)^n - 1]$ to sinking fund + Xr interest on loan.

The annuity repayment method is the most usual. The capital plus interest method results in reducing yearly payments. The second and third are equivalent if the interest received on the sinking fund is the same as that charged on the loan; the third tends to be adopted only if the interest rate on the sinking fund investment is more than the interest charged on the loan.

Public water authorities may be able to borrow from government, from an international lending agency, or may be authorized to borrow from the money market. Private companies usually borrow from the money market. In theory they can also raise more money by offering further shares for sale on the market if their full share capital has not been issued; but this would rarely be adopted for new works. Of course if an undertaking has an excess of income over expenditure, part of the excess can be set aside to build up a fund to help finance new capital works. If a variety of capital projects has to be funded, short-term borrowing from current funds or from the bank may be adopted to fund the capital outlay, until the time is considered appropriate for raising a single long term loan. Much depends on interest rates prevailing and whether these are expected to move up or down in the future.

2.22 DEPRECIATION AND ASSET MANAGEMENT PLANNING

'Depreciation' is an accountancy term for the practice of writing down the initial cost of an asset annually. In company accounts the amount an asset is depreciated is debited against income, with the result that an equivalent amount of money (in cash or securities) is set aside and allocated to a depreciation fund, which would be available later to meet the cost of renewing the asset when it becomes worn out. Depreciation has been defined as representing—'the consumption of assets by current users thereof'. The amount of the depreciation depends on accountancy practice; it needs to be sufficient to write off the whole cost of the asset before it has to be replaced. However, adjustment may be needed to allow for higher cost of replacement due to inflation or 'betterment' for new standards and improved efficiency.

The technique of depreciation is particularly suited to manufacturing companies. Application to a water utility may not produce the desired result with respect to some large, long life assets, such as dams and large pipelines. These may continue in use and to present a liability long after they cease to have any 'book' value. Nevertheless, it is prudent to build up a 'contingency fund' to meet future repair or renewal costs.

Of more value in the financial planning of a water utility, is the 'asset management plan'. This requires assessing the condition of each asset and estimating its remaining life and replacement cost. However, each of these estimates is prone to errors, which may be compounded by difficulty in forecasting future interest rates. All assets should be treated individually; as a result the exercise is major task for an asset rich entity such as a water utility. Usually a long-term plan for asset renewal is developed with a horizon of say 20 years and a 5-year rolling programme of renewals is produced in more detail with estimates of the year-by-year expenditure involved.

2.23 PRIVATE SECTOR PARTICIPATION IN WATER SUPPLY

Since the later 1980s there has been increasing interest in using the private (i.e. commercial) sector to provide some or all of the functions of a water supply company. Whilst the public sector needs always to retain ultimate control of water supply, the advantages of private sector participation can be:

- commercial funds for capital and improvement works may be easier to obtain than government or state funds (at least without adding to government debt);
- specialized technical and managerial skills not available to small water undertakings can be brought in to benefit a water undertaking;
- improved efficiencies of water services can be obtained by setting appropriate targets as contractual obligations on a private company.

Private companies have always been used by publicly owned water utilities to provide such services as the design and construction of new works, repair of burst mains, analysis of water samples and so on. But a wider variety of contractual arrangements for private sector inputs have now been developed, the two main types being 'Build Contracts' which cover the design and construction of new water or sewerage works, and 'Operations Contracts' which deliver operational activities.

2.24 PRIVATE SECTOR BUILD CONTRACTS

The 'Build, Own, Operate and Transfer' (BOOT) contract was developed for the design and construction of new works. A private company, backed by a financial institution e.g. bank, or using a loan provided by its government or a funding agency, finances and builds some new works, operates them for a term of years using revenues to cover operating costs and to repay any loans, and transfers them to the public water authority at the end of the contract. The operation period is typically 20 to 25 years. Construction costs may be repaid by the water authority to a defined schedule and payment for the operation of the works is on some time-related or works output basis.

A 'Build, Operate, and Transfer' (BOT) contract is similar but excludes financing. It is usually used to cover the case where the contractor is paid in stages for design and construction of some works, after which he proceeds to operate the works for a given period on some agreed basis of payment. The contractor is responsible for all repairs and maintenance of the works during the operation period, at the end of which ownership is transferred to the water authority. Appropriate incentives can be included. For example, if the contract includes a fixed sum for maintenance, the contractor will endeavour to build to achieve minimum maintenance.

2.25 PRIVATE SECTOR OPERATION AGREEMENTS

Operations contracts involve a private company taking over and managing some or all of the operational activities associated with a set of assets for a defined period. In practice virtually all operations can be outsourced should the utility so desire. The principal types of operational agreements are as follows.

Contracting out. A private company takes responsibility for a fixed duration for a task such as meter reading; or billing and revenue collection; or operation of a water treatment plant. The responsibility of the private company is limited to its contracted task.

Management Contracts. A private company contracts to provide a service to improve certain water supply operations for a fixed fee, or a fee partly or wholly based on achieving some specified performance target, such as: reduction of leakage, betterment of service levels to consumers, etc. In this case the management contractor is really only an adviser, since the execution of the work is carried out by the water authority's staff or others contracted-in. It differs from a management contract for construction where the contractor lets sub-contracts and takes greater responsibility (Twort, 2004).

Leasing (or Affermage). Under leasing a private company takes over all, or some of a water authority's operations for a fee, or a share of the income collected from consumers. Specific targets are normally set. Leasing does not usually involve the company (i.e. lessee) in financing any new capital works. The lessee is principally responsible for the management and technical aspects of running the works including their routine maintenance. If the lessee's payment comprises a share of the undertaking's income, he has opportunities for increasing his profit by improving the efficiency of the undertaking. The duration of leasing agreements is normally 10 to 15 years.

In many French cities, mainly for the operation of treatment works, leasing (termed 'Affermage') has been widely used for many years for fixed periods. One of the reasons for this is that, whereas in England there are only 10 major water and sewerage companies with under 20 smaller water-only companies, France has over 3000 separate water utilities. Likewise Germany has 1500 and Spain 8000 (CIWEM, 1996). The use of large companies allows a degree of aggregation where many small undertakings exist.

Concession. Under a concessionary agreement a private company takes over full responsibility for a whole system, including planning and funding all capital works needed for rehabilitation and expansion of the system, taking on most risks in return for receiving all, or part of the total income generated from water sales and other charges. This motivates the private company to use its expertise to develop the system in the most efficient manner in order to maximize its income. The duration of such agreements is normally 20 to 30 years.

The drawing up and settlement of any long term contract is a complex matter on which it is essential to take experienced advice. It may take two years or more to produce bid documents, evaluate bids and negotiate a contract. Bidding costs for contractors are high; few bids may be received or bids may only be received for the less onerous and less risky contracts. Regulator and supervisory bodies will need to be set up to monitor performance and adjust tariffs.

With concessions it is difficult to achieve a balance between income and expenditure through the period and to ensure that investment decisions are not distorted by the time remaining under the concession. It is also difficult to assess the risks, misjudgement of which can lead to an unfair financial advantage or penalty for one party or the other, thus leading to disputes. A number of disputes have arisen over the information provided by the utility about the condition of assets (particularly buried assets) at the time of bidding.

2.26 CHARGING FOR PUBLIC WATER SUPPLIES

In some countries water supplies are given free or below cost as a social service to those on very low incomes. Most standpipe supplies in low income countries are given free. The consequent financial cost has to be recovered either by the water supplier charging other consumers more (cross subsidy), or by the state or government meeting the deficit (subsidy). In the 1940s to 1960s in the UK the extension of water mains to rural areas had to be funded by government or local government through a series of Rural Water Supply Acts because the cost of the long water mains required to supply rural dwellers did not give an economic return for the water undertakings and was too expensive for rural dwellers to finance.

Where domestic supplies are unmetered, it is common practice, as in UK, to apply a charge related to the value of a householder's dwelling; this results in a socially equitable structure since most low income householders occupy low value property. Where domestic supplies are metered, a basic quantity can be allowed for each household (per charge period) at a charge sufficiently low for the low income householders to afford. Neither arrangement is perfect since there are some low income householders who occupy large properties because of a large family size and, with metering, if the lowest rate of charge covers basic needs for a large family then it gives a small family a liberal supply at the cheapest rate.

With fixed charges according to property value, it is relatively simple for the water utility to set the charges at a level that will provide the required income because the number of properties in each valuation range is known. However, if all supplies are metered, the setting of an appropriate tariff to produce a given income has to take the elasticity of the demand into account (Section 1.16). This is particularly important because the margin of a utility's annual income over expenditure to pay loan charges and to provide a profit may be heavily dependent on the larger consumptions taken by the wealthier domestic consumers. A graph of the percentage of households taking less than a given amount needs to be produced, as shown in Figure 2.1. The area under this graph represents the quantity of water taken under each charge band. Assuming the elasticity (e) of demand for the lowest band is −0.3, and 31.5 m^3/d is taken per 100 households within that band when the price is £1.00/m^3; then the amount taken (V) when the price is raised by 10% to £1.10/m^3 is given by:

$$V = 31.5 \times [(1.00)/(1.10)]^{0.3} = 30.6 \text{ m}^3/\text{d. per100 households.}$$

With this value of e, a 10% rise in the rate of charge results in a 6.9% increase of income for that band. Proceeding thus one can find the approximate total revised income to be expected from domestic properties, and a similar approach can be used for finding the income to be expected from a charge increase on metered trade and industrial demand. Comments on appropriate values for price elasticity are given in Section 1.16.

FIGURE 2.1

Cumulative frequency graph of consumption per household.

2.27 COMPARISON OF CHARGES FOR WATER AND OTHER DATA

Ofwat publishes annual reports on charges, service levels and other data. In 2005–06 (Ofwat, 2006) 30% of households in England and Wales had metered supplies. However, the proportion varied between suppliers from 8% to 65%. Average household bills in 2006/07 were about the same level in real terms as they were in 1999, but 39% higher than on privatisation in 1989.

Average household bills for water 2006/07 were as follows:

	Unmetered £ per annum		Metered £ per annum	
England and Wales	Median	Range	Median	Range
Water Service Co's. (10 No.)	155	115–207	123	103–144
Water-only Co's. (17 No.)	146	80–200	123	82–152
Scotland	72	66–77	not applicable	

2006/07 tariffs for metered domestic supplies in England and Wales averaged £0.95/m³ (range £0.44–£1.48) plus a standing charge in most cases of between £15 and £60 per annum.

Table 2.3 shows comparisons, between water supply companies in England and Wales and other countries, made by Ofwat for the period 2004/05 (OFWAT, 2007). In this case the costs are internal costs, not tariffs.

The Asian Development bank (ADB) publishes data on water utilities in the region. The data for 1995 (McIntosh, 1997) allow comparisons to be made between the water supplies to key cities as shown in Table 2.4. However, more recent worldwide tariff data are published annually in Global Water Intelligence. Data for 2006 are summarised in Table 2.5 (GWI, 2006).

Table 2.3 Comparison between water companies in Europe, Australia and USA

Region	Companies in sample (number)	Properties served (million)	Population served (million)	Total supply (Ml/d)	Total pipe length (1000 km)	Unit cost (2004/5) (£/m³)[a]
N Europe	10	6.3	13.7	2573	91.7	0.80
USA	9	3.1	10.1	5373	66.9	0.54
Australia	8	5.0	11.9	3785	71.9	0.52
England & Wales	24	26.3	58.0	3848	360.0	0.77
Scotland	1	2.5	4.9	1362	47.0	0.80

[a]Unit costs are average total internal costs in Pounds weighted for volume of total supply in region
Source of information: Ofwat, 2007

Table 2.4 Comparison of city water supplies in Asia for 1996

Region	Cities in sample (number)	Connections (million)	Population served (million)	% of total population	Total supply (Ml/d)	Av. tariff (1995)[a] (US$/m³)
Indian sub-continent	12	4.2	61.0	76	11 333	0.17
Central & E Asia	11	8.0	44.4	99	22 514	0.21
South East Asia	11	4.5	24.7	77	8129	0.34
E Indies	7	1.91	26.3	48	1701	0.34
Small islands	7	0.23	0.45	52	180	0.27

[a]Tariffs are averages weighted for volume of total supply in region
Source of information: McIntosh, 1997

Table 2.5 Tariffs for water in 148 cities of world for 2006[a]

Region	Number in sample	Number making no direct charge	Average tariff (US$/m³)[b]	Tariff range (US$/m³)
Northern Europe	24	2	2.20	1.22–3.14
Southern Europe	14	1	1.17	0.48–2.53
Ex. Soviet Europe	26		0.35	0.04–1.05
Ex. Soviet 'Stans'	8	1	0.12	0.02–0.29
Middle East	13	1	0.69	0.01–1.21
Indian sub-continent	6		0.12	0.01–0.37
China/Mongolia	8		0.31	0.13–0.83
Far East	3		0.79	0.22–1.86
South East Asia	5		0.38	0.11–0.66
Australia	3		1.00	0.94–1.06
US and Canada	19		0.94	0.35–1.80
Central America	4		1.04	0.01–2.75
South America	8		0.47	0.01–1.39
Africa	12		0.48	0.21–1.05

[a] Tariffs do not include connection or indirect charges;
[b] Tariffs are not weighted for volume of total supply in region
Source of information: GWI, 2006

REFERENCES

Anderson, P. (1997). Demands and resources for public water supplies in Scotland from 1991 to 2016. *Journ. CIWEM*. June, p. 164–169.

ADB (1998). *Guidelines for the Economic Analysis of Water Supply Projects*. ADB.

ADB (1998). *Handbook for the Economic Analysis of Water Supply Projects*. ADB.

AWWA (1997). Small Systems Research Committee, Research needs for small systems: a survey. *Journ. AWWA*, January, p. 101–113.

Barker, P. (1991). Agricultural pollution control and abatement in the upper Thames region. *Journ. IWEM*, June, p. 318–325.

CIWEM (1996). Report on Proceedings of Conference on 'Supply and Demand: a Fragile Balance'. *Journ. CIWEM*, August, p. 298.

CN (1994). *Construction News*. 1 Dec.

Eckstein, G. A. (2005). Hydrological Perspective of the Status of Groundwater Resources in the UN Watercourse Convention. *Columbia Journal of Environmental Law*, June.

GWI (2006). Tariff survey. *Global Water Intelligence*, Vol. 7, issue 9, September.

HM Treasury (2003). *Appraisal and Evaluation in Central Government*. Treasury Guidelines. London.

ICE (1969). *An introduction to engineering economics*. ICE.

ILA (1966). *Helsinki Rules on the Uses of Waters of International Rivers and Comments*. Report of the Fifty-Second Conference 484, Article II. International Law Associates.

Isack, F. (1995). Money down the drain, *Construction News*, 31 Aug 1995, pp. 16–17.

Juuti, P. S. and Katko, T. S. (2005). *Water, Time and European Cities. Water Time Project*. EC Contract EVK4-2005-0095.

Kinnersley, D. (1994). *Coming Clean: The politics of Water and the Environment*. Penguin Books.

MacDonald, J. A. et al. (1997). Improving service to small communities. *Journ. AWWA*, January, p. 58–64.

McIntosh, C. A. and Yniguez, C. E. (Eds) (1997). *Second Water Utilities Data Book, Asian pacific Region*. ADB. Manila.

Melosi, M. V. (2000). *The Sanitary City. Urban infrastructure in America from colonial times to the present*. John Hopkins University Press.

NCE (1994). *New Civil Engineer*. 1 Dec 1994 and 9 Feb 1995.

Newman, P. and Thornley, A. (1996). *Urban planning in Europe. International Competition*. National Systems and Planning. Routledge.

Owat (2006). *Water and sewerage charges*. 2006–07 report. Ofwat.

Owat (2007). *International comparison of water and sewerage service. 2007 report*. Covering the period 2005–05. Ofwat.

Okun, D. A. (1995). Addressing the problems of small water systems, *IWSA Conf. Proceedings 1995*, Special Subject 9, p. 439–442.

Pontius, F. W. (1997). Future directions in water quality regulations. *Journ. AWWA*, March, p. 40–48.

Teclaf, L. A. and Utton, A. E. (1981). *International Groundwater Law*. Oceans Publishers Inc.

Twort, A. C. and Rees, J. G. (2004). *Civil Engineering Project Management*. Elsevier.

WCA (1992). *Waterfacts 1992*, p. 52. Water Companies Association.

Water UK (2005). *Towards sustainability 2000–2004*.

Hydrology and Surface Supplies

CHAPTER

3

PART I   HYDROLOGICAL CONSIDERATIONS

3.1 INTRODUCTION

When a new water supply is required, it is necessary to assess whether the desired extra water is available and how its development might impact on existing water users (Section 2.1) and on water quality. In developed countries, reference should be made to existing river basin or water resource plans or, in Europe, to river basin management plans under the Water Framework Directive (Section 2.11). A detailed impact assessment is likely to be required in support of any application for permission to abstract. In developing countries, a new assessment of the water resources may be needed to consider all possible sources and define a suitable development plan. The types of development are as follows:

Surface water sources—(a) direct supply from an impounding reservoir or lake, supplemented, if necessary, by gravity feed from an adjacent catchment or pumped inflow from another source; (b) abstraction from a river or canal, supplemented if necessary by releases from a storage reservoir; (c) collection of rainfall runoff from the roofs of buildings or bare catchments and feeding to storage tanks.

Groundwater sources—(a) springs, wells and boreholes; (b) adits and collecting galleries driven underground; (c) riverside wells or sub-surface extraction wells sunk in the bed of a river course or 'wadi'; (d) artificial recharge of aquifers.

Water reclamation schemes—(a) desalination of brackish water or seawater; (b) re-use of acceptable wastewater discharges by appropriate treatment; (c) demineralisation or other treatment of a minewater, including blending with a freshwater supply.

Other types of scheme—(a) integrated (conjunctive) use of surface water, ground water, or water reclamation schemes; (b) transfer of a bulk supply from another supplier or river basin.

Studies of the hydrology of a catchment should focus on dry years or critical drought periods but the overall available resource is defined by a catchment water balance. This quantifies for a given period of one or more water years:

- rainfall on catchment less losses from evaporation, transpiration and natural surface runoffs (i.e. after correcting for abstractions and discharges);
- resultant percolation into aquifers that results in outflow from the catchment below ground rather than as surface runoff;
- *plus* inflow *less* outflow across underground water boundaries;
- *minus* abstractions from wells and boreholes;
- *plus or minus* change in soil moisture content, underground water storage and water in transition zone;
- balance remaining = water unaccounted for.

If all inflows and outflows have been accounted for; a balance of zero will be obtained. This will not be true where there are over-year storage effects (e.g. in reservoirs or aquifers) or where there are significant errors in the components in the water balance. However, once a reasonably satisfactory balance has been achieved there can be confidence that the hydrology of the catchment has been properly appraised.

3.2 CATCHMENT AREAS

A precise knowledge of the catchment area draining to a surface water source is an essential prerequisite for any hydrological assessment of that source. Where this is not known, it must be determined, usually from an examination of contours on a topographic map.

The catchment boundary is located by defining the direction of downhill flow at right angles to the contour lines. Particular care must be taken in areas of low relief and where water-courses, marked on maps, appear to cross the catchment divide. If not the top pound of a canal, these water-course may well be a contour leat or catchwater designed to augment the flow into a reservoir or to bring water to a mill or mine from an adjacent catchment. Field visits to key sections of the topographic divide may be required to resolve uncertainties, identify apparent cross-boundary channels and check whether they are active or derelict, thereby carrying no flow in or out the catchment.

Large scale maps showing topography in sufficient detail can be hard to obtain in some parts of the world. For very large basins, satellite imagery now widely available on the Internet may prove adequate. Alternatively, 500 000 scale maps produced by the USA Air Force in their 'World Tactical Pilotage' series (obtainable in UK from the Director of Military Survey, Ministry of Defence or major mapsellers) are recommended. Large inaccessible basins can be examined from satellite photographs, libraries of which are now available internationally. However, the best sources are those giving digital descriptions of terrain and rivers, now on CD-ROM; the most detailed global coverage is emerging from ESRI, Redlands, California, the major American geographic information software company.

Catchment areas can be determined quickly from GIS systems and by the use of digital terrain models. These include computer programs that will define a catchment boundary from any gridded data set. One such program, available from the Centre for Ecology and Hydrology (CEH), permits catchment area and numerous other parameters to be computed for any catchment in the UK (NERC, 2006). However, care must be taken on small catchments (<10 km^2) and in areas of low relief where the stream drainage network is poorly defined. Manual adjustments to computed values may be required in some cases.

The catchment contributing to a ground water source may not be readily defined. In many cases, it corresponds closely to the surface water catchment and water table contours reflect surface topography, albeit with a reduced vertical range. However, there are cases where surface and groundwater catchments differ markedly, making precise catchment water balance calculations difficult or impossible. Groundwater catchment divides can migrate seasonally, especially in response to pumped abstractions. Monitoring networks need to have very high densities of observation wells in order to track these variations, but is rarely practicable except under research funding.

3.3 DATA COLLECTION

Good hydrological data forms the basis on which water schemes must be planned and designed. The collection, archiving and dissemination of hydrological data are expensive. It is usually a government funded activity. This funding can be vulnerable when governments try to reduce public spending as the impact is not felt immediately and the benefits of data collection are not always obvious. However, it has been demonstrated that the availability of reliable hydrological data results in benefits an order of magnitude larger than the costs of collection (Simpson, 1987). The need for more data is particularly important where climatic or environmental conditions are changing and water supplies are critical. In Nigeria, for example, it was reported in 1998 that failure of a number of large scale resource developments to achieve their forecasted output was partly due to the absence of accurate data on rainfall and river flows (Tor, 1998). If large capital sums are not to be wasted, the basic hydrological data on which water schemes are designed must be adequate and reliable. Where good hydrological data are available, very large savings can be made.

To assess a potential surface water source the prime need is for long streamflow records. Similarly, for ground water sources, long records of changing aquifer water levels are required. Streamflow records need to be naturalized (Section 3.9) to eliminate the effects of any artificial influences. For this, details are required of all abstractions, effluent returns and reservoir storage changes in the catchment. Similarly, in order to interpret changes in groundwater levels correctly, details should be kept of pumped output from wells together with pumping and rest water levels.

Before embarking on detailed analyses, it is essential to check the validity of all basic data used. This may involve visits to monitoring sites to check their condition and probable degree of accuracy. If relevant data do not exist, temporary gauges or permanent systems must be set up to acquire it. Once data is obtained it should be carefully filed or computer-archived so that it is permanently available for subsequent analyses such as updating yield estimates when catchment or other conditions change.

3.4 STREAMFLOW MEASUREMENT

River or stream flow records taken in the vicinity of an existing or proposed intake or dam site are an invaluable aid to the assessment of the potential yield of a source. The longer the period of record the more reliable any yield estimates are likely to be but even a very short record is often a significant improvement over estimates derived from generalized regional relationships.

Flows are often obtained by measuring the 'stage level' of a river i.e. the elevation at some location of the water surface above an arbitrary zero datum. Continuous or regular measurements of stage are then converted to discharge by means of a rating curve. Increasingly, use is being made of methods that measure flow velocity directly, such as ultrasonic and electromagnetic devices.

The simplest way to measure stage level is by a permanent staff gauge, set so that its zero is well below the lowest possible flow. Although such gauges are simple and inexpensive, they must be read frequently when the water level is changing rapidly to define the shape of the streamflow hydrograph. It is preferable to construct a stilling well to house a data logger or chart recorder connected to a shaft encoder, pressure transducer or float. Electronic data loggers have marked advantages for data processing but the chart record gives an important visual check of existing conditions and may still have a role to play in areas without a power supply or in providing a backup record.

A rating curve for a site can be obtained in a variety of ways of which the two most common are by means of velocity-area methods using a current meter (USGS, 1968), or by means of weirs (Section 12.14) or flumes (Section 12.15) which are mostly permanent structures. The latter include:

- sharp edged plate weirs (BS 3680, 1981);
- broad crested weirs (Bos, 1990);
- triangular profile weirs (Herschy, 1977);
- critical depth flumes (Ackers, 1978).

Dilution gauging (USGS, 1985) and ultrasonic gauging (BS EN 6416: 2005) can also be used to help define the stage/discharge relationship at sites where conditions are difficult for current meter measurements and to check an existing rating curve. The choice of gauging method depends on channel and streamflow characteristics, staff time availability and cost.

Current meter measurements are most often used when large flows have to be measured and the available fall is small. They are also often desirable for smaller rivers with sediment laden flows. Current meter gauging stations are relatively easy to set up since they often require little modification of the existing channel. Since each potential gauging site is unique, each requires a careful pre-assessment of the width and depth of the channel, likely flood velocities and alternative ways of current meter measurement. Wherever possible, current meter gauging stations should be located at accessible sites in straight uniform channel reaches with relatively smooth banks, no obstructions and a stable bed.

Current meter measurements can be made by wading with a current meter attached to a graduated wading rod to measure depth. Wading can usually be carried out safely in water depths are about 1 metre, less if the stream velocity is more than about 1 m/s. Otherwise, current meter observations should be carried out from above the water. Where the current is strong the current meter should be ballasted with a weight; in extreme cases, this may be so heavy that a crane is necessary. Access may be from the deck of a single span bridge, a specially constructed cableway or from a boat. An alternative, which is rapidly gaining popularity, is the acoustic Doppler current profiler (ADCP). This uses ultrasonic technology to measure integrated flow velocity across the river. The ADCP can be mounted on a boat or dragged across the surface of a river from the bank on a small raft. This has several safety and time advantages for the operator.

At current meter stations, the cross-section is divided into vertical sections in which velocity measurements are taken at set depths. The mean velocity in each section is determined and applied to the area it represents. The river discharge is computed from the sum of the individual section values. The number and spacing of the verticals should be such that no section accounts for more than 10% of the river flow. Velocity measurement points are normally located by means of a tagged tape or wire stretched across the river or from graduations painted on the deck of a bridge. Detailed observations for many different types of river have shown that the mean velocity of a vertical can be closely approximated from the mean of two observations made at 0.2 and 0.8 of the depth. If the water depth in the vertical section is less than about 0.6 m, or if time is limited, then one observation at 0.6 of the depth will approximate to the average over the whole depth.

The Water Resources Division of the United States Geological Survey has produced some excellent publications on current meter gauging techniques (USGS, 1968); while standard practice in the United Kingdom is contained in BS EN 748:2007, BS 1100:1998, BS TR 8363:1997 and BS 3680-3Q:2002.

Despite the relatively low capital cost of velocity-area stations, the need for sufficient current meter measurements over a wide range of river levels and for a repeat of the process each time a major flood is thought to have shifted the river bed profile, make heavy demands on staff time. These factors and the difficulty of access and gauging when the river is in flood make completing a rating curve take a long time. The alternative of a standard gauging structure, therefore, is often more attractive to the engineer, particularly for catchments of less than 500 km^2.

Sharp edged plate weirs, of rectangular or vee-shape depending upon the sensitivity required, are generally only suitable for spring flows or for debris-free small streams. The need to keep the weir nappe aerated at all stages limits their use but they are frequently used for low flow surveys as they can be rapidly placed in small channels.

Of all the weirs that designers have tried, perhaps the most successful has been that of Crump (Herschy, 1977) (Section 12.14). It has a simple and efficient flow characteristic while operating up to the total head at which tailwater reaches 75% of that upstream (relative to crest height). Using the crest tapping designed by Crump it is possible to go further and attain reasonable results up to 90% submergence, but this versatility is marred by a tendency for the crest tappings to block in floods carrying sediment. A minimum head of 600 mm on the weir is necessary for accuracy, but with compounding of the weir crest this can be achieved. The flat-vee variant is another possibility.

The accuracy of streamflow data obtained from weirs can often be poorer than generally realised. Laboratory rating conditions rarely appear in real rivers, deterioration of the weir crest and siltation upstream, resulting in non-standard approach conditions, can cause errors in excess of 10%. Drowning out of the weir at high flows can result in gross over-estimation of flood discharges.

Critical depth flumes are appropriate on smaller catchments (area under about 100 km^2), which have a wide flow variation and where sensitive results are required. Such flumes force flow to attain critical depth whatever the upstream head (Section 12.15). A unique upstream head-discharge relationship is then created as for a rectangular throat.

Calibration of an existing sluice structure may be achieved using formulae derived from laboratory model tests but these should be checked by current meter measurements wherever possible. Sometimes long, relatively homogeneous, sluice keeper's records exist (Sargent, 1992), and the calibration of such sites can then produce records of flow of several decades in length for a modest cost.

Dilution gauging (USGS, 1985) is a flow measurement technique that is particularly well suited to small turbulent streams with rocky beds where the shallow depths and high velocities are unsuitable for accurate current meter gaugings. The approach can also be used to calibrate non-standard gauging structures. With dilution gauging, the discharge is measured by adding a chemical solution of known concentration to the flow and measuring the dilution of the solution some distance downstream where the chemical is completely mixed with the stream flow. Sodium dichromate is the most commonly used chemical although dyes such as Rhodomine B have the advantage that they can be easily detected at very low concentrations. With the commonly used 'gulp injection' method a known volume of chemical is added to the stream flow as quickly as possible in a single 'gulp' and downstream samples are used to construct a graph of concentration against time. It follows that if a known volume of chemical V of concentration C_1 is added to a streamflow and the varying downstream concentration C_2 is measured regularly then

$$VC_1 = Q\int_{t_1}^{t_2} C_2\,dt$$

A graph of C_2 against time is drawn and the area below it, between t_1 when the chemical just starts to be detected in the stream and t_2 when it ceases to be detectable, is measured. This gives the integral on the right, hence Q can be found.

Ultrasonic gauging uses the transmission of sound pulses to measure the mean velocity at a prescribed depth across a river channel. Two sets of transmitters/receivers are usually located on either bank of a rectangular channel, offset at an angle of about 45° to the direction of flow. They send ultrasonic pulses through the water, mean water velocity at pulse level being a function of the difference in pulse travel times in upstream and downstream directions. In essence, ultrasonic flow measurement is a velocity-area method requiring a survey of the channel cross section at the gauging site. If used in conjunction with a water level recorder, ultrasonic measurements can provide a complete record of stream flows.

3.5 RAINFALL MEASUREMENT

Precipitation is measured with a rain gauge, the majority of which are little more than standard cylindrical vessels so designed that rainfall is stored within them and does not evaporate before it can be measured (WMO, 1983). In an effort to ensure that consistent measurements of the precipitation reaching the ground are obtained, observers are recommended to use standard instruments, which are set up in a uniform manner in representative locations. Many national meteorological institutions provide pamphlets designed to ensure good standard observation practice and the World Meteorological Organization (WMO) plays an effective co-ordinating role.

The standard daily rain gauge in UK is the Meteorological Office Mark II instrument. It consists of a 127 mm diameter copper cylinder with a chamfered rim made of brass. Precipitation that falls on the rain gauge orifice drains through a funnel into a removable container from which the rain may be poured into a graduated glass measuring cylinder. Monthly storage gauges are designed to measure the rainfall in remoter areas and are invaluable on the higher parts of reservoir catchments. The Seathwaite gauge is a monthly storage gauge developed for use in the Lake District in north-west England.

Ideally, rainfall should be measured at ground level but this gives rise to problems due to rain splashing into the gauge. The higher the rim is placed, the more some rain will be blown away from the gauge orifice and goes unrecorded. All standard storage gauges in the United Kingdom are set into the ground with their rims level and 300 mm above the ground surface, which should be covered by short grass or gravel to prevent any rain splash. Many international gauges are set with their rim one metre high: these can be expected to read 3% lower than the standard British gauge.

In the United Kingdom, daily storage gauges are inspected each day at 09.00 hours and any rainfall collected is attributed to the previous day's date. If the inner container of a rain gauge should overflow as the result of exceptional rainfall, or possibly because of irregular emptying, it is important that the surplus water held in the outer casing should also be recorded. Monthly storage gauges are usually inspected on the first day of each month to measure the previous month's rainfall total. Corrections may need to be made to the measurements taken at any gauges visited later than the standard time during spells of wet weather.

Specialized problems occur in snow prone areas. Small quantities of sleet or snow which fall into a rain gauge will usually melt to yield their water equivalent, but if the snow remains in the collecting funnel it must be melted to combine with any liquid in the gauge. If there is deep fresh snow lying on the ground at the time of measurement, possibly burying the gauge, a core of the snow should be taken on level ground and melted to find the equivalent rainfall. Countries with snow cover throughout the winter months require regular snow course surveys (Hudleston, 1933) to measure precipitation.

Continuously recording rain gauges are invaluable for flood studies. The original type gives a daily chart recording of the accumulated contents of a rain-filled container, which empties by a tilting siphon principle each time 5 mm has collected. A more recent development is the tilting bucket gauge linked to a logger, which runs for at least one month. Each time the bucket tilts to discharge 2 mm the event is recorded in a computer compatible form. A daily or monthly storage gauge is often installed on the same site as a recording gauge to serve as a check gauge and to avoid the possibility of loss of data due to instrument failure.

Particular care must be taken when siting a new rain gauge from which the resulting records are to be published. The gauge should be placed on level ground, ideally in a sheltered location with no ground falling away steeply on the windward side. Obstructions such as trees and buildings, which affect local wind flow, should be a distance away from the gauge of at least twice their height above it. In particularly exposed locations, such as moorlands, it used to be standard British practice to install a turf wall (Hudleston, 1933) around the gauge but it has proved difficult to sustain the level of turf maintenance that is needed.

Rain gauges provide a spot sample of the rain falling over a catchment area. The number of gauges required to give a reliable estimate of catchment rainfall increases where rainfall gradients are marked. A minimum density of 1 per 25 km^2 should be the target, bearing in mind that significant thunderstorm systems may be only about 20 km^2 in size. In hilly country, where orographic effects may lead to large and consistent rainfall variations in short distances, it may be necessary to adopt the high densities suggested in Table 3.1 for the first few years. Thereafter high densities are only required where control accuracy necessitates it.

Table 3.1 Rain gauges required in a hill area (IWE, 1937)				
Catchment area (km²)	4	20	80	160
Number of gauges	6	10	20	30

In large areas of the tropics, there is great variation in rainfall from place to place on any one day but only a relatively small variation in annual totals. In such areas the rain gauge densities of Table 3.1 will be excessive and it is better to concentrate on obtaining homogeneous records of long duration at a few reliable sites.

Measurement of Catchment Rainfall

There are several methods for computing catchment precipitation from rain gauge measurements ranging from simple numerical procedures, interpolation from isohyetal maps or Thiessen polygons and from numerical interpolation procedures of which Kriging (Creutin, 1982) and trend surface are most frequently used. The simplest objective method of calculating the average monthly or annual catchment rainfall is to sum the corresponding measurements at all gauges within or close to the catchment boundaries and to divide the total by the number of gauges. The arithmetic mean provides a reliable estimate provided the whole catchment is of similar topography and the rain gauge stations are fairly evenly distributed. If accurate values of area rainfall are obtained first from a large number of rainfall stations, by one or other of the more time consuming methods described below, the mean of the corresponding measurements from a smaller number of stations may provide equally acceptable results. In the Thames Basin, for example, it was found that the annual catchment rainfall, for the 9980 km^2 area, derived from the arithmetic mean of 24 well distributed representative gauges was within $\pm2\%$ of the value computed by a more elaborate method using measurements from 225 stations.

The isohyetal method is generally considered to be the most accurate method of computing catchment rainfall although a good understanding of the rainfall of a region is needed to ensure its reliability. The monthly or annual rainfall total recorded by each gauge within or close to the catchment boundaries is plotted on a contour base map. Isohyetal lines, i.e. lines joining points of equal rainfall, are then drawn on the map taking into account the likely effects of topography on the rainfall distribution. The total precipitation over the catchment, for the period considered, is obtained by measuring the areas between isohyets, by GIS techniques or by planimeter. The mean catchment rainfall is calculated by summing the products of the areas between each pair of isohyets and the corresponding mean rainfall between them, and then dividing by the total catchment area.

The most popular method of weighting gauge readings objectively by area has been that of Thiessen. An area around each gauge is obtained by drawing a bisecting perpendicular to the lines joining gauges, as shown in Figure 3.1. The portion of each resulting polygon lying within the catchment boundary is measured and the rainfall upon each is assumed to equal the gauge reading.

▲ = rain gauge

FIGURE 3.1

Thiessen's method of estimating general rainfall over an area.

The total precipitation is the weighted average of these values. One drawback is that, if the gauges are altered in number or location, major alterations to the polygonal pattern ensue. To maintain homogeneity it is better to estimate any missing individual gauge values. The gauges must also be reasonably evenly distributed if the results are to lie within a few per cent of the isohyetal method. The approach is not particularly good for mountainous areas because no account is taken of the effects of altitude on rainfall when deriving the Thiessen coefficients for individual polygons.

In mountainous areas, where there may be few stations, the main difficulty is to allow for the influence of topographic. One widely used approach for such areas is to develop a multivariate regression model using parameters such as elevation, orientation, exposure or distance from the sea, and then to use a numerical interpolation procedure such as Kriging (Creutin, 1982) to smooth out residual discrepancies from the regression correlation.

Progress has been made in rainfall estimation by both weather radar (Lau, 2006) and satellite although the establishment and operation of such networks is the domain of national meteorological organisations rather than individual projects. The strength of both lies in the spatial view they afford with the former being particularly useful for flood forecasting. Radar images for low altitude scans are calibrated to ground measurements of actual rainfall and rainfall values assigned to each pixel of the image, normally at 5 or 15 minute intervals. Pixels cover 1 km squares close to the radar increasing to 2 km or 5 km squares further away. GIS techniques can be used to compute catchment rainfall for each time step when digital catchment boundaries are applied to the gridded data. Satellite estimates are far more approximate, being related to cloud top temperature and only indirectly to actual rainfall amount. For large basins in the tropics it is now possible to obtain public domain estimates of $0.5° \times 0.5°$ grid satellite 'monthly rain estimates' from the Climate Analysis Center, Washington D.C.

3.6 EVAPORATION AND TRANSPIRATION MEASUREMENT

Evaporation is a key part of the hydrological cycle in that, globally, about 75% of total annual precipitation is returned to the atmosphere by the processes of evaporation and transpiration. Water evaporates to the air from any open water surface or film of water on soil, vegetation or impervious surfaces such as roads and roofs. The rate of evaporation varies with the colour and reflective properties of the surface (the albedo) and with climatic factors. Energy from solar radiation is the main factor for which air temperature is often used as a proxy measurement. However, windspeed, the relative humidity of the air and the temperature of the water are also important.

Transpiration is the water used by plants. A small part of this water is retained in the plant tissue but most passes through the roots to the stem or trunk and is transpired into the atmosphere through the leaves. As it is almost impossible under field conditions to differentiate between evaporation and transpiration when the ground is covered with vegetation, the amounts of water used by both processes are usually combined and referred to as '*evapotranspiration*'.

Evapotranspiration losses vary with the same meteorological factors as evaporation, but depend additionally on the incidence of the precipitation, the characteristics and stage of development of the vegetation, and the properties of the soil.

Evaporation can be measured directly or estimated indirectly (WMO, 1966), although both have their difficulties. The USA Class A evaporation pan is one of the most commonly used instruments for measuring evaporation directly. It is constructed of galvanized iron or monel metal, 1.21 m in diameter and 255 mm deep, and is set on a standard wooden framework 100 mm above ground level, thus allowing air to circulate all round it. As a result, measured evaporation is higher than that of a natural water surface and a reduction factor must be applied. This is generally taken to be 0.7, but it can vary between 0.35, in areas of low humidity, very strong wind and surrounded by bare soil and 0.85 where high humidity and light winds prevail (Doorenbos, 1976). The British Symons sunken tank is 1.83 m square and 610 mm deep, with the rim 75 mm above ground level. It is more nearly a model of reservoir evaporation but suffers from inconsistent results if it is not in tight contact with the surrounding ground. The heat storage of a small tank is correspondingly small, whereas a large lake takes time to warm up or cool down. As a consequence, tank results do not quite match the evaporation of a nearby lake in regions with strong seasonal temperature variations. Peak lake evaporation rates in the Kempton Park experiment (Lapworth, 1965) occurred up to a month after peak tank measurements; this was explained by heat storage theory. Annual open water evaporation ranges from 700 mm in Northern Europe, through 1500 mm in much of the tropics, to more than 2500 mm in hot arid zones.

Attempts have been made experimentally to measure evapotranspiration by means of percolation gauges (Rodda, 1976) and, on a larger scale, by the use of lysimeters. However, no standard way of measuring it routinely has yet been devised. Indirect techniques therefore have to be relied on.

Penman's formula (Penman, 1963) for predicting water surface evaporation is recognized as the most accurate of these indirect techniques. It is based on physical principles, but involves the use of data which may not always be available e.g. measurements of radiation (or sunshine duration), wind run at 2 m above ground, vapour pressure and air temperature, all of which should be taken at the same site. At altitudes above about 1000 m, McCulloch's fuller version of Penman's equation (McCulloch, 1965) should be used as it makes express allowance for the corresponding pressure drop and adjusts the radiation term for latitude.

Penman also showed how simple coefficients could be applied to his open water evaporation figures to obtain the evapotranspiration rate from a grassed surface. The use of the latter became standardized as being the evapotranspiration (designated 'ET_o') 'from green grass surface cover 80–150 mm high, actively growing'. Crop coefficients k_c could then be applied to give the evapotranspiration of various types of crops at various stages of development and meteorological conditions, e.g. $ET_{(crop)} = k_c.ET_o$.

Thornthwaite's formula for evapotranspiration from short vegetative cover, widely used in USA, is empirical and simpler than Penman's, being dependent on sunshine hours and mean monthly temperature. Thornthwaite's method has been widely used, but it is strictly valid only for climates similar to that of eastern USA where the method was developed. The method tends to give evaporation estimates higher than those produced by the Penman formula, particularly during the summer months.

The most recent guide to assessing the irrigation requirements of growing crops is the FAO's *Irrigation and Drainage Paper 46, CROPWAT: A Computer Program for Irrigation Planning and Management,* 1992. This computes evapotranspiration ET_o values according to the Penman-Monteith method, to which quoted crop coefficients, k_c are then applied. An earlier FAO publication, *Irrigation and Drainage Paper 24', Crop water requirements,* 1976, was based on the Blaney-Criddle formula, which uses temperature and length of daylight hours to give ET_o values, which are then modified by values given for a range of humidities and wind speeds.

Evapotranspiration formulae assume no shortage of water to meet crop growth and potential evaporation. In dry periods when moisture availability becomes limiting, actual evapotranspiration is less than the computed potential. A good explanation of all the foregoing indirect methods of estimating evapotranspiration is given by Wilson (1990).

In the UK, the Meteorological Office (Thompson, 1981) has used an adaptation of the Penman-Monteith equation in their MORECS, and more recently MOSES, evaporation estimation systems. Up-to-date values of actual and potential evaporation are computed and published regularly, as well as precipitation and soil moisture deficits for a network of 40 km × 40 km grid squares covering the whole country. Similar systems are likely to exist in other countries and practitioners will in most cases find it preferable to use published data rather than derive their own evaporation estimates from climate data.

3.7 SOIL MOISTURE MEASUREMENT

Agriculturalists have always been interested in studying soil moisture. More recently, the potential impact of soil moisture content on runoff has been understood by a wider audience. The temporary storage of rainfall in the soil and aquifer layers of a catchment can be significant in the overall catchment water balance. Soil moisture variations occur predominantly in the first metre below the surface. The water content in very sandy soil may vary from 3% to 10% from the driest (Wilting point) condition to the wettest drained state (Field capacity), or from 20% to 40% in a clay soil. Thus, the maximum range of water storage in one metre of soil may be as much as 200 mm. Markedly higher values apply to peat. Additional water may be held under waterlogged conditions whenever the drainage rate is lower than the rainfall intensity.

Knowledge of the types and distributions of soils over the area of interest is an essential prerequisite for the selection of sampling points at which to measure soil moisture. For England and Wales, this information can be obtained from maps published by the Soil Survey and Land Research Centre at Silsoe, while for Scotland soil maps can be obtained from the Macaulay Land Use Institute, Aberdeen. In other countries, similar sources should be consulted but it may be necessary to engage a soil scientist to carry out soil surveys in the catchments where detailed hydrological measurements are required.

A common method of measuring soil moisture is by gravimetric determination, whereby a soil sample of known volume is removed from the ground with a soil auger. It is weighed, dried in a special oven and then reweighed. The method is accurate provided care is taken with the measurements and is often used to calibrate other techniques. However, the method is time consuming and requires laboratory facilities. It is also a destructive process and has obvious limitations where regular sampling is required. A soil capacity probe (Dean, 1987) can be used to provide direct field measurements.

3.8 CATCHMENT LOSSES

A significant proportion of rainfall is lost by immediate evaporation or by the later transpiration of growing vegetation. In some cases, there will also be deep infiltration that eventually emerges in the sea without reappearing on the surface. Catchment losses are best estimated from a water balance conducted over a number of years on the catchment concerned or from one with similar rainfall, geology and land use.

A typical loss rate in England would be 450 mm per annum, with figures significantly above 500 mm per annum generally only occurring where afforestation predominates in a high rainfall area. Table 3.2 gives some idea of the variation in loss in different regions of the world. Although

Table 3.2 Typical catchment losses in various parts of the world

Country	Location	Catchment cover	Annual rainfall (mm)	Annual loss (mm)	Marked seasonal variation
Nigeria	Ibadan	Rain forest	2500	2350	No
Malaysia	Johor basin	Forest/oil palm	2320	1240	No
Sri Lanka	Kirindi Oya	Mixed forest	1650	1230	Yes
Hong Kong	Islands	Grass	2100	1050	No
Zaire	Fimi	Rain forest	1700	1040	No
Thailand	Chau Phraya	Forest/rice paddy	1130	1000	No
Japan	Ota	Conifer forest	1615	890	Yes
Australia	Perth	Mixed grass/forest	875	760	Yes
S. Africa	Transvaal	Mixed grass/forest	870	760	Yes
Kenya	Tana	Forest/savannah	1100	730	Yes
India	Bombay	Rain forest	2550	700	Yes
Zimbabwe	Low Velt	Mixed grass/forest	655	560	Yes
Lesotho	Maseru	Grassland	600	530	Yes
Holland	Castricum	Low vegetation	830	450	Yes
Britain	South England	Pasture/arable	600–900	450–530	Yes
	Midlands	Pasture/arable	650–850	440	Yes
	Central Wales	Moorland/forest	1500–2300	480–530	Yes
	Pennies	Moorland/forest	1150–1800	410–460	Yes
	NE England	Moorland/pasture	700–1250	380	Yes
	S. Scotland	Moorland/pasture	600–1800	360–410	Yes
	N. Scotland	Moorland/pasture	1250–2500	330–380	Yes
Algeria	Hamman Grouz	Scrub	420	400	Yes
Russia	Moscow	Agricultural	525–600	375	Yes
Iraq	Adheim basin	Scrub/grassland	420	350	Yes
S. Korea	Had basin	Forest	1180	320	Yes
Oman	Oman	Rock	160	130	Yes
Iran	Khatunabad	Bare ground	150–550	50–200	Yes

it might be thought that losses would be higher in years hotter than average, this is often more than offset by the concurrent dryness of the weather, which leads to a deficit of moisture in the soil and suppression of transpiration by growing vegetation. As soil dries out and approaches wilting point evapotranspiration rates can drop to only about one-tenth of the potential. Some plants, such as pine trees, are more successful than others in control of their water use in drought conditions.

Losses do not always decrease at cooler altitudes because advected wind energy and higher radiation may offset the lower temperatures. Detailed measurements at experimental catchments by CEH have shown that losses increase with the density and height of natural vegetation growth and crops, particularly in mature coniferous forests. However, predicting forest annual loss compared with that from grazed pasture is still not easy. It has been shown that the forest loss is due to intercepted raindrops being evaporated back into the atmosphere at rates up to five times normal transpiration values for short grass. This is because water laid out in thin films on vegetation can take up available heat in the atmosphere more readily; the sight of a forest steaming gently in a short spell of sunshine between showers is not uncommon. To quantify the extra loss to be expected requires (Calder, 1992) an idea of the depth of water the canopy of vegetation can hold during a shower and the frequency of showers. One point to note is that whilst rain is being evaporated off the outside of leaves, water is not transpired through them. As a result, perhaps only 90% of forest interception losses are in addition to the catchment losses that would prevail anyway.

A review (Bosch, 1982) of 94 catchment experiments in Africa, Asia, Australia and North America concluded that conifer forests could be expected to reduce catchment water yield, on average, by 40 mm per 10% forest cover. More recent studies (Kirby, 1991), at Plynlimon in the UK, found that the magnitude of the reduction in water yield was 29 mm per 10% forest cover, which equates to a 15% reduction in the water yield for a completely forested upland catchment. It should be noted, however, that while heavy afforestation of catchments reduces their overall water yield, the results of the Plynlimon study and a study (Gustard, 1989) of catchment data from 40 European agencies found no statistically significant relationship between the proportion of forest cover and measures of low flow. These findings support the view that low flows in upland headwater catchments are primarily influenced by drainage from minor sources of groundwater, which remain largely unaffected by forest interception losses.

3.9 STREAMFLOW NATURALISATION

In most river basins, very few flow regimes can be considered entirely natural, that is, free of artificial influences such as abstractions, discharges or storage effects from impounding reservoirs. It is therefore not usually sufficient to evaluate a potential resource directly from the 'as-measured' stream flow records. The need to naturalize the available flow records should be considered at the start of any major water resource assessment. This involves considering the extent of artificial influences within a catchment, the scale of their impact on flows and the purpose for which the flow sequences are required.

Flow naturalisation does not normally correct for anthropogenic effects such as urbanisation or land use changes. Except on very small catchments, these effects are not usually significant. The main types of abstraction or artificial losses from a river system are public water supply abstractions; irrigation and other abstractions for agriculture; power station cooling water abstractions; and industrial (non-cooling) water abstractions. The more common gains are from sewage and industrial effluent returns; irrigation return flows; and interbasin transfers.

The need for flow naturalisation arises for three main reasons:

- artificial influences have most impact on low flows which are the focus of water resource assessments;
- the long-term variability of flow records is characterized by comparison with a suitable statistical distribution; natural flow records tend to conform to such distributions, those affected by artificial influences may not;
- options for water resource development are often assessed by means of a computer simulation model; model inflow sequences need to be free of any historic artificial influences embedded in the observed flow record so as not to bias results or lead to the double-counting of influences.

In the UK, all time series of daily river flow data held on the National River Flow Archive have been graded according to the degree of artificial influence at low flows. Elsewhere, review of abstraction licences/permits, where they exist, can often give an initial indication of the scale of abstractions, even though actual abstracted quantities will almost certainly be less than the licensed total. The type of abstraction is also relevant. Power stations are often licensed to abstract very large quantities of river water for cooling, although most of this water is returned to the river a short distance downstream. The net effect on river flow is therefore small except immediately downstream of the abstraction. Similarly, large groundwater abstractions may be from parts of the aquifer not in hydraulic connection with the river and therefore have little or no effect on river flows. Some influences are seasonal so winter abstractions, in temperate climates for example, may have no effect on summer low flows, although they could affect the refilling of storage.

The purpose of flow naturalization should be clearly born in mind and the parameters that are critical to the study in question should be assessed, for example water levels, minimum flows or flow volume over a critical period. It may be considered, for example, that if gauged flows are within 10% of natural flows over the critical period, then naturalisation is not necessary.

If naturalisation is required, the two most commonly used methods are naturalization by decomposition and rainfall-runoff modelling. Naturalisation by decomposition involves breaking the observed flow record down into its component parts. If there are no reservoirs within the catchment then the natural flow is deduced by quantifying all the artificial components in the observed flow record at the appropriate time step (daily or monthly) and using the following equation:

Natural flow = gauged river flow + sum of all upstream abstractions − sum of all upstream discharges and return flows to river

This process is critically dependent on the availability of good quality, complete data for both the observed stream flows and for artificial influences (abstractions and return flows) upstream of the location of interest. In catchments where there are many small abstractions that individually have little or no impact on river flows, these may be filtered so that the smallest are either ignored or combined. It may be adequate to consider only the main public water supply abstractions or sewage treatment works discharges. These are usually the largest artificial influences and are often monitored so that specific time series data are available.

Factors may be applied to certain types of abstraction to account for the proportion of abstracted water that is returned to the river. If time series data are not available, a seasonal profile may be applied to the mean annual abstraction where quantities vary over the year. Factors may also be applied to groundwater abstractions according to their degree of impact on river flows. Impacts may be time-lagged depending on the length of the flow path between river and groundwater table. The results of regional groundwater models may help decide how best to allow for the effect of groundwater abstractions in the naturalisation process. A detailed discussion of these and other aspects is contained in EA Guidelines (EA, 2001) and in Hall (1994).

For catchments that contain a reservoir, allowance needs to be made for the storage and attenuation of flows and for additional evaporation losses from the reservoir surface, which can be significant in warm semi-arid environments. In order to unravel storage effects, time series data are required for changes in reservoir level and/or storage, drawoff to supply and spills/releases to the channel downstream. If data on reservoir spills are not available, the naturalisation can still be carried out on dry season flows, which are usually the most critical from a water resources point of view.

If the available data are inadequate for a decomposition approach, it may be possible to generate natural flow sequences by means of rainfall-runoff modelling. Nevertheless, this approach still requires data to represent rainfall and evaporation at an appropriate time scale over the catchment in question. Ideally, a brief period of observed (natural) river flows is also required with which to calibrate the model, although if this is not available, some spot gaugings at low flow may suffice. Often, the available data permits naturalisation by decomposition for the recent historic period with rainfall-runoff modelling required to extend flow sequences back to earlier years.

3.10 LONG TERM AVERAGE CATCHMENT RUN-OFF

Wherever possible, average catchment runoff should be calculated from a long streamflow record, which, if subject to artificial influences, has been naturalized as described in Section 3.9. Where streamflow records are either short term or non-existent, long term average runoff can be estimated by:

- correlating the brief records available for the study catchment with those of a long record station in a catchment with similar characteristics;
- deducting loss estimates from catchment rainfall figures;
- use of a rainfall/runoff model.

Correlations between long and short record stations are best carried out using monthly data. The use of daily figures is more time consuming and often produces a large scatter while annual values provide too few points. An initial mathematical 'best fit' relationship between the stations may be derived by computer but a graphical plot should always be produced as a check on the mathematical fit. A manual adjustment should be made if, for example, the mathematically computed relationship is unduly influenced by the values for a few high flow months, or if the relationship implies unreal intercept values.

An estimate of average runoff for a catchment for which no streamflow records exist is often obtained by deducting a value for average annual catchment losses from the average annual catchment rainfall for a suitable standard period (e.g. 1961 to 1990). The latter can be obtained from an isohyetal map of the region or from a GIS-based gridded database such as the FEH CD-ROM (NERC, 2006) (Section 3.5). If possible, the value for average annual catchment losses should be based on the typical average annual loss value obtained from similar gauged catchments in the region. However, if no such data exist, estimates of actual evapotranspiration can provide a reasonable measure of catchment losses as long as there are no significant abstractions or groundwater outflows from the catchment. In the UK, the Meteorological Office publishes estimates of actual evapotranspiration for a network of grid squares covering the whole country from its MORECS

Table 3.3 Adjustment factor for estimating actual evaporation in the United Kingdom (Gustard, 1992)

Standard Average Annual Rainfall (mm)	500	600	700	800	900	1000	>1100
Adjustment factor	0.88	0.90	0.92	0.92	0.94	0.96	1.00

(Thompson, 1981) and MOSES evaporation systems. If actual evaporation data are not available, values can be estimated from catchment average annual potential evapotranspiration (Smith, 1975) multiplied by the adjustment factors listed in Table 3.3.

For most regions, rainfall stations are more numerous and have longer records than stream-flow measurement stations. Rainfall/runoff models such as HYSIM (Manley, 1978) or HEC-RAS (USACE, 2000) are therefore commonly used to extend short term stream flow records. Typically, a chosen model is calibrated by adjusting the model parameters so as to produce the best possible match between predicted and measured flows. The calibrated model is then used to extend the short term stream flow data to cover the longer period of rainfall records. One of the usual checks on the synthetic stream flow data generated by rainfall/runoff models is whether they accurately reproduce the long term mean runoff estimated by other means, particularly over a standard period.

3.11 MINIMUM RAINFALLS

Experience of past recorded droughts and low rainfall is an important factor in assessing probable future conditions that may be encountered. In the variable climate of the UK the longest known spell without any recorded rain at all (Holford, 1977) was for 73 days from 4 March 1893 at Mile End in London. At the other extreme, in desert climates many years may have no rainfall at all. At Calama, which is in the Atacama desert of northern Chile, it is believed that virtually no rain fell for 400 years until a sudden storm in 1972.

The most notable droughts in England and Wales in the last 100 years were the following:

- 1921   Southeast England. Annual rainfall lowest in over 100 years; spring sources hard hit as the autumn rainfall was insufficient to prevent flow recession, which began in a dry spring and continued until January 1922 in many parts;
- 1933–34 Wales and mid-England. Two dry summers with a remarkably dry winter intervening;
- 1943–44 Southern England. A similar pattern to 1933–34; very low flows experienced in spring-fed rivers because preceding years were also dryish;
- 1949   Exceptionally low summer rainfall and high temperatures affected sources reliant on river flows or with little storage;
- 1959   Similar to 1949;
- 1975–76 Many low flow records broken because of low summer rainfall;
- 1988–92 South and East England. A succession of dry winters taxed groundwater supplies; runoff deficits in parts of east England were the largest for 150 years;
- 1995–96 Pennines. Two dry summers with a very dry winter intervening;
- 2004–06 Southeast England. Two consecutive dry winters combined with elevated temperatures; steep declines in reservoir storage and groundwater levels.

There was no predictable pattern to these diverse low rainfall events. Hence, when estimating the minimum yield of a source it is necessary to bear in mind the types of drought that past experience shows are possible.

An estimate of drought rainfall for catchment modelling can be made either by using a knowledge of recorded minimum rainfall for a region, expressed as a percentage of the average, or by carrying out a statistical analysis (Tabony, 1977) of available rainfall measurements.

3.12 MINIMUM RATES OF RUN-OFF

In temperate climates with variable rainfall, when minimum runoffs are expressed as rates per unit catchment area, it can often be seen that catchment geology and topography are the major influences, except where human activity has interfered. Clearly the dry weather flow of many small catchments is zero; and bournes, which are streams flowing strongly when the water table is high, dry out gradually from their headwaters as the water table level falls away from the stream bed. In temperate zones where rainfall occurs throughout the year, large rivers do not dry up.

Several studies have attempted to predict minimum flows of specified severity after regional analysis of flow records. Notable examples include those for: Malaysia (Enex, 1976), Europe (Gustard, 1989), New York State (Darmer, 1970) and that produced in 1992 by the IoH, now the Centre for Ecology and Hydrology (CEH) for the UK (Gustard, 1992). The IoH provided formulae for estimating drought flows based on a detailed classification of 29 types of soil systems. A summary national equation for UK, which accounted for 62% of the variations encountered, is:

$$\text{l day mean flow exceeded 95\% of the time, } Q_{95(1)} = 44B^{1.43} \, S^{0.033} \, A^{0.034}$$

where $Q_{95(1)}$ is the 1 day flow, expressed as a percentage of the long-term average daily flow, B is the Base Flow Index (BFI) (Table 3.4); S is the Standard Average Annual Rainfall 1941–70 in mm; A is the catchment area in km^2.

$$\text{Mean annual 7-day minimum flow, } MAM(7) = 6.40 \, Q_{95(1)}^{0.953} \, S^{-0.0342}$$

where $MAM(7)$ is expressed as a percentage of the long-term average daily flow.

The BFI represents the proportion of river flow, which it is estimated to be derived from underground storage. IoH developed a procedure for evaluating BFI from an analysis of a long sequence of daily flows by locating the minima of consecutive non-overlapping 5-day flow totals. The analysis then searches for the 'turning points' in the sequence of minima, connecting them together to form the estimated baseflow hydrograph. Figure 3.2 shows a graphical representation of the procedure. The BFI is the volume of flow below the baseflow hydrograph, divided by the total flow for the same period.

The BFI values for over 1300 gauged catchments in UK are given in the latest issue of the CEH/ British Geological Survey publication *Hydrological Data UK: Hydrometric Register and Statistics*. For any ungauged catchment an estimate of the BFI can be obtained either by interpolation between the values for upstream and downstream gauging stations, quoted in the Hydrometric Register for the same river, or by transposing the value for a gauged catchment with a similar average annual rainfall, surface geology and soil type. The approach has been adopted by a number of countries for analysis of their river low flows, the advantage being that annual values of BFI tend to be more stable than other low flow variables. Details of the method are given in the IoH's Low Flows Study (1980), the predecessor to IoH Report 108 (Gustard, 1992). The range of BFI values typically applying is shown in Table 3.4.

FIGURE 3.2

Derivation of base flow index.

Table 3.4 Typical baseflow indices for various rock types

Dominant characteristics			
Permeability	**Storage**	**Example of rock type**	**Typical BFI range**
Fissured	High storage	Chalk	0.90–0.98
		Oolitic limestones	0.85–0.95
	Low storage	Carboniferous limestone	0.20–0.75
		Millstone Grit	0.35–0.45
Intergranular	High storage	Permo-Triassic sandstone	0.70–0.80
	Low storage	Coal measures	0.40–0.55
		Hastings Beds	0.35–0.50
Impermeable	Low storage at shallow depth	Lias	0.40–0.70
		Old Red Sandstone	0.46–0.54
		Metamorphic-Igneous	0.30–0.50
	No storage	Oxford & London Clay	0.14–0.45

Many of the formulae presented in IoH Report 108 have now been incorporated into the software package 'Low Flow 2000' (Young, 2003), a GIS-based decision support software system that has been adopted by EA and S EPA (the Scottish Environment Protection Agency) as a best practice tool for estimating low flows in ungauged catchments throughout the UK. The system includes extensive data on artificial influences to enable these to be taken into account but, because of the size and complexity of the software, it is not generally available to the average practitioner.

3.13 MAXIMUM RAINFALLS

Figure 3.3 is a plot of the maximum measured rainfalls recorded at individual points in the world. It must be stressed that the highest falls are precipitated only in the most unusual hill areas in certain climatic zones, once the duration exceeds one day. In addition, the lines do not represent any continuous event, except possibly for a storm lasting for up to, say, four hours. The maximum figures for Britain are seen to be about 20–30% of world maxima, with the lowland easterly part of the country suffering less severely in long duration storms. The greatest is the Martinstown, Dorset, storm of 1955 in which 280 mm (11 in) of rain were officially noted in 18.5 hours, with an unofficial estimate at the heart of the storm claiming 350 mm (14 in).

Maximum rainfalls vary with the season of the year because thunderstorm intensities are associated with high sea and air temperatures. However, high rainfalls over a day or so may occur at any time of year, whenever the weather system can bring in moist air steadily and where conditions (often orographic) exist to cause precipitation.

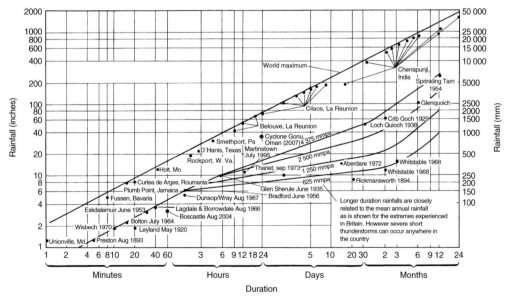

Sources: Paulhus, J. L. H. in Indian Ocean and Taiwan Rainfalls Set New Record. *Monthly Water Review,* **93**(5), 1965, pp. 331–335; Dhar, O. N. and Farooque, S. M. T. in A Study of Rainfalls Recorded at Cherrapunji. *Int Assoc Hydrol Sci,* **XVIII**(4), 1973, pp. 441–450 and Black & Veatch.

FIGURE 3.3

Maximum recorded world and British rainfall.

It is quite possible for many years to elapse without any outstanding maximum rainfall event and then for several unusual maximums to be clustered close together, probably due to high concurrent sea temperatures and dominant weather system movement routes. Many countries have compilations of extreme meteorological events and these should not be overlooked. Many national metereological agencies make available long period rainfall measurements and duration-intensity-frequency estimates. These are often adequate for estimating storm magnitudes that can be expected more frequently than once every 50 years; but possible rarer storms need to be investigated by thorough regional studies. In the UK the most up-date information is contained in Volume 2 of the Flood Estimation Handbook (FEH, 1999) and shows how rainfall frequency calculations can be applied to UK catchments, leading to estimation of either the rainfall depth for a given return period and duration, or of the return period corresponding to a given depth and duration of rainfall. The parameters of the rainfall frequency model are provided digitally on FEH CD-ROM to a 1 km grid.

Improved rainfall depth-duration-frequency values for use in UK flood studies should become available from the Defra project "WS194/2/29, Reservoir Safety–Long Return Period Rainfall", due for completion in late 2008. A project report by Svensson (2006) provides a review of rainfall frequency and probable maximum precipitation estimation methods, both in the UK and several other countries.

3.14 MAXIMUM RUN-OFFS

Peak flood discharges are notoriously difficult to measure directly because of their transient nature, high velocity, large debris content and the difficulty in gaining access to the river during times of flood. The problem is exacerbated when flows come out of bank and significant discharge occurs across the flood plain. Peak discharges can, however, often be estimated indirectly by means of hydraulic calculations using sediment or debris marks left behind by the flood to indicate peak water level, cross section area and water surface slope.

Figure 3.4 gives some recorded maximum runoff data experienced in UK. In lowland areas, peak runoff rates are normally much lower than Figure 3.4 suggests because of the temporary storage available in side channels and drains or in adjacent low lying land and the greater likelihood of drier soil conditions before a storm. For example in the fenland area of eastern England entire catchments may be drained quite adequately by pumping stations capable of pumping no more than 13 mm of runoff from their catchment per day. This is very much less than the maximum precipitation rates shown in Figure 3.3 where rainfalls up to about 250 mm in 24 hours can occur in Eastern England, as shown by the plot for Thanet. These figures are not however, large compared with experience elsewhere in the world where ten times higher rates can be experienced.

There are four methods of estimating maximum runoffs; the fourth applies specifically to the UK.

1. ***Probability analysis of existing flood records.*** This method plots recorded peak floods against a probability distribution, as described for the analysis of droughts in Sections 3.18 and 3.19, in order to estimate the probability of occurrence of a flood of given magnitude. The type of probability distribution used is that which gives a best straight line plot for the recorded floods. Usually the peak flood for each year of record (the AMAX series) is plotted; but sometimes all peak flows above a given level are plotted (a peaks-over-a-threshold analysis). In the latter case care has to be taken to ensure that the events plotted are truly independent and that they are counted per 'water year' (i.e. summer plus winter) and not per calendar year.

Source of information: ICE (1960) and others.

FIGURE 3.4

Flood data in Britain in and since the ICE 1933 report.

There are limitations to the value of this method because a probability plot of past annual maximum floods can only be extrapolated reliably to estimate return periods up to about twice the length of the period of record. Thus, a 30-year record of annual maximum floods cannot be safely used to estimate the magnitude of a 100 year return period flood. In addition, the accuracy of past flood records cannot be checked and catchment conditions may have altered since records were taken or may alter in the future.

2. *Use of regional flood probability curves.* A different approach, applicable worldwide, is to use 'regional flood probability curves' which are published for many regions (Fig. 3.5). The mean annual flood for a given catchment, termed 'the index flood', is obtained from the period of historic record. A 'growth factor' (i.e. multiplier) taken from the appropriate regional flood probability curve is then applied to the index flood to give the probable magnitude of a flood of given return period T. Published flood probability curves are derived from analyses of flood magnitude frequencies for catchments in the same region possessing similar characteristics. In the absence of an adequate historic record, the mean annual flood may be estimated using formulae based on regression analyses with various catchment characteristics such as catchment area, average annual rainfall, soil type or average catchment slope.

   In the UK, the latest approach is to adopt the median annual flood as the 'index flood', rather than the mean. The FEH (1999) also introduces the concept of a 'pooling group' of catchments with similar characteristics to replace geographical regions. Full details of the method are given in the Volume 3 of the handbook.

3. *Flood estimation derived from maximum precipitation.* In this method the maximum precipitation that has occurred, or may occur, on a catchment is converted to consequent flood flow using either a unit hydrograph derived from the catchment, or a 'synthetic' hydrograph

Sources: UK—Law, F. Inst. of Hydrology. Elsewhere—Meigh, J. R., Farquharson, F. A. K. and Sutcliffe, J. V. *Hydrological Sciences Journal,* 42(2), April 1997, pp. 225–244.

FIGURE 3.5

Examples of regional dimensionless flood frequency curves.

whose derivation assumes the most unfavourable catchment conditions that are likely to pertain during a probable maximum flood event. The method is of wide application because use of the unit hydrograph permits the calculation of flood flows from a wide range of rainstorms. Both approaches are similar but as the second is principally used for designing impounding reservoir overflows, it is described in Chapter 5, Section 5.6.

4. ***Flood estimation in UK.*** The principal guidance for estimating flood magnitudes in UK and Ireland is now Volumes 3 and 4 of the FEH (1999). Alternative methods are presented for flood frequency estimation on any catchment, gauged or ungauged. Volume 3 provides statistical procedures intended principally for use for return periods between 2 and 200 years. However, these statistical procedures should not be used for assessing the required design capacity of impounding reservoir flood overflow works, except where no loss of life can be foreseen as a result of a dam breach and very limited additional flood damage would be caused (Section 5.6). Volume 4 uses a unit hydrograph rainfall-runoff method and is applicable to a wider range of return periods than the statistical approach given in Volume 3. The rainfall-runoff method uses Volume 2 procedures for estimating depth and duration of rainfall for a given return period, or vice versa. This permits the estimation of floods for return periods between 2 and 10 000 years and derivation of the probable maximum flood from estimates of the probable maximum precipitation. In addition, up-dated summaries of flood peak data and flood event data are presented in Appendices to these volumes.

The procedures set out in Volumes 3 and 4 are relatively complex by reason of the number of influencing and limiting factors that need to be taken into account. Although the FEH provides much useful guidance on the choice and use of method, the application of the procedures to any particular case in the UK or Ireland is best undertaken by an experienced hydrologist. Issues identified with the FEH are mentioned in Section 5.17.

PART II   YIELD OF SURFACE SOURCES

3.15 INTRODUCTION, DEFINITIONS AND CONCEPTS

The term 'yield' does not have a precise meaning. Its definition varies according to the context, which is important to understand when interpreting yield values quoted by others. Similarly, it is essential to state clearly the basis of yield calculations when quoting a yield figure to others.

The following terms are used to define yield more precisely or to indicate the way in which yield has been calculated:

- hydrological yield
- source yield
- average yield
- probability yield
- operational yield
- system yield
- historic yield
- failure yield

Terms such as 'deployable output' or 'water available for use' (WAFU) are sometimes understood to be general yield terms whereas in fact they have precise meanings in legal or regulatory contexts. The meaning of all of these terms is discussed in the following paragraphs while examples of their use are given later in the chapter.

Understanding the difference between terms in the first two rows of the above list is fundamental to appreciating the subtleties in more precise definitions. For surface water sources, *hydrological yield* is limited by the overall availability of rainfall and river flow, perhaps boosted by storage to iron out short-term shortages. However, the water available to satisfy demand may be less than that available in purely hydrological terms. It may be limited by hydraulic constraints such as pump, intake or pipe capacity or by the terms of an abstraction licence or agreement. Hydraulic yield so constrained is referred to as *operational yield*. In 1996, the UK Government published its *Agenda for Action* which introduced a new framework for the assessment of water resources in the newly privatized UK water industry. The paper assigned fairly precise definitions within the regulatory framework to terms like *deployable output* and *water available for use*. Their relationship to other terms like 'hydrological yield' is illustrated in Plate 1(a).

Deployable output (DO) is the constant rate of supply that can be maintained from a water resources system except during periods of restriction. DO may be constrained by the specified 'Level of Service'(LoS); the historic period for which data are available; the physical capacity of the supply system (pipes, pumps, intakes etc.); abstraction licences or permits; and water quality and environmental constraints (UKWIR/EA, 2000).

From the above, it can be seen that quoting a DO figure without defining the constraints under which it was calculated is usually rather meaningless. Levels of service are discussed below under 'source reliability' while other constraints are discussed in sections 3.18 to 3.24 in relation to specific types of sources.

The supply figure that must be relied on is *Water Available for Use* (WAFU). This can be compromised by Outage, which is the temporary loss of DO due to planned events, such as maintenance of source works, or unplanned events such as power failure, system failure and unacceptably high levels of pollution; turbidity, nitrates and algae. Outage must therefore be added to WAFU to get DO.

More often than not, the yield quoted for a source is its *drought yield*; the amount of water the source can produce in times of drought. However, for much of the time the source can provide more than this. When water is plentiful, higher rates of output can support higher rates of demand or may enable other sources to be rested. As rainfall and river flows decrease in dry periods, output is reduced to ensure that supplies can be maintained throughout a period of drought. The long-term output of the source through wet and dry periods is called the *average yield*. This is a good indicator of source potential and its overall contribution to satisfying demand. However, drought yields are the figures most often quoted since they can be relied on most of the time.

Droughts vary enormously in their severity and duration so when quoting a drought yield, the drought parameters must be defined. Drought severity can be considered in terms of the quantity of rainfall or runoff that occurs in a given period and the frequency with which this happens. The definition of yield that incorporates this combination of quantity and frequency is called the *probability yield*; the source output that can be maintained throughout a drought of given severity or frequency.

In sparsely populated or remote areas with scattered communities, a single water supply source is often adequate to supply the needs of the local population. The yield of this source is referred to as *source yield*. As the population grows and communities merge into towns and conurbations, these single sources are often connected by distribution systems so that any one community may be served by more than one source. The yield of these linked systems is termed *system yield*.

Linked systems can also realise benefits from the *conjunctive use* of different types of sources. Most surface water sources face a reduction in water availability in the summer or dry season. The early summer is a time when groundwater storage is normally at its maximum following winter recharge. The linking of these different types of source may increase security of supply by taking advantage of their differing yield characteristics. This introduces the concept of *critical drought period*.

In essence, the yield of any source is defined by:

$$\text{Yield} = \frac{\text{inflow over the critical period} + \text{any storage that is available}}{\text{Length of the critical period}} \tag{3.1}$$

At a run-of-river intake, there may be no storage available to buffer brief periods of minimum flow. The critical period in this case would be zero and yield is effectively the minimum river flow of specified frequency. Impounding reservoirs provide storage, which can maintain supplies above the rate of minimum flow for a certain period. How long this period is depends on how much storage is available and how long flows remain below the target rate of drawoff; i.e. drought duration (Plate 1B and Section 3.19).

The period over which available storage is drawn down is critical to determining the size of the yield, hence *critical period*. The concept of *critical drought period* is fundamental to both yield assessment and water supply system design. The critical period of a single source system is relatively easy to determine. However, once a number of sources of different types are linked together and subject to complex operating rules, the critical period is more difficult, if not impossible, to define as it varies with system configuration and with different operating rules. This is one of the main reasons why yield assessments in the UK are generally required to use system simulation modelling (Section 3.24).

All water utility managers have to be aware that almost every source has a quantifiable risk of partial failure. However, *source reliability* can be viewed from two perspectives:

- climatic or hydrological variability;
- the capacity or constraints of the water supply system.

The method used to quantify yield depends on which of these perspectives is the focus of concern.

Hydrological parameters such as rainfall and river flow have a natural variability and ultimately their availability will limit the yield attainable. Individual sources may be developed to abstract all of the water available in times of drought. Their output is therefore related closely to hydrological variability and attempts are made to describe this variability by comparing it to a statistical or probability distribution (Section 3.18).

In regions where water is relatively plentiful and living standards are high, supply systems are developed in stages in response to the growing demand of the population and the needs of the environment. As demand rises, the capacity of source works is increased or new sources are developed to satisfy that demand for a given level of reliability. The required reliability may be specified by a central planning authority as one of the criteria in system design. In a privatized water industry, a reliability standard may be set or imposed by regulation, the standard being determined by a combination of cost, customer preference and environmental implications. Typically, such a standard would be one of a number of *levels of service* to which the utility managed its supply. Relevant typical levels of service that have been used in the UK are:

- a hosepipe ban not more than once in 10 years on average;
- a need for a major public campaign requesting voluntary savings of water not more than once in 20 years on average;
- Drought Orders imposing restrictions on non-essential use not more frequently than one year in 50 on average;
- Drought Orders authorising standpipes or rota cuts not more frequently than one year in 100 on average.

Such levels of service are not statutory requirements but are in essence a contract between customer and supplier setting out the standards of service that customers can expect to receive. Reliability involves a balance between the needs of customers and the needs of the environment and as such, it may change with time. This affects the deployable output of the system. If DO is set too high, cut-backs in supply will be too frequent and the levels of service will not be met; the DO then has to be reduced until the levels of service are achieved.

Hydrological yield is usually calculated on the assumption of drawoff to supply at a constant target rate. However, it has always been recognized that water supply sources are not operated in an "on/off" way. If the target rate cannot be maintained, a lesser rate almost certainly can. In reality, water resources managers do not wait until supplies are about to run out before initiating cut-backs. System operation rules normally specify that target drawoffs are reduced when reservoir storage or river flows fall below pre-defined thresholds. Such thresholds may be stipulated in the terms of an abstraction licence or agreement. System operating rules are often linked to levels of service.

The types of analyses that are involved in the 'hydrological' and 'operational' approach to yield estimation are discussed in sections 3.18–3.24.

Based on information from: *Water Resources & Supply: Agenda for Action* (DoE, 1996)

(a) Relationship between yield-related definitions and terms.

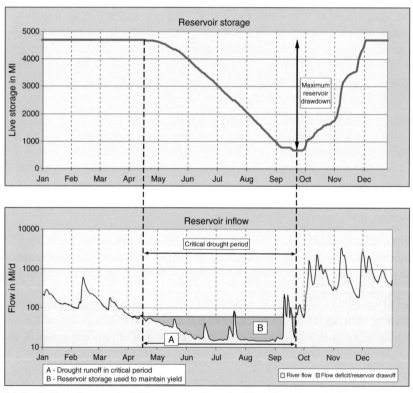

(b) Illustration of reservoir yield components and critical drought period.

PLATE 1

(a) A 450 mm percussion chisel.

(b) A 'down-the-hole' compressed air operated hammer drill.

(c) A roller rock bit with reverse circulation water supply.

PLATE 2

3.16 HISTORY OF YIELD ESTIMATION IN THE UK

In England and Wales, the original approach to yield estimation was to adopt historic yields based on experience in major historic droughts. These yields normally referred to single sources or small groups of sources.

In the late 1960s and 1970s there was a change towards probability yields based on statistical analyses of historic drought flow sequences. These analyses allowed synthetic flow sequences to be constructed to represent droughts of varying duration, severity or return period.

As individual sources increasingly became merged into linked systems for which complex operating rules were sometimes devised, the yield of individual sources became less relevant and the difficulty of calculating yield for such systems became more apparent. In 1995–96 there was a notable drought across much of the country and the Secretary of State for the Environment initiated a review of the lessons to be learned for water resources and supply arrangements, both for the ongoing drought and for the longer term. The Government consulted widely amongst interested parties and the results of the review were published in a paper entitled *Water Resources and Supply: Agenda for Action* (DoE, 1996). Annex E of the paper outlines the framework in which the yield of surface water sources for public water supply are to be assessed in the UK on a consistent basis.

The framework embodies two main principals:

- yield is to be assessed by simulation of the realistic operation of the water resources system in question; and
- yield is defined as the supply that can be met with a given level of service.

Effectively, this means that a computer simulation model of the water resources system is required (Section 3.24). However, not all water supply sources occur in integrated water supply systems, particularly in sparsely populated or developing countries. Different source types have different characteristics and hence methods of estimating their yield vary.

3.17 METHODS OF YIELD ESTIMATION—GENERAL

Equation (3.1) describes the basic yield calculation for all single sources. Any yield assessment relies on historical data sufficient for the purpose. However, it must be accepted that available data are unlikely to reveal a drought that would be critical to yield determinations. In arid or semi-arid regions, rivers frequently dry up altogether for at least part of the year. Without storage, the 'historic yield' of such streams is zero. However, they may not dry up every year and so some water is available in certain years. The more water that is required, the lower the frequency at which it is available. This combination of quantity and frequency is called the *probability yield*.

3.18 RIVER INTAKE YIELDS

By far the most reliable estimate of river intake yield is obtained by analysing a long, reliable record of river flows at or near the point of abstraction. If there are no restrictions on the quantity of water that can be abstracted, yield for a given level of reliability (i.e. a probability yield) can be obtained from a statistical analysis of the available flow record. It is important that the flows are natural (i.e. free of any artificial influences such as upstream abstractions, discharges or river regulation effects) so that the observed variability in the flow record can be assumed to conform to an appropriate statistical distribution. This is likely to be true if the flow variability is natural but it is very unlikely to be true if the record contains artificial influences. If such influences are present in the flow record, the record must be naturalized (Section 3.9). Statistical analyses can then be performed on the natural flows to derive estimates of flow for a given frequency/level of reliability. Alternatively, a rough assessment may be made by ranking the low flows and plotting them on probability paper (Fig. 3.6). If appropriate, estimates of the artificial flow component are then added back in to give a total flow figure. This is important because in some rivers effluent returns from industry or treatment works are a major component of river flow in the dry season. If of sufficient quality, these effluent returns may be a vital resource and so cannot be ignored in any yield assessment.

As with all yield assessments, the question of critical period must be considered. For a water supply system fed from a river intake that has no storage, the critical period is zero. In practice, most water supply systems have at least some storage (e.g. bankside storage at the intake or service reservoirs within the distribution system) that can provide a buffer to maintain supplies through brief shortages. Bankside storage reservoirs may typically store the equivalent of 7 days supply to safeguard the system against pollution incidents in the river when the intake may have to be closed.

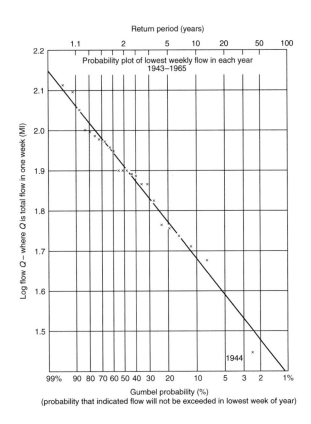

FIGURE 3.6

Naturalized dry weather flow probability plot.

The critical period for the source would then be 7 days and source yield would be determined by the average flow (of appropriate frequency) over a 7 day period. This is likely to be a much more stable value than daily flow, which can be subject to erratic changes due to gate operation or other artificial influences upstream. The shorter the critical period, the more important it is to base the analysis on daily flows. If only monthly data are available for the location in question, a scaling factor should be applied to the monthly values to reduce them to the relevant shorter period. An appropriate factor can be obtained from an analysis of daily low flow sequences from a flow record on a similar or nearby river in the region.

The above discussion relates to the determination of the *hydrological yield* of a river intake i.e. water availability is limited only by climatic variability and hydrological factors. In reality, the yield of river intakes may be constrained by a number of additional factors including the following:

- Licence or permit constraints: abstraction limits linked to river flows; compensation flows or environmental sweetening flows; and the rights of other users;
- Hydraulic capacity of the intake;
- Water quality constraints: tidal effects and saline intrusion; pollution etc.

In modern UK parlance, this constrained figure tends to be referred to as the 'deployable output' of a source.

If the *hydrological yield* of the river is greater than the physical capacity of the intake and any pumps involved, these would constrain source yield. Such factors could be inherent in the design or have resulted from post-construction morphological changes in the river channel (erosion, deposition or changes in river course).

Many river intakes are located near the tidal limit of rivers in order to be able to exploit the maximum potential hydrological yield of the catchment. Even where water levels at the intake exhibit tidal influence (e.g. a twice-daily cycle) the water itself may remain fresh in all but the most extreme circumstances. These usually occur when extreme high tides coincide with low river flows allowing saline water to intrude further up the estuary than normal and perhaps reach the intake. The saline water (intrusion) travels up the estuary as a wedge (due to its greater density) with freshwater on top. The depth from which water is drawn into the intake may therefore determine the increase in salinity. The frequency of such an event determines whether it affects yield. Rare events are likely to be treated as *source outages*, in which case they would have no effect on quoted yields or deployable output. Frequent events may mean that for high levels of reliability (low probability of failure) the source has no reliable yield at all. This situation might arise due to continuing increases in upstream abstractions.

Water quality problems at intakes are not restricted to saline intrusion. A serious pollution incident in the river upstream could cause the intake to be closed for a time. As mentioned earlier, some intakes have a bankside storage provision to provide supplies during the period of closure. As with saline intrusion, intake yields may become affected if the water quality problems become more frequent or protracted. Background levels of pollution by chemicals such as nitrates (from fertilizers) are one example. High nitrate levels are often associated with the first period of significant runoff after a long dry spell when nitrate build-up in the soil is washed into rivers. It may be possible to model this effect using long sequences of river flow record if the relationship between nitrates, flows and dry periods can be adequately defined. The risk of a pollution incident affects yield if the river intake is a single source; it is less likely to affect the yield if the source is used conjunctively with other sources.

3.19 YIELD OF DIRECT SUPPLY IMPOUNDING RESERVOIRS

The yield of a direct supply impounding reservoir is defined by the following equation:

$$\text{Yield} = \frac{\text{Storage} + \text{inflow over the critical drawdown period}}{\text{Length of the critical drawdown period}} \qquad (3.2)$$

The critical period (Section 3.15) is the time from when the reservoir first starts to be drawn down to the time when it is at its maximum drawdown. It is the period over which storage is used to supplement river flows to maintain a reliable yield, the larger the available storage, the longer the critical period. The other main factors determining yield are the length of the dry season, over which flows have to be boosted, and the volume of reservoir inflows during this period. If winter rains are regular and the reservoir fills every year, the critical period is short, perhaps 6–10 months. If the storage is small, it may be even shorter. If winter rains occasionally fail, droughts may extend over 2 years or even longer. In this case, storage has to be eked out over a longer period making the critical period longer. Plate 1B illustrates the principles involved. The volume of water that supports yield is A (the volume of inflow over the period) plus B (the volume of storage used to maintain the required supply over the same period). So if the volume is X million litres and the critical period is Y days the reservoir yield is X/Y = V Ml/d.

The historic yield is calculated by identifying the longest period in the historic period that has the smallest volume of inflow (per unit of time). To get a robust yield figure you need the longest period of record you can in order to ensure it contains a severe drought that is unlikely to be exceeded often. This may require the reservoir inflow record to be extended back in time (by flow correlation or rainfall/runoff modelling) if the direct record is not long enough. Figures 3.7 shows

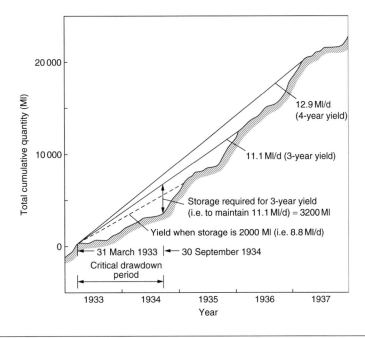

FIGURE 3.7

Five-year mass flow diagram.

a method for assessing historic yield graphically; cumulative catchment run-off is plotted over the period of record.

If a *probability yield* is required, the drought of required duration (the critical period) with that probability of recurrence (return period) must be identified in the historic record. The chances of finding one are slight so a 'synthetic drought' is sometimes constructed from available runoff records. For the design of a new reservoir, many of the variables such as storage volume, compensation flow etc. may be undecided and a number of options may need to be assessed. Each option may have a different critical period. It is very unlikely that data will be available for a historic drought for which all the critical periods of interest have the same return period. One way of getting round this is to construct a nested synthetic drought in which runoff of all durations has the same return period so that, in say a 36 month sequence of river flows, the driest 6, 7, 8, 15, 18, 24, etc. periods all have the same probability of recurrence; say 2% (1 in 50 years). An illustration of this if shown in Figure 3.8. Another graphical method, illustrated in Figure 3.9, is to express the minimum yield as a proportion of the mean catchment flow over the period of record and the reservoir storage as a proportion of the mean annual flow volume.

Most of the commercial software available to simulate complex water resources systems summarized in Section 3.24 can also be used to simulate individual reservoirs. Alternatively, a spreadsheet may be used with parameters such as inflow, storage, outflow etc. arranged in columns across the page and values for each time step arranged in rows down the page. The elevation/storage and/or elevation/surface area relationships (Section 5.18) can be represented as lookup tables. As the time step for such calculations is rarely less than a day and often as long as a month, reservoir spills,

FIGURE 3.8

Minimum runoff diagram.

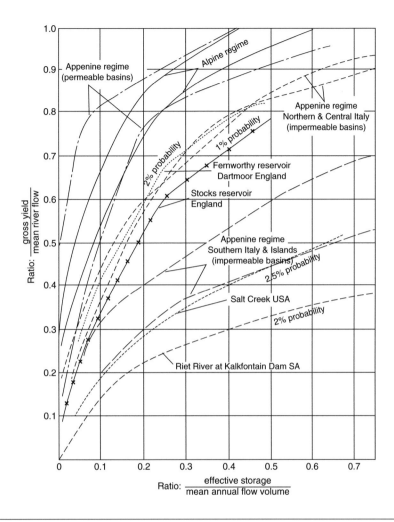

FIGURE 3.9

Typical yield/storage relationships for direct supply reservoirs.

if any, are simply the balance of inflows minus outflows when the storage is full. There are no hydraulic considerations involved.

Evaporation from the surface of a water body, such as a reservoir, is greater than from the equivalent area of land surface. Potential evaporation is higher because the albedo (reflectivity) of the reservoir surface is lower and more of the sun's energy is available to drive evaporation. However, the ratio of actual to potential evaporation is also higher because of the availability of water to evaporate. In temperate regions, the increased losses from the surface of the reservoir may not be significant, especially in the case of a relatively small reservoir in a large catchment. However, in arid or semi-arid zones, such losses can be very large and should be allowed for in the yield calculation. In the simplest calculations, daily or monthly evaporation figures can be applied to the average surface area of the reservoir. However, if a spreadsheet model is used, it is straightforward to apply evaporation figures to the area of the reservoir at each time step, computed using a storage/elevation/area relationship.

Other factors that should be taken into account in a reservoir simulation include drawoff, licence stipulations, compensation flows and emergency storage allowance.

3.20 YIELD OF A PUMPED STORAGE RESERVOIR

To estimate the yield of a pumped storage reservoir it is more accurate to work on the basis of daily flows. The use of mean monthly flows is inexact and necessitates reducing the computed yield by an arbitrary amount to allow for part of the flows in excess of the mean being uncollectable. If some daily flow records are available, these may assist in estimating what allowance should be made; otherwise a frequent practice is to assume that only 90% of the potential abstraction is possible. The better option is to develop a series of daily flows that represent flows during a period of minimum runoff. The method involves use of minimum monthly flows of a suitable probability (say 2%) and applying to them a factor derived from a relevant, albeit short, record of daily flow for one of the months.

With daily flows available, the basic calculations are as shown in Table 3.5. However, the sub-calculation required to calculate potential abstraction depends on: (a) the rule(s) laying down the abstraction conditions; (b) the assumed maximum pumping capacity; and (c) the time lag between a change of flow and the consequent change of pumping rate.

Factors (b) and (c) need careful consideration before starting the calculation. With appropriate control equipment and variable speed pumps, the abstraction may closely follow flow variations; but other arrangements are often adopted. A range of fixed speed pumps may be started and stopped automatically to give outputs such as *0.5Q, 1.0Q, 1.5Q* or *2.0Q* according to their combination. In addition, if pump output is manually controlled, the time lag between change of flow and change of abstraction will depend on the manning pattern and on whether variable or fixed speed pumps are used.

A simplistic calculation can be based on assuming an unlimited volume of reservoir storage is available so the maximum drawdowns occurring each year can be calculated and plotted on probability paper to find the drawdown probable once in, say, 50 years. However, this does not solve the problem of what yield is available with a reservoir of a lesser capacity than the maximum drawdown. Calculations then have to proceed for a range of drawoffs to produce a yield-storage

Table 3.5 Pumped storage reservoir calculation in 10^3 m³/day (Abstraction condition—two-thirds flow above 30×10^3 m³/day)

Day	Flow to intake	Quantity available at intake	90% pumped to storage	Amount supplied ex storage	Reservoir Contents	
					Net change of storage	In store end of day
15.8	42.0	8.0	7.2	5.5	+1.7	151.7
16.8	39.0	6.0	5.4	5.5	−0.1	151.6
17.8	38.0	5.0	4.5	5.6	−1.1	150.5

relationship with an associated risk in order to be able to determine the yield for a given storage for any degree of risk. This process is difficult to carry through successfully unless the flow record is long enough to make the probability plot one for interpretation only. Points from an unnatural distribution do not permit confident extrapolation.

In view of these difficulties, it is more satisfactory to work directly on an estimate of the daily flows likely during, say, a drought of 2% risk. It is then possible to compute the yield of 2% risk according to the size of storage reservoir adopted and the three factors mentioned above. A further limiting consideration is the need to ensure refilling of the reservoir after some maximum drawdown. The most secure provision is to ensure refilling of the reservoir in any single wet season following a dry season, but this is not always possible. Some pumped storage schemes accept that refilling will only occur once every few years, but they should be tested to make sure they can achieve an initial filling sufficient to meet the anticipated initial demand.

All such problems can be dealt with by applying appropriate computer calculations to the record of daily flows. Among the most useful result is that of finding the most economic size of pumps to install. Increasing pump capacity beyond a certain level may increase the yield by only a small amount when flows fall rapidly and critical drawdown periods are short. On the other hand, the need to ensure refilling of the reservoir during a wet period may be an over-riding factor determining maximum pumping capacity.

3.21 YIELD OF REGULATING RESERVOIRS

A regulating reservoir (Fig. 3.10) impounds water from a catchment A, and releases water to support an abstraction at some location B downstream, when flows at B are not sufficient to meet the required abstraction. This means that the yield obtainable is greater than that provided by catchment A alone. Usually some compensation water has to be released from A to maintain a flow in the stream below it; and at B various abstraction conditions may apply in order to preserve the natural low flow regime of the river. This may require either:

■ maintenance of a given flow continuously below the intake B; or
■ abstraction at B to be fully supported by equivalent releases from A until the natural flow at B reaches a certain figure; thereafter, as flows at B continue to rise, releases from A are cut back until the natural flow at B is sufficient to support the whole abstraction.

FIGURE 3.10

Regulating reservoir.

As with pumped storage schemes it is best to calculate minimum yield by first producing daily flows for, say, a 2% drought period for the natural flows at the intake point B. The proportion of this flow diverted into the regulating reservoir A has to be assessed and probably bears a varying relationship to the natural flow at B. The calculation can then proceed according to the rules laid down for compensation releases at A and the abstraction conditions at B. An example is given in Table 3.6.

Two points need to be borne in mind. Releases from the reservoir may need to include an allowance for evapotranspiration and other losses en route to the abstraction site (hence the 10% addition to Col. 4 in Col. 5 of Table 3.6). Secondly, the time taken for released water to travel to the abstraction point must be taken into account. Depending on the distance involved, the time lag may range from several hours to 1 or 2 days or more. This can mean that an increase of release in expectation that flow at the intake will decline may be wasted if rain should come and increase the flow. Hence, a further allowance of 10% or 15% may have to be added to the releases to cover discrepancies between actual and theoretical release requirements. A dry weather recession curve for the natural flow at the intake, converted into a guiding rule, can be used to aid release decisions. Use of such a decision rule in UK has shown that actual releases tend to be up to 20% more than the theoretical requirement in wet years and about 3% more in dry years.

Table 3.6 Regulating reservoir calculations. Conditions: abstraction required 28 Ml/day. Reservoir compensation release 2.0 Ml/day. Residual flow below intake 10 Ml/day

Day	(1) Natural flow at intake 1000 m³	(2) Reservoir inflow 1000 m³	(3) Net flow at intake 1000 m³	(4) Intake flow deficiency 1000 m³	(5) Reservoir release 1000 m³	(6) Reservoir change/day 1000 m³	Cum 1000 m³
1	45.0	3.3	41.7	Nil	2.0	+1.3	+1.3
2	41.0	3.0	38.0	Nil	2.0	+1.0	+2.3
3	38.0	2.8	35.2	0.8	2.9	−0.1	+2.2
4	35.5	2.6	32.9	3.1	5.4	−2.8	−0.6
5	33.0	2.4	30.6	5.4	7.9	−5.5	−6.1
6	etc.						

Col (3)	= Col (1) − Col (2)
Col (4)	= Abstraction + Residual flow − Col (3) − Compensation release
Col (5)	= Col (4) × 110% + Compensation release
Col (6)	= Col (2) − Col (5)

3.22 **YIELD OF CATCHWATERS**

A catchwater is usually a channel that leads water from some remote catchment into an impounding reservoir. The remote catchment would not otherwise contribute any flow to the reservoir so the flow of the catchwater increases the yield of the reservoir. The yield of the catchwater is the amount of water the catchwater can provide with specified frequency or reliability over a given period. This period would normally be the critical drawdown period of the reservoir so catchwater yield is effectively the amount by which the catchwater raises the yield of the reservoir.

It is very unusual for flows in catchwaters to be measured directly. The contribution made by the catchwater therefore has to be estimated. Unless it is extremely large, the catchwater will not be able to divert all of the runoff reaching it. Flows in excess of catchwater capacity will be lost (unless the overflow occurs within the catchment of the reservoir). Catchwaters often originate at an intake on another stream, or capture flows from streams they cross along their route. Intake capacity and compensation flow requirements may therefore be other limiting factors. In view of the above, catchwater yield requires the evaluation of two main parameters:

a. the volume of water reaching the catchwater in the critical period; and
b. the proportion of total flow that the catchwater can capture.

Parameter (a) depends in turn on the capacity of the catchwater and the variability of potential inflows. One way of determining catchwater capacity is to install a temporary gauge and measure it directly during a period of heavy rainfall. If this is not feasible, the capacity may have to be estimated by hydraulic analysis (according to channel dimensions and gradient).

If the catchwater is to be incorporated into a simulation model of the overall system, parameter (a) must be provided as a long, historic flow series while parameter (b) is calculated by the model for each time step. The time step must be short enough (15 minutes to an hour) to represent the variability of flows in the inflow sequence. If the time step is too long, average flow over the time step may be less than the capacity of the catchwater, suggesting that all of the runoff can be captured.

If the catchwater is not to be modelled using a long sequence of historic flows, the volume of water reaching the catchwater and the proportion of runoff captured have to be calculated separately. The former can be obtained from a flow duration curve (Figure 3.11) for an appropriate period; e.g. an average year or a 2% drought year. The proportion of runoff captured can then be obtained from a catchwater transfer curve such as that shown in Figure 3.12. This shows a relationship between catchwater size and percentage of runoff collectable, derived from hydrographs observed on small British catchments with average rainfalls of 1500 mm or more. However, the relationship was found to hold true against data from a tropical catchment, such as one in Singapore. Although Figure 3.12 strictly applies to average annual flows, it can be used without too much error to assess yield during a given season of the year, provided the appropriate average daily runoff for that season is used in place of the average annual flow.

The catchwater contributions to the reservoir are added to its direct catchment inflow, so that the minimum yields for various consecutive months can be plotted on a diagram such as Figure 3.13. This shows the extra yield a catchwater provides for a given reservoir storage and the change of critical drawdown period.

Since many catchwaters are simply open unlined channels cut to a gentle gradient in a hillside, they often contribute little or no inflow during the dry season, and do not contribute flow to

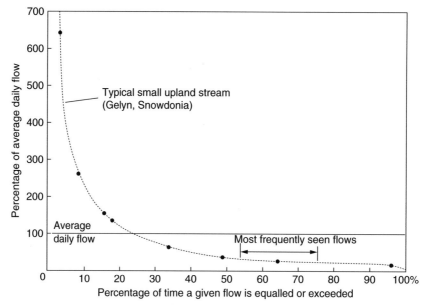

Note: The curve is derived by measuring the number of hours the stream flow exceeds given level of flow.

FIGURE 3.11

Flow duration curve.

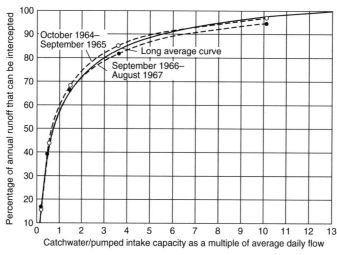

Note: The design curve is based on hourly flow duration data from small mountain catchments in England and Wales.

FIGURE 3.12

Catchment transfer curve.

Note: The figure by each straight line denotes the length of the critical period in months.

FIGURE 3.13

Catchwater yield /storage diagram.

the reservoir until initial precipitation is sufficient to wet their bed and banks to saturation level. Therefore, their main contribution to the reservoir is during the wet season and their impact on yield is more on average yield rather than drought yield.

3.23 COMPENSATION WATER

Compensation water is the flow that must be discharged below an impounding reservoir to maintain the water rights of riparian owners and other abstractors downstream. Each country tends to have its own water law to preserve water rights and setting quantities of compensation water can involve extended legal dispute. In Britain, the compensation water from most impounding reservoirs is set by some Parliamentary Act. In the early 1900s compensation water was often set at one-third of the gross yield of the reservoir, but this proportion tended to reduce to one-quarter in later years (Gustard, 1987). Nowadays compensation water is often required to be varied seasonally and extra discharges as 'spates' may be stipulated at certain times of the year to meet fishing interests.

The discharge of a fixed amount every day has been criticized on environmental grounds as being 'unnatural' and not conducive to the maintenance of fauna and flora, which need periods of varying flow. Considerable progress has been made in quantifying the water requirements at different stages in the life cycle of fish, invertebrates and macrophytic vegetation (Bullock, 1991). American studies of physical habitats have been followed in France, Norway, Australia and UK as a means of defining environmentally acceptable flows. The software calculations with the PHABSIM program, available in the public domain, depend on field measurement of river velocity, depth, substrate and tree cover. They determine ecological preferences and hence seasonal variation of compensation water but, inevitably, not all the requirements can be met if a reasonably economic yield is to result. Hence, some compromise solution has to be found. Nevertheless, the technique gives a satisfactory means of engineering water resource developments to achieve minimum environmental damage.

3.24 YIELD OF WATER RESOURCES SYSTEMS

In relatively sparsely populated regions, water supplies are typically drawn from single sources, particularly where settlements are separated by large distances and it is impractical to link them. However, in more densely populated regions, water supply systems are generally supplied from multiple sources of varying types. There are numerous advantages in having integrated water supply systems: individual sources can be rested or taken out of commission for maintenance without undue interruptions to supply; and water can be transferred from a part of the system with a surplus to an area experiencing a drought. The lack of a fully integrated system contributed to the shortages that were felt in the Yorkshire region of the UK in 1995. However, estimating the yield of an integrated system is more complicated than for a single source, largely as a result of the differing critical drought periods for each source type; the only practical way of calculating the yield of an integrated system is by computer simulation.

In UK Environment Agency Guidelines (EA, 2007) for the estimation of the yield of surface water and conjunctive use schemes, the recommended approach includes behavioural analysis; a model is used to simulate the realistic operation of a water resource system as currently configured using historic data corrected for current catchment conditions. The method further requires:

- the derivation of long, naturalized historic sequences for all inflows to the water resource system being modelled;
- system control rules linked to levels of service which allow for the introduction of demand management practices to cut supplies during periods of drought;
- yield estimates to allow for the provision of emergency storage to accommodate the operational uncertainty regarding the duration of a particular drought.

The hydraulics of water movement around the system is not normally simulated by the model, but realistic physical limitations such as pump and mains capacity are taken into account, as are the limits and conditions imposed by licence agreements.

Inflows to the system need to be for as long a period as possible in order to represent the full range of hydrological variability in the modelled sequence and to allow the company's target levels of service to be tested. If necessary, naturalized historical flow records need to be extended back in time, either by correlation with other long term records or by the use of rainfall-runoff modelling.

Demand is defined at model nodes, which coincide with major demand centres. Base demand is the average demand over the year, to which a seasonal demand profile is applied.

The deployable output of the system is calculated as the output to supply which can be met over the whole period of the simulation with the required levels of service. It is defined by system performance throughout the worst drought in the record and takes account of the varying critical drought periods for different sources and parts of the system.

Simple water resource simulation models can be developed using spreadsheets. They are quick and easy to develop and have the advantage that numerical and graphical output can be tailored to the user's requirements. Commercial programs are available including:

- WRMM Alberta Environment, Calgary, Alberta
- HEC-ResSim *www.hec.usace.army.mil/*

- Aquator *http://www.oxscisoft.com/aquator/index*
- MIKE-Basin *www.dhi.di/mikebasin*
- Miser *www.tymemarch.co.uk*

Some of these, such as WRMM and HEC-ResSim, are freely available while others are designed for use by water companies with large, permanent water supply systems and may include automatic source optimisation routines.

3.25 CONJUNCTIVE USE AND OPERATION RULES

When a water utility has several sources, conjunctive or integrated use of them may be a means of improving the total yield or of reducing costs, or both. Thus extra water from an underground source when the water table is high, or from a river in flood, may permit a cut-back in the supply from an impounding reservoir with a 2-year critical drawdown period, enabling it to store more water. Similarly, it may be possible to keep storages with short critical periods in continuous full use to avoid overspills and so maximize their supply, at the same time reducing drawoff from a larger reservoir with a longer critical period, thereby gaining larger reserves to meet critical drawdown conditions. Similarly, it may be possible to reduce costs if the source producing the cheapest water can be over-run for part of the year, whilst dearer sources are cut back (Lambert, 1992: Parr, 1992).

However, there may be physical conditions that limit possibilities for conjunctive use, such as:

- isolation of sources and their supply areas;
- supply areas at different elevations;
- incompatibility of one source water with another;
- the need of certain manufacturers to use only one type of water.

It is not always advisable to change frequently from one type of water to another, particularly if one is a 'hard' water from underground and the other a 'soft' river or impounding reservoir supply (Section 14.8).

To ascertain potentials for conjunctive use the whole system of sources needs drawing out in diagrammatic form, showing:

- source outputs (average day critical yield; maximum day plant output);
- impounding or pumped storage capacity; length of critical drawdown period;
- area served, line of trunk feeders, key service reservoirs fed;
- elevation of supply at sources; high ground areas in the supply area;
- any legal or other restrictions on source outputs.

It is helpful to allocate a different colour for each source and its associated works. The possibilities can then be examined for conjunctive use. Key factors will probably be the need for major interconnecting pipelines and extension of treatment works capacity. The cost of these must be roughly assessed to see whether they are likely to be worthwhile having regard to the possible gain in yield. Once a possible scheme for conjunctive use has been clarified, this can be tested by computer calculations on a month-by-month basis to check the combined yield during a chosen critical dry period.

Operating rules can be developed to assist in judging when storage reservoirs can supply more than their minimum yield for a given risk. Monthly reservoir drawdowns over a long period of simulated inflows can be used to develop a control curve as shown in Figure 3.14. This shows the

Note: The control line represents the levels to which the reservoir could have been drawn down during any month in the period 1910–64 (55 years) and still maintain a total outflow of 46 Ml/d to supply and 13 Ml/d to compensation. When the contents are above the control line the draw off may exceed the 46 Ml/d to supply (up to the limit of the treatment works capacity). If the contents fall to or below the control line the draw off must be limited to 46 Ml/d to supply.

FIGURE 3.14

Reservoir control rule.

minimum storage required at the beginning of every month to ensure maintenance of a given supply rate. To produce a control curve of this type involves calculating backwards in time, from an assumed zero storage at the end of each month of the year. By applying this process to droughts of every duration, it is possible to locate the maximum storage required at any time of the year to ensure the reservoir never quite empties at the assumed abstraction rate. Different abstraction rates will require different control curves. A family of such curves is therefore produced, each for a different level of supply. Hence, reference to the curves and a storage/water level chart, can show whether the water level in the reservoir permits an increased abstraction or not.

Reservoir control curves have to be based on the most severe droughts historically recorded, or on a 'design' drought of specified probability. Neither can forecast the magnitude of some drought. Hence a control curve tends to be of more practical use in permitting extra water to be supplied when storage is high, than when storage is low. If a reservoir is three parts empty with the dry season not yet ended, most engineers would attempt to restrain demand, in preference to relying solely on a control curve, which poses a significant risk that it might not apply to what the future may bring.

3.26 RAINWATER COLLECTION SYSTEMS

Rainwater tanks collecting runoff from roofs or impervious surfaces form a useful source of drinking water where daily rainfall is frequent, as in equatorial climates and in the monsoon periods of monsoon climates. The supply is particularly useful if local sources are polluted, since only simple precautions are necessary to keep the rainwater free of pollution.

For individual house rainwater collecting tanks the principal constraint is usually the size of tank, which can be afforded or which it is practicable to install. Tanks up to 600 or 800 litres are generally formed of one piece of material and are transportable. Tanks of 1000 litre capacity are usually more economical if constructed in situ, for example in reinforced concrete. Such tanks have the advantage they are repairable if they leak. Tanks made of plastic plates bolted together tend to fracture under the repeated bending caused by changing water levels and the fracture is usually unrepairable. Steel tanks made of plates bolted together tend to rust at the joints.

Roof areas are not usually a principal constraint provided houses are one storied. There is usually an area of roof, which can discharge to guttering along one side, perhaps with a return along another side. Traditional roofing for low income communities is galvanized iron, but corrugated asbestos cement roofing, clay tiling and asphalt may be found. With house occupancies varying from six to 12 people, about 12–22 m² of developable roof area per person is usually available.

Daily rainfall records need to be tabulated. The daily runoff to the collecting tank is usually calculated according to the roof area available per person. The runoff is taken as *95% × (daily rainfall − 1 mm)*, the 1 mm deduction being for initial evaporative loss on the roof, and the 95% allowing for guttering overspill. Some field tests need to be undertaken to assess the size of guttering required; where it should be located; and what allowance should be made for overspill during intense rainfall. Local practice and experience, where available, can act as a guide. Although UK rainfall conditions are not likely to apply overseas where roof rainfall collection systems are mainly used, some UK publications can provide guidance (BS EN 12056:2000: Sturgeon, 1983). If large roofs, e.g. of commercial premises, etc. are used, the use of siphonic outlets to guttering may achieve economy in downpipe sizing. The siphons become primed by air-entrainment of initial flows, thus making the whole head between gutter outlet and the bottom of the downpipe available for the discharge, reducing the size of downpipe required (May, 1982; May, 1997).

Calculations proceed by using trial abstractions over typical recorded dry periods, commencing when the collecting tank can be assumed full. It is convenient to work on 'units of roof area' and 'storage available per person', sizing this up later according to the average number of people per household for which the design should cater. Operating rules for householders should be simple, e.g. either a fixed amount X per person per day, reduced to $0.5 \times X$ when the tank is half empty. An appropriate minimum drawoff rate would be 8 litres per person per day, the basic minimum requirement for direct consumption (drinking and cooking purposes) (Section 1.4). It is necessary to add an allowance for the *bottom water* and *top water* to the calculated theoretical tank capacity. The top water allowance, about 150 mm, means the householder does not have to restrict his or her take until the water level is that much below overspill level. The level can be marked clearly inside the tank. The advised abstraction X has to be expressed in terms of commonly available vessels. A standby supply must also be available to afford householders a supply when longer dry periods occur than assumed for the calculation and also for rescuing householders who run out of water for any reason.

Although house tanks will not give a sustainable amount in dry seasons of the year, they will be well used in the wet season because they relieve the householder of having to carry water from a distance. The rainwater in the tanks does not need to be chlorinated, but tanks do need cleaning out annually. Mosquito breeding in tanks can prove a nuisance in some climates and specialist advice may need to be sought as to how to control it. Water quality and treatment aspects of rainwater collection are discussed in Section 7.24.

3.27 THE LIKELY EFFECTS OF CLIMATE CHANGE

Analyses of hydrological variability inevitably have to assume climatic *stationarity*; i.e. wet years, dry years, floods and droughts all varying about a constant long term average. This assumption is necessary in order to describe the variability in terms of a statistical or probability distribution.

This enables estimates/predictions of extreme events (e.g. the 1 in 100 year drought) that perhaps have not yet been observed in climatic records.Climate change is an additional factor that has to be considered. The term climate change is used imprecisely. It is essential, therefore, to understand the difference between *climate change* and *climate variability*.

Climate variability is the natural variation of climatic parameters such as temperature or rainfall about the long term average. Large variations in these parameters occur, both seasonally and from year to year. Without climate change, these variations occur more or less symmetrically about the long-term mean, which is effectively constant. The words 'long-term' are important here because clusters of hot or cold, wet or dry years might produce short-term trends or cycles. These do not provide evidence of climate change.

The climate can be said to be changing when there is a clear trend in the long-term average of one or more climatic parameters over a long period. Trends might be seen in maximum and minimum values as well as the mean, or may be observed in event frequency such as the period between floods or droughts of a given magnitude.

Few hydrological records extend back more than 50 years and so most are too short to identify long term trends with much confidence. Where trends in river flows can be identified, it is difficult to separate the effects of climate change from changes in land use, farming practice or urbanisation. Nevertheless, conventional wisdom states that our present climate is changing and the possible impact of this change on future source yields needs to be considered.

Climate change predictions come from the extrapolation of observed values by global climate models, of which there are many. Model predictions vary widely although there is more consensus on rising global temperatures than there is on changes to rainfall or river flow. As a result of this uncertainty, local or Government policy on how to take climate change into account may vary. Guidance may also change as the results of new research become available. It is therefore advisable to check the publications and websites of national and international organisations for the latest advice or policy on allowing for the impact of climate change on yield. The situation in the UK is described below by way of example.

The UK Climate Impacts Programme (UKCIP) coordinates research and provides scenarios that show how our climate might change. A new set of scenarios (UKCIP08) has recently been published but until these are incorporated into climate change methodologies, most of the guidance is based on the previous UKCIP02 scenarios. These present predicted changes relative to a 1961–90 baseline in monthly, seasonal and annual averages for 15 climatic variables including temperature, relative humidity, wind and precipitation. The predictions were derived using the climate model developed at the UK Hadley Centre and were made for three 30-year time slices up to the year 2100 assuming four scenarios for future greenhouse gas emissions. The overall conclusions were that:

- the UK will continue to get warmer;
- summers will continue to get hotter and drier;
- winters will continue to get milder and wetter;
- some weather extremes will become more common while others will become less common.

Further details and up to date information may be found on the UKCIP website (www.ukcip.org.uk).

The UKCIP scenarios indicate likely future impacts on precipitation (rainfall and snow) and evaporation but do not directly predict the effects on river flows and groundwater recharge that are needed to determine future yield. These are provided by the results of separate research projects (UKWIR, 2002; Arnell, 2007; UKWIR; 2006; UKWIR; 2007). Factors are presented which enable historic records of river flow to be perturbed in order to represent future climatic conditions. If rainfall/runoff models are available for the catchments of interest, these can be used to derive flow sequences from perturbed rainfall data. Similarly, soil moisture models can be used to estimate groundwater recharge under future climatic conditions. Wetter winters may increase recharge while drier summers are likely to shorten the recharge period and may give rise to lower groundwater levels at the start of the recharge season. The overall impact on recharge therefore may be hard to predict and require detailed modelling to derive credible scenarios.

If rainfall-runoff or recharge models are not available, the guidelines provide factors to perturb historic flow and recharge sequences directly. These perturbed sequences are then used as input to groundwater or surface water system simulation models to determine yield at specified future time horizons. The impact on yield in intervening years is determined by interpolation between these time horizons and the present-day baseline.

REFERENCES

Ackers, P., White, W. R., Perkins, J. and Harrison, A. J. M. (1978). *Weirs and Flumes for Flow Measurement.* John Wiley & Sons.

Arnell, Prof. N. and Reynard, N. (2007). *Climate change implications in estimates of water resource zone deployable output. Supplementary guidance to Chapter 8 of the Environment Agency's Water Resources Planning Guidelines.* EA.

Bos, M. G. (1990). *Discharge measurement structures.* ILRI publication 20, 3rd Edn.

Bosch, J. M. and Hewlett, J. D. (1982). A review of catchment experiments to determine the effects of vegetation changes on water yield and evapotranspiration. *J. Hydrology*, **55**, No. 1/4, p. 3–23.

BS EN 748:2007. *Hydrometry. Measurements of liquid flow in open channels using current-meters or floats.* BSI.

BS 1100:1998. *Measurement of liquid flow in open channels. Determination of stage-discharge relation.* BSI.

BS 3680. *Measurement of liquid flow in open channels. Part 3Q:2002: Guidelines for safe working practices in river flow measurements, and Part 4A:1981: Method using thin plate weirs.* BSI.

BS EN 6416:2005. Hydrometry. *Measurement of discharge by the ultrasonic (acoustic) method.* BSI.

BS TR 8363:1997. *Measurement of liquid flow in open channels. General guidelines for the selection of method.* BSI.

BS EN 12056:2000. *Part 3: Gravity drainage systems inside buildings: Roof drainage, layout and calculation.* BSI.

Bullock, A., Gustard, A. and Grainger, E. S. (1991). *Instream flow requirements of aquatic ecology in two British rivers, Report 115*, Inst. Hydrology.

Calder, I. R. (1992). *Evaporation in the Uplands.* John Wiley & Sons.

Creutin, J. D. and Obled, C. (1982). Objective analysis and mapping techniques for rainfall fields: an objective comparison. *Water Resources Research*, **18 (2)**, p. 413–431.

Darmer, K. I. (1970). *A proposed streamflow data program for New York*, USGS Water Resources Division, Open File Reports, Albany, New York.

Dean, T. J., Bell, J. P. and Baty, A. J. B. (1987). Soil moisture measurement by an improved capacitance technique. Part 1. Sensor design and performance. *J. Hydrology*, **93**, p. 67–78.

DoE (1996). *Water Resources and Supply: Agenda for Action.* (including Annexes D and E contributed by EA). HMSO.

Doorenbos, J. and Pruitt, W. O. (1976). *Crop water requirements. Irrigation and Drainage Paper 24.* FAO.

EA (2001). *Good Practice in Flow Naturalisation by Decomposition*, National Hydrology Group, Version 2.0. EA.

EA (2007). *Water resources planning guideline.* EA.

Enex (1976). *Magnitude and frequency of low flows in Peninsular Malaysia, Hydrological Procedure No. 12*, Enex/ Drainage and Irrigation Dept, Malaysia.

FEH (1999). *Flood Estimation Handbook, Inst Hydrology 1999. Vol 1—Overview; Vol 2—Rainfall frequency estimation; Vol 3—Flood frequency estimation: statistical procedures; Vol 4—Flood estimation: FSR rainfall-runoff method; Vol 5—Catchment description.* IoH.

Gustard, A. et al, (1987). *A study of compensation flows in UK, Report 99*, IoH.

Gustard, A. and Gross, R. (1989). Low flow regimes of Northern and Western Europe, *Proc. FRIENDS in Hydrology Conf.* IAHS Publ, 187, p. 205.

Gustard, A., Roald, L. A., Demuth, S., Lumadjeng, H. S. and Gross, R. (1989). *Flow regimes from experimental and network data (FREND) I and II.* IoH.

Gustard, A., Bullock, A. and Dixon, J. M. (1992). *Low flow estimation in the United Kingdom. Report No. 108*, IoH.

Hall, J. and Nott, M. (1994). "The naturalisation of flow records by decomposition", *presentation at the National Hydrology Meeting of the British Hydrological Society. London, 17th March.*

Herschy, R. W., White, W. R. and Whitehead, E. (1977). *The Design of Crump Weirs.* Technical Memo No. 8, DoE Water Data Unit.

Holford, I. (1977). *The Guiness Book of Weather Facts and Feats*, Guiness Superlatives Ltd.

Hudleston, F. (1933). *A summary of seven years experiments with rain gauge shields in exposed positions 1926–1932 at Hutton John, Penrith.* British Rainfall, p. 274–293.

IoH (1980). Low *flow studies report*, IoH.

IWE (1937). Report of Joint Committee to consider methods of determining general rainfall over any area. *Trans. IWE XLII*, p. 231–299.

Kirby, C., Newsom, M. D. and Gilman, K. (1991). *Plynlimon research: the first two decades. Report No. 109*, Inst. Hydrology.

Lambert, O. A. (1992). *An introduction to operational control rules using the ten component method, Occasional Paper No. 1, British Hydrological Society 1990. Parr N. M. et al (eds.) Water resources and reservoir engineering*, British Dam Society, Thos. Telford, p. 11–40

Lapworth, C. F. (1965). Evaporation from a reservoir near London. *J.IWE*, **19**, No.2, p. 163–181.

Lau, J., Onof, C. and Kapeta, L. (2006). Comparisons between calibration of urban drainage models using point rain-gauge measurements and spatially varying radar-rainfall data, *7th International Workshop on Precipitation in Urban Areas, 7 December, Switzerland.*

Manley, R. E. (1978). Simulation of flows in ungauged basins, *Hydrological Sciences Bulletin, 3*, p. 85–101. (See also Hysim user manual, Manley R. E. and Water Resources Association Ltd, 2006).

May R. W. P. (1982). *Report IT 205, Design of gutters and gutter outlets*, Hydraulics Research Station.

May, R. W. P. (1997). The design of conventional and siphonic roof drainage systems. *Journ. CIWEM*, p. 56–60.

McCulloch, J. S. G. (1965). Tables for the rapid computation of the Penman estimate of evaporation. *East African Agricultural and Forestry Journal*, No. 3.

NERC (2006). FEH CD-ROM, Version 2.0, Centre for Ecology & Hydrology, UK.

Parr, N. M. (1992). *Water resources and reservoir engineering*, British Dams Society, Thomas Telford, pp. 11–40.

Penman, H. L. (1963). *Vegetation and hydrology. Technical Communication No.53.* Commonwealth Bureau of Soils.

Rodda, J., Downing R. A., and Law. F. M. (1976). *Systematic Hydrology.* Newnes-Butterworth, p. 88–89. NERC (2006), *FEH CD-ROM, Version 2.0*, Centre for Ecology & Hydrology.

Sargent, R. J. and Ledger. D. C. (1992). Derivation of a 130 year run-off record from sluice records for the Loch Leven catchment, south-east Scotland. *Proc. ICE Water, Maritime and Energy,* **96** No. 2, p. 71–80.

Simpson, R. P. and Cordery, I. (1987). A review of methodology for estimating the value of streamflow data. *Inst. Engrs. Aust. Civil Eng. Trans,* **24** p. 79–84.

Smith L. P. and Trafford. B. D. (1975). *Technical Bulletin 34, Climate and drainage,* MAFF. HMSO.

Sturgeon C. G. (Ed.) (1983). *Plumbing engineering services design guide.* Inst. of Plumbing.

Tabony, R. C. (1977). *The variability of long duration rainfall over Great Britain, Scientific Paper 37,* Meteorological Office.

Svensson, C., Jones, D. A., Dent, E. D., Collier, C. G. (2006). *Review of Rainfall Frequency and Probable Maximum Precipitation Methods, Report No. 2 of Defra project number WS194/2/39 Reservoir Safety–Long Return Period Rainfall.* Defra. UK.

Thompson N., Barrie, I. and Ayles, M. (1981). *The Meteorological Office rainfall and evaporation calculation system:* MORECS. The Meteorological Office.

Tor, S. M. (1998). Strategies towards sustainable development in Nigeria's semi-arid regions, *J.CIWEM,* p. 212–215.

UKWIR/EA (2000). *Report 00/WR/18/2, A unified methodology for the determination of deployable output from water sources, Volume 2,* UKWIR (EA R&D Technical Report Ref. no. W258, 2000).

UKWIR (2002). *Report 03/CL/04/2 Effect of Climate Change on River Flows and Groundwater Recharge, UKCIP 02 Scenarios.* UKWIR.

UKWIR (2006). *Report 05/CL/04/6. Effect of Climate Change on River Flows and Groundwater Recharge, a Practical Methodology: A Strategy for Evaluating Uncertainty in Assessing the Impacts of Climate Change on Water Resources.* UKWIR.

UKWIR (2007). *Report CL04. Effects of climate change on river flows and groundwater recharge: a practical methodology for draft recharge and groundwater level impact assessment.* UKWIR.

USACE (2000). *Hydrologic Engineering Center Hydrologic Modeling System, HEC-HMS, Technical Reference Manual,* USACE.

USGS (1968). *Techniques of Water Resources Investigations of the USGS, Book 3. Chapters A6–A8 on stream gauging procedure.* United States Geological Survey. US Government Printing Office.

USGS (1985). *Techniques of Water Resources Investigations of the USGS. Book 3, Chapter A16, Measurement of discharge using tracers.* United States Geological Survey. US Government Printing Office.

Wilson, E. M. (1990). *Engineering Hydrology,* 4th. Edn. Macmillan.

WMO (1966). Measurement and estimation of evaporation and evapotranspiration. Note No. 83. World Metereological Organisation (WMO).

WMO (1983). *Guide to meteorological instruments and observing practices* No. 8, WMO.

Young, A. R., Grew, R. and Holmes, M. G. R. (2003). Low Flows 2000: a national water resources assessment and decision support tool. *Water Science & Technology.* Vol. **48**, No. 10, 00, p. 119–26.

Groundwater Supplies

4

4.1 GROUNDWATER, AQUIFERS AND THEIR MANAGEMENT

The most prolific sources of underground water are the sedimentary rocks, sandstones and limestones, the latter including the chalk. They have good water storage and transmissivity, cover large areas with extensive outcrops for receiving recharge by rainfall, and have considerable thickness but are accessible by boreholes and wells of no great depth.

The porosity of a rock is not an indicator of its ability to give a good water yield. Clays and silts have a porosity of 30 percent or more but their low permeability, due to their fine grained nature, makes them unable to yield much water. Solid chalk has a similar high porosity and low permeability but is a prolific yielder of water because it has an extensive network of fissures and open bedding planes, which store large quantities of water and readily release it to a well, borehole or adit. Limestone, on the other hand, is so free draining that, though it may yield large quantities of water in wet weather, the aquifer rapidly drains away in dry weather and so has low yield.

In England the main aquifers used for public supply (Rodda, 1976) comprise:

1. The Chalk and to a lesser extent the Upper and Lower Greensands below it, both of the Cretaceous Period.
2. The Sherwood Sandstones of the Triassic-Permian Period, previously named the Bunter and Keuper Sandstones.
3. The Magnesian and Oolitic Limestones of the Jurassic Period and, to a lesser extent, the Carboniferous Limestones and Millstone Grits of the Carboniferous Period.

The chalk and greensands, and the limestones of the Jurassic Period, are widely spread over southern and eastern areas of England; the Triassic sandstones and Carboniferous limestones occur in the Midlands. Similar formations occur in northern France and across the lowland northern plains of Europe where they give good yields.

Some of the largest aquifers worldwide are listed in Table 4.1. The formations most extensively used for water supply are the following:

4. The shallow alluvial strata, sands and gravels of Tertiary or Recent age, which are so widespread that, despite their varying yields, they form the principal source of supply in many parts of the world because of the large areas they cover and ease of access.

Table 4.1 Major aquifer systems worldwide

		Estimated volumes			
	Strata	**Basin area km²**	**Reserves MCM**	**Recharge MCM/yr**	**Use MCM/yr**
Nubian Aquifer System[1,2] Egypt, Libya, Chad, Sudan	Cambrian & Tertiary	2.0 m	150 m	Small	460
Great Artesian Basin Australia[3,4]	Triassic—Cretaceous	1.7 m	20 m	1,100	600 (1975)
Hebei Plain China[3]	Quaternary alluvium	0.13 m	−0.75 m	−35 000	10 000
Algeria, Tunisia North Sahara[3,5]	Lower—Upper Cretaceous alluvium	0.95 m	very Large	very Small	> recharge
Libya[6]	Cambrian—Cretaceous sandstones	1.8 m	24 m	Small	> recharge
Ogallalah Aquifer USA[7,8]	Jurassic sediments	0.075 m	0.35 m	60	3400
Dakota Sandstone Aquifer USA[8,9]	Upper Mesozoic sandstones	>0.4 m	>4.0 m	>315	725
Ummer Radhuma Aquifer Saudi Arabia[10]	Tertiary—Palaeocene sediments	>0.25 m	0.025 m	1048	n.a.

Sources: [1]Idris, H. and Nour, S. (1990). Present groundwater status in Egypt and the environmental impacts. *Env. Geol.Wat. Sci.*, 16.3, pp. 171–177.
[2]Lamoreaux, P. E. et al. (1985). Groundwater development, Kharga Oasis Western Desert of Egypt: a long term environmental concern. *Env. Geol.Wat.Sci.* 7.3, pp. 129–149.
[3]Margat, J. and Saad, K. F. (1984). Deep-lying aquifers; water mines under the desert. *Nature & Resources* 20.2, pp. 7–13.
[4]Habermehl, M. A. (1980). The Great Artesian Basin, Australia. *B.R.MJ.Austr.Geol. Geophys.*, 51.9.
[5]De Marsily, G. et al. (1978). Modelling of large multi-layered aquifer systems: theory & applications. *J. Hydrol.* 36, pp. 1–34.
[6]Mayne, D. (1991). *The Libyan Pipeline Experience*, Brown & Root Ltd.
[7]ASCE (1972). *Groundwater Management.* ASCE.
[8]Johnson, R. H. (1997). *Sources of water supply pumpage from regional aquifer systems. Hydrogeology J.* 5.2, pp. 54–63.
[9]Helgeson, J. O. et al, (1982). Regional Study of the Dakota Aquifer, *Ground Water* 20.4, pp. 410–414.
[10]Bakiewicz, V. et al, (1982). Hydrogeology of the Umm er Rhaduma aquifer Saudi Arabia, with reference to fossil gradients, *Q. J.Geol.* 15, pp. 105–126.

5. The many areas of Mesozoic to Carboniferous age sedimentary rocks—chalk, limestones and sandstones that mostly give good yields, subject to adequate rainfall.

6. The hard rock areas of Paleozoic to Cambrian or pre-Cambrian age. This includes igneous and metamorphic rocks which are relatively poor yielders of water unless well fractured and fissured, but which have to be used because better sources are scarce in many parts of the world.

Over vast areas of Africa and India, reliance has to be placed on the relatively small yields that can be drawn from the ancient hard rocks under (6) above. The Deccan Traps of India, for example, cover almost 0.5 million km² (Singhail, 1997) and are used by a large rural population. Almost one half of the African continent is underlain by hard basement rocks. Although these provide relatively poor yields, they are the main source of underground water for rural populations (Wright, 1992; MacDonald, 2000). These formations, typically yielding 10–100 m³/day per borehole, are sufficient for basic domestic use by small populations but are normally inadequate for any large scale agricultural or manufacturing development.

The advantages of groundwater are substantial. The wide area typically occupied by aquifers makes it possible to procure water close to where it is required. Many aquifers provide water that requires no treatment other than precautionary disinfection though in many developing countries even disinfection is not adopted. The cost of borehole installation and associated pumping is relatively modest if the water depth is not great; and the supply can be increased if necessary by drilling additional boreholes subject to the availability of groundwater resources.

However, this ease of exploitation often leads to failure to conserve and protect underground supplies. For example, the large cities of Bangkok, Jakarta, Calcutta and Manila were initially able to gain supplies by drilling and pumping from boreholes close to, or even within, their urban areas but yields have reduced due to spread of paved and built-on areas. Subsidence can result from over-pumping; intensive pumping of groundwater from the thick series of alluvial aquifers beneath Bangkok has resulted in their partial dewatering and consolidation and consequential ground subsidence at the surface.

There is often a failure to monitor a groundwater resource and the effects of overdraw may remain unseen until the resource is seriously depleted. The European Environment Agency estimates that about 60% of European cities with more than 100 000 inhabitants are located in or near areas with groundwater over-exploitation, as shown by recent severe supply problems in parts of Spain and Greece (Stanners, 1995). There is also commonly a failure to understand the importance of aquifer protection, which is particularly important in urban areas overlying an aquifer where poor sanitation methods or badly maintained sewerage systems result in shallow aquifers becoming polluted.

Sometimes the physical or chemical quality of a groundwater can act as a constraint on its use. Proximity to the sea poses a salinity hazard affecting boreholes in coastal areas and on oceanic islands. Development of freshwater resources then requires great care because, if once seawater is drawn in, it may prove difficult or impossible to reverse the process (Section 4.15). In areas where water circulates to great depth, the groundwater that eventually emerges at the surface may be warm or even hot. Whilst air cooling methods can be applied, the warm or hot groundwater may have taken up an undesirably high concentration of minerals or gases that may be difficult to deal with.

Cessation of large scale abstractions of groundwater can cause groundwater levels to rise. In Paris, London, Birmingham and Liverpool, rising groundwater levels are occurring because major water-consuming industries, that historically abstracted water from on-site boreholes, are being replaced by other lower water demand activities. These rising groundwater levels threaten the flooding of tunnels and deep basements, can cause chemical attack on the structural foundations of buildings, and decrease the ability of drainage systems to dewater surface areas. Similarly, rising groundwater levels can occur where excessive irrigation is applied or where large scale river impounding schemes for hydro-power development or irrigation raise water levels upstream, causing soil salinization and sometimes land slope instability.

If groundwater is to provide sustainable resources for the future, aquifers have to be managed. The techniques described in this chapter allow aquifer behaviour to be assessed and thereby managed. However, conservation and pollution control require a suitable legal and regulatory framework (Chapter 2).

4.2 YIELD UNCERTAINTIES AND TYPES OF ABSTRACTION WORKS

The yield of a well, borehole or adit is dependent on the following:

- aquifer properties of the strata from which the water is drawn, the thickness and extent of the aquifer and area of its outcrop;
- extent to which the water storing and transmitting network of fissures, cracks and open bedding planes in the aquifer are intercepted;
- depth, diameter and construction details of a borehole or well, and the type of screen or gravel packing;
- effect of abstraction on other users of water from the same aquifer, be it subterranean abstraction, springs or discharges to water courses.

Uncertainty arises because a borehole intercepts only a small volume of the strata. In fissured rock, it may not intersect fissures or planes large enough to give a good supply. Many cases can be quoted where a boring in one location gives a poor yield, when a second boring only a few feet away provides a yield four or five times as much. Clearly the more fissures or cracks there are, the more likely it is that a reasonable yield will be obtained. Thus, boreholes in the UK are favoured in the well-fissured layers of the Upper Chalk or in the looser Pebble Beds of the Bunter Sandstone. In alluvial gravels and sands there is less uncertainty with yield but lenses of low permeability can reduce transmissivity.

Estimation of the probable yield of a proposed underground development is difficult. Knowledge of the hydrogeology of the area and records of the yield of other boreholes or wells in a similar formation can be of help. However, where little is known about the hydrogeology of a proposed borehole site, it is advisable to sink a 'pilot hole', usually about 150 mm diameter, to gather information from the samples withdrawn. If the results seem promising and it is decided to adopt the site, the pilot hole can be reamed out to form a borehole large enough to accommodate a pump. However, it must be borne in mind that, if the small pilot bore appears to give a good flow for its size, this may not be a reliable indicator that a larger yield could be obtainable from a larger hole in the same or at a nearby location. A method of approach Sander (1997) for locating a borehole for optimum yield uses mapping data from lineations, bedrock geology, vegetation and drainage to produce probabilities of yield. The method is used to assist rural groundwater development in poor aquifers. A recent practical guidance for rural water supply in sub-Saharan Africa is available in MacDonald (2000). The guidance is based on research carried out by BGS with funding from the UK DFID. The manual presents a number of methods for quick assessment of groundwater resources.

Need for Hydrogeological Survey

Where an aquifer is already supporting abstractions and feeds rivers and streams flowing through environmentally sensitive and recreationally important areas, it is essential to conduct a full hydrogeological investigation of the aquifer before new abstraction is proposed. An environmental assessment of the possible effect of a proposed new abstraction for public supply is often a requirement of an abstraction licensing authority or government department. Section 4.19 lists the key steps required in the UK. A hydrogeological investigation may also be needed to make sure possible sources of pollution on the catchment, such as leachates from solid waste tips or abandoned contaminated industrial sites, will not render water from a proposed abstraction unsuitable for public supply. The work involved in such an investigation may be extensive and is best carried out by an experienced hydrogeologist who is familiar with the range of geophysical and other techniques available (Section 4.7).

Types of Abstraction Works

The range of abstraction works which are used for various different ground conditions include the following:

1. Boreholes can be shallow or deep, sunk by different methods, and have many different methods of screening to exclude fine material and keep the bore stable, or to draw upon the best quality or quantity of water available.
2. Large diameter wells, 15 to 25 m deep, can have a borehole sunk from their base to considerable depth to intercept some main water-bearing strata, the well lining being sealed to prevent entry of surface waters.
3. Although not built now, many older wells have one or more adits driven from them. An adit is an unlined tunnel approximately 1 m wide by 2 m high driven at some level below ground surface where it is expected that more water bearing fissures or bedding planes will be intercepted. Some adits are several kilometres long.
4. A 'well field' can be adopted by sinking several moderately sized boreholes, spaced apart in some pattern, their yields being collected together.
5. Collector wells and galleries, using porous or unjointed pipes installed in river bed or in river bankside deposits of sand and gravels, can be used to abstract shallow groundwater; or galleries can be driven into a hillside to tap the water table (Section 4.16).

These methods of development are described in detail in the following Sections.

4.3 POTENTIAL YIELD OF AN AQUIFER

It is often necessary to quantify the limiting yield of an aquifer for water supply purposes. Formerly this used to be taken as equal to the 'long-term' average recharge from rainfall percolation, provided the storage capacity of the aquifer is large enough to even-out year to year variations of the recharge. However, further studies of climate variability have shown there can be long runs of years of below-average recharge so that aquifer yields rarely exceed 90% of the 'long term' average recharge. Additionally, allowances may have to be made to prevent the reduction of aquifer-fed spring flows that could result in environmental damage and to limit saline intrusion. As rough guide, therefore, abstraction of about 75% of the average 'long-term' recharge is a safer estimate of the likely maximum sustainable development.

To assess the potential yield of an aquifer, the groundwater catchment must first be defined from the contours of the water table. It is usually assumed that the water table reflects the ground surface to a reduced scale, but variations from this can occur in asymmetric scenery containing features such as escarpments, or where one valley has cut down deeper than its neighbours. It is important to check this, because the underground water table catchment 'divide' may not coincide with the topographical surface catchment divide above. It has to be assumed that underground flow is in the direction of the major slope of the water table. This may not always be true, for instance, where there are karst limestone fissures, but it is best to follow the general rule in the absence of any evidence to the contrary.

Assessing the amount of recharge. Sometimes the aquifer catchment outcrop may be remote from the point at which the wells tap the strata concerned. The average recharge is estimated as average rainfall minus evapotranspiration over the outcrop area (Sections 3.6–3.8). Recharge can also occur as leakage where the catchment includes clay covered areas, surface runoff subsequently leaking directly into the aquifer through river beds. This amount is difficult to estimate and can be near guesswork until there is a long record of successful pumping to confirm any mathematical modelling (Section 4.5) of the aquifer hydrology. Recharge from river bed leakage is usually only significant in arid areas. In temperate wetter areas the hydraulic gradient between the water level in the river and the adjacent groundwater table level is usually too small to cause significant river bed leakage.

Percolation formulae (Summers, 1987) can be of use to assess aquifer recharge, but they need confirming before being used outside the area for which they were derived. Formulae for semi-arid areas like Jordan or Western Australia are based on a simple percentage of 'long-term' average rainfall, say 3% or 5%. Research by tracking the tritium contents of chalk-limestone pore water and fissure water has demonstrated that, whereas water passing through fissures may travel down to the water table at more than 0.3 m/day, the pore water recharge front may move down at only 1 to 2 m/year. Distinguishing between the volumes travelling by these alternative routes is fraught with difficulty because it depends on the size of the recharge event and the possibility that pore water will drain out through fissures during any major drawdown of the water table. A more helpful analytical approach lies in recession analysis of dry-weather flows from groundwater catchments.

River flow recession curves can be used to determine that part of river flow, termed 'base flow' which is fed from underground aquifer storage (Section 3.12). In prolonged dry weather the natural flow of a river comprises only aquifer drainage through springs. At any instant of time, spring flow Q is related to the volume of stored water S in the aquifer by the relationship:

$$Q = kS$$

If Q_o is the spring flow at time $t = 0$, and Q_t is the flow at later time, t,

$$Q_t = Q_o e^{-kt}$$

where k has the unit day^{-1} if t is in days. Typical values for k lie between 0.01 per day in a good aquifer, to 0.10 per day in a relatively impermeable aquifer.

Consequently if log Q_t is plotted against t (days) for a prolonged dry period, this should give a straight line of slope k as shown in Figure 4.1. If a single long period free of rain is not available for analysis, it is possible to link together shorter dry period flow recessions which, plotted as shown in Figure 4.2, will be asymptotic to the natural recession curve. Section 3.12 shows how this recession curve can be used to estimate the base flow for a given period of record.

Base flow is percolation routed through storage. Hence, by summing the baseflows over a period and adding a correction for the change of storage between the beginning and end of the period the percolation for the period can be estimated. The correction for change of storage is obtained by taking the change of water table level between the beginning and end of the period and multiplying this by the aquifer storage coefficient (Section 4.4). Ranked and plotted on probability paper (Figure 4.3) these seasonal percolation values can be used to estimate recharge probabilities.

Percolation gauge or lysimeter drainage readings can also be used, but it is difficult to be sure that they represent actual average catchment conditions. Their rims prevent surface runoff and, being on level ground, local runoff does not occur; therefore, the measured percolate may be an optimistic estimate of the amount that reaches the water table under natural conditions. Practical difficulties occur when such gauges are kept in use for a long period. They can become moss covered or the soil may shrink away from the edge of the container. Alternatively, percolation can be estimated by a soil moisture storage balance method of the type demonstrated by Headworth (1970). Once the readily available moisture in the root zone is used up by evapotranspiration in a dry period, the subsequent build-up of soil moisture deficit must be made up before excess rainfall can percolate to the aquifer once more. The average percolation produced by this method depends strongly on the amount of moisture stored within the root zone. Storage or 'root constant' values range from 25 mm for short-rooted grassland to over 200 mm for woodland and need to be calculated for each specific climatic regime, preferably with daily rainfall data.

Yield constraints. Having found the potential yield of a groundwater catchment, it is possible to compare it with the total authorized (or licensed) existing groundwater abstractions. This gives the

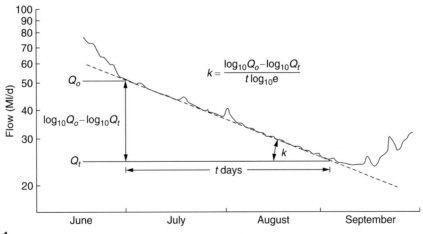

FIGURE 4.1

Groundwater recession graph.

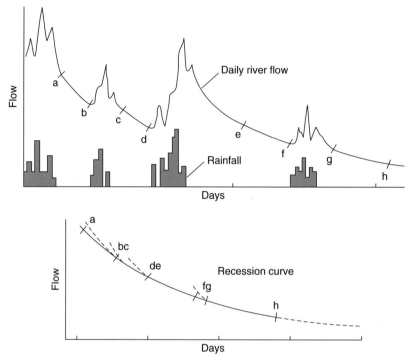

FIGURE 4.2

Recession curve drawn from parts of daily flow record.

upper limit to what extra yield may be obtained or, if too much is already being taken, the extent of groundwater mining that must be taking place. It should be noted that mining can occur in one part of an aquifer whilst elsewhere springflows may indicate no mining is apparent. This indicates low transmissivity within the catchment or parts of it, suggesting a better siting of abstraction wells or a different approach to estimating the maximum yield.

Where a group of wells exists within a catchment, the average maximum drawdown they can sustain without interfering with their output can be assessed. Using the average storage coefficient S' for the aquifer supplying the wells, the amount of storage that can be draw upon to produce this drawdown can be estimated. From this it is possible to calculate the minimum yield for various time periods for a given severity of drought, i.e.

Abstraction possible = storage available for given pumping level + recharge during the given time period − loss from springflows.

This calculation would be carried out for say a 2% drought (50 year return period) for periods of 6–7 months (1st summer); 18 months (2 summers + 1 winter); and so on. This can reveal what is the length of the dry period that causes the lowest yield and whether pumps are sited low enough to average out percolation fluctuations.

Seawater intrusion into wells close to the sea may limit abstraction. The pumping has to be kept low enough to maintain a positive gradient of the water table to the mean sea level. However, since seawater is 1.025 times denser than freshwater, under equilibrium conditions the freshwater-saltwater interface is 40 h below sea level, where h is the difference in height between the

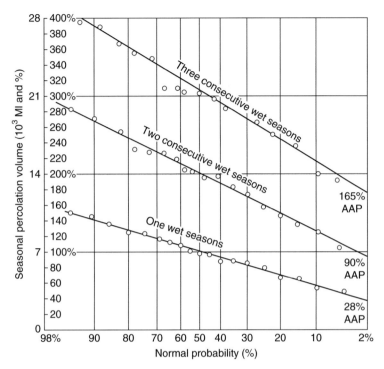

FIGURE 4.3
Seasonal percolation probability plot.

freshwater surface level and the sea level. Thus if the rest water level in a well is 0.15 m above mean sea level, the saline water interface will be about 6.0 m lower, or 5.85 m below mean sea level. (In practice, the interface will be a zone of transition from fresh to salt water rather than a strict boundary.) This means a pump suction can be sited slightly below mean sea level without necessarily drawing in seawater; but this is heavily dependent on the aquifer's local characteristics, because a stable interface may only be formed in certain types of ground formations. A more reliable policy is to use coastal wells conjunctively with inland wells. In the wet season the coastal wells are used when the hydraulic gradient of the water table towards the sea is steepest, reducing the risk of drawing in seawater. In the dry season the inland wells are used. By this method a higher proportion of the water that would otherwise flow to the sea is utilized than if only the coastal wells were used. Small low-lying oceanic islands face special seawater intrusion problems described in Section 4.15.

4.4 ASSESSMENT OF AQUIFER CHARACTERISTICS

The two principal characteristics of an aquifer are its horizontal transmissivity T, which is the product of its permeability times its wetted depth; and the storage coefficient S'. Transmissivity T is the flow through unit width of the aquifer under unit hydraulic gradient. Its units are therefore m^3/m per day, often abbreviated to m^2/d.

The storage coefficient S' is defined as the amount of water released from an aquifer when unit fall in the water table occurs. Where free water table conditions occur it is the volume of water that will drain from a unit volume of an aquifer (expressed as a percentage of the latter) by gravity with a unit fall of the water table level; sometimes known as Specific Yield S_y. However, when the aquifer is confined under pressure because of some impervious layer above, it is the percentage of unit aquifer volume that must be drained off to reduce the piezometric head by unit depth. The difference between these two meanings, although subtle, is vital. Whereas the former may be in the range 0.1% to 10.0%, the latter may be 1000 times smaller (demonstrating the incompressibility of water).

By considering a well as a mathematical 'sink' which creates a cone of depression in the water table, it has been shown by Theis (1935) that drawdown in a homogeneous aquifer due to a constant discharge Q initiated at time $t = 0$ is:

$$h_o - h = \frac{Q}{4\pi T}\left(-0.5772 - \log_e u + u - \frac{u^2}{2.2!} + \frac{u^3}{3.3!} + \frac{u^4}{4.4!}\Lambda\right)$$

$$h_o - h = \frac{Q}{4\pi T}W(u) \tag{4.1}$$

where $u = r^2 S/(4Tt)$, h_o is the initial level and h is the level after time t in a well distance r from the pumped well. Any consistent set of units can be used. For example, if Q is in m³/d and T is in m²/d, then h_o and h must be given in metres. S is a fraction. $W(u)$, the 'well function' of the Theis equation, can be obtained from tables (Walton, 1970). Although both this equation and the Jacob (1950) simplification of it (see below) are derived for a homogeneous aquifer, they are found to work in well-fissured strata, such as Upper and Middle Chalk. The essential need is to be able to assume reasonably uniform horizontal flow on the spatial scale being considered, together with the absence of any impervious layer that interferes with drawdown.

In the Theis method of solution a type curve of $W(u)$ against u is overlaid on a plot of the pump test drawdowns versus values of log (r^2/t). Where a portion of the type curve matches the observed curve, coordinates of a point on this curve are recorded. With these matchpoint values the equations can be solved for S and T.

However, Jacob's less exact method is easier to apply and meets most situations that confront an engineer. He pointed out that if time t is large, as in most major pumping tests, then u is small, (say less than 0.01); therefore, the series in the Theis equation can be shortened to:

$$h_o - h = \frac{2.30Q}{4\pi T}\log_{10}\left(\frac{2.25Tt}{r^2 S}\right) \tag{4.2}$$

Plotting drawdown against time at an observation borehole within the cone of depression thus produces a straight line (Fig. 4.4).

If the drawdown for one logarithmic cycle of time is read off, the value of $\log_{10}\left(2.25Tt/r^2 S\right) = 1.0$ in equation (4.2). Hence, where $h_o - h$ is in metres and Q is in m³/d:

$$h_o - h = \frac{(2.30Q)}{(4\pi T)}$$

i.e. $T = (2.30Q)/4\pi(h_o - h)$

Note: Data plotted from observation well 200 m from test well.

FIGURE 4.4

Jacob's pump test analysis for test pumping of a well at 4000 m³/day.

Reading off the time intercept, t_o days, for zero drawdown,

$$S = 2.25T \ t/r^2, \text{ where } T \text{ is in m}^2/\text{d and } r \text{ is in metres.}$$

The following qualifications apply.

1. If the regional water table rises or falls as a whole during the test then the drawdown should be adjusted by the equivalent amount.
2. Early time data should not be used because there will be substantial initial vertical flow as the storage is evacuated. Boulton (1963) suggests the necessary horizontal flow conditions exist when $r > 0.2d$ and $t > 5dS/K_t$ days; where d is the wetted aquifer depth and K_t is the vertical hydraulic conductivity which can be taken as T/d if K_t is not otherwise known. (In horizontally layered strata, K_t may be one-third or less of the horizontal conductivity.) Some reiteration with values of S and T is required to use this guide.
3. Where the pumped well only partially penetrates the aquifer (Kruesman, 1990), it is necessary to adjust the drawdown values for the resulting non-standard flow lines unless the observation well is at least 1.5 times the aquifer depth away from the source. Rather than make complex adjustments it is preferable to use the Jacob rather than the Theis method on a long term pumping test.
4. Where drawdown is large compared with the aquifer thickness and the aquifer is unconfined, the measured drawdowns should be corrected (Jacob, 1950) by subtracting from them a correction for large drawdown:

$$\text{Correction} = (\text{Drawdown})^2/2 \times \text{wetted aquifer depth}$$

Data observed during the recovery part of a test can also be analysed to derive S and T. If only levels in the pumped well are obtainable, the information that can be gained is usually limited to an estimate of T on recovery (Driscoll, 1986).

Once S and T are established, it is possible to predict with the above equations what drawdown below current rest water level will result at different pumping rates, different times and other distances. Where more than one well can create a drawdown at a point of interest, the total effect can be calculated by the principle of superposition, i.e. the drawdowns due to individual well effects can simply be added. Analytical solutions (Walton, 1970; Kruesman, 1990) exist for many aquifer conditions, including boundary effects from impermeable faults and recharge streams. Care is needed to adopt the solution appropriate to the lithology and recharge boundary.

It will be appreciated that the engineer has no control over the values of S and T found at a well site. The water drawn from the hole may come from local aquifer storage after a long residence time. However, resiting, deepening and duplicating a bore are all options that may be called upon once S and T are known.

4.5 GROUNDWATER MODELLING

The use of numerical models in hydrogeology has increased dramatically in the last twenty years with the easy availability of computing power and groundwater modelling packages. The development of a useful groundwater model is a multistage process. Anderson and Woessner provide an excellent review of groundwater modelling techniques and in their introduction, they give a concise guide to modelling protocol (Anderson, 1991).

Early groundwater movement models used an electrical analogue of Darcy's law of groundwater flow, which states:

$$\text{Flow } Q = Tiw$$

where T is transmissivity, and i is hydraulic gradient through an aquifer cross-section of width w. Transmissivity was defined in Section 4.4 as flow per m width i.e. m^3/m per day under unit hydraulic gradient. With the introduction of computers, mathematical modelling of groundwater flows became possible. The most common models use the finite difference or finite element approach. Under the former, and assuming flow is near enough horizontal, a grid is superimposed on a plan of the aquifer to divide it into 'nodes'. Between nodes the flow is related to the hydraulic gradient and the transmissivity of the aquifer in directions 'x' and 'y' (the transmissivity sometimes being taken the same in each direction). The finite difference method adjusts the calculations so that 'boundary conditions' between each flow stream and its neighbours match. Finite difference models in general require fewer input data to construct the finite difference grid. The finite element method is better able to approximate irregularly shaped boundaries and it is easier to adjust the size of individual elements as well as the location of boundaries, making the method easier to test the effect of nodal spacing on the solution. Finite elements are also better able to handle internal boundaries and can simulate point sources and sinks, seepage faces and moving water tables better than finite differences.

Equations (with many terms) can therefore be set up connecting node-to-node flows and head changes, and to which overall limiting boundary conditions apply, such as the assumed (or known) upstream initial head applying, the lateral boundaries of the aquifer, and the downstream conditions applying, such as the outcrop of the aquifer. The equations can then be solved by the computer to match the boundary constraints by reiterative methods or matrix formulation, to a specified degree of accuracy, resulting in the computer providing water table contours and field flows. Three dimensional flows and both transient and steady-state condition of flows can be dealt with, together with such matters as consequent aquifer-fed spring flows, the effect of pumped abstractions and so on.

There are many computer models available for simulating groundwater flows; a resume of their characteristics and capacities is given by Maidment (1992). MODFLOW developed by US Geological Survey in 1988 is quoted as 'popular and versatile'. It can deal with two- and three-dimensional flows and incorporates numerous ancillary facilities. Another useful finite difference model is PLASM, originally written for the Illinois State Water Survey and described as 'recommended as a first code for inexperienced modellers because it is interactive and easiest to operate'. Groundwater modelling requires a sound knowledge of hydrogeology and practical experience of modelling techniques. A hydrogeological investigation of the aquifer is also essential to derive the data for the model (Sections 4.2 and 4.7).

Data Requirements

Setting up and running a model requires a large amount of data. The advantages numerical models have over analytical models are that they allow for spatial and temporal variations in many parameters. The disadvantage is that values for these parameters must be specified by the modeller. Most of the parameters listed below must be defined for every element of a numerical model. Generally the area modelled is divided into smaller regions, which are allocated the same values, but this can lead to loss of flexibility.

- *Boundaries* must be defined for all models, the area modelled being completely surrounded, so that the flow equations can be solved; for resource modelling it is sufficient to ensure that the boundaries are far enough removed from the area of interest in the model that they do not significantly affect the solution but this is not true for protection zone modelling;
- *Recharge rates* must be specified for every node and, since they are critical to model credibility, they should allow for spatial variation due to differences in cover, soil type, vegetation, rain shadows caused by hills and leakage from distribution systems;
- *Hydraulic Conductivity* or, in some cases transmissivity, must be defined for every node;
- *Storage Parameters* are required for each formation in transient models and in models used for particle tracking;
- *Transport Parameters* are difficult to obtain but, if required for a particular model for contaminant transport (as opposed to particle tracking), should include dispersion coefficients, adsorption isotherms and decay coefficients;
- *Interaction Parameters* for defining aquifer relationships with rivers are important for regional hydrogeological models; for rivers two parameters are, in most cases, needed: river elevation (easily ascertained) for each node affected and river bed permeability (usually not well known);
- *Leakance Parameters* are needed where the aquifer level is known, or expected, not to be the same (in hydraulic continuity) as the surface level in a connected water body and where the connection is categorised as 'leaky'; leakance can be determined by careful head measurements but is usually obtained during model calibration and can, therefore, be a source of error; it varies significantly along a river.

Construction of a model is often an iterative procedure. It is unusual for all of the parameters required for the model to be well quantified at the start. The development of a useful and useable model involves fine tuning of the data in order to achieve the desired results.

Model Calibration

Once the data sets are acquired and the model run, it is necessary to assess if the parameter set and the conceptualisation reproduce the observed performance of the aquifer. The output from the model is compared with measured data, generally of water levels from boreholes and flow in rivers. The model parameters are varied until a good correlation is achieved between the model and measured data. To verify interaction, measurements of base flow are often compared with model estimates of river flows to calibrate the model; therefore it is important that the rivers are correctly represented. However, there is no unique solution to the flow equations. Similar head patterns can be achieved with quite different combinations of input parameters. This is especially true for regional models where there are large numbers of parameters that can be varied.

4.6 TEST PUMPING OF BOREHOLES AND WELLS

In most countries concerns, about the need to preserve the natural aquatic environment and the rights of existing abstractors, mean that authorisation to abstract more underground water will only be obtained if the engineer is able to show, from records taken during test pumping, that no unacceptably harmful effect will be caused (Boak, 2007). Therefore, the work of test pumping a borehole and monitoring its impact on the aquifer and the catchment area is essential. The aim of test pumping is to find how much water a well or borehole will presently yield and to determine:

- the effect, if any, of the pumping on adjacent well levels, spring and surface flows;
- the sustainable amount that can be abstracted during dry periods of different severities;
- the drawdown/output relationship whilst pumping in order to decide the characteristics required for the permanent pumps installed;
- the quality of the water abstracted;
- data sufficient to derive estimates of the key characteristics of the aquifer penetrated, i.e. its transmissivity and storage coefficient.

An initial problem is to decide what size of pump to use for test pumping. If the pump's maximum output is much less than the well is capable of yielding, the test will not prove how much more water the well could give and what effect this would have on adjacent sources. On the other hand, a test pump with an output larger than the well can yield, means a waste of money. A major cost of test pumping is the temporary discharge pipeline required. For a large output the discharge line may need to be laid a considerable distance to ensure there is no possibility of the discharged water returning to the aquifer and affecting the water levels in the test well and in other wells used for observation.

The test pump size has to be estimated from experience in drilling the well and on what other holes in similar formations have given. Water level recovery rates after bailing out (Section 4.9) and on stopping drilling are important indicators of possible yield; the drill cores can indicate where well fissured formations have been encountered, and an experienced driller should be able to notice at what drilling level there have been signs of a good ingress of water to the hole. Electrical submersible pumps are most commonly used for test pumping; output being adjusted by valve throttling. In rare instances a suction pump (i.e. one above ground having its suction in the well) may be used if the water level in the well is very near ground level; and for small holes an airlift pump is possible and cheap although rarely used. The discharge must be accurately measured and continuously recorded. A venturi meter, orifice meter, or vane meter can be used for measuring discharge or, for best accuracy, the discharge may be turned into a stilling tank equipped with a V-notch (section 12.14) measuring weir outlet with a float recorder to log the water level over the notch. The advantage of a stilling tank and weir is that it provides a nearly constant outlet head for all outputs and is a positive method of measurement. If meters in the discharge pipeline are used, the pump outlet head will vary with the flow making it more difficult to maintain a constant test pump output, and the meter accuracy must be checked before test pumping and at least once during the test.

Test Pumping Regimes

Every effort should be made to carry out the test pumping when groundwater levels are near their lowest seasonal decline towards the end of the dry season. Whilst dry weather during the test is an advantage, it cannot be guaranteed in a variable climate, and some rainfall during the test period does not invalidate the observations taken, provided it is not extreme. Test pumping outside the dry season when groundwater levels are high should be avoided if at all possible, because it gives uncertain estimates of the yield during the dry season.

The test pumping usually runs for at least 3 weeks, with three stages of increasing output, such as 1/3rd, 2/3rds and full output without stoppage, each for at least 24 hours. The pump is then stopped and the recovery rate of the well water level is carefully measured. When sufficient of the recovery curve against time has been obtained to show its trend towards starting rest level, the pump should be restarted at maximum output and kept at that rate for 14 days, the recovery again being measured when pumping is stopped.

If the maximum pump output is greater than the well can yield, there will be a continuous dropping water level, and the output must be throttled back in an endeavour to achieve a steady drawdown. In other cases the output, after causing an initial drop of water level, may either show a continuous slow decline of water level as pumping continues, or a diminishing decline asymtotic to some maximum drawdown for the given output. It is not always possible to reach stable conditions in a 14 day pumping test.

Problems can occur. If the pump fails in the middle of the test, it is important to record the well water level recovery. After re-starting the pump, the flow rate should be adjusted to that previously. It is not easy to keep the pump output constant because it reduces as drawdown increases, and fluctuations in the voltage of the electrical supply to the pump can cause quite marked changes of pump output. Adjustment of the pump output by valve operation has therefore to be done carefully, to avoid over-adjustment and re-correction, which produce readings that are difficult to interpret. Measuring the water level accurately in a pumped borehole can present difficulties. Using an electrical contact device to detect the water level must be done manually. An air pressure measuring device sited below the water level is not very accurate. A submerged electronic pressure transducer can record levels, but the instrument must be calibrated and readings corrected for barometric variation.

The pumping water levels and pump outputs should be plotted as shown in Figure 4.5 to derive a 'type curve' of output against drawdown can. It is also useful to replot the stepped drawdown results as a Bruin and Hudson curve (Bruin, 1955) as shown in Figure 4.5. This can reveal the proportion of the drawdown due to the characteristics of the aquifer, and the proportion due to hydraulic characteristics of the well or borehole, the latter being distinguished by the turbulent flow losses, the former by the laminar flow losses. Examination of the turbulent loss coefficient c can sometimes suggest high entry losses to the well due to poor design of the well screens or gravel packing. Good (i.e. low) figures for c in cQ^2, where Q is in m^3/min, are below 0.5. The laminar loss coefficient b describes the relative permeability of the aquifer(s) feeding the well.

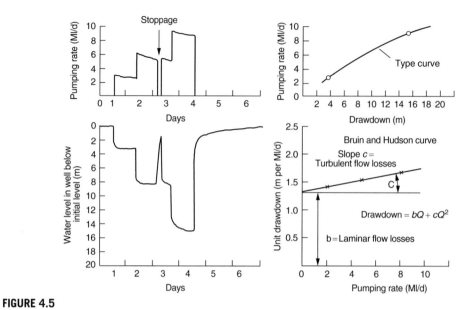

FIGURE 4.5

Step-drawdown test results. (Bruin, 1955)

Use of observation wells and monitoring

To determine aquifer characteristics (Section 4.4) it is desirable to have two observation boreholes sited within the likely cone of depression of the water table around the borehole when test pumping takes place. The radial distance of these holes from the well to be tested should be such that the drawdown they experience is large enough to give a reasonably accurate measure of the drawdown but small relative to the depth of saturated aquifer which contributes flow to the pumped well, so that the flow through the aquifer at the observation bore site is essentially horizontal. Experience has shown that these requirements frequently result in observation holes being sited between 50 m and 200 m of the pumped well. The larger this distance is the greater is the 'slice' of aquifer brought under observation. Similarly, more information will be gained if the two observation holes are on radials 90° apart. The radial distance of each bore from the well should be measured.

Water levels in the observation boreholes should be measured regularly before during and after test pumping, the frequency being increased after starting, stopping or altering the pump output sufficient to log rapid changes of water level. Preferably, a water level recorder should be used. An existing well, such as a household well, can be used for observation purposes, provided the quantity of water being drawn is too small to affect its standing water level. Usually, however, no conveniently placed existing well will be available and, if observation bores are required, they will have to be drilled. Observation boreholes are often omitted because of their installation cost, with reliance being placed solely on results from the pumped borehole and the effect, if any, on existing boreholes and wells nearest to the pumped site.

It is also essential to monitor water levels in all wells and ponds, and the flows from springs and in streams, that could conceivably be affected by the test. The area and features to be monitored is a matter of judgment according to the particular circumstances, but would often be about 2 km radius from the test pumping site. However, all important abstractions that could be sensitive to flow diminution, such as watercress beds and fisheries, should be monitored even if they appear to lie outside any possible range of influence of the proposed abstraction. The paramount need is to have proof that the features were not affected by the test pumping. The measurements of key indicator wells and springs may be started 6 to 9 months before the planned timing of the test pumping. Temporary weirs may be needed to measure springs and stream flows. On streams or rivers, measurements of the flow both upstream and downstream of the test pump site may be necessary. The aim is to get a sufficiently long record of water table levels and stream flows that their trend before and after the test pumping can show whether there is any interruption or change caused by the test pumping.

There is also a need to check that the underground water table 'catchment area', which contributes to the test pumping, is the same as the topographical catchment. Increased abstraction from one underground catchment area can cause the water table divide to migrate outwards, reducing the area of an adjacent catchment feeding other sources, which would, therefore, at least notionally suffer a reduction of potential yield.

Daily rainfall measurements must be taken also, if there are no existing raingauge stations to give the catchment rainfall.

4.7 GEOPHYSICAL AND OTHER INVESTIGATION METHODS

Samples of strata will be retrieved during the sinking of a borehole. However they may not be properly representative of aquifer water quality because inevitably the samples will be disturbed and the groundwater quality will have been disrupted also, being only a mixture of all the flows entering the bore. More precise information can be derived from a wide variety of subsurface 'down-hole' geophysical methods. Direct physical observation or measurement can be obtained by use of:

- a television camera to view the walls of a boring;
- a calipering device to measure and record bore diameter with depth;
- recording instruments to obtain profiles of water temperature, pH and conductivity with depth;
- small sensitive current meters to detect differences of vertical flow rates in the borehole, thus indicating changes in inflow or outflow rate from strata.

Electrical and nuclear instruments can provide information concerning the strata penetrated. Resistivity measurements taken down the hole, using an applied electrical potential through probes, can locate the boundaries of formations having different resistivities. This can aid identification of the type of strata penetrated and distinguish between fresh and saline waters in the formation. Measurement of the 'self' or 'spontaneous' electrical potential existing between strata at different depths of the formation can assist in detecting permeable parts of the aquifer. Nuclear downhole tools used in aquifers are mainly (in water) of three types:

- natural gamma detectors which pick up natural radioactivity from potassium in clays (so correlating with clay content);
- gamma-gamma tools which use a gamma radiation source to bombard the formation and measure back-scattered radiation produced, which is inversely proportional to the formation density and can be used to indicate degree of cementation or clay content also;
- neutron tools which are used to give an indication of water content and porosity.

All these measures, co-ordinated and supplementing each other, can give additional information about an aquifer. In particular they can indicate the best locations for inserting well screens or gravel packs (Section 4.8). More information is available in IWES *Manual No. 5* (1986).

Many aquifer systems comprise different aquifer formations at different levels. These can be investigated by using packers down the borehole to isolate individual aquifers and, for each, measure inflow and piezometric head and take water samples for analysis. Various multi-tube assemblies can be used to withdraw samples of water from the isolated aquifers and, if inserted in an observation borehole, can remain in place for subsequent monitoring purposes. The production bore can, of course, be similarly investigated before a pump is installed. The method can reveal whether it is advisable to seal off inflows of undesirable quality (including signs of contamination), or which come from formations whose catchment outcrop is known to include potential sources of pollution best avoided.

4.8 BOREHOLE LININGS, SCREENS AND GRAVEL PACKS

The components of a borehole are illustrated in Figure 4.6.

The upper part of a borehole is usually lined with solid casing, which is concreted into the surrounding ground. This is to prevent surface water entering and contaminating the hole and to seal off water in the upper part of the formation, which may be of lower quality than that within the main aquifer.

FIGURE 4.6

Components of a borehole.

Where a borehole encounters weakly cemented or uncemented sands and gravels, a perforated or slotted lining may be needed to support such formations, while permitting fine particles adjacent the screen to be washed out and improve the yield of the borehole. There are many different types of screen; the simplest comprise slots cut in borehole casing, whilst the most sophisticated are stainless steel, wedge wire-wound screens with accurately set apertures. In non-corrosive waters screens are usually made of steel; in corrosive waters, plastic coated steel, phosphor bronze, glass reinforced plastic or rigid PVC slotted or perforated screens may be used. Figure 4.7 shows the aperture and percentage open area for some typical 300 mm diameter screens. The screen aperture required has to be decided according to the size of particles in the formation (see below). Because of unavoidable partial blocking of the screen by sand particles, the effective area of a screen is usually estimated at less than half its initial open area. The water entry velocity through the apertures needs to be limited to about 30 mm/s to avoid high turbulence losses. Working from the required yield, the open screen area and the amount of blockage that might occur over time, the length of screen necessary can be calculated.

If the apertures or slots of a well screen are suitably sized, withdrawal of the fines from the formation should result in the coarser material forming a 'naturally graded pack' against the screen that prevents further withdrawal of fines when the borehole is pumped. A formation is considered capable of forming a natural pack if the D_{10} size (aperture size through which 10% of the material by weight passes) is greater than 0.25 mm, and the D_{90} size (size through which 90% passes) is less than 2.5 mm, and the Uniformity Coefficient (UC) is between 3 and 10.

$$UC = D_{60}/D_{10},$$

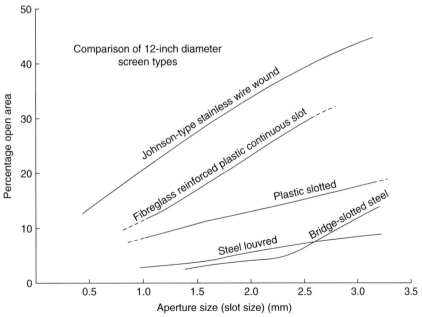

FIGURE 4.7

Comparison of screen open areas.

where D_{60} is the aperture size through which 60% of material passes (by weight) and D_{10} is the aperture size through which 10% of material passes.

During test pumping and development of the borehole, abstraction rates should be higher than will occur during the permanent production pumping so that no fines are withdrawn in production.

Where the grading of the formation material is unlikely to form a 'natural pack', a gravel pack should be placed outside the screen. The screen is suspended centrally in the borehole by means of lugs fitted at intervals on the outside and gravel is tremied into the annular space. The gravel pack can be of uniform sized gravel or graded gravel. Uniform sized packs are suitable when the Uniformity Coefficient of the formation material is less than 2.5, in which case the 50% size of the gravel pack should be 4 to 5 times the 50% size of the formation material (Terzaghi, 1943). Otherwise, a graded gravel pack should be used, its grading being selected to parallel the grading curve of the formation material so that the Uniformity Coefficients are similar; the D_{15} size of the pack should be 4 times the D_{15} size of the coarsest sample from the formation, with the D_{85} size of the pack less than 4 times the D_{85} size of the finest formation material sampled.

The minimum thickness of the gravel pack should be 75 mm. Installing a thinner pack becomes uncertain because of imperfect verticality of the boring, variations in its diameter, and the problem of placing the gravel evenly in the annulus at depth. Even with the use of a tremie pipe there may be zones where the pack is virtually absent and others where the thickness is greater than necessary. Part of the development of a newly sunk borehole may involve the need to remove 'mud cake' from the formation face, and a thicker pack tends to reduce the ability to remove the cake. There are many methods for selecting an appropriate gravel pack (Monkhouse, 1974; Campbell, 1973) each usually for a particular type of aquifer formation.

In loose sandy formations such as Greensand, the need for a gravel pack has to be taken into account when deciding on the starting diameter for a boring. The expected water pumping level is also an important factor because it determines where the pump is to be sited, and consequently the diameter required down to pump level to ensure the pump can be accommodated with reasonable flow conditions to its suction. Although it is possible to ream out a borehole to a larger diameter in some cases, it may not be practicable in a loose formation where solid or slotted lining has to be driven down closely following the boring tools to prevent collapse of the hole.

4.9 CONSTRUCTION OF BOREHOLES AND WELLS

Boreholes for public water supply are usually of large diameter; methods for sinking them differ from the smaller diameter holes sunk for oil drilling or for the water supply to a few houses. The percussive method is widely used. A heavy chisel (Plate 2(a)) is cable suspended and given a reciprocating motion on the bottom of the hole by means of a 'spudding beam', which alternately shortens the cable and releases it at a rate that varies with the size and weight of chisel used. Considerable skill is required by the operator to adjust the rate of reciprocation to synchronize with the motion of the heavy string of tools down the hole and avoid any violent snatch on the cable. This cable has to be paid out gradually and kept exactly at the right length for the chisel to give a sharp clean blow to the base of the hole. The wire cable is left hand lay and, under the weight of the string of tools tends to unwind clockwise, thus rotating the chisel fractionally with each blow. After a while the unwinding builds up a torque so that, when the weight is off the cable just after a blow on the base of the hole, the torque breaks a friction grip device in the rope socket attachment and the cable returns to its natural lay. Thus slow rotation of the chisel continues. Every so often the string of tools must be withdrawn and a bailer—a tube with a flap valve at the bottom—is lowered to clear the hole of slurried rock chippings. Progress is therefore slow and, depending on the hardness of the formation, the drill chisel may require resharpening from time to time.

Another percussive method is use of a 'down the hole' hammer operated by compressed air. This is up to ten times faster in hard rock than a cable-operated chisel, but it is unsuitable for soft formations and cannot be used in substantial depths of water. The chippings have to be ejected to the surface by the exhaust air from the hammer passing through the annular space between the drill rods and the borehole. Unless the boring is of small diameter there may be difficulty in raising the chippings, in which case an open topped collector tube may be positioned immediately above the tool to collect the chippings. The down-the-hole hammer is predominantly used on small boreholes up to about 300 mm diameter (Plate 2(b)).

For rotary drilling of large holes in hard material a roller rock bit is used, equipped with toothed cutters of hard steel that rotate and break up the formation (Plate 2(c)). Water is fed to the cutters down the drill rods and, rising upwards through the annular space, brings the rock chippings with it. In the 'reverse circulation' method the water feed passes down through the annular space and up through hollow drill rods. This achieves a higher upward water velocity making it easier to bring chippings to the surface. Air is sometimes used instead of water, but the efficiency falls off as depth below water increases. Heavy drill collars may have to be used to give sufficient weight on the rock cutters. In addition, a large quantity of water may be required for the drilling of a large diameter hole. Rotary rock bit drilling is widely used by the oil industry, usually for holes of substantially smaller diameter than those needed for public water supply. Clay or bentonite drilling fluids can be used to support unconsolidated formations and assist in raising chippings. However, the use of such suspensions is not always advisable in water well drilling in case they seal off water bearing formations. Instead polymer based, low solids, biodegradable muds are preferred since these are not so likely to seal off water bearing formations.

Diamond core drilling is seldom used for sinking water supply holes except for small diameter holes in hard formations or for trial holes where rock cores are needed.

An alternative method of rotary core drilling for large diameter holes is to use chilled shot fed down to the bottom of a heavy core-cutting barrel. The barrel has a thickened bottom edge under which some of the chilled shot becomes trapped and exerts a strong point load on the formation causing it to break up. A small amount of water is added for lubrication and further chilled shot is added so that eventually, with the annular cut deepening, a core of rock enters the barrel. To break off the core and bring it to the surface, some sharp pea gravel or rock chippings are sent down, causing the core to be gripped by the barrel, rotation of which causes the core to be broken off. Progress in hard formations is slow and, if fissures are encountered, problems of loss of shot or water can occur.

When soft or loose material is encountered, a solid casing or slotted lining may have to be pushed down to the level required as the boring proceeds, or after withdrawal of the chisel. Once a lining is inserted, the hole may have to be extended in smaller diameter. Several such 'step-ins' of diameter may be required if the formation has several separated layers of loose material. Biodegradable muds may be used with rotary drilling to keep the hole open, provided the mud does not seal off water bearing formations or is not lost through fissures; in which case it may not be necessary to line the hole until it is completed to full depth. However, rotary drill rigs may have difficulty in penetrating ground containing hard boulders and rotary drilling is considerably more expensive than percussion drilling.

A typical specification for verticality is that the hole should be no more than 100 mm off vertical in 30 m of depth but it is not necessary to insist on this at depths below the possible siting for the pump. It is especially difficult to keep a boring vertical when steeply inclined hard strata are encountered. Verticality of a borehole is not as important as are 'kinks' in the boring down to the proposed level for the permanent pump. Kinks above pump level may throw the pump to one side of the boring, making poor pump suction entry conditions and causing difficulties in lowering or removing the pump. If correction is essential it may be necessary to ream out a hole to larger diameter to get the remainder of the boring on line.

The duties of the borehole driller are:

a. not to lose drilling tools down the hole;
b. to note and log down every change of drilling conditions such as—increased or decreased speed of drilling, change of tools necessary, quick or sluggish recovery of water level after bailing out, etc, and
c. to note and log down the lengths of different casings used, the nature and depth of all cores and chippings produced from the boring and to keep all cores and chipping samples.

Tools lost down a hole, such as the drill bit which is attached to the drilling rods, will cause days of delay in attempts to recover them and, if 'fishing' (with a wide variety of special tools) is unsuccessful it may be necessary to ream out a hole to larger diameter down to the tool to recover it. Sometimes tool recovery may defeat all efforts and the hole has to be abandoned if it has not been driven to the required depth. When a water borehole is being sunk it is particularly important to watch for signs of large fissures, indicated, for example, by a sudden drop of the chisel or drill, increased flow of water into the hole or fast recovery of water level when bailing out. The engineer needs to see that borehole progress logs are meticulously kept. Core samples should be stored in wooden core boxes, their depths being clearly marked. Samples of chippings and soft material should be stored in plastic bags properly labelled. The samples need to be examined by an experienced geotechnical engineer or hydrogeologist following the identification and classification guidelines set out in BS EN 14688 and BS EN 14689.

Wells up to about 1.2 m diameter can be sunk in hard material by rotary drilling using a heavy core barrel and chilled shot as previously described. In less hard materials, such as chalk, a large percussion chisel may be used to chop up the formation, the excavated material being removed by suction pump or by grab. In very soft or loose material, wells may be sunk kentledge fashion, a concrete caisson of the required well diameter, with a lower cutting edge, is sunk into the ground by hand excavation of the core material below the cutting edge. As excavation deepens, further rings are added to the top of the caisson. The excavated material is removed by skips or by a crane grab. In extremely soft material, excavation by grab alone may be sufficient for the caisson to descend by its own weight until harder material is encountered. Early wells were hand dug with linings of brick; later wells had linings of cast iron segments bolted together. Either interlocking precast concrete rings or concrete segments bolted together are now used. The upper part of a well is generally sealed by cement grouting on the outside to prevent ingress of surface water, except where a well is sunk in river bed deposits.

4.10 DEVELOPMENT AND REFURBISHING BOREHOLES AND WELLS

On completion of a borehole or well it should be 'developed' to maximize its yield. The objective is to remove the clay or finer sand particles from the natural formation surrounding the slotted linings or well screens to improve the flow of water into the boring and to ensure that any gravel packing is properly compacted. 'Surging' is the most common development technique. It consists of pumping from the well at maximum rate, and then suddenly stopping the pump thereby causing flow to wash in and out of fissures in the formation thus dislodging silt and fine sand therein. The piston effect of using a bailer can also be used, but must not be too energetic or screens might be damaged or an uncased hole collapse. 'Swabbing' and 'surging' with a piston can be used, a swab having a valve in it which allows water upflow. Both tools promote energetic flow through a screen during their up and down movement, the surge plunger in particular causes strong agitation in a gravel pack, encouraging rearrangement of particles in the pack and removal of fines. However, these tools cannot be used with wire wound screens, which are supported by internal vertical bars; instead they have to be operated in the casing above the screen where they may still be effective but considerably less so.

Air lifting, although an inefficient form of water pumping, can be useful for the development of a hole carrying sand-charged water, which would be highly abrasive to normal centrifugal pumps. There are several variations on the use of an air-lift for this purpose, which depend on the vertical movement of the air pipe up or down inside the eductor pipe. It is also possible to seal the top of a boring and inject compressed air at the top, driving the water down into the formation, prior to release of the pressure and starting the air lift.

Jetting can be used. Depending on whether the gravel pack has to be agitated, or the screen or formation cleaned, either low or high pressure can be used. A jetting head with horizontal water jet nozzles is suspended on the end of a drill pipe and slowly rotated whilst jetting. Care has to be taken with non-metallic screens or where the screen may have been weakened by corrosion.

Other methods sometimes adopted comprise chemical treatment or acidisation but it is strongly advisable that these should only be undertaken by an engineer experienced in such techniques. The most common chemical treatment uses a dispersant, such as sodium hexametaphosphate, e.g. Calgon, to assist in the removal of fine material. Acidisation is used in calcareous strata to enhance the size of fissures in chalk or limestone in the immediate vicinity of the borehole. Concentrated hydrochloric acid is usually used, together with an inhibitor to minimize corrosion of any mild steel casing. The acid is applied by pipe below the water level. If the top of the borehole is closed the pressure of the evolved carbon dioxide can enhance penetration of the acid into fissures. However, the handling of acids is dangerous and should be subject to safety procedures. In addition, the carbon dioxide produced can be hazardous unless special attention is paid to ventilation.

The performance and condition of water supply boreholes will change with time and performance monitoring is a core activity in good groundwater supply management strategy. Comprehensive data gathering and interpretation of the results enable the optimum maintenance and rehabilitation strategy and implementation plan to be adopted. Best practice guidance on monitoring, maintenance and rehabilitation can be found in a CIRIA report on the subject (Howsam, 1995).

4.11 POLLUTION PROTECTIVE MEASURES: MONITORING AND SAMPLING

The dangers of groundwater contamination are widespread and varied. Industry, agriculture and urbanisation generate large and varied suites of groundwater pollutants, many of which are of public supply significance. The characteristics of the various pollutants vary in time and location as Table 4.2 indicates. All represent the background addition of some substances to groundwater, which may be controllable somewhere in the water supply cycle, but prevention can not be managed necessarily within the catchment. The implications for the public supply are very high. Treatment becomes necessary initially, and the resources may then become difficult to manage to avoid inducing or accelerating further contamination. Groundwater pollution can sometimes be 'cleaned up'; many methods have been used in the last twenty years, mainly in the USA where funds have been provided for the purpose but the technology for restoration of contaminated groundwater and aquifers remains costly and imprecise. Remedial action may be impracticable where predicted clean-up time is decades or longer. The aquifer may become completely unusable so that substitute resources, such as imported supplies or treated surface water, are needed and a valuable and inexpensive resource is lost.

The concept of evaluating the vulnerability of groundwater sources to pollution is based on consideration of the lithology and thickness of the strata above the aquifer and the surface soil leaching properties. From this, the size of protection zone required around a borehole or well may be derived, based on the estimated 'travel times' of potential pollutants within the saturated zone to the abstraction point (Adams, 1992). Work initiated by the National Rivers Authority and later taken over by the EA, suggested three zones of protection should be investigated for protection of underground abstraction works:

- *Inner Zone I* defined by a 50-day travel time from any point below the water table to the source, based principally on biological decay criteria;
- *Outer Zone II* defined by a 400-day travel time, based on the minimum time required to provide dilution and attenuation of slowly degrading pollutants;
- *Source catchment Zone III* defined as the area needed to support the protected yield from long-term groundwater recharge from effective rainfall.

Table 4.2 Major sources of potential groundwater pollution

Occurrence	Local/linear mode	Distributed mode
Seasonal/periodic	Road salting Rail and road verge herbicides Silage Combined sewer overflows	Agricultural fertilizers, herbicides and pesticides Sewage sludge spreading
Continuous	Road drainage Cesspool overflows Septic tank effluent disposal to land Solid waste tip and landfill leachates Contaminated abandoned industrial land Influent polluted rivers Runoff from grazing land (cryptosporidium)	Industrial atmospheric discharges
Random	Road/rail tanker spills Pipeline and sewer breakages Fires Defective storage of industrial or agricultural chemicals	Nuclear and industrial accident fallout

To delineate such zones about any particular source, a series of field investigations are necessary; initially to create a conceptual model which later, as more data is obtained, is sufficient to produce a calibrated model of the catchment so that the zones can be more accurately defined. In the UK, zones are used as a guide for the control of catchment activities by water utilities but elsewhere, as in Germany, the protection zones can be statutory, being set up under local or Federal legislation.

Continual vigilance over the surface catchment to an underground source should be exercised. All potential sources of pollution (cesspools, septic tanks, farm wastes, industrial waste and solid waste tips, farm or industrial storages of chemicals etc.) should be identified and recorded on a map of the catchment area. All these risks should be regularly monitored to ensure they are properly controlled. Poorly functioning household sewage disposal works may have to be replaced by more efficient plant and farmers may have to be advised of improvements to their waste disposal practices.

Monitoring water quality will identify signs of contamination and whether it is a recent and local incident or more likely to be from a distant part of the catchment. The chemical constituents of the water can show evidence of the types of pollution listed in Table 4.2. If a pumped borehole penetrates several water bearing strata, depth specific sampling devices can be used below the pump with isolating inflatable packers. This can be useful for ascertaining at what level a contaminant is entering the borehole. Gas operated ejection devices or small diameter piston pumps can be used to withdraw samples from sections of the borehole using the isolating packers. Otherwise samples from the pumped water will be a composite sample of the water entering the hole from the full thickness of the formation in contact with the well.

Groundwater sampling needs a slightly different approach to surface water sampling. The chemistry of groundwater tends to be relatively stable, but dissolved oxygen and redox (Section 7.1) potential may be more important than for surface waters. Quality changes during sampling have to be considered, mitigated as much as possible by taking measurements of parameters such as dissolved oxygen, pH and conductivity in the field. The groundwater temperature may also be a useful indicator, because surface percolation to the upper levels of an aquifer will show a substantially greater seasonal temperature variation than water drawn from depth. Long term groundwater abstraction may cause an increase in dissolved solids, which may imply that abstraction is drawing water from more distant parts of the aquifer, possibly indicating that the catchment area monitored may need to be extended.

4.12 RIVER FLOW AUGMENTATION BY GROUNDWATER PUMPING

Regulating river flow by pumping from groundwater is practised, although not currently favoured. Water is pumped from an aquifer and discharged to a river to augment its flow during a low flow period. Usually the abstraction points lie within the river catchment basin. The Shropshire Groundwater Scheme is an example in the UK where water is abstracted from the sandstone aquifer underlying North Shropshire and discharged into the River Severn to meet peak dry weather demands. To be successful river regulation by groundwater pumping has to meet certain basic criteria:

- the scheme should provide a satisfactory increase in yield above that which could be obtained by direct abstraction from the aquifer;
- discharge of the pumped groundwater to the river should result in economies in pipeline costs;

- the abstraction pumping capacity must be not less than the desired gain in river flow during the design drought period, plus any reduction of springflows feeding the river caused by the pumping;
- if recirculation losses through the river bed occur, it must be practicable to adopt equivalent extra abstraction capacity.

Prolonged groundwater abstraction will almost certainly reduce some springflows contributing to the river flow. To avoid iterative calculations it may be better, in a simple approach, to assume that all springflows within a given distance from the groundwater abstraction points will be reduced to zero by the pumping during a prolonged drought. Pilot well tests and a detailed knowledge of the catchment hydrogeology are necessary to assess which springflows will be affected. A better approach is to adopt groundwater mathematical modelling. This permits the cumulative effect of pumping to be traced for droughts of different severity and length. The method can reveal the effect on local wells used for private supplies, farming and other interests, whether adjacent catchments are affected; and to check if percolation is sufficient to restore the aquifer storage during later wet periods.

In general river regulation by groundwater pumping is most successful where the pumping has a delayed, minimal, short-lived effect on natural river flows. To achieve this, several carefully sited abstraction points, pumped according to a specific programme, may be necessary. Where a regulated flow substantially in excess of the natural low river flow is required, prolonged pumping tests at the proposed abstraction points during a dry period are essential. This can be an expensive operation. In addition, caution is required in estimating the possible yield during some extreme drought conditions not previously experienced, because aquifer drawdown conditions during some future critical event cannot be known with any accuracy.

Augmenting river baseflows as a consequence of groundwater recharge has the flowing advantages:

- allows better management of water resources by recharging with surplus water during high flow periods;
- raises low flows in the summer, provided that the timing and movement of recharge water is appropriate, with consequent environmental benefits.
- However, the disadvantages include:
  - only applicable where recharging an unconfined aquifer in hydraulic continuity with the surface water system (springs or river channel inflow);
  - additional studies and trials are required to determine feasibility and possible effects.

4.13 ARTIFICIAL RECHARGE AND AQUIFER STORAGE RECHARGE

Artificial recharge (AR) supplements the natural infiltration of water into the ground to augment groundwater yields. AR can be used strategically where an aquifer is overexploited such that no further abstraction would be allowed or where lack of natural recharge prevents its utilisation. The water abstracted is not necessarily the same water that was recharged and abstraction boreholes may be some distance apart. It is often practised for water treatment reasons, as in the Netherlands where the dune sands are used to store potable water (Jones, 1999).

The two basic methods adopted are the use of spreading areas or pits and injection through boreholes. The former technique has dominated because of its simplicity and the ease with which clogging problems can be overcome. The relatively unknown performance of wells and the need to pass only pretreated water down them limits their usefulness except where land for recharge pits is at a premium. Untoward events can happen during recharge operations: at an Israeli site (Sternau, 1967) the injection of water down a well caused unconsolidated sands around it to settle and, in less than two days after pumping ceased, the borehole tubes and surface pump house sank below ground level. In a British experiment (Marshall, 1968) even the use of a city drinking water supply for recharge did not prevent the injection well lining slots from becoming constricted with growths of iron bacteria.

At times when existing surface water treatment works are working below capacity, e.g. when there is a seasonal fall in demand, it may be possible to make water available for well recharge. This is being practised in the Lee Valley (O'shea, 1995) north of London. Complete efficiency cannot be expected because of the relatively uncontrollable and 'leaky' nature of aquifer storage. Because the recharged source may not be used for some months, there is a loss of resources as the 'recharge mound' decays outwards and down the hydraulic gradient. Even where abstraction facilities have been located specifically to minimize this, the losses are considerable. Special artificial recharge operations in coastal areas have been used to prevent saline intrusion from the sea, as for example in coastal Israel (Harpaz, 1971), (Aberbach, 1967) and southern California (Bruington, 1965).

Recharge as a way of improving wastewater quality has been used in the Netherlands, Germany and Scandinavia for many years (IAHS, 1970). Alluvial sand aquifers of the coastal Netherlands are recharged with heavily contaminated River Maas water after primary filtration. The abstracted water still has to be treated using activated carbon to remove heavy metals but can be used for public supply. The strata that provide the natural treatment during infiltration have to be 'rested' and allowed time to recover and re-oxygenate. At Atlantis in south west Africa a successful artificial recharge scheme is part of the local water resources management system. Treated domestic sewage effluent and urban storm runoff from the 67 000 population is recharged through lagoon systems and re-abstracted from specially sited production wellfields (Wright, 1996).

Recharge boreholes are vulnerable to various types of clogging due to accumulation of suspended solids, gas bubbles that come out of solution, and microbial growth filling interstices or screen apertures. Borehole recharge systems using polluted water are normally not acceptable. Hence, only recharge pits are considered below.

Although dimensions vary, a recharge pit bears resemblance to a slow sand filtration bed because replaceable filter media normally covers the base. The rating of a pit depends on the rate at which the raw water will pass through the filter media and the rate at which the underlying aquifer will accept it. The former depends upon raw water quality, any pretreatment of the water and the depth of water kept in the pit. Little, if anything, can done to improve the rate at which the aquifer accepts the filtrate, but a decline in the infiltration rate must be guarded against by tests to ensure the filter media is working satisfactorily. Published results for infiltration rates vary widely.

Pilot tests are always required at a new site, as are initial investigation bores to ensure the pit floor will be above the water table by a margin as big as possible to give the best opportunity for water quality improvement and to allow for increased groundwater storage and its 'mounding' below the recharge pit. Iron-pan layers and other impedances to vertical flow should be avoided.

Aquifer storage and recovery (ASR) is a subset of AR and comprises one or more of the following:

- storage of water in a suitable aquifer (through wells) during times when water is available and recovery of water from the same well when needed;
- normally achieved by using dual purpose boreholes;

- injection of water into an aquifer containing non potable water;
- abstraction of the same water that was injected;
- utilisation of a confined aquifer to minimize potential environmental impacts.

The water is stored in the aquifer locally to the borehole; it improves resource management and has operational advantages. Water is recharged into the aquifer during periods of low demand and then recovered during periods of high demand.

ASR could be considered rather than AR in parts of confined aquifers that have not been used for productive water supply because of poor quality (Jones, 1999). The use of ASR in areas of outcrop of potable aquifers is less likely as the aquifer may be fully licensed, natural recharge will occur and the groundwater-surface water interaction will be more immediate and hence have environmental impacts (Williams, 2001).

The key constraints on the development of ASR identified by research (Jones, 1999) include:

- recovery efficiency, a measure of how much is recovered against how much is put in, is less significant where both recharged and native waters are of similar quality;
- clogging issues including: air entrainment, suspended sediment, bacterial growth, chemical reactions, gas production and compaction of clogging layer;
- water quality changes: chemical components of the recharged water reacting with groundwater and aquifer, particularly the organochlorine compounds in the treated recharge water;
- hydraulic properties of aquifers;
- operational issues including: variability in volume and quality of water available for recharge, less significant when treated effluent is used; variation in daily and weekly demand in relation to average monthly or annual demand; and site selection;
- regulatory issues.

Of particular environmental concern is the possible impact on water levels in adjacent aquifers or in the target aquifer. ASR schemes are designed to result in no net change in abstraction from the aquifer. In the long term there may be some net change but this should be small in relation to the overall resource. A scheme that injects and then abstracts the same volume of water, as a long-term steady state average, will have no impact on local or regional groundwater levels. However, the seasonality of the scheme means that local water levels will first be increased and then reduced. The timing and absolute value of these changes are important in assessing the eventual impact and will need to be determined by site specific model studies or field trials.

4.14 GROUNDWATER MINING

In many countries of the Middle East and North Africa groundwater is being used at a rate greatly in excess of the rate at which the resources are being replenished. This tends to occur either because there is no institutional framework for controlling development or because demands for water cannot be ignored while adequate resources appear to exist. Climatic patterns have changed during the last few tens of thousands of years; areas, which are now arid, were once comparatively humid. In that earlier period, high rates of recharge applied where none now takes place, as in the Libyan Desert and much of Saudi Arabia. In both regions there are groundwater developments on a vast scale that 'mine' this water. Many thousands of cubic kilometres of groundwater can be developed, though the cost of doing this and taking the water to where it is needed are high.

The Great Man-Made River Project in Libya (Table 4.1) is a modern example, where the capacity now exists to pump many millions of cubic metres of groundwater through 4 m diameter concrete pipelines for over 600 km from inland desert wellfields to the Mediterranean coast. In Saudi Arabia 'fossil' groundwater is being mined for public supply and here, as in some other countries, groundwater exhaustion problems have only been managed through large scale adoption of seawater desalination.

The use of water resources without knowledge of their sustainability, or where it is known that mining is taking place, may be looked upon as an irresponsible course of action in most cases. However, often the use of technology to minimize demand or maximize efficiency in the use of water is too costly to be politically acceptable, especially in respect of agriculture, which remains the greatest and least efficient user of water resources. It is estimated that a 10% improvement in the efficiency of agricultural use would double the resources available for public water supply.

4.15 ISLAND WATER SUPPLIES

Large numbers of islands abound in all the major oceans, particularly off South East Asia and in the Pacific. In the Philippines alone there are over 7000 islands and Indonesia comprises over 13 000 mainly small islands. The Republic of the Maldives in the Indian Ocean is a nation comprised entirely of about 1300 coral atoll islands, of which about 200 are populated. Geologically the islands range from the Maldive type of atoll islands which are less than 2 m above high tide level, to larger islands with a rocky core surrounded by a rim of sediment and reef. Many of these islands are isolated, lack natural resources and are short of water. Such freshwater as exists often occurs as an extremely fragile lens, floating on saline water below. Population densities are high. On Malé island, capital of the Maldives, a population of over 83 000 live on 200 ha of land. The fresh groundwater therefore not only became polluted with sewage but eventually was virtually consumed by over abstraction so that dependence on desalinated water and external resources is now near total. Similar problems can occur in other places where tourism increases water demand beyond the ability of local resources to supply.

On the larger rocky islands, surface water resources can sometimes be developed from perennial streams or springs, but stream flow is ephemeral on all but the larger islands. Usually there is heavy dependence on groundwater, which exists as a reserve of freshwater, usually described as forming 'a lens' of freshwater within the strata forming the island, below which lies denser saline water infiltrated below from the sea. The lens is not usually regular, but strongly distorted by complications of geology that influence fresh groundwater flow to the sea. A lens thickness of 20 to 30 m is common, with a transition zone between the fresh and saline water, the thickness of which is a function of the aquifer properties, fluctuations in the tidal range and due to variation in the rate of rainfall recharge. The amount of water represented by a given lens depends on the specific yield of the strata; a coral sand may have a high specific yield (20% or more) whereas a limestone or volcanic rock may have a much lower capacity to store water (less than 1%).

The simplified elements of an island water balance are illustrated in Figure 4.8 including the transition zone between fresh and saline waters. The dynamic balance between percolation from rainfall and outflow of freshwater to the sea determines the dimensions of the lens of freshwater, mainly its depth because the width is limited. This is the 'classical view' of the lens configuration, in which the height of the lens surface above mean sea level is 1/40th of the depth of the lens below mean sea level (Section 4.3). However, observations of many islands show that a more realistic model

FIGURE 4.8

Diagram of lens of freshwater floating on seawater on a coral island.

incorporates vertical and horizontal flow due to the effect of a typical two-layered sand overlying a reef limestone system (Wheatcroft, 1981; Ayers, 1986; Falkland, 1993). The transmission of the tidal fluctuations to the lens is not necessarily proportional to the distance from the shore, but rather to the depth of the sand aquifer and the transmissivity of the underlying reef limestone. Dispersion is also incorporated in the model to allow for the thickness of the transition zone between the fresh and saline water, which is typically of the order of 3 to 5 m.

However, other than rainfall, assigning a quantity to all the components in Figure 4.8 is very difficult. For instance the groundwater outflow is usually a significant proportion of the water balance, but it cannot be directly measured; it can only be roughly estimated from the groundwater gradient and the hydraulic conductivity of the aquifer. Similarly evapotranspiration use by vegetation is difficult to estimate because of the unusual situation that such vegetation can find freshwater continuously available at shallow depth. Hence a water balance estimate can only reveal whether the likely replenishment of the lens is obviously less (or more) than the estimated losses from outflow and abstraction. Consequently the only safe way to evaluate the situation is to measure the thickness (and extent) of the lens and to keep it monitored.

Measuring the electrical conductivity at the base of the freshwater lens and across the transition zone is the most sensitive method of monitoring freshwater reserves. Over this zone, small changes in the elevation of the lens are reflected in large changes of electrical conductivity. Figure 4.9 shows measurements taken on Malé Island between 1983 and 1988 at a number of depths. The figure illustrates the virtual demise of the freshwater reserves during the period.

Controlled development of island freshwater is best achieved by shallow abstraction, using groups of shallow wells or collector systems, or even infiltration trenches. The use of boreholes is usually not the best approach because drawdown of the freshwater surface during pumping tends to cause upward movement of the saline water below. The rate of 'up-coning' of saline water depends on the geology, rate of pumping and depth of the borehole; but once contaminated with saline water the aquifer may take many weeks or longer to recover a usable quality of freshwater. The proportion of rainfall that recharges groundwater is subject to many factors. A relationship between rainfall and recharge derived by Falkland (1993) for atolls and larger topographically low islands is shown in Figure 4.10 and is a useful guide for preliminary use.

However, the difficulties of preventing overdraw from an island aquifer has to be recognized. The local population will have been able to use house wells for their supply. However, because the level of the water in wells remains virtually constant (even though over-abstraction is causing the freshwater

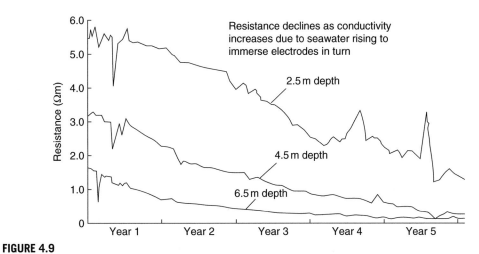

FIGURE 4.9

Graph indicating rise of freshwater/seawater interface as lens of freshwater thins due to over-abstraction—
Malé 1983–1988.

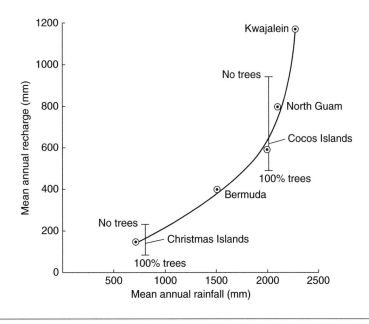

FIGURE 4.10

Relationship between mean annual rainfall and estimated mean annual groundwater recharge.

lens to thin), householders will see no evidence to suggest they should reduce their take. The technical solution required to preserve the lens is to close all the private wells and install a public surface
abstraction system, which takes only an amount equal to the average replenishment of the aquifer,
supplying householders with a rationed supply via standpipes or metered connections. Such a policy
would be difficult to implement because householders may not be persuaded of the technical need.

4.16 COLLECTOR WELLS AND OTHER UNDERGROUND WATER DEVELOPMENTS

Collector wells sunk in the river bed are widely used in alluvial deposits in valleys in arid areas where the dry weather flow of the river is underground. They take the form of large diameter concrete caissons sunk in the river bed, the caisson having to be of sturdy construction to withstand the force of the flood flow in wet weather. Access by bridge is necessary, the bridge and the top of the caisson being sited above maximum flood level. From the base of the caisson, collector pipes are laid out horizontally usually upstream, though sometimes across the stream bed or fanwise upstream. The collectors usually comprise porous, perforated or unjointed concrete pipes, 200 to 300 mm diameter, laid in gravel filled trenches cut in the river bed sediments, connected to the wet well. It is important to lay the collectors at sufficient depth to avoid their destruction during flood events during which some of the stream bed can mobilize into suspension. Where the bed is narrow with deep deposits a simple large diameter well may suffice to intercept flow down the valley. The yield of such wells is limited by replenishment flow down the valley. If ordinary centrifugal pumps (which are the cheapest) are used, they have to be sited above maximum flood level with their suctions dipping into the water, which limits the range of water levels from which water can be pumped. Use of submersible pumps can overcome this difficulty.

The *patented Ranney well* is an early form of collector well, usually made of cast iron and sunk in river bankside gravels or alluvial deposits fed from the river, with perforated collector tubes jacked out horizontally from the base in the most appropriate configuration, often parallel to the river. This type of well uses the bankside deposits to act as a filter. Consequently, the well may have a short productive life if suspended sediment is drawn into the bankside deposits; but back-flushing of the collectors can prolong the useful life, sometimes very effectively. The advantage of a Ranney well is its relative cheapness.

Galleries are similar to the pipes of collector wells and are common in some areas of the world where seasonal watercourses cross thick and extensive alluvial strata that offer large storage potential. Galleries are comparatively large diameter perforated collector pipes buried 3 to 5 m depth or below the minimum dry season groundwater level, and surrounded by a filter medium. It is essential to locate galleries to avoid those parts of a watercourse where active erosion of the bed could occur when the river is in spate.

Qanats are believed to be of ancient Persian origin and are widespread on the flanks of mountain ranges in Iran. They are also widespread in the Maghreb where they are known as 'foggaras', in the Arabian Peninsula where they are known as 'aflaj' (singular 'falaj'), and are also found in Afghanistan and China. The source or 'mother well' is dug to some depth below the water table relatively high on the flanks of a mountainous area (Fig. 4.11). A series of additional wells is dug in a downhill line towards the area where the water is needed and the wells are connected by tunnels below the water table so that the groundwater is drained down-gradient. The tunnel gradient is slightly less than the groundwater gradient and much less than the topographical gradient; therefore the tunnel eventually emerges above the water table and subsequently at the surface where it acts as a canal to deliver the water where required.

The systems were hand dug and hence tunnels are comparatively large for ease of construction and to allow regular maintenance. The tunnel section from its point of emergence from the water table has to be clay-lined to minimize losses. In an increasing number of cases, competing large scale pumped groundwater abstraction is causing serious interference with these ancient gravity

FIGURE 4.11

Section through typical Qanat system.

systems. As a result they are slowly going out of use through neglect as their discharge declines and the necessary maintenance is no longer done.

4.17 BOREHOLE AND WELL LAYOUTS

For most public supply purposes, a borehole capable of giving upwards of 5 Ml/d is required and, for this, a boring of at least 300 mm diameter is necessary to accommodate the pump. In practice a more usual size of borehole would be 450 or 600 mm diameter because of the need for good flow characteristics in the boring and through the rising main, a possible need to reduce the diameter of the boring with depth or for installing screens, and to give allowance for any lack of verticality in the hole. If a boring is to be sunk at a site where there are no boreholes in the vicinity to provide information concerning the likely nature of the strata to be penetrated it is advisable to sink a pilot borehole first. Choice of the size of this boring can present a problem. If it is small, starting say 150 mm and reducing to 100 mm, it will be cheap and give the strata information required, but should it strike a good yield of water it may turn out too small for insertion of a pump to develop its yield. It can, perhaps, be reamed out to a larger diameter or a larger borehole may be sunk nearby to accommodate the pumps. However, there is no certainty in fissured formations such as chalk and sandstone that a nearby hole will yield as much, or more. In one case where two 300 mm diameter borings were sunk in the chalk of southern England, the first hole gave just over 3 Ml/d, which was considered not enough and a second hole was sunk only 5 m away and gave more than 18 Ml/d.

A single boring of 450–600 mm diameter will permit only one pump to be installed and this may not adequately safeguard a public water supply. Even though the modern practice is to use submersible pumps, the time taken to remove a defective submersible pump from a boring and to substitute another may cause an interruption to the supply longer than can be tolerated. If the pumping level is not too far below ground level, say about 20 m, it may be economic to construct a well sufficiently large to accommodate two pumps. The well can be sunk over a trial boring, if the latter gives a good yield; or a second boring can be sunk a few feet away and a connection between the two can be 'blown through' by using a directional charge (Fig. 4.12). If, however, the trial borehole fails to give a satisfactory yield, but it is nevertheless thought the site should be capable of giving more, a well may be sunk elsewhere on the site, a second boring being sunk from its base. By use of a pump in this second boring, the well may be dewatered so that an adit can be driven towards the trial boring (also dewatered) connecting the two bores together. This makes it possible to utilize the combined output of the two borings.

Adits are seldom constructed now, but in the past, many miles of adits were driven from wells sunk in chalk or sandstone formations. Their purpose was to increase the yield of a well by cutting more water bearing fissures: this was sometimes successful. The adits were generally unlined, about 2 m high by 1.2 m wide, driven to a slight upward grade from the well and with a channel on one side to drain incoming water to the well shaft when driving. The risk of finding ever present possibility of meeting a large water-bearing fissure meant that quick, reliable escape means had to be assured for the miners. Dual power supplies for the dewatering pumps and for the crane operating the access bucket were therefore essential; and adit cutting was slow due to lack of space for muck haulage. The consequent high cost of adit driving and improvements in developing yields from boreholes has therefore effectively ended adit construction today.

A hand pump is the simplest device to lift groundwater to the surface. Hand operated village pumps are considered the most suitable low cost option for safe water supply in resource poor

FIGURE 4.12

Layout using trial borehole, deep well and adits.

settings common in rural areas in developing countries. The hand pump gives access to deeper groundwater up to 100 m that is not polluted and also improves the safety of a well protecting the water source from contaminated buckets.

During the International Drinking Water Supply and Sanitation decade (1981–90) boreholes, hand dug wells and tubewells were constructed and water pumps provided to developing countries. Unfortunately this top down approach led to the installation of pumps that were difficult to maintain. Studies showed that about 90% of most hand pumps break down within 3 years due to worn out or broken components. A *Village Level Operation and Maintenance* (VLOM) approach, also referred to as *Management of Maintenance* (VLOMM), was introduced and led to the development of hand pumps that required minimal maintenance. These 'VLOM pumps' allow the villagers in remote locations to maintain the pumps themselves and are part of a larger strategy to reduce dependency of villages on government and donor agencies. The World Bank completed a technical review of a large number of hand pumps in 1987 and provided guidelines for the selection of water supply technology and systems that best meets the needs of a community (WB, 1987). The main types of traditional hand pumps are the India Mark III and the Afridev deep well pumps. Hand pumps that have minimum moving parts, such as the Afripump, have now been developed.

4.18 CHOICE OF PUMPING PLANT FOR WELLS AND BOREHOLES

From the test pumping of the completed well the yield-drawdown characteristic curve for the source will be obtained, such as curve A shown in Figure 4.13. It shows the pumping water level expected when the rest level is 'average'. However, allowance has to be made for the fluctuation of the water table level during the wet and dry seasons of the year and the lowest rest level that might occur in an extreme dry weather period. There may also be a deterioration of the yield with time due to such unpredictable matters as silting of fissures, change of aquifer recharge due to changed conditions on the outcrop, or increased abstraction from the aquifer elsewhere.

The engineer needs to know what duty range to specify for the pump(s) to be inserted in the well, and what will be the range against which the pumps will most times be operating. The normal seasonal fluctuation of the water table level should be known from data collected before the pumping test takes place and, if historical records are available for existing wells in the same underground catchment, these may indicate the lowest water table level likely in an exceptionally prolonged dry season. The second problem is what to allow for deterioration of the yield with time. Experience of boreholes installation in the same area or in similar formations elsewhere, can act as a guide. Where necessary, a typical arbitrary allowance can be made such that future yield will decline by 33% for a given drawdown level or, alternatively, that drawdown will increase by 33% for a given pumping rate.

If, an arbitrary reduction of yield is assumed as shown by Curve B on Figure 4.13 curves bb' and cc' show its variation in the wet season and dry season respectively. The characteristic curve of a possible pump has also to be added onto the diagram, shown as curve x-x' for a given fixed speed pump. The curve x-x' applies to the head necessary to lift the water to the top of the well and to any additional lift to some storage tank. The pump characteristic has to be such that it is capable of giving the maximum output required (10 Ml/d in Figure 4.13) if the well characteristics should decline to curve c-c'. However, if the well yield-drawdown characteristics remain as Curve A, the pump would have to be partially throttled by part closing of the delivery valve to keep the flow to 10 Ml/d. Use of a motor with a variable speed drive (Section 17.22) would allow the pump

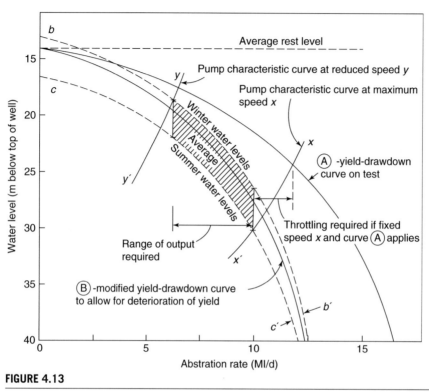

FIGURE 4.13

Yield-drawdown curves for a well with pump characteristics curves superimposed.

to be driven at a lower speed and produce correspondingly lower output (Section 17.14). The curve y-y' represents the pump output at a lower speed of rotation, when only about 6 Ml/d, say, is the required output and the pumping water level in the well is higher. Another alternative would be to use a pump with a dummy stage, into which an impeller would be fitted later when drawdown increases (Section 17.3).

The choice of pumping arrangement depends on cost. Throttling a fixed speed pump wastes energy. A pump with a variable speed drive will be more expensive than a fixed speed pump and its average efficiency over the range of operation may not be as high as that of a fixed speed pump, which can spend most of its time running at its design duty. The best procedure for the engineer is to define the range of expected operational conditions and discuss the most suitable arrangement with the pump manufacturer. The data provided should include:

- the well yield-drawdown characteristics;
- the additional lift required from the well head (and its characteristics);
- the expected time periods per annum for which the drawdown will be in the upper, middle and lower range of values;
- what increase of drawdown should be assumed for the first 10 or 15 years;
- the average, maximum and minimum rates of pumping that will be required.

An important point to note is that the licensed abstraction is often stated as a quantity that must not be exceeded in any 24 hour period, whereas generally the actual pumping hours will be less, for example, 22 hours to allow for shutdowns to attend to routine maintenance matters. Similarly demands less than the licensed abstraction would normally be met by pumping for reduced hours, which implies that storage must be available to maintain the supply to consumers when the pump is not operating.

Generally pumps for well and borehole pumping are fixed speed submersible pumps (Section 12.3). Vertical spindle centrifugal pumps would only be installed for large outputs because of their much greater cost. A mixed flow, fixed speed submersible pump (Fig. 12.5) with a fairly flat efficiency curve about the design duty and a reasonable slope to the head-output curve may be the most suitable for well pumping if the drawdown range is not excessive.

4.19 ENVIRONMENTAL IMPACT ASSESSMENTS

An assessment of the hydrogeological impact of groundwater abstractions is required for licence applications in UK under the Water Act 2003. A methodology for hydrological impact assessment (HIA) is outlined by the Environmental Agency in line with their abstraction licence process (Boak, 2007). The methodology identifies 14 steps necessary in carrying out an assessment. They are:

1. Establish the regional water resource status.
2. Develop conceptual model for the abstraction and the surrounding area.
3. Identify all potential water features that are susceptible to flow impacts.
4. Apportion the likely flow impacts to the water features.
5. Allow for mitigating effects of any discharges, to establish net flow impacts.
6. Assess the significance of the net flow impacts.
7. Define the search area for drawdown impacts.
8. Identify all features in the search area that could be impacted by drawdown.
9. For all these features, predict the likely drawdown impacts.
10. Allow for the effects of measures taken to mitigate the drawdown impacts.
11. Assess the significance of the net drawdown impacts.
12. Assess the water quality impacts.
13. If necessary, redesign the mitigation measures to minimize the impacts.
14. Develop a monitoring strategy.

These steps are not intended to be prescriptive; the detail required for each step should match the situation. The methodology depends on the development of a good conceptual model of the aquifer and the abstraction itself. Advice is also given on how to undertake an HIA in karstic aquifers and fractured crystalline rocks.

Several other pieces of legislation and regulatory regimes remain highly relevant to the assessment of the impacts of groundwater abstraction on water resources and the water-related environment. These include the Habitats Directive, the Water Framework Directive (Section 2.11), and Catchment Abstraction Management Strategies (CAMS).

REFERENCES

Aberbach, S. H. and Sellinger, A. (1967). *A review of artificial groundwater recharge in the coastal plain of Israel*, Bull. Int. Assoc. Sc. Hydrology, **12**(1), pp. 75–77.

Adams, B. and Foster, S. S. D. (1992). Land-surface zoning for groundwater protection. *J. IWEM*, June pp. 312–320.

Anderson, M. P. and Woessner, W. W. (1991). Applied Groundwater Modeling: Simulation of Flow and Advective Transport, Academic Press Inc.

Ayers, J. F. and Vacher, H. L. (1986). Hydrogeology of an atoll island; a conceptual model from a detailed study of a Micronesian example, *Ground Water* **24**(2), pp. 185–198.

Boak, R. and Johnson, D. (2007). *Science Report—SC040020/SR2, Hydrogeological impact appraisal for groundwater abstractions*. EA.

Boulton, N. S. (1963). Analysis of data from non-equilibrium pumping tests allowing for delayed yield from storage, *P. ICE* **26** pp. 469–482.

Bruin, J. and Hudson, H. E. (1955). *Selected Methods for Pumping Test Analysis, Report 25*, Illinois State Water Supply.

Bruington, A. E and Seares, F. D. (1965). Operating a seawater barrier project, *J Irrigation. & Drainage Division*, *ASCE*, **91** (IR-1 pt.1), pp. 117–140.

BS EN 14688. *Geotechnical investigation and testing. Identification and classification of soil. Part 1:2002 Identification and description, Part 2:2004 Principles of classification*. BSI.

BS EN 14689. *Geotechnical investigation and testing. Identification and classification of rock. Part 1:2003 Identification and description*. BSI.

Campbell, M. D. and Lehr, J. H. (1973). *Water Well Technology*. McGraw Hill.

Driscoll, F. G. (1986). *Groundwater and Wells*. Edward E. Johnson Inc.

Falkland, A. (1993). Hydrology and water management on small tropical islands, *Yokohama Symposium, Hydrology of warm humid regions*. IAHS Publication No. 216 pp. 263–303.

Headworth, H. G. (1970). The selection of root constants for the calculation of actual evaporation and infiltration for chalk catchments, *J. IWE* **24**, p. 431.

Harpaz, Y. (1971). Artificial ground-water recharge by means of wells in Israel. *Hydraulics Division, Proc. ACE.* HY12. pp. 1947–1964.

Howsam, P., Misstear, B. and Jones, C. (1995). *Monitoring, maintenance and rehabilitation of water supply bore-holes, Report 137* CIRIA.

IAHS (1970). *International Survey on Existing Water Recharge Facilities*, Publication No. 87, IAHS.

IWES Water Practice Manual No.5 (1986). *Chapter 9, Downhole Geophysics, Groundwater Occurrence, Development and Protection*. IWES.

Jacob, C. E. (1950). *Flow of Groundwater*, John Wiley & Sons.

Jones, H. K., Gaus, I., Williams, A. T., Shand, P. and Gale, I. N. (1999). *A review of the status of research and investigations, Report WD/99/54*. ASR-UK, British Geological Survey,

Kruesman, G. P. and de Ridder, N. A. (1990). *Analysis and Evaluation of Pumping Test Data, Int. Inst. for Land Reclamation and Improvement, Publication 47*. Wageningen.

MacDonald, A. M. and Davies, J. (2000). *A brief review of groundwater for rural water supply in sub-Saharan Africa, BGS Technical Report WC/00/33*. British Geological Survey, NERC.

Maidment, D. R. (Ed). (1992). *Handbook of Hydrology, Ch.22, Computer models for sub-surface water*, McGraw-Hill Inc.

Marshall, J. K., Saravanapavan, A. and Spiegel, Z. (1968). Operation of a recharge borehole, *P. ICE.* **41**, pp. 447–473.

Monkhouse, R. A. (1974). *The use of sand screens and filter packs for abstraction wells*. Water Services, 78.

O'Shea, M. J., Baxter, K. M. and Charalambous, A. N. (1995). The hydrology of the Enfield-Haringey artificial recharge scheme, north London, *QJEG, 28*, pp. 115–129. The Geological Society.

Rodda, J. C., Downing, R. A. and Law, F. M. (1976). *Systematic Hydrology*, Newnes-Butterworth, Table 3.10.

Sander, P. (1997). Water well siting in hard rock areas; identifying probable targets using a probabilistic approach. *Hydrogeology Journal*. **5**(3), pp. 32–43.

Singhail, B. B. S. (1997). The hydrogeological characteristics of the Deccan Trap formation in India, Proc. *Rabat Symposium*, Hard rock hydrosystems, *IASH Publications No. 241*, pp. 75–80.

Stanners, D. and Bourdeau, P. (Eds) (1995). *Europe's Environment, Chapter 5, Inland Waters.* European Environment Agency, pp. 57–108,

Sternau, R. (1967). Artificial recharge of water through wells: experience and techniques, *P Haifa Symposium*, Artificial Recharge and Management of Aquifers, IAHS.

Summers, I. (1987). *Estimation of Natural Groundwater Recharge*, Reidel Publishing Co. p. 449.

Terzaghi, K. (1943). *Theoretical Soil Mechanics.* John Wiley & Sons.

Theis, C. V. (1935). The relation between lowering the piezometric surface and the rate and duration of discharge of a well using groundwater storage, *Trans. American Geophysical Union*, **16**, pp. 519–614.

Walton, W. C. (1970). *Groundwater Resource Evaluation*, McGraw-Hill.

WB (1987). Community Water Supply. The Hand Pump Option. The World Bank.

Wheatcroft, S. W. and Buddemeier, R. W. (1981) Atoll hydrology, *Ground Water* **19**(3), pp. 311–320.

Williams, A. T., Barker, J. A. and Griffiths, K. J. (2001). *Assessment of the Environmental impacts of ASR Schemes. Report CR/01/153*, BGS, NERC.

Wright, A. and du Toit, I. (1996). Artificial recharge of urban wastewater; the primary component in the development of an industrial town on the arid west coast of South Africa, *Hydrogeol. Journ.* **4**(1), pp. 118–129.

Wright, E. P. and Burgess, W. G. (1992). *The hydrogeology of crystalline basement aquifers in Africa, Special Publication 66*, Geological Society.

Dams, Reservoirs and River Intakes

5

5.1 INTRODUCTION

Reservoirs are provided to regulate supply and may be required to store rainfall in wet periods for use in dry periods. Such storage typically requires an impounding dam across a natural river valley. Service reservoirs provide a similar function but related to treated water (Chapter 18). A dam across a natural river valley is sometimes called an impounding dam, as it impounds a natural watercourse.

The earliest dam Smith was able to report in his book. *A history of dams* (Smith, 1971) was the 37 feet high Sadd el-Kafara dam, built between 2950 and 2750 BC, the remains of which lie 20 miles south of Cairo. It had upstream and downstream walls of rubble masonry each 24 m thick at the base, with a 36 m wide, gravel-filled space between; it appeared to have had a short life because it suffered from the two principal defects that continued to plague many dams for the next 4500 years—it leaked and was probably overtopped. However, the nearby Ma'la dam, located in Giza, operated successfully for over 3000 years. The remains of the dam can still be seen today. The dam was constructed using a greater volume of material than that contained in the Great Pyramid at Giza; the irrigation water storage that it provided was one of the keys to Egypt's prosperity.

There are many materials of which a dam can be made—earth, concrete, masonry or rockfill. The choice depends on the geology of the dam site and what construction materials are nearest to hand. Concrete and masonry dams require hard rock foundations; rockfill dams are normally built on rock, but have been built on alluvial deposits; earth dams can be built on rock and also on softer, weaker formations such as firm clays or shales.

Masonry dams are still built in developing countries where labour costs are low. Where labour costs are high, masonry has generally been replaced by mass concrete compacted by using immersion vibrators or, in recent years, by roller compaction. It was not until the middle of the nineteenth century that concrete and masonry dams began to be designed according to mathematical analysis of the internal and external forces coming upon them; nor until the first quarter of the twentieth century was the behaviour of earth dams sufficiently understood for mathematical design procedures to be applied to them also. Recent advances in dam design are such that only the salient principles involved can be given here.

5.2 ESSENTIAL RESERVOIR CONDITIONS

For a successful reservoir the following conditions need to be fulfilled:

1. The valley sides of the proposed reservoir must be adequately watertight to the intended top water level of the reservoir and must be stable at that level.
2. Both the dam and its foundations must be sufficiently watertight to prevent dangerous or uneconomic leakage passing through or under the dam.
3. The dam and its foundations must withstand all forces coming on them.
4. The dam and all its appurtenant works must be constructed of durable materials.
5. Provision must be made to pass all floodwaters safely past the dam.
6. Provision must be made to draw water from the reservoir for supply and compensation purposes and for lowering the reservoir water level in emergencies.

To fulfil these conditions, any proposed reservoir site must be subjected to detailed geological investigation to ensure that there are no fault zones, hidden valleys or permeable strata through which unacceptable leakage would take place. In addition, the nature and stability of the hillsides surrounding the reservoir must be investigated. At high reservoir levels areas of hillside will become saturated. Any rapid lowering of the reservoir may leave such areas with high internal pore pressures, higher (saturated) unit weights and reduced cohesion. Should this reduce stability sufficient to cause a large volume to slide into the reservoir surge waves would result, threatening the dam and also any people on the lake shore.

5.3 WATERTIGHTNESS

All dams leak to some extent: but risk of that leakage carrying with it material from the dam or its foundations must be avoided. Hence, as much as possible of all such leakage should be collected by an underdrainage system and delivered to a collecting basin where it can be continuously measured to reveal any signs of increase and inspected to see it is not carrying material with it.

To reduce seepage below the foundations of a dam, two methods are in common use:

- the construction of a 'cutoff' across the valley below the dam; and
- grouting the foundations beneath the dam to reduce their permeability.

A *concrete cutoff*, typical of those used prior to about 1960, is illustrated in Figure 5.1. It is taken down sufficiently far to connect into sound rock or clay at the base and is usually filled with concrete, the trench being about 2 m wide. Many earlier dams had the cutoff backfilled with puddle-clay.

Where a deep cutoff is necessary, it is now more common to use a plastic concrete diaphragm wall. The trench is excavated under bentonite slurry using a grab in soils or a 'hydrophraise' machine in rock. After excavation, the slurry is replaced by a plastic concrete mix, the slurry being de-sanded and recirculated for further use. Alternatively, particularly for temporary works such as cofferdams, a self-setting slurry can be used for both trench support and provision of the cutoff.

Whatever form of deep cutoff is used, the junction of the top of the cutoff with the core of the dam requires the most careful design, to ensure that the differential stiffness does not allow settlement to open up a leakage path around the top of the cutoff, or act as a propagator for cracks.

FIGURE 5.1

Deep concrete cut-off at Taf Fechan dam, Wales. (Engineer: Binnie & Partners).

A wide, shallow cutoff is shown in Figure 5.2 and this is used where a sound foundation material exists not far below ground surface. The corewall material—in this case 'rolled clay', which is a mixture of clay and coarser materials—is taken right down to the bottom of the cutoff.

Some open excavation cutoffs have been extraordinarily deep and therefore difficult and expensive to construct. The classic case is the cutoff for the Silent Valley dam for Belfast (McIldowie, 1934–35), which reached a maximum depth of 84 m before sound rock was met. Diaphragm wall cutoffs have been taken down to depths of more than 120 m, for example Mud Mountain dam in the USA.

As a 'rule of thumb', cutoffs are generally taken down to between 50% and 100% of the depth of retained water above. Two-thirds the impounded water depth is an often used, but may need to be adjusted locally for particular geology. It is also important to extend the cutoff wall into the abutments of the dam, to reduce seepage around the ends of the dam.

Other forms of cutoff have been used in poor ground conditions; for example Linggiu dam in Malaysia has two rows of small-diameter secant piles through the permeable alluvial deposits, with jet grouting of the soils between them.

Diaphragm walls have been used in several cases for the repair of leaking dams. Diaphragm walls were constructed within the puddle-clay corewalls of Lluest Wen dam in South Wales and Balderhead dam in Teesdale.

Grouting is normally adopted to improve the watertightness of a rock which is already basically sound, reasonably impermeable, and not liable to decompose, even with some leakage through it. In practice, it is difficult and generally uneconomic to grout rock where the packer permeability test results show a value of less than 5 Lugeon units (about 5×10^{-7} m/s). In exceptional cases it may be economically justified to attempt permeation grouting of materials other than fissured rock, such as sands, gravel, silts, clay, and mixtures of these materials. These are often difficult to grout, despite use of a variety of methods; complete success is not ensured and additional precautions may be needed in the design of the dam to allow for unavoidable leakage (Ischy, 1962; Geddes, 1972). Guidance on grouting is given in the CIRIA guide (Rawlings, 2000).

After completing foundation grouting it is usual to ensure that the contact zone between the dam material and the excavated ground surface is sealed by carrying out 'contact' or 'blanket' grouting, through a line of grout pipes with packers set above and below foundation level. After sufficient

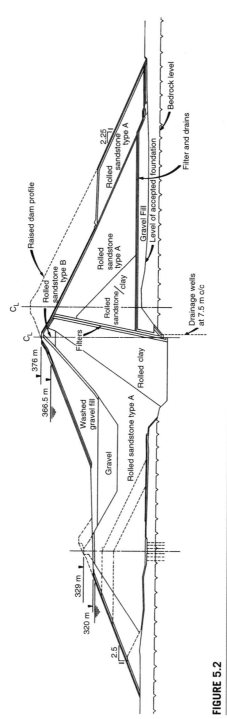

FIGURE 5.2

Shallow, wide clay cut-off at Mangla dam, Pakistan. (Engineer: Binnie & Partners).

height of the dam has been constructed to provide weight, grout is injected into the pipes between the packers.

Grouting can be highly successful, but tales are legion concerning the damage that can be done if the procedure is not carefully controlled. Risks of causing heave and un-wanted 'hydrofracture' are great.

A fully impervious cutoff is often not possible, but water pressures beneath the dam need to be controlled for stability and to prevent erosion. This can often be achieved using a 'partial' cutoff, together with pressure-relief wells downstream. Partial cutoffs do not significantly reduce flow volumes; many dams have been built with both a full cutoff and pressure-relief wells.

5.4 STRENGTH AND DURABILITY OF A DAM

Every dam must be secure against failure by sliding along its base; this applies whether the dam is of concrete, masonry, rock or earth. Resistance to failure against shearing along any surface is another criterion to be applied. Sliding is resisted by the weight of the dam acting in combination with friction and cohesion along its base contact with the foundations. Usually the resistance of rockfill and earthfill embankments dams against sliding or direct shear is ample, on account of their large widths. Stability checks on such dams tend to focus instead on internal failure mechanisms, such as the development of slip circle instabilities or erosion and piping failures.

Sliding is a more crucial failure mechanism to examine when designing a masonry or concrete dam; these, being constructed to much steeper slopes than earth dams, should also be checked against overturning. Where concrete dams are arched, the abutments must be strong enough to take the end thrusts of the arch, and the foundations must also be strong, as they are more highly stressed than would be the case for an equivalent height concrete gravity dam.

Durability is of the utmost importance since dams are one of the few structures built to last almost indefinitely. Their design life is sometimes described as 'monumental'. The durability of an earth dam is primarily dependent upon the continuance of its ability to cope with drainage requirements. It has to contend with seepage, rainfall percolation and changing reservoir levels. The movement of these waters must not dislodge or take away the materials of the dam; otherwise its condition and stability will deteriorate at a progressively increasing rate. It must perform this function despite long-term settlement of itself and its foundations and may be required to withstand seismic movements.

5.5 TYPES OF DAM

The most economic forms of dam construction are those that make maximum use of natural materials available nearby. In the lower reaches of a river there may be an abundance of soils such as clay, sand and gravels. Use of such materials would usually be economically and environmentally preferred and the resulting dam would be compatible with the foundations, which are likely to be of similar materials. The upper reaches of a river may be in much steeper and more rocky terrain. There may even be a dearth of soils. In such a terrain a rockfill or concrete dam is likely to more appropriate.

Dams constructed of earthfill or rockfill are termed embankment dams. The simplest example would be a homogeneous embankment made entirely of the same material, which would need to be sufficiently impermeable to retain the reservoir without excessive leakage, while being sufficiently strong so as to be self-supporting. More advanced designs might feature a central core of clay or silt as the impervious element, with supporting shells, or shoulders, of stronger soils or even rockfill. In between the clay and shell materials, graded sand and gravel filters are provided as a transition between the different materials. Clay-core earth embankments and clay-core rockfill embankments are amongst the most common forms of dam construction.

Where clay is not economically available, the central waterproofing element may be of another material, such as asphalt. A number of rockfill dams with a central asphalt core have been constructed in Norway. In the case of rockfill embankments, it is also possible to make the dam watertight by providing a suitable upstream facing, instead of an impermeable central core. Such upstream facings are usually either of asphalt or reinforced concrete. The latter have grown in popularity in recent years.

Where the foundations are good and significant flood passage structures are required, it may be economical to form the whole dam of concrete. The various forms of concrete dam are generally named after their geometry or function. A simple concrete dam, approximately triangular in cross-section and relying on its own dead weight for stability, is termed a concrete gravity dam. Similar stability can result by providing an upstream wall, generally sloped in order to attract a vertical component of the water load, supported on the downstream side by a series of propping walls or buttresses. This arrangement is termed a buttress dam. Concrete arch dams take the form of a shell spanning across a narrow valley, such that the load is conveyed to the valley sides as well as to the valley base.

There are many variants to those described above. The adaptation of local materials to provide a specific function remains an area where dam engineers can use their experience and talents to be imaginative and creative. In some cases, it might be advantageous to combine two different types of dam in one location. The main types of dam and some considerations in their design and construction are discussed in the following sections.

EMBANKMENT DAMS

5.6 TYPES OF DESIGN

To avoid a misunderstanding, it is necessary to mention that the term 'earth' does not mean that tillable soil can be used in an earth dam. All such 'soil' must be excluded, because it contains vegetable matter which is weak and, in its decomposition, could leave passages for water percolation. The first operation when undertaking earth dam construction is therefore to strip off all surface soil containing vegetable matter. The dam can be constructed, according to the decision of the designers, of any material or combination of materials such as clay, silt, sand, gravel, cobbles and rock.

Early designs of earth dams usually had a central core of impermeable 'puddle clay', supported on either side by one or two zones of less watertight but stronger material (Fig. 5.1). The central puddle-clay core is impermeable but its structural strength is low. This clay core must join to a cutoff in the ground below. The earliest dams had clay-filled cutoff trenches but use was soon made of concrete to fill the cutoff trench, although the corewall continued to be made of clay. The inner zones of the shoulders of the dam contained a mixture of clay and stones, so was impermeable to some extent, but the stony material added to its strength. The outer zones would contain less clay again and more stones, perhaps boulders and gravel. The main purpose of the shoulders was to hold the inner core of clay and add strength to the dam. Boulder clay (a glacier-deposited mixture of clay, silt and stones with boulders) was a favourite material for this outer zone in the UK, because it had

FIGURE 5.3

Reinforced concrete core wall in Gyobyu dam, Rangoon, Myanmar. (Engineer: Binnie & Partners).

the right characteristics and there were extensive deposits of it in those areas where impounding schemes developed.

Many such dams have been and continue to be successful. Their outer slopes were decided by accumulated experience and were often 3:1 upstream and 2:1 or 2.5:1 downstream, with flatter slopes for the higher dams. There were few measurement tests that could be applied. Although many new kinds of tests are now available, earth dam construction still involves judgment from experience and constant supervision as the work proceeds to ensure sound construction. Only a minute proportion of the variable earth materials in a dam can be tested, so construction must always come under continuous detailed supervision by an experienced engineering team.

In modern dams, a puddle-clay core is no longer used. Instead the rolled-clay core is made of a mixture of compacted clay and coarser materials. The core usually has a base width of 25–50% of the height of the dam, to limit the hydraulic gradient across it and reduce the risk of leakage at foundation level. A large dam of this sort is shown in Figure 5.2.

Sometimes, a dam may consist wholly of rolled clay with only an outer protective zone of coarser material on the water face for protection against wave action. However, dams with the same core and shoulder materials normally include internal drainage to improve stability and reduce the risk of erosion.

Although clays of moderate to high plasticity are preferred for core material, dams have been constructed successfully using clays of very low plasticity and silt for the core. In the latter case, the filters protecting the core against internal erosion are of great importance. An alternative used is the concrete corewall; Figure 5.3 shows one of the earliest examples. Generally a concrete corewall is limited to use with stiffer shoulder materials such as sand, gravel or rockfill and to good foundations.

The principal aim is to use as much locally available material as possible, because this keeps the cost of the dam as low as possible. The problem for the designer is how to incorporate the available materials in the dam in a manner which uses their different characteristics efficiently and safely.

5.7 PORE PRESSURE AND INSTRUMENTATION IN EARTH DAMS

An earth dam is built in layers of material, 150 to 450 mm thick, which are compacted with vibrating or heavy machinery. The material has to have a moisture content close to its 'optimum' value to achieve maximum in-situ density. Soils that need to be flexible, such as the core material are usually compacted slightly wet of the optimum. If the soil is too wet, the compaction is ineffective and the plant cannot operate. Soil used in the shoulders is often stiffer and, to achieve this, the placement water content is at or just below the optimum value. At the other end of the range, if the soil is too dry, effective compaction cannot be achieved. Control testing of the fill materials, both at the borrow pit and after placement is necessary to ensure compliance.

The water within the soil ('porewater') is affected by the construction process. When the soil is compressed by the compaction process and the added load, water pressure in the pores increases. The air in the soil is either pushed out or goes into solution, increasing the degree of saturation. As the effective strength of the soil is reduced by high porewater pressure, the dam can become unstable. Clays of low to medium plasticity such as Glacial Till (Boulder Clay) tend to compress when loaded, and high construction pore pressures can often occur. Heavily over-consolidated clay, such as the Lias Clay or Oxford Clay dilate when excavated and at low stresses (less than 10 to 15 m of fill) the compaction porewater pressures are often low or even negative (suction).

The maximum allowable construction porewater pressures to ensure an acceptable factor of safety against failure can be calculated. During construction, these porewater pressures can be monitored using piezometers. Some examples are shown in Figure 5.4. The Casagrande type is suitable for dam foundations and the Bishop type suitable for installation in the embankment fill. Other types of piezometers using different methods of measuring the water pressure at the tip are available. These include vibrating wire and pneumatic systems. Details of such instruments are given in Dunnicliff (1993). In unsaturated fill materials only piezometers that can be de-aired, such as the Bishop type, are likely to give reliable long term readings. The piezometer readings can be used to control the rate of construction.

During first impounding, piezometers placed in the upstream shoulder of an earth dam can register how the water level in the shoulder responds to the reservoir water level; piezometers in the foundation and downstream shoulder can check the performance of the core and cutoff. During

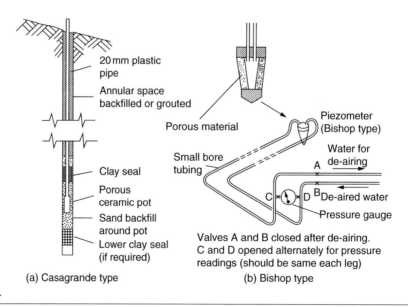

(a) Casagrande type   (b) Bishop type

FIGURE 5.4

Pore water pressure sensing devices.

operation, piezometers can show how fast the water drains out of the upstream shoulder with a fall in reservoir water level. This is important in order to avoid a slip failure due to 'rapid draw-down'. Piezometers at low level just downstream of the corewall can show pressures which indicate leakage through the corewall, especially so if their pressure rises and falls with reservoir water level. It is thus possible to monitor the behaviour of an earth dam through the use of piezometers, both during and immediately post construction and long-term operation. In many soils, the end-of-construction porewater pressures are a worst case, but in some—particularly the high plasticity clays—the dam stability deteriorates as the pore pressures come to equilibrium. This process may take tens of years.

Settlement gauges, extensometers and inclinometers can measure the deformation of a dam both during and after construction. Surface markers permit movement and settlement to be detected by precise surveying, which must be based on benchmarks established outside the limits of any land settlement caused by the presence of the dam and by the weight of water in the reservoir.

5.8 STABILITY ANALYSIS IN DAM DESIGN

The stability of the earth dam against slope failure is an essential element of the design. There are many limit equilibrium slope stability analysis methods and a significant number of commercial slope stability computer packages are available. For the majority of embankment dams, a circular slip analysis based on Bishop's rigorous method (Bishop, 1955) is satisfactory. However, the geology of the site and the shape of the dam may indicate that a non-circular slip surface may be more critical. Simple non-circular methods include two-part and three-part wedge analyses; more sophisticated methods include Morganstern and Price's method (Morganstern, 1965).

Computer packages may offer a number of different analyses methods but some methods are not reliable for embankment dam design, particularly where there are a variety of materials in the dam and its foundation. The Ordinary or Fellenius method, which ignores interslice forces, should not be used and methods which do not fully define the interslice forces such as Janbu's method or Bishop's simplified method should only be used with caution.

The dam needs to be checked for a number of different modes of slope failure. The most important of these situations modes are:

- the end of construction (prior to impounding for the upstream slope and with the reservoir impounded for the downstream slope);
- long term, with steady-state seepage conditions established within the dam;
- after the rapid drawdown of the reservoir, which can destabilize the upstream slope; and
- in most countries including Britain, seismic conditions.

In practice, the required factor against failure (factor of safety) used in design is related to the risks and consequences of a failure, so differs for the various modes. Natural materials such as soils are inherently variable and, even with good quality ground investigation and strength testing, there will always be some uncertainty over the choice of suitable design parameters. Although a moderately conservative set of design parameters should be chosen from the test results, the factor of safety still needs to cover these uncertainties.

Figure 5.5 shows the output from a rapid drawdown analysis for Mangla Dam. This figure shows the reservoir not fully drawn down. There is normally a critical pool level for which the minimum factor of safety is lower than that for complete drawdown.

Finite element analyses of embankment dams can be very helpful, particularly for those embankments that have wide intermediate berms. The strains required to mobilize the resistance in some parts of a dam may be sufficient to cause failure elsewhere. Elasto-plastic or strain softening soil models are necessary to obtain realistic results. Dynamic finite element analyses are also used to assess changes in pore water pressures and deformations under seismic conditions.

Failures can occur on pre-existing sheared surfaces. These are usually in the foundation soils and are a result of tectonic shearing or periglacial solifluction. An example is the failure of Carsington Dam during construction, where some sheared foundation soil was not removed (Fig. 5.6(a)). There are a few cases of failure within the embankment where shear surfaces were formed by previous movement and not removed. Figure 5.6(b) shows a section of the failure at Tittesworth Dam.

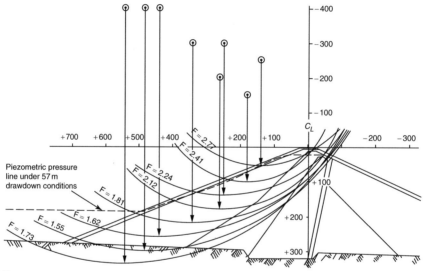

FIGURE 5.5

Slip-circle stability analysis results for upstream slope for Mangla dam for 57 m drawdown and static conditions. Lowest factor of safety is 1.55.

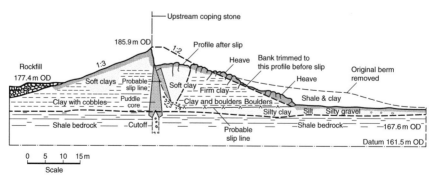

FIGURE 5.6(a)

Failure of Tittesworth dam 1960, UK. Removal of part of small berm to downstream shoulder caused wedge-type failure through puddle clay core. Later investigations showed that the embankment had failed in the same location but had not been properly rebuilt.

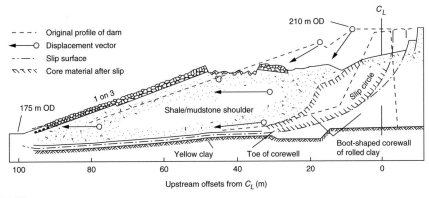

FIGURE 5.6(b)

Failure of Carsington dam 1984, UK. Overstressing of material near toe of core, as construction was nearing completion, caused progressive slip circle failure.

5.9 DRAINAGE REQUIREMENTS FOR AN EARTH DAM

Much emphasis has been put on the slope stability aspects of dam design but dam failure due to inadequate assessment of slope instability is extremely rare. Failures are more likely to be due to poorly designed drainage or to poor detailing of the contact between the dam and its foundation.

The preceding sections illustrate that an earth dam is subject to continuous water movements within it. Water levels in the upstream shoulder rise and fall with reservoir level; the downstream shoulder has a moisture content and often a positive porewater pressure which rise and fall with rainfall, and it may have to accept some inflow from seepage through the core and from springs discharging to its base and abutments. The principal requirement is that all these water movements must not dislodge or carry away any material of the fill.

Drains are therefore necessary within a dam to accommodate the water movements that occur. However, a drainage layer of coarse material, such as gravel, cannot be placed directly against fine material such as clay, or water movement would carry clay particles into and through the drainage layer. The removal of the clay would progressively enlarge leakage paths until there is a serious danger of disruption of the fill and breakthrough of water from the reservoir. Every drainage layer of coarse material has therefore to be protected by a 'filter'. The filter material must have a particle size small enough to prevent the fine material moving into it, and coarse enough to prevent its own movement into the larger material of the drain. Sometimes two filters have to be placed in succession when there is a large difference in size between the fine material to be drained and the coarse material of the drainage layer. All such drainage layers must convey their water individually to inspection traps where the flows should be regularly measured and inspected to make sure they are not carrying suspended material from the dam. Any flow of cloudy drainage water from a dam should be tested for the amount and size of suspended solids it carries.

The literature includes a number of sets of filter rules to design the grading of filter materials to protect soils from internal erosion. The work by Sherard and Dunnigan (1989) gives simple and effective rules for most soil types. Special care needs to be exercised for gap-graded and widely graded soils that are not internally stable.

An example of a drainage layout is shown in Figure 5.7. The main drainage system comprises a 'chimney' drain, against the downstream slope of the core, to transfer seepage flows to a 'drainage

FIGURE 5.7

Chimney and blanket drains in 58 m high Upper Muar dam, Malaysia, constructed of residual clay soils on weathered rock foundations. (Engineer: Binnie & Partners, in association with SMHB).

mattress' laid at formation level, discharging to an outlet chamber at the downstream toe of the dam. Drainage 'blankets' are laid at intervals in the upstream shoulder to relieve pore pressures that build up during construction and for drainage of the upstream shoulder during rapid drawdown. In some embankment dams drainage galleries have been used, but they are principally included for interception of pressure relief wells in the dam foundations.

5.10 SURFACE PROTECTION OF EARTH DAMS

The upstream face of an earth dam must be protected against erosion by water and wave action. Common forms of protection are:

- stone pitching;
- riprap;
- concrete blockwork;
- concrete slabs; and
- asphaltic concrete.

Stone pitching, either grouted or open-jointed, comprises stones of fairly uniform thickness and size, generally about cuboid, which are laid by hand with relatively close jointing. Pitching is generally found on older dams. Riprap is formed by the random packing of pieces of angular rock with a range of sizes and shapes, which are generally placed by machine; although it is generally advantageous to hand-place small pieces of rock between the large pieces to assist in locking them together. Beneath the rock must lie layers of material which act as filters between the large rock and the finer material of the dam that must be prevented from being drawn through the rock surface by wave action.

Where large waves are anticipated and large rock is unavailable, concrete blocks can be used. These are shaped to dissipate the wave energy and a number of proprietary designs are available, such as tetrapods and dolosse. When concrete slabs are used for protection they need to be placed without open joints to avoid dislodgement by wave action, but drainage of the fill through them should be possible. They should be laid on one or two graded gravel filter layers.

For many years slope protection of dams has been designed empirically, with satisfactory results in most cases (Johnston, 1999). A conventional rule used by many engineers required the depth of riprap or concrete blockwork on dams in the UK to be not less than one sixth of the 'significant' wave height in the design event. Experience at a number of reservoirs suggests that this is not always adequate, although a factor in some cases has apparently been the adoption of an insufficiently rare windstorm event for the design of the wave protection works.

The key current reference on the stability of riprap under wave attack (and on the appropriate detailing and underlayers) is CIRIA *Rock manual* (CIRIA, 2007). For blockwork and slabbing, a design procedure was recommended by Yarde *et al* (1996), based on physical model studies undertaken at HR Wallingford. Adoption of this procedure, however, may be hampered by limitations in the ranges of experimental parameters used to derive the equations. A experience-based approach using data provided by Herbert et al (1995) has been proposed (Besley, 1999). Further information on the subject of wave protection at UK embankment dams is given by Johnston *et al* (1999).

Erosion of the downstream face of a dam by rainfall runoff must be prevented. It is usual to achieve this by turfing the surface of the dam or by soiling and seeding it with a short-bladed strongly rooting grass. If the dam is more than about 15 m high then, according to the type of rainfall climate, berms should be constructed on the downstream slope, each berm having a collector drain along it so that surface runoff is collected from the area above and is not discharged in large amounts to cause erosion of the slope below. Cobbles or gravel are sometimes used to surface the downstream slope, particularly where rainfall is insufficient to maintain a good grass growth throughout the year. Surface drains on the downstream slope should discharge separately from the internal drains to the shoulder.

It is unwise to allow any tree growth on, or at the foot of, the downstream slope of an earth dam. Some tree roots can penetrate deeply in their search for water, and drainage layers could be penetrated by such roots. The presence of trees may obscure proper observation of the condition of the slope. A damp patch, or development of a depression on the slope, may indicate internal drainage systems not working properly or some leakage through the core. Any signs of this sort need to be investigated. It is best to treat the downstream slope uniformly, with the same type of covering (gravel or grass) throughout, for example Jari Dam, Mangla (Plate 3(a)), since this helps to disclose any discrepancies that might need investigation.

5.11 ROCKFILL AND COMPOSITE DAMS

Rockfill dams are appropriate for construction at locations where suitable rock can be quarried at or near the dam site, and where the foundations will not be subject to material settlement due to loading or to erosion from any seepage through or under the dam. The design must, of necessity, incorporate a watertight membrane, which is generally sited either centrally in the dam or in the form of an upstream facing.

Early concrete faced dams of dumped rockfill featured embankment slopes of about 1 on 1.5 and were satisfactory in service with heights up to about 75 m. Above that height face cracks and excessive leakage tended to occur due to the compressibility of the dumped rockfill. This problem was alleviated somewhat when it was discovered that sluicing rockfill with water improved compaction, but the modern rockfill dam really dates from the 1960s, when vibrating rollers became standard equipment for compaction (Plate 3(b)).

(a) Jari 42 m high earth dam, Mangla, Pakistan. (Engineer: Binnie & Partners)

(b) Llyn Brianne 91 m high rockfill dam, Wales; side spillway: capacity 850 m³/s.
    (Courtesy of Aled Hughes) (Engineer: Binnie & Partners)

PLATE 3

(a) Olivenhain dam under construction, USA. RCC being placed at a rate of about 750 m³/hour. (Courtesy of Dr. M. R. H. Dunstan)

(b) Kariba 128 m high double curvature concrete arch dam, 1320 MW hydro. (Engineer: Coyne & Bellier) (Courtesy of Dr. W. R. White)

(c) Nant-y-Moch 53 m high buttress dam, Wales. (Courtesy of John Sawyer)

PLATE 4

A variety of rocks can be used with modern methods of compaction; even relatively weak rocks such as sandstones, siltstones, schists and argillites have been used. It is usual to examine all possible sources of locally available material, to carry out laboratory tests on samples and pilot construction fills and to base construction procedures and zoning of materials on the results of such tests. The specification for construction should set out the required layer thickness, normally 1 to 2 m for sound rock and 0.6 to 1.2 m for weak rock, and the number of passes required by a 10 tonne vibratory roller to achieve adequate rockfill breakdown and strength (Cooke, 1990).

Centrally located waterproofing can typically comprise a core of clay, silt, asphalt or concrete. Central cores of clay or silt are thicker than asphalt or concrete cores. An earth core would be protected by zones of transitional material between the clay and the rockfill (Section 5.9). Usually two or more layers of transitional material are necessary, grading through from clayey sand against the corewall, to a crushed rock with fines against the rockfill.

Where upstream facings are provided, these are generally of either reinforced concrete or asphalt. In both cases the final facing is screeded up the slope over one or more layers of fine bedding material, compacted both horizontally and up the slope. However, a more recent development has been to place the rockfill against upstream kerbs of roller compacted concrete, as construction proceeds. This produces a relatively impermeable face, which is a useful safeguard against construction floods. It also provides a good base on which to cast the subsequent waterproofing membrane. Watertightness of a concrete face is achieved by sealing vertical joints with waterbars. Horizontal joints are not required, construction joints being formed with reinforcing steel passing through them.

In the case of an upstream facing, the associated cutoff trench (or plinth) must necessarily follow the curve of the upstream toe of the dam in order to connect with it. Such upstream membranes need to be sufficiently flexible to be able to accommodate long term settlement of the rockfill without unacceptable leakage. Concrete-faced rockfill dams are now among some of the highest dams in the work. Their essential simplicity and construction economy has made them very attractive where good rockfill is plentiful. However problems with a few such dams occurred in 2006. These were high dams, greater than 140 m, in steeply sided valleys, where vertical expansion joints in the facing had been omitted and the rockfill was basalt. In these cases, settlement deformations caused central shearing, or crushing, failures of the concrete facings over almost their entire heights. The leakages produced were easily accommodated by the internal rockfill and so the dams were not in structural danger, however the matter illustrated the need for care when design concepts are taken beyond the ranges of previous experience.

The rockfill body of an upstream faced rockfill dam is inherently stable. The water load is applied more vertically downwards than horizontally, the rockfill is totally drained and sliding resistance is over the whole of the base. As a result no concrete-faced rockfill dams have experienced structural failure and stability analyses on assumed failure planes are not generally done. Instead the upstream and downstream slopes of the dam are decided based on the quality of the rockfill and on precedent. Slip circle analyses are, however, necessary when an earth core is incorporated, to determine the extent to which the core strength and pore pressures affect the dam's stability. In the case of a central-core dam the water load is applied in a mainly horizontal direction and only the downstream shoulder resists sliding forces.

An asphalt or bituminous concrete facing can also be used. It has an inherent flexibility and may be able to absorb larger deformations than a reinforced concrete facing without cracking. However, it can still be vulnerable to differential settlement at its connection with the plinth. Cold weather affects the flexibility of bituminous materials, hot weather causes creep, frequent variations in temperature can cause fatigue; and ultraviolet radiation can damage a coating that is not submerged.

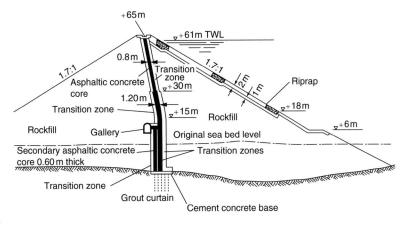

FIGURE 5.8

Rockfill shoulders with asphaltic concrete core in High Island West dam, Hong Kong, China. (Engineer: Binnie & Partners).

Hence the standard practice is to provide a protection layer and a membrane sealer. Modern practice is to use a single 100 to 200 mm thick impervious bituminous concrete layer spread by a vibratory screed and then rolled by a winched vibratory roller.

A centrally installed bituminous diaphragm wall avoids problems with extreme temperatures and ultraviolet radiation, is less vulnerable to damage, and permits the cutoff to be constructed along the centreline of the dam. An example of an internal asphaltic core to a rockfill dam is shown in Figure 5.8. At 107 m high, this is one of two dams at High Island, Hong Kong (Vail, 1975), which are, unusually, founded below the sea bed across a sea inlet. The core consists of 1.2 m thick asphaltic concrete, made of 19 mm aggregate mixed with cement and water, preheated and mixed with about 6% of hot bitumen. Below a certain level, the core is duplicated and above a certain level it is reduced to 0.8 m thick.

CONCRETE AND MASONRY DAMS

5.12 GRAVITY DAM DESIGN

The weight of a gravity dam (and the width of its base) prevents it from being overturned when subjected to the thrust of impounded water (Kennard, 1995). To prevent sliding the contact with the foundation and the foundation itself must have appreciable shear strength. In carrying out stability calculations for any proposed section of a dam, the 'uplift' must be taken into account. Uplift is the vertical force exerted by seepage water which passes below a dam or which penetrates cracks in the body of the dam. Maximum or '100% uplift' on any section through a dam would be a triangular pressure diagram as shown in Figure 5.9; this assumes the upstream value equals the pressure exerted by the maximum water level during flood conditions above the section. Below the base of a dam with no cutoff, the pressure diagram is trapezoidal if, at the downstream toe, there is a potential uplift from any tailwater level.

Taking these three forces into account—the uplift, the water thrust and the weight of the dam— the accepted rule is that the resultant of these forces should pass within the middle third of the

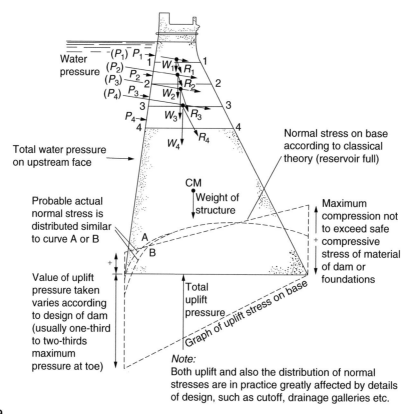

FIGURE 5.9

Simple stability analysis of gravity dam showing effect of uplift on base of dam and at other sections.

section being analysed. This ensures that no tension is developed at the upstream face of the dam. At first sight the foregoing appears illogical since the assumption of 100% uplift assumes a crack exists, which the design should prevent. However, perfect construction everywhere is not likely to be achieved and cracks might progressively develop if there is any tension at the face. Even if no crack exists, there can be uplift pressure in the pores of concrete of the dam due to the passage of seepage. Figure 5.10 shows pore pressures measured inside the concrete of the Altnaheglish dam in Northern Ireland, where concrete deterioration had occurred, before remedial work was undertaken. This pore pressure acts, of course, only on the proportion of the concrete which consists of voids, so that the total uplift is only some proportion of the concrete area multiplied by the pore pressure.

In order to reduce uplift from seepage below a dam, a concrete-filled cutoff trench may be sunk at the upstream toe, as shown in Figure 5.10. Even if this cutoff does not make the foundation rock wholly watertight, it lengthens any seepage path and thus reduces uplift below the base. Relief wells or drainage layers can be inserted at intervals below the downstream half of the dam base; these also reduce the total uplift. With these provisions the amount of uplift assumed in the design can be less than '100% uplift'.

Ice thrust may have to be taken into account in the design of gravity dams in cold climates. Estimates (Davis, 1969) of the thrust force vary according to anticipated ice thickness, and range from 2.4×10^2 to 14.4×10^5 N/m^2 of contact with the vertical face of the dam. Seismic forces from

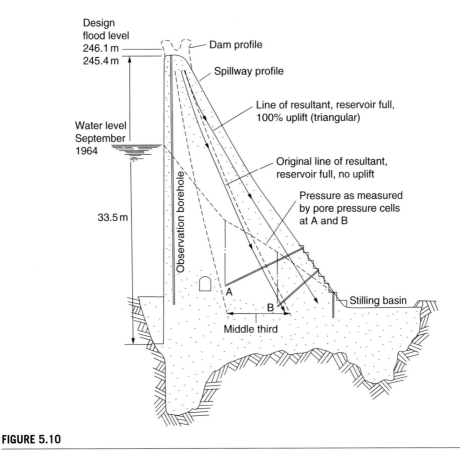

Design
flood level
246.1 m
245.4 m

Dam profile

Spillway profile

Line of resultant, reservoir full,
100% uplift (triangular)

Water level
September
1964

Original line of resultant,
reservoir full, no uplift

Pressure as measured
by pore pressure cells
at A and B

33.5 m

Observation borehole

A

B

Stilling basin

Middle third

FIGURE 5.10

Effect of uplift on 'resultant' at Alnaheglish dam, Northern Ireland.

earthquakes also have to be considered. It is now accepted that minor seismic damage can occur in areas, such as the UK, previously thought to be free of earthquake risk (Section 5.22).

The design of gravity dams appears deceptively simple, if only the overall principles described above are considered. In fact, controversy raged for more than half a century over the subject of how stresses are distributed within a gravity dam, until computer programs permitted computation of stresses using two- or three-dimensional finite element analysis. The method determines displacements at each node, and stresses within each element of the structure, the latter being considered for analysis as an assemblage of discrete elements connected at their corners.

5.13 GRAVITY DAM CONSTRUCTION

Concrete. The construction of concrete gravity dams is relatively simple. Plate 5(c) shows the 55 m high Tai Lam Chung concrete gravity dam. Most of the remarks concerning the preparation of foundations and cutoff trenches for earth dams apply also to gravity dams. The key problem with mass concrete dams is to reduce the amount of shrinkage that occurs when large masses of concrete cool off. Heat is generated within the concrete as the cement hydrates and as this heat dissipates, which may take many months, the concrete cools and shrinks. In order to reduce the effects of this shrinkage the concrete is placed in isolated blocks and left to cool as long as possible before adjacent blocks are concreted. 'Low heat' cement is frequently used to reduce shrinkage, or sometimes a blend of cement and pozzolan. Pozzolans, such as pulverized fuel ash, also have cementitious properties, but a much slower and lower rate of heat development. Shrinkage is also reduced if the content of cementitious materials is kept as low as possible and if the aggregate size and proportions are as high as possible. Ice flakes may be added to water used in the concrete mix, while in the larger installations, water cooling pipes may be laid within the concrete to draw off excess heat as it is produced. Concrete dams are best constructed in areas where rock suitable for the making of concrete aggregates is abundant.

Some concrete dams have been raised or strengthened by installing post-tensioned cables through the concrete mass near the upstream face, and grouting them into the foundation rock to provide additional vertical forces to counterbalance the overturning moments introduced by increased water loads.

Masonry. The construction of masonry dams is expensive because of the large amount of labour required to cut and trim the masonry blocks. In the UK, masonry dams are no longer built, but elsewhere dams have had masonry facing and concrete hearting, primarily to improve the appearance of the dam and also to avoid the need to provide shuttering for the concrete. Overseas, particularly in India, construction of masonry gravity dams continued, but seldom for large structures, due to the unpredictability of masonry behaviour under seismic forces.

Great care must be taken with masonry dams to fill all the joints and beds completely with a watertight mortar mix and to pay special attention to the quality and watertightness of the work on the upstream face. It was Dr G F Deacon's insistence on this when constructing the 44 m high masonry Vrynwy dam for Liverpool in 1881 that has left the dam still in excellent condition over 120 years later. Up to then, he remarked 'there was probably no high masonry dam in Europe so far watertight that an English engineer would take credit for it' (Deacon, 1895–96).

5.14 ROLLER-COMPACTED CONCRETE DAMS

Roller-compacted concrete (RCC) dams were developed to combine the best attributes of a fill dam (offering a plant-intensive method of construction) and a concrete dam (offering a small volume of erosion-resistant material). Although there are reported to have been a number of RCC dams constructed in the 1960s and 70s, it was not really until the early 1980s that the development of the methodology really started. A sound foundation similar to that required for a traditional concrete dam is needed, and then RCC can be placed in successive layers to form a monolithic dam structure. One of the main advantages of the construction methodology is the very rapid rate of placement. RCC dam stability requirements are similar to those adopted for traditional concrete dams. The main differences between the two types are the concrete mix design and method of placing and the design of spillways, outlet works, drainage and inspection galleries to facilitate RCC construction.

Initially, different approaches were made to RCC dams (Dunstan, 1994). One approach was to use a relatively lean and porous mix for the body of the dam, relying on upstream waterproofing for watertightness. This was usually achieved by using a thickness of richer, impermeable concrete on the water face. The other approach was to use a 'high-paste' concrete containing more cementitious material (that is cement and pozzolan) that is sufficiently impermeable not to need upstream waterproofing. Its greater tensile strength permits economy in the profile of the dam by the use of steeper side slopes. The lean-mix dam can suffer seepage if the upstream waterproofing is not entirely effective whereas the high-paste RCC dam has a greater propensity to suffer thermal cracking, so the thermal conditions in the dam have to be studied carefully. In recent years, some low-paste RCC dams have been constructed using a geo-membrane to ensure a waterproof upstream face.

The Japanese have developed their own form of RCC dams with thick upstream and downstream faces of traditional concrete protecting the RCC in the interior of the dam. The latter has a cementitious content of 120 or 130 kg/m^3 depending on the height of the dam. This form of RCC dam has not been found to be economic outside Japan. Another interesting development is the faced symmetrical hardfill dam (FSHD) (Londe, 1992). This has a low-strength body made of cement-stabilized fill but with a waterproof reinforced concrete upstream face. Both upstream and downstream faces are sloped, with the upstream slope therefore attracting vertical water load. The FSHD is generally economic where there is a lower quality foundation, a high dynamic loading and/or a large design flood.

Recent trends have been towards use of high-paste RCC that has been found to be the economic solution under most conditions. A further factor is the increasing size of RCC dams; during the 1980s the average height was only 40 m, whereas in the first decade of the 21st century the average height of RCC dams is approaching 100 m. Plate 4(a) shows the RCC being placed at Olivenhain dam in the USA. The main part of this 97 m high dam was placed in only six months. Several RCC dams over 200 m in height were under construction at the end of 2007.

5.15 ARCH DAM DESIGN

The principle of design of an arch dam is greatly different from that of a gravity dam. The majority of the strength required to resist the water thrust is obtained by arching the dam upstream and transferring the load to the abutments. The abutments must therefore be completely sound. The theory of design is complex, with the dam resisting the water thrust partly by cantilever action from the base and partly by arching action from abutment to abutment. Hence, an arch dam can be much thinner than a gravity dam—for example Mudhiq (Plate 5(b)) and Kariba (Plate 4(b)). Early designs were based on the 'trial load' procedure. The dam was assumed to consist of unit width cantilevers one way and unit width arches the other. The water load at each point was then divided between the 'cantilevers' and 'arches' so that their deflections at every point matched. The modern method is to construct a three-dimensional finite element model of the dam to evaluate stresses under various loadings. Physical models are also used to measure strains and are useful, partly as a check on the mathematical calculations and partly as an aid to them by giving a first approximation to the likely distribution of stresses.

There are many variations from the simple uniform arch shape; the most economic section is curved both vertically and horizontally and results in the horizontal arches varying in radii with level. A dam of this kind is called a double curvature dam and is especially economical in the use of concrete.

River and flood diversions are usually taken in tunnels through the abutments; flood overspill may be passed over a central spillway. Arch concrete dams are among the highest in the world and are inherently stable when the foundations and abutments are solid and watertight. However, the stresses in the concrete of the dam and in the foundations and abutments can be very high, so that the utmost care needs to be taken in the site investigations and in the design and the construction. Whatever may be the results of the theoretical analyses of forces and stresses on the dam, sufficient reserve strength must be included in the design to meet unknown weaknesses.

5.16 BUTTRESS OR MULTIPLE ARCH DAMS

Where a valley in rock is too wide for a single arch dam, a multiple arch dam may be used. This comprises a series of arches between buttresses, as shown in Plate 4(c) for Nant-y-Moch dam. Each section of the dam, consisting of a single arch and its buttresses, may be considered as achieving stability in the same manner as a gravity dam. A similar design uses a 'diamond head' at the upstream end of each buttress. In this case there is no arch action and seals are needed between each pair of 'diamond heads'. The buttress dam provides considerable saving of concrete compared with a mass concrete dam; on the other hand this saving may be offset by the extra cost of the more complicated shuttering required and the more extensive surface areas requiring a good finish for appearance. A buttress dam has the advantage that uplift is negligible, because the space between the abutments allows uplift pressure to dissipate. Occasionally a buttress dam is advantageous where the depth to sound rock would make a mass concrete gravity dam more expensive.

Buttress dams (or barrages) may also be used in connection with river control and irrigation works, where many gates have to be incorporated to control the flow. In this case there are no arches: the buttresses are to gravity design and have to take the extra load from the gates when closed.

FLOOD AND DISCHARGE PROVISION

5.17 DESIGN FLOOD ESTIMATION

The assessment of flood risk is a vital element in the safe design, maintenance and operation of impounding reservoirs. Earth dams are inherently erodible and uncontrolled overtopping can lead to catastrophic failure. Overtopping of a rockfill, masonry or concrete dam needs to be avoided except where the design specifically provides for this. It is therefore necessary to specify a design flood, in combination with wave action, which the dam must be capable of withstanding. Greater security is required against dam failure where there is a major risk of loss of life and extensive damage and a lower security where the threat is less severe.

A wide range of methods are used for computing reservoir design floods. These methods include:

- empirical and regional formulae;
- envelope curves;
- flood frequency analysis;
- various types of rainfall-runoff and losses models, including the unit hydrograph, the US Soil Conservation Service, the transfer function and the Gradex methods.

The Design Flood (Guidelines) produced by the International Commission on Large Dams (Committee on Design Flood) contains useful summaries of the more commonly used methods and their limitations. The guidelines also contain general guidance and recommendations about reservoir design flood estimation, some of which are the following:

1. The determination of design flood is a complex problem requiring the contributions of specialist engineers, hydrologists and meteorologists whose involvement must be sought through the whole process.
2. The choice of the design flood involves too many and varied phenomena for a single method to be able to interpret all of them.

3. All methods available are based on meteorological and hydrological records both for the river basin under investigation and often for other comparable areas. The results derived from the application of a particular method will essentially depend on the reliability and applicability of the data adopted.
4. The exceptional circumstances causing extreme floods are often such that the records of historical moderate floods may provide only a very poor indication of conditions during a Probable Maximum Flood (PMF) or a 10 000 year event.

For dams in the UK the key advisory document is the third edition of *Floods and Reservoir Safety* (ICE, 1996) which recommends the levels of protection for different categories of dam. The recommended floods standards in the UK are shown in Table 5.1. Chapter 3 of *Floods and Reservoir Safety* guides the engineer on the use of the 1975 Flood Studies Report (FSR) (NERC, 1975) unit hydrograph rainfall-runoff and losses model to derive reservoir design flood inflows, while Chapter 5 contains a method for estimating wave surcharge allowances. The *Flood Studies Report (FSR)* was superseded by the *Flood Estimation Handbook* (FEH, 1999); of which volume 4

Table 5.1 Flood standards by dam category according to *floods and reservoir safety*, ICE, 1996

| | | **Reservoir design flood inflow** | | |
Dam category	Potential effect of a dam breach	Initial reservoir condition standard	General	Minimum standard if overtopping is tolerable
A.	Where a breach could endanger lives in a community	Spilling long-term average inflow	Probable maximum flood (PMF)	10 000-year flood
B.	Where a breach (1) could endanger lives not in a community or (2) could result in extensive damage	Just full (i.e. no spill)	10 000-year flood	1000-year flood
C.	Where a breach would pose negligible risk to life and cause limited damage	Just full (i.e. no spill)	1000-year flood	150-year flood
D.	Special cases where no loss of life can be foreseen as a result of a breach and very limited additional flood damage would be caused	Spilling long-term average inflow	150-year flood	Not applicable

Note: Where reservoir control procedures require, and discharge capacities permit operation at or below specified levels defined throughout the year, these specified initial levels may be adopted providing they are stated in the statutory certificates and/or reports for the dam.

presents an updated version of the *FSR* rainfall-runoff model. MacDonald and Scott (MacDonald, 2000) outlined the key issues surrounding the use of this updated model to derive design flood estimates for UK reservoirs, and also how the flood estimates might be improved. To this end the Defra research project *"Reservoir Safety—Long Return Period Rainfall"*, due for completion in late 2008, should provide improved rainfall depth-duration-frequency values for use in UK flood studies. A project report by Svensson et al. (Svensson, 2006) provides a review of current rainfall frequency and probable maximum precipitation estimation methods, both in the UK and several other countries.

The statistical analysis of flood events has a very limited role in reservoir design flood estimation in the UK. The reason for this is that extrapolation of statistical flood estimates to the high return periods relevant to freeboard and spillway design may lead to gross under- or over-design, given the relatively short period for which flood data are typically available. Flood estimates become increasingly unreliable when the return period of the design event is greater than about twice the length of the record. It is not possible here to provide a guide to each of the methods of flood estimation currently employed. The following paragraphs, however, provide some basic guidance on the unit hydrograph rainfall-runoff method which has been used for design flood estimation in many countries.

Unit hydrograph approach

There are numerous different forms of unit hydrograph and loss models, but all are similar in that they convert a rainfall input to a flow output using a deterministic model of catchment response. These deterministic models have three common elements:

- the unit hydrograph itself which often has a simple triangular shape defined by time to peak;
- a losses model which defines the amount of storm rainfall which directly contributes to the flow in the river; and
- baseflow (i.e. the flow in a river prior to the event).

Where possible the model parameters should be derived from observed rainfall runoff events. If no records exist the model parameters may be estimated from catchment characteristics.

A unit hydrograph for a given catchment shows the flow resulting from unit effective rainfall in unit time on the catchment, e.g. the flow following, say, 10 mm effective rainfall falling in 1 hour. It assumes the rainfall is uniform over the catchment and that runoff increases linearly with effective rainfall. Thus the runoff from 20 mm of effective rainfall in one hour is taken as double that due to 10 mm and so on, and the ordinates of the hydrograph are doubled. Similarly if rainfall continues beyond the first hour, the principle of superposition can be used, the resulting hydrograph being the sum of the ordinates of the hydrographs for the first, second, third, and so on, hours.

Due to the complexities involved, the derivation of a catchment unit hydrograph from storm rainfall and runoff records is best carried out by an experienced hydrologist. Losses by evaporation, interception and infiltration have to be deducted from rainfall, and the baseflow has to be deducted from the measured streamflow. A complication is that rainfall events are seldom of unit duration, so further analysis is required to produce a unit hydrograph for unit time and unit rainfall. It is usual to derive several unit hydrographs from separate rainfall events for comparative purposes and, where a catchment is too large for uniform rainfall intensity over the whole of it to be assumed, it is necessary to treat individual tributaries separately and assess their combined effect.

One relatively straightforward method of obtaining an adequately severe unit hydrograph is suggested by the US Bureau of Reclamation (USBR, 1977) as follows:

(1) Time T_c of concentration is given by:

in US units	in Metric units
$T_c = (11.9\ L^3/H)^{0.385}$ hours	$T_c = (0.87\ L^3/H)^{0.385}$ hours

where: L = length of longest tributary (miles—km); H = fall of this tributary (feet—metres); T_c = time elapsing between onset of storm and time when all parts of the catchment begin contributing to flow to measuring point.

(2) Time to peak T_p for the hydrograph for one inch of rainfall over the catchment is given by:

$$T_p = 0.5\ (\text{rainfall duration}) + 0.6T_c$$

where T_p, T_c and rainfall duration are in hours.

(3) Peak runoff rate R_p:

in US units	in Metric units
$R_p = 484\ AQ/T_p$ ft³/s (cusec)	$R_p = 0.2083\ AQ/T_p$ m³/s (cumec)

where A is catchment area (sq. miles—km²) and Q is the unit rainfall (inch—mm). If A is in acres and Q is in inches, the peak runoff in US units is given by $R_p = 0.756\ AQ/T_p$ acre-feet.

Use of this formula tends to over-estimate flood peaks in temperate, flat and permeable areas; for these, a simple triangular unit hydrograph, with a base of 2.5 times the time to its peak, should suffice. Its peak flow is $2.2/T_p$ m³/s for each km² of catchment for every 10 mm of effective rainfall. The value of T_p often lies in the range $1.3(\text{area})^{0.25}$ to $2.2(\text{area})^{0.25}$ where the area is in km².

A variety of alternative loss models can be incorporated into the unit hydrograph approach. The two most common are to:

- derive a single percentage runoff value applicable throughout the whole storm; or
- adopt an initial loss x mm at the start of the design storm followed by continuing losses of y mm/hour throughout the event.

Once each of the model elements has been defined for a catchment, the unit hydrograph method may be used to estimate the total runoff from any rainfall event. The design rainfall is in the form of a hyetograph defined by a depth/duration characteristic of the area and arranged into a selected storm profile. A 'bell-shaped' profile is most commonly used for storm events up to about 24 hours duration, whereas the recorded profiles in severe historic events are used to define more realistic design storm profiles for longer duration events.

The design storm rainfall may be a statistically derived design event to produce a flood of specific return period (the T-year event), or may be a probable maximum precipitation (PMP) to produce a probable maximum flood (PMF). Rainfall depth–duration–frequency values, sometimes including estimates of the PMP, are available for many regions. Table 5.2 gives some examples of the PMP estimates adopted for the areas draining to a number of major dam sites; Figure 3.3 also gives information on maximum precipitations that have been recorded.

Table 5.2 Some probable maximum precipitation (PMP) estimates

Location	Country	PMP in mm				
		20 min	1 hour	6 hours	15 hours	24 hours
South Ontario	Canada	—	—	410	—	445
Guma	Sierra Leone	81	183	—	580	630
Selangor	Malaysia	—	160	300	—	460
Shek Pik	Hong Kong	101	220	—	915	1200
Garinono	Sabah	81	162	420	620	675
Brenig	Wales					
	May–Sept	74	109	183	—	254
	Oct–Apr	38	72	165	—	272
Tigris (50 000 km²)	Iraq	—	—	60	—	167
Jhelum, Mangla	Pakistan					
(2500 km² sub-area)	Dec–May	—	—	185	—	295
	Jan–Nov	—	—	365	—	575

5.18 SPILLWAY FLOOD ROUTING

If a reservoir is full to top water level, a flood inflow causes the reservoir water level to rise and this causes increasing rates of discharge over the spillway. In a conventional ungated spillway, with a fixed weir crest, the temporary ponding of water in the reservoir results in a maximum spillway discharge that is less than the maximum rate of flood inflow to the reservoir, with the peak outflow occurring after the peak inflow, as illustrated in Figure 5.11. (It may be noted that, when the outflow is at its maximum, it is equal to the inflow.)

This phenomenon is referred to as 'flood attenuation' and the calculation to process the inflow hydrograph and produce an outflow hydrograph is termed 'flood routing'. In the past, graphical approaches were sometimes adopted to assist in flood routing but these are no longer appropriate, as computer programs or spreadsheets can be used much more conveniently. In most practical circumstances the reservoir water surface can be treated as horizontal at each timestep and the flood routing approach is therefore described as 'level pond'.

Suitable software for undertaking reservoir flood routing is included in industry-standard open-channel flood modelling software. The principal data required to carry out a reservoir flood routing are:

- the inflow hydrograph, at a suitable timestep (preferably no longer than about a tenth of the time to the peak of the incoming flood);
- the spillway rating (that is the relationship between discharge and reservoir level); and
- the relationship between the water level and surface area in the reservoir.

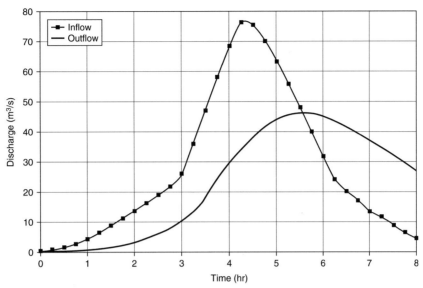

FIGURE 5.11

Example flood routing through a reservoir.

In practice, a storage volume relationship for the reservoir may be available, rather than a surface area relationship, and some software may accept the volume relationship, rather than require an area relationship to be provided. However, it is always best practice to convert a volume relationship to the corresponding area relationship, to check that it is consistent with the known surface area of the reservoir and increases appropriately as the water level rises.

The user needs to define the initial water level in the reservoir, for example whether it is assumed that the reservoir is initially just full, or is discharging the mean discharge or some other amount. There may be further complexities, such as the presence of a bywash, which intercepts part of the incoming flow and carries it past the reservoir. Such cases such should be evaluated by an experienced hydraulics engineer.

Level-pond flood routing is undertaken for each timestep equating the difference between the inflow and outflow to the change in the volume of water contained. The basic flow balance equation (normally in metre-second units) is:

$$\frac{I_1 + I_2}{2} - \frac{O_1 + O_2}{2} = (H_2 - H_1)\frac{(A_1 + A_2)}{2\Delta t}$$

where I is the incoming discharge, O is the outflow, A is the reservoir surface area and H is the water level in the reservoir, and the subscripts 1 and 2 refer to the beginning and end of the timestep, Δt. Theoretically it is necessary to iterate to a solution, and this is what the best software does. However, if a short enough timestep is adopted, it can generally be assumed that, within a timestep, $O_2 = O_1$ and $A_2 = A_1$, in which case the basic equation simplifies to:

$$H_2 = H_1 + \frac{(I_1 - O_1)\Delta t}{A_1}$$

In this case, the values of O_2 and A_2 are calculated only after H_2 has been determined, ready to be used as the values of O_1 and A_1 in the next timestep. If this simplified approach is adopted, the user should beware of using too small a timestep Δt. If the calculations proceed with values of ΔH that are too small to be evaluated accurately within the available precision of the calculator or computer, the accuracy of the simulation could be severely prejudiced.

5.19 DIVERSION DURING CONSTRUCTION

When a dam is constructed in a natural river valley, some means of temporarily diverting the river flow is needed in order for the dam foundations to be properly prepared and for the construction of the works to proceed. Where flows are seasonal, advantage can be taken of this, ensuring that especially critical periods of construction are carried out in low-flow periods. Plate 5(a) shows closure of the River Chira at Poechos Dam, Peru.

In some cases, it may be possible to divert the river via tunnels in the abutments and to work within an effectively dry river bed. In wide valleys it may be more economic to divert the river onto one side of the valley while starting construction on the other side, then to divert the river though temporary openings in the partially completed structure and start construction on the remaining side. In exceptional cases, where rivers are subject to regular high floods, it may be necessary to allow the partially completed dam to periodically over-top and to design the works accordingly to accommodate this.

Diverting through a temporary opening left in the main dam body or allowing the partially completed structure to be overtopped by floodwater are techniques especially suitable for concrete or masonry dams. These are essentially non-erodible and therefore can be easily recovered after inundation.

It is not possible to follow this procedure with an earth dam which, if overtopped at any stage during construction, would probably be destroyed. In such cases diversion is generally via a tunnel or culvert. Culverts and tunnels used initially for river diversion can later be used for housing the permanent drawoff pipes from the reservoir. Care needs to be taken with culverts to ensure that they do not cause differential stress concentrations in the earthfill lying above. Unless an earth dam is founded on sound rock (which is often not the case), it is undesirable to have a central culvert left through the body of an earth dam. If a culvert is unavoidable it is better to recess the culvert as much as possible into a trench, to minimize the tendency for it to generate stress differentials in the main dam body. Differential settlement can also cause the culvert to fracture or open up a path for impounded water to leak through the junction where the culvert passes through the corewall of the dam (see Appendix).

The flood capacity of temporary diversion works may well be laid down in the construction specification but is also of key importance to the contractor building the works, since he would normally have to shoulder any risks associated with an event of greater severity occurring during the construction of the scheme. The flood discharge capacity of the diversion works required during construction generally reflects the consequences of damage should a larger flood occur. In the early stages of construction the consequences of under-capacity may be slight. Later, insufficient capacity may entail loss of much of the dam. It is not uncommon therefore to see the design return period for diversion increase as construction develops. Diversion works may initially be required to accommodate the 1 in 5 year flood, but eventually be required to take the 1 in 50 or 1 in 100 year event immediately prior to dam closure.

Each case has to be considered individually, taking all the relevant circumstances into account. These include the nature of the formation, the type of dam and the proposed method of construction. In planning a diversion scheme, rainfall, runoff and flood routing studies need to be carried out to find the critical circumstances applying at each stage of the construction.

Some risk is unavoidable and should be the subject of insurance provision. In the case of the Oros dam in Brazil (W&WE, 1960), a diversion tunnel of 450 m³/s capacity had been built, but 600 m of rainfall fell on the catchment in a week. Despite all efforts to raise the dam ahead of the rising waters, it was overtopped on 25 March 1960 and required major re-construction.

5.20 FLOOD SPILLWAYS

Spillway works can generally be classified into three types:

- integral overflows that convey floodwater over the crest of the dam;
- open 'chute' spillways, which convey floodwater via a channel constructed on natural ground at a dam abutment or at a low point on the reservoir perimeter; and
- structures that discharge floodwater via an opening or culvert through or under a dam, or via a tunnel driven through an abutment.

The crest of a dam has generally to be high enough above the highest water level in the reservoir under flood conditions to prevent waves and spray passing onto the crest and the downstream face of the dam. For an earth dam, the usual practice in the UK results in the top of the embankment being up to about 2m higher than the crest of the overflow weir, with a substantial wavewall capable of protecting the bank against the runup from wave action also being provided. For dams in other parts of the world, where reservoirs and floods can be of much greater magnitude than in the UK substantially higher freeboards may be necessary as for Mangla dam (Fig. 5.2), where almost 10m of freeboard was provided for flood surcharge and wave runup (Binnie, 1968).

Integral Spillway

Direct discharges are permissible over concrete (including RCC) and masonry dams and some rock-fill dams, if provided with a suitably designed spillway. The practice can be economical, because very little alteration to the profile of such a dam is necessary to accommodate the overflow section. Typical examples are shown in Plate 5(b) and 5(c). The water usually cascades down a smooth or stepped face of the dam. At the foot of the dam the water must be turned into a stilling basin to disperse some of its energy, because the main danger with this type of overflow is scour at the toe of the dam during an extreme flood. Sometimes the water is ejected off the face of the dam by a ski-jump or other means (Plate 5(b)), which throws the water some distance away from the toe of the dam, thereby lessening the danger to the toe. It is essential that all the construction is massive and soundly founded on solid rock. The design is normally not appropriate to a valley where good hard rock does not appear in the river bed at the dam and for some distance downstream.

Sometimes an additional feature of a concrete or masonry dam is the inclusion of automatic crest gates at the overflow weir. The gates permit water to be stored above the level of the overflow weir crest but, after a certain level is reached, the gates are opened automatically and permit progressively increasing discharges to pass over the weir. This permits increased storage without raising the flood overflow level. Except in the case of some concrete dams on hydro-electric schemes in Scotland, gates have seldom been used on dams in the UK, because storm conditions might interfere with the power supplies for operating the gates at the crucial time.

Another method is the use of a siphon hood over the weir crest, in order to achieve a greater discharge per unit length of crest than would occur for the same water head over a fixed weir. Unit discharges of between 6 and 20 m²/s have been achieved for design heads, relative to the spill level, of between 0.12 m and 0.9 m (Ackers and Thomas, 1975). Modern spillway siphons are usually of the air-regulated type, being designed to provide a progressive increase in discharge at increasing heads, in comparison with older 'make-and-break' designs that tend to cause sudden fluctuations in discharge and hence unstable conditions both upstream and downstream. Ultimately, if the upstream water level continues to rise and the discharge increases, an air-regulated siphon reaches 'blackwater' flow, in which there is little or no air entrained in the flow. Beyond this point, the siphon behaves like a short culvert, with the discharge being approximately proportional to the square root of the operating head, so further increases in discharge capacity require substantial rises in the upstream pool level. For this reason, spillway siphons should normally be designed with a generous factor of safety on discharge, to allow for the uncertainties inherent in the determination of design floods.

Siphons require careful design, either adopting a previously proven design or verification by model testing. A key design consideration is the avoidance of cavitation, and this limits the operating head to about 8 m. Siphons are not normally suitable for use in cold climates, where ice formation in winter might block or decrease the capacity of the siphon. Booms may be needed to protect syphon inlets from trash.

Chute Spillway

At most major reservoirs in the UK floodwaters are discharged via a spillway channel around the end of the dam (Plate 6(a)). Such spillways should be constructed on natural ground and not on the embankment dam itself where settlement will occur. The construction has usually to be massive, of concrete or masonry, because of the need to prevent dislodgement of any part under flood conditions, when significant pressure fluctuations can occur in the flow. Plates 7(a) and 7(b) show larger examples where the spillway gated structure is sited on original ground between massive monoliths against which the dam embankments abut.

At the upstream end of the channel there is an overflow weir, often with a rounded crest to increase the discharge coefficient and minimize damage by floating debris. A common configuration is a 'side-channel' (Plate 6(a)), in which the overflow weir forms one side of a channel that runs along the reservoir shoreline for a short distance upstream of the dam abutment. The channel downstream of the weir must be designed to take the water away without any backing up of water on the weir. Flow here is usually subcritical and the channel gradient is gentle, until it reaches a point near the abutment where it starts to fall more steeply and critical flow occurs (Sections 12.8–12.13). This acts as a control point for the flow of water in the spillway and determines whether the flow conditions upstream allow the overflow to perform up to its full discharge capacity, or is 'drowned out' and behaves as a submerged weir. In many cases, flows in this area are further constricted by the presence of a bridge, affording access to the dam crest; this can create a 'control' based on orifice-type flow conditions.

A common design problem is that the floor of the channel upstream of the corewall of the dam is subject to uplift when the reservoir is full, so must be designed accordingly. The channel must also be effectively sealed against the corewall of the dam, to prevent leakage passing below the floor or behind the walls of the spillway.

Downstream of the control point near the abutment of the dam, the overflowing water becomes supercritical and passes at increasing speed down the steeper part of the channel. Attempts have sometimes been made to destroy some of its energy by constructing piers or steps in the channel, but these are normally not particularly effective and can cause problems with air bulking of the flow and with spray, which can erode soft material adjacent to the walls of the channel. At the end of the spillway channel a 'stilling basin' is normally constructed to dissipate some of the energy of the fast-moving water. The most usual method is to induce the water to form a hydraulic jump within the stilling basin (Section 12.12).

Historically, spillway channels have often been curved, especially in the vicinity of the abutment, where the channel steepens and when the channel reaches the base of the valley, approaching the stilling basin. It was also common to accompany the steepening gradient with a reduction in channel width. These curves and transitions generally result in complex patterns of cross-waves in supercritical flow and often result in overtopping of the channel walls at locations that were not anticipated by the designer. Wherever possible, new spillways should be designed so that they are straight throughout sections where supercritical flow occurs. Unless designed in a manner that ensures virtual certainty of the flow conditions, the designs of new channel spillways should normally be subjected to model testing.

Bellmouth Spillway

A bellmouth (or morning-glory) spillway (Plate 6(b)) may be adopted where the expense of cutting a spillway channel is great, or where a tunnel has to be driven through one of the abutments for the diversion of the river during construction of the dam. The bellmouth must be constructed on firm ground, preferably either clear or almost clear of the toe of the dam. If built within the body of the dam, differential settlement of the bellmouth and embankment fill may cause disruption of the fill, or the different levels of fill against it might cause it to tilt. The bellmouth shaft may be vertical or sloping and it may join the tunnel with a smooth or sharp bend.

Three primary flow modes are usually possible with a bellmouth spillway:

- crest control, where the rim of the bellmouth acts as a weir;
- orifice control, in which the throat of the shaft (normally at the base of the bellmouth, where the diameter of the shaft becomes uniform) acts as a control and drowns out the crest;
- tunnel control, in which the shaft and tunnel flow full and the discharge is a function of hydraulic friction and form losses.

These three flow modes would usually apply in the above sequence, at increasing upstream pool levels but in some cases there may be a direct transition from crest control to tunnel control, without any region of orifice control. A potential disadvantage of the bellmouth spillway is that there is little increase in discharge capacity once the weir is drowned and orifice or tunnel flow control applies. It is therefore normally best practice to design a bellmouth spillway so that crest control applies over the entire discharge range, with a generous margin provided before either orifice control or tunnel control would occur.

The profile adopted for the rim and interior of the bellmouth is normally similar to that for an ogee type of spillway. *Design of small dams*, Section 212 (USBR, 1977) gives profiles and discharge relationships for bellmouth spillways, based on research by the US Bureau of Reclamation in the 1950s. See also *Morning Glory Spillways* (USACE, 1987) for a simpler comprehensive presentation.

(a) Chira River closure and diversion for Poechos Dam, Peru. (Courtesy of R. L Brown)

(b) Mudhiq concrete arch dam, 73 m high, Saudi Arabia. (Engineer: Binnie & Partners)

(c) Waterfall masonry faced concrete gravity dam, 55 m high, Tai Lam Chung, Hong Kong. (Engineer: Binnie & Partners)

PLATE 5

(a) Lower Lliw dam side spillway, Wales. (Courtesy of Welsh Water).

(b) Taf Fechan dam drawoff tower and bellmouth spillway, Wales. (Engineer: Sir Alex Binnie, Son & Deacon)

PLATE 6

(a) Main spillway of 50 m high earth dam at Poechos, Peru—capacity 7200 m³/s.

(b) Main spillway of 138 m high earth dam at Mangla, Pakistan—capacity 25 000 m³/s.
(Engineer: Binnie & Partners) (Courtesy of O J Berthelsen)

PLATE 7

(a) 14 000 m³/s capacity fuse plug emergency spillway at Poechos Dam, Peru.

(b) Concrete tipping gates at dam in UK. (Courtesy of John Ackers)

(c) Reinforced concrete labyrinth fusegates at Terminus Dam, California. Capacity 7545 m³/s after tipping of all gates. (Courtesy of Hydroplus, Inc, VA)

PLATE 8

Radial vanes are generally incorporated in the top of a bellmouth shaft to minimize the occurrence of vortex action, which can otherwise reduce the discharge capacity and cause problematic flow conditions in the shaft and tunnel. If a bellmouth spillway is located close to the reservoir shoreline or the upstream face of the dam, the limited depth and width of the approach on that side of the rim can lead to 'starvation' of that part of the weir, reducing the overall discharge capacity. Normally, the bellmouth should be positioned so that there is a minimal starvation effect.

A key requirement in the design of bellmouth spillways is to avoid (or severely limit) the occurrence of syphonic conditions, which can occur at the base of the shaft and upstream end of the tunnel when tunnel control applies. This is primarily to avoid cavitation; this is the formation of pockets of water vapour when pressure falls in usually fast flow to below the vapour pressure of water and their subsequent collapse which, if at a surface, can fast erode that surface (Section 13.12). The damage done by cavitation can be catastrophic. At the Tarbela dam in Pakistan in 1974 (Kenn, 1981) a thickness of over 3 m of concrete tunnel lining was eroded within about 24 hours. This damage is believed to have been caused by erosion due to the collapse of vapour pockets against the concrete surface after being generated at the interface between the very high velocity jet emerging from the part open gate and adjacent slower moving water. Great care should be taken in the interpretation of model test results in cases where subatmospheric pressures occur.

The junction of the base of the shaft with the tunnel requires careful design, as the falling water brings with it appreciable quantities of air. The design should be such that this air can be safely evacuated, avoiding a phenomenon known as 'blow-back', in which pockets of air force their way up the shaft, temporarily reducing discharge.

An advantage of the bellmouth overflow and tunnel, apart from use of the tunnel for river diversion during construction, is that it may also accommodate the drawoff functions. The supply pipe and compensation water pipe may be led through the tunnel, either fully encased in concrete or passing through a separate compartment, whilst the scour valve may discharge directly to the base of the bellmouth shaft.

Bellmouth spillways are not suitable in climates where substantial ice may form on the reservoir during winter. Also, even for small reservoirs, the size of the bellmouth and its throat must be large enough to allow passage of any debris that might be brought down by the river. The bellmouth presents a danger to sailing and fishing boats when the reservoir water level is high, so guards must be fixed around its perimeter; these must also be designed not to hold back floating debris that could restrict the bellmouth discharge capacity.

Emergency Spillways

It is sometimes not feasible to provide sufficient capacity at a single spillway to accommodate the whole design flood flow. This may be for reasons such as space limitation or cost. In any case, if the main spillway is gated, it may be held that there should some back-up spillway provision in case the main spillway gates cannot be opened.

Operation of an emergency spillway should be automatic and reliable and so should not be reliant on any human intervention or any systems. It will be required to operate in the most extreme circumstances when normal operation procedures are unlikely to be possible. Often limited space means that a simple weir cannot be used, so designs usually involve triggering the opening of ample water passages. Several arrangements are commonly used.

Fuseplug embankments. These consist of erodible embankments constructed on a non-erodible (concrete or rock) base, often in bays with different crest levels to allow increments in discharge.

The embankments consist of a clay or silt core sloping downstream and supported by shoulders of non-cohesive sandy gravel. Sand filters are used each side of the core for the usual reasons (Section 5.9) and as their presence downstream of the core facilitates erosion. The embankment must be designed and constructed as a dam but the grading of the downstream shoulder material is critical. Plate 8(a) shows a 4 bay fuse plug spillway of capacity 14 000 m³/s for Poechos Dam, Peru.

Arrangements in which a whole fuse plug section is designed to fail at once may not be suitable where there is concern about the effects of the consequent significant flood wave passing downstream. In such cases a pilot channel may be used. This is a section of width about half the height of the fuse plug and with crest level about 1 m below the main plug crest (for plug height of 3 to 9 m). Model tests carried out by USBR (1985) showed that the pilot channel encourages the initial breach and allows the plug to be eroded away laterally at a rate of between 1.4 and 2.8 m/hr for heights of 3 and 9 m respectively. A 100 m wide fuse plug 6 m high with a central pilot channel would thus take about 25 minutes to erode away fully, thus reducing the steepness of the flood wave considerably.

Tipping gates. These may be of two types, both allowing progressive flow increase as gates tip:

- reusable 'folding' steel or concrete walls 'hinged' at the base and designed to topple into a recess in the spillway channel floor; Plate 8(b) shows concrete tipping gates at a reservoir in UK;
- disposable steel or concrete fabrications designed to topple and be carried away by the resulting flow; these may be made with straight or labyrinth weir sections (Plate 8(c)) such as those designed by Hydroplus—used on over 30 dams worldwide since 1990.

Another arrangement sometimes used, where appropriate resources (perhaps military) are assured locally, is to arm sections of the dam embankment with explosives or facilities for their quick installation.

Operation of emergency spillways is intended to be rare, if at all, but some arrangements produce a sudden large discharge—even if the full flow is delivered in steps. The consequences of the sudden increases in flow downstream should be carefully considered and controls may have to be put in place to avoid the worst damage or loss of life that sudden flood flows could cause.

5.21 DRAW-OFF ARRANGEMENTS

Draw off works at water supply reservoirs can be for water supply, reservoir draw-down and for providing downstream river compensation flows. A summary of requirements and UK practice is provided in the paper by Scott (2000).

Except when a reservoir is shallow, water supply drawoff pipes are usually designed for withdrawing water at several different levels from within the reservoir. A common provision may be simply upper, middle and lower, although more may be used in deep reservoirs. The choice of levels depends on the depth/volume relationship of the reservoir and the expected variation of water quality (and perhaps temperature) with depth, which may change seasonally (Section 7.1). The upper drawoff must be a sufficient distance below top water level to avoid constant changing from it to mid-level drawoff with normal reservoir level fluctuations. The bottom drawoff level may need to allow for sediment.

Such arrangements are normally sited in drawoff towers. Wet-well towers feature drawoffs discharging into a flooded tower and a single extraction point from the tower at low level. Such towers are less common nowadays owing to the difficulty of maintaining the underwater valves and pipework. A more common option is the dry-well tower, in which all extracted water is contained by valved pipework accessible by means of internal ladders and platforms. At each drawoff there are usually two valves, or else an outer sluice-gate and an inner valve. The inner valve is used for normal operation, the outer valve or sluice-gate is used to permit the inner valve to be maintained.

The drawoff tower at an embankment dam would typically be sited close to the upstream toe of the embankment with access by footbridge from the dam crest or shoreline (Plate 6(b)). In the case of concrete dams they are more economically tied to the main dam or sited within it. In some cases drawoff arrangements have been located in sloping galleries on the reservoir abutments. This can be a cheaper and less prominent option as well as attracting less seismic loading. The drawoffs transfer water to one or more outlet pipes. It is not good practice to site such pipes directly in an embankment, but rather they should be sited in culverts or tunnels and with some means of upstream isolation. They can, of course, be directly embedded in the body of a concrete dam.

A scour pipe is required for a reservoir to ensure the approach channel to the lowest draw-off is kept free of silt, and to make it possible to lower the water level in the reservoir at a reasonably fast rate in case of emergency. Recommended drawdown rates for reservoirs are not standardized and are to a large extent site-specific. Typical criteria can be a drop of a quarter of the reservoir height in 14 to 21 days in conjunction with an average wet month inflow or, alternatively, half of the reservoir volume in 30 days in conjunction with mean inflow. In both cases the rate of reservoir drop should not be so as to threaten the stability of the dam or reservoir margins and the rate of release should not be so great as to cause excessive distress downstream.

Those pipes that release flow to the downstream watercourse (including the scour pipe, any scour branch off the main drawoff pipes and the outlet for compensation water) normally discharge either into a stilling basin, designed to dissipate the energy, or via a disperser valve that turns the flow into a spray.

5.22 SEISMIC CONSIDERATIONS

Earthquakes are severe and common in some countries, especially those bordering the Pacific Ocean. In other countries, such as the UK, they are less common and less severe. Nevertheless, dams are major items of infrastructure with the potential to cause considerable damage should they fail. In view of this, just as dam and reservoir works are designed to accommodate rare flood events, they have also to be designed to be able to accommodate very rare seismic events.

Seismic loadings need to be considered in the design of both new dams and the appraisal of old dams. In the UK, the key references are *An engineering guide to seismic risk to dams in the United Kingdom* (Charles, 1991) and an 'application note' covering aspects of the use of the guide (ICE, 1998). These documents provide a structured framework for the consideration of seismic risk similar to that adopted for floods.

DAM REGULATION, SUPERVISION AND INSPECTION

5.23 STATUTORY CONTROL OVER DAM SAFETY

Regulations are adopted in many countries to ensure that dams are regularly inspected and are constructed or altered only under the charge of properly qualified engineers. Reservoir safety in the UK is governed primarily by the Reservoirs Act 1975 (ICE, 2000) which was brought into effect in

1986, replacing the earlier Reservoirs (Safety Provisions) Act 1930. Reservoir safety regulations in a number of other countries follow the model of the UK Reservoirs Act 1975.

The 1975 Act applies to 'large raised reservoirs' which are defined as having a capacity of 25 000 m^3 or more above the level of any part of the adjacent land. Specifically excluded from the 1975 Act are lagoons covered by the Mines and Quarries (Tips) Act 1969 (together with the corresponding 1971 Regulations and the Quarries Regulations 1999) and navigation canals.

The 1975 Act includes roles for the Secretary of State, for the 'undertaker' (the owner and/or operator of the reservoir), for the 'enforcement authority' and for 'qualified civil engineers'. A series of 'panels' of such engineers have been set up to perform the particular functions under the Act, namely:

Construction Engineer	Responsible for supervising the design and construction or the enlargement of a reservoir.
Inspecting Engineer	Carries out periodic inspections, normally at intervals of ten years.
Supervising Engineer	Has a continuous appointment to 'supervise' the reservoir, watching out for problems and keeping the undertaker informed.

Three panels have been set up to cover the duties of Construction Engineer and Inspecting Engineer, the panels being defined according to the type of reservoir on which the engineer is qualified to exercise their duties, as follows:

All Reservoirs Panel	All reservoirs covered by the Act
Non-impounding Reservoirs Panel	All reservoirs, except impounding reservoirs
Service Reservoirs Panel	Service reservoirs only

All engineers on the above panels are also entitled to act as Supervising Engineer for all reservoirs covered by the Act. In addition, the Supervising Engineers Panel covers those qualified to act only as Supervising Engineers. Panel appointments are made for a period of five years and are renewable.

The 1975 Act contains provisions for the registration of reservoirs and for the enforcement of the Act, both of which functions were transferred to the Environment Agency in October 2004. The enforcement of the Act is facilitated by requirements for the issue of certificates by the qualified civil engineers under specific circumstances, in particular in association with the construction or enlargement of a reservoir, the issue of inspection reports and the implementation of 'measures taken in the interests of safety'. Appointments of qualified civil engineers by the undertakers also have to be notified to the enforcement authority.

The Water Act 2000 contained a number of changes to the Reservoirs Act 1975, including transferring the role of the enforcement authority to the EA and creating a new power to direct an undertaker to prepare a 'flood plan' for a large raised reservoir. The flood plan is intended primarily to cover any escape of water from the reservoir in terms of the predicted extent and depths of inundation, the risks to life and property and arrangements for issuing warnings.

A guide to the Reservoirs Act 1975 (ICE, 2000) contains the full text of the Act and the relevant regulations up to that date, a commentary on the application of the Act and flowcharts to illustrate the duties of the various parties. Anyone engaged on work involving reservoirs in the UK should refer to this guide and to the current guidance on subsequent developments given on the Defra website, http://www.defra.gov.uk/environment/water/rs/index.htm.

Particular points worth noting are:

- responsibility for the construction or enlargement of a reservoir remains with the Construction Engineer for between three and five years from first filling;
- a Supervising Engineer is not required until the end of the Construction Engineer's involvement, upon issue of their final certificate;
- the first inspection under Section 10 of the Reservoirs Act 1975 is required within two years of the issue of the final certificate; and
- the Inspecting Engineer must be 'independent' of the Construction Engineer and the undertaker.

The provisions of this Act and the earlier Act of 1930 have worked well, in that they have assisted in preventing any dam failures involving a loss of life in the UK for three-quarters of a century. A particular feature of the 1975 Act is that it imposes a personal responsibility on the engineer who issues a certificate. Such responsibility can only be effectively exercised by an engineer having adequate experience of dam design and construction and who has access to the specialist services frequently required in making a proper inspection. Many dams in the UK are 100 to 150 years old and require careful attention to ensure their continued safety.

5.24 DAM DETERIORATION SIGNS

Routine observations to ensure a dam remains in good condition are too numerous to list here. Reference can be made to guides on the safety of embankment dams (Johnston, 1999) and concrete and masonry dams (Kennard, 1995), both of which provide detailed checklists, but each dam requires its own specific programme of monitoring. All instrumentation of a dam, such as settlement or tilt gauges and instrument for measuring porewater pressures and underdrain flows need regular monitoring and checking for accuracy. Only some matters of importance, principally related to signs of leakage, can be mentioned here.

Good access to a dam is important and should be suitable for heavy construction plant that might be needed for repairs. The situation arising if emergency work should be required has to be envisaged: night work may be necessary under heavy rainfall and in high winds and with a flood overflow. All gates, valves and other mechanical controls should be in easily operational condition, and access to shafts, galleries and inspection pits should be safe, properly ventilated and lit to safeguard against accidents to personnel adding to the troubles of an emergency.

The catchment needs monitoring for important changes of use and for any evidence of hillside movement that could indicate instability.

For earth dams leakage may be evidenced by damp patches on the downstream face, by areas with unusually luxuriant plant growth or by increased underdrain flows or porewater pressures. However, leakage may also cause settlement. When properly built, an earth dam should have smoothly regular upstream and downstream slopes; the crest should either be straight or to a preformed curve. Normally the dam crest should initially show an even rise towards the centre or highest part of the dam because a settlement allowance should have been incorporated during construction, usually of about 1% of the dam height. Irregularities of line and level of the crest that are not explained by the anticipated post-construction settlement may be evidence of erosion due to leakage, as could a localized depression in the embankment.

The upstream facing of the dam, whether of pitching, riprap, concrete blockwork or slabbing, should be checked for slope movement, sometimes visible from distortions in the water line. Concrete slabbing of the upstream face should be checked for signs of settlement or damage caused by wave action pulling out the supporting material.

The ground at the toe of an earth dam is often wet, sometimes supporting a growth of rushes. This is a natural collecting point for surface run-off but undue wetness may be caused by seepage due to a hidden defect such as cracking, erosion or settlement of the corewall. All underdrain inspection pits and drainage water should be examined for evidence of silt or clay being carried out of the dam. The puddle-clay corewalls of old dams are particularly vulnerable to erosion through leakage, and thought must be given to the possibility that the dam has suffered leakage or settlement, in the past, which has not been effectively dealt with.

Concrete dams can show signs of leakage by damp patches on the downstream face or by the growth of moss and lichen at joints. Signs of stress or movement in a concrete dam are spalling of concrete at joints, opening up of joints and cracks, or displacement irregularities, both on the surface of the dam and in any internal gallery or shaft. Since displacement of shuttering when the dam was constructed, or initial settlement which has ceased, can cause irregularities, it is important to log them and the areas where no irregularities occur, so that any new signs of movement can be detected. In masonry or masonry faced dams, leakage will be through joints; cracking or fallout of pointing needs investigation to see whether it is caused by weathering, mortar softening, increased stress, or possibly acid water seepage attack.

The flood discharge works to any dam need regular inspection for signs of settlement and for any other irregularities which could induce scour that might damage the works during a flood overflow.

A 'trained eye' is necessary for the inspection of dams, especially for old earth dams, where a small surface defect may eventually prove to be related to a dangerous condition of the dam developing internally, hence the need to investigate the cause of anything that seems to be 'not quite right' or 'not as it should be'.

5.25 RESERVOIR SEDIMENTATION

There is a vast range in the degree to which reservoirs worldwide are affected by sedimentation. Figure 5.12 shows the range of suspended sediment loads experienced in rivers of different size, as reported by Fleming (1965). However, the design and operational practices at the reservoir can significantly affect the degree to which the incoming sediment load is trapped. Morris and Fan (1998) cite a reference to an average worldwide rate of loss of storage through sedimentation of up to 1%, and White (2001) quotes the case of a reservoir in Austria that was completely filled with sediment within one year of commissioning. On the other hand, a survey of 95 reservoirs in the UK found an 'almost negligible loss of storage', of about 0.1% per annum.

The most comprehensive current source of information specifically applicable to water supply reservoirs in the UK is by Halcrow Water (Halcrow, 2001). This points out that, although sedimentation rates in the UK are generally not high by global standards, they can be significant in some locations. Historic approaches to sediment management for British reservoirs include:

- provision of a silt trap ('residuum lodge') where the main watercourse enters at the upstream end of the reservoir; and
- provision of a bypass channel ('bywash') to divert the most sediment-laden flows.

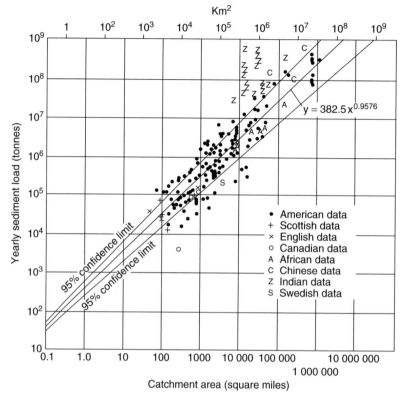

Based on information from Fleming (1965).
FIGURE 5.12

Average suspended sediment concentrations.

These measures have generally been successful, although there can be problems in the disposal of trapped sediment, including the operation of bottom outlets to release sediment accumulated in the main reservoir basin into the downstream watercourse.

In many parts of the world, the sustainability of reservoirs is threatened by high sediment loads in the incoming rivers. These loads may be the result of many factors, such as:

- natural erodibility of the catchment, for example during rainfall runoff;
- landslides, occurring naturally, or triggered by man's activities;
- earthquakes;
- deforestation, agricultural and urban development; and
- fallout of volcanic dust and ash.

In recent years, attention has been focussed on how to design and operate reservoirs in those parts of the world that are subject to high sediment loads. White (2001) concentrates on the feasibility of regular flushing of reservoirs to pass the sediment downriver and minimize the loss of useful storage capacity. Information is given on the factors that make sediment flushing practicable, which include:

- the shape of the reservoir basin—flushing is more effective in comparatively narrow steep-sided reservoir basins;

- the ability to lower the water level in the reservoir, which depends on the discharge capacity of the low-level outlets provided;
- the availability of sufficient water for flushing, which depends mainly on the reliability of seasonal patterns of inflow to the reservoir and the ease of refilling after flushing;
- the capacity of the reservoir in relation to the annual inflow; and
- the mobility of the deposited sediments.

Morris and Fan (1998) provide many case studies and examples of techniques that have been deployed to manage sediments successfully, including the removal of sediments that have been deposited over many years.

5.26 ENVIRONMENTAL CONSIDERATIONS AND FISHPASSES

Water storage reservoirs represent major investments, so are constructed only where there is a strong, demonstrable need. However, the construction of such major works and the blockage of natural rivers, inevitably have environmental impacts and some of these can be seen as negative. Typical impacts include the need to relocate people displaced by the reservoir, loss of heritage features, destruction or change of local flora and fauna and changes to local water quality and groundwater regimes.

No such reservoir project can now proceed unless environmental issues, such as the ones above, are addressed at the planning stage. All major funding agencies and most countries now have rules and/or laws to deal with associated environmental impacts. At the planning stage, the benefits of the project are assessed as well as any negative impacts. A focus at this stage is on trying to eliminate negative aspects or where elimination is not possible, to mitigate the effects. Projects can only proceed if positive outcomes exceed any negative impacts; on occasion a negative impact may be so severe—for example the loss of a unique species—that it may be sufficient to stop a project proceeding.

An overview of the environmental issues involved with respect to dams and reservoirs is given by Carpenter (2001). The issues are too numerous to be dealt with here, but a special mention is made of fish passage, as fishpass structures are often incorporated into dams, so that fish migration can continue along rivers after the dam and reservoir have been constructed.

The primary emphasis for the provision of fishpasses in the past was to allow the upstream passage of migratory salmonid species (salmon and sea trout). These species are diadromous, meaning that their life cycle involves both the sea and freshwater. Historically little or no attention has been paid to other species or indeed to downstream fish migration, except with regard to their exclusion from water intakes. The safe passage of freshwater fish and diadromous species such as eel and shad is now also considered.

There are a number of approaches to the design of fishpasses, which are mostly variations on the themes of steps, slopes and lifts. The 'step' approach splits the height difference into a series of smaller drops, with various forms of traverse separating pools in which the fish can rest. In the 'slope' approach, the water passes down a relatively steep slope that contains various forms of baffle to dissipate the energy and limit the water velocity, the best-known versions of these devices being due to Denil (1908) and Lariner and Miralles (1981). In a 'lift', the fish are attracted into a confined space and then lifted either mechanically or hydraulically before being released upstream.

The design of fish passage arrangements for adult migratory salmonids has been developed over many years but knowledge is still being gained on the needs and appropriate passage arrangements for other species. The design of fish passage arrangements is a specialist matter, requiring appropriate expert advice. In the UK, the key reference on the subject is the EA's *Fish pass manual* (Armstrong, 2004).

5.27 STATUTORY CONSENTS AND REQUIREMENTS

Reservoirs and associated structures require the consent of the national and local planning authorities. Regulations vary from country to country but, in the UK, reference must be made under the Town and Country Planning Act 1990 for consent. This action normally attracts comment from various statutory bodies, local organisations, environmental groups and other stakeholders depending on the sensitivity of the location. Early consultation with these bodies is necessary to avoid delay or refusal of the planning application.

In the UK, where nationally or internationally designated nature conservation sites could be affected, consultation with Natural England, the Countryside Commission for Wales, the Environment and Heritage Service (Northern Ireland), or Scottish Natural Heritage is necessary. This may lead to the modification of the reservoir's design, method of construction, or operation to ensure that its impacts are moderated and concerns allayed. The requirements are set out in the Wildlife and Countryside Act 1981, Countryside and Rights of Way Act 2000, and the Conservation (Natural Habitats, &c) Regulations 1994 (as amended).

If other designated features, such as archaeological sites, recreational assets or important landscapes might be affected, there would be similar requirements under other legislation to consult with the relevant statutory authority and potentially modify the proposals to accommodate their concerns.

Larger reservoirs—or those likely to have a significant effect on the natural or human environment—require the preparation of a formal environmental impact assessment (EIA) under the Town and Country Planning (Environmental Impact Assessment) (England and Wales) Regulations 1999 (or the equivalent legislation in Scotland and Northern Ireland). Typically, the EIA draws together, in a systematic way, an assessment of the project's likely significant environmental effects. This helps to ensure that the importance of the predicted effects, and the scope for reducing any adverse effects, are properly understood by the public and the relevant regulators. Dams and other installations designed to impound or permanently store water always require an EIA if the new or additional amount of water impounded or stored exceeds 10 million cubic metres. Smaller structures likely to have an adverse effect on sensitive sites also require an EIA.

Similar requirements for an EIA apply to major reservoir projects outside the UK, either under the relevant national law or to meet the requirements of international funding agencies.

5.28 RIVER INTAKES

The design of a river intake requires consideration of a number of issues including:

- any need for a structure to control the river levels, particularly at low flows;
- potential problems of saline intrusion if abstraction is from a river subjected to tidal influences;

- the exclusion of as much sediment and trash as possible and the potential need for de-silting basins;
- the provision or otherwise of pumping;
- any need for bankside storage to guard against closure of the intake in the event of a contamination spillage into the river upstream;
- the construction and operation of the facility built, as is almost certainly the case, in the flood plain of the river.

Intake designs must be chosen to suit the individual site, the characteristics of the river and the relative magnitudes of river flow and abstraction requirement. No hard and fast rules or criteria are possible—there are too many variables—and each intake must be designed individually. The comments below are intended purely as general guidance and to raise awareness of some of the issues.

The need for a river control structure needs careful assessment as it can be an expensive requirement. On a large river with a relatively small offtake requirement then the intake can probably be designed with sufficient depth at low river flows that the offtake supply can be assured even under severe drought conditions. This may involve training works to encourage low flows towards the intake. However, if the abstracted flows are a significant proportion of the river low flows then it is much more likely that a control structure will be needed. If possible, the intake should be sited to take advantage of any existing control structure but this may not be possible and the designer must then consider the type of structure required. The simplest option is a weir with fixed crest. Figure 5.13 shows such a structure with a gated intake upstream—but the 'backwater' effect on

FIGURE 5.13

Annalong River intake, Northern Ireland. (Engineer: Binnie & Partners).

water levels upstream should be checked. The alternative is a gated barrage, which allows better control of water levels and the opening of the gates under flood conditions to reduce any backwater effects. A gated barrage is, however, a costly solution and is likely to be justified only if the intake is supplying a major demand. In both cases, if the river is navigable, the costs of providing a lock to afford passage past a control structure could be prohibitive.

The design of a river control structure can be a major undertaking and is beyond the scope of this book. Some of the main design issues are the need for energy dissipation, the potential for scour downstream and how to limit and control that scour and, for a structure built on alluvial foundations, the development of potential uplift pressures under the structure, which affect not only the overall stability but also the local variation of loadings. The designer will be fortunate if the chosen location allows the structure to be founded on rock.

Intake designs have been developed for specific applications (Avery, 1989; Bouvard, 1992). On steep mountain streams, there are likely to be high sediment loads with the movement under flood conditions of gravels cobbles and even boulders. Figure 5.14 shows one intake arrangement designed for such conditions in Cyprus and hence known as the 'Cyprus intake'. Another type of intake for such conditions is the 'Tyrolean' or bottom rack intake in which the flow is abstracted through a bottom rack or screen into a channel running across the river below the bed. The concept is that cobbles, boulders and trash are carried right over the rack and the abstracted flow is relatively clean.

More commonly on lowland rivers the design of an intake with a river control structure is with the gated intake on one bank with a gate-controlled sediment flushing channel through the weir immediately in front of the offtake. This flushing arrangement may be in the form of a tunnel or culvert with the roof slab below the intake level so that the heavier sediment-laden bottom water is sluiced through the weir with the tunnel roof slab acting to separate the cleaner surface flow and allow it to be drawn off. The need to minimize sediment abstraction is important on many rivers with heavy sediment loads and much advice regarding the location and orientation of intakes is centred on the sediment movement associated with flow around bends. Spiral flow is set up naturally in such

FIGURE 5.14

Groyne intake for small abstraction on a 'flashy' river, Cyprus.

instances and tends to carry sediment across the bed towards the inside of the bend whilst the cleaner surface flow is directed towards the outside of the bend. If the designer is fortunate he may be able to position the intake at a point on the outside of a natural bend in the river. Abstracting surface flows at this location, with the intake gate facing at an angle upstream, will minimize sediment intake. This is also likely to be the location where the river is deepest and where low flows are concentrated. If this is not practicable then it may be possible to achieve curving flow approaching the intake artificially with river training works or with a suitable layout of the intake flushing channel. One interesting design concept (Avery, 1989) involves a tapered, curved flushing channel which enhances the movement of sediment away from the intake.

Figures 5.15 and 5.16 show designs for typical uncontrolled river intakes. That in Figure 5.15 is for a major intake on a large river. There are bankside screening facilities comprising both coarse screens and finer band screens. A low level tunnel feeds back to a deep pump sump with the pump house containing the electrical motors and associated equipment above maximum flood level. The second figure shows a similar concept but a much smaller intake with submersible pumps located in the same structure as the coarse screens. Similar designs with the inclusion of finer band or drum screens have also been employed. In the design illustrated a gravel trap has been included upstream of the pumps and provision needs to be made in the maintenance schedules for regular clearing of this area. The trap will collect the heavier material drawn in from the river but much of the sand-sized material will pass through to the pumps and consideration needs to be given with both designs as to whether a bankside settling basin should be included. The advantage of removing sand-sized material before pumping up to the treatment works is that it may be possible to flush the material back into the river; the flushing flows do not need to be pumped up to the same head but the arrangement may require double pumping. One arrangement that has been utilized on many intakes is to have a channel leading from the river to the intake set back from the bank and hence to form a forebay that can also act as a settling basin. On large intakes the removal of deposited sediment from this area has been achieved with a small floating dredger.

FIGURE 5.15

Intake on River Severn. (Engineer: Binnie & Partners).

FIGURE 5.16

A river side channel intake for a flow of 600 m³/h.

Infiltration galleries and bankside collector pipes (Section 4.16) allow abstraction of water from a river without the need for river control works or intake structures. In some cases it may be practical to suspend submersible pumps from a piled crib or jetty directly into the river if the local channel is deep and stable enough. A crib intake is potentially relatively cheap but it may need to be of sturdy construction if there is boat traffic on the river and, depending on the width of flood plain, the provision of access above flood levels may be costly. Other solutions adopted have included, on rivers with large water level variations, pontoon-mounted pumps with flexible connection to the bank. However, the flexible connection is a point of weakness and needs secure anchoring.

APPENDIX—DAM INCIDENTS

Dams are monumental structures, designed to rigorous standards and for a prolonged, some might say indefinite, life. Many thousands of dams have been built over the years, the majority in the last 100 or so years, and the vast majority have served society well. However, very occasionally failures have occurred and it is useful for practicing engineers to learn from such events as are described below.

Inundation of hillside materials increases their unit weight and decreases their cohesion, so that, if the water level in the reservoir should be rapidly lowered, the loosened wet material may slide into the reservoir. This happened in the Vaiont dam disaster of 9 October 1963 in Italy (Jaeger, 1965) when a landslide of gigantic proportions fell into the reservoir, causing a 100 m high floodwave to pass over the crest of the 206 m high arch dam causing the deaths of 3000 people in the valley below. The dam was not destroyed but the reservoir was afterwards abandoned.

The Dolgarrog disaster (ENR, 1926) in Wales on 25 November 1925 (the last to cause any loss of life in UK) was caused by continuing leakage below a low concrete wall only 3 m high that had been used to heighten the level of water in Lake Eigiau. At one point the wall had been taken only 0.5 m deep into clay foundations and leakage at this point so widened its passage that there was ultimately a sudden breakthrough of the lake waters. The wave of water destroyed another small dam below and engulfed the village of Dolgarrog, causing sixteen deaths.

An early failure in 1884 was that of the Bouzey Dam in France (ICE, 1896), a 19 m high masonry dam which moved 0.35 m downstream under the static force of the water. It was cemented back on its foundations, but 11 years later in 1895, it split horizontally about the middle. The 60 m high concrete St Francis Dam feeding Los Angeles failed on 13 March 1928 (ENR, 1928). It was placed on weak foundation material, so that a section of the dam broke out and a floodwave 40 m high travelled down the valley at 65 km/h, causing 426 lives to be lost.

The failure of the Malpasset Dam in France (Jaeger, 1963) on 2 December 1959 illustrates the catastrophic nature of an abutment failure. The arched concrete dam was 60 m high but only 6.5 m thick at the base. It was judged afterwards that, owing to rock joint pressurisation, the left abutment moved out ultimately as much as 2 m, causing rupture of the arch in a few seconds, instantaneously releasing the whole contents of the reservoir which engulfed the Riviera town of Fréjus 4 km downstream. The dam was almost completely swept away.

In Italy, the multiple-arch 43 m high Gleno Dam suddenly failed on 1 December 1923 (ENR, 1924) when one of the buttresses cracked and burst in a matter of a few minutes. The cause was attributed to poor quality workmanship: the concrete in the arches was poor and inadequately reinforced with scrap netting that had been used as hand grenade protection in the Great War, and there was evidence of lack of bond with the foundations.

The near failure of Lluest Wen dam in South Wales in 1970 (Twort, 1977) illustrates the dangers resulting from placing a culvert through the centre of an earth dam, even when the foundation is rock. The dam was built in 1896 in a coal mining area. A 'pillar' of un-mined seams of coal was left below the dam, but land settlement occurred outside this pillar and this, combined with the weight of the upstream drawoff tower, caused the plug of concrete in the culvert to fracture within the zone of the puddle-clay corewall. A small 150 mm drain pipe through the plug was fractured and seepage occurred into this, bringing with it clay from the corewall. After an unknown length of time, probably several years, a 2 m deep hole appeared in the crest of the dam 20 m above and, shortly afterwards, clay slurry began emerging from the drain pipe. The matter was rectified by lowering the reservoir and inserting a concrete cutoff wall through the clay corewall into bedrock below.

REFERENCES

Ackers, P. and Thomas, A. R. (1975). Design and operation of air-regulated siphons for reservoir and head-water control. *Symposium on design and operation of siphons and siphon spillways.* May. London. BHRA.

Armstrong, G., Aprahamian, M., Fewings, A. et al. (2004). *Fish passes—Guidance notes on the legislation, selection and approval of fish passes in England and Wales,* Version 1.1. EA.

Avery, P. (Ed.) (1989). *Sediment Control at Intakes: A Design Guide.* BHRA.

Besley, P., Allsop, N. W. H., Ackers, J. C. et al. (1999). Waves on reservoirs and their effects on dam protection. *Dams & Reservoirs*, Vol 9, No 3, December.

Binnie, G. M., Gerrard, R. T., Eldridge, J. G. et al. (1968). Mangla. Engineering of Mangla. *Proc. ICE*, 1967, 38 (Nov) 337–576 and 1968, 41 (Sept) 119–203. ICE.

Binnie, W. J. E. and Gourley, H. J. F. (1939). The Gorge Dam, *Journ.* ICE March p. 174.

Bishop, A. W. (1995). The use of the Slip Circle in the Stability Analysis of Slopes. *Geotecnique*, ICE. March p. 7.

Bouvard, M. (1992). *Mobile Barrages and Intakes on Sediment Transporting Rivers*, IAHR. Monograph Series. Balkema, Rotterdam.

Carpenter, T. G. (Ed.) (2001). *Environmental, Construction and Sustainable Development*, Vol 2, Sustainable Civil Engineering. John Wiley & Sons Ltd.

Charles, J. A, Abbiss, C. P., Gosschalk, E. M. and Hinks, J. L. (1991). *An Engineering Guide to Seismic Risk to Dams in UK*, ICE.

Cooke, J. B. (1990). Rockfill and Rockfill Dams, S*tan Wilson Memorial Lecture*, Univ. of Washington, USA.

CIRIA, CUR, CETMEF. (2007). *The rock manual—The use of rock in hydraulic engineering* (2nd edition). C683. CIRIA. London.

Davis, C. V. (1969). *Handbook of Applied Hydraulics*, McGraw-Hill.

Deacon, G. F. (1895–96). The Vyrnwy Works for the Water Supply to Liverpool, ICE *Trans* CXXVI Pt. IV pp. 24–125.

Denil, G. (1909). *Les echelles a poissons et leur application aux barrages de Meuse et d'Ourthe.* (Fish ladders and their use at the Meuse and Ourthe dams). Bull. Acad. Belg. pp. 1221–1224.

Dunnicliff, J. (1993). *Geotechnical Instrumentation for monitoring field performance*. John Wiley & Sons Inc. New York.

Dunstan, M. R. H. (1994). The state-of-the-art of RCC dams, *International Journal of Hydropower and Dams*. March.

ENR. (1924). *Engineering News Record*. 7 August, pp. 213–215.

ENR. (1926). *Engineering News Record*, 7 Jan 1926 and 25 Nov.

ENR. (1928). *Engineering News Record*, 29 March p. 517, and 12 April p. 596.

FEH. (1999). *Flood Estimation Handbook*, 5 Volumes, IoH.

Fleming, G. (1965). Design curves for suspended load estimation, *Proc. ICE*, 43, pp. 1–9.

Geddes, W. E., Rocke, S. and Scrimgeour, J. (1972). The Backwater Dam. *Proc. ICE*, 51, March p. 433.

Halcrow Water (2001). *Sedimentation in storage reservoirs*. Final Report. DETR. London.

Herbert, D. M. Lovenbury, H. T. Allsop, N. W. H. and Reader, R. A. (1995). *Performance of blockwork and slabbing protection for dam faces*. HR Wallingford report SR 345.

ICE (1896). The Failure of the Bouzey dam (Abstract of Commission Report). *Proc. ICE* CXXV, p. 461.

ICE. (1996). *Floods and Reservoir Safety: An Engineering Guide*, 3rd Edn. ICE.

ICE. (1998). *Application Note to 'An Engineering Guide to Seismic Risk to Dams in UK'*, ICE.

ICE. (2000). *A guide to the Reservoirs Act 1975*. ICE. Thomas Telford.

Ischy, E. and Glossop, R. (1962). An Introduction to Alluvial Grouting. *Proc. ICE* 21 March p. 449.

Jaeger, C. (1963). The Malpasset Report. *Water Power*, 15 Feb pp. 55–61.

Jaeger, C. (1965). The Vaiont Rock Slide, *Water Power*, 17 March pp. 110–111 and 18 April pp. 142–144.

Johnston, T. A. Millmore, J. P. Charles, J. A. and Tedd, P. (1999). *An Engineering Guide to the Safety of Embankment Dams in the United Kingdom*, Report 171, Building Research Establishment, 1990 new Edn.

Kenn, M. J. and Garrod, A. D. (1981). Cavitation Damage and the Tarbela Tunnel Collapse of 1974, *Proc. ICE* Feb. p. 65.

Kennard, M. F., Owens, C. C. and Reader, R. A. (1995). *An Engineering Guide to the Safety of Concrete and Masonry Dams in the UK*. Report 148, CIRIA.

Lariner, M. and Miralles, A. (1981). *Etude hydraulique des passes a ralentisseurs* (Hydraulic study of Denil fishways). Unpublished report, CEMAGREF.

Londe, P. and Lino, M. (1992). The faced symmetrical hardfill dam: a new concept for RCC, *Water Power and Dam Construction*. February.

MacDonald, D. E. and Scott, C. W. (2000). Design floods for UK reservoirs—personal view of current issues. *Dams 2000*. Proc. British Dam Society conference at University of Bath, July.

McIldowie, G. (1934–35). The Construction of the Silent Valley Reservoir. *Trans. ICE* 239, p. 465.

Morganstern, N. R. and Price, V. E. (1965). The analysis of the stability of general slip surfaces, *Geotechnique, Vol. 15*, pp. 79–93.

Morris, G. L. and Fan, J. H. (1998). *Reservoir sedimentation handbook: Design and management of dams, reservoirs, and watersheds for sustainable use*. McGraw-Hill. New York.

NERC. (1975). *Flood Studies Report*. NERC, London.

Rawlings, C. G., Hellawell, E. E. and Killkenny, W. M. (2000). *Grouting for ground engineering*. CIRIA C514, London.

Scott, C. W. (2000). Reservoir outlet works for water supply. *Proc. ICE*, Water & Maritime Engineering, Vol 142, December.

Sherard, J. L. and Dunnigan, L. P. (1989). Critical filters for impervious soils. *Journal for Geotechnical Engineering, ASCE*, 115(7), pp. 927–947.

Smith, N. (1971). *A history of dams*. Peter Davies. London.

Svensson, C., Jones, D. A., Dent, E. D. et al. (2006). *Review of Rainfall Frequency and Probable Maximum Precipitation Methods*. Report No. 2 of Defra project number WS194/2/39 *Reservoir Safety—Long Return Period Rainfall*. October.

Twort, A. C. (1977). The repair of Lluest Wen dam, *J IWES* July p. 269.

USACE (1987). Hydraulic Design Criteria. Morning Glory Spillways. HDC 140-1 to 140-1/8. Publication of the Headquarters. USACE.

USBR. (1977). *Design of Small Dams*, US Bureau of Reclamation, US Govt Printing Office, Washington.

USBR (1985). *Hydraulic Model Studies of Fuse Plug Embankments*. REC-ERC-85-7. Engineering and Research Center, Denver. USBR.

Vail, A. J. (1975). The High Island Water Scheme for Hong Kong, *Water Power & Dam Construction*, Jan. p. 15.

White, W. R. (2001). *Evacuation of sediment from reservoirs*. Thomas Telford.

W&WE. (1960).The Breaching of the Oros Dam, North East Brazil, *Water & Water Engineering*, August p. 351.

Yarde, A. J., Banyard, L. S. and Allsop, N. W. H. (1996). *Reservoir dams: wave conditions, wave overtopping and slab protection*. HR Wallingford Report SR 459.

Chemistry, Microbiology and Biology of Water

6.1 INTRODUCTION

This chapter is divided into six parts:

- Part I lists alphabetically 51 parameters or groups of parameters that make up some of the more usual physical and chemical characteristics of water and describes their significance. Some of these parameters constitute a risk to human health, others affect the aesthetic quality of the water supplied and others relate to treatment issues.
- Part II looks at the standards relating to the above parameters and their derivation. It also looks at the levels of monitoring and the analytical requirements.
- Part III looks at the microbiology of water and the most common waterborne diseases; the associated standards; and the testing of water for pathogenic organisms.
- Part IV looks at water biology in terms of the significance of macro organisms on water quality.
- Part V looks at new areas of concern in respect of drinking water quality.
- Part VI looks at Drinking Water Safety Plans, which are based on risk assessment and management of water supplies from source to tap.

The various Guideline Values and Standards associated with the various physical, chemical and microbiological parameters discussed in Parts I and III can be found in Tables 6.1(A)–(E) and 6.2.

PART I SIGNIFICANT CHEMICAL AND PHYSICO-CHEMICAL PARAMETERS IN WATER

6.2 ACIDITY

An acid water is one which has a pH value of less than 7.0 (Section 6.39). The acidity of many raw waters is due to natural constituents, such as dissolved carbon dioxide or organic acids derived from peat or soil humus. These are unlikely to lead to pH values much below 5.5. Some apparently unpolluted moorland waters may have pH values below 4.5 due to 'acid rain' which is formed when atmospheric sulphur dioxide, derived from the burning of fossil fuels, combines with water vapour to form dilute sulphuric acid.

Surface waters sometimes become contaminated with acidic industrial effluents. Acidic wastes from disused mines can also provide a significant source of acid contamination in some parts of the world. One of the worst incidents in the UK in recent years occurred in 1992 when 50 million litres of acidic, metal laden water was accidentally released from the Wheal Jane tin mine into the River Carnon in Cornwall.

Acidity is an important factor in water treatment, especially when optimizing the coagulation process and in ensuring that the treated water entering supply is non-corrosive.

6.3 ACRYLAMIDE

Polyacrylamides are used extensively in water treatment as coagulant aids, filter conditioners and in sludge treatment. There is a potential health risk from free residual acrylamide monomer in the treated water but this is controlled by dose and product specification rather than direct measurement. The maximum dose should not exceed 0.50 mg/l of the active ingredient and each batch should contain less than 0.02% of free acrylamide monomer, based on the active ingredient content. The presence of residual monomer in the recycled flows from sludge treatment has to be taken into account when calculating the overall monomer concentration in treated water.

6.4 ALGAL TOXINS

Algae are discussed in detail in Part IV.

Intense algal blooms can occur in lakes, reservoirs and rivers under certain conditions and it is possible that climate change will increase the frequency of such blooms unless catchment management reduces the potential for eutrophication (Section 6.75). As the plants die off their decay products tend to form scums along the margins of the water body. The scums associated with blue-green species of algae (commonly referred to as Cyanobacteria) contain toxins which, if concentrated by reason of the scum collecting in large masses at the shallow margins of a lake or reservoir, may prove fatal to fish or to animals drinking at the water's edge (Carmichael, 1993). Similar occurrences can also occur in the marine environment, with several beaches in Italy being closed in 2006 due to the levels of cyanotoxins in the sea water.

Some algae such as *Microcystis, Oscillatoria* and *Anabaena* produce hepatotoxins, the best studied of which is Microcystin LR. Other algae, including *Oscillatoria* and *Anabaena,* produce neurotoxins.

Activated carbon has been used successfully to remove the toxins from raw waters and ozone has also been shown to be effective at breaking down the toxins into less toxic by-products (Falconer, 1989; Lahti, 1989).

6.5 ALKALINITY

In a general sense 'alkalinity' is taken to mean the opposite of 'acidity' i.e. as the pH value increases (Section 6.39) alkalinity increases. More accurately, the alkalinity of a water principally comprises the sum of the bicarbonates, carbonates and hydroxides of calcium, magnesium, sodium and potassium. Calcium and magnesium bicarbonates predominate in waters that are associated with chalk or limestone and comprise the temporary hardness of a water (Section 6.30). Where the alkalinity is less than the total hardness, the excess hardness is termed permanent hardness. Conversely where the alkalinity is greater than the total hardness, the excess alkalinity is usually due to the presence of sodium bicarbonate, which does not affect the hardness of the water. Because bicarbonate ions can exist at pH values below pH 7.0, a measurable alkalinity is still obtained with 'acidic' waters down to pH values of 4.5.

Alkalinity provides a buffering effect on pH, which is an important factor in many water treatment processes. It is also a key factor in determining the corrosive nature of a water.

6.6 ALUMINIUM

Aluminium can occur in detectable amounts in many natural waters as a consequence of leaching from the sub strata. It is also found in the run-off from newly afforested areas. However the most usual source of aluminium in public water supplies comes from incomplete removal of aluminium based coagulants during treatment (Section 7.21). It should be possible, by taking a holistic approach to optimising the treatment processes, to keep the residual aluminium in water going into supply after clarification and filtration to below 0.05 mg/l (or 50 μg/l) as Al. A percentage of this aluminium is still likely to settle out in the distribution system and will tend to accumulate as flocculant material, especially in areas where the flow is low. Inadequate treatment or deficiencies in process control can result in much higher concentrations of aluminium in the water leaving the treatment works. This is also likely to deposit as flocculant material within the distribution system and any disturbance of these sediments, either through flow reversals or changes in flow, may result in consumer complaints of 'dirty' water.

Standards relating to aluminium are based on the aesthetic quality of the water supplied. However, public concerns have been expressed about the possible neurotoxic effects of aluminium in drinking water. It has been established that the aluminium content of water used for renal dialysis should be no greater than 0.01 mg/l to avoid neurological problems in dialysis patients. Concerns have also been expressed that aluminium in drinking water might be a risk factor in the development and acceleration of the onset of Alzheimer's disease. This was initially examined in 1985 by the then Committee on Medical Aspects of the Contamination of Air, Soil and Water (CASW) at the UK Department of Health, without establishing a relationship. Further studies were carried out following the Lowermoor incident in 1988, when 20 tonnes of aluminium sulphate were inadvertently placed in the treated water tank at an unmanned treatment works in Cornwall. This resulted in contaminated water being supplied to a large number of consumers in the surrounding area. The findings of the most recent of these studies, carried out by a sub group of the Committee on Toxicity of Chemicals in Food, Consumer Products and the Environment at the Department of Health, were reported in February 2005 (DoH, 2005). The study concluded that it was unlikely that there would have been any persistent or delayed harm to health from any of the contaminants involved in the incident, but that further studies were needed as a precautionary measure including studies into the toxicity of aluminium.

6.7 AMMONIACAL COMPOUNDS

Ammonia is one of the forms of nitrogen found in water (see also Section 6.36). It exists in water as ammonium hydroxide (NH_4OH) or as the ammonium ion (NH_4^+), depending on the pH value, and is usually expressed in terms of mg/l 'free' ammonia (or 'free and saline ammonia'). 'Albuminoid ammonia' relates to the additional fraction of ammonia liberated from any organic material present in the water by strong chemical oxidation. 'Kjeldahl nitrogen' is a measure of the total concentration of inorganic and organic nitrogen present in water.

Ammoniacal compounds are found in most natural waters. They originate from various sources, the most important being decomposing plant and animal matter. Increased levels of free ammonia in surface waters may be an indicator of recent pollution by sewage, agriculture or industrial effluent. However, some deep borehole waters, which are of excellent organic quality, may contain high levels of ammonia as a result of the biological reduction of nitrates. The source of any substantial amount of ammonia in a raw water should always be investigated, especially if it is associated with excessive bacterial pollution.

The level of free ammonia in a raw water is important in determining the chlorine dose required for disinfection. Chlorine reacts with any ammonia present in the water to form monochloramine, dichloramine and finally nitrogen dichloride. This reaction has to be completed before a free chlorine residual can be achieved (Section 11.8). Therefore, if the ammonia concentrations create a high chlorine demand, ammonia removal may have to be considered (Sections 10.28 and 10.29).

There is no health risk associated with the levels of ammonia typically found in drinking water, although the water may become unacceptable on taste and odour grounds if high concentrations of dichloramine or nitrogen trichloride are allowed to develop during chlorination.

6.8 ANTIMONY

Trace amounts of antimony may be found in some raw waters but the main source in drinking water is likely to be from the corrosion of brass tap fittings and solders.

6.9 ARSENIC

Arsenic is toxic to humans and, if detected in water, the origins should always be investigated. It is a natural water contaminant in many parts of the world, particularly in areas of geothermal activity. Arsenic can also be found in surface runoff from mining waste tips or in areas where there are certain types of metalliferous ore. Its presence in water may also be the result of pollution from weedkillers and pesticides containing arsenic. Significant health problems have been caused by high levels of arsenic in the drinking water supplies in parts of Bangladesh since the 1990s (Smith, 2000). Arsenic has also been found in a number of groundwater sources in the UK, with some water companies having to install treatment.

6.10 ASBESTOS

The widespread use of asbestos cement pipes in distribution systems has raised concerns that the fibres found in water may be a danger to health. A study carried out in the UK found that some drinking waters can contain up to 1 million asbestos fibres per litre, with more than 95% of these being less than 2 microns in length (WRc, 1984). The UK Department of Health CASW Committee reported in 1986 that there was substantial evidence to show that asbestos, as found in drinking water, did not represent a hazard to health (DoE, 1990a). WHO likewise considers that there is no consistent evidence to show that ingested asbestos is hazardous to health. Few countries continue to install asbestos cement pipes, mainly because of the inhalation risks associated with working with the dry material. In 2000 the use of new asbestos cement pipes was banned in the UK because of this health and safety risk.

6.11 BIOCHEMICAL OXYGEN DEMAND (BOD)

The BOD test gives an indication of the oxygen required to degrade biochemically any organic matter present in a water, as well as the oxygen needed to oxidize inorganic materials such as sulphides. The test provides an empirical comparison of the relative oxygen requirements of surface waters, wastewaters and effluents. For example, if a sewage effluent with a high BOD is discharged into a stream, the oxygen required by organisms to break down the organic matter in the effluent is taken from the overall oxygen content of the receiving water. This depletion could potentially destroy fish and plant life.

Nearly 100 years ago the UK Royal Commission on Sewage Disposal observed that unpolluted rivers rarely had BOD values of more than 2 mg/l and could accept added pollution up to a total BOD value of 4 mg/l with no apparent detriment. This implies a maximum permissible BOD value of 20 mg/l in a sewage effluent entering a watercourse, assuming a 8:1 dilution with freshwater. Today discharge standards are related to the quality of the receiving waters and could be more stringent (Sections 2.11 and 2.12).

6.12 BORON

The main sources of boron in the aquatic environment are from industrial discharges or detergents in treated sewage effluents. The concentrations found in drinking water tend to be very low. Seawater contains 4 to 6 mg/l of boron and is not normally rejected in reverse osmosis desalination unless the process is operated at high pH values (Section 10.47).

6.13 BROMIDE AND IODIDE

Seawater contains 50 to 60 mg/l bromide, so the presence of bromide in well or borehole sources near the coast could be evidence of seawater intrusion. Ozone treatment of water containing bromide may result in the formation of bromate as a disinfection by-product (Section 6.25).

Many natural waters contain trace amounts of iodide, usually at levels less than 10 μg/l, although higher concentrations are found in brines and brackish water. The levels found in drinking water are too low to contribute significantly to dietary requirements.

6.14 CADMIUM

The main sources of cadmium in the aquatic environment are from discharges or diffuse pollution. It is used for electroplating of metal fastenings and can be an impurity in the zinc used to galvanize pipes or other metal fittings. It is occasionally detected in drinking water.

6.15 CALCIUM

Calcium is found in most waters, the level depending on the type of rock through which the water has permeated. It is usually present as calcium carbonate or bicarbonate, especially in waters that are associated with chalk or limestone, and as calcium sulphate. Calcium chloride and nitrate may also be found in waters of higher salinity. Calcium bicarbonate forms temporary hardness; the other salts being linked to the permanent hardness (Section 6.30).

Calcium is an essential part of human diet, but the nutritional value from water is likely to be minimal when compared to the intake from food. The main limitation of calcium in drinking water is the potential for excessive scale formation.

6.16 CARBON DIOXIDE

Free carbon dioxide in a water (as distinct from that existing as carbonate and bicarbonate) depends on the alkalinity and pH value of the water. It is an important factor in determining the corrosive properties of a water (Section 10.41).

Surface waters usually contain less than 10 mg/l free CO_2 but some groundwaters from deep boreholes may contain more than 100 mg/l. A simple means of reducing free CO_2 in a water is by aeration (Section 10.18).

6.17 CHLORIDE

Chloride is found in nearly all waters and comes from a number of sources, including natural mineral deposits; seawater intrusion or airborne sea spray; agricultural or irrigation discharges; urban runoff due to the use of deicing salts; or from sewage and industrial effluents. It is usually combined with sodium and to a lesser extent with potassium, calcium and magnesium, which makes chloride one of the most stable components in water.

Most rivers and lakes have chloride concentrations of less than 50 mg/l Cl and any marked increase may be indicative of sewage pollution or, if the increase is seasonal, urban runoff. The chloride content of a sewage effluent under dry weather flow could increase the chloride content of the receiving water by as much as 70 mg/l. Brackish or estuarine waters can contain several hundred mg/l of chloride, with seawater typically containing around 20 000 mg/l.

Excessive chlorides can give rise to corrosion and also to taste problems. High concentrations of chloride tend to enhance the corrosion rates of iron, steel, and plumbing metals, especially when coupled with low alkalinity. A sensitive palate can detect chlorides in drinking water at as low a level as 150 mg/l and concentrations above 250 mg/l may impart a distinctly salty taste (Section 6.50). However in arid or semi- arid areas, people may have to drink water containing much higher levels of chloride if no alternative supply is available.

Conventional water treatment processes do not remove chlorides. If the chloride content of a water has to be reduced then some form of desalination has to be applied (Section 10.44).

6.18 CHLORINATED HYDROCARBONS

The term chlorinated hydrocarbons covers a wide range of volatile organic chemicals used as solvents, metal cleaners, paint thinners, dry cleaning fluids etc. and in the synthesis of other organic chemicals. They tend to be found as micro contaminants in groundwaters that have been subjected to industrial pollution and can remain long after the original source of pollution has been removed. Some may also be found in surface waters as a result of effluent discharges.

The three most common groups to be detected in drinking water are the chlorinated alkanes, which include tetrachloromethane (or carbon tetrachloride); the chlorinated ethenes which include trichloroethene (TCE) and tetrachloroethene (PCE); and the chlorinated benzenes. Many chlorinated solvents are considered to be potential carcinogens and some, such as TCE and PCE, may degrade in anaerobic groundwaters to produce more toxic substances such as vinyl chloride. There is little commonality between the recommended standards and, in the case of the chlorinated benzenes, the taste and/or odour thresholds are likely to be well below any health risk levels.

6.19 CHLORINE RESIDUAL

Chlorine remains the principal biocide and disinfectant used in water treatment in most countries (Chapter 11). Chlorine gas tends to be used at major water treatment works, although many works in urban areas now use on site generation by electrolysis of salt on health and safety grounds. Chlorine containing powders, crystals and solutions tend to be used at smaller works or at sites where the procurement of chlorine gas presents difficulties. The effectiveness of chlorine as a disinfectant arises from its high chemical reactivity. However, it is recognized that the use of chlorine for disinfection can result in the formation of a wide range of undesirable chlorinated compounds, known collectively as disinfection by-products (Section 6.25). Although usually found at very low concentrations, some of these compounds are potentially hazardous to health; others produce objectionable tastes and odours. Notwithstanding these problems, which can be minimized by appropriate treatment control (Section 11.7), the use of a suitable disinfectant such as chlorine is essential for ensuring that a water is bacteriologically safe to drink, unless other reliable means of disinfection (e.g. boiling) can be used.

There is no evidence that the levels of chlorine residual normally found in drinking water are harmful to health. Most consumers become familiar with the levels that are normal for their local water supply. However sudden increases are likely to be noticed and generate taste complaints, particularly from consumers with sensitive palates. Taste and odour complaints can also arise from the reaction of chlorine with other trace substances present in the water (Section 11.11) or with certain plumbing materials (Section 6.50).

WHO recommends that there should be a free residual chlorine of ≥0.5 mg/l after at least 30 minutes contact time at a pH of less than 8.0 to achieve effective disinfection. The UK Regulations require that all water leaving a water treatment works be disinfected, without specifying the type of disinfectant or level of residual. In the USA, public water supplies have to comply with the Stage 1 Disinfectants/Disinfection By-products Rule.

6.20 COLOUR

The colour of a water is usually expressed in Hazen units, which are the same as TCU (true colour units) or mg/l on the platinum cobalt (Pt-Co) scale. Water can often appear coloured because of colloidal or other material present in suspension, which means that true colour must be determined only after filtration, usually through a 0.45 micron filter. The colour in unpolluted surface waters is caused by the presence of humic and fulvic acids, which are derived from peat and soil humus. In some waters the brown colour is enhanced by the presence of iron and manganese, which is often organically bound. Waters subject to industrial pollution may also contain a wide variety of coloured materials.

The level at which colour becomes unacceptable depends largely upon consumer perception, with most consumers noticing a colour of 15 TCU in a glass of water.

6.21 COPPER

Copper is rarely found in unpolluted waters, although trace amounts can sometimes be found in very soft, acid moorland waters. The most usual source of copper in drinking water is from corrosion of copper and copper containing alloys used in domestic plumbing systems.

Newly installed copper pipework may give rise to 'blue' water, especially if there is prolonged stagnation in the domestic distribution system. Water containing as little as 1 mg/l of copper can cause blue/green stains on sanitary fittings. Much lower concentrations can accelerate corrosion of other metals in the same system (Cambell, 1970) making it inadvisable to use galvanized steel piping or storage tanks downstream of copper piping.

Although copper is an essential element in the human diet, concentrations above 2.5 mg/l can impart an unpleasant and astringent taste to the water and some individuals may suffer acute gastric irritation at concentrations above 3 mg/l.

6.22 CORROSIVE QUALITY

Ideally a treated water entering supply should not be corrosive or aggressive to plumbing materials or to concrete. Many factors determine whether a water will be corrosive but three specific characteristics are important:

- a low pH value, i.e. acidity;
- a high free carbon dioxide (CO_2) content; and
- an absence or low amount of alkalinity.

Free chlorine residual is also a factor in the dezincification of some brasses (those with more than 15% zinc).

Waters that tend to be corrosive include soft moorland waters; shallow well waters of low pH with low temporary hardness but high permanent hardness; waters from greensands and other iron bearing formations; chalk and limestone waters with a high CO_2 content; waters from coal measures; and waters with a high chloride content. Desalinated water is also very corrosive unless suitably treated.

One measure of assessing the corrosive nature of a water is by placing the water in contact with powdered chalk or marble for a defined period. The final pH value is referred to as the stability or saturation pH (pHs) and if this is greater than the initial pH value (i.e. a negative value), the water is likely to be corrosive to iron, steel and cement. If the pHs value is less than the initial pH value, then the water is likely to deposit calcium carbonate as a protective layer on the interior of metal pipework. If the two pH values are the same, the water is said to be in equilibrium. The stability pH value can also be calculated using the Langelier formula (Section 10.41) (Langelier, 1936). There are also a number of other indices with wider capabilities that can be used to calculate the corrosion potential of a water.

Another aspect of corrosion potential is whether a water will dissolve lead or copper (i.e. be plumbo-solvent or cupro-solvent) (Section 6.33). Such waters may require pH and alkalinity adjustment as well as the addition of orthophosphate or silicate, either individually, or in various combinations, to reduce the corrosion potential.

6.23 CYANIDE

Cyanide and cyanide complexes are only found in waters polluted by effluents from industrial or mining processes involving use of cyanide. Most cyanides are biodegradable and should be removed by chemical treatment before an effluent is discharged into a receiving water. Chlorination to a free chlorine residual under neutral or alkaline conditions effectively decomposes any remaining cyanide that may be present in a raw water.

6.24 DETERGENTS

There are several substances that can cause foaming in water, the largest group being synthetic detergents or surfactants. In recent years the increased use of biodegradable detergents, which can be removed by normal sewage purification processes, has led to a reduction in the residual detergent discharged in sewage effluent. The main limitation on detergents is to prevent foaming in drinking water, although some components of anionic surfactants are toxic to aquatic life. Methylene blue active substances (MBAS) relate to the more common anionic surfactants found in detergents, which react with methylene blue. Many surface waters downstream of urban areas contain detergent.

6.25 DISINFECTION BY-PRODUCTS

Chemical oxidants, such as chlorine and ozone, are traditionally used as disinfectants to control pathogenic organisms present in raw waters (Chapter 11). They are also used to assist in the oxidation of iron and manganese and to break down taste and odour forming compounds. Ozone is used extensively to break down pesticides. Side reactions with organic and inorganic constituents present in a raw water can give rise to low concentrations of a number of compounds, collectively known as disinfection by-products or DBPs. In the case of organic constituents, the rate of DBP formation can be significantly reduced by optimising the treatment process to remove or substantially reduce these precursors prior to the final disinfection stage. However the efficacy of the disinfection process must never be compromised by the need to reduce DBP levels. In 1991 a major cholera epidemic occurred in Peru after water officials were put under pressure to reduce chlorination because of a perceived hypothetical cancer risk from chlorination by-products.

Over the last 25–30 years much research has been carried out into the significance and control of DPBs, especially those associated with chlorination. The more commonly monitored DBPs are discussed in more detail below.

Bromate. The presence of bromate in drinking water mainly tends to be associated with disinfection. However, early in 2000, a UK water company carrying out checks in advance of the new Drinking Water Directive requirements for bromate found unacceptable levels at one of its groundwater sources. The site of a former chemical works was identified as the source of contamination. The affected source had to be taken out of supply and there were increasing concerns for other sources in the area as the plume of contamination moved across the aquifer.

In terms of DBP formation, ozone reacts with any bromide ion present in a raw water to form bromate (DoE, 1993a). The rate of bromate formation in drinking water depends on the amount of natural organic matter present in the water, the alkalinity and the ozone dose applied (Siddiqui, 1995). Bromate may also be present in commercially available sodium hypochlorite or be formed during on site electrolytic generation of sodium hypochlorite (Section 11.15) if bromide is present in the brine. The concentrations found in hypochlorite solution can range from 2.8 to more than 20 mg/l (DoE, 1993b). In March 2004 the producer of a bottled table water in the UK had to withdraw supplies because of elevated bromate levels. The bottled water was produced from tap water which was first desalinated. Calcium chloride was then added to meet a minimum standard for calcium and the product water disinfected with ozone. Traces of bromide present in the calcium chloride were oxidized to bromate, giving concentrations of up to 22 µg/l.

Chloramines. The process of chloramination involves dosing a controlled amount of ammonia to chlorinated water (Section 11.8). The resultant monochloramine has disinfectant properties. Although less powerful than free chlorine, it provides a more stable residual within the distribution system. Dichloramine and nitrogen trichloride can also be formed if the process is poorly controlled, with resultant taste and odour problems. Similar problems arise if chloraminated water is mixed with water containing a free chlorine residual.

Chlorate and chlorite. Although not strictly DBPs, chlorate and chlorite can be found in water that has been treated with chlorine dioxide (Section 11.17). Chlorate, like bromate, may also be present in commercially available sodium hypochlorite and in sodium hypochlorite produced by electrolysis.

The UK Regulations specify a maximum concentration of chlorate of 0.7 mg/l as ClO_3 in the treated water as a condition of use for plants producing sodium hypochlorite by electrolysis. The UK Regulations also apply British Standards BS EN criteria for the use of chlorine dioxide and sodium chlorite, such that the combined concentration of chlorine dioxide, chlorite and chlorate must not exceed 0.5 mg/l as chlorine dioxide in the water entering supply.

Chlorophenols may be formed from the chlorination of trace levels of phenolic compounds present in the raw water or as degradation products from the breakdown of phenoxy acid herbicides. They have very low organoleptic thresholds and, if present, are immediately noticed by consumers as an antiseptic taste. The taste thresholds for the most commonly found chlorophenols in drinking water are well below any health related value.

Haloacetic acids (chloro- and bromoacetic acids) and haloacetates may be formed when surface waters are chlorinated, depending on the organic precursors present. Trichloroacetic acid can be used as a herbicide and has been found, in this context, in detectable concentrations in some raw waters.

Halogenated acetonitriles Chloro- and bromo-halogenated acetonitriles may be formed when surface waters containing algae are chlorinated or chloraminated. Dichloroacetonitrile is the most commonly found.

Sodium Dichloroisocyanurate Sodium dichloroisocyanurate, which is widely used as a disinfectant for swimming pool water, can also be used as an emergency disinfectant of drinking water. It may also be used as a source of chlorine for point-of-use water treatment. Although not strictly a DBP, WHO sets guideline values for both residual sodium dichloroisocyanurate and associated cyanuric acid.

Trihalomethanes (THMs) may be formed when surface waters containing naturally occurring organic compounds, such as humic and fulvic acids, are chlorinated. They are also formed by the reaction of chlorine with some algal derivatives (Section 11.7). The most commonly found members of the group are trichloromethane or chloroform, bromodichloromethane, dibromochloro-methane, and tribromomethane or bromoform; and they are best controlled by removing as much of the organic precursors as possible before the water is chlorinated. Some epidemiological studies have suggested weak correlation between incidences of certain cancers with consumption of water containing THMs, but the level of exposure has not always been adequately characterized and, in some cases, no account has been taken of other cancer risk factors or other routes of exposure. Likewise recent studies in the US claim a correlation between THM intake and early pregnancy terminations and low birth weight. A study into the relationship of THM concentrations in public water supplies to stillbirth and birth weight in three water regions in England suggested a significant association of stillbirths with maternal residence in areas with high total THM exposure but it also concluded that further research was needed (Toledano, 2005).

WHO sets guideline values for each of the four individual THMs and uses a fractionation approach (the sum of the ratios of the concentration of each to its respective guideline value should be equal to or less than 1) to derive a standard for total THMs.

Nitrites (see also Section 6.36) can be formed as a DBP where ultra violet radiation is used to disinfect waters containing moderate to high concentrations of nitrate (Section 11.27). Undesirable levels of nitrite can also be formed at the end of long and complex distribution systems where disinfection by chloramination is practiced. (Section 11.8)

6.26 ELECTRICAL CONDUCTIVITY AND DISSOLVED SOLIDS

Conductivity measures the ability of a solution to carry electrical current. This depends on the presence of ions in solution and therefore provides a useful indication of the total dissolved solids, or salts, present in a water. For most waters a factor in the range 0.55–0.70 multiplied by the conductivity gives a close approximation to the dissolved solids in mg/l. The factor may be lower than 0.55 for waters containing free acid and greater than 0.70 for highly saline waters. Conductivity is temperature dependent and a reference temperature (usually 20 °C or 25 °C) is used when expressing the result. One of the advantages of the conductivity determination is that it can be easily measured in the field or used in continuous monitoring.

High levels of dissolved salts can result in taste complaints, as well as causing excessive scaling to domestic and industrial water systems. Water with low dissolved salts is desirable for many industrial processes but may be unacceptable to consumers, again on taste grounds, and may also be corrosive to domestic plumbing.

Water with a dissolved solids content of more than 1000 mg/l could be unpalatable, although many natural mineral waters exceed this level.

6.27 ENDOCRINE DISRUPTING SUBSTANCES

It has been long recognized that exposure to certain chemicals such as polychlorinated biphenyls, tributyl tin, certain pesticide residues and some pharmaceutical residues can have an adverse effect on the endocrine (or reproductive) system of animals in the aquatic environment. Some of these substances are already subject to environmental quality standards. The full impact of others, including steroids such as oestrogen, are still being researched. Most of these substances enter the aquatic environment through industrial discharges and sewage effluents, or via diffuse pollution.

The UK EA has investigated the impact of synthetic and natural oestrogen residues on the aquatic environment as part of its environmental strategy especially in terms of the feminisation of male fish (EA, 1998).

The impact of such residues tends to be more of an environmental issue than a drinking water one. There is usually significant dilution before downstream water is abstracted for drinking water and the advanced treatment processes installed for pesticide removal are effective at removing any remaining residues. Traces of oestrogens can also be removed by chlorination at concentrations typically used for final disinfection.

6.28 EPICHLOROHYDRIN

Epichlorohydrin is a chemical intermediate used in the manufacture of polyamine coagulants and also in some ion exchange resins used in water treatment and softening. Traces may be found in drinking water but concentrations tend to be controlled by product specification and use.

6.29 FLUORIDE

Naturally occurring fluoride is found in varying concentrations in most drinking waters. Some deep groundwaters in the UK contain between 2 and 5 mg/l of fluoride and much higher concentrations are found in other parts of the world, especially in areas associated with fluoride-containing minerals.

The levels of fluoride in drinking water have to be closely controlled as excessive amounts can lead to fluorosis, with resultant mottling of the teeth, and in extreme cases even skeletal damage. The maximum concentration also has to be related to climatic conditions and the amount of water likely to be consumed. Specialized treatment has to be used to remove excess fluoride from a water but it is costly to operate (Section 10.16). Blending is always the preferred option provided there is sufficient low fluoride water available.

Low levels of fluoride are dosed to public supplies in the UK (MoH, 1969), the USA, Australia and elsewhere as an effective means of reducing dental caries. The decision to fluoridate water supplies has to be taken by the relevant health authority and must be evidence based. The greatest reduction of dental decay occurs if fluoridated water is drunk in childhood during the period of tooth formation. Although consumers accept the presence of naturally occurring fluoride in the drinking water, many object to the deliberate addition of fluoride to public water supplies in spite of the stringent controls that are attached to the process.

6.30 HARDNESS

The original definition of 'hardness' was the extent of scum or curd formation when a typical hard water reacts with ordinary soap. Hardness also relates to the scale that precipitates in kettles and utensils when water is boiled. The precipitate formed on heating is the temporary, or carbonate, hardness consisting of the bicarbonates of calcium and magnesium. Permanent, or non-carbonate, hardness (which is not precipitated by heating) is due to other salts of calcium and magnesium present in the water, usually in lesser quantities than the bicarbonates. Hardness, like alkalinity

(Section 6.5), is usually expressed in mg/l as $CaCO_3$ although other notations, such as French or German degrees, may be found in association with domestic appliances such as dishwashers.

The descriptive terms commonly applied are as follows:

Hardness description	Hardness as $CaCO_3$ mg/l
Soft	0–50
Moderately soft	50–100
Slightly hard	100–150
Moderately hard	150–200
Hard	>200
Very hard	>300

Problems caused by excessive hardness mainly relate to the formation of scale in boilers and hot water systems. Consumers in hard water areas also complain of scale deposition on kitchen utensils and increased soap usage, with associated scum formation. Conversely, waters containing less than 30–50 mg/l total hardness tend to be corrosive and may need additional treatment to reduce the risk of plumbo- and cupro-solvency. Desalinated water has virtually zero hardness and is highly corrosive, requiring treatment to render it non-aggressive to metallic plumbing materials.

A number of studies have been carried out into relationships between the hardness of water and the incidence of cardiovascular disease. In 2005 the UK Defra commissioned a literature review, which concluded that there was evidence that hard water gave a protective effect (Defra, 2005). However, many other characteristics vary with the hardness of a water and so it is not clear whether the relationship observed might be due to a protective factor in hard water or a harmful factor in soft water. Very few UK water companies now soften public water supplies following advice issued by the DHSS (1971).

6.31 HYDROCARBONS

Hydrocarbons include petroleum, mineral oils, coal and coal tar products, and many of their derivatives such as benzene and styrene, which are produced by the petro-chemical industry for industrial processes. In oil producing parts of the world, background concentrations of hydrocarbons may be present in surface and groundwaters from natural sources. However their presence is more usually the result of pollution.

On rare occasions, traces of benzene and other solvents have been reported in domestic water supplies. This has usually occurred where there has been localized permeation of plastic water pipes from contaminated ground conditions or other external sources of hydrocarbons such as petroleum spillages.

A number of hydrocarbons have health related standards, but most are well above the taste or odour thresholds.

6.32 IRON

Iron is found in most natural waters and can be present in true solution, or in suspension as a colloid, or as a complex with other mineral or organic substances. Iron in surface waters is usually in the ferric (Fe^{3+}) form, but the more soluble ferrous (Fe^{2+}) form can be found in deoxygenated conditions that may occur in some deep boreholes or in the bottom waters of lakes and reservoirs. On exposure to air, such waters rapidly become discoloured as the iron oxidizes to the ferric form and precipitates out. Severe corrosion can occur where unlined iron is used in rising mains and pumps at boreholes that contain aggressive water. This gives rise to very iron rich water, especially when there are long periods of stagnation between pumpings.

Iron salts are extensively used as coagulants in water treatment (Section 7.22). With good process control, it should be possible to keep the residual concentration of iron in the water entering supply to less than 0.05 mg/l. However incomplete removal during treatment can lead over a period of time to significant deposits of iron in the distribution system. Corrosion products from unlined cast iron mains also contribute to the build up of ferruginous material in a distribution system. Regular flushing, providing it is adequately controlled, can help to control this build up of deposits which otherwise gives rise to discoloured water problems when disturbed. The deposits can also contribute to the growth of iron bacteria, which in turn can cause further water quality deterioration by producing slimes or objectionable odours.

Iron is an essential element in the human diet. There is no health related standard for iron in drinking water but consumers are likely to reject discoloured water on the grounds of appearance. Excess iron can cause brown stains on laundry and plumbing fixtures and can also impart a bitter taste to the water when present above 1 mg/l.

6.33 LEAD

Lead is a cumulative poison and the hazards of exposure to lead in the environment have been well documented over the years. Recently there has been growing concern about the possibility that quite low levels of exposure can affect learning ability and behavioural problems in children. With the decrease in levels of lead in atmospheric emissions and with reduced use of paint containing lead, lead in drinking water is now considered to be the main source of controllable exposure.

Lead is rarely found in detectable concentrations in natural waters, except in areas where soft acidic waters come into contact with galena or other lead ores. The main source of lead in drinking water is the dissolution of lead service pipes and internal domestic plumbing which may still be present in older properties in the UK and elsewhere. Although plumbosolvency tends to be associated with soft, acidic waters from upland and moorland catchments, it can also occur in hard water areas, especially where the hardness is mainly non-carbonate. Thus the concentrations of lead in drinking water should always be monitored closely in areas where lead pipes are known to be in use.

Unacceptable levels of lead have also been found with some new unplasticized PVC pipes, where lead compounds used as stabilizers in the pipe manufacture are leached out. Although the use of solder with a high lead content for copper pipe joints is now banned in the UK, occasional high levels of lead are still found where the wrong type of solder has been used.

Treatment for plumbosolvency (Section 10.14) by stabilizing the final water pH value or orthophosphate dosing has been shown to be very effective in many parts of the UK and USA. Since 1990 in England and Wales there has been a gradual but significant increase in the number of samples taken with lead concentrations of less than 10 µg/l. In 2006 97.9% of regulatory samples taken contained less than 10 µg/l of lead and 99.7% of samples contained less than 25 µg/l (DWI, 2006). Although these results are very promising, it is recognized that in some areas the remaining failures are due to particulate lead, which sloughs off the interior of the pipe. In such situations the only viable options are pipe replacement or, if the pipe is still structurally sound, possibly lining with a thin plastic liner. Lead pipe replacement is the only permanent solution but this may be difficult where the water company does not own the pipework. In the UK the owner of the property is responsible for the part of the service pipe within the curtilage of the property and for internal domestic plumbing. However, under the UK Regulations, if the property owner replaces his part of the pipework, then the water company is required to replace its part of the lead service pipe.

6.34 MANGANESE

Manganese can be found in detectable concentrations in both surface and groundwaters. The concentration of manganese in solution rarely exceeds 1.0 mg/l in a well aerated surface water but much higher concentrations can occur in groundwaters subject to anaerobic conditions. Manganese can re-dissolve from the bottom sediments in impounding reservoirs if the bottom water becomes deoxygenated. This leads to an increase in the overall manganese content of the water when the reservoir water 'turns over' (Section 7.1).

Manganese is an essential element in the human diet but levels as low as 0.1 mg/l can cause staining of laundry and sanitary ware. It is also undesirable even in small quantities in water supplies as it can precipitate out in the presence of oxygen or after chlorination. The precipitate forms a black slime on internal surfaces in the distribution system and, if disturbed, gives rise to justifiable consumer complaints. The tolerable concentrations of manganese in a distribution system are generally lower than those for iron for although deposition of manganese is slow, it is continuous. Thus the onset of serious trouble may not become apparent for some 10 to 15 years after putting a manganese rich supply into service.

6.35 NICKEL

Nickel occurs widely in the environment and is occasionally found in drinking water. Increased concentrations may occur when the plating on nickel- or chrome-plated taps becomes damaged or starts to break down. Increased concentrations may also be found in the water boiled in electric kettles with exposed elements. Nickel is known to cause skin sensitisation and the ingestion of high levels via food or drinking water may cause dermatitis in some, but not all, nickel sensitive individuals.

6.36 NITRATE AND NITRITE

These parameters are considered together because both are oxidation states of ammonia and conversion from one form to the other can occur naturally within the aquatic environment.

Nitrate

Nitrate is the final stage of oxidation of ammonia and the mineralisation of nitrogen from organic matter. Most of this oxidation in soil and water is achieved by nitrifying bacteria and can only occur in a well oxygenated environment. The same bacteria are also active in percolating filters at sewage treatment works, resulting in large amounts of nitrate being discharged in sewage effluents from such works. The use of nitrogenous fertilizers on the land can also give rise to increased nitrate concentrations in both surface and underground waters. Nitrate levels in surface waters often show marked seasonal fluctuations, with higher concentrations occurring in winter when runoff increases due to winter rains at a time of reduced biological activity. During summer the nitrate levels are likely to be reduced by biochemical mechanisms and by algal assimilation in reservoirs. Bacterial denitrification and anaerobic reduction to nitrogen at the mud interface can, in addition, substantially reduce the nitrate levels in reservoirs. The EC Nitrates Directive of 1991 (CEU, 1991) required Member States to take measures to control farming practices that are likely to cause increased nitrates in surface and groundwaters. In the UK a designated action programme required Nitrate Vulnerable Zones to be set up and the promotion of best practice to farmers in the use and storage of fertilizers and manure. By 2002 55% of England was designated a Nitrate Vulnerable Zone, with similar designations in Scotland, Wales and Northern Ireland. The focus now is on diffuse pollution, with farmers being encouraged to adopt catchment sensitive farming. Implementation of the 1998 EC Directive on Urban Waste Water Treatment (CEU, 1998b) has also resulted in the setting up of Sensitive Areas across many parts of England. These are intended to reduce the level of eutrophication in inland or coastal waters and to protect surface waters used for the abstraction of drinking water by reducing the levels of phosphorus (Section 6.41) and nitrates in discharges from sewage treatment works.

Waters containing high nitrate concentrations are thought to be potentially harmful to infants. At the neutral pH of the infant stomach, nitrate can undergo bacterial reduction to nitrite, which is then absorbed into the bloodstream and converts the oxygen-carrying haemoglobin into methaemoglobin. Whilst the methaemoglobin itself is not toxic, the effects of reduced oxygen-carrying capacity in the blood can be serious, especially for infants having a high fluid intake relative to body weight. However, it has become apparent that methaemoglobinaemia is a problem associated with rural shallow wells subject to microbial contamination, rather than public water supplies. There are many examples of public water supplies with relatively high nitrate levels in Europe and North America where no problems with methaemoglobinaemia have been reported. In 1997, an epidemiological study carried out by Leeds University indicated that there was a small but statistically significant correlation between the incidence of childhood diabetes and nitrate in drinking water. The study was restricted to a very small geographical area and the correlation could have arisen by chance. A follow up study, commissioned by the then DETR examined the incidence of childhood diabetes over large areas of England and Wales and the results failed to demonstrate any correlation with nitrate concentrations in drinking water (DETR, 1999a).

Nitrite

Nitrite is the intermediate oxidation state between ammonia and nitrate and can be formed by the reduction of nitrates under conditions where there is a deficit of oxygen. Surface waters, unless badly polluted with sewage effluent, seldom contain more than 0.1 mg/l nitrite as N. Thus the presence of nitrites in surface waters in conjunction with high ammonia levels indicates pollution from sewage or sewage effluent. The presence of nitrites in groundwater may also be a sign of sewage pollution. Conversely it may have no hygienic significance as nitrates in good quality groundwaters can be reduced to nitrite under anaerobic conditions, especially in areas of ferruginous sands. New brickwork in wells is known to have a similar effect.

Ingested nitrites, some of which may be formed by bacterial reduction of nitrates, can react with secondary and tertiary amines found in certain foods to give nitrosamines. This has given rise to concern as nitrosamines are potentially carcinogenic. However, there is no epidemiological evidence of an association between nitrite levels in drinking water and cancer incidence.

There are no simple methods for treating a water to reduce its nitrate-nitrite concentration, other than by blending it with another supply with low or negligible concentrations of the same. Treatment processes currently used for nitrate removal include ion exchange or desalination by membranes, and biological removal under controlled conditions using denitrifying bacteria (Sections 10.24–10.27). Improved catchment management is the most sustainable approach in dealing with point source or diffuse pollution of groundwaters.

6.37 ORGANIC MATTER AND CHEMICAL OXYGEN DEMAND (COD)

The organic matter found in water can come from a variety of sources such as plant and animal material, including partially treated domestic waste and industrial effluents. The total organics present in a water can be estimated from the chemical oxygen demand (COD), the oxygen absorbed permanganate value (PV) or from the total organic carbon content (TOC). The oxygen demand tests are indirect measures of the total organic carbon by determining its oxidisability. All are gross measures of the total concentration of organic substances and are usually dominated by the concentration of natural organics e.g. humic and fulvic acids. (Humic and fulvic acids are complex molecular structures resulting from the decay of humus, itself the product of decay of plant, animal and microbial matter.)

Organic Micropollutants

The application of gas chromatography and mass spectrometry to water analysis since the 1970s has revealed the presence of many hundreds of different organic compounds. These are derived from naturally occurring substances in the environment, as well as from materials produced, used or discarded by industry and agriculture and, when present, are usually at very low concentrations of less than 1 µg/l. Some are known to be toxic or carcinogenic to animals at concentrations far higher than detected in drinking water; others are known to be mutagenic (i.e. capable of making heritable changes to living cells) under laboratory testing. A number have yet to be identified fully.

Assessing the risk that these compounds might present to human health is difficult, given the very low levels found in drinking water. Different methods of extrapolating data obtained from observing the effect of high concentrations on animals, in order to deduce the risk to human health of ingesting far smaller quantities, have given results in some cases varying by several orders of magnitude (Hunt, 1987). Epidemiological studies to discover some statistical relationship between cancer mortality and water type (e.g. comparing the effect of using surface waters likely to contain organic matter, with the effect of using groundwater likely to contain less) show "some association" in some studies, whilst other studies have given inconsistent results. Thus any risk to health due to the presence of organic micropollutants in water is likely to be very small, difficult to measure, and to be manifest only after a long period of exposure. However, the potential risk has led to the view that, where possible, the concentration of certain pollutants should be kept as low as possible in drinking water.

6.38 PESTICIDES

Pesticides cover a wide range of compounds used as insecticides, herbicides, fungicides, and algicides. The term can also refer to chemicals with other uses such as wood preservation, public hygiene, industrial pest control, soil sterilisation, plant growth regulation, masonry biocides, bird and animal repellents and anti-fouling paints.

Pesticides find their way into natural waters from direct application for aquatic plant and insect control, from percolation and runoff from agricultural land, from aerial drift in land application, and from industrial discharges. Organic pesticides are often toxic to aquatic life even in trace amounts and some, particularly the organo-chlorine compounds, are very resistant to chemical and biochemical degradation. Where pesticides and algicides are used for aquatic control they can cause deoxygenation as a result of the decomposition of the treated vegetation. This in turn can cause other problems such as the dissolution of iron and manganese and the production of tastes and odours.

Even with controlled application, traces of pesticides are still found in sewage effluents and in a number of groundwater and river sources used for public water supplies. Pesticides used in non-agricultural situations, particularly on hard surfaces, pose a higher risk of contaminating water sources. Accidental discharges of pesticides in bulk to watercourses occasionally occur and can have serious implications, causing fish death and making it necessary for a temporary shutdown of any water intakes. In the UK the Pesticides Safety Directive approves and regulates the use of pesticides; similar arrangements exist in the USA. Even so some pesticides that have been banned or restricted for many years remain persistent in the environment and have very occasionally been found in groundwater sources long after the original contamination occurred.

Information on the pesticides most widely used and likely to be detected in water supplies in the UK can be found in the former Department of the Environment publication *Guidance on Safeguarding the Quality of Public Water* Supplies (DoE, 1990a), which is still relevant today. Many pesticides are insoluble in water or have limited solubility; others degrade readily, depending on the soil type. Many are rapidly adsorbed onto sediment or suspended material and this property can be utilized in treatment processes involving coagulation followed by sedimentation and filtration. Oxidation using ozone, chlorine, chlorine dioxide or potassium permanganate effectively breaks down organic pesticides; although in some cases the use of chlorine or ozone may result in degradation products, which are more toxic than the original pesticide, or give rise to odorous compounds (Section 6.50). Ultra violet radiation, when used on good quality groundwaters, has also been shown to effectively degrade low concentrations of atrazine (Bourgine, 1995). For more effective pesticide removal, adsorption onto activated carbon should be adopted either with or without a pre-oxidation stage (Sections 10.35–10.37). A preferable policy is to restrict the levels of pesticides in the raw water prior to treatment, rather than having to remove them in the treatment process.

In certain situations selected pesticides may actually be applied to sources of drinking water to control disease-carrying insects. Such pesticides have to be approved for use by the national authorities and applied under very strictly controlled conditions.

Overall the amounts of pesticides which are likely to be consumed from drinking water tend to be a very small fraction of those likely to be consumed from foodstuffs. WHO and US EPA set health related standards for a number of individual pesticides. However, the EC Directive applies a single standard of 0.1 µg/l to most pesticides (and relevant metabolites and degradation products), which does not take into account of the wide variation in toxicity of the various pesticides that might be detected in drinking water.

6.39 pH VALUE OR HYDROGEN ION

The pH value, or hydrogen ion concentration, determines the acidity of a water. It is one of the most important determinations in water chemistry as many of the processes involved in water treatment are pH dependent. Pure water is very slightly ionized into positive hydrogen (H^+) ions and negative hydroxyl (OH^-) ions. In very general terms a solution is said to be neutral when the numbers of hydrogen ions and hydroxyl ions are equal, each corresponding to an approximate concentration of 10^{-7} moles/l. This neutral point is temperature dependent and occurs at pH 7.0 at 25 °C. When the concentration of hydrogen ions exceeds that of the hydroxyl ions (i.e. at pH values less than 7.0) the water has acid characteristics. Conversely when there is an excess of hydroxyl ions (i.e. the pH value is greater than 7.0) the water has basic characteristics and is described as being on the alkaline side of neutrality.

The pH value of unpolluted water is mainly determined by the inter-relationship between free carbon dioxide and the amounts of carbonate and bicarbonate present (Section 10.41). The pH values of most natural waters are in the range 4 to 9, with soft acidic waters from moorland areas generally having lower pH values and hard waters which have percolated through chalk or limestone generally having higher pH values.

Most water treatment processes, but particularly clarification and disinfection, require careful pH control to fully optimize the efficacy of the process. The pH of the water entering distribution must also be controlled to minimize the corrosion potential of the water (Section 6.22).

6.40 PHENOLS

Phenolic compounds found in surface waters are usually a result of pollution from trade wastes such as petrochemicals, washings from tarmac roads, gas liquors and creosoted surfaces. Decaying algae or higher vegetation can also release natural phenols into the aquatic environment, whilst traces of phenols and other phenol-like compounds may be found in good quality groundwaters, especially in areas with coal or oil-bearing strata. Most phenols, even in minute concentrations, produce chlorophenols on chlorination (Section 6.25). Even trace amounts can render the water unacceptable to consumers because of objectionable taste and/or odour (Section 10.33).

Phenols can be effectively removed by superchlorination, whereby the excess chlorine chemically decomposes the phenol; by oxidation with ozone; or by adsorption onto activated carbon. However, it is preferable to seek to eliminate the source wherever possible.

6.41 PHOSPHATES

Phosphates in surface waters mainly originate from sewage effluents containing phosphate-based synthetic detergents, from industrial effluents, or from agricultural runoff following the use of inorganic fertilizers. Groundwaters usually contain insignificant concentrations of phosphates, unless they become polluted. Phosphorous is one of the essential nutrients for algal growth and can contribute significantly to eutrophication of lakes and reservoirs (Section 6.75).

Orthophosphates may be added during water treatment for plumbosolvency control (Section 6.33). The applied dose is initially around 1 mg/l as P, gradually decreasing to around 0.6 mg/l as P as the treatment takes effect and the system becomes optimized. Although phosphate treatment is effective for lead, phosphate dosing of waters with high alkalinity and low pH may slightly increase cuprosolvency. Orthophosphates, along with polyphosphates, can also be used as corrosion inhibitors to reduce the risk of supplying 'red water' from corroding iron pipes.

Phosphate dosing has played an important role in meeting the tighter lead standards. However concerns have been expressed about the impact of such dosing on the environment. Typically up to 40% of the phosphate in sewage effluent comes from detergents and about 50% from human excretion, with the phosphate added to water supplies to control plumbosolvency contributing no more than 10%. In areas of low intensity agriculture, sewage effluents represent the main source of phosphate in the aquatic environment but the impact from plumbosolvency treatment is still very low. High intensity agriculture contributes significantly more amounts of phosphates to the environment, making the contribution from plumbosolvency treatment even lower. Thus the overall impact of phosphate dosing is likely to be insignificant, even in situations where it is deemed necessary to reduce phosphate levels in order to further protect the aquatic environment.

6.42 POLYNUCLEAR AROMATIC HYDROCARBONS (PAHs)

PAHs are a group of organic compounds that occur widely in the environment as the result of incomplete combustion of organic material. Trace amounts of PAHs have been found in industrial and domestic effluents. Their solubility in water is very low but can be enhanced by detergents and by other organic solvents which may be present. Although PAHs are not very biodegradable, they tend to be taken out of solution by adsorption onto particulate matter. If present in raw waters, they are usually removed during coagulation, sedimentation and filtration. However, they can be re-introduced in the distribution system from mains that have been lined with coal tar pitch. Until the 1970's coal tar, which can contain up to 50% of PAHs, was used to line iron water mains to prevent rusting. In some situations this lining may eventually break down, releasing PAHs in solution and as particulates into the water. There tends to be a seasonal variation in the concentrations found, with solubility linked to water temperature. There is also evidence that chlorine dioxide, when used as a disinfectant or for taste and odour control, can result in elevated concentrations of PAHs at consumers' taps (DoE, 1990a).

Several PAHs are known to be carcinogenic at concentrations considerably higher than those found in drinking water, with the main routes of exposure being from food and cigarette smoke. Drinking water is normally monitored for five indicator parameters, namely benzo(b)fluoranthene, benzo(k)fluoranthene, benzo(a)pyrene, benzo(ghi)perylene and indeno (1,2,3-cd)pyrene, of which benzo(a)pyrene is considered to be the most harmful and has a separate standard.

6.43 RADIOACTIVE SUBSTANCES

Many water sources contain very low levels of radioactive substances, which are mainly naturally occurring radionuclides in the uranium and thorium decay series. These tend to be alpha particle emitters, although some (e.g. radium-228 and potassium-40) are beta particle emitters. Other radionuclides may be found in water sources as a result of pollution from human activities, e.g. the nuclear fuel and power industries, or where radioactive isotopes are used in medicine or industry, or other similar activities. Most of these man-made radionuclides are beta emitters, including tritium which is usually man-made but may also occur naturally. The impact of man-made radionuclides on the environment is subject to tight regulatory control. If there is any likelihood of a water source or a drinking water supply becoming contaminated then these regulatory controls would be used to ensure that remedial action was taken.

The contribution of drinking water to total radiological exposure is generally extremely small. The International Commission on Radiological Protection (ICRP) provides detailed advice and recommendations on the control of exposure to radiation. WHO uses these recommendations, along with advice and recommendations from the International Atomic Energy Agency (IAEA), in formulating its guidelines on radionuclides in drinking water. These guidelines are based on a recommended reference dose level (RDL) of the committed effective dose, equal to 0.1 mSv from 1 year's consumption of drinking water and the dose coefficients for adults; mSv (milli Sievert) is the 'effective dose equivalent'. It is a measure of the effect produced on a person by different types of radiation, taking into account the nature of the radiation and the organs exposed. This reference dose level represents less than 5% of the average effective dose each year from natural background radiation. These recommendations apply only to existing operational water supplies and to new supplies. Other advice would apply in emergency situations where radionuclides have been accidentally or deliberately released into the environment.

It is neither feasible nor practical to routinely monitor water supplies for all individual radionuclides. WHO recommends a screening approach, with no further action being required if the gross alpha activity is ≤0.5 Bq/l and the gross beta activity is ≤1.0 Bq/l. Bq (Becquerel) corresponds to one nuclear transformation per second; 1 curie (Ci) = 37×10^9 Bq. Specific radionuclides need only be identified if either of these screening levels is exceeded. In the event of this happening, a dose estimate has to be made for each radionuclide likely to be present, using activity to dose conversion factors (based on 1 year's consumption of 2 litres of water per day). The sum of the individual dose estimates determines whether the RDL (or total indicative dose) is likely to be exceeded. If the screening relates to a single sample, the RDL would only be exceeded if consumers were exposed to the measured concentrations for a full year. A single exceedence does not, therefore, necessarily mean that the water supply is unsuitable for consumption, but signals the need for further investigations. In the event of the RDL being exceeded, appropriate medical advice should be sought and remedial measures taken to reduce the level of exposure.

The more commonly found nuclides in drinking water include:

- Potassium-40, which occurs naturally in a fixed ratio to stable potassium;
- Radium 226 and 228;
- Radon, a radioactive gas released as a decay product of radium;
- Tritium, which occurs naturally at low levels but at higher concentrations, is indicative of man-made pollution, thereby providing an indication of other potentially more harmful radionuclides; and
- Uranium.

For several years Defra routinely monitored some 30 surface and groundwater sources in England and Wales for gross alpha and gross beta activities and for selected specific radionuclide concentrations. The monitored sources supply about one-third of the population. The results for 1995 to 2001 are available as part of the environmental statistics published annually by Defra (2007a). Most samples have been below the gross alpha and gross beta screening levels and any exceedences of these values have been marginal and transient. Many of the radionuclides detected are of natural origin, but traces of some artificial radionuclides have been found. Similar monitoring has been carried out in Scotland and Northern Ireland.

6.44 SELENIUM

Selenium is naturally present in soil in many parts of the world but may have increased in some areas through air pollution from coal-fired installations. It is also present in water, particularly in areas of geothermal activity. In some areas the concentrations of selenium are increased significantly through irrigation returns.

Selenium is an essential element at low levels. It is present in foods (fish, cereals, eggs and Brazil nuts) and is included in some mineral supplements—particularly in Europe where selenium levels in soil are low. However, excess selenium is toxic and cumulative in the body.

6.45 SILICA

Silica can be found in water in several forms, caused by the degradation of silica-containing rocks such as quartz and sandstone. Natural waters can contain between 1 mg/l of silica in the case of soft moorland waters and up to about 40 mg/l in some hard waters. Much higher levels are found in waters from volcanic or geothermal areas.

The levels of silica in drinking water can be important in a number of industrial processes because it forms a very hard scale which is difficult to remove.

6.46 SILVER

Trace amounts of silver are occasionally found in natural waters but it is rarely found in detectable concentrations in drinking water. However silver can be used as a disinfectant in domestic water treatment units or point of use devices (e.g. ceramic filter candles or granular activated carbon impregnated with silver) and this may result in elevated levels in the treated water (Butkus, 2003).

6.47 SODIUM

Sodium compounds are very abundant in the environment and are also very soluble. The element is present in most natural waters at levels ranging from less than 1 mg/l to several thousand mg/l in brines. The threshold taste for sodium in drinking water depends on several factors, such as the predominant anion present and the water temperature. The threshold taste as sodium chloride is around 150 mg/l as Na, whereas the threshold taste as sodium sulphate is higher at around 220 mg/l as Na (WHO, 1996).

The use of base-exchange or lime-soda processes to soften hard waters can lead to a significant increase in the sodium concentration of the softened water. This situation also applies to domestic softeners and ideally, when a domestic softener is installed, there should always be a separate tap available for drinking water that is supplied directly off the mains supply. It is advisable that consumers on low sodium diets do not drink point of use softened water and the water should not be used for making up baby food.

6.48 SULPHATES

The concentration of sulphate in natural waters can vary over a wide range from a few mg/l to several thousand mg/l in brackish waters and brines. Sulphates come from several sources such as the dissolution of gypsum and other mineral deposits containing sulphates; seawater intrusion; the oxidation of sulphides, sulphites and thiosulphates in well aerated surface waters; and from industrial effluents where sulphates or sulphuric acid have been used in processes such as tanning and pulp paper manufacturing. Sulphurous flue gases discharged to atmosphere in industrial areas often result in acid rain water containing appreciable levels of sulphate.

High levels of sulphate in water can impart taste and, when combined with magnesium or sodium, can have a laxative effect (e.g. Epsom salts). Consumers tend to become acclimatized to high sulphate waters and, in some parts of the world, waters with very high sulphate contents have to be used if no alternative supplies are available. WHO recommends that health authorities be notified if concentrations exceed 500 mg/l.

Bacterial reduction of sulphates under anaerobic conditions can produce hydrogen sulphide, which is an objectionable gas smelling of bad eggs. This can occur in deep well waters but the odour rapidly disappears with effective aeration. It can also occur if there is seawater intrusion to a shallow aquifer which is polluted by sewage.

6.49 SUSPENDED SOLIDS

The suspended solids content or filter residue of a water quantifies the amount of particulate material present and includes both organic and inorganic matter such as plankton, clay and silt. The suspended solids content of a surface water can vary widely depending on flow and season, with some rivers under flood conditions having several thousand mg/l of material in suspension.

The measurement of suspended solids is usually on a weight-volume basis and gives no indication as to the type of material in suspension, the particle size distribution or the settling characteristics. However, it is usually a key condition in effluent discharge consents. In terms of drinking water quality, it is important that suspended solids are adequately removed from the raw water prior to final disinfection so that the efficacy of the final disinfection process is not impaired.

6.50 TASTE AND ODOUR

There are four basic taste sensations, namely sweet, sour, salt and bitter. What is regarded as taste is in fact a combination of these sensations with the sensation of smell. In examining water samples, the odour rather than the taste of a sample is often evaluated as it avoids putting a possibly suspect sample into the mouth. However, it is still desirable to check the taste of a final treated water. A subjective or qualitative assessment of the taste or odour is often carried out at the time of sampling. This can then be supplemented by a quantitative measurement, reported as the Dilution Number (or the Threshold Number), which is carried out under controlled laboratory conditions. The quantitative test is based on the number of dilutions of the sample with taste and odour-free water necessary to eliminate the taste or odour.

Many tastes and odours in drinking water are caused by natural contaminants such as extra-cellular and decomposition products of plants, algae and micro fungi. Certain of the blue-green algae and actinomycetes, when present in a raw water, can give rise to very distinctive earthy and musty tastes and odours (Section 6.73). Raw waters contaminated by agricultural and industrial discharges may also give rise to serious taste and odour problems, which are often exacerbated by the use of chlorine for disinfection. Chlorine itself can give rise to extensive taste and odour complaints, especially if the applied dose has been increased suddenly for operational reasons. However some tastes and odours may be caused by the condition of the domestic plumbing system e.g. chlorophenolic tastes associated with old washing machine hoses; astringent tastes from the dissolution of plumbing materials such as zinc or copper; and poorly sited pipes resulting in the build-up of biofilms.

Taste and odour of drinking water tends to be very subjective, with consumers becoming familiar with the taste or odour associated with their local water supply. Thus any complaint of an unusual taste or odour should always be investigated immediately, in case it relates to more serious changes.

6.51 TURBIDITY

The measurement of turbidity, although not quantitatively precise, is a simple and useful indicator of the condition of a water. Turbidity is defined as the optical property that causes light to be scattered and absorbed rather than transmitted in straight lines through a sample. Although turbidity is caused by material in suspension, it is difficult to correlate it with the quantitative measurement of suspended solids in a sample, as the shape, size, and refractive indices of the particles in suspension all affect their light-scattering properties. For the same reason turbidity measurements can vary according to the type of instrument used. Nephelometers measure the intensity of light scattered in one particular direction and are highly sensitive for measuring low turbidity. Other instruments measure the amount of light absorbed by particles when light is passed through a water sample. Early measurements were made using the Jackson Candle Turbidimeter, with results reported in Jackson units (JTU). However, this was a fairly crude form of measurement and not suitable for measuring low turbidities. Current methods use a primary standard based on formazin which, if used with nephelometric instruments, gives an equivalent turbidity measurement in FTU (i.e. FTU = NTU).

Raw water turbidity can vary over a very wide range, from virtually zero to several thousand NTU. Effective treatment should be able to consistently produce final waters with turbidity levels of less than 1 NTU. Turbidity meters are therefore an essential tool in optimizing and controlling water treatment processes, although particle size counters may eventually take over from turbidity meters as the optimum analytical tool for monitoring the performance of rapid gravity filters, particularly at the start of a filter run. WHO suggests a median turbidity of <0.1 NTU is necessary for effective disinfection. The EC Directive has a treatment standard of 1 NTU for the turbidity of final waters at works treating surface waters. This is applied at all treatment works in the UK regardless of the source. US EPA considers turbidity under the Long Term 1 Enhanced Surface Water Treatment Rule.

6.52 ZINC

Zinc tends to be found only in trace amounts in unpolluted surface waters and groundwaters. However, it is often found in the water at consumers' taps as a result of corrosion of galvanized iron piping or tanks or dezincification of brass fittings.

Zinc is beneficial in moderation in diets but is harmful in excess, producing copper deficiency and other adverse effects. There are no health related standards but zinc can cause an astringent taste and also cause opalescence in some waters at concentrations of more than 3 mg/l.

PART II   WATER QUALITY STANDARDS FOR CHEMICAL AND PHYSICAL PARAMETERS

6.53 DRINKING WATER STANDARDS (CHEMICAL AND PHYSICAL)

Standards apply to many of the parameters discussed in Part I and these are listed in Tables 6.1(A)–(E). Reference is made to the following documents; the units and notations quoted are as they appear in the respective document. Care should be taken when comparing the standards as the units of measurement and notations for some parameters are not consistent between authorities.

The WHO Guidelines for Drinking-water Quality

Third edition, including addenda published in 2006 and 2008 (WHO, 2008).

The third edition of the WHO *Guidelines for Drinking-water Quality* focuses on the public health aspects of drinking water and provides health related guideline values that can be applied worldwide. Many countries therefore use the WHO guideline values as a basis for setting national standards. The guideline values for chemical parameters are based on the potential for such parameters or substances to cause adverse health effects after long periods of exposure, either as a cumulative toxin such as lead, or as a possible carcinogen in the case of some organic contaminants. Some parameters have provisional guideline values on the grounds that available data on health effects are either limited or at present poorly defined. The Guidelines also provide guidance on the aesthetic quality of drinking water, whereby a water may be unacceptable to consumers on the grounds of appearance, or taste and odour, but there is no associated health risk. Unlike previous editions, the third edition provides guidance on the application of the Guidelines in certain circumstances and places more emphasis on drinking water safety and the need for a risk management approach, based on sound science. The Guidelines are subject to an ongoing rolling revision; the 2008 third edition includes addenda published in 2006 and 2008.

The European Commission Directive on the quality of water intended for human consumption (CEU, 1998a)

The 1998 Drinking Water Directive updates the original Directive of 1980 (CEU, 1980). It is a legal document that applies only to Member States and is intended to enable equal water qualities and obligations to be achieved throughout the European Union. Member States were required to adopt the Directive into national law within two years of it coming into force and to achieve compliance with most of the standards specified by 25 December 2003. These standards are, for the most part, based on the 2008 WHO guideline values. The Directive contains 43 numerical standards for chemical and radiological parameters, of which 30 are mandatory on the grounds of health risks. The remaining 13 standards, along with five parameters that do not have numerical values, are termed indicator parameters. Most of the indicator parameters relate to the aesthetic and organoleptic characteristics of a water and have no associated health risk. However failure to meet a parametric value

Table 6.1(A) Inorganic chemical parameters of health significance

	WHO Guidelines 3rd Ed. Volume 1, with addenda. Health related guideline value	UK Water Supply (Water Quality) Regulations. Prescribed conc. or value	EC Directive 98/83/ EC November 1998. Parametric value	US EPA Regulations under Safe Drinking Water Act. Maximum contaminant level (MCL)
Antimony (Sb)	0.02 mg/l	5 µg/l	5 µg/l	0.006 mg/l
Arsenic (As)	0.01 mg/l[P]	10 µg/l	10 µg/l	0.010 mg/l[1]
Asbestos (fibres >10 µm in length)	No consistent evidence			7×10^6 fibres/l
Barium (Ba)	0.7 mg/l			2 mg/l
Beryllium (Be)				0.004 mg/l
Boron (B)	0.5 mg/l[P]	1.0 mg/l	1.0 mg/l	
Bromate (BrO₃)*	0.01 mg/l[P]	10 µg/l	25 µg/l to Dec 2008 10 µg/l thereafter	0.010 mg/l annual average[2]
Cadmium (Cd)	0.003 mg/l	5 µg/l	5 µg/l	0.005 mg/l
Chloramines (Cl₂)*	Mono- 3 mg/l			4 mg/l[2]
Chlorate (ClO₃)*	0.7 mg/l[P]	0.7 mg/l[3]		
Chlorine (Cl₂)	5 mg/l[C]			4 mg/l[2]
Chlorite (ClO₂)*	0.7 mg/l[P]	0.5 mg/l[4]		1.0 mg/l monthly average[2]
Chromium (Cr)	0.05 mg/l[P]	50 µg/l	50 µg/l	0.1 mg/l
Copper (Cu)	2 mg/l	2.0 mg/l	2.0 mg/l	1.3 mg/l[5]
Cyanide (CN)	0.07 mg/l	50 µg/l	50 µg/l	0.2 mg/l free cyanide
Fluoride (F)	1.5 mg/l	1.5 mg/l	1.5 mg/l	4 mg/l
Lead (Pb)	0.01 mg/l	25 µg/l 10 µg/l from 25 Dec 2013	25 µg/l[6] 10 µg/l from 25 Dec 2013	0.015 mg/l[5]
Manganese(Mn)	0.4 mg/l[C]			
Mercury (Hg)	0.006 mg/l	1.0 µg/l	1.0 µg/l	0.002 mg/l
Molybdenum (Mo)	0.07 mg/l			
Níkel (Ni)	0.07 mg/l	20 µg/l	20 µg/l	
Nitrate	50 mg/l as NO₃	50 mg/l as NO₃	50 mg/l as NO₃	10 mg/l as N
Nitrite	3 mg/l as NO₂ 0.2 mg/l[P,7]	0.50 mg/l as NO₂ 0.10 mg/l ex works	0.50 mg/l as NO₂ 0.10 mg/l ex works	1 mg/l as N
Nitrate + Nitrite	NO₃/50 + NO₂/3 ≤1	NO₃/50 + NO₂/3 ≤1	NO₃/50 + NO₂/3 ≤1	
Selenium (Se)	0.01 mg/l	10 µg/l	10 µg/l	0.05 mg/l
Sodium dichloro- isocyanurate	50 mg/l as sodium dichloro- isocyanurate, 40 mg/l as cyanuric acid			

(*continued*)

Table 6.1(A) (*Continued*)

	WHO Guidelines 3rd Ed. Volume 1, with addenda. Health related guideline value	UK Water Supply (Water Quality) Regulations. Prescribed conc. or value	EC Directive 98/83/ EC November 1998. Parametric value	US EPA Regulations under Safe Drinking Water Act. Maximum contaminant level (MCL)
Thallium (Tl)				0.002 mg/l
Turbidity – see also Table 6.1(E)	<0.1 NTU[8]			1 NTU to 5 NTU[9]

Notes:
[P]Provisional guideline value
* Usually present in drinking water as a disinfection by-product.
[C]Consumers may reject water with concentrations at or below the health based guideline value on aesthetic grounds
[1]The Arsenic and Clarifications to Compliance and New Source Monitoring Rule.
[2]Stage 1 Disinfectants and Disinfection Byproducts Rule—maximum residual level on an annual average.
[3]Set as a condition of use under Section 31(4) of the Regulations for on-site electrolytic generation of chlorine.
[4]Set as a condition of use under Section 31(4) of the Regulations where chlorine dioxide is used in the treatment. The combined concentration of chlorine dioxide, chlorite and chlorate must not exceed 0.5 mg/l as chlorine dioxide in the water entering supply.
[5]The Lead and Copper Rule—Remedial action required if the concentration in more than 10% of tap water samples collected during any monitoring period is greater than the value stated.
[6]Based on a weekly average value ingested by consumers.
[7]For long term exposure.
[8]Not health based but ideally the median value for effective disinfection.
[9]Long Term 1 Enhanced Surface Water Treatment Rule—performance standards for works treating surface waters or groundwaters under the direct influence of surface waters. Conventional and direct filtration—the maximum turbidity of the combined filtrate should be <1 NTU, with turbidities of ≤0.3 NTU in at least 95% of measurements taken each month. Slow sand and diatomaceous earth filtration—the maximum turbidity of the combined filtrate should be ≤5 NTU, with turbidities of ≤1 NTU in at least 95% of measurements taken each month. Individual State-set limits, based on those set for slow sand filters, apply to alternative technologies for filtration.

could indicate problems or potential problems with the treatment or distribution of the water. Member States are allowed to set values for additional parameters not included in the Directive. They can also adopt more stringent standards than those specified, although the Commission has to be notified. Any failure of a standard has to be investigated but the water supplier may then be granted a special dispensation, which permits non-compliant water to be supplied for up to three years. Such dispensations or 'derogations' are time limited and can only be authorized if there is no associated risk to human health in continuing to supply the water. They are also conditional on appropriate remedial action being taken. There is a requirement in the Directive to review the standards at least every five years, 'in the light of scientific and technical progress'. It seems unlikely that any revisions to the 1998 Drinking Water Directive will be enacted before 2012.

The UK Water Supply (Water Quality) Regulations

Water supply arrangements are different in England and Wales compared to Scotland and Northern Ireland, as is the role of the regulatory authorities. The requirements of the EC Directive were transposed into the respective national laws by way of the Water Supply (Water Quality) Regulations. This was done in England in 2000 (UK, 2000), with subsequent amending Regulations in 2001 (UK, 2001a); in Wales in 2001 (UK, 2001b); in Scotland in 2001, with subsequent amending

Table 6.1(B) Organic chemical parameters of health significance

	WHO Guidelines 3rd Ed. Volume 1, with addenda. Health related guideline value	UK Water supply (Water quality) Regulations. Prescribed conc. or value	EC directive 98/83/EC November 1998. Parametric value	US EPA Regulations under Safe Drinking Water Act. Maximum contaminant level (MCL)
Aromatic hydrocarbons				
Benzene	0.01 mg/l	1.0 µg/l	1.0 µg/l	0.005 mg/l
Ethylbenzene	0.3 mg/l[c]			0.7 mg/l
Styrene	0.02 mg/l[c]			0.1 mg/l
Toluene	0.7 mg/l[c]			1 mg/l
Xylenes	0.5 mg/l[c]			10 mg/l total
Chlorinated alkanes				
Tetrachloromethane (Carbon tetrachloride)	0.004 mg/l	3.0 µg/l		0.005 mg/l
Dichloromethane	0.02 mg/l			0.005 mg/l
1,2-dichloroethane	0.03 mg/l	3.0 µg/l	3.0 µg/l	0.005 mg/l
1,1,1-trichloroethane				0.2 mg/l
1,1,2-trichloroethane				0.005 mg/l
Chlorinated benzenes				
Monochlorobenzene				0.1 mg/l
1,2-dichlorobenzene	1 mg/l[c]			0.6 mg/l
1,4-dichlorobenzene	0.30 mg/l[c]			0.075 mg/l
1,2,4-Trichlorobenzene				0.07 mg/l
Chlorinated ethenes				
1,1-dichloroethene				0.007 mg/l
1,2-dichloroethene	0.05 mg/l			*cis* 0.07 mg/l *trans* 0.1 mg/l
Tetrachloroethene (PCE) and	0.04 mg/l	)	)	0.005 mg/l
		) Sum 10 µg/l	) Sum 10 µg/l	
Trichloroethene (TCE)	0.02 mg/l[P]	)	)	0.005 mg/l
Pesticides				
Total pesticides		0.50 µg/l[1]	0.50 µg/l[1]	
Individual pesticides *except for:*	See Table 6.1(C)	0.10 µg/l[1]	0.10 µg/l[1]	See Table 6.1(C)
- Aldrin	Sum of aldrin	0.030 µg/l	0.030 µg/l	
- Dieldrin	+ dieldrin = 0.00003 mg/l	0.030 µg/l	0.030 µg/l	
- Heptachlor		0.030 µg/l	0.030 µg/l	0.0004 mg/l
- Heptachlor epoxide		0.030 µg/l	0.030 µg/l	0.0002 mg/l

(*continued*)

Table 6.1(B) (*Continued*)

	WHO Guidelines 3rd Ed. Volume 1, with addenda. Health related guideline value	UK Water Supply (Water Quality) Regulations. Prescribed conc. or value	EC Directive 98/83/EC November 1998. Parametric value	US EPA Regulations under Safe Drinking Water Act. Maximum contaminant level (MCL)
Polycyclic aromatic hydrocarbons (PAH)		0.10 µg/l[2]	0.10 µg/l[2]	
Benzo(a)pyrene	0.0007 mg/l	0.01 µg/l	0.01 µg/l	0.0002 mg/l
Other organic compounds				
Acrylamide	0.0005 mg/l	0.10 µg/l[3]	0.10 µg/l[3]	TT
Di(2-ethylhexyl) adipate (DEHA)				0.4 mg/l
Di(2-ethylhexyl) phthalate (DEHP)	0.008 mg/l			0.006 mg/l
1, 4 Dioxane	0.05 mg/l			
Dioxin				3×10^{-8} mg/l
Edetic acid (EDTA)	0.6 mg/l (free acid)			
Epichlorohydrin	0.0004 mg/l[P]	0.10 µg/l[3]	0.10 µg/l[3]	TT
Microcystin-LR	0.001 mg/l[4]			
Nitrilotriacetic acid (NTA)	0. 2 mg/l			
Polychlorinated biphenyls (PCBs)				0.0005 mg/l
Vinyl chloride	0.0003 mg/l	0.50 µg/l	0.50 µg/l	0.002 mg/l
Disinfection by-products - see also Table 6.1(A).				
Haloacetic acids:				0.060 mg/l as a total of 5 acids
- monochloroacetate	0.02 mg/l			
- dichloroacetate	0.05 mg/l[P]			
- trichloroacetate	0.2 mg/l			
*Chlorophenols**				
- 2,4,6-trichlorophenol	0.2 mg/l[c]			
Cyanogen chloride	0.07 mg/l as CN			
Halogenated acetonitriles				
- dichloroacetonitrile	0.02 mg/l[P]			
- dibromoacetonitrile	0.07 mg/l			
Trihalomethanes:				

(*continued*)

Table 6.1(B) (*Continued*)

	WHO Guidelines 3rd Ed. Volume 1, with addenda. Health related guideline value	UK Water Supply (Water Quality) Regulations. Prescribed conc. or value	EC Directive 98/83/EC November 1998. Parametric value	US EPA Regulations under Safe Drinking Water Act. Maximum contaminant level (MCL)
- bromoform	0.1 mg/l[6]	)	) 150 µg/l sum of	)0.080 mg/l total[5]
- chloroform	0.3 mg/l[6]	) 100 µg/1 sum of	) concentrations;	)
- dibromochloro-methane	0.1 mg/l[6]	) concentrations[7]	) 100 µg/l after 25	)
- bromodichloro-methane	0.06 mg/l[6]	)	) December 2008[7]	)
N-Nitrosodium-methylamine (NDMA)	0.0001 mg/l			

Notes:
[c]Threshold taste and/or odour concentration considerably lower than any associated health risk
[p]Provisional or proposed
[TT]Treatment standard based on dose and monomer level, when used in drinking water treatment
  Acylamide - 0.05% dosed at 1 mg/l (or equivalent)
  Epichlorohydrin - 0.01% dosed at 20 mg/l (or equivalent)
[1]'Pesticides' means organic insecticides, herbicides, fungicides, nematocides, acaricides, algicides, rodenticides, slimicides and related products (inter alia, growth regulators) and their related metabolites, degradation and reaction products. 'Total pesticides' means the sum of all individual pesticides detected and quantified in the monitoring procedure.
[2]Total PAHs—Sum of concentrations of benzo(b)fluoranthene, benzo(k)fluoranthene, benzo(ghi)perylene, indeno(1,2,3-ed)pyrene and benzo(a) pyrene.
[3]Residual monomer concentration calculated according to the specification of the maximum release from the polymer in contact with the water. (Acrylamide derives from use of polyacrylamide flocculants in water treatment; epichlorohydrin from its use in the manufacture of water treatment resins.)
[4]Total microcystin-LR (free plus cell-bound)
[5]Stage 1 Disinfectants and Disinfection Byproducts Rule—based on annual average
[6]The sum of the ratios of concentration for each to their respective GV not to exceed 1.
  (Note: bromoform is tribromomethane; chloroform is trichloromethane)
[7]A lower value should be aimed for but without compromising disinfection.

Regulations (UK, 2001c and d); and in Northern Ireland in 2002 (UK, 2002), with subsequent amending Regulations in 2003 (UK, 2003). In 2007 new Regulations were introduced in Northern Ireland (UK, 2007a) to reflect changes in the delivery of water services and the English and Welsh Regulations were subject to further Amendment Regulations (UK, 2007b and c). The Regulations contain a number of national standards, most of which are indicator parameters under the Directive. They are also statutory instruments, with a legal obligation for water suppliers to supply wholesome water as defined under the Regulations.

The US EPA National Primary Drinking Water Regulations

As specified in the 1996 amendments to the Safe Drinking Water Act (US EPA, 1996).

In the USA, drinking water quality regulations were mandated by law under the Safe Drinking Water Act of 1974 and are issued by US EPA. There are two categories of standards, the National

Primary Drinking Water Regulations (NPDWR), which are health related standards, and National Secondary Drinking Water Regulations, which are non-enforceable guidelines relating mainly to aesthetic parameters. Maximum Contaminant Levels (MCLs) are specified for a wide range of parameters under the primary standards, along with Maximum Residual Disinfectant Levels where appropriate. Several standards are linked to treatment or monitoring requirements, which have been promulgated as a series of Rules. Full details of these are available on the US EPA website (*http://www.epa.gov/safewater*). The Regulations apply to public water supply systems and the primary standards are enforceable by US EPA and are mandatory on all public water suppliers. However, US EPA is required to carry out a cost benefit analysis on each standard and may, if necessary, adjust the standard for a particular supply to a level that 'maximizes health risk reduction benefits at a cost that is justified by the benefits'. The Safe Drinking Water Act requires US EPA to review and revise, as appropriate, each of the NPDWR at least every six years. US EPA is also required to identify and prioritize new contaminants that may become subject to regulation in the future. This is done by way of the Drinking Water Contaminant Candidate List, which was last published in February 2005. Some, but not all of the contaminants listed are also included in the WHO forward work programme. Good accounts of the Regulations are given by Pontius (Pontius, 1998; 2002; 2003).

6.54 COMMENT ON THE APPLICATION OF HEALTH RELATED STANDARDS

Drinking water could be one of several sources of exposure for the health related standards listed in Tables 6.1(A) to (D). In some cases it may be a very minor source, with little impact on the overall exposure. In other cases, such as lead, drinking water may play a significant role in the overall level of exposure.

Occasionally it may be possible to determine impact of exposure and the associated health effects from studies involving humans. More usually studies are carried out on laboratory animals (e.g. rats and mice), especially if carcinogenic or potentially carcinogenic substances are involved. The dose at which there is no observed adverse effect (NOAEL) is determined and 'uncertainty factors' varying from 10^{-2} to 10^{-4} are applied to take account of species differences, nature and severity of adverse effects, and the quality of the data. Further factors are then applied to give a tolerable daily intake (TDI), which is the amount that can be ingested without any appreciable risk to health over an average human life span (usually taken to be 70 years), adjusted for the average human body weight (usually taken to be 60 to 70 kg). Consideration has to be given to the contribution of drinking water to the overall exposure, compared to the intake from food and other sources, and the volume of water consumed (usually taken to be an average of 2 litres per day). Account also has to be taken of the impact on infants and children who, because of their higher intake and lower body weight, may be more susceptible than adults to exposure from certain substances. In the case of carcinogens or potential carcinogens, mathematical models may be used to extrapolate the effect to dose levels low enough to reduce the risk of cancer development in human cells to an 'acceptable level', such as 1 case per lifetime per 100 000 individuals. Problems may arise over interpretation whichever procedure is used, for example: limited term tests at high dosage rates may not be representative of life-duration ingestion of much smaller quantities; the extrapolation relationship may be assumed wrongly; and different effects may arise with combinations of substances or different forms of a substance.

Table 6.1(C) Pesticides—parameters of health significance

	WHO Guidelines 3rd Ed. Volume 1, with addenda. Health related guideline value	US EPA Regulations under Safe Drinking Water Act. Maximum contaminant level	UK DoE Guidance on Safeguarding the Quality of Public Water Supplies 1989. Advisory value
PESTICIDES—see also Table 6.1(B)			
Alachlor	0.02 mg/l	0.002 mg/l	
Aldicarb	0.01 mg/l		
Atrazine	0.002 mg/l	0.003 mg/l	2 µg/l
Carborfuran	0.007 mg/l	0.04 mg/l	
Chlordane	0.0002 mg/l	0.002 mg/l	0.1 µg/l - total isomers
Chlortoluron	0.03 mg/l		80 µg/l
Chlorpyrifos	0.03 mg/l		
Cyanazine	0.0006 mg/l		
DDT+ metabolites	0.001 mg/l		7 µg/l - total isomers
2,4-Dichlorophenoxy-acetic Acid (2,4-D)	0.03 mg/l	0.07 mg/l	1000 µg/l
2,4-DB	0.09 mg/l		
2,4,5-T	0.009 mg/l		
1,2-dibromoethane (Ethylene dibromide)	0.0004 mg/l^P	0.00005 mg/l	
1,2-dibromo-3-chloropropane (DBCP)	0.001 mg/l	0.0002 mg/l	
1,2-dichloropropane (1,2-DCP)	0.04 mg/l^P	0.005 mg/l	
1,3-dichloropropene	0.02 mg/l		
Dalapon		0.2 mg/l	
Dichlorprop	0.1 mg/l		40 µg/l
Dimethoate	0.006 mg/l		3 µg/l
Dinoseb		0.007 mg/l	
Diquat		0.02 mg/l	
Endothall		0.1 mg/l	
Endrin	0.0006 mg/l	0.002 mg/l	
Fenoprop (2,4,5-TP)	0.009 mg/l	0.05 mg/l	
Glyphosate		0.7 mg/l	1000 µg/l
Hexachlorobenzene (HCB)		0.001 mg/l	0.2 µg/l
Hexachlorobutadiene (HCBD)	0.0006 mg/l		
Isoproturon	0.009 mg/l		4 µg/l
Lindane	0.002 mg/l	0.0002 mg/l	
MCPA	0.002 mg/l		0.5 µg/l

(*continued*)

Table 6.1(C) (*Continued*)

	WHO Guidelines 3rd Ed. Volume 1, with addenda. Health related guideline value	US EPA Regulations under Safe Drinking Water Act. Maximum contaminant level	UK DoE Guidance on Safeguarding the Quality of Public Water Supplies 1989. Advisory value
MCPB			0.5 µg/l
Mecoprop (MCPP)	0.1 mg/l		10 µg/l
Methoxychlor	0.02 mg/l	0.04 mg/l	30 µg/l
Metolachlor	0.01 mg/l		
Molinate	0.006 mg/l		
Oxamyl (Vydate)		0.2 mg/l	
Pendimethalin	0.02 mg/l		
Pentachlorophenol	0.009 mg/l[P]	0.001 mg/l	
Permethrin	0.3 mg/l[a]		
Picloram		0.5 mg/l	
Pyriproxyfen	0.3 mg/l[b]		
Simazine	0.002 mg/l	0.004 mg/l	10 µg/l
Terbuthylazine (TBA)	0.007 mg/l		
Toxaphene		0.003 mg/l	
Trifluralin	0.02 mg/l		

Notes: [P]Provisional guideline value; [a]When used as a larvicide; [b]Guideline value not intended when used as a vector control agent in drinking water.

The above is a very simplified overview of the way in which health related drinking water quality standards are developed and there are numerous papers that give a more scientific account (e.g. Fawell, 1991). Suffice to say that all the health related chemical standards listed in Tables 6.1(A)–(D) have large safety margins built in to ensure that the results achieved are no worse than those estimated on the basis of an 'acceptable degree of risk'. Failure to meet a health related standard means that the water is unwholesome but it does not necessarily mean that the water is unsuitable or unfit for human consumption.

6.55 RAW WATER QUALITY

WHO no longer provides specific guidance on raw water classification. Instead it recommends effective catchment management and source protection as part of the water safety plan approach (see Part VI). However the relationship between treatment and the levels of faecal contamination developed by WHO in 1993 (1993), summarized in Table 6.2, remains relevant today, as does the conclusion that adequate disinfection should produce at least 99.99% reduction of any enteric viruses present in the raw water.

Although such raw water classifications provide useful guidelines in assessing whether a source of water is suitable for public supply, in many parts of the world it may be necessary to take account

Table 6.1(D) Radioactive substances—parameters of health significance

	WHO Guidelines 3rd Ed. Volume 1, with addenda. Health related guideline value	UK Water Supply (Water Quality) Regulations. Prescribed conc. or value	EC Directive 98/83/EC November 1998. Parametric value	US EPA Regulations under Safe Drinking Water Act. Maximum contaminant level (MCL)
Reference dose level/total indicative dose	0.10 mSv/year	0.10 mSv/year [#,1]	0.10 mSv/year [#,1]	
Screening levels –				
Gross alpha activity	≤0.5 Bq/l[2]	0.1 Bq/l		15 pCi/l[3]
Gross beta activity	≤1 Bq/l[2]	1 Bq/l		4 mrem/year[4]
Radium 226 + 228				5 pCi/l
Radon	>100 Bq/l			
Tritium[5]		100 Bq/l[#]	100 Bq/l[#]	
Uranium (U)	0.015 mg/l[P]			30 µg/l

Notes:
[#] Indicator parameter, with standard set for monitoring purposes only.
[P] Provisional
[1] Total indicative dose excludes tritium, potassium-40, radon and radon decay products.
[2] Further action required above these levels.
[3] 1 pCi/l = 0.037 Bq/l.
[4] Beta particles and photon emitters. 1 mrem/year = 0.01 mSv/year.
[5] Effectively a screening parameter for the presence of artificial radionuclides. Monitoring only required if there is a source of tritium in the catchment and it cannot be shown by other means that the level of tritium is well below 100 Bq/l.

of other circumstances. Consideration should always be given to the unlisted or unquantifiable risks that apply in a catchment, such as the desirability of choosing a source that is least likely to be affected by domestic and industrial wastes, or subject to dangerous accidental pollution. Account also needs to be taken of physical and financial constraints that may make it impossible for the water supplier to provide water that complies fully with the quality requirements of the WHO Guidelines or other standards. For example, problems may arise in maintaining consistent treatment and disinfection because of lack of resources, inadequate supplies of materials, or lack of skills in the local labour force. However, every effort should be made to provide consumers with a palatable and aesthetically pleasing water supply that is free of bacteria and objectionable tastes or odours, especially if the alternative is an untreated source of doubtful quality.

The 1975 EC Directive on the quality required of surface water intended for the abstraction of drinking water (CEU, 1975) was repealed at the end of 2007 as part of the implementation of the Water Framework Directive (CEU, 2000), although the recommended steps to prevent deterioration of raw water quality are likely to remain relevant for some time to come. The Water Framework Directive sets out a long-term perspective for the management and protection of the aquatic environment by Member States and one of the aims is to achieve 'good status' for all water bodies by 2015.

Table 6.1(E) Physical characteristics and substances undesirable in excess

	WHO Guidelines 3rd Ed. Volume 1, 2004, with addenda. Values based on aesthetic quality and acceptability to consumers	UK Water Supply (Water Quality) Regulations. Indicator parameters specification concentration or value	EC Directive 98/83/EC November 1998 indicator parameters - parametric value	US EPA Regulations under Safe Drinking Water Act. National Secondary Drinking Water Regulations. Secondary standards
Aluminium (Al)	0.2 mg/l	200 µg/l*	200 µg/l	0.05–2.0 mg/l
Ammonia (NH_4)		0.50 mg/l	0.50 mg/l	
Chloride (Cl)	250 mg/l	250 mg/l	250 mg/l	250 mg/l
Colour	15 TCU	20 mg/l Pt/Co scale*	Acceptable to consumers and no abnormal change	15 colour units
Conductivity		2500 µS/cm at 20°C	2500 µS/cm at 20°C	
Copper (Cu)	1 mg/l			1 mg/l
Dissolved solids	1000 mg/l			500 mg/l
Fluoride (F)				2 mg/l
Hydrogen sulphide	0.05 mg/l as H_2S			
Iron (Fe)	0.3 mg/l	200 µg/l*	200 µg/l	0.3 mg/l
Manganese (Mn)	0.1 mg/l	50 µg/l*	50 µg/l	0.05 mg/l
Oxidisability (O_2)			5 mg/l[1]	
pH (Hydrogen ion)		6.5–9.5*	≥6.5 and ≤9.5[2]	6.5–8.5
Silver (Ag)				0.10 mg/l
Sodium (Na)	200 mg/l	200 mg/l*	200 mg/l	
Sulphate (SO_4)	250 mg/l	250 mg/l	250 mg/l	250 mg/l
Surfactants (Foaming agents)				0.5 mg/l
Taste/Odour		Acceptable to consumers and no abnormal change*	Acceptable to consumers and no abnormal change	3 as Threshold Odour Number
Total organic carbon		No abnormal change	No abnormal change	
Turbidity - See also Table 6.1(A)	5 NTU	1 NTU ex works 4 NTU at the consumer's tap*	Acceptable to consumers and no abnormal change <1 NTU ex works[3]	
Zinc	3.0 mg/l			5 mg/l

Notes:
* National standards with prescribed concentrations or values.
[1] Does not have to be measured if TOC is being measured.
[2] Should not be aggressive.
[3] For works treating surface waters.

Table 6.2 Classification of water sources according to bacterial quality and the recommended level of treatment

Source	Level of contamination	Treatment
Ground Water[a]		
Deep protected wells[b]	Free of faecal contamination *E. Coli* 0/100 ml	Disinfection[c] for distribution purposes only
	Evidence of faecal contamination *E. Coli* ≤20/100 ml	Disinfection[c]
Unprotected ground water e.g. shallow wells	Faecal contamination *E. Coli* ≤2000/100 ml	Filtration[d] and disinfection[c]
	Evidence of faecal contamination *E. Coli* >2000/100 ml	Not recommended as a source of drinking water[e]
Surface Waters[a]		
Protected impounded waters	Essentially free of faecal contamination *E. Coli* ≤20/100 ml	Disinfection[c]
Unprotected impounded upland water or upland river	Faecal contamination *E. Coli* 20–2000/100 ml	Filtration[d] and disinfection[c]
Unprotected lowland river	Faecal contamination *E. Coli* 200–20 000/100 ml	Long term storage or pre-disinfection, filtration[d], additional treatment[f] and disinfection[c]

Notes:
[a]If the sources are contaminated with *Giardia* cysts or *Cryptosporidium* cysts they must be treated by processes additional to disinfection (Sections 8.19 and 8.20).
[b]Water must comply with the WHO guideline criteria for pH, turbidity, bacteriological and parasitological quality.
[c]WHO conditions for final disinfection must be satisfied (Section 11.5).
[d]Filtration must be either rapid gravity (or pressure) preceded by coagulation-flocculation and where necessary clarification or slow-sand filtration. The degree of virus reduction must be >90%.v
[e]Water from these sources should be used only if no higher quality source is available. Drinking water from such sources carries a risk of inadequate virological quality.
[f]Additional treatment may consist of slow sand filtration, ozonation with granular activated carbon absorption or other processes demonstrated to achieve >99% virus reduction.

Annex VIII of the Directive sets out an indicative list of the main chemical pollutants that need to be assessed and in July 2006 the Commission set out a proposal for a Directive on related environmental quality standards (CEU, 2006). There are 33 priority substances listed, along with eight other pollutants that have to be considered.

The Water Framework Directive also requires Member States to establish a register or registers of areas where special protection of groundwaters or surface waters is needed. This includes bodies of water that are used for the abstraction of water for human consumption and provide more than an average of 10 m³/day in total, or serve more than 50 consumers. It also includes bodies of water intended for such level of use in the future. The main objectives of these Drinking Water Protected Areas are to ensure that, with the correct treatment regime in place, the treated water entering supply will meet the requirements of the Drinking Water Directive and that the raw water quality is unlikely to deteriorate to the point where additional treatment is required. US EPA likewise advocates identifying and controlling potential contaminants in the catchment and the need for source protection.

6.56 SAMPLING FOR PHYSICAL AND CHEMICAL PARAMETERS

Sampling Frequencies to WHO, EC, UK and US EPA Requirements

WHO provides guidance both on operational monitoring and on the level of sampling needed to confirm the quality of the water supplied. Rather than specifying sampling frequencies, emphasis is placed on the need for clearly defined objectives that take account of the local conditions, including facilities available for sampling and analysis, and the variability and likely levels of contamination that may be present in the water.

The EC Drinking Water Directive sets out a minimum frequency for sampling and analysis based on the volume of water distributed each day within a water supply zone. Where water is supplied via a distribution system, the point of compliance is deemed to be the point in a building where drinking water is normally made available to consumers. There are two levels of monitoring under the Directive, audit monitoring and check monitoring. Audit monitoring is carried out for the health related parameters that have mandatory standards, whereas check monitoring is intended to provide regular information on the general microbiological quality of the water and the overall effectiveness of the treatment process. Check monitoring, which relates mainly to the indicator parameters, is carried out at higher sampling frequencies than those required for audit monitoring. For a typical zone supply receiving 10 000 m^3/day and serving a population of around 45 000 consumers, 34 samples a year would be required for parameters on check monitoring and four samples a year would be required for parameters on audit monitoring. In all cases the samples have to be representative of the quality of the water supplied throughout the year in each water supply zone.

The UK Regulations follow the same pattern of check monitoring and audit monitoring, taking into account the need for both check and audit monitoring for those parameters that are also national standards. Sampling frequencies at treatment works are based on the volume supplied, whereas sampling frequencies at consumers' taps are based on zonal population, with a maximum allowable population of 100 000 per water supply zone. Zonal samples are usually taken at randomly selected properties, although some parameters may be sampled from a fixed point within the supply system, such as the outlet of a treatment works, provided the water is representative of the zone as a whole. Fixed points can only be used for certain parameters, such as pesticides, that remain at the same concentration or value throughout the distribution system. Samples for parameters such as lead, copper and trihalomethanes have to be taken from consumers' taps, with the metal samples always being taken as 'first-draw' samples.

US EPA sampling requirements are complex. They vary with the parameter to be monitored, the type of source used and whether the source is defined as 'vulnerable' or 'non vulnerable' to specific contaminants. Parameters under the National Secondary Drinking Water Regulations are required to be monitored at intervals no less frequently than those for inorganic chemical contaminants. Monitoring is also required for parameters on the Chemical Contaminants List, for which final standards have yet to be set.

Minimum Sampling Requirements Where no Regulations Apply

In countries where there are no legal requirements for sampling, the following regime is likely to provide an adequate minimum level of monitoring, linked with the priorities suggested above.

Simple chemical tests should be carried out daily on the raw water and on the treated water leaving the treatment works. Samples should also be taken at least weekly from the associated distribution system and preferably from consumers' taps. The tests should be for the more easily measurable, but important parameters, such as colour, taste, odour, turbidity, pH, conductivity and chlorine residual in the case of treated waters. Other parameters might be included to meet the requirements of a particular source or situation, for example, chloride if salinity is a problem; nitrate and ammonia as simple pollution indicators; iron, manganese and lead as necessary; and residual coagulant metal and possibly hardness to check treatment performance.

Full chemical analyses should be carried out, including tests for toxic substances, on any new raw water source before it is put into supply, whenever treatment processes are being significantly altered or if new sources of pollution are suspected. Routine samples for full chemical analysis of water in the distribution system should be taken quarterly, six-monthly, or yearly, depending on the type of water and the size of the population supplied. If there is any cause for concern then more frequent checks should be carried out for parameters of health significance, for example trihalomethanes, pesticides, PAH and the heavy metals. However, this should be done on a risk management basis, especially where local laboratory capabilities are limited.

Sampling Techniques for Physical and Chemical Parameters

It is of paramount importance that correct procedures are always followed whenever water samples are taken, to ensure that the samples are representative of the water being supplied. Whenever possible samples should be taken by trained and experienced personnel, using dedicated sampling bottles and equipment.

Methods of sampling for chemical parameters are fully documented in a series of UK publications under the title of Methods for the Examination of Waters and Associated Materials (EA, 1976). Individual water companies in the UK have developed their own sampling manuals based on these publications. Another useful publication is Standard Methods for the Examination of Water and Wastewaters (AWWA, 2005) jointly produced by AWWA, the American Public Health Association and the American Water Pollution Control Federation. ISO also provides guidance on design of sampling programmes and sampling techniques (ISO, 2006).

On-site Testing and Field Analysis

It is desirable that the analysis of some parameters, such as temperature, pH and residual disinfectant, is carried out at the time of sampling as significant changes in concentration or value may occur even over a short period of time. Other parameters requiring on-site measurement include redox potential, dissolved oxygen and carbon dioxide. These are particularly important for groundwater samples where there can be a rapid change in concentration (or value) due to pressure changes as the sample is taken. The loss of carbon dioxide from a deep borehole sample may result in an increase of pH and, if present, soluble iron and manganese can precipitate out as the water comes into contact with atmospheric oxygen. Special sampling techniques are required in such situations.

Under some circumstances it may be necessary to carry out a more comprehensive analysis in the field, especially if the site is remote and without ready access to a laboratory. There are numerous test kits available for field analysis, covering a wide range of parameters, and varying greatly in complexity and accuracy. The simplest are 'test strips' which are dipped into the water sample. The intensity of the subsequent colour development is then compared against a strip of standard colours for specific concentrations of the parameter under test. These provide a fairly crude but objective result. There is a wide range of pre-calibrated test discs also available for monitoring most of the common water quality parameters. Reagents, usually in tablet form, are added to a standard volume of sample contained in a glass sample cell. The resultant colour development is compared visually against a blank sample, using a coloured glass test disc in the appropriate range of concentration. Most of these test kits are user friendly and can be used with minimal training but the results should always be treated with a degree of caution, as they involve visual comparisons. It is also advisable to introduce an element of analytical quality control, as far as is practicable, to such systems, especially if the results are intended for regulatory reporting.

A broad range of parameters can be monitored in the field by means of electronic meters, for example, pH, redox, dissolved oxygen, turbidity and temperature. Multi-parameter meters are also available which can measure a number of parameters with a single instrument. Such instruments are available with an integral data logging system and are extremely useful for continuous monitoring surveys. Comprehensive 'field laboratories' are also commercially available. These typically provide reagents and apparatus for measuring single parameters, such as residual chlorine, or multiple parameters within a single unit. A pre-calibrated spectrophotometer is used for colorimetric tests, with electronic meters for pH and conductivity. A digital titration system may also be included for parameters such as hardness and chloride. A higher level of skill is required to operate such instruments and a higher level of analytical quality control should be applied to ensure the validity of the results.

Water Quality Monitoring at Treatment Works

On-line monitoring systems can be used to check the quality of the raw water and the final treated water as it leaves a water treatment works. Adequate monitoring of the raw water forms an essential part of process optimisation, especially if the source is of variable quality or is subject to sudden and dramatic changes in quality. Devising a reliable system can prove difficult as it is impossible to monitor for all the potentially harmful substances that may be present. However, any raw water that is abstracted directly from a lowland river or canal should be monitored for pH, turbidity, dissolved oxygen, temperature and conductivity as a minimum, with visual checks being carried out at the intake at regular intervals. Surrogate monitoring systems such as fish monitors have been used with some success. The simplest form is a tank, containing a number of suitably sensitive fish such as salmonids, which is supplied continuously with the water being monitored. Any fish deaths may indicate the presence of a pollutant that needs to be investigated further. A more sophisticated version involves continuously monitoring the gill movement or electrical responses of a number of fish. Organic monitors that detect changes in the absorbance of the raw water at a specific ultra violet frequency provide another alternative. In an ideal situation the raw water quality should be monitored upstream of the intake, with alarm systems connected either to the treatment works control room or to a central control room that is manned permanently, and with sufficient time built in to shut the intake in the event of a major pollution incident. Bankside storage helps to even out fluctuations in raw water quality and also helps to maintain supplies if the intake has to be closed for a short period to allow a plume of pollution to pass by. Ideally such storage should be divided into at least two separate compartments, so that the raw water in each can be tested alternately before being used.

On-line monitoring systems for pH and turbidity are widely used to check the efficacy of the clarification and filtration processes and for checking the quality of the treated water entering supply. Similar systems are used for monitoring the residual disinfectant level at various stages during the disinfection process and also in the treated water entering supply (Section 11.16).

6.57 PRIORITIES IN WATER QUALITY CONTROL

In some parts of the world, the availability of well equipped laboratories and resources for water quality testing may be limited. The level of testing in such circumstances must concentrate on the most essential parameters. The following is a suggested list of priorities in testing which, for the sake of completeness, includes microbiological testing as defined in Section 6.69.

Simple Checks at Source Works

If the source is a river or canal, visual checks of the raw water should be carried out twice daily at the inlet to the works. Daily measurements should also be carried out on samples of the raw water for turbidity, colour, odour, conductivity and pH. Where coagulation, clarification and filtration are applied coagulant dosage and the pH and turbidity of the water ex clarifiers and ex filters should be checked daily.

Twice daily checks should be carried out on chlorine dosage rate and on residual chlorine content of water entering supply, along with daily checks for turbidity, colour, taste, odour, conductivity and pH. All results should be recorded on a day sheet and reviewed at regular intervals by a suitably qualified person. Any trends should be identified and acted upon.

Microbiological Testing (See also Sections 6.69 and 6.70)

Where possible, analysis for coliform bacteria and *Escherichia coli* should be carried out at least weekly on samples of treated water leaving a water treatment works. Samples of raw water and samples from the distribution system should be analysed at least monthly, taking into account the level of risk and population factors as appropriate. Every effort should be made to achieve these minimum frequencies but, where they cannot be met, resources should be directed towards determining coliform counts in the treated water leaving the treatment works. If no suitable laboratory is available for microbiological analyses, consideration should be given to on-site testing by a trained operative visiting the works on a routine basis. Portable test kits are available using filter membrane techniques, as are a number of test kits which determine whether an indicator organism is simply present or absent in 100 ml of sample.

Chemical Testing

At all sources, chemical analyses covering the most important parameters of a given raw water should be conducted at least twice a year. Samples should be taken at appropriate times, e.g. following the onset of heavy rainfall after prolonged dry weather, along with samples for microbiological analyses.

Source Watch (see also Part VI)

The catchment to a new source should be surveyed and mapped fully, with any potential sources of pollution noted, before the source is brought into use. Subsequent catchment surveys should be carried out at regular intervals and any changes noted. Where chemical tests show that toxic substances are present in a raw water, their source should be traced and if possible eliminated. If this is not possible, additional monitoring must be set up for such substances in both the raw and the treated water. The use of waters having consistently high coliform and *E. coli* counts, or dangerously sited with respect to any waste discharge, should be avoided if at all possible. Attempts should be made to divert any sources of pollution that are close to an intake. If this is not possible then they should be kept under frequent observation.

In general a given raw water tends to have characteristics that fall within a certain range, often according to rainfall or other seasonal conditions. As soon as sufficient microbiological and chemical data have been obtained to establish this range and after appropriate treatment has been adopted, routine testing can concentrate on monitoring the parameter or parameters which are most likely to indicate any abnormal change of quality. However, routine microbiological testing of the treated water leaving the treatment works should always be carried out at a frequency that demonstrates that the treatment processes and, in particular, disinfection have not been compromised.

6.58 METHODS OF CHEMICAL ANALYSIS

It is important that methods of analysis be standardized as far as possible in order to achieve comparability of results. In 1972 in the UK, the then DoE established the Standing Committee of Analysts to set up working groups to produce suitable methods for water analysis. The committee, which now comes under aegis of the EA, represents a wide range of interests in the water industry and has produced detailed guidance in a series of publications under the title of Methods for the Examination of Waters and Associated Materials (EA, 1976). Each publication looks at a single analytical method or linked group of methods. Another useful publication is Standard Methods for the Physical and Chemical Examination of Water and Wastewaters (AWWA, 2005). This comprehensive book, which is updated at regular intervals, provides a valuable reference and the methods described in it have had a wide influence on standards adopted in other countries. Field Testing of Water in Developing Countries (Hutton, 1983) provides further useful reference.

All analytical methods for chemical parameters should be fully evaluated and validated by the laboratory carrying out the tests before a method is adopted for routine use. The increasing use of sophisticated analytical equipment means that many methods have to be adapted for use with a particular instrument. The initial performance tests should demonstrate that the analytical system is capable of establishing, within acceptable limits of deviation and detection, whether any sample contains the parameter under analysis at concentrations likely to contravene the required standard. Such performance testing should cover the entire analytical procedure, including sample preparation and any concentration steps. The precision and accuracy of the test in terms of maximum tolerable values for total error and systematic error and limit of detection should be ascertained, along with checks for recovery and resilience against possible interferences. The required performance characteristics for chemical parameters are specified both in the EC Drinking Water Directive and in the US NPDWR. In the UK, guidance on analytical systems was issued in the DoE publication on Guidance on Safeguarding the Quality of Public Water Supplies (DoE, 1990a) and in subsequent information letters sent by the DWI to water companies in England and Wales. These can be found on the DWI website (*http://www.dwi.gov.uk*). However it is recognized that the required level of performance cannot always be met with the current methods of analysis available, especially for some of the more obscure organic parameters.

Each laboratory should also have established and documented procedures for routine analytical quality control as applied to each validated method. External quality control schemes or inter-laboratory proficiency testing schemes, where available, provide further useful information on a laboratory's capabilities for carrying out an acceptable level of analysis. Guidance on setting up such schemes is available from the ISO (ISO, 2007).

Under the US NPDWR all analyses, other than turbidity, chlorine residual, temperature and pH, have to be carried out by a certified laboratory. The UK Regulations requires that samples be analysed by or under the supervision of a person who is competent to perform the task. The DWI has issued guidance on the level of training considered necessary to meet these requirements. Furthermore, in England and Wales, water companies are expected to either use suitably third party accredited laboratories or have their laboratories inspected by DWI.

6.59 QUALITY ASSURANCE OF WATER TREATMENT CHEMICALS AND MATERIALS IN CONTACT WITH DRINKING WATER

Some water treatment chemicals and materials that come into contact with drinking water may have an adverse impact on the quality of the water supplied. WHO refers to the need for verification protocols as part of its Water Safety Plan approach to ensuring the safety of drinking water supplies (Section 6.83). The EC Drinking Water Directive requires Member States to take appropriate measures to ensure that such chemicals and materials or associated impurities do not pose a risk to human health. The majority of chemicals and filter materials used by Members States in water treatment are covered by appropriate European Standards; the European Commission's (DG Enterprise) Expert Group—Construction Products Drinking Water is providing advice on the development of a revised European Acceptance Scheme for drinking water construction products. Under the UK Water Supply (Water Quality) Regulations all chemicals used in the treatment of drinking water and materials which come in contact with drinking water (such as pipe linings etc.) have to bear an appropriate CE marking, or conform to an appropriate harmonized standard or European technical approval or an appropriate British Standard. All other products and materials have to be approved by the appropriate authorities. For England and Wales the DWI publishes an electronic list of products for which the Secretary of State's approval has been granted, refused, revoked or modified or for which use has been prohibited. This is available on the DWI website (*http://www.dwi.gov.uk*) and covers items such as: water treatment chemicals; filtration, including membrane filtration and electrodialysis systems, and ion exchange media; systems used to generate disinfectants *in-situ*; construction products used in water treatment processes, including pipework and storage installations; *in-situ* applied repair material; construction products and coatings used in raw water and treated water installations; and water retaining vessels and pipework used for the provision of drinking water in emergencies. Maximum permissible dosages are stipulated for some proprietary products, together with other conditions of use as appropriate (Sections 6.3, 6.25 and 6.28). If a water company fails to meet these requirements, then it is guilty of an offence. Similar arrangements apply in Scotland and Northern Ireland.

PART III WATER MICROBIOLOGY

6.60 DISEASES IN MAN THAT MAY BE CAUSED BY WATER-BORNE BACTERIA AND OTHER ORGANISMS

There are a number of diseases in humans, which can be water-borne and may be caused by the presence of pathogenic bacteria and other organisms such as protozoa and viruses in drinking water supplies, in water used for bathing or immersion sports, or via other routes. It has long been recognized that the ingestion of water contaminated with excrement can result in the spread of diseases such as cholera and typhoid and that adequate treatment and other measures are required to prevent outbreaks of such diseases. The WHO Guidelines state that 'The most common and widespread health risk associated with drinking-water is microbial contamination, the consequences of which mean that its control must always be of paramount importance.'

The following sections look at some of the intestinal diseases which have commonly, although not invariably, been water-borne. In all cases the associated organisms are present in large numbers in the excreta of an infected host and are relatively resistant to environmental decay. Many can cause illness even when ingested in small numbers. Schistosomiasis and other parasitic diseases of the tropics are reviewed as part of water biology in Part IV.

6.61 BACTERIAL DISEASES

Cholera is caused by the bacterium *Vibrio cholerae* and is characterized by acute diarrhoea and dehydration. Many strains have been identified and those in serogroups O1 and O139, which produce cholera toxin, are associated with outbreaks of epidemic cholera. Infection is usually contracted by ingestion of water contaminated by infected human faecal material, but contaminated food and person to person contact may also be sources. In recent years cholera has moved from the Far East to the Near East, Africa and southern Europe and has the potential to enter Britain and other European countries through carriers or by individuals who are in the incubation period of the disease. In 1991 the seventh cholera pandemic caused by the so called "El Tor" biotype which started in Indonesia in 1961 appeared in South America causing 750 000 cases and 6500 deaths in one year. The bacteria are short lived in water and have low resistance to chlorine so it is unlikely to spread in communities with controlled water supplies and effective sewage disposal.

Typhoid fever is caused by the bacterium *Salmonella* Typhi. Infection is usually contracted by ingestion of material contaminated by human faeces or urine, including water and food (e.g. milk, shellfish). *Salm.* Typhi occasionally continues to proliferate in the gall bladder of a few patients who have recovered from the primary infection; these carriers continue to excrete the organisms in their faeces or, occasionally, in their urine for long periods, even for life. In 1937 a large water-borne outbreak of typhoid fever occurred in the town of Croydon, killing 43 people. Investigations revealed that it was caused by a combination of circumstances including a person who was a carrier of *Salm.* Typhi working down a well, which was pumping into supply, coincidental with the filtration and chlorination plants being bypassed (Suckling, 1943). There have been a number of more recent outbreaks, which are believed to have been due to water contamination coinciding with inadequate disinfection as the bacteria are moderately persistent in water but have low resistance to chlorine. *Paratyphoid fevers* are also caused by *Salmonella,* in this case Paratyphi *A, B,* or *C* and infection may exceptionally be via contaminated water.

Bacillary dysentery is caused by bacteria of the genus *Shigella*, with *Sh. dysenteriae* 1, *Sh. flexneri*, *Sh. boydii* and *Sh. sonnei* being the four recognized species. Infection can occasionally be contracted via water contaminated by human faeces but more commonly is due to ingestion of food contaminated by flies or by unhygienic food handlers who are carriers. *Traveller's diarrhoea* is a term applied generally and may have several potential causes, but it may be that some forms of pathogenic *Escherichia coli* or, rarely, *Shigella* are responsible. It is probably transmitted in the same way as bacillary dysentery and water may sometimes be the vehicle. However the bacteria are short lived in water and have a low resistance to chlorine. The problems associated with enterovirulent *E. coli* are reviewed in Section 6.81.

Leptospirosis is caused by strains of *Leptospira interrogans* of which there are 20 serogroups with numerous serovars. Leptospires are motile, spiral-shaped organisms and pathogenic strains can cause symptoms ranging from mild fever to severe jaundice in the case of Weil's disease. The organisms are shed in the urine of infected rats, dogs, pigs and other vertebrates, and are often present in ponds and slow-flowing streams visited by such animals. People who bathe in, fish in or sail on these waters are at risk, becoming infected via the mouth, nasal passages, conjunctiva, or skin abrasions. The bacteria are very sensitive to disinfectants and are eliminated by normal water treatment processes but sewer workers remain at risk.

Legionnaires' disease is an acute form of pneumonia caused by the bacterium *Legionella pneumophila*. More than 45 species of the genus *Legionella* have been identified and at least 18 have been implicated in infections of humans, causing a range of infections generally termed Legionellosis (EA, 2005). The bacterium *L. pneumophila* is regarded as the most dangerous, having been identified in all outbreaks of Legionnaires' disease. (The name comes from an outbreak of the disease causing high mortality in war veterans attending a Legion Convention in Philadelphia, USA in 1976). Outbreaks of the disease can be sudden and result in mortality rates of around 13%. People who smoke or drink heavily, or are stressed through other illness, are particularly susceptible and there is a distinct age and sex distribution, with older males being most at risk. Other forms of Legionellosis such as Pontiac fever are less severe and exhibit no particular age distribution. *Legionella* bacteria are widely present in low numbers in surface waters and possibly in some groundwaters. They are persistent in water and will multiply, but have a low resistance to chlorine. However, if present within protozoa they can survive conventional water treatment, including disinfection with chlorine, and retain an ability to colonize water systems such as cooling towers, evaporative condensers and even domestic hot water tanks, showers and other plumbing system components. They are thermo-tolerant and ideally suited to grow in the warm water systems of buildings at 30–55 °C. However, they do not survive sustained temperatures above 60 °C. Infection occurs from the transport of the bacterium by aerosols or air-borne water droplets, which are inhaled. Several of the outbreaks of Legionnaires' disease have been caused by aerosols blown from exposed cooling towers of air-conditioning plants at hotels, hospitals and similar large buildings. Whirlpools and Jacuzzis, where recirculated warm water is sprayed, have also been implicated. Official advice is available (HSE, 2000) on preventive measures, including designing and maintaining water systems in buildings to minimize the risk of colonisation; minimising the accumulation of sediments and slimes; and maintaining hot water systems above 60 °C and cold water systems below 20 °C. Biocides are effective at controlling *Legionella* in air conditioning systems using wet evaporative cooling towers but are generally less effective in water distribution systems within buildings. There is no evidence of transmission by ingestion, so infection is not attributable directly to drinking water supplies.

Campylobacteriosis is caused by bacteria of the genus *Campylobacter*, which includes 14 species. It is the most common bacterial enteric pathogen in England and Wales with *Campylobacter jejuni* causing many cases of food poisoning every year. These bacteria are associated with the faeces of a wide range of wild and domesticated animals and birds including poultry and gulls. *Campylobacters* are widespread in the environment and frequently found in sewage. They have been detected in surface waters, where they can survive for several weeks at cold temperatures. They are sensitive to disinfectants used in water treatment and conventional water treatment should prevent their presence in water supplies. However, there have been a number of reported outbreaks associated with unchlorinated or inadequately chlorinated surface water supplies or with contaminated storage facilities. In May 2000 a major drinking water quality incident occurred in Walkerton, Ontario, which affected more than 2300 consumers and caused seven deaths. *Campylobacter* and enterovirulent *E.coli* (Section 6.81) were both implicated. Problems can also occur with private water supplies where there is no treatment.

6.62 OTHER BACTERIA

The *Pseudomonas* group is commonly found throughout the environment. Some species may be present in human and animal excrement and many can multiply in water containing suitable nutrients. Their subsequent growth, if present in drinking water, may result in an overall deterioration in the microbiological quality and lead to consumer complaints of taste and odour. *Ps.aeroginosa* is an opportunist pathogen; it can cause infection particularly in people whose natural defence mechanisms may be impaired, e.g. the very old, the very young or the immuno-suppressed. Other pseudomonads may produce undesirable slime layers or biofilms within the distribution system.

The *Aeromonas* group is also naturally present in the aquatic environment. Their presence in drinking water does not necessarily indicate faecal pollution but highlights possible inadequacies in the treatment process or ingress within the distribution system. The number of such organisms likely to be found in a distribution system depends on the residence time of the water, its organic content and the residual disinfectant level.

6.63 PROTOZOAL DISEASES

Amoebic dysentery is caused by the microscopic parasite *Entamoeba histolytica.* The parasite is distributed throughout the world and exists in two stages, only one of which, the cyst, is infective. The parasite infects mainly primates and, following infection, resides in the large intestine in humans where it reproduces and generates further cysts, which are passed in the faeces. Infection takes place by ingestion of these cysts, which range in size from 10 to 20 μm. They can survive for several days in water at temperatures of up to 30 °C and are relatively resistant to disinfection. Coagulation, clarification and filtration are the most commonly employed methods for physical removal of parasites, while ozone or ultraviolet light are used for inactivation. Combinations of treatment technologies can result in parasite removal/inactivation greater than 6-log, resulting in reliable public health protection. Outbreaks can occur if water supplies are contaminated with domestic sewage containing viable cysts. More commonly the disease is transmitted by person-to-person contact or via food contaminated by carriers.

Cryptosporidiosis is an acute self-limiting diarrhoeal disease caused by the parasite *Cryptosporidium parvum.* Two genotypes are recognized, one of which infects domestic and farm animals and the other, *C.hominis*, is more specific to humans. The faecal-oral route is the usual means of exposure, either by direct contact with infected animals or humans or via contaminated water or food. Not all infected persons necessarily develop the symptoms of the disease but the illness is likely to be serious or even life threatening for patients who are immunologically compromised (Smith, 1992).

The parasite has a complex life cycle, which takes place within the body of the host and can include repeated cycles of autoinfection. Infective oocysts, which are 4–6 μm in diameter, are then shed in vast numbers in the faeces of infected animals and persons. These oocysts are often found in surface waters, particularly in areas associated with intensive animal grazing, and are also found occasionally in some groundwater sources. Treated sewage effluent can, on occasion, also contain large numbers. The oocysts can remain infective in water and moist environments for several months and are resistant to high concentrations of chlorine (Section 11.6). The latter, coupled with the small size of the oocysts, can result in low numbers of oocysts occasionally penetrating conventional water treatment processes. In recent years there have been an increasing number of outbreaks of cryptosporidiosis in both the UK and North America. Following an outbreak in Swindon and Oxfordshire during the winter of 1988-9, an expert committee was set up by the UK Government under Sir John Badenoch, to examine the problem. The Badenoch Report (DoE, 1990b) made many recommendations concerning water treatment practices, monitoring and the role of various authorities in the event of a suspected waterborne outbreak. Five years later a second report (DoE, 1995) was issued providing further guidance. A third report was produced after the Group was re-convened under Professor Ian Bouchier (DETR, 1998), following an outbreak of cryptosporidiosis in North London in which groundwater supplies were implicated. Knowledge has moved on significantly in the intervening years, with ongoing research into methods for assessing oocyst viability and infectivity; methodology for the detection of oocysts in water. Both ozone and ultra violet irradiation have been shown to inactivate the oocysts (Sections 8.19, 11.22 and 11.25), with suitable membrane systems being used to remove them.

One of the largest outbreaks of cryptosporidiosis associated with drinking water occurred in the City of Milwaukee in 1993, when more than 400 000 consumers became ill and a number died. Changes in the seasonal incidence of cryptosporidiosis in the UK point to the effectiveness of focusing on treatment measures. Recent trends show a sustained reduction in the number of cryptosporidiosis cases reported during spring, coincidental with an increased occurrence in farm animals. However, an autumn peak continues to be reported across all regions of the country. Although other factors such as overseas holidays and swimming pools may play an important part in this peak, there is no room for complacency (Sopwith, 2005; Goh, 2005).

Giardiasis is also a diarrhoeal disease and is caused by the protozoan parasite, *Giardia duodenalis*. Like cryptosporidiosis, the disease is usually self-limiting and is caused by the ingestion of cysts by a susceptible host. The faecal-oral route is the usual means of exposure, either by direct contact with infected animals or humans, or via contaminated water or food. Infected animals can contaminate surface waters and, in North America, beavers are frequently blamed for associated outbreaks. The cysts are larger than *Cryptosporidium* oocysts, being 7–10 μm wide and 8–12 μm long, and can survive for many days in a cool aqueous environment. Being larger, the cysts are effectively removed from drinking water by physical methods of treatment, such as filtration. They are also much more susceptible to disinfection than *Cryptosporidium*.

Giardiasis is a worldwide disease and there have been a number of reported waterborne outbreaks in the USA. Evidence of waterborne infection in the UK has been confined to situations where there has been direct faecal contamination of water used for drinking, usually with untreated private supplies.

6.64 VIRAL DISEASES

Viruses differ from bacteria in that they are very much smaller and can multiply only within suitable host cells, in which they produce changes that give rise to a range of diseases. More than one hundred different types of virus have been identified in faeces and the main sources of human enteric viruses in the aquatic environment are sewage discharges. Little direct information is available on the removal of viruses by water treatment processes but information gained using cultured viruses indicates that effective treatment, particularly disinfection, will, if applied properly, produce effectively virus free drinking water. The main areas of risk then become the use of sewage polluted waters for recreational purposes, or the recycling of wastewater for domestic use without adequate treatment and disinfection.

Human enteroviruses, such as poliovirus which causes *Poliomyelitis*, can be found in untreated sewage and even in the effluent from sewage treatment works. The world-wide vaccination programme means that the polioviruses now found in sewage effluents are largely vaccine strains, so infection tends to be very limited and restricted to a few countries in the developing world. Like other viruses, enteroviruses do not multiply in the absence of living cells.

The viruses associated with *Hepatitis A and E* have been detected in sewage and polluted rivers. Contaminated water has been associated with outbreaks of *Hepatitis* in the Far East, either as a result of treatment failure, or where the distribution system has been compromised, or where badly constructed wells have been contaminated from adjacent cesspits, or as a result of heavy rainfall.

Other enteric viruses that are emerging as important risks in the transmission of waterborne viral diseases are reviewed in Part V.

6.65 MICROBIOLOGICAL STANDARDS FOR DRINKING WATER

Pathogenic bacteria and other organisms are often difficult to detect in water that has been treated effectively. If present, their numbers are likely to be very small. Their presence even in sewage effluent or polluted river waters may be only infrequent or at irregular intervals, depending on the level and source of contamination. Analysing directly for pathogenic bacteria is not therefore a practical safeguard for a water supply and, indeed, routine monitoring for process control purposes would be both impracticable and unnecessary. Instead, evidence of any pollution with excreta from man or animals should be sought using simpler and more accessible tests. If the evidence is positive, it should be assumed the water may also contain pathogenic bacteria and must therefore be regarded as unsuitable for supply purposes.

Table 6.3 sets out the WHO guidelines, the EC Drinking Water Directive standards, the current UK standards and US EPA standards for the microbiological quality of drinking water. Coliform bacteria should not be detected in the water leaving a water treatment works provided the treatment processes, particularly disinfection, are adequate. The principal requirements common to all the standards for drinking water at the point of supply, or the consumer's tap, are:

- no *Escherichia coli* (faecal coliforms) detected in 100 ml of sample; and
- at least 95% of 100 ml samples must not show the presence of coliform organisms.

Standards for both parameters should be rigorously adhered to. Depending on the quality of the source water and the type of treatment adopted, biofilms may develop in the distribution system and give rise to occasional failures for coliform bacteria at service reservoirs and at consumer's taps. If a

Table 6.3 Microbiological standards

	Colonies/ml at 22°C	Colonies/ml at 37°C	Total coliforms	E.coli (faecal coliform)	Enterococci (faecal streptococci)	Clostridium perfringens	Cryptosporidium
WHO Guidelines 3rd Edition Vol 1 2004 with addenda							
Treated water entering the distribution system				ND in any 100 ml sample[a]			
Treated water in the distribution system				ND in any 100 ml sample[a]			
All water directly intended for drinking				ND in any 100 ml sample[a]			
EC Directive 1998							
The point where water emerges from taps that are normally used for human consumption for water supplied from a distribution network	No abnormal change[b]		0/100 ml[b]	0/100 ml	0/100 ml	0/100 ml[b]	
UK Water Supply (Water Quality) Regulations							
Water leaving a water treatment works	No abnormal change[b]	No abnormal change[b]	0/100 ml	0/100 ml			
Water in service reservoirs	No abnormal change[b]	No abnormal change[b]	0/100 ml[c]	0/100 ml			
Water at consumers' taps	No abnormal change[b]	No abnormal change[b]	0/100 ml[b]	0/100 ml	0/100 ml	0/100 ml[b]	
US EPA							
At sites representative of the water throughout the distribution system <500 colonies/ml	Treatment requirement for surface water <500 colonies/ml	Treatment requirement for surface water <500 colonies/ml	Treatment requirement for surface water MCL <5% positive samples in a month[d, e] MCLG—zero				MCL-99% removal MCLG—zero

Notes:
ND—Not detectable.
MCL—Maximum contaminant level (US, mandatory standard).
MCLG—Maximum contaminant level Goal (US, non-mandatory).
[a] *E.coli* or thermotolerant coliform bacteria.
[b] Indicator parameter.
[c] 95% of the last 50 samples taken must meet the standard.
[d] Including faecal coliform and *E.coli*.
[e] No more than one sample should be positive for water supply systems that take less than 40 samples per month.

positive result is obtained, the number should always be very low and *E.coli* should not be present. All such failures should be investigated immediately and the need for corrective action, such as increasing the level of disinfection by booster chlorination, assessed. Immediate action should also be taken in the event of even a single *E. coli* (faecal coliform) being detected in a 100 ml sample. Action should always be taken on the presumptive result, even if this does not subsequently confirm, and repeat samples should be taken from the tap giving the original failure and from at least two other taps in adjacent properties and on the same supply.

In addition to specific microbiological standards, the EC Drinking Water Directive requires that water intended for human consumption should not contain any micro-organisms and parasites in numbers that constitute a potential danger to human health. The Directive also sets microbiological standards for water offered for sale in bottles and containers. These are more stringent than the standards that apply at consumers' taps and include additional numerical standards for colony counts and *Pseudomonas aeroginosa*.

There are maximum contaminant level goals of zero for *Cryptosporidium, Giardia* and enteric viruses in public water supplies in the USA. The Surface Water Rules, which apply to systems treating surface water sources or groundwater sources under the direct influence of surface water, enable US EPA to issue a system of 'log removal credits' for *Cryptosporidium, Giardia* and viruses. Two-log or 99% removal is required for *Cryptosporidium*, at works with filtration, with a catchment control programme for works without filtration. Three-log or 99.9% inactivation is required for *Giardia* and four-log or 99.99% inactivation is required for viruses, either by disinfection alone or by a combination of filtration and disinfection. Under the Groundwater Rule, one of the options for corrective action at sources identified as being at risk is to install treatment to achieve four log or 99.99% inactivation or removal of viruses.

Standards of microbiological testing can vary in different parts of the world. Terminology is often inconsistent and confusing, with the same term being used to cover one test procedure in one country and a different test procedure in another. In hot climates many waters give total bacterial counts substantially higher than in temperate climates, and certain types of organisms may be more abundant. It can be difficult for the engineer to interpret bacterial test results and the advice of an experienced water microbiologist should always be sought in obtaining a definitive interpretation.

6.66 USE OF COLIFORMS AS AN INDICATOR OF MICROBIOLOGICAL POLLUTION

Coliform bacteria are widespread throughout the environment and have long been used as indicator organisms by water microbiologists because tests for them are relatively simple and they can be detected in low numbers. The coliform group contains many species which can multiply in water and are not of faecal origin. Some coliforms are able to grow at higher temperatures, giving rise to the terms 'thermotolerant' and 'faecal' coliforms'. Emphasis is usually placed on hygienic importance, for which the occurrence in faeces is the most significant.

Current standards tend to be more specifically based, focusing on *Escherichia coli*. This is a thermotolerant coliform and is consistently present in very large numbers in the faeces of warm-blooded animals, including man, where it is a natural inhabitant of the intestine. Many coliform bacteria, including *E.coli*, can survive for a considerable time in water, making them a good indicator for the presence of many pathogenic bacteria. Thus the detection of *E. coli* in drinking water supplies provides clear evidence of faecal contamination. If coliform bacteria are detected, but no *E. coli*, it is likely that the contamination may be from soil or vegetation, or it may provide a warning that more serious contamination could follow, especially after heavy rain. However the presence of any coliform bacteria in treated water indicates either deficiencies in the treatment process or some form of post treatment contamination and the circumstances should always be investigated immediately.

Although the absence of coliform bacteria, and more particularly *E. coli*, implies that the water is unlikely to be contaminated, it cannot be guaranteed that other intestinal pathogens are absent. This is because other pathogens such as viruses and protozoa, although less likely to be present, may be more resistant to disinfection.

6.67 FREQUENCY OF SAMPLING FOR MICROBIOLOGICAL PARAMETERS

The WHO Guidelines do not specify sampling frequencies. The EC Drinking Water Directive sets minimum frequencies for samples taken from consumers' taps, based on the volume of water supplied within a water supply zone. Colony counts at 22 °C, coliform bacteria, *E. coli* and *Clostridium perfringens* are all subject to check monitoring, with samples for Clostridium *perfringens* only being required if the water originates from or is influenced by surface water. For a zone supply of up to 1000 m³/day the minimum number of samples required per year is four, the number increasing by three samples for every 1000 m³/day or part thereof for larger volumes. Enterococci are monitored at the audit frequency, which is one sample per year for volumes of up to 1000 m³/day, increasing by one sample for each 3300 m³/day or part thereof up to 10 000 m³/day, then three samples plus one sample for each 10 000 m³/day or part thereof up to 100 000 m³/day, and 10 samples plus one sample for each 25 000 m³/day or part thereof up to volumes greater than 100 000 m³/day.

Similar minimum frequencies are specified in the UK Regulations, with additional samples being required to meet the national standards for coliform bacteria and *E.coli* in the water leaving water treatment works and in service reservoirs. US EPA sets out routine monitoring requirements for coliform bacteria and *E.coli* in the Total Coliform Rule. The minimum number of samples is based on the population supplied and range from one sample per month for a population of up to 1000 to 480 samples per month for populations greater than 3 960 000.

Where untreated water is supplied without disinfection, for example with non-piped or community supplies, a minimum sampling frequency should be established based on local conditions.

There are no set frequencies for taking microbiological samples from raw water sources. A degree of monitoring is required for abstraction points within drinking water protected areas in the UK but this is linked to risk assessments of the catchment and the water treatment requirements to meet the Regulation. US EPA requires an average *Cryptosporidium* concentration to be calculated in source waters under the Long Term 2 Enhanced Surface Water Treatment Rule and source water monitoring for *E.coli*, Enterococci or coliphages under the Groundwater Rule.

6.68 SAMPLING FOR ROUTINE MICROBIOLOGICAL PARAMETERS

The sample should be representative of the water in supply and care should always be taken to avoid accidental contamination either during or after sampling. Personnel taking microbiological samples should be adequately trained and aware of the responsibilities of their role. They should be aware of the need to avoid cross contamination between raw and treated water samples. Microbiological samples should be transferred immediately to dark storage conditions and kept at temperatures between 2 and 8 °C during transit to the analysing laboratory. They should be analysed as soon as practicable on the day of collection. In exceptional circumstances, commencement of the analysis may be delayed by up to 24 hours after a sample is taken but only if the sample has been stored under the above conditions.

Raw water samples should be taken at the inlet to the treatment works and preferably from a dedicated sampling point. Unless there is no alternative, dip samples should be avoided.

Water leaving a water treatment works should be sampled from a point that is representative of the water entering supply, which may mean having more than one sample point. Samples should be taken from a dedicated sample tap that should not be used for any other purposes. The tap should be of an approved design and made of metal. It should be kept clean and also be prominently labelled with a dedicated reference number. Delivery pipework to the tap should be as short as possible, with the pipe made of a suitable material and opaque. It is generally not advisable to take microbiological samples from constantly running taps as the action of turning the tap off to disinfect it and then turning it on again to run to waste can dislodge particulates or biofilm from the sample line.

Sample taps and pipework at service reservoirs should be as described above. The system should be designed to ensure that water sampled is representative, as far as possible, of the reservoir as a whole (Section 18.6).

Sampling from consumers' taps should ideally be carried out at a tap which is in good repair and free of attachments. The tap should be supplied direct from the service pipe so as to be representative of the water in supply. Many of the mixer taps currently available in developed countries have components or are made of materials that make them very difficult to disinfect. Extra care is therefore always needed to ensure that the tap is adequately prepared before being sampled.

Any microbiological failure either in the water leaving a water treatment works or within the distribution system must be investigated immediately, with additional samples being taken for testing. Such samples should also be tested for Enterococci and *Clostridia perfringens*. It is also advisable to take additional samples following any interruptions to supplies or any repair work that might have compromised the integrity of the distribution system. In the event of a suspected water-borne outbreak of illness, sampling for a variety of other microbiological parameters may be appropriate.

Method of Sampling

The order of sampling should always be physico-chemical samples and then samples for microbiological analysis. This is because some chemical tests on samples taken at consumers' taps have to be carried out in the first litre of water drawn from the tap, before flushing (e.g. copper, zinc and lead), and samples for PAH should always be taken before the tap is disinfected. As soon as any chemical samples have been taken, the sample tap should then be disinfected using a blow torch in the case of a metal tap, or by swabbing the outside and as much of the inside of the tap as possible with a chemical disinfectant such as sodium hypochlorite solution. A few minutes contact time should be allowed for the disinfectant to work before the tap is run to waste until the water is cool or all the chemical disinfectant has been removed. During running to waste and sampling the flow rate of the water should remain steady to reduce the risk of any biofilm being dislodged into the sample.

Only sterilized sample bottles should be used and, if the water being sampled contains residual chlorine, the bottle should contain sufficient 1.8% w/v sodium thiosulphate to dechlorinate the sample. The sample bottle should not be rinsed but filled in one action, with the cap or stopper being removed for the minimum time possible. The lip of the bottle should not be allowed to come into contact with the tap and the bottle should be filled without splashing, leaving a small air space below the cap or stopper. Care should always be taken not to contaminate the cap or stopper and if accidental contamination is suspected the sample should be discarded and taken again in a new bottle.

In some cases it may be necessary to take samples from hydrants or standpipes, particularly for new and repaired mains. The hydrant box should always be cleared of any accumulated debris and water and the outlet should be dosed with sodium hypochlorite before the standpipe, which should be kept in a clean condition, is attached. The hydrant should then be cracked open to fill its outlet and the standpipe and allowed to stand for at least 5 minutes before flushing. Flushing should continue until the residual chlorine level is that of the mains supply and the microbiological sample should be carefully taken without turning the water off.

Occasionally dip samples have to be taken for investigational purposes. Special sterilisable sampling cans can be obtained for this purpose; otherwise wide mouthed sterile sample bottles securely attached to sterile wire can be used for taking the sample. This requires special care to avoid contamination of the water being sampled.

6.69 ROUTINE TESTS FOR BACTERIAL CONTAMINATION OF WATER

The rationale for routine microbiological monitoring usually includes the following tests:

Colony counts at 20–22 °C and 37 °C. Large numbers of micro-organisms occur naturally in both ground and surface waters, many of which are associated with soil and vegetation and can survive for long periods in the environment. Counts of such organisms, grown as colonies on or in nutrient agar, provide a useful means of assessing the general bacterial content of a water. The colony count, or plate count, following incubation at 20–22 °C gives an indication of the diversity of bacteria present at normal environmental temperatures. Although the result does not have any direct health significance, it provides a useful means of assessing the efficacy of the various water treatment processes in terms of overall bacterial removal. It also gives an indication of the general microbiological state of a given distribution system. Incubation at the higher 37 °C temperature encourages the growth of bacteria that can thrive at body temperature and which, therefore, may be of animal origin. The main value of both tests is to provide a background level or reference for a particular source water, treatment works or distribution system. A sudden marked increase, particularly in the 37 °C count, could be indicative of treatment deficiencies or of a more serious problem developing. Marked changes over and above the normal seasonal trends for colony counts at both temperatures could indicate longer term changes in the microbiological quality of the water. The counts are also of value where the water is used in the manufacture of food and drink as they could be an indication of a potential spoilage problem.

Coliform bacteria count. As discussed in Section 6.66, tests for coliform bacteria make it easy to detect and enumerate these bacteria when present in water. The term 'coliforms' traditionally referred to bacteria capable of growing at 37 °C in the presence of bile salts and of fermenting lactose at this temperature, producing acid and gas after 24–48 hours incubation. They are also gram and oxidase-negative and non-spore forming. Improved understanding of the coliform group has enabled more rapid and direct test methods to be developed. These are based on the expression of the ß-galactosidase gene, without the requirement for gas production. Methods using membrane filtration provide at least a 'presumptive' result under normal laboratory conditions after 18 hours incubation. Further tests are then carried out to confirm the result.

Escherichia coli (Faecal coliform) count. Traditionally coliforms of faecal origin, characterized by *E. coli*, have been considered capable of growth and of expression of their fermentation properties at the higher temperature of 44 °C. This has resulted in them sometimes being referred to as 'thermotolerant'. Standards now tend to require the application of a stricter taxonomic definition that is based on the detection of *E.coli*, and recognizing that strains that are not thermotolerant have the same sanitary significance. As with coliform bacteria, a result that is at least 'presumptive' is available after 18 hours of incubation. Methods incorporating expression of the ß-galactosidase gene, which is highly specific for *E. coli*, are now available. These can provide a confirmed result after 18 hours, otherwise specific confirmation based on production of indole from tryptophan is required. A number of confirmatory test kits are commercially available.

The detection of *E. coli* provides a reliable indicator of recent faecal contamination. The presence of other 'thermotolerant' coliform bacteria, particularly when detected in warmer tropical or sub-tropical waters, provides a less reliable indication of such contamination. However for many routine monitoring purposes an acceptable correlation between *E. coli* and faecal coliforms, as defined by thermotolerance, is often assumed.

Enterococci and Clostridium perfringens. Tests for the presence of Enterococci and *Clostridium perfringens* are required under the EC Directive to verify the quality of the water at the tap and the effectiveness of treatment of surface derived sources. These bacteria are commonly used as secondary indicators of faecal pollution and can be used to extend the scope of the testing, especially if there is a need to confirm whether there is a problem.

The test for Enterococci can be used to assess the significance of coliform organisms in the absence of confirmed *E. coli*, as they are more persistent indicators of faecal contamination than *E. coli*. The ratio between the numbers of *E. coli* and Enterococci present in a sample may in some circumstances provide some indication as to whether the source of the contamination is human or animal. However, careful interpretation of any such ratios is necessary and even then the outcome may be unreliable.

The presence of spore-forming, sulphite-reducing anaerobes, such as *Clostridium perfringens* is also associated with faecal contamination. The presence of such bacteria, especially in well or borehole supplies, can indicate remote or intermittent contamination. During treatment the presence of C. *perfringens* in the filtered water and/or the final water may indicate deficiencies in the filtration or disinfection processes. At some works this may correlate with a potential for the breakthrough of protozoan cysts such as *Cryptosporidium*, although this has been the subject of debate and is not universally true.

6.70 METHODOLOGY FOR MICROBIOLOGICAL EXAMINATION

The methods for the routine microbiological examination of water are well documented in a series of booklets published by the EA under the auspices of the Standing Committee of Analysts (EA, 2002a, b, c and d). Equivalent test procedures are given in the American Standard Methods (AWWA, 2005) and also in ISO procedures published by ISO (ISO, 1985). Established methods should always be used for routine analyses and good laboratory practice should be adopted at all times, with special precautions being taken to avoid accidental contamination of samples once they are in the laboratory. Appropriate quality control procedures should be used at all stages of the analysis and laboratories should partake in external quality control schemes, such as inter laboratory tests, where these are available. The choice of method for coliform analysis depends to a certain extent on the number of organisms likely to be present. Methods based on membrane filtration and colony counting and Most Probable Number, such as the defined substrate 'Colilert' method, are widely used for the analysis of drinking water. Membrane filtration is a good approach for treated waters but it may not be suitable for highly turbid waters, or for waters containing only a small number of indicator organisms in the presence of large numbers of other bacteria capable of growing on the media used. Multiple tube or most probable number methods can also provide good results for treated waters and in some instances may be more suitable for highly contaminated samples or samples containing appreciable quantities of sediment or particulate matter.

Present—absence tests for coliform bacteria. A modification of the traditional multiple tube procedure, based on a single 100 ml volume of appropriate medium, such as minerals modified glutamate, instead of a series of tubes of different volumes can be used to determine whether coliform bacteria are present in a sample. There are now a variety of proprietary formulations, such as 'Colilert', which can be used in this format. These proprietary formulations have additional advantages in being easier to set up than the traditional method, with confirmed results including the presence of *E. coli* available after incubation at 37 °C for 18 hours, rather than 18–24 hours. However the drawback of such tests is that a positive sample gives no indication of the relative number of organisms present in the sample.

Field test kits. There are a number of field test kits available. Most are based on membrane filtration or presence-absence testing and consist of a small portable incubator, which can if necessary be plugged into a car battery, and a means of aseptically preparing and processing the sample.

The most simplistic form consists of a 'dip slide' with a small area of sterile agar medium, which is selective for coliforms, at the end of a stick attached to a screw cap for a dedicated tube or container. The 'slide' is dipped in the water sample and incubated in its tube. The result is essentially a presence-absence test, although a very crude semi-quantitative estimate of numbers can be obtained. However, this does not relate to a specific volume of sample, only to the magnitude of contamination. This would not be applicable to treated drinking water but can assist in initial assessments of raw water. The more sophisticated test kits provide the equivalent of a portable laboratory. Some test kits offer rapid initial results, where significant numbers of *E. coli* may be present, after only 8–12 hours incubation. However, the full 18 hours incubation is essential to provide reliable results.

Considerable care has to be taken with all field test kits to ensure that samples do not become contaminated during analysis and that adequate sample dilutions have been prepared to cover the expected concentration ranges. Colony identification and counting must be carried out by an experienced person, familiar with techniques of identifying coliform bacteria, since non-coliform organisms may also be visible on the membrane. This is particularly so in tropical or sub-tropical climates. A basic level of analytical quality control should always be adopted with these test kits, to include positive and negative control samples and a record of the incubator temperature.

6.71 PROTOZOAL EXAMINATION

Sampling methods for *Cryptosporidium* and *Giardia* include continuous capture by filtration, large volume grab samples or composite samples. For general surveillance purposes, a large volume of water sampled through a filter over a long period of time, usually 1 m³ over 24 hours, is likely to give the best result. Although this works well for treated waters filtration may not be possible for raw waters that are very turbid or contain a lot of algae. Dedicated sampling equipment is required for continuous filtration, whilst grab samples, usually 10–20 litres, can provide a more accessible means of obtaining operational samples. Grab samples will always give a less representative result and liaison with the analysing laboratory is required to avoid overloading analytical capacity.

Standard methods of analysis are available for both protozoa. Although greatly improved in recent years, the biggest problem associated with methods for *Cryptosporidium* is still the maintenance of analytical consistency through a multi-stage procedure to ensure acceptable percentage recovery efficiencies. In addition, the method detects both live and dead oocysts, resulting in problems in the application of treatment technologies that do not physically remove the oocysts, e.g. ozone or ultra violet irradiation (Sections 11.22 and 11.25). Research is ongoing to further improve methodology and performance criteria.

6.72 VIROLOGICAL EXAMINATION

Human enteric viruses constitute a diverse group. Although usually many times less numerous than *E.coli* in domestic sewage, they occur widely in surface waters and are also found in some ground waters that have been subject to contamination. They are obligate parasites and are usually quite specific in their ability to infect, so that viruses affecting animals do not tend to infect humans.

Routine testing of water for viruses is not recommended, as it would require specialist laboratory facilities and trained virologists. Large volume samples in excess of 50 litres are often required, depending on the type of water, and the analytical procedures tend to be time-consuming and complicated. Conventional treatment processes, particularly disinfection, when properly applied are usually regarded as being effective in addressing any risk posed by viruses in drinking water.

Bacteriophages (or coliphages), which are viruses that can infect bacterial cells, have been proposed as possible viral indicators in drinking water, since it has been found that some bacteriophages are inactivated at similar rates to enteroviruses during treatment. Furthermore the isolation of bacteriophages is relatively straightforward and rapid, thereby providing an effective means of monitoring treatment processes for the removal and inactivation of enteroviruses.

6.73 NUISANCE ORGANISMS

Iron bacteria. There are several groups of iron bacteria, all of which are capable of abstracting and oxidising any ferrous and manganous ions present in a water. The process is continuous with a large accumulation of rust coloured or black deposits developing over time. These deposits tend to accumulate in storage tanks and on the walls of pipes in low flow areas of the distribution system. Pipes may become blocked or the flow seriously impaired and any disturbance of the deposits results in badly discoloured water.

The growth of iron bacteria also results in an increase in the organic content of the water that could in turn encourage the growth of other nuisance organisms. In combination with sulphur bacteria, they also contribute to the corrosion of iron and steel pipelines.

The development of the organisms can only occur where there is sufficient iron or manganese present in a water at the ideal oxidation-reduction potential. Thus iron bacteria are widely found in the bottom sediments of raw water reservoirs where the depletion of oxygen provides adequate ferrous ion. They can also be associated with ferruginous groundwaters containing high levels of free carbon dioxide and low levels of oxygen.

Sulphur bacteria. There are two groups of sulphur bacteria with implications for the water industry. Sulphate reducing bacteria grow in anaerobic conditions and reduce any sulphate present in the water to hydrogen sulphide. They contribute to galvanic corrosion of water mains and can cause taste and odour problems. Sulphur oxidising bacteria grow in aerobic conditions and produce sulphuric acid from any sulphides present, for example in sewers.

Nematodes Nematodes are small wormlike organisms and a number of non-pathogenic free-living varieties are occasionally found in drinking water. Their presence does not necessarily indicate a health risk, rather they are regarded more as an aesthetic problem in terms of consumer acceptability. They are more likely to be found in supplies derived from nutrient rich surface waters and in the 'dead zones' of distribution systems. If large numbers are detected then some form of remedial action is required.

Actinomycetes are a diverse group of bacteria which, together with some genera of microfungi, can give rise to earthy or musty tastes and odours, particularly in water derived from nutrient rich lowland sources. Actinomycetes occur particularly at the margins of surface water bodies wherever decomposing organic material is present. Two compounds commonly formed as metabolites by actinomycetes are geosmin and 2-methylisoborneol, both of which have very low threshold taste/odour values.

PART IV   WATER BIOLOGY

6.74 INTRODUCTION

Rivers and lakes can support a wide range of plants and animals. These living organisms (the biota) form ecosystems which are in balance with important variables such as climate, water quality and human uses. The biota usually contribute beneficially to water quality of surface waters but may, particularly when not in balance, also have deleterious effects. A large number of organisms have been reported as causing problems in water supply systems. Examples include blue green algae resulting in blooms, toxins and scum and Chironomid midge swarms (affecting rivers and reservoirs); sponges, barnacles, mussels and pipe mosses (reducing hydraulic transfer capacity); algae (causing taste and odour and blocking filters and clarifier launders); and animals such as water fleas, *Rotifers*, *Cyclops*, *Asellus*, *Gammarus*, *Daphnia* and Chironomid (leading to contamination and tastes and odours in the distribution system and service reservoirs). Most organisms can readily be removed during treatment but some are difficult to prevent from passing through to the distribution system and to the consumer.

Consumers regard the presence of living organisms in potable water as aesthetically unacceptable, as they indicate an impure product. Apart from their mere presence, such biota can cause deteriorating water quality by introducing turbidity, tastes and odours as a result of their metabolism, and greatly increasing the chlorine demand.

Only one animal, whose adult life is free living, has been positively identified as a health hazard in drinking water. This is a water flea, which is the host for a human parasite, the guinea worm, which can infest man if ingested. There are other animals, which are parasitic in man and which produce life stages that are free living and could be transmitted if not killed or removed by treatment. Finally there are animals which if not removed would proliferate in the distribution systems where their activities could facilitate the survival of disease organisms such as bacteria or viruses.

6.75 SOURCE WATER AND STORAGE RESERVOIRS

There are two main groups of plants in rivers and standing waters, algae that are largely microscopic and the larger, readily visible, macrophytes—the plants commonly named water weeds. Algae are simple green plants, most of which are free floating (planktonic) forms ranging in size from single celled species of 2 to 5 microns diameter to larger colonial forms up to several millimetres. There are, however, some species of algae which grow to macroscopic size as attached fronds in rivers or as upright forms on lake beds. The number of algae and the species found depend on environmental conditions such as temperature, the concentrations of dissolved salts and particularly the available nutrients such as nitrogen and phosphorus which the plants, with energy from sunlight, build into their biomass.

Given the right conditions, plants will grow until the available nutrients are exhausted, and if not controlled may cause a variety of nuisances. For example water weeds or larger attached algae can cause major engineering problems such as mechanical breakages to control structures and even river channel blockages. Surface 'blooms' of floating algae can affect slow moving and standing waters, causing fish kills through de-oxygenating the water and the sliming of fish gills as the algae die and decay. In 1998 a major fish kill of over 150 tonnes of trout and coarse fish occurred in a fish farm fed from the Kennet and Avon canal in the UK due to this effect.

In recent years many rivers have shown an increase in nutrients, a feature which also occurs as rivers flow downstream. Starting as clean upland streams they often become polluted lowland waters from direct and indirect releases of wastes. This tendency to nutrient enrichment has become known as eutrophication and is outlined in Chapter 7. Whilst additional nutrients may improve plant and animal growth up to a certain point, beyond that point the excess nutrients can have a deleterious biological effect. This is because the growth of rapid growing biota is favoured and these species tend to be tolerant of most forms of pollution. They take over or dominate the other biota at the expense of the less tolerant forms, a loss of biodiversity. Many of these less tolerant forms are clean water biota and are highly valued for attracting animals such as salmon and trout, dragonflies and mayflies. By comparison the more tolerant species include nuisance organisms such as midges or some of the bloom forming algae, few of which are regarded as attractive.

The macrophytes, the larger aquatic plants, may be floating plants or grow as emergent or submerged waterweeds. These plants maintain healthy ecosystems by providing food and shelter for numerous animals. They also create water quality improvements by allowing settlement of particulates in quiescent areas, by incorporation of nutrients from the water and by oxygenation. When in excess macrophytes can prove troublesome by reducing the carrying capacity of water courses, blocking intakes and water control structures, whilst sudden die off can lead to foul water.

Although lakes and reservoirs vary considerably in size and other conditions, the range of species of algae is small and there is a characteristic make-up of species of algae that is found in similar water bodies anywhere in the world. As the waters leave the uplands and enter lowland plains there is an increase in dissolved solids, in temperature, and, in many developed countries, in nutrients from sewage discharges. There is commonly a transition in the algae from diatoms that tend to dominate in cold upland waters to green and to blue-green algae that thrive in warm, often shallow and nutrient rich lowland waters.

Nutrient poor upland waters are known as oligotrophic (poorly fed) whilst the richer nutrient waters are eutrophic (well fed) waters. Oligotrophic waters tend to have a few species of flagellate algae such as the Chrysophyte genera *Synura*, *Uroglena* and in the spring there may be blooms of diatoms such as *Cyclotella*, *Tabellaria* or *Asterionella*. Green algae are common in many ponds and shallow water bodies, common species being *Chlamydamonas, Chlorella, Euglena, Chlorococcus, Coelastrum, Cosmarium, Oocystis, Pediastrum, Scenedesmus* and *Staurastrum*. The term blue-green algae is retained here for convenience, although now becoming known as Cyanobacteria, being more closely related to bacteria than algae. Many blue-green algae are colonial and able to grow very rapidly in warm waters. A number of the commonest blue-green have flotation 'devices' such as gas vacuoles in the cells or mucus which binds the colonies into large rafts; examples are *Anabaena, Microcystis, Aphanizomenon, and Oscillatoria.* As already noted in Section 6.4, some of the blue-green algae also produce toxins present in the mucopolysaccharides that make up the mucus released by the cells. The toxins are not always present and their presence cannot currently be predicted except that high concentrations of the named blue-green species give rise to a higher risk. There have been numerous accounts of the ecology and biology of algae but the works of Palmer (1977) in the USA and Bellinger (1992) in the UK provide a good general introduction. Illustrations of algae genera of significance in water supply taken from Bellinger (1992) are shown in Figure 6.1.

There has been a great deal of research and development of methods to reduce the adverse effects of eutrophication, varying from control of nutrient contribution from agriculture and from sewage works using a range of phosphorus and nitrogen removal technologies (Anon, 1997) to in-lake methods of lake management. The traditional method, of direct intervention by dosing copper sulphate (IWEM, 1969) direct to the water at times of blooms to kill the algae, is not now acceptable to most authorities due to the toxicity of copper to other biota including humans. The reduction by precipitation of soluble phosphates by dosing aluminium or iron salts in a contained volume at the

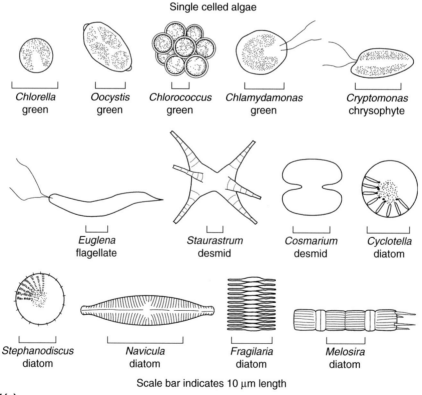

FIGURE 6.1(a)

Algae genera of significance in water supply—single celled algae.

Colonial algae

Scale bar indicates 10 μm length

FIGURE 6.1(b)

Algae genera of significance in water supply—colonial algae.

entry to the water body has been used successfully (Croll, 1992). Typically 1 to 3 moles of Fe or Al to 1 mole of phosphate is required, with a pH range for precipitation from 5.5 to 7.0 depending on the metal salt used. The phosphate concentration can be reduced to <10 μg/l. The phosphate precipitated should be regularly removed by dredging. Introducing species or biological communities as a form of bio-manipulation or adding barley straw, which decays in the water releasing an algal toxin, have been employed and, whilst attractive in large water supply reservoirs, could also prove useful and cost effective in small impoundments (Garbett, 2005).

The use of ultrasound for disruption of algal cells is being developed (Lee, 2002) and has been used in UK where units of modest power have achieved worthwhile operational gains (Dynamco, 2004). Units available in 2006 operate at frequencies of 15 kHz to 10 MHz and have an effective operating radius of 200 m. They can be arranged to traverse the reservoir or fixed by mooring and can be operated from solar panels thereby giving greater flexibility in their deployment. It is critical to keep the transducer face clean for effective performance. The use of ultrasound in this context is reported as being harmless to fish and other aquatic life. This book cannot cover all the literature on the control of algae and other biota in eutrophic waters but the reader may find reviews by Sutcliff (1992) and Moss (1996) useful.

6.76 TRANSFER STAGES

Plants can seriously reduce the cross section of open channels and also increase the frictional resistance to flow. This is usually a minor problem in UK but can be an important issue in tropical countries where plant growth can be substantial. It is also more often a problem associated with bulk water transfers as in irrigation canals, in which velocities are usually low.

Plant growth in water courses, whilst maintaining a good aquatic habitat, can at the same time promote other problems for the water manager. In the Ely Ouse scheme in UK, the good habitat provided by plants has encouraged coarse fish. These shoal and block the intake screens for this inter-river transfer scheme. In many tropical countries snails, carrying the intermediate vector for schistosomiasis (bilharzia), thrive in clean water canals and can create a major health hazard. Molluscicides are needed to control the snail, combined with adequate education of the local population with regard to sensible personal hygiene practice.

Plants do not in themselves cause problems in pipelines, as without light they cannot grow. However, other organisms may affect raw water transfer pipelines. These tend to be filter feeding animals such as sponges, moss animalcules (these are colonial forms resembling the common hydroid *Hydra*), mussels and, where sediments build up in major transfer systems, even cockles can be found within the sediments. These forms thrive upon algae and other fine plant debris carried into the pipelines and can result in increased headloss in the system, tastes and odours imparted to the water, and blockages when dead animals fall from the pipe walls. Water louse, shrimps and fish also occur in low velocity bulk water transfer systems.

In an unusual instance in Hong Kong, animals have given water quality benefits. 'Moss animalcules', which grow in a layer resembling a moss, grow naturally in a number of the extensive raw water tunnels. The presence of copious nitrifying bacteria in the matrix of fibres making up the colony of animals is able to strip any ammonia from the water in mains or tunnels that carry this biota. The improvement in water quality, and the reduced chlorine demand for pre-chlorination, has encouraged promotion of this situation in other pipelines.

6.77 TREATMENT STAGES

The seasonality and speciation of algae has become an important issue in water supply as many of the algae have highly specific effects on the treatment process. In investigating problems with algae in the treatment process, it is important to be able to measure algae and the specific removal rates accurately. This has been difficult as the techniques of counting or measuring the plants have to vary to suit the organisms being counted. The most standard measurement in the past has been direct counting under a light microscope with numbers being reported as cells or organisms or colonies per unit volume, with some agencies preferring to report the plan area or even the volume of the cells per unit volume. The cost and subjectivity of counting has led to the use of chemical determinands such as the measurement of particulate organic carbon (POC) or the green pigment chlorophyll a, which is extracted and measured or measured fluorimetrically. Particle counting using modified optical Coulter counter techniques has been used more recently in special studies.

As noted in the introductory section of this part there are no limits set for animals or plants in drinking waters though there is a presumption that they should be absent. In practice it is difficult to prevent some carry over of algae and small animal components from treatment processes though these are killed by the final chlorination before leaving the treatment works. Values for an algal content of between 100 to 1000 cells/ml, which is a range close to the lowest discernible limit for direct optical counting, have been used as a guide over the years. The lower number (100 cells/ml) has been advocated as a practical limit for the treated water prior to final chlorination (Mouchet, 1984).

6.78 SERVICE RESERVOIRS AND DISTRIBUTION SYSTEMS

Water in service reservoirs and distribution systems will be chlorinated and, in the majority of cases, kept in the dark as most service reservoirs are covered to avoid photo-oxidation of the chlorine residual or aerial contamination of the water entering supply zones. It is therefore often surprising how many living organisms are found in such areas. Plants are seldom found in such systems as light levels are too low, but where brick or concrete conduits have been laid too near the surface the roots of trees can penetrate and cause blockage and a pathway for other organisms to enter.

Service reservoirs should have midge and mosquito proof mesh fitted to all the essential air vents provided near the roof of the reservoir as such gaps have been shown to allow ready colonisation by the flying adults which lay their eggs on cool damp surfaces.

Apart from contamination at the surface contact points, the main source of animals found in the distribution system is the treatment process and, in a number of cases in the past, it has been found that the source may be not simply the by-passing of the clarification and filter stages by animals in the raw water, but rather that animal communities build up in the sedimentation tank and filters and provide a regular slippage of animals to the final water. These animals include a wide variety of freshwater forms including nematode worms, water fleas, the water louse and Chironomid midge larvae. Pre-treatment oxidation by chlorine or ozone of raw water from a eutrophic water body and, where necessary, occasional oxidation of process stages have been shown to be effective in controlling such nuisance animals.

PART V   NEW AND EMERGING ISSUES

6.79 INTRODUCTION

Under the forward work programme for the ongoing revision of the WHO Guidelines on Drinking-water Quality, addenda will be issued as guideline values are set following the evaluation of new contaminants, or as new scientific information becomes available for existing guidelines. Similar processes of review and revision are built into the EC Directive and the US EPA National Primary Drinking Water Regulations. WHO and US EPA collaborate on a number of emerging issues, especially those relating to water and infectious disease (WHO/US EPA, 2003). However, a number of topics, both chemical and microbiological, have recently emerged and have caused consumer concern. Some of these issues are discussed below.

6.80 CHEMICAL ISSUES

Arsenic (Section 6.9) is included in the WHO programme of work for the rolling revision of the Guidelines. Recent epidemiological studies in Bangladesh, India and Chile (Ahsan, 2006; von Ehrenstein, 2006; Hopenhayn, 2006) indicate correlations between arsenic in drinking water and skin cancer, reproductive effects and anaemia.

Disinfection by-products. A number of disinfection by-products are discussed in Section 6.25. Epidemiological studies continue to suggest associations between adverse health effects and exposure for several disinfection by-products that do not currently have health related standards. Concerns have been raised over halo-acetic acids and also the short term and long term effects of cyanogen chloride, which may be present as a by-product of chloramination and is metabolized to cyanide in the body.

Endocrine disrupting chemicals and pharmaceutical residues. Endocrine disrupting chemicals are discussed in Section 6.27. They may disrupt human or animal endocrine systems and can either be oestrogenic or androgenic. It has been claimed that they may be responsible for increases in breast and testicular cancer and for decreases in sperm counts. Media interest relates mainly to hormone residues associated with the use of contraceptive pills. However, recent studies in the UK and New Zealand indicate that hormones present in animal wastes from livestock farms may also be having an impact on the aquatic environment (Matthiessen, 2006; Sarmah, 2006). A study commissioned by the UK DETR (1999b) investigated a number of approved water supply products as potential sources of endocrine disrupters. Transient and low levels of leaching were observed in some construction materials. Higher and more persistent levels were noted in some *in-situ* glues and additives but exposure is likely to be very low. These products were all used within buildings rather than the public water supply system. Research is continuing on other substances such as bisphenol a, nonyl phenol and nonyl ethoxylates, which have been shown in bio-assays to possess endocrine disrupting properties. Analytical methods have been developed to determine the presence of oestrone, 17α-ethynl oestradiol and 17 β-oestradiol at the levels of interest and further research is ongoing.

Surveys on the levels of pharmaceutical residues in treated sewage effluent and the associated receiving waters have been carried out by the EA. A recent study of the lower River Tyne (Roberts, 2006) identified traces of a number of pharmaceutical compounds. A desk based review carried out for Defra (2007b) found reports of several compounds being detected in drinking water in different parts of the world. The structure and nature of the individual compounds were key to the efficacy of removal during treatment. Reverse osmosis was found to be very effective at removing a wide range of pharmaceuticals but advanced treatment processes such as ozone and activated carbon gave removal rates of greater than 90% in many cases. There is very limited data on the concentrations of pharmaceuticals in UK drinking waters but levels of up to 0.1 µg/l have been found in some treated waters in Europe and the USA.

Metaldehyde is a molluscicide that is widely for slug control both by the domestic market and by agriculture. Typically between 6000 and 10 000 tonnes of slug pellets are applied annually in the UK and, given the current climatic conditions, increasing levels of metaldehyde are being found in surface water sources. A number of UK water companies have detected metaldehyde in concentrations above the standard for individual pesticides in the treated water leaving their works. Research has shown that the conventional advanced oxidation processes for pesticide removal are ineffective at removing metaldehyde and that currently the only viable treatment is reverse osmosis.

Methyl tertiary butyl ether (MTBE) is the main fuel oxygenate for unleaded petrol. It is highly soluble, highly mobile in the aquatic environment, not significantly absorbed and not significantly attenuated or biodegraded. There is a possibility of it contaminating groundwaters via leaking storage tanks and pipes. MTBE is considered to be more of a problem in the USA and is included on the US EPA Drinking Water Contaminant Candidate List. Monitoring has been carried out since 2001. In the UK, water companies carry out some limited operational monitoring at sites considered to be at risk. If present in drinking water, it would be detected by taste/odour at levels much lower than any perceived health risk.

N-Nitrosdimethylamine (NDMA). This potent carcinogenic compound is formed by the reaction of dimethylamine (DMA) with monochloramine or hypochlorous acid (HOCl) in the presence of ammonia or nitrite (WHO, 2002). DMA can be present in surface waters (some pesticides, industrial or sewage effluent) or can be introduced into drinking water if polyDADMAC (polydiallyldimethylammonium chloride) or Epi-DMA products are used as coagulants or coagulant aids. NDMA has been detected following the recycle of backwash supernatant containing high levels of DMA and also when ion exchange resins containing the quaternary ammonium group come into contact with chlorinated water (Wilczak, 2003).

Perchlorates can be used in the manufacture of explosives but are more generally used as an additive to petroleum products to increase the octane number. They have been detected in groundwater sources in California and other parts of the US. Although a number of epidemiological studies have been carried out, there remain significant gaps in knowledge of health effects, levels in drinking water and effective removal during treatment. Perchlorate is on the US EPA Contaminant Candidate List.

Perfluorooctanesulfonate (PFOS) was a key ingredient in the manufacture of 'Scotchguard' and is also a component of fire fighting foams. It is a toxic substance and, although produced and used in very limited quantities these days, it remains persistent in the environment. It excited interest in the UK following the Buncefield oil depot explosion and fire in December 2005, when there were concerns that the local aquifer had become contaminated. Trace levels of PFOS and other perfluoralkyl surfactants have been detected in a number of surface waters in the USA.

Uranium is widespread throughout the environment. In some countries it is naturally present in drinking water and drinking water can constitute the major source of intake. Although the radiological aspects of uranium are widely recognized, little information is available on chronic health effects caused by uranium as a chemical. The BGS has carried out a survey of uranium in British groundwaters (Defra, 2006) which found concentrations of <0.02 to 48.0 µg/l. Much higher concentrations have been found in groundwaters in other parts of the world and it is of increasing concern in some Eastern European countries.

6.81 MICROBIOLOGICAL ISSUES

Aeromonas. There are now fifteen named species of *Aeromonas*, of which some have sub-species/biotypes. The organisms are widely found in the aquatic environment and have been associated with regrowth and biofilm problems in water distribution systems. Various methods of detection are available. Most treatment processes effectively remove Aeromonads and free chlorine residuals of 0.2–0.5 mg/l are considered sufficient to control the organisms in distribution systems. The main health effects are the causation of wound infections and septicaemia. Aeromonads have also been associated with diarrhoeal illness.

Arcobacter. Four species of *Arcobacter*, which are gram-negative, spore forming rods, are generally recognized. The organisms have been associated with diarrhoeal illness particularly through contact with raw meat but there have been isolated incidents when water supplies might have been implicated as a potential source of infection. However, the epidemiology is not well understood and there is little evidence indicating causation. No data are currently available on removal efficiency during treatment but, as *Arcobacter* species are similar to *Campylobacter*, they are likely to be inactivated by disinfectants such as chlorine and ozone.

Enterovirulent Escherichia coli. Certain strains of *E. coli* are an important cause of diarrhoeal illness and have been termed enterovirulent *E. coli*. Several classes have been defined, based on the possession of distinct virulence factors. These are the enteropathogenic, enterotoxigenic, enteroinvasive and Vero cytotoxigenic (VTEC) *E. coli*. Most data currently available relates to *E. coli* 0157, one of the VTEC strains that does not ferment sorbitol. The organisms can cause a wide range of symptoms including vomiting, fever, bloody and mucoid stools and, in severe cases, acute renal failure. Most outbreaks have been associated with the consumption of contaminated food. However, depending upon the source of contamination, these organisms may be found in water sources and specific techniques exist to detect them in water. Water treatment processes provide effective removal at the same level as for normal *E. coli*., A number of minor outbreaks have been linked to small private water supplies, usually associated with farms, with no treatment, or to recreational water use. In May 2000 a large outbreak occurred in Walkerton, Ontario, and involved a municipal water supply. The cause of the outbreak was contamination of a well with cattle excreta, following heavy rain, coupled with inadequate treatment and inadequate disinfection. *Campylobacter* was also implicated and there were more than 2300 reported cases of illness and seven deaths.

Mycobacterium avium subsp. Paratuberculosis (MAP) is one of several related species of bacteria collectively referred to as the *M. avium* complex (MAC). It is the causative agent of Johne's disease, a chronic inflammatory disease of the intestines of herbivores, monkeys and other wild animals. This has resulted in suggestions that MAP could be the cause of Crohn's disease, a similar inflammatory bowel disease in humans. The organism is excreted in large numbers by infected animals and is common in the environment. It is likely that human exposure is common, and transmission is thought to be through the faecal-oral route. However other risk factors, such as immunologic and genetic, are involved in the establishment of disease. Detection of MAP in water samples is a difficult process, as it is a very slow growing and fastidious organism requiring special culture media for isolation. Progress has been made in identification using Polymerase Chain Reaction (PCR) techniques. Other members of the MAC, but not MAP, were isolated from drinking water samples in a recent study (Defra, 2003). MAP should be removed by conventional water treatment processes, although it has been shown to be comparatively resistant to chlorine. Conditions in distribution systems are not considered favourable to its proliferation.

Microsporidia are single cell protozoan which are obligate intercellular parasites. They proliferate inside the cell and are released in the faeces as spores. They are common in nature and rarely cause infection in humans, although they do represent a significant risk for immuno-deficient individuals. They present a potential challenge for removal by conventional treatment because of their very small size.

Cyclospora are also protozoan parasites similar but larger, at 8–10μ, to *Cryptosporidium*. They cause diarrhoeal illness and infection is understood to be through consumption of contaminated food or water. Water associated outbreaks have been reported in some parts of the world. Treatment processes effective at removing *Cryptosporidium* will also remove Cyclospora, although disinfection has little effect.

Adenoviruses. Human adenoviruses belong to the *Mastadenovirus* genus in the family *Adenoviridae*. There are 51 serotypes, grouped into six sub-groups, which can cause upper respiratory infections, conjunctivitis and febrile illness. Traditionally identified by electron microscopy there are now cell culture techniques being developed which are capable of detecting some sub-groups from water samples. They can be transmitted by water but, while they are reported to be more resistant than some viruses, they should be inactivated by conventional water treatment processes, particularly disinfection.

Norovirus. Formerly known as 'Norwalk Like Virus', Norovirus is the most common cause of viral gastroenteritis in England and Wales. Molecular characterisation of the RNA genome has demonstrated the Norovirus to be a genus of the *Calicivirus* family, distinct from classic members. There are two genogroups, of which genogroup II is commonly reported from clinical specimens. Multiple strains are recognized and different strains can be circulating in the community at any one time. These viruses can be detected by electron microscopy, using polymerase chain reaction and ELISA techniques in clinical samples but detection in environmental samples is more difficult. They are responsible for nausea and diarrhoea and are most commonly spread from human-to-human but can also be food or waterborne. They often cause major problems on cruise liners.

Other Enteroviruses. The term enterovirus encompasses a range of different virus groups broadly related by their ability to infect the intestinal tract, causing a wide range of symptoms often including fever, meningitis and abdominal pain. They are usually present in large numbers in the faeces of infected individuals and, therefore, in sewage. Research in recent years has lead to an improved understanding of these viruses, methods of detection and differentiation. Prominent among the enteroviruses are Coxsackieviruses, Caliciviruses and Echoviruses. Normal water treatment processes, including disinfection, should remove or inactivate such viruses.

Biofilms and associated problems. The ability of an increasing number of bacteria to be associated with biofilms and to survive in otherwise adverse environments, for example the presence of chlorine, by virtue of being internalized by protozoans such as Acanthamoeba spp. provides a potential challenge in distribution and plumbing systems. Among the bacteria potentially implicated in this respect are *Helicobacter pylori*, a cause of gastritis and duodenal ulcers, and *Legionella pneumophila*.

Cyanobacteria, commonly referred to as blue-green algae, are discussed in detail in Section 6.4.

Climate change and the increase in travel and international mobility present unknown challenges for the future as ambient temperatures and weather patterns may provide conditions for the emergence of novel infections and exposure of new populations to infections to which they have little immunity.

PART VI   WATER SAFETY PLANS

6.82 INTRODUCTION

It has long been recognized that effective management of the water supply chain is an essential part of safeguarding public health. It is now usual to have some form of continuous monitoring of the raw water entering larger water treatment works to provide early warning of potential contamination. Continuous monitoring is also used for process control, with automatic shutdown often available if results do not fall within preset limits. Even so, most operational and regulatory water quality monitoring is based on routine samples taken at intermittent intervals, where the results are often not available until long after the body of water that they represent has left the distribution system. Test results showing the presence of contaminants requiring immediate action would be too late to be of benefit.

In recognizing the problem, experts drafting the WHO *Guidelines for Drinking-water Quality*, Third edition (WHO, 2008) developed a holistic approach to the management of the supply of safe drinking water, which could be applied to all drinking water supplies regardless of size, type or location. The approach is based on carrying out a systematic risk assessment of the drinking water supply chain and looking at ways in which to manage the risks identified. WHO has drawn on a number of risk management processes in formulating its approach, including those used for food safety.

6.83 WATER SAFETY PLANS

The concept of drinking water safety plans goes beyond the need for source water protection and good water supply management. It builds on good practice, but also takes a comprehensive risk assessment and risk management approach to the whole water supply chain from catchment to consumer. The key objectives are to:

- minimize the contamination of source waters;
- reduce or remove any contamination by using appropriate treatment; and
- prevent contamination as the water passes through the distribution system to the point of supply.

Although the water supplier has a key role to play in the development and implementation of such plans, other parties such as drinking water quality regulators, authorities responsible for monitoring public health and the environment, and consumers also have an important role to play.

The key components of a water safety plan, and the steps involved in developing one, are set out in Chapter 4 of the WHO *Guidelines for Drinking-water Quality* (WHO, 2008). The Guidelines also refer to a series of supporting documents, which are available on the WHO website (*http://www. who.int/water_sanitation_health/dwq/en/*). A further supporting document looking specifically at the chemical safety of drinking water was published by WHO in 2007 (WHO, 2007).

6.84 DEVELOPMENT OF WATER SAFETY PLANS

A water safety plan is essentially made up of three components, namely:

- system *assessment* of the whole water supply chain to determine whether water of a quality required to meet the health-based targets can be delivered up to the point of consumption; this includes the assessment of design criteria for new plants;
- operational monitoring to ensure that the *control measures* are effective and ensure that health-based criteria are met and that deviations from the required system performance are detected in a timely manner; and
- *management and communication* procedures documenting the system assessment, the level of monitoring and the actions required under normal operational and incident conditions.

The system assessment is intended to identify potential hazards in every part of the supply system from source to tap. A level of risk is then assessed for each hazard, along with the measures needed to control that risk to ensure that the water as supplied is safe to drink and that the required standards are being met. The likelihood of occurrence of each hazard and severity of the consequences determines the level of risk, which can then be ranked in priority for action. The control measures in turn collectively determine the nature and frequency of the operational monitoring. Management procedures should include appropriate actions for investigating a failure to meet a standard or an operational control; proposed remedial action following such a failure; and appropriate levels of communication and reporting both internally and to external stakeholders. Validation processes also need to be developed to ensure that the plan is effective, based on sound science and technical information and benchmarked against other similar systems utilities. Plans need to be supported by training programmes and the development of standard operational and maintenance procedures and processes for ensuring timely corrective actions if necessary.

Ideally a water supplier should develop a water safety plan for each individual water system in its area of supply. In the case of very small supply systems, this may not be practicable and a generic model may be developed for supplies of a similar nature and with similar levels of risk.

Although the water supplier is ultimately responsible for preparing and implementing its water safety plans, a multidisciplinary team of experts on drinking water supply should ideally be convened, with representatives from the authorities responsible for the protection of public health in the area of supply. The team would then draw up the plan, with each step being documented, for that supply chain. Once agreed, the plan should be implemented as soon as possible. However, it should be periodically reviewed to ensure that all components are still relevant. It should also be reviewed if any one component changes significantly or if there has been a problem with the quality of the water supplied.

The three key components to be covered by a water safety plan are the catchment (resource and source protection), treatment and piped distribution systems. These are discussed in more detail below.

Catchment

The factors that could impact the quality of the source water include: type of catchment, in terms of its geology and hydrology; the impact of local weather patterns; the nature of the land within the catchment and how it is used; any existing catchment controls or source protection; other water uses (e.g. irrigation) and how they could impact the raw water; and any other known or planned activities within the catchment. The source of water could be a deep borehole, a shallow well, an upland catchment or a lowland surface water. Each will have different hazards that need to be identified. Control measures could be applied to the catchment, or at the point of abstraction. A water supplier with little or no influence over the catchment may look to treatment as the best means of controlling the risk. On the other hand, if the source is a transboundary river that flows through more than one country, there is a need for good communication and co-operation between the various governments involved.

Treatment

The team assessing the treatment processes should have a complete understanding of the processes involved and their capability to remove the contaminants or potential contaminants that might be present in the raw water. The efficacy and reliability of each stage of process needs to be assessed, with particular focus on disinfection. The appropriateness of the treatment chemicals used should be included in the assessment, along with the levels of process monitoring and control. Consideration also needs to be given to any identified hazards that can not be controlled in the catchment and that may not be removed or acceptably reduced by the existing treatment. Suitable control measures should be identified. These may include further optimization of the treatment processes, with new or upgraded processes being installed as necessary, or stopping or restricting abstraction during periods when the raw water quality is seriously compromised.

Piped Distribution Systems

Once the water leaves the treatment works its quality deteriorates; the rate of deterioration depending on the condition of the distribution system and the residence time in each component. Distribution systems are likely to include storage facilities (such as service reservoirs), pumping stations and pipe networks, with either constant or intermittent supplies. The hazards associated with the condition of the network and its performance and the relationship with the associated operation and maintenance strategies and practices need to be understood in order to identify suitable control measures.

Non-piped, community and Household Systems

Intermittent supplies and non-piped supplies significantly increase the complexity of the problem and risks to public health, over which the water supplier may have little control. Similarly the supplier is unlikely to have any control over the point of supply in terms of hygiene, or the condition of the plumbing materials used. In such situations the relevant authorities need to work with the supplier to develop surveillance and education programmes, and to provide advice, so that consumers are aware of the risks involved.

Plumbing within buildings can represent a significant potential source of deterioration in both the microbiological and the chemical quality of the water supplied. Risks arise through poor design and materials or from later alterations to the plumbing system. There is also a risk that back-flow contamination could affect the public supply. The control of hazards at consumer premises requires the involvement of other stakeholders and the development of appropriate supporting programmes including building regulations and water fittings regulations. However, control measures can only be as effective as the policing by the enforcement agencies.

To date water safety plans are not universally available for public water supplies. However, it is important to recognize that each utility will be producing their plans from different starting points. National drinking water standards apply in many countries and utilities in the USA, Western Europe and some Australasian countries tend to have comprehensive management practices that can be used as a basis for developing such plans. In the UK the Drinking Water Inspectorate has issued guidance on water safety plans (DWI, 2005) and expects water companies in England and Wales to develop plans accordingly. Future drinking water legislation in Europe is also likely to incorporate the principles of water safety plans. However, in countries where there are no national standards or trade associations through which consistency can be achieved, individual suppliers will need to draft their standards, procedures and documentation based on their individual circumstances and resources, using the guidance provided by WHO. For developing countries, plans should be appropriate to the situation and be attainable but, at the same time, should provide for continuous improvement. For example, it would not be cost effective to attempt to produce very high quality water with sophisticated treatment processes only to distribute it into a distribution system in poor condition and with an intermittent supply to consumers. The risk to public health would be better managed by first improving the reliability of supply and then by addressing the treatment processes.

REFERENCES

Ahsan, H., Chen, Y., Parvez, F. et al. (2006). Arsenic Exposure from Drinking Water and Risk of Premalignant Skin Lesions in Bangladesh: Baseline Results from the Health Effects of Arsenic Longitudinal Study. *American Journal of Epidemiology*, 163(12).

Anon. (1997). Biogrowth, *Water Quality International*, March/April

AWWA (2005). *Standard Methods for the Examination of Water and Wastewaters*, 21st Edn.

Bellinger, E. G. (1992). *A Key to Common Algae: Freshwater, Estuarine and Some Coastal Species*, 4th Edn. IWEM.

Bourgine, F. P., Chapman, J. I., Kerai, H. et al. (1995). The Degradation of Atrazine and Other Pesticides by Photolysis. *J CIWEM,* 9.

Butkus, M. A., Edling, L., Labare, M. P. et al. (2003). The Efficacy of Silver as a Bactericidal Agent: Advantages, Limitations and Considerations for Future Use. *J WSRT-Aqua*, 52(6).

Cambell, H. S. (1970). Corrosion, Water Composition and Water Treatment. *J SWTE*, 19.

Carmichael, W. W. (1993). Harmful Algal Blooms: a Global Phenomenon. *Proceedings of the 1st International Symposium on Detection Methods for Cyanobacterial (Blue-Green Algal) Toxins.* Royal Society of Chemistry. Bath.

CEU (1975). Directive on the quality of surface water intended for the abstraction of drinking water. 75/440/EC, *EC Official Journal L194.*

CEU (1980). Directive on the quality of water intended for human consumption. 80/778, *EC Official Journal L229/11.*

CEU (1991). Directive concerning protection of waters against pollution caused by nitrates from agricultural sources. 91/676/EC. *EC Official Journal.*

CEU (1998a). Directive on the quality of water intended for human consumption. Directive 98/83. *EC Official Journal L330/41.*

CEU (1998b). Directive concerning urban waste water treatment. 91/271/EC, *EC Official Journal L135: as* amended by Directive 98/15/EC, *CELEX-EUR, EC Official Journal L67.*

CEU (2000). Directive establishing a framework for the community action in the field of water policy. 2000/60/EC. *EC Official Journal L327/1.*

CEU (2006). Proposal for a Directive of the European Parliament and of the Council on Environmental Quality Standards in the Field of Water Policy and amending Directive 2000/60/EC. 2006/0129(COD).

Croll, B. T. (1992). *Phosphate reduction in reservoirs.* IWEM Scientific Section, Huntingdon.

Defra (2003). *Further Studies on the Incidence of Mycobacterium avium Complex (MAC) in Drinking Water Supplies (Including the Detection of Helicobacter pylori in Water and Biofilm Samples).* FWR. DWI contract 0833.

Defra (2005). *Review of Evidence for Relationship Between Incidence of Cardiovascular Disease and Water Hardness.* DWI contract 70/2/176.

Defra (2006). Uranium Occurrence and Behaviour in British Groundwater. *British Geological Society Commissioned Report CR/06/050N.*

Defra (2007a). *The Environment in Your Pocket.*

Defra (2007b). *Desk Based Review of Current Knowledge on Pharmaceuticals in Drinking Water and Estimation of Potential Levels.* Watts and Crane Associates. DWI contract 70/2/213.

DETR (1998). Cryptosporidium in Water Supplies. *Third report by Group of Experts.*

DETR (1999a). *Nitrate in Drinking Water and Childhood-Onset Insulin-Dependent Diabetes Mellitus in Scotland and Central England.* DWI contract 0801.

DETR (1999b). *Exposure to Endocrine Disrupters via Materials in Contact with Drinking Water.* DWI contract 0809.

DoE (1990a). *Guidance on Safeguarding the Quality of Public Water Supplies.* HMSO.

DoE (1990b). Cryptosporidium in Water Supplies, *Report by the Group of Experts.* HMSO.

DoE (1993a). Formation of Bromate during Drinking Water Disinfection. *Report No DWI0137.* FWR, DWI contract 0137.

DoE (1993b). Formation of Bromate during Electrolytic Generation of Chlorine. *Report No DWI0136.* FWR, DWI contract 0136.

DoE (1995). Cryptosporidium in Water Supplies, *Second Report by the Group of Experts.* HMSO.

DHSS (1971). *Circ.71/150.* 17 August.

DoH (2005). *Lowermoor Water Pollution Incident.* Committee on Toxicity of Chemicals in Food. Consumer Products and the Environment.

DWI (2005). *A Brief Guide to Drinking Water Safety Plans.* October.

DWI (2006). *Drinking Water 2006—A Report by the Chief Inspector.* (on line), http://www.dwi.gov.uk (February 2008).

Dynamco (2004). Ultrasonic Algae Control at Barcombe Reservoir. *Dynamco Research Report Ref 85358.*

EA (1976). *Standing Committee of Analysts—Methods for the Examination of Waters and Associated Materials—* (a series of booklets 1976 onwards).

EA (1998). *Endocrine-Disrupting Substances in the Environment: What Should be Done?* EA.

EA (2002a). The Microbiology of Drinking Water: Part1-Water Quality and Public Health. *The Standing Committee of Analysts—Methods for the Examination of Waters and Associated Materials.* EA.

EA (2002b). The Microbiology of Drinking Water: Part2—Practices and Procedures for Sampling. *The Standing Committee of Analysts—Methods for the Examination of Waters and Associated Materials.* EA.

EA (2002c). The Microbiology of Drinking Water: Part3—Practices and procedures for laboratories. *The Standing Committee of Analysts—Methods for the Examination of Waters and Associated Materials.* EA.

EA (2002d). The Microbiology of Drinking Water: Part 4—Methods for the Isolation and Enumeration of Coliform Bacteria and Escherichia coli (including E. coli 0157:H7). *The Standing Committee of Analysts—Methods for the Examination of Waters and Associated Materials.* EA.

EA (2005). The determination of Legionella bacteria in waters and other environmental samples: Part 1 Rationale of surveying and sampling. *The Standing Committee of Analysts—Methods for the Examination of Waters and Associated Materials.* EA.

Falconer, I. R., Runnegar, M.T.C., Buckley, T. et al. (1989). Using Activated Carbon to Remove Toxicity from Drinking Water Containing Cyanobacterial Blooms. *J AWWA*, 18.

Fawell, J. K. (1991). Developments in Health-Related Standards for Chemicals in Drinking Water. *J IWEM*, 5.

Garbett, P. (2005). An investigation into the Application of Floating Reed Beds and Barley Straw Techniques for the Remediation of Eutrophic Waters. *Water and Environment. J,* 19(3).

Goh, S., Reacher, M., Casemore, D. et al. (2005). Sporadic Cryptosporidiosis Decline after Membrane Filtration of Public Water Supplies, England, 1996–2002. *Emerging Infectious Disease*, 11(2).

HSE (2000). Legionnaire's Disease: The Control of Legionella Bacteria in Water Systems. *Approved Code of Practice and Guidance L8.* HSE.

Hopenhayn, C., Bush, H., Bingkang, A. et al. (2006). Association Between Arsenic Exposure from Drinking Water and Anaemia during Pregnancy. *Journal of Occupational & Environmental Medicine*, 48(6).

Hunt, S. M. and Fawell, J. K. (1987). The Toxicology of Micropollutants in Drinking Water; Estimating the Risk, *J IWES*, 41.

Hutton, L. G. (1983). *Field Testing of Water in Developing Countries*, WRc.

ISO (1985). 07.100.20 Microbiology of water. ISO.

ISO (2006). Water quality—Sampling—Part 1:Guidance on the Design of Sampling Programmes and Sampling Techniques, ISO 5667-1:2006. ISO.

ISO (2007). Water quality—Interlaboratory Comparisons for Proficiency Testing of Analytical Chemical Laboratories, ISO/TS 20612:2007. ISO.

IWEM (1969). *Manual of British Water Engineering Practice*, 4th Edn. IWEM.

Lahti, K. and Hissvirta, L. (1989). Removal of Cynaobacterial Toxins in Water Treatment Processes: Review of Studies Conducted in Finland. *Water Supply* 7.

Langelier, W. F. (1936). Analytical Control of Anti-Corrosion Water Treatment. *JAWWA*, 28(10).

Lee, T-J., Nakano, K. and Matsumura, M. A. (2002). Novel Strategy for Cyanobacterial Bloom Control by Ultrasonic Irradiation. *Water Science and Technology*, 46(6–7).

Matthiessen, P., Arnold, D., Johnson, A. C. et al. (2006). Contamination of Headwater Streams in the United Kingdom by Oestrogenic Hormones from Livestock farms. *Science of the Total Environment* 367.

MoH (1969). Fluoridation Studies in UK and Results Achieved After Eleven Years, *Report of Research Committee of the MoH*, No. 22. HMSO.

Moss, B., Madgwick, F. J. and Phillips, G. L. (1996). *A Guide to the Restoration of Nutrient-Enriched Shallow Lakes.* Environment Agency and Broads Authority.

Mouchet, P. (1984). Potable Water Treatment in Tropical Countries: Recent Experiences and Some Technical Trends. *Aqua*, 3.

Palmer, C. M. (1977). *Algae and Water Pollution.* EPA-600/9-77-036.

Pontius, F. W. (1998). New Horizons in Federal Regulation. *J AWWA*, 90(3).

Pontius, F. W. (2002). Regulatory Compliance Planning to Ensure Water Supply Safety. *J AWWA*, 94(3).

Pontius, F. W. (2003). Update on USEPA's Drinking Water Regulations. *J AWWA*, 95(3).

Roberts, P. H. and Thomas, K. V. (2006). The Occurrence of Selected Pharmaceuticals in Wastewater Effluent and Surface Waters of the Lower Tyne Catchment. *Science of the Total Environment*, 356(1–3).

Siddiqui, M. S., Amy, G. L. and Rice, R. G., (1995). Bromate Formation: a Critical Review. *J AWWA*, 87(10).

Sarmah, A. K., Northcott, G. L., Leusch, F. D. L. et al. (2006). A Survey of Endocrine Disrupting Chemicals (EDCs) in Municipal Sewage and Animal Waste Effluents in the Waikato Region of New Zealand. *Science of the Total Environment*, 355(1–3).

Smith, A., Lingas, E. and Rahman, M. (2000). Contamination of Drinking-Water by Arsenic in Bangladesh: a Public Health Emergency. *Bulletin of the World Health Organisation*, 78(9).

Smith, H. V. (1992). Cryptosporidium and Water: A Review. *J IWEM*, 6(5).

Sopwith, W., Osborn, K., Chalmers, R. et al. (2005). The Changing Epidemiology of Cryptosporidiosis in North West England. *Epidemiology and Infection*, 133(5).

Sutcliffe, D. W. and Jones, J. G. (1992). *Eutrophication: Research and Application to Water Supply.* Freshwater Biological Association.

Suckling, E. V. (1943). *The Examination of Water and Water Supplies*, 5th Edn. Churchill.

Toledano, M. B., Nieuwenhuijsen, M. J., Best, N. et al. (2005). Relation of Individual Trihalomethane Concentrations in Public Water Supplies to Stillbirths and Birth Weight in Three Water Regions in England. *Environmental Health Perspective*, 113(2).

UK (1996). The Surface Waters (Abstraction for Drinking Water) (Classification) Regulations 1996. Statutory Instrument 1996/3001.

UK (2000). The Water Supply (Water Quality) Regulations 2000, Statutory Instrument 2000 No.3184.

UK (2001a). The Water Supply (Water Quality) (Amendment) Regulations 2001, Statutory Instrument 2001 No.2885.

UK (2001b). The Water Supply (Water Quality) Regulations 2001, Welsh Statutory Instrument 2001 No.3911.

UK (2001c and d).The Water Supply (Water Quality) (Scotland) Regulations 2001, Scottish Statutory Instrument 2001 No.207 and Amendment Regulations 2001, Scottish Statutory Instrument 2001 No.238.

UK (2002). The Water Supply (Water Quality) Regulations (Northern Ireland) 2002, Statutory Rule 2002 No.331.

UK (2003). The Water Supply (Water Quality) (Amendment) Regulations (Northern Ireland) 2003, Statutory Rule 2003 No.369.

UK (2007a). The Water Supply (Water Quality) (Northern Ireland) Regulations 2007, Statutory Rule 2007 No.147.

UK (2007b). The Water Supply (Water Quality) Regulations 2000 (Amendment) Regulations 2007, Statutory Instrument 2007 No.2734.

UK (2007c). The Water Supply (Water Quality) Regulations 2001 (Amendment) Regulations 2007, Welsh Statutory Instrument 2007 No.3374.

US EPA (1996) *Public Law 93-523*, Safe Drinking Water Act (December 1974) and amendments under the Safe Drinking Water Act 1996. USA.

von Ehrenstein, O. S., Guha Mazumber, D. N., Hira-Smith, M. et al. (2006). Pregnancy Outcomes, Infant Mortality and Arsenic in Drinking Water in West Bengal, India. *American Journal of Epidemiology*, 163(7).

White, G. C. (1992). *Handbook of Chlorination*. 2nd Edn. Van Nostrand Reinhold.

WHO (1993). *Guidelines for drinking-water quality. Volume 1, Recommendations.* 2nd Edn. WHO.

WHO (1996). *Guidelines for drinking water-quality, Volume 2, Health Criteria and other supporting information.* 2nd Edn. WHO.

WHO (2002). N-Nitrosodimetheylamine. *Concise International Chemical Assessment Document 38*. WHO.

WHO/US EPA (2003). *Emerging issues in water and infectious disease*. WHO.

WHO (2008), *Guidelines for Drinking water Quality*, 3rd Edn, Incorporating the first and second addenda, Volume 1; Recommendations. WHO.

WHO (2007). *Chemical safety of drinking-water: Assessing priorities for risk management*. WHO.

Wilczak, A. J., Assadi-Rad, A., Lai, H. H. et al. (2003). Formation of NDMA in chloraminated water coagulated with DADMAC Cationic Polymer, *J AWWA*, 95(9).

WRc (1984). Asbestos in drinking water. *Technical Report TR 202*. WRc.

Storage, Clarification and Chemical Treatment

7.1 RAW WATER STORAGE

This may be regarded as a first stage in treatment as it may involve a complex combination of physical, chemical and biological changes. Raw water storage has been regarded as a 'first line of defence' against the transmission of water-borne diseases; this aspect is still of major importance if the unstored water is liable to excessive bacterial pollution from sewage, even though such pollution may only occur occasionally, e.g. if storm-water sewage overflows discharge into a river. A few days storage improves the physical and microbiological characteristics of a surface water through the effect of a combination of actions including sedimentation, natural coagulation and chemical interactions, the bactericidal action of ultraviolet radiation near the water surface and numerous biotic pathways which help to reduce enteric micro-organisms (Sykes, 1971). Storage in a reservoir for a period from one to several months produces a substantial decrease in the numbers of bacteria of intestinal origin; the specific organisms of typhoid and cholera also disappear. The die-off rate for enteric coliforms, designated here as the time to achieve a 90% loss of bacteria or T_{90}, in lakes and other open waters varies from 2–3 hours in strong sunlight in clear waters, to 10 hours in more turbid waters (Kay, 1993). An additional benefit of short-term storage is that it allows a river intake to be shut down to avoid or investigate any pollution which might, for example, be indicated by the death of fish or by other information (Young, 1972) such as changes to physical chemical characteristics of the water.

The UK DoE recommended in circular No. 22/72 (DoE, 1971) that water supplies at risk from accidental spillages of industrial chemicals on roads and manufacturing sites should be protected by at least seven days' storage. This was to allow closure of the intake until the pollution risk was over, to dilute any polluted water entering the intake with clean stored water and to allow further self purification to take place. By 1979 rather more than two thirds of river derived supplies in England and Wales had received some storage in a reservoir prior to treatment (DoE, 1979). Such buffer storage is still desirable but alternative strategies are now considered, particularly when storage is neither economic nor practical. Such strategies (Chapter 6, Part VI) might include catchment management, regulations to avoid or reduce the risk of industrial chemicals being spilt in locations where they might enter an aquifer or surface source, or protection of abstraction sites by continuous water quality monitoring systems coupled with the provision of an alternative source of supply (DoE, 1993). The Water Framework Directive (Section 2.7) will have an impact on the need for management of raw water storage.

Instruments for water quality parameters such as temperature, pH, conductivity, ammonia or dissolved oxygen, have been available since the 1960s and have been in use by some undertakings from that time (Ward, 1998). Incidents caused by accidental spillage of compounds such as phenols, which cause taste and odour problems at very low concentrations and were able to pass through conventional process works and enter distribution before being detected, have led to the development of increasingly sophisticated raw water monitoring (Palmer, 1993). Often there is a need for monitoring but no one compound could be generally said to be the main risk, as for example at an intake downstream of a roadbridge at which accidental spillage might occur. In such instances fish monitors have been in use for many years; behavioural and metabolic activity monitoring of sensitive fish such as trout has been devised and improved by rigorous research and development (Harman, 1998). There are several instrumental packages capable of measuring very low concentrations of soluble organic substances at very low concentrations such as UV absorption or total organic carbon (TOC); these can be used as general monitors to raise alarms when the background levels of such substances are suddenly raised. Manufacturers of liquid and gas chromatography have since been able to produce reliable, stand-alone, instrumentation which, together with facilities for alarms and computer-driven libraries of traces, allows notice to be given of sudden peaks and identification of the compounds within 10–20 minutes. This is usually sufficient time to close down an intake before a water supply is threatened by irretrievable contamination (Fowlis, 1996).

Potential Problems in Raw Water Storage

There are potential disadvantages in the prolonged storage of raw waters, which should be taken into account when considering adoption of storage and its management. The most obvious of these is the likelihood of growth of various forms of plants, either rooted aquatic types (macrophytes) which may choke shallow waters, or free floating water weeds such as duckweed and, in the tropics, water hyacinth or salvinia as well as planktonic types such as algae (phytoplankton), which may increase the difficulties of treatment. The main issues for water supply raised by the presence of these plants and also animals are discussed in Part IV of Chapter 6. Storage reservoirs that are less than about 10 m deep can allow light to reach the bottom; this may encourage the growth of rooted plants unless the stored waters are sufficiently turbid to reduce light penetration. Shallow reservoirs are therefore generally avoided if there is any likelihood that plant growth could be high.

Waters which contain sufficient nutrient materials to support prolific growths of aquatic plants are usually described as eutrophic. Lakes and water bodies exhibit a range of concentrations from low nutrient conditions in upland lakes on igneous rocks (oligotrophic) through moderate or mesotrophic lakes where a balanced ecology with some shoreline macrophytes and a wide range of planktonic algae occur, through to lowland water bodies which tend to be eutrophic and even hypertrophic waters where prolific growths of plants commonly occur due to enrichment by sewage and where in temperate climates there are usually seasonal peaks. In most reservoirs that are more than about 10 metres deep, and many are designed in this way to avoid excessive plant growth, an additional complication is that thermal layering or stratification may occur on a seasonal basis in temperate climates. As temperature rises in the spring warmer water tends to remain at the surface due to its lower density. In the absence of any strong wind induced circulation the colder and now denser water below remains and ceases to mix with the surface water. The upper and lower layers are known respectively as the epilimnion and hypolimnion; in between there is a zone known as the thermocline in which there is a relatively steep change with depth. In reservoirs averaging less than 10 m deep thermal stratification in temperate lands is often only temporary, lasting for a few days at most. In colder climates there may be winter stratification due to ice formation which ensures the water body and its biological life is protected from lethal frosts (Moss, 1988).

Thermal stratification can clearly affect retention time of incoming water and is often of major importance with reference to water quality. In many large reservoirs, there are facilities for withdrawing the water for treatment at several different levels which can be chosen as circumstances dictate. Multiple drawoff facilities are discussed in Section 5.21. In the case of eutrophic reservoirs the ability to avoid drawing from surface water with high concentrations of algae is particularly useful. The tower can be sited some way from the shore in order to avoid the build up of surface aggregations of algae that onshore winds can cause.

In some large reservoirs the water in the hypolimnion is of a high standard of purity, as well as being cool, e.g. in some places in Scandinavian countries, in the Lake District of England (Lund, 1963), and in Lake Constance (Bodensee). However, in eutrophic reservoirs organic impurities in the incoming water or released by leaching from bottom muds and inundated soils may accumulate in the bottom water. As a result of plant and animal respiration and bacterial activity, the concentrations of dissolved oxygen fall and may approach zero. Under such conditions major chemical changes take place in the transition from anoxic conditions, when only combined oxygen compounds such as sulphate (SO_4^{2-}) and nitrate (NO_3^-) are present, to full anaerobic conditions when even these compounds have been reduced to sulphide and nitrogen. This transition can be measured by redox potential which falls from +200 mV when free oxygen is present to −200 mV in anaerobiosis. Redox potential or oxidation–reduction potential (ORP) is the potential developed in a cell between a metal electrode (e.g. platinum) and a reference electrode during an oxidation–reduction reaction. It is reported with respect to the potential of the hydrogen electrode which is zero. The typical range is −1000 mV to +1000 mV. Oxidising agents (e.g. Cl_2, O_2) increase the ORP, whereas reducing agents (e.g. sulphites) lower the ORP. Therefore, it is used to determine the oxidising or reducing characteristics of a solution. At high ORP ions such as sulphate, nitrate or ferric predominate whilst the presence of sulphites, ammonia or ferrous ions in significant concentration are an indication of low ORP. In aquatic redox reactions ORP is related to bacterial activity (e.g. oxidation of iron in biological filters and reduction of ferric hydroxides or oxides in lakes). The important consequence for water quality is that under negative redox conditions numerous other chemical compounds can re-dissolve from the sediments into the overlying water. Iron, which is present as insoluble oxide or hydroxide, is usually the most prominent of these substances but if manganese is present (usually as insoluble oxides), it may also become soluble. Both iron and manganese are often in combination with organic colouring matter. The concentrations of plant nutrients, phosphates as well as ammonia commonly also increase in bottom water close to anaerobic sediment. Actinomycete fungi often proliferate in these conditions (Burman, 1965) and lead to tastes and odours in the raw water. Under such circumstances abstraction from the hypolimnion for water treatment should be avoided. Some authorities prefer to abstract this water for compensation flows to 'bleed' the higher dissolved contaminants from the stored water.

When the surface water cools down in the autumn and wind action becomes effective, the reservoir water mixes and, under normal circumstances, water from mid depth or deeper is of good quality. Rapid mixing caused by strong autumnal winds can cause sudden 'turnover' resulting in rapid deterioration in the quality of the water at the surface in respect of colour, iron and manganese. A more gradual effect of the resumed mixing is that the plant nutrients carried up from the bottom may increase the growth of algae in the water and cause an autumnal bloom of algae. This is not uncommon in eutrophic lakes and reservoirs and may render the surface water difficult to treat.

Reservoir management to control stratification has been established practice since the 1960s. Mechanical pumps have been used to induce mixing artificially but with varying degrees of success. The systems have usually been expensive to install and operate. They are intended both to oxygenate the bottom layers of a reservoir and to control stratification. Pioneering studies in UK (Pastorok, 1980) showed that algal populations as well as water quality could be managed by water movement control. This approach led to arrangements for carefully controlled raising of cool low oxygen water from below the thermocline using jetted inlets in those reservoirs, which had pumped inflows. Jets entrain the bottom waters (also low in algae) in the flows induced to the surface and increase the amount of satisfactory water above the thermocline. The system is suitable for large flat-bottomed reservoirs where the bottom water volume is large relative to the surface epilimnion, but is not effective in narrow natural valley impoundments where the hypolimnion is small and where more vertically oriented mixing is needed. Air lift pumping systems such as the 'Bubble Gun' (Henderson-Sellers, 1984) and 'Helixor' (Ridley, 1966) which made use of rising air bubbles in tubes to entrain bottom water so that it rose to the surface whilst being aerated by the bubbles were shown to be more effective than mechanical pumps. Later in the 1970s simpler and more economical methods were developed such as the use of compressed air plumes from perforated airlines laid along the bottom of the reservoir (Davis, 1980; Lorenzen, 1977). An alternative air injection system used a number of ceramic domes to produce fine air bubbles (Speece, 1994). Recently the increased effectiveness of fine bubbles has been incorporated into a design comparable to the 'Helixor' design and reported to show improvements in its performance (HRW, 2003).

It may be beneficial to have an emergency bypass so that water may be taken directly from a river instead of from a reservoir. This could be used in the case of exceptional high algal growth in the water or of pollution having occurred, or being suspected in water in the reservoir. However, this action can have operational and hence economic consequences in that waters exhibiting eutrophication are very likely to be at risk of carrying oocysts of *Cryptosporidium* and appropriate processes need to be included in the treatment plant.

Growth of plants can deplete nutrients from the water particularly if the growth is not limited by light or grazing. Management of stored water to make use of this characteristic has been carried out by TWUL for many years (Steel, 1975). Water from the River Thames in autumn and winter frequently has nitrate nitrogen concentrations, which exceed the EC Directive value (Section 6.36) of 50 mg/l as NO_3 due to pollution from agricultural runoff. Toms (1981) showed that an empirical relationship could be used to predict the reduction of nitrate by storage and that long-term stored water with low nitrate could be used to reduce nitrate concentrations entering supply from other sources by blending. TWUL use storage in one of their larger reservoirs for prolonged periods to reduce nitrates by the combination of algal growth and bacterial denitrification.

SCREENING

7.2 PASSIVE SCREENS

Passive screens are static screening devices which have no moving parts in water and are used for off-shore or shoreline applications. The passive screens are generally used at locations where it is necessary to keep aquatic life from entering the screened water and may not require further fine screening. Therefore they avoid the need and cost of handling trash.

Passive screens, usually vertical, are installed completely below the water surface; they depend on having low velocity (0.1–0.15 m/s) through the screen bars to prevent fish impingement. They are best suited for areas where an ambient cross-flow current is present. The screen bar spacing is generally in the range 0.5–10 mm depending on application. Some designs use perforations, up to 12 mm diameter, in plates instead of screen bars. Combination of low velocity and slot width has reduced the entrainment of eggs, larvae and juvenile fish (Stefan, 1986). Periodic air sparging with bursts of air, about 3 times the volume of screen units at about 10 bar pressure, is used to keep the screen surface clean. This type of screen is only available from a small number of suppliers and is of proprietary design (Gille, 2003). Fouling of the screens by bacteriological slime, algae, mussels, sponges, barnacles and other marine organisms can be a problem as chlorination of the water outside the screens is normally not acceptable on environmental grounds. Cupro-nickel alloy is suggested by the manufacturers of the screen to minimize the fouling as copper ions produced are toxic to biological organisms. However, the effect of copper is reduced by cathodic protection and by the presence of ions, which form insoluble copper compounds such as sulphide and carbonate.

7.3 ACTIVE SCREENS

Active screens are either moving or require moving devices for cleaning and are used for off-shore or shoreline applications. Active screening is traditionally used to keep trash and floating debris out of the screened water. Active screens include coarse bar screens, which may have clear spaces between bars from as little as 10 mm to 150 mm or more to suit the application. The aim of the screening facility is to remove trash and protect downstream equipment such as pumps from large objects.

Where it is necessary to remove finer material from the water, band screens or drum screens are widely used. Another type of screen that uses revolving screen discs, of non metallic construction, is also available for fine screening, particularly for applications where large quantities of water need to be screened (Peltier, 2004).

Bar screens. Practically all intakes are screened, even though the screens may be of the simplest type of bar grille (Plate 9(a)). The bars are up to about 25 mm thickness and are normally spaced at 75–100 mm centres. The design velocity is in the range 200–600 mm/sec. The bars are best inclined for ease of cleaning by hand rake for occasional manual cleaning or by mechanical rake (automated or manual) for more frequent cleaning. Should a smaller mesh be necessary, it is best to group bars into frames so that each frame can be lifted out of the water, cleaned and lowered back into position. To prevent unscreened water from passing through the intake when screens are lifted, they should be provided in duplicate or provision should be made for stop log insertion upstream.

Bar screens generally having larger spacing would not arrest smaller size debris. Where this could adversely affect downstream plant it is advisable to install a robust band or drum screen downstream of the protective bar screen.

Band and drum screens. If fine screening is adopted, some means must be found for continuously cleaning the screens or they rapidly become clogged. Therefore, fine screens are usually arranged as endless bands of wire mesh panels or rotating drums of material perforated with holes of openings of about 6–9.5 mm or finer mesh screens with openings 0.5–5 mm. Plates 9(b) and (c) show a cup screen and a band screen. Captured debris is lifted by the moving screen and then washed off into a trough as it passes over water jets. Screen units of sizes up to about 100 000 m^3/hr are available.

A supply of clean water under pressure is needed for the wash water jets; this may have to be pumped from the strained water. The total amount of water required for washing may be of the order of 1% of the throughput depending on the amount of material retained. Fine screening must always be preceded by a coarse screen. Band screens in general have a small foot print and require smaller civil works but have more mechanical parts and need more maintenance. Drum screens in general require a larger foot print and usually require a larger structure but can deal with larger flows and have lower maintenance costs.

7.4 MICROSTRAINERS

These are revolving drums mounted in open tanks with a straining medium which is usually a stainless steel wire fabric of a very fine mesh, fitted to the periphery of the drum. The drum is submerged for about 75% of its diameter (66% of area) and rotates at about 0.5–5 rpm (peripheral drum speeds of 3–50 m/min). Water to be treated enters the drum axially under gravity and flows out radially through the fabric, depositing particulate matter. Cleaning is accomplished by a row of water jets along the full length of the drum operating at about 2.5 bar pressure. Particulate matter intercepted by the fabric rotates to the top of the drum where it is backwashed into a hopper running the full length of the drum and conveyed by a pipe which also acts as the axle for the drum assembly, to a point outside. Water jets use about 1–1.5% of the total quantity of water strained but this washwater should be filtered and chlorinated.

Total headloss through a microstrainer unit including inlets and outlets varies from about 150–200 mm. Single units have capacities of 10 m³/h to a maximum of 4000 m³/h for a 3.2 m diameter × 5 m wide drum.

Microstrainers bring about an improvement in the physical quality of a water but there is no change in the chemical characteristics of the water. The ideal water for microstraining is a lake or large reservoir supply which does not contain a large amount of suspended matter but which contains moderate quantities of zooplankton, algae and other microscopic-sized particles; total algae removal ranges from 50 to 75% depending on the size. When applied for the removal of zooplankton microstrainers are located either at the beginning or end of the treatment process. Fabrics commonly used with stored waters are made of woven stainless steel wires of 0.05 mm diameter with apertures of 23 and 35 µm. However, a coarser mesh at 200 µm aperture is sometimes used after granular activated carbon (GAC) filters to remove eroding particles of the carbon and any bacterial flora or zooplanktons that sometimes develop in GAC filters. Attempts to use plastic mesh materials have not been successful in water supply.

Microstrainer operation is fully automatic. A microstrainer screen can easily be damaged if too great a loading is placed upon it; hence head across the screen should be monitored and an alarm initiated if it approaches the maximum desirable value. The rotational speed of the drum assembly can be adjusted so that the optimum differential headloss across the fabric can be maintained to achieve maximum removal efficiency irrespective of the raw water flow or quality. Most installations have an automatic fail-safe bypass weir which diverts unstrained water when the screens become overloaded. This causes deterioration of the treated water at times of peak loading and illustrates why microstrainers are not generally acceptable as an alternative to media filtration for potable waters. Their most extensive and successful use in water supply has been to lighten the loading upon rapid or slow sand filters so that the length of run of these filters between cleaning is extended, thereby increasing their output by as much as 50%. In comparison with roughing filters, microstrainers (which have surface loading rates up to 80 m³/h.m²) require less space and produce lower headloss; their capital and running costs are lower.

SEDIMENTATION AND SETTLING TANKS

7.5 GENERAL DESIGN CONSIDERATIONS

Sedimentation tanks are designed to reduce the velocity of water so as to permit suspended solids to settle out of the water by gravity. There are many different designs of tanks and most are empirical. A design which may be very successful on one kind of water may perform poorly with another. The success of a design may be judged on its ability to maintain the claimed throughput and the agreed effluent water quality under adverse raw water quality conditions. The effluent quality for suspended solids or turbidity depends on the design and can be in the range 1–5 mg/l and 1–5 NTU, respectively. The amount, size, shape, density and nature of suspended solids in a water, the water temperature and the extent of clarification required all influence the performance of a tank design. Laboratory jar tests are performed on samples of raw water and a suitable tank design has to be selected. Designs which have been successful before under similar conditions are a useful start but alternatives that could provide a more economical and efficient solution should be considered.

7.6 PLAIN SETTLING

In plain settling (or sedimentation), suspended solids in a water are permitted to settle out by gravity alone; no chemicals are used. For this purpose the water can be left to stand in a tank, although with continuous supply at least two such tanks have to be used alternately. Such fill-and-draw tanks are seldom used in modern plants, except for filter washwater recovery. Instead plain sedimentation tanks are designed for continuous throughput, the velocity of flow through the tank being sufficiently low to permit gravitational settlement of a portion of the suspended solids to occur. In practice the application of plain sedimentation in waterworks is very restricted because impurities such as algae, aquatic plant debris and finely divided mineral matter do not settle at a rate sufficient for a tank of reasonable size to be utilized. Plain settling is most frequently used as a preliminary treatment for fast flowing river waters carrying much suspended solids as occur in the tropics or at intake works on water transfer schemes in order to minimize the amount of suspended material passing into the system. Under most circumstances chemically assisted sedimentation, which is a more complex process is adopted (Section 7.9).

The velocity with which a particle in water falls under gravity depends on the horizontal water velocity, the size, relative density and shape of the particle and the temperature of the water. The theoretical velocity V (mm/s) of falling spherical particles in slowly moving water (Reynolds numbers of less than 0.5) is given by:

$$V_c = \frac{g}{1.8 \times 10^4}(r-1)\frac{d^2}{\gamma} \tag{7.1}$$

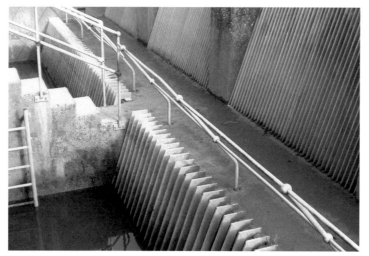

(a) Manually raked bar screens at a river intake in UK.

(b) Large diameter cup screen.

(c) Band screen.

PLATE 9

(a) 2.8 m diameter static mixer for pH control using 98% sulphuric acid.
   (Manufacturer: Statiflo, UK)

(b) Schematic and inset photograph of static mixer with chemical dosing pipes installed in
    a channel. (Manufacturer: Statiflo, UK)

PLATE 10

where $g = 9.81$ m/s^2, r is the relative density of the particles, d is the diameter of the particles in mm and γ is the kinematic viscosity of water in m^2/s, which varies with the temperature of the water as given in Table 7.1 (Camp, 1946). The coefficient of kinematic viscosity (m^2/s) = coefficient of absolute viscosity (Ns/m^2) divided by density (kg/m^3) where N (Newton) is kg.m/s^2.

A number of different (mainly empirical) formulae have been given for the settlement of sand and soil particles in still water; some of the values derived are given in Table 7.2.

Aluminium and iron flocs have a specific gravity of about 1.002, particle size as large as 1 mm and a settling velocity (at 10 °C) of about 0.8 mm/s (Fair, 1968). Clay particles generally have a grain diameter of 0.01 mm to less than 0.001 mm (1 µm) so that it is impracticable to remove them from a water by simple sedimentation, or even by filtration, without prior chemical coagulation treatment (Sections 7.11–7.18).

Table 7.1 Kinematic viscosity of water

Temperature (°C)	0	5	10	15	20	25
Value γ (m^2/s) $\times$ 10^{-6}	1.79	1.52	1.31	1.15	1.01	0.90

Table 7.2 Settling speeds of particles of relative density r

Diameter of particle (mm)	Settling speed (mm/s)[1] – sand[a] ($r = 2.65$ at 10 °C)	Settling speed (mm/s)[2] – sand[a] ($r = 2.65$ at 10 °C)	Settling speed (mm/s)[2] – coal ($r = 1.5$ at 10 °C)	Settling speed (mm/s)[2] – sewage solids ($r = 1.2$ at 10 °C)
1.0	100	140	40	30
0.6	63	—	—	—
0.5	—	70	20	17
0.4	42	—	—	—
0.2	21	22	7	5
0.1	8	6.7	—	—
0.06	3.8	—	—	—
0.05	—	1.7	0.4	0.3
0.04	2.1	—	—	—
0.02	0.62	—	—	—
0.01	0.15	0.08	0.02	0.008

Note: [a] Particle sizes 0.1 mm and below are classed as silt
Sources: [1] AWWA (1969); [2] Imhoff (1971)

Maximum velocity to prevent bed uplift or scour. Apart from the settling rate in still water it is, of course, essential that once a particle has reached the base of the tank it shall not be resuspended by the velocity of flow of water over the bed. Camp (1946) gives the channel velocity V_c (m/s) required to start motion of particles of diameter d (mm) as:

$$V_c = \left(\frac{8\beta g}{10^3 f} (r - 1)d \right)^{1/2} \tag{7.2}$$

where r is the relative density, f is the friction factor in the Darcy–Weisbach formula, head loss = $(4flv^2/2gd)$, β is in the range 0.04–0.06 for sticky flocculent materials and 0.10–0.25 for sand and $g = 9.81$ m/s^2.

Maximum horizontal velocity of flow. A third flow measure which must be taken into account is that the horizontal velocity of flow must not be so great as to prevent, by turbulence, the settling of particles under gravity. There is general agreement that this velocity should not be more than 0.3 m/s to allow sand grains to settle. This is, of course, too high a velocity for the settling of particles of light relative density (1.20 and less), but this is the figure normally used for sewage grit chambers where the heavier material is to be deposited and the lighter material left to carry over. At 0.2 m/s faecal matter, i.e. organic matter, begins to settle. These velocities are in contrast to 1.0 to 1.5 m/s to minimize suspended solids settlement in pipelines.

7.7 THEORY OF DESIGN OF TANKS

The flow Q through a rectangular sedimentation tank of length l, water depth d and width b is shown in Figure 7.1. The time of fall (d/V) of a particle of silt of vertical falling speed V from entry to the tank to reach the bottom before the water leaves the tank must equal the time of horizontal flow (Lbd/Q). Therefore $V = Q/A$, where A (= Lb) is the surface area of the tank. Q/A is known as the surface loading rate and is expressed as m^3/h.m^2, m/h or mm/s.

In a tank with uniform flow distribution all particles with a falling speed greater than Q/A reach the bottom before the outlet end of the tank. Particles with a speed less than Q/A are removed in the same proportion as their speed bears to Q/A, e.g. if the speed V is only half Q/A then only half the

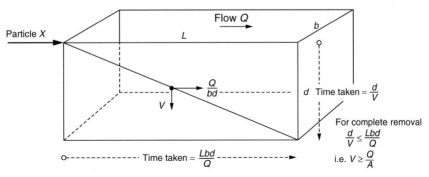

FIGURE 7.1

Theoretical flow in a horizontal flow rectangular tank.

particles falling at this speed reach the bottom. Q/A is thus a measure of the effective removal of the particles in any tank. For example, in a tank of 300 m^2 surface area with inflow of 1.2 m^3/s, Q/A = 14.4 m^3/h.m^2 (4 mm/s) so that, theoretically, all particles with V of 4 mm/s or more are be removed, 50% of those having V of 2 mm/s, 25% of those with V of 1 mm/s, and so on. Therefore, the performance of a tank is independent of depth and retention time. This concept is the basis for the design of multi-tray horizontal flow tanks and inclined plate and lamella settlers (Sections 7.14 and 7.17).

The foregoing theory, however, assumes that the falling particles do not hinder each other; but McLaughlin (1959) showed in laboratory experiments with clay and aluminium sulphate that the faster particles, settling through the slower ones, gather some of the latter up and drag them out of suspension. This is the case with flocculated suspensions where, during settlement, agglomeration of the particles takes place and so particles settle much faster. The performance of a tank in which this phenomenon occurs is related not so much to its surface loading rate as to the time of residence. These findings relate to still water. Thus, dependent upon the nature and size of the settling particles, the range of sizes, the degree of concentration of the suspension and the amount of turbulence, the performance of a sedimentation tank may relate to its surface loading rate or to its residence time, or partly to both.

7.8 GRIT TANKS

In waterworks, grit primarily constitutes sand, gravel and other abrasive material and is present mostly in water abstracted from rivers. If allowed to enter the works it could damage intake pumps and settle in raw water pipelines and inlet process units. Intakes should therefore be sited and designed to minimize the uptake of grit (Section 5.22) and tanks should be provided upstream of pumps to trap the grit. Grit tanks are also used as traps for sand, anthracite, granular activated carbon and other filter media carried over in used filter washwater to protect any pumping equipment downstream.

Grit tanks operate as plain settling tanks (unaided by coagulants) and are sized to capture particles with diameter larger than 0.1 mm. From Table 7.2 a silt particle of 0.1 mm falls in still water at 8 mm/s. The residence time in the tank should be at least 1.5 times the time taken for 0.1 mm particles to settle to the floor of the tank. Grit tanks for sand removal usually have surface loading rates of 10–25 m^3/h.m^2, a water depth of 3–4 m and length to width ratio of at least 4:1 (Kawamura, 2000). To prevent scour of settled material the horizontal velocity must be smaller than the scour velocity calculated from the formula in Section 7.7. This is usually about 0.3 m/s for sand.

CHEMICALLY ASSISTED SEDIMENTATION OR CLARIFICATION

7.9 CHEMICALLY ASSISTED SEDIMENTATION

This comprises several separate processes of treatment which together make up the complete system known as 'clarification'. This system is designed to remove from a water suspended materials, colour and other soluble material such as those of organic origin and soluble metals such as iron, manganese and aluminium ahead of filtration processes. It is a delicate and chemically complex phenomenon having three stages: (1) the addition of measured quantities of chemicals to water and their thorough mixing, (2) coagulation and flocculation, or the formation of a precipitate which coalesces and forms a floc and, (3) sedimentation.

7.10 CHEMICAL MIXING

Most chemical reactions in water treatment applications are completed within 5 seconds and therefore the principal objective in chemical mixing is to obtain rapid and uniform dispersion of the chemical in the main flow of water to ensure that chemical reactions are completed in the shortest possible time. Inadequate mixing of a coagulant such as aluminium sulphate can impair the formation of a good floc and would result in poor plant performance which can only be corrected by using excess of the chemical. The addition and mixing of chemicals to the main flow of water is a continuous process and is frequently described as either rapid or flash mixing. The design of mixers is often based on the concept of velocity gradient, which was first developed by Camp and Stein (1943) for flocculation. It is an inadequate parameter for design of mixers but, in the absence of a better design approach, it is still being used for mixer design and its value is used to express the degree of mixing at any point in the liquid system. The velocity gradient G (s^{-1}) is defined in terms of power input by the following relationship developed by Camp and Stein for flocculation:

$$G = \left(\frac{P}{\mu V}\right)^{1/2} \tag{7.3}$$

where P is the useful power input (W), V is the volume (m^3) and μ is the absolute viscosity (Ns/m^2 or kg/m.s) at the water temperature. V can be taken as the flow rate multiplied by the residence time in the mixer. Mixing efficiency is directly related to the local flow turbulence created and should give a high degree of chemical-in-water homogeneity within a short time, with low absorption of power. The methods used for mixing can be hydraulic or mechanical. In hydraulic mixers all elements of liquid within the mixer are subjected to the same retention time, akin to plug flow and they are suitable for many mixing applications. Mechanical mixers are mostly backmix devices and when applied to a continuous flow system the elements of liquid within them have a distribution of residence time. This is considered unsuitable for coagulant mixing. They require long residence time to make allowance for short-circuiting and the head loss across the mixing chamber is as much as that required for hydraulic mixing.

Hydraulic mixing makes use of the turbulence created due to the loss of head across an obstruction to flow such as an orifice plate, pipe expansion or valve or by the sudden drop in water level when water flows over a weir or hydraulic jump. The latter is usually formed at a flume (Chow, 1959) (Section 12.12) in a channel with a local width constriction and change in floor level designed to produce supercritical flow under all operating flows. The ratio of the upstream depth d_1 (m) to the downstream depth d_2 (m) of such a device is given by the following equation (Fair, 1968):

$$\frac{d_2}{d_1} = \frac{1}{2}\left[(1 + 8F^2)^{1/2} - 1\right] \tag{7.4}$$

where: F (Froude number) $= V_1/(gd_1)^{1/2}$, V_1 is the velocity upstream of the jump (m/s) and g is the acceleration due to gravity (m/s^2). For the hydraulic jump to form the ratio (d_2/d_1) should be >2.4

and therefore $F > 2$. For F between 4 and 9 about 40–70% of energy is available for mixing (Schulz, 1968). Hydraulic mixer efficiency is flow dependent. Mixers should be designed for the operating range of plant flow rates, if necessary with facilities for taking sections of the mixer out of service at low flow rates.

When pipelines are used for mixing, a pipe of equivalent hydraulic length of at least 20 pipe diameters should be allowed. Static mixers are pipeline hydraulic mixers with radial mixing employing stationary, shaped diverters which force the liquids to mix themselves through a progression of divisions and recombinations until uniformity is achieved. These mixers comprise fixed mixing elements installed in a housing of diameter the same or larger than the pipe diameter. The most efficient designs provide the shortest mixing length with lowest headloss and should be used where mixing times required are short, e.g. coagulation. Plate 10(a) shows a staic mixer for a pipeline. In raw water applications, debris such as weeds and twigs could obstruct this type of unit, which may not be suitable for unscreened river abstraction. If instantaneous mixing is not achieved when mixing alkalis (e.g. lime, caustic soda) in waters with alkalinity greater than 10 mg/l as $CaCO_3$ calcium carbonate scale may form on the elements due to localized softening; also ferric salts could form hydroxide precipitate on the elements. Gases should ideally be injected as solutions using a static mixer in a side stream of flow 1–2% of the plant flow.

Chemicals are injected at about twice the velocity of pipe flow at a point within 1 m upstream of the mixer and samples of the mixed solution are taken at up to three pipe diameters downstream after normal flow patterns have been re-established. Location of a static mixer in a pipe should be discussed with the supplier as with some designs clear straight lengths are required both upstream and downstream of the mixer. The ratio of pipe flow to the chemical flow should be in the range 10^2–10^4 depending on the design of the injection system. Mixer units should be designed for easy removal for cleaning although a bypass is considered unnecessary. Static mixers can also be installed in channels (Plate 10(b)) where the headloss required is in the range 10–150 mm. Lower water depth with reduced flow has little or no effect on the performance of the mixer.

The useful power input of hydraulic mixers is related to headloss by the equation:

$$P = Q\rho gh \qquad (7.5)$$

where P is the useful power input (W), Q is the flow (m^3/s), h is the headloss (m), ρ is the density of water (kg/m^3) and g is the acceleration due to gravity (m/s^2). It is generally found that adequate mixing is obtainable in a free fall weir with a headloss of between 300 and 400 mm. The power input for a headloss of 350 mm is about 3.5 kW per m^3/s of water flow and, with a mixing time of 5 seconds, gives a G value of about 700 s^{-1}. For mixing polyelectrolytes (Section 7.23) as a coagulant aid following coagulant mixing the headloss should be kept to about 150 mm and G value in the range 300–500 s^{-1} to minimize the damage to the floc already formed. Static mixers require a headloss in the range 100–1000 mm although headloss as much as 1500 mm is not uncommon where a high turndown in water flow is required. The residence time is typically 2–3 seconds and G value could be as high as 5000 s^{-1}. The mixing efficiency of all types of mixer is defined by coefficient of variation (C_{ov}) which, for good mixing, should be less than 0.05. $C_{ov} = \sigma/x$ where σ is the standard deviation of the tracer concentration in each sample and x is the mean concentration of the tracer. For example for C_{ov} of 0.05 (that is '95% mixed') there is a 99.9% probability that all samples taken downstream will be between $\pm 3\sigma$ of the mean mixed value. This is rarely achieved except in static mixers.

Hydraulic mixers are usually simple and particularly suitable where some headloss can be tolerated. They have no moving parts or direct power consumption so maintenance is negligible. A disadvantage is that the efficiency of mixing suffers if works throughput is lowered outside the operating range of the mixing device. Mechanical mixing is achieved in purpose built chambers equipped with mechanical rotary impellers such as radial flow turbines or axial flow propellers. Ideally mixer manufacturers should be consulted for guidance on sizing. Typical residence times are of the order of 15–30 seconds and the velocity gradient value G varies between about 300 and 600 s^{-1}. This gives a power input range equivalent of about 4–10 kW per m^3/s of water flow at 20 °C and the value selected depends upon the raw water quality, chemical to be mixed and degree of short-circuiting in the chamber. In recirculation pump jet mixing, about 2.5–5% of the plant flow is drawn upstream of the mixing chamber and returned in a pipe through an orifice plate on to a plate in the mixing chamber (Skeat, 1969). The residence time and the power input are the same as for the impeller type mixers. This concept has been extended to pipelines where the water is returned into the pipe against the direction of plant flow through an injection nozzle selected to give a full cone spray (Kawamura, 2000). Due to a tendency for nozzle blockage in raw water mixing applications, clarified, filtered or plant service water should be used as the motive water for this type of mixer. Mixing time and velocity gradient are similar to those for hydraulic mixers. In both types of pumped jet mixers the chemical may be injected into the return pipe or at the point of turbulence. Mechanical mixers have the advantage that they are not affected by flow variations. To maintain a uniform velocity gradient at varying works throughputs the power input could be varied by fitting the mixer with a variable speed motor.

Pumps, in particular the centrifugal type, are good mixers. The lower the efficiency of the pump the better the mixing; for example, in a pump of 75% efficiency, a significant part of the 25% energy lost is due to turbulence (Section 17.2) which can be used for mixing. However, consideration should be given to possible volatilisation of the chemical due to low pressure on the suction side and hence corrosion of pump internals by the concentrated chemical. For this reason gases such as chlorine in solution should not be mixed in pumps and even for other chemicals, materials of wetted parts should be carefully selected. If the alkalinity of the water is greater than 10 mg/l as $CaCO_3$, alkalis should not be dosed into pumps because of the potential to form calcium carbonate scale in the pump. When chemicals are injected into pumps, the manufacturer must be consulted to verify the suitability of the pump materials. Mixing is rapid when viscosities of the chemical and the receiving water are similar and therefore it is improved by applying the chemical in a diluted form (Section 7.20).

The method of injecting chemicals also contributes to the performance of the mixer. For dosing chemicals into pipelines, 1, 2 or 4 injection tubes of length 33%, 15% and 15%D should be used for up to 600, 1000 and 2500 mm pipeline diameters (D) respectively. Injection tubes used for slurries or with chemicals likely to form precipitates in contact with water (e.g. alkalis) should be withdrawable. The injection tube could be a pipe fitted with a nozzle or a perforated diffuser. The nozzle velocity should be about 0.75 m/s or 50% of the velocity of flow in the large pipe, whichever is the greater. In all cases the injection tubes are mounted in a plane perpendicular to the direction of flow and in the case of horizontally laid pipes the tubes are inserted with their axes at 45° to the horizontal. For static mixers the injection criteria applied vary according to the mixer manufacturer.

Injection of chemicals at weirs or flumes is best achieved via perforated pipes (for clear solutions) or channels (for slurries) running the full length of the weir about 50 mm above the water surface and just upstream of the point of turbulence. Submerged perforated pipes similarly located are used for gases such as chlorine in solution.

7.11 CHEMICAL COAGULATION AND FLOCCULATION

Coagulation and flocculation are essential processes in the treatment of most surface waters; one exception being the application of slow sand filtration (Section 8.12). The two processes operating in conjunction with solid–liquid separation processes, remove turbidity, colour, cysts and oocysts, bacteria, biological matter, viruses and many other organic substances of natural and industrial origin. Some are removed directly and some indirectly through attachment or adsorption onto particulate matter.

The terms coagulation and flocculation are two separate processes, contrary to common usage. In coagulation the coagulant containing the aluminium or iron salt is mixed thoroughly with the water and various species of positively charged aluminium or iron hydroxide complexes are formed. These positively charged particles adsorb onto negatively charged colloids such as colour, clay, turbidity and other particles through a process of charge neutralisation. Flocculation is the process in which the destabilized particles are bound together by hydrogen bonding or Van der Waal's forces to form larger particle flocs during which further particulate removal takes place by entrapment. Flocculation is usually achieved by a continuous but much slower process of gentle mixing of the floc with the water in one of the many possible types of plant. In the theory of flocculation the rate at which it takes place is directly proportional to the velocity gradient (Camp, 1943) and the same equation as used in Section 7.12 for mixing is used for determining the velocity gradient G for flocculation. If the residence time in a flocculation chamber is t seconds then the extent of flocculation which takes place, or the number of particle collisions which occur, is a function of the dimensionless expression Gt which is given by the equation:

$$Gt = \frac{1}{Q}\left(\frac{PV}{\mu}\right)^{1/2} \tag{7.6}$$

where the symbols have the same meaning as in Section 7.12.

For the common coagulants of aluminium and iron salts the value of G for flocculation is usually in the range 20–100 s^{-1} with the residence times in flocculation chambers varying from 10 to 40 minutes. However, there are cases where flocculation times approaching two hours have been necessary for waters of extremely high colour and temperature close to 0 °C. The value of Gt would be in the range from 20 000 to 200 000. The values of G and t depend on the raw water quality (e.g. colour, turbidity, algae) and the required floc size, which in turn depends on the downstream treatment process (clarifiers, 2–5 mm; filters in direct filtration, pore space in the media up to 0.2 mm; on the type of clarifier: e.g. dissolved air flotation, similar to bubble size 10–100 μm with a mean of 50 μm (Edzwald, 1990a)) and on water temperature. Therefore, each application should be individually evaluated by pilot trials unless adequate information is available for almost identical conditions. In direct filtration where the intention is to form a micro-floc, G values of the order of 100 s^{-1} and a residence time of about 10 minutes are used. For dissolved air flotation the values used are: G about 50–70 s^{-1} and residence time about 15 minutes for algal laden water and 20 minutes for waters with colour. Flocculation for sedimentation would require G value in the range 30–70 s^{-1} and residence time between 20 and 40 minutes. For optimum flocculation the coagulated water should be subjected to a decreasing level of energy with time; the so-called 'tapered energy' flocculation provided in two or three equal size compartments. The G values quoted above are the mean values for two or three stage flocculators. The G value applied in the last stage is about 10–30% of the first stage G value. For example, in dissolved air flotation the first and last stage G values could be 100 and 25 s^{-1}. The G and residence time values in the high rate clarifier 'Actiflo', for flocculation of micro-sand and the coagulant are 150–300 s^{-1} and 4–8 minutes respectively and are usually applied in one stage.

7.12 TYPES OF FLOCCULATORS

The agitation required for flocculation is usually provided by either hydraulic or mechanical means. The most common hydraulic flocculator is the baffled basin in which a sinuous channel is equipped with either around-the-end or over-and-under baffles. The flocculation energy is derived primarily from the 180° change in direction of flow at each baffle. For the around-the-end type, which is preferred for the ease of cleaning, the minimum water depth is 1 m and the head loss across the flocculator is in the range 500 mm to 1 m. The residence time is 20–25 minutes. The distance between baffles should be at least 500 mm and that between the end of each baffle and the wall should be about 1½ times the distance between the baffles. The baffle spacing should be increased gradually with channel length to achieve tapered flocculation. The floor of the channel should slope towards the outlet. The baffled basin has no mechanical or moving equipment and produces near plug flow with low short-circuiting. The disadvantages are that most headloss occurs at the 180° bends and therefore the value may be too high at the bends and inadequate in the straight channels, G value varies as the flow varies (but this could be partly overcome by providing removable baffles) and settlement of suspended solids in the channel. When designing baffled sinuous channels some allowance must therefore be made for water quality, suggested velocities to minimize settlement are 0.40, 0.30 and 0.25 m/s for high, moderate and low turbidity waters respectively.

Other hydraulic flocculators are helicoidal-flow, staircase-flow, gravel-bed and Alabama types (Schulz, 1968). Hydraulic flocculation is also used in sludge blanket clarifiers (Section 7.16).

There are several types of mechanical device for flocculation, common designs being the paddle type stirrers, mounted either horizontally or vertically in the flocculating chamber, or axial flow impeller type stirrers. The latter is also the design used for high energy flocculation. The power term in the velocity gradient expression for a paddle type is given by the equation:

$$P = F_D V = \frac{1}{2} C_D \rho A V^3 \tag{7.7}$$

where P is the power input (W), F_D is the drag force (m kg/s^2), C_D is the drag coefficient, A is the submerged area of the paddles (m^2) and V is the relative velocity of the paddles (m/s) with respect to water, ρ is the density of water (kg/m^3), C_D is about 1.8 for a paddle stirrer. V may be approximated to 0.75 times the peripheral velocity of the paddle or equal to 1.5 πrn, where r is the effective radius of the paddle (m) and n is the number of revolutions (s^{-1}). The speed of rotation of the paddles varies from 2 to 15 rpm and the peripheral velocity of the paddles ranges from 0.3 to 1.2 m/s. For one, two or three stage flocculators the peripheral velocity would be 0.6, 1.2/0.6 and 1.2/0.6/0.3 m/s respectively. For high energy flocculation the paddle tip speed could be as much as 3 m/s.

Tapered flocculation chambers are usually made up of two or three equal size compartments in series to minimize short-circuiting and the stirrers should be counter-rotating. In order to optimize the velocity gradient the stirrer in each compartment should if possible be fitted with variable speed motors with that in the last compartment having infinitely variable speed facilities. The compartments should be separated by either around-the-end or over-and-under baffles arranged to give diagonal flow in the compartment; the headloss across the baffle walls produces a G value of up to 20 s^{-1}. Flow velocity should be limited to 0.25 m/s, depending on the floc characteristics to minimize floc shear. The tank dimensions vary according to the type. For the horizontal shaft type, the tank should be long and narrow (L:W of at least 4:1 with a square cross-section perpendicular to the direction of flow) and with depth of about 3 m. The tank for the vertical shaft type must have square plan (up to 6 × 6 m); the depth is not critical provided the stirrer can be supported from the gear box without having to use bottom bearings. The paddle type is therefore limited to a depth of about 4–5 m whereas the turbine type could be as deep as 7 m; the clearance to the floor should not be less than 500 mm. The vertical shaft paddle type stirrers should have a diameter and paddle height greater than 2/3rd of a plan dimension and should have 500 mm clearance to the walls. For good performance the horizontal shaft type should have at least three paddles on each diametrical arm mounted in sinuous channels since short circuiting is reduced. When flocculation tanks are dedicated to individual clarifiers, measures should be taken (such as a free fall outlet) to prevent flocculator flow patterns being transmitted to the downstream clarifier and the transfer velocity should be less than 0.1 m/s.

Performance Considerations

Impurities which require coagulation and flocculation before they can be removed by solid-liquid separation processes can broadly be classified as either inorganic or organic material. The more usual type of inorganic material encountered in water treatment settles easily, especially when it is of particulate size; when chemically unaided settling is used one or two hours retention usually removes at least 40–50% of particulate matter. A primary coagulant, sometimes assisted by a coagulant aid, usually removes 98% or more of the suspended particulate matter for lightly loaded water with suspended solids not exceeding about 300 mg/l, but such a removal rate may not be adequate for those river waters typical of the regions where suspended solids are regularly measured in excess of 1000 mg/l and over 10 000 mg/l is not uncommon. In such places it may be necessary to introduce a settlement stage unaided by chemicals before chemical sedimentation or, alternatively, to have two separate stages of chemical treatment and settlement in series because the quantity of solids in suspension and the sludge formed by the addition of chemicals are too much to remove in one stage of treatment.

Not all inorganic matter settles out quickly in plain settlement and in some waters, such as are found in Central Africa, it may be weeks or even months before any significant settlement of colloidal material takes place. With such waters there is no option but to use two clarification stages each with chemical coagulation. Organic colour and finely divided mineral matter, including various forms of clay, comes in this category. Clay in water is a hydrophobic colloidal suspension, or 'sol', in which the surfaces of the particles are considered to have a negative charge which contributes to the stability of the sol by helping to prevent the particles coalescing into larger particles which would have a relatively rapid rate of settling.

The excellent coagulating effect of aluminium or ferric salts may be due to the triple positive charge on the trivalent aluminium or ferric ion neutralising the negative charges on the clay particles. Literature on the coagulation of clay and similar materials is reviewed by Packham (1962 and 1963) who describes his own investigations which have developed the theories in a somewhat different direction. His experiments confirmed that it is not primarily the aluminium or ferric ions in solution which react with the clay particles, but it is the mass of rapidly precipitating hydroxides of aluminium or ferric which enmesh them. In subsequent work by other researchers this is identified as one of two mechanisms in coagulation and is called 'sweep coagulation'. It requires excess coagulant and occurs in about 1–7 seconds. The other mechanism is one of charge neutralisation of negatively charged colloids by positively charged hydrolysis products which are soluble hydrated hydroxide complexes of aluminium or ferric salts. The reaction is completed within a second (Amirtharajah, 1978). The presence of some anions such as sulphate (SO_4^{2-}) helps the coagulation process (Fair, 1968). Sweep coagulation uses more chemicals and produces more sludge than for charge-neutralisation, but removal of trace contaminants is better.

pH is an important factor in coagulation. The optimum pH range corresponds to that over which minimum solubility of the hydrolysed coagulant products occurs and maximum turbidity and colour removal is achieved. Typically for ferric salts the coagulation pH is greater than 5 and for aluminium salts it is between 6.5 and 7.2. For coagulation of natural organic matter such as colour, coagulation pH values for ferric and aluminium salts are about 4.5 and 5 respectively, whilst optimal turbidity removal typically occurs at pH 6–7.5.

US EPA (1999) includes in the Disinfection By-Products Rule a requirement for 'enhanced coagulation' for greater removal of natural organic matter to minimize disinfection byproduct formation. It is defined as the addition of excess coagulant, a change in coagulant type, or a change in coagulation pH for improved total organic carbon (TOC) removal. US EPA proposes that enhanced coagulation is applied to surface waters treated by conventional treatment unless the concentration of TOC in the treated water is <2 mg/l, the concentration of TOC in the raw water is <4 mg/l and the alkalinity is >60 mg/l as $CaCO_3$ or the concentrations of chlorine by-products in the distribution system are below 50% of the maximum contaminant levels (MCLs) when chlorine is the disinfectant.

Polyelectrolytes (Section 7.23) can be used as coagulants for waters containing high turbidities (and for sludge conditioning). For example in the 1365 Ml/day Al Karkh water treatment works for Baghdad a cationic polyacrylamide was used as a coagulant to reduce the raw water suspended solids from 30 000 mg/l to about 500 mg/l in pre-settlement tanks. In the subsequent clarification stage the same polyelectrolyte was used as the coagulant aid to aluminium sulphate. For waters of low turbidity polyelectrolytes are ineffective as coagulants because, at the very low dosages applied, the residence time provided for flocculation is insufficient to produce large floc. They can, however, be made effective by increasing the number of particles by combining their use with aluminium or ferric salts. In this application, polyelectrolytes are called coagulant aids and are added between 60 seconds (warm water) and 5 minutes (cold water) after the mixing of the coagulant, preferably closer to the clarifier to minimize the damage to the floc. This time delay is shown to be more important for ferric than aluminium coagulants (Brejchova, 1992) and found to be applicable in particular to polyacrylamide type polyelectrolytes. Coagulation by polyelectrolytes follows charge neutralisation or bridging or both. Therefore they would be effective even when the polyelectrolyte carries a charge of the same sign as that of the particles to be coagulated. Hence, the best type of polyelectrolyte for an application should be determined by the laboratory jar tests.

7.13 EFFECT OF ORGANIC CONTENT AND ALGAE

The variety of impurities present in raw waters and their varying concentrations explain the dictum that all waters are different and that water treatment is an 'art' as well as a science. This particularly applies to waters whose primary natural impurity is organic in origin, as these are very often the most difficult to treat. Miscellaneous fragments of animal and vegetable matter contribute towards the organic content, as does the organic colouring matter derived from peat and similar sources consisting largely of humic and fulvic acids and of more complex compounds, partly in true solution and partly in colloidal form. The optimum chemical treatment for such waters is sometimes very difficult to achieve, particularly when it is necessary to remove dissolved iron or manganese. Laboratory jar tests are an essential tool in the formulation of treatment for all waters. Where possible these should be followed by testing at pilot scale.

Planktonic animals and plants, particularly the plants (phytoplankton) which are algae, can cause a variety of problems in treatment. Their removal in microstrainers has been mentioned in Section 7.4. Table 7.3 shows a short list of some commonly occurring algae and their implications for treatment in temperate climates are given. They often occur in large quantities in lowland and in eutrophic waters. Their removal by flocculation, sedimentation and filtration is basically similar to, but more variable than, that of other forms of suspended material because of the widely diverse types, densities and shapes of plankton. Some algae, for example diatoms which have silica in the cell walls, are notably denser than water. These algae and many other algae with large or simple shaped cells are removed effectively by chemical treatment. On the whole, however, plankton do not settle very readily when growing rapidly, as they are buoyant either because of oxygen gas in their cells or as they actively swim, and the coagulant dose usually has to be increased to be effective. Many small celled types and those which by their motile flagellae, are often not well retained in floc, are liable to pass through the subsequent rapid filters. Conversely problems often arise because of the tendency of some algae to form dense blooms in lake waters; blue–green algae (*Cyanobacteria*) are typical of this group. Such blooms can cause a rapidly increasing loss of head in the filters. The removal of planktons by coagulation and settlement can be improved by the use of a polyelectrolyte as a coagulant aid and killing them or simply inactivating them using pre-treatment with chlorine, ozone (Bauer, 1997) or an algicide (Parr, 1991). However, these methods have problems due to toxicity of the residues and, in the case of chlorine, the formation of disinfection byproducts or other side effects such as taste and odour formation (Section 10.33). Without killing algae their removal by sedimentation is about 50–75% (Markham, 1997). Some algae such as blue-green algae are especially adapted to live in the top layers of water; the cells of these forms contain minute air vacuoles and are particularly buoyant and difficult to settle. The dissolved air flotation process is therefore particularly suited to blue-green algae removal because of the natural tendency for the algae to rise to the surface (Bare, 1975; Edzwald, 1990b). The process can also be effective with other less buoyant algae, such as some diatoms and is described in more detail in Section 7.18.

Micro-organisms such as cysts and oocysts are readily removed by coagulation and flocculation followed by filtration with or without clarification (Section 8.19); a removal of 99.9% should be feasible for bacteria (WHO, 1996), although it is reported that such removal efficiencies are achieved mostly in the summer months, reducing to 70% in the winter months (O'Conner, 1997). Removal mechanisms include direct removal by charge neutralisation and entrapment in floc and through removal of particles to which bacteria are attached.

Of the other impurities which are encountered in raw waters, it is far less easy to be precise as to removal efficiency by chemical coagulation. In many lowland waters there are to be found small but appreciable concentrations of substances in the run-off from farmland, drainage from agricultural land and sewage and trade effluent discharges. Many of these are organic substances and are usually classified as volatile or non-volatile with many more subdivisions, such as biodegradability. Whilst it is known that chemical coagulation, flocculation, clarification and rapid filtration are at

Table 7.3 Organisms reported as having caused difficulties in british waterworks

Group	Genus	Difficulty experienced (✓)					
		Blocking of filters	**Penetration of filters**	**Taste or odour**	**Growth in pipes, channels**	**Unsightly scums**	**Potentially toxic substances**
Bacillario-phyceae (Diatoms)	Asterionella	✓					
	Cyclotella	✓					
	Fragilaria	✓					
	Synedra	✓		✓			
	Nitzschia	✓					
	Stephanodis-cus	✓	✓				
	Melosira	✓					
	Diatoma	✓					
Chlorophyta	green unicells		✓				
	chlamydomo-nas		✓				
	Cladophora				✓	✓	
	Spirogyra	✓					
Cyanobacte-ria 'Blue-green algae'	Anabaena			✓		✓	✓
	Aphani-zomenon	✓				✓	
	Lyngbya	✓	✓				
	Microcystis			✓		✓	✓
	Oscillatoria	✓	✓	✓	✓	✓	✓
Xanthophy-ceae	Tribonema	✓					
Chrysophy-ceae	Synura		✓	✓			
	Dinobryon	✓					
	Mallomonas	✓	✓				
Euglenophyta	Euglena		✓	✓			
Dinophyceae	Peridinium	✓					
Rotifers	Diglena		✓				

(continued)

Table 7.3 (*continued*)

Group	Genus	Difficulty experienced (✓)					
		Blocking of filters	**Penetration of filters**	**Taste or odour**	**Growth in pipes, channels**	**Unsightly scums**	**Potentially toxic substances**
Crustacea	*Cyclops*	✓	✓				
	Diaptomus		✓				
	Daphnia	✓	✓				
	Asellus				✓		
	Bosmina	✓					
Nematodes					✓		
Oligochaetes	*Nais*				✓		
Insects	*Chironomus*		✓		✓		
	Tanypus		✓		✓		
Filamentous bacteria	*Sphaerotilus* (*Cladothrix*)				✓	✓	
Iron bacteria	*Crenothrix*				✓		
Polyzoa	*Plumatella*				✓		
	Cristatella				✓		
Porifera (sponges)	*Ephydatia*				✓		
	Spongilla				✓		
Molluscs	*Dreissena*				✓		

least partially effective in their removal, in particular those proportions adsorbed on to particles, different processes can offer a higher degree of removal of organic substances. These are described in Chapter 10.

CLARIFIERS

7.14 HORIZONTAL FLOW CLARIFIERS

Some simple types of settlement tank are in use for the clarification of flocculated waters. For large volumes of water containing a relatively heavy load of suspended solids a relatively dense floc is formed which settles easily and in warm climates, where the viscosity of the water is lower thereby permitting more rapid settlement of floc, the large horizontal flow sedimentation tank can be an economical solution for clarification. Although a large tank is necessary in order to keep velocities sufficiently low to permit settlement of floc to the base of the tank, its construction is simple because it need not be very deep and few internal walls are required. In the simplest design of horizontal flow tank, floc is allowed to accumulate on the floor of the tank until such time as the increasing velocity of water above the accumulated sludge begins to stir it up, thereby affecting the clarity of the effluent. When this occurs the tank should be cleaned out. Alternatively, moving scrapers operated continuously or intermittently add to the overall efficiency by pushing the settled floc to outlets in the base of the tank where it may be drawn off as sludge.

Accepting their initial low rating and correspondingly high civil construction cost, horizontal tanks are nevertheless versatile clarifiers; modifications to an original simple design to meet changing conditions of raw water can be accommodated if provision has been made at the design stage.

FIGURE 7.2

Circular Centrifloc® clarifier, 51 m diameter, unit flow 76 Ml/d @ 1.7 m³/h.m², at Al Karkh for treatment of R. Tigris water for Baghdad. (Engineer: Binnie & Partners; Contractor: Paterson Candy Ltd.).

For example, rotating flocculators and sludge scrapers can be added and some tanks have had their rate of flow increased by the addition of inclined tube or plate modules (Section 7.17). Since the depth of a tank does not influence its performance some horizontal flow tank designs use the principle of shallow depth sedimentation, with a sloping tray so that the direction of flow of the water reverses up and over the tray and exits near the inlet. Yet another variation is where the depth of the rectangular tank is divided by as many as four inclined trays and flow takes place in parallel streams between the trays. In this way the capacity of a tank can be increased 3 to 4-fold on the same foot print. Circular tanks have centre feed and, after chemical flocculation which usually takes place in a central compartment, water flows radial and upwards to peripheral collecting launders or to a combination of peripheral and radial launders. Circular sedimentation tanks are fitted with rotating sludge scrapers, either with the drive mechanism mounted on a central platform or bridge or, with the drive unit mounted on the outside wall and the scraper bridge pivoted on a central support.

In the design of rectangular or circular tanks for treating heavily silted water, the feed water should be conveyed in channels rather than pipes to permit easy access for cleaning of settled solids. The advantages of these tanks are: greater tolerance to hydraulic and quality changes, ideal for stop/start operation; infinite turn down; simplicity of operation; suitable for water containing high silt loads; and performance is not appreciably affected by diurnal temperature change. The primary drawback is their low surface loading rate and hence the large footprint and associated capital costs. Compared to rectangular tanks, circular tanks do not lend themselves to a compact layout. One such circular tank design known as Centrifloc® is used for the treatment of R. Tigris water at the 1365 Ml/day Al Karkh water treatment works in Baghdad and is shown in Figure 7.2.

7.15 DESIGN CRITERIA

Without the aid of a coagulant a rectangular or circular tank only works at a surface loading up to about 1/3 the rate shown by jar tests (33% efficiency). With coagulant assisted sedimentation, the actual efficiency is generally much closer to the results obtained by laboratory jar tests. However, theoretical settling rates cannot be directly translated to tank design because of short-circuiting in the tank. The safety factor can be limited to between 1.3 and 1.5 by careful design of the inlet and outlet arrangements. These design features include multiple inlets at about 1.5 m centres sized to give an inlet velocity of about 0.5 m/s and perforated baffles, with orifice diameters of 100–200 mm to give a head loss of less than 10 mm to minimize floc shear, at the inlet and outlet ends across the whole cross section of the tank to create a more uniform flow pattern through the tank. Whilst these measures help, they may have little effect when the temperature of the incoming water differs from that in the tank and so creates turbulence which must subside before the particles can resume settling under quiescent conditions.

In horizontal flow tanks the horizontal flow of water should occur under laminar conditions (Reynolds number <2000) and short-circuiting and instability of flow should be minimum (Froude number >10^{-5}). For tanks of dimensions l, w and d (m), surface loading rate S (m^3/h.m^2) and flow Q (m^3/s), Reynolds and Froude numbers are given by:

$$\text{Re} = \frac{Q}{w+2d}\frac{1}{\gamma} \quad and \quad F = \frac{S^2}{g}\frac{l^2(w+2d)}{1.3 \times 10^7 \, wd^3} \tag{7.8}$$

where γ is kinematic viscosity (m^2/s) and $g = 9.81$ m/s^2.

Chemically assisted sedimentation tanks for the removal of organic matter and light flocculated particles, have length to width ratios greater than or equal to 4:1 and surface loading rates between 0.75 and 1.75 $m^3/h.m^2$, which, with a coagulant aid, may increase to 2.5 $m^3/h.m^2$. The mean horizontal flow velocity should be in the range 0.25–1.0 m/min. The flow over the outlet weir should be maintained below 50 m^3/h per m of weir length and should preferably be about a quarter of this value. Alternatively, 90° V-notches with a water depth of about 75 mm and spacing less than 0.6 m may be used. Submerged orifice outlets are useful in minimising the passage of floating material to filters or where freezing is likely. Orifices should be sized to give a flow velocity of about 0.6–0.7 m/s with an orifice diameter greater than 30 mm and head loss 35–40 mm. They should be about 75 mm below the water surface or more (0.1–0.2 m) below the water surface where freezing is a problem, depending on the thickness of the ice layer. To achieve the outlet flow, double sided launders are usually used. These are placed about 0.5 m from the end wall (rectangular tanks) or peripheral wall of circular tanks. Alternative arrangements include 'finger' launders of length <20% of the tank length (rectangular tanks) or radial launders (circular tanks) spaced at less than 8 m on the circumference feeding a collector launder. The depth of the rectangular tanks should be adequate for sludge deposits and storage; a minimum of 3 m is recommended for scraped tanks and 5 m recommended for manually cleaned tanks (Section 7.19). Circular tanks typically have a side water depth of 4.5 m. Detention times for particles of low relative density, i.e. the theoretical time of travel of water in the tank, varies from a minimum of 1½ hours to an average of 4 hours, but is more usually about 2½–3 hours.

Surface loading rates in clarifiers are often obtained by dividing the daily or hourly flow by the gross surface area of the tank. This can be misleading as it may not take into account the area occupied by the mixing and flocculation compartments (where these are provided), and the effluent launders, which usually account for at least 12% of the gross area. A better guide to the true loading rate is to take the horizontal clarification area available at a depth of about 1.25 m below the surface.

7.16 SLUDGE BLANKET OR SOLIDS CONTACT CLARIFIERS

Probably the earliest type of sludge blanket tank is the so called 'hopper bottomed tank'. The sludge blanket effect is obtained simply by allowing the chemically treated water to flow upwards in an inverted pyramid type of tank with the angle of slope usually 60° to the horizontal and terminating in a vertical section of 1.5 m water depth. They are usually square in plan (occasionally circular), although rectangular multiple hopper tank variations have been constructed for large capacity works. As the water rises in the tank of increasing area, its velocity progressively decreases until, at a given level, the reduced upward drag on the particles counterbalances the weight of the particles, leaving them suspended in the water. Their presence forms a kind of blanket in the water in which chemical and physiochemical reactions can be completed and in which a straining action to remove some of the finer particles may also take place. While the top of the blanket is usually well defined there is no distinguishable bottom to the blanket because of the varying density of the particles forming the blanket.

Flocculation takes place within the clarifier and is hydraulically induced as the coagulant dosed water enters downwards and flow reverses upwards at the base of the clarifier. Flocculation is completed within the sludge blanket through a process known as contact flocculation. The concentration of the sludge blanket is controlled by allowing it to bleed off via suitably placed hoppers. Sludge is also removed from the bottom of the hopper. The hopper bottomed tank is not cheap to construct as the total water depth, governed by geometry, is usually about the same as a dimension of the plan.

During construction care must be taken to ensure that the inverted pyramid is reasonably accurate and the inlet pipe discharges at the geometric centre; if it does not, streaming of the water to one side can occur and upset the stability of the sludge blanket. The surface loading of hopper bottomed tanks is within the range 2 to 3.5 $m^3/h.m^2$, the lower velocity may sometimes be needed for floc formed in the removal of colour from a soft reservoir water.

In order to reduce construction costs there has been a continual striving towards simplification of the tank shape, and the most recent addition to the sludge blanket family is a flat-bottomed variety, usually rectangular in plan, as illustrated in Figure 7.3. In this type of design chemically treated water is directed downwards onto the base of the tank through suspended pipe 'tridents' before passing upwards through the blanket to the surface collecting launders. Flocculation is hydraulically induced at the trident outlet. With cold waters, however, this should to be supplemented with external flocculation. The sludge is collected in hoppers with their lips placed at the top level of the blanket and is removed under hydrostatic head.

A proprietary design of upward flow flat bottomed sludge blanket tank is the Pulsator® (Fig. 7.4) where a proportion of the incoming chemically dosed water is lifted into a chamber built onto the main inlet channel by applying a vacuum of about 0.65 m water gauge using a centrifugal fan and released into the tank. This creates a pulsing effect in the blanket. The frequency and duration of pulsation is varied according to the flow and water quality and typically would be of the order of 30–50 seconds and 8–10 seconds, respectively. These would normally be set to achieve a loading during the pulse in the tank of about 7–8 $m^3/h.m^2$ for waters containing low settleable solids and 10–12 $m^3/h.m^2$ for waters containing high settleable solids. Flocculation in flat bottomed tanks is induced hydraulically as the water enters via downward facing orifices in the distribution pipes and turns upwards and is completed within the blanket by contact flocculation.

Sludge blanket clarifiers are suitable for many types of waters, including turbid ones, provided the particulate matter is of low density. A safe upper limit for turbidity is about 500 NTU but, depending on the nature of the particulate matter, much higher (up to 1000 NTU) peaks can be accommodated. Heavy particulate matter tends to settle on the bottom of the clarifier. When used in such applications these clarifiers should be provided with scrapers or other methods for intermittent removal of bottom sludge. Alternatively they should be drained down and cleaned after the rainy season, or about once or twice a year. In this respect, hopper bottomed tanks are more appropriate for highly turbid waters. For heavily silt laden waters sludge blanket clarifiers should be preceded by simple rectangular or circular sedimentation basins as described in Section 7.14. The surface loading rates of flat bottomed sludge blanket clarifiers typically vary from 2 to 5 $m^3/h.m^2$. They have a side water depth of about 4.5–5 m, which is typically made up of a bottom distribution zone of 0.6–1.0 m, sludge blanket of 2.15–2.25 m and clarified water depth of 1.75 m. In some designs the clarified water depth could be as low as 1.0 m. In sludge blanket tanks, the blanket concentration is typically 20–25% $^v/v$ (after 10 minutes settlement in a 250 ml cylinder) and 0.1–0.2% $^w/v$, although with some waters containing high settleable solids due to high colour and turbidity it could be as much as 30–35% $^v/v$ and 0.25–0.5% $^w/v$.

The operation of sludge blanket tanks is somewhat sensitive to sudden changes in flow and raw water quality and requires greater operator skill. They are not suitable for stop/start operation. Restart following a lengthy shut down may take more than 24 hours. However this could be reduced if sludge from a similar clarifier is available for seeding the blanket. Stoppages of 3–6 hours, can however be accommodated. There is also a constraint on flow turn-down, which is normally limited to about 2/3rd of the maximum flow. The Pulsator clarifier, because of its high intermittent flow, can be operated down to about 1/3rd the maximum flow.

FIGURE 7.3

Flat-bottomed sludge blanket clarifier with Gravilectric® sludge cone. (Paterson Candy Ltd).

First half-cycle: Air valve A closed and water rises in vacuum chamber C. Water in clarifier D at rest and sludge settles.

Second half-cycle: Water in C rises to upper contact and air valve A opens. Water in C falls and enters D raising sludge which enters concentrator B. When water falls to lower contact, air valve A closes.

FIGURE 7.4

Pulsator® clarifier. (Degremont, France).

The performance of sludge blanket clarifiers (and horizontal flow settling tanks) is known to be influenced by temperature (Hudson, 1981). The temperature effect is normally diurnal and is caused by the creation of thermal gradients within the clarifier due to the walls of the tank being heated by the sun or by warmer water entering from an open raw water storage tank or an exposed raw water main, giving rise to density currents within the clarifier. The result is disturbance of the blanket and carry-over of floc towards the evening. Sometimes similar effects have been attributed to the release of gases due to bacterial activity in sludge. Measures taken to minimize carry-over of floc include: use of polyelectrolyte as a coagulant aid; inclusion of tube modules in the clarified water zone; or perhaps both techniques. Intermittent chlorination would help to overcome bacterial activity in sludge. So long as their sensitivity is appreciated and they are operated intelligently sludge blanket clarifiers will produce a good quality effluent (turbidity of about 1 NTU) and are very tolerant to changing conditions of raw water quality which would be detrimental to the operation of many other types of clarifiers.

There are many other designs which endeavour to achieve the same high level of performance as sludge blanket tanks by providing extra mixing energy for flocculation or by the recirculation of sludge, or by a combination of both methods. Nearly all such variations have circular configurations and are usually equipped with bottom sludge scrapers, but operate as solids contact clarifiers. The essential feature is that the settled floc is used to seed the incoming dosed water, thereby accelerating the flocculation process. This is achieved internally (i.e. Accentrifloc clarifier) by feeding the chemically dosed raw water into the flocculation zone where it is mixed with sludge recirculated by the aid of a rotor impeller or externally (i.e. Pre-treator clarifier or Densadeg® clarifer) where a pump is used to recycle sludge to mix with dosed water. Clarifiers of these designs can be operated at surface loading rates about two to three times the loading of those without recirculation. Densadeg®, which incorporates tube modules, can operate at much higher ratings. Optimisation of performance requires some skill as there are several operating variables such as: rotor impeller speed (peripheral speed 0.5–1.5 m/s); recirculation flow (up to 20% of tank throughput for external recirculation or up to 5 times or 15 times for softening the tank throughput for internal circulation); blanket depth; and sludge withdrawal rate. Such clarfiers have a high mechanical plant content and are comparatively high in capital and operating costs.

7.17 HIGH RATE CLARIFIERS

Commercial competition between treatment plant manufacturers and limited space for construction of treatment facilities has provided the main incentive for the development of new techniques for obtaining higher flow ratings in clarifiers. The maximum rate for a sludge blanket tank using coagulants is about 2 m³/h.m², the upper limit with polyelectrolytes is about 5 m³/h.m². In proposals for clarification it is important to check the probable performance at the lowest water temperatures likely to apply, when the viscosity of the water is highest. This can be the limiting criterion for performance.

Tube or Plate Settlers

Some success with increasing flow rates through existing clarifiers has been achieved by the use of tube modules or plates, which make use of the principle established long ago by Hazen (1904), that a settling tank should be as shallow as possible in order to shorten the falling distance for particles. The use of all-plastic tube module or plates is a logical development of shallow depth settlement by multiple trays (Section 7.14), which helps to lessen the sludge removal problem of wide shallow trays. The 'tubes' are circular, hexagonal or square in cross-section, of hydraulic diameters in the range 50–80 mm and made of plastic, usually polystyrene loaded with carbon black to protect against ultra violet radiation. The best shape for a tube of a given area is a shape with the largest perimeter. The order of preference based on projected area is hexagonal (regular or chevron), then square followed by circular (Degremont, 2007). The tubes have a large wetted perimeter relative to the cross section area and thereby provide laminar flow conditions which theoretically offer optimum conditions for sedimentation. Laminar flow is not achieved immediately when water enters the tubes, but after a transition length L_t (m) given by the Schiller formula (Coulson, 1999), $L_t = 0.0288\, Re.D$, where Re is the Reynolds number which should be less than 280 and D is the hydraulic diameter (m) which is equal to $4s/p$, where s is the cross-section area (m²) of the tube and p its wetted perimeter (m). In the design calculations this transition length must be deducted. Sedimentation takes place in the length following the transition length and it will retain all particles with a settling velocity less than V_s (m³/h.m²) (Yao, 1973) which is given by:

$$V_s = \frac{V \cdot k}{\sin \theta + \dfrac{L_s}{d} \cos \theta} \qquad (7.9)$$

where, V is average velocity of flow in the tubes (m³/h.m²) which is equal to $U \sin \theta$ where U is the average upward velocity (m³/h.m²) which is the surface loading rate of the settling tank (i.e. rate of flow ÷ plan area); L_s is settling length (m) which is equal to $(L - L_t)$ where L is the total tube length (m); k is a coefficient (1.0 for parallel plates, 1.33 for circular and hexagonal tubes and 1.38 for square tubes); θ is angle of inclination of tubes to the horizontal; and d is depth of water in a tube at right angle to the direction of flow (m).

Flow in the tubes or plates should be laminar with Re <200, but <50 is preferred. Froude number should be >10⁻⁵. V should be <10 m/h, detention time in the tubes should be <10 minutes and in the plates should be <20 minutes and surface loading rate for the tank should be <7.5 m³/h per m² plan area of the tank depending on the raw water quality. The clear water depth above the plates or tube modules should not be less than 300 mm. The vertical depth occupied by the tubes or plates is 500–1000 mm. The velocity under tubes or plates should be less than 10 mm/s and the clarified water removal rate should not be greater than 15 m³/h per m equivalent launder length. In tube (or plate) settler design at least one-third of the tank length should remain tube (or plate) free. When the angle of inclination of the tubes to the horizontal is between 55–60° the solids settled on the inclined wall of the tube slide downwards under gravitational forces along its lower side, the clarified water flowing in the counter-current direction. Most tube settlers have been used in circumstances where it was desired to increase the flow through or improve the performance of an existing clarifier, particularly horizontal and radial flow clarifiers and some sludge blanket clarifiers, so that the increased clarification rates achieved were still in the range 2–5 m³/h.m² (Galvin, 1992).

Sludge blanket clarifiers are uprated by incorporating tube or plate modules in the clear water zone. The clear distance between the tube pack and the top water level and the sludge layer below is usually maintained at about 500 mm in each direction. The vertical depth occupied by the tube pack is about 650–750 mm. An example of a sludge blanket clarifier with tube modules in the

Pack of plates, tubes or modules

FIGURE 7.5

Pulsatube Pulsator® clarifier. (Degremont, France).

clarified water zone is illustrated in Figure 7.5. The use of tube packs allows up to 2-fold increase in the surface loading rate; in essence the tube packs, which provide a much larger settling area than the clarifier plan area, trap the floc carried over from the blanket when subjected to such high rates. Major disadvantages with tube settlers particularly those in sludge blanket clarifiers are that the floc carried over to the filters tends to be fine and may not be retained well in the filters and that the clear water depth of only about 500 mm encourages algal growth on the tube modules which, along with slime growth if allowed to form, would partially clog them and affect performance. It is desirable to use anthracite-sand filters downstream of these clarifiers. However, the use of the tubes helps to minimize the effect of thermal 'boiling' and wind on the clarifier performance. There are few examples of proprietary clarifiers where tube packs are incorporated into the design of new tanks.

An example of a sludge blanket clarifier which has made practical use of the plate system within the sludge blanket is the Super Pulsator®. The plates are about 300 mm apart, at an angle of 60° to the horizontal and perpendicular to the sludge collection hoppers; coagulated water passes upwards and sludge travels between the plates into the hoppers. It is reported to achieve clarification rates in the range 5–10 m/h and sludge concentration about twice that in a Pulsator operating at the same upward flow rate.

Lamella Clarifiers

The principle of shallow depth sedimentation has been extended to the design of a parallel plate system, sometimes referred to as lamella clarifiers. Clarifiers using the plate system are usually purpose built to take advantage of the high settlement rates which can be obtained and the greater density of sludge provided. They require an efficient flocculation stage which is critical for successful operation. The flocculated water enters at the base of lamella plates and travels upward between the plates counter current to settled sludge moving down (Plate 11(a)). In some designs the uneven distribution to the inlet of the lamellas is corrected by introducing the flocculated water flow individually into each lamella space via slotted openings in the side walls of channels running on both sides of the plate pack along the length of the tank. Each space between the lamella plates therefore acts as an independent settling module. The lamella plates extend the full depth of the tank and rise about 125 mm above the top water level. Clarified water is collected in decanting launders running along each side of the plate pack by submerged orifices or V-notches one between each pair of plates (Plate 11(b)). About 1.5 m is allowed in the bottom of the tank for the collection of sludge which is removed by a circular scraper or a scraper of the chain and flight or reciprocating type to a central hopper or a series of small hoppers at one end of the tank. For small tanks sludge could be collected in hoppers placed underneath the plate pack. The plates are inclined at 55–60° to the horizontal. The turbidity of the clarified water is about 1–2 NTU.

   With this arrangement, the settling area available is equal to the sum of the projections of plates in a horizontal plane. Thus the settling area is very large on account of the overlapping of plates but occupies a relatively small plan area. The total settling area is equal to $(n-1)\,LW\cos\theta$ where n is the number of plates, L the plate length in water (m) after deducting the transition length, W the plate width (m) and θ the angle of inclination of the plates to the horizontal. The value of n should be determined taking plate thickness and spacing between plates into consideration. The plates should be flat and not corrugated and they are usually made of stainless steel but sometimes of plastic. Plate width is about 1.25 to 1.5 m and plate length is about 2.5 to 3.25 m including the length of 125 mm above the normal water surface; plate thickness is usually 0.7 mm for stainless steel. The horizontal spacing between plates is varied according to the application and is normally in the range 50–80 mm. Depending on the settling velocity of the particles (hence quality of the raw water) the lamella clarifiers could be operated at about 0.8–1.5 m/h settling rate (also known as the Hazen velocity). The surface loading rates are about 10–25 m³/h.m² and can therefore give a much reduced surface area (up to 95%) compared with more conventional horizontal flow clarifiers. This also means that the retention time within the clarifier is low, sometimes 20 minutes or less, so that control of chemical treatment becomes more exacting. The tendency for algae to grow on the plates is a problem with lamella clarifiers. Such clarifiers should therefore be enclosed in a building. The use of chlorine for algae control could corrode the length of the stainless steel plates above water (due to air above the water containing moist chlorine) and should therefore be protected by a plastic laminate down to about 300 mm below the water surface.

Other High Rate Clarifiers

In ballasted flocculation clarifiers (e.g. the Actiflo® process) a suspension of fine quartz sand (effective size 100–150 μm and uniformity coefficient of 1.6) is used as a ballasting agent to form a weighted floc of density in excess of 2.5 kg/l, resulting in very high settling velocities. In the process coagulant is first mixed in the water using turbine stirrers (mixing time 1–2 minutes), followed by sand in a similar mixer (1–2 minutes) and high energy flocculation (4–8 minutes). Polyelectrolyte is added to any one of the three tanks depending on the water quality. Settlement takes place in flat bottomed or hopper bottomed tanks fitted with inclined parallel plates or tubes (at 60° and vertical height about 1.25 m) operating at rates in the range of 25–60 m³/h.m² (Cailleaux, 1992). The sludge containing the sand falls to the bottom and in the case of flat bottomed tanks, is removed by scrapers to a series of hoppers located at one end. The sludge is then recycled at 3–6% of works throughput via hydrocyclones to separate the sand from the sludge. Hydrocyclones are operated at constant rate and multiple units are required for clarifier output turndown, which may be over 5:1. The sludge discharge from the hydrocyclones (overflow) is about 80% of the recycle rate. The recovered sand (underflow 20%) is then made up with fresh sand to account for losses in the sludge stream and recycled to 'seed' the incoming water again. The process requires about 3–6 g/l of sand which is equivalent to about 0.08% of the volume of the settling tank. The make up sand is of the order of 1–2 mg/l and is injected either continuously or in larger quantities intermittently (about once a week). The process removes *Cryptosporidium* oocysts and *Giardia* cysts like any other clarification process, which follows coagulation and flocculation. There is however the potential risk of returning some oocysts and cysts back to the process along with the recycled sand. This aspect should be investigated by pilot trials. The process is highly dependent on polyelectrolyte and the control of the dose is paramount as over dosing could harm the downstream filters. The proportion of water lost as sludge can be about 2.5% of works throughput consisting about 0.1% ^{w}/v solids. The energy consumption of the process can be in the range 0.01–0.02 kWh/m^3 of water treated.

In the 'Sirofloc' process magnetite in fine particulate form is used as a ballasting agent to achieve high settling rates. The magnetite is 'activated' with sodium hydroxide solution then introduced as a slurry and, after a period of contact during which the magnetite adsorbs destabilized colour, it is flocculated by passing it between the poles of a permanent magnet and the floc thus formed settles readily in the clarifiers. Magnetite is recovered by re-treating the sludge drawn from the clarifiers with sodium hydroxide and returned with a make up dose to the main stream. The effluent, which contains very high colour (600–850° Hazen for raw water colour of 50–90° Hazen) and pH (about 12), can be discharged to a sewer or treated by conventional treatment (Kolarik, 1994).

The principal advantage of high rate clarifiers is that they occupy a small foot print, but they need close attention and optimisation of chemical treatment. Most of them depend on polyelectrolyte dosing. The retention time in some of them is very short and therefore tolerance to changing water quality is significantly reduced and the sensitivity to optimum operating parameters is significantly increased.

7.18 DISSOLVED AIR FLOTATION

Dissolved air flotation operates on the principal of the transfer of floc to the surface of water through attachment of air bubbles to the floc. The floc accumulated on the surface, known as the 'float', is skimmed off as sludge (Section 7.19). The clarified water is removed from the bottom and is sometimes called the subnatant or 'floated' water. Since rain, snow, wind, freezing could cause problems with the float, flotation tanks must be fully enclosed in a building; some users enclose the flocculation tanks as well. The process is particularly suited to treatment of eutrophic, stored lowland or otherwise algae laden waters and soft, low alkalinity upland coloured waters (Longhurst, 1987; Rees, 1979). Like all clarification processes flotation performance depends on the effectiveness of coagulation and flocculation. Polyelectrolyte dosing is often included to compensate for reduced performance at low water temperature or if the floc is fragile. Although the process has been successfully used for some directly abstracted waters other clarification methods tend to be more suitable for treatment of such waters especially when the turbidity consistently exceeds about 100 NTU (Gregory, 1999). Table 7.4 shows some typical results when treating algal laden waters.

There is, however, some experience with eutrophic waters with very high counts of algae where dissolved air flotation has not been successful, so that caution is necessary when choosing

Table 7.4 Comparison of algal cells in the raw water and % removal after coagulation by a ferric salt and sludge blanket clarifiers at 1 m/h or flotation at 12 m/h (Parr, 1991)

Alga	Raw water	Sedimentation	Flotation
Aphanizomenon	179 000	87%	98%
Microcystis	102 000	76.5%	98%
Stephanodiscus	53 000	58.7%	82.8%
Chlorella	23 000	84.3%	90.4%

the process. It should be noted that sedimentation can achieve degrees of removal comparable to flotation, if algae are first inactivated by chlorination. This would however result in the formation of DPBs by the action of chlorine on algal metabolic products.

Flotation is preceded by a flocculation stage of the hydraulic or mechanical type usually dedicated to each flotation cell. The flocculation tank should have at least two compartments in series (Section 7.12). Flotation is normally carried out in rectangular tanks designed with surface loading rates between 8–12 $m^3/h.m^2$ but rates as low as 5 $m^3/h.m^2$ or as high as 15–20 $m^3/h.m^2$ have been used on some plants (Pfeifer, 1997; Nickols, 1997). With such high rates there is a risk of air entrainment in the clarified water causing problems such as negative head due to air binding in downstream filtration processes (Section 8.2). This can be overcome by installing lamellas in the clarified water section as in DAFRapide®. The use of the lamellas enhances the physical separation air bubbles (Edzwald, 2007). A similar effect can be achieved by minimising the high velocity that can cause bubble entrainment at the DAF outlet.

In flotation the solids loading can vary in the range 4–15 kg dry solids/h.m². Typical tank depth is 2–3 m and the preferred length:width ratio is 1.33–2.5:1 with lengths up to 15 m using end-feed of air or 20 m with centre-feed of air. Width is limited to about 6 m for scarped tanks. The retention time in the flotation tank is between 10–20 minutes. The velocity in the subnatent opening should not exceed 0.05 m/s. The flow over the clarified water discharge weir should be less than 100 m^3/h per m of weir length.

For effective flotation the quantity of air required is about 6–10 g/m^3 or 4–6 l/m^3 of water treated and requires a recycle flow rate of about 6–15% (typically 8–10%) depending on temperature and dissolved oxygen concentration of the incoming water (Edzwald, 1992). The recycle flow should be included in the flow used for the computation of the rates for the flotation unit and downstream filters. Recycle water should preferably be filtered water. Clarified water if used should be strained to prevent recycle nozzle blockage. Oil-free compressors are preferred but not essential for the air supply. Air is dissolved in recycle water under pressure either in pressure vessels equipped with an eductor on the inlet side for adding air or in a packed column; the operating pressures of the two respective saturator systems are 6–7 bar and 3.5–6 bar. In packed columns a packing depth of 0.8 to 1.2 m of 25–37.5 mm Pall or Rashig rings of polypropylene (unsuitable for chlorinated water) or PVDF are used. The hydraulic loading rate of the air dissolving units lies in the range 50–90 $m^3/h.m^2$. Saturator efficiency for packed column type is about 90–95% whilst that for unpacked type is about 65–75% (Amato, 1997). Saturator efficiency is 100 times the amount of air measured in the recycle water divided by the amount of air that could be dissolved theoretically. Air saturated water is returned to the flotation tank

(a) Schematic of lamella sedimentation clarifier. (Purac, Sweden)

(b) Lamella clarifier, capacity 3155 m³/d at the 472,000 m³/d Water District #1 of Johnson County, Kansas, pre-sedimentation facility. (Engineer: Black & Veatch; manufacturer: Parkson Corporation—formerly Purac)

PLATE 11

(a) Schematic of AquaDAF® flotation clarifier. (Courtesy of Infilco Degremont Inc. VA USA)

(b) Interior view of a CoCo-DAFF (counter-current dissolved air flotation filtration) clarifier unit.

PLATE 12

through a series of nozzles or needle valves to give a sudden reduction in pressure and release of air bubbles in a white water curtain. Typically bubble size ranges from 10 to 100 μm with a mean diameter of 40 μm (Zabel, 1984). The outlets are usually spaced at 0.3–0.6 m for needle valves and 0.1 to 0.3 m for nozzles (Dhalquist, 1997). A typical nozzle density is about 10 per m^2 provided in 2 or 3 manifolds which could be isolated independently to facilitate greater turndown of recycle flow without loss of pressure. The contact time in the riser section should be about 100–120 seconds.

In plants where there is a need for raw water ozonation and flotation, the two processes could be combined with air in the flotation process being replaced by an ozone–air or ozone–oxygen mixture (Boisdon, 1994).

High rate flotation processes are finding application as they require smaller footprint. These include proprietary designs DAFRapide® (see above) AquaDAF® (Plate 12(a)) and Clari-DAF®. AquaDAF comprises a pre-fabricated perforated false floor with distribution of holes of different sizes across the floor, designed for uniform withdrawal of flow over the whole area of the tank which is said also to maintain a deeper bubble blanket throughout the float area. The holes are also thought to act as bubble collectors and allow for bubble coalescence preventing carry over to the filters. The combination of these effects is believed to provide performance comparable to conventional flotation (where clarified water is collected at one end and the bubble blanket is concentrated at the inlet end and grows shallower along the length of the tank) but at much higher rates. The surface loading rates are between 25–50 m^3/h.m^2. The width of the tank is greater than the length in the ratio 1.5–2:1 and the depth is about 4 m. The other design parameters (such as flocculation requirements, bubble size, recycle ratio and air dose) are similar to conventional flotation. Clari-DAF tank geometry is similar to conventional flotation design, but deeper and clarified water is removed through a pipe lateral system located on the floor of the tank. The surface loading rates up to 50 m^3/h.m^2 are claimed.

Since the clarified water is taken from the bottom of the tank in the flotation process it could be combined with rapid gravity filtration in the same tank with the filtration section placed underneath (DAFF) e.g. Flofilter®. Therefore the surface loading rates of the two processes need to be the same and should include the recycle flow. COCO DAFF® (counter-current dissolved air flotation filtration) is an innovative combined flotation–filtration design in which air and water flow counter-current as against co-current in the conventional dissolved air flotation process (Fig. 7.6 and Plate 12(b)). Air is introduced with recycle water across the total tank sectional area below the flotation zone and therefore only the filter surface loading rate should include the recycle flow. COCO DAFF gives more efficient particle–bubble interaction, and therefore increase in turbidity during desludging is minimized. (Officer, 2001) The process combines flotation and gravity filtration in one tank and uses a group of flocculation tanks common to all of the flotation cells. Flocculation is usually hydraulic and continues within the bubble blanket. Since the recycle flow is dissipated into the clarified water and not to the flocculated water as in conventional DAF, floc damage is minimized. The process requires far fewer recycle nozzles.

The flotation process is suitable for stop/start operation and has a flow turndown of about 2:1 or greater depending on the design of aeration manifolds. The former is one of its advantages when dealing with a water subject to high algal loadings; a plant can be 'switched in' as and when needed and will give a steady quality treated water within 45 minutes (Rees, 1979). Apart from the drawbacks common to all high rate clarifiers, the flotation process has high energy requirements (about 0.05–0.075 kWh/m^3 of water treated).

FIGURE 7.6

Typical arrangement of a COCO DAFF tank. (Black & Veatch).

7.19 SLUDGE REMOVAL FROM CLARIFIERS

Effective removal of sludge is very important for the efficient operation of clarifiers. With a raw water having suspended solids less than about 250 mg/l (most waters used in many countries fall in this category) the sludge volume to be removed from the tank should not exceed about 2.5% of inflow. For raw waters having high suspended solids (>500 mg/l) the sludge may be as high as 5–10% by volume of inflow; for suspended solids 1000 mg/l the sludge may have to be removed continuously in order to keep the tanks in operation, even at a reduced output, and to maintain an acceptable water quality. Under such conditions output may have to be reduced and the sludge volume can be as much as 20–30% of inflow. Special measures for sludge removal must obviously be provided for heavily silted waters and, for those having above 1000 mg/l solids or possibly less, depending on the nature of the solids, it is necessary to provide scraping equipment for all designs if throughput and quality are to be maintained. Scrapers move sludge to a series of hoppers located at the inlet end of rectangular tanks or usually in the centre or periphery of circular tanks. The floor of the tank should have a slope. Hoppers are of an inverted pyramid shape with an included angle of about 60°. Sludge is removed from hoppers individually under hydrostatic head using the full water depth in the tank.

Rectangular tank scraper design depends on the tank geometry; they are either of the travelling bridge type with or without suction headers (for tanks up to 25 × 75 m) or of the chain and flight or cable hauled type (for tanks up to 6 × 50 m). Bridge scrapers with suction headers (speeds varying from 1.0 to 2.0 m/min) and chain and flight type (speeds less than 0.5 m/min) are suitable for tanks treating heavily silted waters. The speed of bridge scrapers without suction headers is 0.5–1.0 m/min for scraping and about 2.5 m/min for the return.

Circular tanks have radial or diametrical scrapers with the bridge supported from the centre and driven with a central or peripheral drive unit. In larger tanks, support is also provided by a travelling wheel on the outside wall. The peripheral speed of the scraper is about 1.0–2.0 m/min and should not exceed 3 m/min. In some square tanks, corners are curved on the bottom and the scraper arm is equipped with a pivot type corner blade extension, which reaches out to the corners and then folds back on itself when traversing the four sides (pantograph mechanism).

Some circular tank designs include a suction header similar to the rectangular tanks. Tanks employing suction headers draw sludge at about 2 l/s.m of tank width or radius from points just in front of the scraper blades or squeegees using pumps mounted on the bridge; others are aided by submersible pumps or down pipes with eductors. The floors of scraped circular tanks have slopes of about 1:10–1:20 and those of scraped rectangular tanks have slopes of about 1:300–1:500. Rectangular tanks equipped with suction headers do not require a slope except for drainage. Unscraped rectangular tanks usually have a cross fall of about 1:10 to a central channel running the length of the tank and a longitudinal fall of 1:200. By including high pressure water jets (at 3.5–4 bar) for cleaning the slope can be reduced to about 1:250. Valves for sludge removal are always better placed outside the walls of the tank. Both valves and pipework should be adequately sized to pass the maximum sludge withdrawal rate, which can be 400% or 500% greater than the average rate. The valves should be of the full bore type like plug valves (Section 16.12).

Sludge removal from blanket tanks is generally easier than with other designs, although the same rules for valve and pipe sizing must be used. The positioning of sludge hoppers is less critical than with other designs as a sludge blanket is in continual movement and migrates towards the space left by evacuated sludge. Hoppers usually occupy about 10–15% of the total settling area of the tank. The hydrostatic head available for sludge removal depends on the discharge level of the sludge pipe. Sludge removal is usually operated by an automatic system having adjustable timers for varying the duration of opening of sludge discharge valves at pre-selected but adjustable time intervals. When suspended solids are high continuous removal of sludge may be necessary. Sludge should be withdrawn from hoppers individually; manifolding hopper outlets to allow for simultaneous withdrawal is not recommended.

A method for initiating sludge removal, which has met with considerable success is a design that relies on sensing the differential weight of concentrated sludge in water. The equipment is shown in Figure 7.3 and consists of several flexible sludge cones suspended in water and one of which (called the pilot cone) is connected by a cable to a load cell. The load cell is sufficiently sensitive that when the weight of the sludge reaches a pre-set value (usually when the cone is about two-thirds full of sludge) the load cell initiates the opening of the desludging valve on all the cone outlets.

In flotation tanks sludge or the float collects on the water surface and is removed by mechanical or hydraulic means or a combination of the two. For highly coloured waters, the float should not be allowed to accumulate in the tank for more than 30 minutes, whereas much longer periods can be tolerated for algal laden or high turbid waters. Mechanical units are usually scrapers of the travelling type e.g. chain and flight, reciprocating or bridge and the choice is made primarily on the tank dimensions. The flight speed should be about 0.5 m/min or greater (up to 2 m/min) if assisted by hydraulic desludging. Another design is a scraper with number of blades fixed to a cylinder which rotates over the beach. It produces a thin sludge. All types are known to cause 'knock down' of float solids to some degree; a process where clarified water is contaminated with the sludge as a result of deaeration of the float due to the disturbances caused by the scraper. In hydraulic desludging, the clarified water draw-off is restricted intermittently to raise the water level in the cell until the sludge layer overflows into a collection trough. The overflow rate should be greater than 6 l/s per m weir length. In COCO DAFF where there is no natural cross-flow velocity, filtered water is pumped to a side channel (central channel in duplex units) and overflows a ski-jumped shaped weir. This method produces sludge of low concentration compared to mechanical methods but at reduced 'knock down'.

In COCO DAFF with the full-depth air blanket across the full process area, there is a greater chance of any sludge knockdown being re-floated. Water spray bars should be used to reduce friction at the walls and float area should be free of any apparatus that would restrict the movement of the float.

7.20 CHEMICAL DOSING EQUIPMENT AND PLANT LAYOUT

Chemical dosing plant comprises: storage facilities, solution or slurry preparation tanks and chemical metering and conveying systems. It is usual for storage facilities to be sized for 28 days' demand at average dose and normal flow rate, or the size of one consignment plus the demand for the period between placing the order and receiving a delivery allowing for public holidays. Longer storage may be required for locations where access is affected by bad weather or where chemicals have to be imported. Properties of some of the commonly used chemicals in water treatment are given in Table 7.5.

Chemicals are delivered as liquids or solids. Most chemicals are made up into solutions of known concentration or suspensions in batches; at least two batching tanks are required for each chemical in order to maintain continuity of dosing; additional tanks would allow maintenance and cleaning without interruption to dosing. Each tank is normally sized so that one or two batches are prepared in a work shift. Accurate batching and dilution, with proper mixing, is required to maintain uniform concentrations as in most cases metering is volumetric. For soluble solids the solution strength should be well below the solubility at the lowest water temperature. For powders such as lime and powdered activated carbon (PAC) suspensions need to be maintained at a value of less than 10% $^w/_v$ and must be continuously stirred. The concentration of batches must be checked for accuracy by hydrometer, conductivity meter or chemical analysis. The solutions and slurries should be further diluted after metering. Liquids are normally metered as delivered product and diluted downstream.

The use of lime as a slurry for final pH correction increases the turbidity of the filtered water depending on the lime dose, the proportion of impurities in lime and the formation of calcium carbonate precipitate due to localized softening caused by poor mixing. This can be overcome either by using caustic soda or a saturated solution of lime (lime water). Lime water is usually prepared in continuous flow saturators which are upward flow hopper bottomed tanks comprising a bed of lime (similar to hopper bottomed sludge blanket clarifiers—Section 7.16) charged with a lime slurry (5–10% $^w/_v$). Water is fed from the bottom through the bed of lime with lime water drawn from the top as a saturated solution. Surface loading rates range from 1 to 1.2 $m^3/h.m^2$. In some designs surface loading rates up to 2.0 $m^3/h.m^2$ are achieved by using a turbine mixer to improve contact between lime and water. Adding a polyelectrolyte to improve settling rate or by lamella plates in the clear solution zone (to increase settling area) would also help to increase the loading rate. Saturators convert only about 80% of the lime in the feed and an allowance for this should therefore be made in sizing the saturators. They are also useful to produce a clear solution of lime when its purity is low. Unconverted lime, calcium carbonate formed by softening and grit are removed from the bottom by regular desludging.

Chemical dosing must be accurate and related to the flow of water to be treated. Positive displacement pumps of the reciprocating type with mechanical or hydraulic diaphragm heads are most frequently used for injection but for lime and PAC suspensions or viscous solutions such as polyelectrolytes, progressive cavity type positive displacement pumps are sometimes used. For suspensions, peristaltic pumps also find application. Pumps should be provided with a calibration vessel on the suction side, a pressure relief valve (venting to waste), pulsation dampener, and a back pressure

Table 7.5 Properties of some chemicals commonly used in water treatment (Table 7.6 for coagulants and chapter 11 for disinfectants)

Chemical	Function	Form	Density (for solids: bulk density)	Materials	Freezing point/ solubility	Storage	Dosing concentration
Hydrated lime 96% w/w[a] $Ca(OH)_2$ (BS EN 12518: 2000)	pH correction	White fine powder	[b]480 kg/m^3 [c]400 kg/m^3 [d]1.81 m^3/t	Steels, thermoplastics (*Al, tin, Zn, brass, galvanized steel)	1.76 g/l at 10°C, 1.65 g/l at 20°C, 1.53 g/l at 30°C	Bags (25 kg, 50 kg) on pallets, steel silos	>2.5% w/v[a] <10% w/v[e]
Hydrated lime 17% w/w $Ca(OH)_2$	pH correction	Milky white liquid	1.11 g/ml	(As for hydrated lime)	0°C	Vertical steel or thermoplastic tanks with mixers or recirculation pumps	2.5% w/v to neat
Quicklime[f] 95% w/w CaO (BS EN 12518: 2000)	pH correction	Hygroscopic powder	[b]1230 kg/m^3	(As for hydrated lime)	Highly reactive with water	Bags (25 kg, 50 kg) on pallets, steel silos	Slaked to form hydrated lime >2.5% w/v <10% w/v
Powdered activated carbon (BS EN 12903: 2003)	Organics removal dechlorination	Powder	[b]410–600 kg/m^3 (depending on the grade) [c]375–500 kg/m^3	Stainless steel (304, 306) rubber/mild steel (for slurry), thermoplastics	—	Bags (25 kg, 50 kg) on pallets, 450 kg bags, 1000 kg bins or steel silos (epoxy paint coated)	<10% w/v
Sodium carbonate (light grade) 95% w/w Na_2CO_3 (BS EN 897: 2005)	pH correction	Anhydrous crystalline powder	[b]550 kg/m^3	(As for hydrated lime)	120 g/l at 10°C, 210 g/l at 20°C	Bags (25 kg, 50 kg) on pallets or steel silos	5% w/v (temperate) 15% w/v (tropics)
Potassium permanganate (BS EN 12672: 2000)	Oxidation	Granular	1600 kg/m^3	Steels, thermoplastics (*Zn, Cu, Al, galvanized steel, rubber)	44 g/l at 10°C; 65 g/l at 20°C	Kegs (50 kg), 150 kg drums	1.5 to 3% w/v;
Sulphuric acid (i) 98% H_2SO_4 w/w[g] (ii) 96% w/w H_2SO_4[g] (BS EN 899: 2003)	pH correction	Corrosive liquid	1.84 g/ml at 20°C	Steels, PTFE (*most other metals)	(i) 3°C (ii) −14°C	Carboys (45 l), steel horizontal pressure vessels or steel vertical tanks (lagged for 98% w/w as applicable)	Neat or 10 w/w H_2SO_4
Sulphuric acid 50% w/w H_2SO_4 (BS EN 899: 2003)	pH correction	Corrosive liquid	1.4 g/ml at 15.5°C	Thermoplastics, rubber/steel	−37°C	Carboys (45 l), PVC/GRP, PP/GRP, HDPE or rubber/mild steel vertical tanks	10% w/w to neat
Caustic soda 47% w/w NaOH (BS EN 896: 2005)	pH correction	Corrosive liquid	1.497 g/ml at 20°C	Steels, thermoplastics, rubber, Ni and Ni alloys (T < 150°C) (*Al, Tin, Zn, galvanized steel, brass)	8°C (−25°C when diluted to 20% w/w NaOH)	Carboys (45 l) steel horizontal pressure vessel or steel or thermoplastic, PVC/GRP vertical tanks (heated and lagged as applicable)	Neat or 20% w/w NaOH

Chemical	Application	Form	Density	Materials of construction	Crystallization point / Solubility	Storage	Dosing
Hexafluorosilicic acid 20% $^{w}/_{w}$ H_2SiF_6 (15.8% $^{w}/_{w}$ F) (BS EN 12175: 2001)	Fluoridation	Corrosive liquid. Highly toxic	1.18 g/ml at 20°C	Thermoplastics (PE, PP, uPVC), Neoprene rubber/steel (*Glass, stainless steel (304, 316), Al, brass, bronze, carbon steel)	−11.6°C	Horizontal or vertical rubber/mild steel or thermoplastic vertical tanks	Neat or diluted to suit
Sodium silicofluoride 98% $^{w}/_{w}$ Na_2SiF_6 (59.4% $^{w}/_{w}$ F)	Fluoridation	Crystalline powder. Highly toxic	1400 kg/m³	Thermoplastics, rubber lined carbon steel	5.5 g/l at 10°C, 6.4 g/l at 20°C	Bags (20 kg, 50 kg) on pallets, steel silos	0.2% $^{w}/_{h}$
Orthophos-phoric acid 75% $^{w}/_{w}$ H_3PO_4 (24% $^{w}/_{w}$ P) (BS EN 974: 2003)	Plumbo-solvency control	Corrosive liquid	1.57 g/ml at 20°C	Stainless steel (316), thermoplastics (*carbon steel, cast iron, Al, Al-alloys, brasses, tinned or galvanized)	−18°C	Lined steel drums (45 l, 200 l), horizontal or vertical stainless steel or rubber/steel or vertical HDPE or PVC/GRP tanks	Neat or diluted to suit^h
Orthophosphates (i) mono sodium (20% $^{w}/_{w}$ P); (ii) di sodium (17% $^{w}/_{w}$ P); (iii) tri sodium (8% $^{w}/_{w}$ P)	Plumbo-solvency control	Crystalline powders	(i) 1200 kg/m³ (ii) 1200 kg/m³ (iii) 900 kg/m³	Thermoplastics, stainless steel (304, 316), rubber/steel (*carbon steel, Al)	(i) 650 g/l at 10°C, 850 g/l at 20°C; (ii) 38 g/l at 10°C, 78 g/l at 20°C; (iii) 125 g/l at 10°C, 215 g/l at 20°C	Bags (50 kg)	(i) 40% $^{w}/_{v}$^h (ii) 15% $^{w}/_{v}$^h (iii) 20% $^{w}/_{v}$^h
Sodium hypochlorite 15.5% $^{w}/_{w}$ NaOCl (15% $^{w}/_{w}$ Cl_2) (BS EN 901: 2000)	Disinfection oxidation	Hazardous liquid	1.27 g/ml at 20°C	Thermoplastics (PE, PVC, HDPE), rubber/steel (*carbon steel, Al, Zn, Cu and their alloys, PP)	−17°C	Carboys (45 litres), horizontal or vertical rubber/steel or PP/GRP, PVC/GRP or HDPE tanks	Neat or diluted to suit^h
Sodium bisulphite 40% $^{w}/_{w}$ $NaHSO_3$ (25% $^{w}/_{w}$ SO_2) (BS EN 12120: 2005)	Dechlorina-tion Deoxy-genation	Hazardous liquid	1.35 g/ml at 15.5°C	Thermoplastics (PP, PVC, GRP, stainless steel (304, 316) (*carbon steel)	10°C	Dums (45 litres, 210 litres), stainless steel (316) or PP/GRP, PVC/GRP, HDPE or rubber/mild steel vertical tanks (heated and lagged as applicable)	Neat or diluted to suit
Sodium metabisulphite 96% $^{w}/_{w}$ $Na_2S_2O_5$ (65% $^{w}/_{w}$ SO_2)	Dechlorina-tion Deoxy-genation	Granulated powder	1.48 g/ml at 20°C	Thermoplastics (PP, PVC) GRP, stainless steel (304, 316) (*carbon steel)	400 g/l at 20°C	25 kg PP bags with PVC liner	10% $^{w}/_{w}$

(continued)

Table 7.5 (continued)

Chemical	Function	Form	Density (for solids: bulk density)	Materials	Freezing point/ solubility	Storage	Dosing concentration
Hydrogen peroxide 35% $^w/_w$ H_2O_2 (BS EN 902: 2000)	Oxidation	Hazardous liquid	1.130 g/ml at 20°C	Aluminium (99.5%), Al-Mg alloys, stainless steel (304, 316), HDPE, PVC (*Fe, Cu, Ni, Cr, brass)	−33°C	PE carboys (50 kg), stainless steel or Al horizontal or vertical tanks or HDPE or PP/GRP, PVC/GRP	Neat
Sodium chloride (Pure Dried Vacuum Grade) 100% $^w/_w$ NaCl (BS EN 973: 2000)	Regeneration of ion exchange resin. On-site generation of sodium hypochlorite	Crystalline powder	1200–1360 kg/m³	Thermoplastics, rubber/carbon steel, stainless steel (316), Aluminium alloy NS4 (*stainless steel, carbon steel for moist or salt solutions)	358 g/l at 10°C; 360 g/l at 20°C	PE bags (25 kg), 1 t containers, saturators of reinforced concrete of rich mix (1:1.5:3) with 40 mm cover or GRP	Saturated solution or diluted to suit
Ammonium sulphate 25% $^w/_w$ NH_3 (BS EN 12123: 2005)	Chloramination	Crystalline powder	1120 kg/m³	Thermoplastics, stainless steel (304, 316) (*iron, Cu, Zn, Tin and their alloys)	727 g/l at 10°C; 754 g/l at 20°C	Bags (20 kg, 50 kg)	10% $^w/_v$
Ammonium hydroxide (25% $^w/_w$ NH_3) (BS EN 12122: 2005)	Chloramination	Hazardous liquid	0.9 g/l at 20°C	Carbon steel, stainless steel, Al,	−55°C	Vertical or horizontal tanks fitted with safety relief valve and vacuum breaker and earthed	Neat or diluted to suit[h]
Sodium chlorite (i) 26% $^w/_w$ $NaClO_2$ (ii) 80% $^w/_w$ $NaClO_2$ (BS EN 938: 2000)	Chlorine dioxide generation	(i) Hazardous liquid; (ii) Hazardous powder	(i) 1.27 g/ml at 20°C; (ii) 1105 kg/m³	Thermoplastics (PE, HDPE, PVC) GRP (*Zn and combustibles)	(i) −7°C (ii) 400 g/l at 20°C	PE kegs (50 kg, 70 kg), Steel drums with PE lining, HDPE vertical tanks	(i) 12.5–20% $^w/_v$ (ii) 25–30% $^w/_v$[h]

Notes:

PTFE—Polytetrafluoroethylene; HDPE—High density polyethylene; PVC—Polyvinyl chloride; GRP—Glass reinforced plastic; PE—Polyethylene;
PP—Polypropylene; PVC/GRP—PVC lined GRP; PP/PRP—PP lined GRP; Rubber/steel—rubber lined carbon steel; All stainless steel grades are to BS 970 or 1449.

* Unsuitable materials.

[a] $x\%$ $^w/_w$ is x percent weight per weight = x grammes of the chemical in 100 g of the product.
[b] For calculating silo capacity.
[c] When aerated during bulk delivery.
[d] When stacked in bags.

[e] $y\%$ $^w/_v$ is y percent weight per volume = y grammes of the substance in 100 ml of solution containing the substance.
[f] Quicklime gives off considerable amount of heat (1.14 × 10⁶ J/kg) during slaking.
[g] Sulphuric acid gives off considerable amount of heat. Therefore when diluting, acid should be added to a large quantity of water.
[h] Softened water should be used for solution preparation and dilution to prevent scaling.
BS EN refers to British European standards, current edition.

valve on the delivery side. All chemical dosing pumps must be of materials appropriate for the chemical handled. The maximum stroking speed (spm) of reciprocating pumps should be about 120 spm, in particular for viscous or abrasive chemicals. The motor speed of progressive cavity and peristaltic pumps should be kept to less than 500 and 50 rpm, respectively.

With the reciprocating pump, dosage adjustment is achieved by altering the pump stroke length. Where plant throughput is variable (greater than ±5%) the pump motor speed is automatically controlled in proportion to the flow rate measured near the chemical injection point. This type of control is called 'open-loop'; it has no feed-back or corrective action and the applied dose rate is strictly proportional to the flow. In a 'closed-loop' system, the pump output is corrected to maintain a given water quality value (such as pH) over a narrow pre-set band, measured downstream of the injection point after the chemical has been well mixed with the water ('feed-back' control). A process controller, working in conjunction with an appropriate water quality measuring instrument, sends a 4–20 mA signal (Section 17.34) back to the pump to adjust its stroke length. In some instances this water quality signal is combined with the rate of flow signal and used to control the pump motor speed. This latter method is used on pumps which have no stroke adjustment (e.g. progressive cavity and peristaltic types), or on chlorinators and similar equipment to control the orifice positioner so as to maintain a pre-set residual chlorine concentration in the water (Section 11.12).

In some plants it is necessary to apply the same dose to two or more equal streams, e.g. dosing to individual clarifiers; it is then vital to ensure equal division of the metered chemical flow and this is economically achieved by use of a splitter box with equally set V-notch weirs.

In remote parts of developing countries chemical dosing systems are kept simple, with manual preparation of solutions and slurries and the use of constant head solution feeders for dosing chemicals (WHO, 1977).

Plant Layouts

Among the most important considerations when planning a layout are the following:

1. The flow through the works should be gravitational: it is inadvisable to re-pump water between clarifiers and filters as this would break up floc. Hence a site having a gentle gradient of 1 in 10 to 1 in 15 is most favourable. The typical head loss across a treatment plant comprising clarifiers and filters and assuming mixing is by hydraulic means (inlet chamber to the treated water reservoir) could be in the range 5.5–6 m of which filters (underdrain, clean media and an allowance for clogging) accounting for 2–2.5 m. Inter-stage pumping may be unavoidable when there are two filtration stages in series for which an additional head of about 2.5–3 m may be needed.

2. When siting works adjacent to a river it is important to avoid siting any structure below highest flood level because of the difficulty and cost of countering uplift problems. No electrical or chemical plant should be put in a basement which could flood owing to a burst pipe.

3. The works should be provided with means to safely evacuate overflow caused by fault or mal-operation. Typical locations are inlets to the works, filters and disinfection contact tank and/or treated water reservoir.

4. All structures conveying and retaining water downstream of filters should be sealed to prevent contamination. All water retaining structures should be provided with means for dewatering.

5. Easy access, including turning-circles, should be provided for chemical delivery vehicles. Access for large equipment for plant repair or replacement should be possible to all buildings and process units.

6. It is preferable to provide chemical dosing lines in duplicate (one duty, one standby). Chemical lines should not be laid in positions where any leakage could damage other lines or cause injury to personnel (e.g. over access ways). Toxic gas under pressure or solution lines should be laid outside buildings in separate ducting. When possible toxic gases should be conveyed under vacuum and mixed with water just before injection. Delivery lines for slurries, such as lime, are difficult to keep clean and should be of the flexible hose type and laid flat. Water-flushing of dosing lines should be provided. Chemical pipes should be laid in trenches in the ground provided with removable covers for better access in preference to buried ducts.

7. All chemicals where possible should be diluted in-line after metering. For lime and PAC, the lower the dilution the better the mixing and typically dilution down to 1% $^w/v$ (10 g/l) is used. If the receiving water contains an alkalinity greater than 10 mg/l as $CaCO_3$ water for in-line dilution (and solution preparation) of those chemicals which react with alkalinity to form $CaCO_3$ scale (e.g. caustic soda, sodium hypochlorite) should be softened by base exchange (Section 10.7) to a hardness less than 25 mg/l as $CaCO_3$. Dilution water for lime should have an alkalinity less than 10 mg/l as $CaCO_3$. This is achieved by de-alkalisation (Section 10.9) or by treating the dilution water with hydrochloric acid (HCl) to destroy the alkalinity followed by degassing to remove the carbon dioxide produced in the reaction; 1 mg/l of alkalinity as $CaCO_3$ requires 0.73 mg/l 100% HCl and produces 0.88 mg/l carbon dioxide. Alternatively lime dilution should be in tanks with at least 15 minutes residence time to allow any softening reactions to reach completion. For coagulants the diluted concentration is usually in the range 1 to 5% $^w/v$ depending on the alkalinity and should not neutralize more than about 2–2.5% of the coagulant. Polyelectrolyte solutions which are viscous should be diluted to less than 0.025% $^w/v$. Dilution is usually carried out downstream of the dosing pump and in the dosing line except for polyelectrolytes where a static mixer is desirable.

8. Instrumentation and electrical cables and sampling and dosing lines form a complex network of cabling inside buildings and on the site. In the early stages of planning the building layout, allowance should be made for chemical pipes, cable trays, ventilation ducts and other services such as site water supply. High tension electrical cables should be separately ducted. Ducts for chemical delivery lines can be adjacent, but must be separate. All ducts must have drainage outlets.

9. Liquid chemical storage vessels need bunding dedicated to each chemical (Cassie, 2003). A bund should be designed to hold 110% of the contents of the largest tank. Dosing pumps and similar apparatus should also be surrounded by a low bund wall. Dust-nuisance chemicals, e.g. lime and PAC, should be fully segregated. PAC is an electrical conductor and should not be allowed to accumulate as dust on open electrical circuits. Toxic gas facilities (chlorine, sulphur dioxide, ammonia and carbon dioxide) should be located in fully segregated buildings, or rooms. Toxic and suffocating gas storage buildings should be separated from each other and other habited buildings by a distance of at least 25 m. All rooms in gas handling facilities should be physically separated from each other (Section 11.12).

10. Substances and products used in the works, which may come in contact with the water, which is to be supplied for drinking or cooking, should not contain any matter, which could impart taste, odour, colour or toxicity to the water or otherwise be objectionable on health grounds.

11. All chemical drainage including that from bunded and hardstanding areas should be collected, neutralized and disposed of separately and should not be allowed to contaminate water courses.

12. Safety precautions for operational staff should receive careful attention. This should include the provision of safety signs, safety showers, eye baths, first aid boxes, protective goggles and clothing and breathing apparatus. Safety screens should be provided around pumps used for hazardous chemicals such as sulphuric acid.

COAGULANTS AND COAGULANT AIDS

7.21 ALUMINIUM COAGULANTS

Aluminium sulphate is the most widely used aluminium coagulant. It is available in a number of solid grades such as block, kibbled or ground and is also available as a solution. In waterworks practice aluminium sulphate is frequently but incorrectly referred to as 'alum'. The solid form has the composition $Al_2(SO_4)_3 \cdot xH_2O$ where x may range from 14 to 21 containing 14–18% $^w/w$ Al_2O_3 (alumina) or 7.5–9% $^w/w$ Al (aluminium), depending on the number of molecules of water (x). The liquid form contains 8% $^w/w$ Al_2O_3 or 4.2% $^w/w$ Al. The amount of Al_2O_3 or Al in any solid grade of aluminium sulphate containing x moles of water is given by $y\%$ $^w/w$ $Al_2O_3 = [5.67 \div (19 + x)] \times 100$ and $z\%$ $^w/w$ Al $= [3 \div (19 + x)] \times 100$, respectively.

The aluminium sulphate dose is therefore normally expressed in mg/l as $y\%$ $^w/w$ Al_2O_3 depending on the grade of aluminium sulphate, or more usefully in mg/l as Al; 1 mg/l as $y\%$ $^w/w$ Al_2O_3 is equal to $5.29\,y \times 10^{-3}$ mg/l as Al. Aqueous solutions of the solid grades are usually prepared in concrete tanks suitably lined and equipped with a collector system of perforated pipe laterals in a bed of gravel. The tanks are usually built below ground level so that the material delivered in bulk can be tipped directly into the tanks. A saturated solution is prepared which contains about 653 g/l (10 °C), 710 g/l (20 °C) or 788 g/l (30 °C) and is subsequently diluted about four- to six-fold in stock tanks before dosing. When the solid grade is delivered in bags, a 200–300 g/l solution is prepared in tanks containing two compartments separated by a timber grid to prevent solid in one compartment damaging the top-entry turbine mixers in the other. The liquid grade containing 8% $^w/w$ Al_2O_3 is usually stored in vertical tanks and dosed by pump in the delivered form; after metering it should preferably be diluted.

When dosed into water, the formation of an aluminium hydroxide floc is the result of the reaction between the acidic coagulant and the natural alkalinity of the water, which usually consists of calcium bicarbonate. A dose of 1 mg/l of aluminium sulphate as Al reacts with 5.55 mg/l of alkalinity expressed as $CaCO_3$ and increases the CO_2 content by 4.9 mg/l. Thus if no alkali is added the alkalinity would be reduced by this amount with a consequent reduction in pH. If a water has insufficient alkalinity or 'buffering' capacity, additional alkali such as hydrated lime, sodium hydroxide, or sodium carbonate must therefore be added; the alkalinity expressed as $CaCO_3$ produced by 1 mg/l of each chemical (100% purity) is 1.35, 2.5 and 0.94 mg/l, respectively. The aluminium hydroxide floc is insoluble over relatively narrow bands of pH, which may vary with the source of the raw water. Therefore pH control is important in coagulation, not only in the removal of turbidity and colour but also to maintain satisfactory minimum levels of dissolved residual aluminium in the clarified water. The optimum pH for coagulation of lowland surface waters is usually in the range of 6.5–7.2, whereas for more highly coloured upland waters a lower pH range, typically 5–6, is necessary.

Table 7.6 Properties of coagulants

	Aluminium sulphate	Aluminium sulphate	Polyaluminium chloride	Polyaluminium um chloride	Polyaluminium chlorosulphate	Polyaluminium silicate sulphate	Ferric sulphate	Ferric chloride	Polymeric ferric sulphate
Physical form	Solid	Liquid	Liquid	Liquid	Liquid	Liquid	Liquid	Liquid	Liquid
Chemical formula	$Al_2(SO_4)_3.xH_2O$ $x = 14-21$	$Al_2(SO_4)_3$	$Al_x(OH)_yCl_z$	$Al_x(OH)_yCl_z$	$Al_2(SO_4)_x.Cl_y.(OH)_z$	$Al_w(OH)_x(SO_4)_y(SiO_2)_z$	$Fe_2(SO_4)_3$	$FeCl_3$	$Fe_2(SO_4)_3$
Typical commercial grade	Blocks 14% $^{w/w}$ Al_2O_3	8% $^{w/w}$ Al_2O_3	10% $^{w/w}$ Al_2O_3	18% $^{w/w}$ Al_2O_3	8.3% $^{w/w}$ Al_2O_3	8% $^{w/w}$ Al_2O_3	40–42% $^{w/w}$ $Fe_2(SO_4)_3$	40–42% $^{w/w}$ $FeCl_3$	48–50% $^{w/w}$ $Fe_2(SO_4)_3$
Fe/Al content (% $^{w/w}$) of the commercial product	300/(19 + x)	4.2	5.3	9.0	4.4	4.4	12	14–14.5	13.5–14
pH	1.5 for a saturated solution (670 g/l of water at 20°C)	1.3	2.3–2.9	1.0	2.8–3.0	3.6–3.8	<1.0	<1.0	1.0
Specific gravity	1 to 1.4 t/m³ (bulk density)	1.32 at 15°C	1.20 at 20°C	1.37	1.16 at 20°C	1.28 at 15°C	1.52 at 15°C	1.45 at 15°C	1.58–1.63 at 15°C
Freezing point	—	−15°C	−12°C	−20°C	−12°C	0	−15°C	−2°C	−20°C
Viscosity at 20°C	—	20 m.Pa.s	3.5–4.5 m.Pa.s	25–35 m.Pa.s	4.5 m.Pa.s	11 m.Pa.s	30 m.Pa.s	7.5 m.Pa.s	55 m.Pa.s
Coagulation pH range	5.5–7.5	5.5–7.5	6–9	6–9	6.5–7.8	6.5–8	4.0–9.0	4.0–9.0	>4.5
BS EN	878: 2004	878: 2004	883: 2004	883: 2004	883: 2004	885: 2004	889: 2004	888: 2004	—

Notes:
All liquid coagulants are dosed neat and solid coagulants are dosed as saturated solutions. They can be diluted to suit (typically 20% $^w/_v$ for metering), but any further in-line dilution following metering should be limited to a level that does not neutralize more than 2.5% of the coagulant. Diluted solutions of some polymerized coagulants gradually hydrolyse with time with subsequent loss of effectiveness.

Suitable materials of construction are thermoplastic material such as polyvinylchloride (PVC), glass reinforced plastic (GRP), polyethylene, HDPE, polypropylene (PP), PVC lined GRP, PP lined GRP, rubber lined mild steel, stainless steel (316) except for those containing chlorides and concrete with suitable linings, e.g. acid resistant bricks, fibre glass or resin coated. Unsuitable materials are mild steel and most common metals such as Al, Zn, Cu and their alloys and concrete.

Lowland waters usually contain higher concentrations of dissolved salts, including alkalinity and may therefore require the addition of an acid in excess of that provided by the coagulant. Under these circumstances it is usually more economic to add sulphuric acid rather than excess aluminium sulphate to obtain the optimum coagulation pH value. The use of coagulant to depress the pH would result in more sludge production.

There are some coagulants, which are polymerized inorganic aluminium salts such as polyaluminium chloride (PACl), aluminium chlorohydrate (ACH), polyaluminium chlorosulphate (PACS) and polyaluminium silicate sulphate (PASS) and are formulated to contain high basicity, a measure of hydroxyl ions present in the coagulant. A definition of basicity is given by Letterman (1999). Aluminium sulphate has zero basicity whereas that of PACl varies from about 50% for $Al(OH)_3Cl_3$ (i.e. 10% $^w/_w$ Al_2O_3) to about 85% for ACH (24% $^w/_w$ Al_2O_3). The H^+ ions produced by aluminium sulphate, PACl (10% $^w/_w$ Al_2O_3) and ACH are 6, 3 and 1 respectively. High basicity coagulants therefore depress the pH of the treated water less than aluminium sulphate, thereby minimising the need for coagulation pH adjustment and reducing the alkali dose required for subsequent final pH correction; these salts also limit the aluminium residuals whilst maintaining optimum coagulation properties and produce stronger and more readily settleable floc than aluminium sulphate, thus reducing the need for polyelectrolytes as coagulant aids; coagulation is less affected by low temperature and produces less sludge than aluminium sulphate. In some waters such salts can be used in lower doses than aluminium sulphate and over a broader optimum pH range (6–9). There are several grades of PACl containing 10, 18 or 24% $^w/_w$ Al_2O_3; the 10% and 18% $^w/_w$ grades being the most commonly available. The other polymeric aluminium salts PACS and PASS although less common, behave in a similar manner to PACl. The properties of the most commonly used aluminium coagulants are summarized in Table 7.6.

It has been suggested that aluminium in drinking water may be associated with neurological disorders and Alzheimer's disease (Packham, 1992). In addition, its presence in filtered water can be harmful to users of renal dialysis (Section 6.6). For these reasons some water undertakings have changed from aluminium to iron coagulants (Carroll, 1991). When making such a change care should be taken to clean process units free of all accumulated floc, which would otherwise dissolve and increase the aluminium concentration in the water if the ferric coagulant is used outside the optimum pH range for the aluminium coagulant.

7.22 IRON COAGULANTS

Iron coagulants in the ferric form behave similarly to aluminium sulphate and form ferric hydroxide floc in the presence of bicarbonate alkalinity. A dose of 1 mg/l of ferric sulphate or chloride as Fe neutralizes 2.7 mg/l alkalinity expressed as $CaCO_3$ and increases the CO_2 content by 2.36 mg/l. The ferric hydroxide floc is insoluble over a much broader pH range (4–10) than aluminium sulphate. The lower end of the pH range (4–5.5) is useful for treating highly coloured moorland waters.

Iron coagulants are available as ferric sulphate, ferric chloride and ferrous sulphate. Ferric salts are very corrosive acidic liquids. Ferric sulphate is usually preferred to ferric chloride since the introduction of chloride ions may increase the corrosivity of a water. Ferrous sulphate, traditionally referred to in its hydrated form ($FeSO_4.7H_2O$) as 'copperas', is used as a coagulant usually in conjunction with chlorine; in practice excess chlorine is used. Chlorinated ferrous sulphate is not used due to the potential risk of DPB formation by the action of excess chlorine with DPB precursors in the raw water. Ferrous sulphate on its own is used as a coagulant in processes utilising high pH values such as lime softening (pH 10–11) and manganese removal (pH 9). Iron coagulants have the advantage of producing a denser floc than that produced by aluminium sulphate thereby producing improved settlement characteristics but at the expense of about a 40% increase in the weight of hydroxide sludge when compared to aluminium coagulants.

Polymeric ferric sulphates are now available; they contain about 12.5% ^{w}/w Fe and are claimed to perform better and at lower doses than ferric sulphate (Jia-Quian, 1996). There are ferric-aluminium sulphate coagulants; one such product contains approximately 8% ^{w}/w of the metal oxides made up of 6% ^{w}/w Al_2O_3 and 2% ^{w}/w Fe_2O_3. The properties of commonly used of iron coagulants are summarized in Table 7.6.

Many iron coagulants contain approximately 2–6 g of manganese per kg of iron as an impurity depending on the product specification. This contributes to the manganese concentration in the water.

7.23 COAGULANT AIDS AND POLYELECTROLYTES

Coagulant aids are used to improve the settling characteristics of floc produced by aluminium or iron coagulants. The coagulant aid most used for a number of years was activated silica; other aids included sodium alginates and some soluble starch products which are still in use. These substances had the advantage of being well-known materials already used in connection with the food industry and were thus recognized as harmless in the treatment of water. Polyelectrolytes came later into use and were more effective. They now comprise of numerous synthetic products: long chain organic chemicals, which may be cationic, anionic or non-ionic. The theory of their action has been reviewed by Packham (1967).

Polyacrylamides are the most effective of the synthetic group of polyelectrolytes, but for their safe use the toxic acrylamide monomer residue (the raw materials used in their manufacture) which is not adsorbed by the floc, should be virtually absent from the product. In the UK Water Supply (Water Quality) Regulations (UK, 2003) states that no batch may contain more than 0.02% ^{w}/w of free acrylamide monomer based on the active polymer content; the dose used must average no more than 0.25 mg/l and never exceed 0.5 mg/l. US EPA allows a maximum dose of 1 mg/l on an assumed acrylamide content of 0.05% by weight.

Polyelectrolyte doses used are very small in relation to the dose of the primary coagulant. Natural polyelectrolyte (starch based) doses vary between 0.5–2.5 mg/l whereas polyacrylamide doses vary between 0.05–0.25 mg/l. Polyelectrolytes are added as a coagulant for turbid waters or after the primary coagulant as a coagulant aid (Section 7.10). Sometimes they are added just prior to filtration in very small doses (about 0.01 mg/l) to flocculate micro-floc particles carried over from the clarifiers and filter passing algae; care in control of the dose is necessary because excess polyelectrolyte could result in 'mud ball' formation and other problems in the filters. The cationic polyelctrolyte PolyDADMAC (polydyallyldimethylammonium chloride) is known to reduce the dose of inorganic coagulants when the two are used together. This has been observed in waters of high colour with both low and high alkalinity and conductivity. The reduction in the aluminium sulphate dose could be as much as 50% (Gebbie, 2005). The effect this reduction in the coagulant dose has on the TOC removal should be tested in the laboratory. PolyDADMAC could form N-nitrosodimethylamine (NDMA) (Section 6.80).

Most polyelectrolytes are powders and a solution must be prepared for dosing. For successful preparation the powder must be wetted properly by using a high energy water spray before dissolving; the solution should be allowed to age for about an hour in cold water or 30 minutes in warm water conditions before use. For polyacrylamide the solution should be prepared at about 2.5 g/l, whereas for natural polyelectrolytes the solution concentration could be as high as 25 g/l. Following metering the solution should be diluted ten-fold to assist transfer in the pipe and dispersion at the point of application. Once a batch of stock solution is prepared it should be used preferably within about 24 hours.

The practical effect of introducing polyelectrolytes in many existing waterworks has been to increase the settling rate and hence allow substantially greater output through the clarifiers; an additional use is to assist in the recovery of used filter washwater and thickening and dewatering of sludges (Chapter 9).

7.24 RAINWATER HARVESTING

Rainwater is typically devoid of all dissolved solids but contains dissolved gases (oxides of carbon, nitrogen and sulphur) which result in pH values of about 5.5 or lower. In coastal areas rainwater may also contain up to 15 mg/l of sodium chloride from sea-spray. Rainwater is therefore devoid of alkalinity; it is acidic (low pH), low in mineral content and aggressive towards calcium bearing materials such as concrete and some of the metals typically used in domestic plumbing. The presence of chlorides exacerbates corrosivity. Rainwater is also unpalatable in taste, again due to the low solids content (Meera, 2006).

Rainwater is harvested from roofs via guttering into storage tanks and may be contaminated, particularly in the 'first flush' after a dry period, by detritus that has collected on the roof, for instance leaves, insects or more importantly bird faeces. In coastal locations, salt deposits from sea spray may also be present.

Water falling on roofs or other surfaces designated for the purpose of collection is directed, via screens to remove leaves, bird feathers and other coarse debris, to a collection tank. The first flush, typically about 1–0.5 l per m² of collection area (Krishna, 2005), is discharged to waste by means of 'stand pipes' or other proprietary designs and may be used for non potable purposes such as gardening or toilet flushing.

The roof surfaces over which the rain is collected should be made of materials from which no undesirable compounds may be dissolved or leached if rainwater is intended for human consumption; examples are lead flashings and bituminous roof material.

REFERENCES

Amato, T. (1997). DAF: Its Place in an Integrated Plant Design, *Proc International Conference on Dissolved Air Flotation*, CIWEM, London.

Amirtharajah, A. (1978). Design of Rapid Mix Units, *Water Treatment Plant Design* (Ed. Sanks, R. L.). Ann Arbor Science.

AWWA (1969). *Water Treatment Plant Design*.

Bare, W. F. R., Jones, N. B. and Middlebrooka, E. (1975). Algal removal Using Dissolved Air Flotation. *JWPCF*, **47**(1).

Bauer, M. J. Bayley, R. and Chipps, M. J. E. A. et al (1998). Enhanced Rapid Gravity Filtration and Dissolved Air Pretreatment of River Thames Reservoir Water. *Water Science and Technology*, 37(2).

Brejchova, D. and Wiesner, M. R. (1992). Effect of Delaying the Addition of Polymeric Coagulant Aid on Settled Water Turbidity. *Water Science Technology*, 26(9–11).

Boisdon, V., Bourbigot, M. M. and Nogueria, F. et al (1994). Combining of Ozone and Flotation to Remove Algae. *Water Supply*, 12(3/4).

Burman, N. P. (1965). Taste and Odour Due to Stagnation and Local Warming in Long Lengths of Piping. *Proc SWTE*, 14.

Cailleaux, C., Pujol, E. and De Dianous, F. et al (1992). Study of Weighted Flocculation in View of a New Type of Clarifier. *Aqua*, 41(1).

Carroll, B. A. and Hawkes, J. M. (1991). Operational Experience of Converting From Aluminium to Iron Coagulants at a Water Supply Treatment Works. *Water Supply*, 9.

Camp, T. R. (1946). Sedimentation and the Design of Settling Tanks. *Trans ASCE*, III.

Camp, T. R. and Stein, P. C. (1943). Velocity Gradients and Internal Work in Fluid Motion. *J. Boston Soc Civil Eng*, 30.

Chow, V. T (1959). *Open Channel Hydraulics*. McGraw Hill. New York.

Cassie, S. and Seale, L. (2003). *Chemical Storage Tank System—Good Practice*. CIRIA.

Coulson, J. M. and Richardson, J. F. (1999). *Chemical Engineering*, Vol. 1, 6th Edn. Pergamon Press.

Dahlquist, J. (1997). State of DAF Development and Application to Water Treatment in Scandinavia. *Proc International Conference on Dissolved Air Flotation*, CIWEM, London.

Davis, J. M. (1980). Destratification of Reservoirs, A Design Approach. *Water Services*, August.

Degremont (2007). *Water Treatment Handbook*, Vol. 1. 7th Edn.

DoE (1971). *First Annual Report of the Steering Committee on Water Quality*. Circular No.22/72.

DoE (1979). *Second Biennial Report: February 1977–February 1979*.

DoE (1993). *Bankside Storage and Infiltration Systems*. UK Water Research Centre Report.

Edzwald, J. K., Walsh, J. P. and Kaminski, O. O. et al (1992). Flocculation and air requirement for dissolved air flotation. *JAWWA*, 84(3).

Edzwald, J. K. and Wingler, B. J. (1990b). Chemical and Physical Aspects of Dissolved Air Flotation. *J. WSRT-Aqua*, 39(1).

Edzwald, J. K. (2007). Developments of High Rate Dissolved Air Flotation for Drinking Water Treatment. *J WSRT-Aqua*, 56(6–7).

Edzwald, J. K., Malley, J. P. and Yu, C. (1990a). Conceptual Model for Dissolved Air Flotation in Water Treatment. *Water Supply*, 8.

Fair, G. M., Geyer, J. and Okun, D. A. (1968). *Water and Wastewater Engineering*, Vol. 1. John Wiley.

Fowlis, I. A. (1996). On-line Analysis for Trace Levels of Phenolic Compounds. *Proc. Brighton Crop Protection Conference*.

Galvin, R. M. (1992). Lamella Clarification in Floc Blanket Decanters: Case Study. *J. W SRT-Aqua*, 41(1).

Gebbie, P. (2005). Dummy's Guide to Coagulants. *68th Annual Water Industry Engineers and Operators' Conference*, Bendigo.

Gregory, R., Zabel, F. F. and Edzwald, J. K. (1999). Sedimentation and Flotation, *Water Quality and Treatment*. 5th Edn. McGraw-Hill.

Gille, D. (2003). Seawater Intakes for Desalination Plants, *Desalination*, 156(1).

Harman, M. M. I. and Baldwin, I. G. (1998). *Intake Protection Systems*. LR 45-M. WRc.

Hazen, A. (1904). On Sedimentation. *Trans ASCE*, 53.

Henderson-Sellers, B. (1984). *Engineers Limnology*. Pitman Advanced Publishing Program. Pitman.

HRW (2003). *Aquaeration Improved reservoir water destratification*. Hydraulic Research. Wallingford Report EX4808.

Hudson Jr, H. E. (1981). *Water Clarification Processes*. Van Nostrand Reinhold.

Imhoff, K. (1971). *Disposal of Sewage.* Butterworth.

Jiang, J-Q., Graham, N. J. D. and Harward, C. (1996). Coagulation of Upland Coloured Water with Polyferric Sulphate Compared to Conventional Coagulants. *J WSRT-Aqua*, **45**(3).

Kawamura, S (2000). *Integrated Design of Water Treatment Facilities*. John Wiley.

Kay, D. and Hanbury, R. (1993). *Recreational Water Quality Management*, Vol. 2, Ellis Horwood.

Kolarik, L. O. (1994). Magnetic Microparticles in Water Treatment. *Water Supply*, **12**(3/4).

Krishna, J. H. (2005). *The Texas Manual on Rainwater Harvesting*. Texas Water development Board, 3rd Edn.

Letterman, R. D., Amirtharajah, A. and O'Melia, C. R. (1999). *Coagulation and Flocculation, Water Quality and Treatment*. 5th Edn. Mc Graw Hill.

Longhurst, S. J. and Graham, N. J. D. (1987). Dissolved Air Flotation for Potable Water Supply. *Public Health Engineer*, **14**(6).

Lorenzen, M. W. and Fast, A. W. (1977). *A Guide to Aeration/Circulation Techniques for Lake Management.* EPA-600/3-77-004.

Lund, J. W. G., Mackereth, F. J. H. and Mortimer, C. H. (1963). *Phil Trans Roy Soc*, B246.

Markham, L., Porter, M. and Schofield, T. (1997). Algal and Zooplankton Removal by Dissolved Air Flotation at Severn Trent Ltd. *Proc Int Conf on Dissolved Air Flotation*, London.

McLaughlin, R. T. (1959). The Settling Properties of Suspensions. *J ASCE*, HY12.

Meera, V. and Ahammed, M. M. (2006). Water Quality of Rooftop Rainwater Harvesting Systems; a Review, *JWSRT-Aqua*, 55(4), pp. 257–268.

Moss, B. (1988). *Ecology of Freshwaters*. Blackwell Scientific.

Nickols, D. and Crossley, I. A. (1997). The Current State of Dissolved Air Flotation in the USA. *Proc International Conference on Dissolved Air Flotation*. CIWEM, London.

O'Conner, J. T. and Brazos B. J. (1997). Evaluation of Rapid Sand Filters for Control of Micro-Organisms in Drinking Water. *Public Works*, **128**(3), pp. 52–56.

Officer, J., Ostrowski, J. A. and Woollard, P. J. (2001). The Design and Operation of Conventional and Novel Flotation System on a Number of Impounded Water types. Water Science and Technology: *Water Supply*, **1**(1).

Packham, R. F. (1962). The Theory of Coagulation Process. (1) The Stability of Colloids. *J SWTE*, **11**(1).

Packham, R. F. (1963). The Theory of Coagulation Process. (2) Coagulation as a Water Treatment Process. *J SWTE*, **12**(1).

Packham, R. F. (1967). Polyelectrolytes in Water Clarification. *J SWTE*, **16**(2).

Packham, R. F. and Ratnayaka, D. D. (1992). Water Clarification with Aluminium Coagulants in the UK. *Proc International Workshop on Aluminium in Drinking Water*, IWSA, Hong Kong.

Palmer, D. (1993). *River Water Quality Monitoring Equipment Annual Handbook*. CIWEM.

Parr, W., Clarke, S. J. and Gourlay, D. L. (1991). *Method of Controlling Phytoplankton*. UM 1223. WRc.

Pastorok, R. A. (1980). Review of Aeration/Circulation for Lake Management in Restoration of Lakes and Inland Waters. *Proc International Conference*, Portland, Maine.

Peltier, R (2004). Multidisc Passes Screen Test. *Power*, **148**(4).

Pfeifer, B. J., Harris, D. I. and Adkins, M. F. (1997). DAF and Catalytic Filtration for Treatment of a Challenging Water Supply in a Small Northern Community. *Proc AWWA Annual Conference*, Atlanta, Georgia.

Rees, A. J., Rodman, D. J. and T. Zabel. (1979). Water Clarification by Flotation. *TR 114*, WRc.

Ridley, J. E., Cooley, P. and Steel, J. A. P. (1966). Control of Thermal Stratification in Thames Valley Reservoirs. *J SWTE*, **15**.

Schulz, C. R. and Okun, D. A (1992). *Surface Water Treatment for Communities in Developing Countries.* John Wiley.

Skeat, W. O. and Dangerfield, B. J. (Ed.) (1969). *Manual of British Water Engineering Practice*, Vol. III, 4th Edn. W. Heffer & Sons.

Speece, R. E. (1994). Lateral Thinking Solves Stratification Problems. *Water Quality International*, 3.

Steel, J. A. P. (1975). 1975 Management of Thames Valley Reservoirs, in: *Effects of Storage on Water Quality.* WRc.

Stefan, H. G., Dahlin, W., Winterstein, T. et al (1986). Passive Screen Water Intake Design Studies. *J ASCE*, **112**(2).

Sykes, G. and Skinner, F. A. (Eds) (1971). *Microbial Aspects of Pollution.* Academic Press.

Toms, I. P. (1981). Reservoir Management. *Proc Water Industry '81*, CEP Consultants.

UK (2003). The Water Supply (Water Quality) (Amendment) Regulations (Northern Ireland). Statutory Rule 2003 No. 369.

US EPA (1999). *Microbial and Disinfection Byproduct Rules Simultaneous Compliance Guidance Manual*, EPA-815-R-99-015.

Ward, R. C. and Loftis, J. C. (1998). Monitoring Systems for Water Quality. *Water Resources Bulletin*, **19**

WHO (1996). *Guidelines for Drinking Water Quality*, Vol. 2, 2nd Edn.

WHO (1997). *Practical Solutions in Drinking Water Supply and Waste Disposal for Developing Countries*, International Reference Centre for Community Supply.

Yao, K. M. (1973). Design of high rate settlers, *J ASCE Env Eng Div*, 99(EE5).

Young, E. F., Wallingford, F. E. and Smith, A. J. E. (1972). Raw Water Storage. *J SWTE*, **21**(2).

Zabel, T. (1984). Flotation in Water Treatment, in: *Scientific Basis of Flotation*, (Ed. Ives, K. J.). NATO ASI Series, Boston.

Water Filtration Granular Media Filtration

8.1 RAPID FILTRATION—MECHANISM

Clarification is usually followed in a waterworks by solid-liquid separation processes which usually include rapid filtration; the basic principles determining the removal of particles by filtration are discussed below. Usually rapid filtration is preceded by chemical treatment of the water; rapid filtration without chemical treatment is used in 'primary' filtration before slow sand filtration.

In rapid filtration the removal of particles is largely by physical action, although physicochemical processes may also occur. The size of grain of the filter media, usually sand, is normally within the range of 0.4–1.5 mm whereas particles which may be removed by simple filtration, e.g. mineral particles or diatoms, may be at least twenty times smaller. In fact a proportion of particles even several hundred times smaller than the size of the sand grains may be removed. In achieving effective removal of the smaller particles, the addition of a coagulant to form a floc, containing aluminium or iron hydroxides, is usually necessary; but even the floc particles may be very small compared with the size of a grain of the filter media. Therefore filtration is more than a simple straining action such as that of microstrainers (Section 7.4). There may be some straining action due to a coating on the surface of a filter but in general filtration is a process in which some depth of the filter media is utilised. During filtration the flow within the filter bed is laminar; the loss of head through the media is proportional to the velocity of flow of water.

Extensive research has been carried out to determine the mechanisms of removing relatively small particles in filtration. Some of this work has been carried out and summarised by Ives (1967; 1969; 1971). The general conclusion is that the principal mechanisms of filtration are physical and they may be considered under the headings of gravity (or sedimentation), interception, hydro-dynamic diffusion, attraction and repulsion. The contribution of each mechanism depends on the nature of the water and its chemical treatment.

Research on filtration has also considered mechanisms of attraction (or repulsion) between particles and filter grains; in the absence of any attraction some particles would tend to become detached. Such considerations are, however, complex and any clear conclusions are of limited application.

Van der Waals forces are well known as attractive forces between molecules and theoretically they apply to nearly all materials in water, but their range is usually limited to minute distances of less than 0.05 μm.

Finally it should be mentioned that in some water treatment processes physicochemical or chemical reactions occur in contact with filter grains. One example is the deposition of calcium carbonate from waters having a positive calcium carbonate saturation index. Another is the chemical oxidation and deposition of compounds of iron and manganese. After such reactions have commenced the coated grains may provide active surfaces as a catalyst for their continuance (Section 10.11). In some instances the biological oxidation of ammonia, iron or manganese or other biochemical action takes place when passing a water through a rapid filter, due to the development of the necessary bacterial flora in organic impurities on the filter grains (Sections 10.12 and 10.29).

8.2 DESIGN AND CONSTRUCTION OF RAPID GRAVITY FILTERS

The part of a rapid gravity filter that removes the solids from the incoming water is the filtering medium which is usually sand. Rapid gravity filters usually constitute the last solid–liquid separation stage in a treatment cycle for drinking water. The objective with all designs of filters is to reduce the solids content measured as turbidity to less than 0.3 NTU with an upper limit of 1 NTU, these being the usual target values. More recently however, the need to ensure the removal of *Giardia* cysts and *Cryptosporidium* oocysts has led to turbidity targets of less than 0.1 NTU (95% ile value) with a maximum of 0.3 NTU being applied to filtered waters.

To achieve these standards of purity there is some limited but nevertheless important scope for varying the basic design of a rapid gravity sand filter: for instance the sand size can either be (nominally) constant, i.e. monograde sand, or it can vary from fine to coarse i.e. graded sand; the depth of the sand can be shallow or deep; the direction of flow of water is usually downflow but could be upflow as well. A further variation in design is to use three or more layers of sand and pebbles or, alternatively, to use other materials, having differing grain size and specific gravity. When a fine sand is used, the collection of solids during filtration, and hence the build-up of headloss, tends to be in the top layers. In coarser sands the solids penetrate to a greater depth. So long as there is an adequate factor of safety in bed depth against complete dirt penetration it makes good sense to utilise at least some of the bed depth for solids capture provided the backwashing system can be relied on to remove accumulated solids and achieve thorough cleansing of the sand before its next working cycle.

Filter Media

The sand filter designs use either graded sand (fine to coarse or heterogeneous) or coarse monograde sand (uniform size or homogeneous). There is no single media specification (size and depth) that can be applied universally for all waters. The choice depends on the water quality and upstream processes, filtered water quality objectives, cleaning method, filtration rate and length of filter runs. In graded sand filters the bed depth typically comprises 0.7 m of 0.6–1.18 mm fine sand (effective size 0.75 mm), 0.1 m of 1.18–2.8 mm coarse sand, 0.1 m of 2.36–4.75 mm fine gravel and 0.15 m of 6.7–13.2 mm coarse gravel. For applications requiring a finer sand the two upper layers are changed to 0.7 m of 0.5–1.0 mm sand (effective size 0.55 mm) and 0.1 m of 1.0–2.0 mm coarse sand, the gravel layers remaining the same. Effective size = size of aperture through which 10% by weight of sand passes (D_{10}). Depending on the slot size of the nozzles the bottom gravel layer can be omitted and replaced by more of the adjoining media. The homogeneous sand filter has a 0.9–1 m deep bed and typically of 0.85–1.7 mm of sand (effective size 0.9 mm) placed on a 50 mm layer of 4–8 mm or 75 mm of 6.7–13.2 mm gravel. Homogeneous sand of effective size up to 1.3 mm has also been used. The stated size ranges for sand and gravel are generally 5 and 95 percentiles. For estimating the sand depth some employ the rule that the depth of sand should be ≥1000 times its effective size (Kawamura, 2000). Some filter plant designers use the term 'hydraulic size' in place of effective size (Stevenson, 1994). It is defined as the size particles would have to be, if all were the same size, in order to match the surface area of a sample covering a range of sizes. For media with size range 1:2 hydraulic size is approximately 1.36 × the lower size in the range, for example for 0.85–1.7 mm sand it is 1.16 mm.

Other filter media such as anthracite (Section 8.6), granular activated carbon (Section 8.8), garnet, pumice (Farizoglu, 2003), expanded clay or glass are used in filtration application. Garnet is a dense (s.g. 3.8–4.2) medium which is used as the bottom layer of multimedia filters containing anthracite and sand. It occupies about 15% of the bed depth and the effective size could be as low as 0.35 mm. Being dense, it requires about 3 times the wash rate as anthracite to give the same bed expansion. Pumice and expanded clay are porous media and could be used in biological filtration (Sections 10.12 and 10.29). Glass is a suitable filter medium of similar specific gravity to sand.

The sand should be of the quartz grade with a specific gravity in the range 2.6–2.7. The uniformity coefficient (UC) should be less than 1.6 and usually lies between 1.3 and 1.5. Bulk density is about 1.56 g/cc.

$$UC = D_{60}/D_{10}$$

where D_{60} is the size of aperture through which 60% of sand passes and D_{10} is the size of aperture through which 10% of sand passes. Lower UC values would make the medium costly as a high proportion of fine and coarse medium is discarded and higher values would reduce the voidage. Typically sand has a voidage of 37–40%. Voidage = 100 × (particle density—bulk density)/particle density. Loss in weight on ignition at 450°C should be <2% and the loss in weight on acid washing (20% ᵛ/v hydrochloric acid for 24 hours at 20°C) should be <2%. The sand should not be too friable to ensure that washing operations do not produce fines. It should therefore be tested for friability (BW, 1996).

Underdrain Systems

The filter media is placed on a system to collect water from the underside of the bed in an even manner and to spread air and water uniformly through the bed during cleaning. There is a choice of collector system design. One system comprises nozzles set in PVC pipe laterals, the spacing between laterals being infilled with concrete (Fig. 8.1). The design is commonly adapted to apply air and water separately during cleaning and therefore finds application in graded media, dual or triple media or granular activated carbon filters. In another design nozzles are set in a reinforced concrete false floor with a plenum (space below) (Fig. 8.2). The floor is either constructed *in situ* on plastic formwork (Plate 13) or made up of pre-cast concrete slabs supported on concrete sills. The design allows water and air to be applied simultaneously or separately and the system provides for better distribution of air and water than the pipe lateral systems. In separate air-water design the supporting layers underneath are used to prevent sand from penetrating down to the floor as nozzles have slot sizes larger than the sand size. In combined air-water washing the gravel layer limits erosion of the nozzle dome caused otherwise by the localised fluidised sand; the nozzles which

FIGURE 8.1

Pipe lateral filter floor arrangement. (Paterson Candy Ltd.).

FIGURE 8.2

Plenum floor. (Paterson Candy Ltd.).

usually have slot sizes in the range 0.25–0.5 mm also minimise the risk of sand penetration. Nozzle density depends on the type of nozzles and is greater than about 35 nozzles/m^2. There are several other underdrain systems, mostly of proprietary designs, successfully used in many parts of the world. An example is the design by Leopold which comprises underdrain blocks, formed from high-density polyethylene, which clip together to form longer laterals. The blocks incorporate a dual lateral design with a water recovery channel that ensures uniform distribution of air and washwater even over long lateral lengths. The floor can be fitted with a porous plate that helps to eliminate the need for support gravel and is ideal for GAC media. Existing graded media filters with pipe lateral underdrains designed for separate air and water washing can be converted to monograde media filters with combined air-water washing by adopting the Leopold system or by placing separate air laterals, usually at right angles to the water laterals, above the gravel layer with orifices covered in gauze and pointing downwards.

Filter Configuration

The overall number and size of filters vary. The number of filters is selected to minimise the sudden increase in the filtration rate when removing a filter from service for washing as this would dislodge retained deposits and increase the filtered water turbidity in the other filters. Ideally it should be possible to take two filters out of service simultaneously (one draining down and one washing) although it would be adequate to design for one filter out for washing. The more filters the better but a minimum of six filters is desirable; four filters may be used with at least 3 operating in filtration mode provided plant throughput is reduced during maintenance of another filter. Up to five filters could be arranged in one bank and six filters or more even numbers of filters are usually arranged in two banks. The limiting factors for size are the uniform collection of filtered water, even distribution of washwater and air and the travel length of washwater to the collection channel during washing. Usually filter sizes vary from 25 to 100 m^2 with lengths in the range 8–20 m and widths 3–5 m. The washwater collection channel is located on one side along the length of the filter. Filter beds up to twice these sizes can be constructed by providing two identical beds separated by the washwater collection channel in the middle, thus limiting the travel length to 5 m. For some applications, such as combined air-water backwashing of anthracite-sand filters, washwater collecting troughs are placed above the filter bed (Section 8.6).

Filtration Rates

Filtration rates are selected on the basis of the application. Filters with deep bed homogeneous sand for iron removal are rated at 6–7.5 m^3/h.m^2 and for manganese removal at about 15–18 m^3/h.m^2. When used downstream of clarifiers homogeneous sand filters are rated at about 6–12 m^3/h.m^2 with the higher rate being used when water upstream of the clarifiers is treated by a combination of a coagulant and a polyelectrolyte. At filtration rates above 15 m^3/h.m^2 the quality of filtrate tends to deteriorate and at rates in excess of about 20 m^3/h.m^2 the rate of headloss development becomes too rapid. Shallow bed, graded sand filters are usually rated at about 75% of the rates for deep bed, homogeneous sand filters. The rates achieved with multi-media filters are similar to those achieved with deep bed homogeneous sand filters (Section 8.6). Where there is concern over *Cryptosporidium* oocysts and *Giardia* cysts in raw water supplies filtration rates are limited to about 6–7 m^3/h.m^2 to minimise the risk of particulate breakthrough.

Head Loss, Air Binding and Negative Head

In the downflow filter design with upflow washing it is usual for the filter to operate with about 1.5–2 m or more of water depth over the bed. However, there are some proprietary designs which operate with a much smaller depth of water (down to 0.5 m) and even with negative head conditions, but the latter is not regarded as good practice as difficulties can occur with cracking and mud balling in the filter bed and air binding (also called air blinding), especially with high filtration rates. The pressure distribution in a filter bed is illustrated in Figure 8.3. Negative head can occur when 'clogging head' (total head loss less the clean media loss) at any depth exceeds the static head (water depth) at that point. The point of negative head development varies with the filter media; it is nearer the media surface for graded sand, about one-third way down the media in coarse homogeneous sand and just below the anthracite layer in dual media filters. Under negative head conditions dissolved gases in water are released into the space between sand grains restricting the water flow, increasing the head loss and prematurely terminating filter runs (Scardina, 2004). It can also result in 'mud-ball' formation and poor filtrate quality when air binding is restricted to part of the filter and due to channels formed by escaping gases. Negative head is particularly a problem in filters treating cold waters because of greater solubility of gases in water at low temperature, surface waters or well-aerated ground waters. It could also be a problem when filters are preceded by dissolved air flotation clarifiers operating at high rates typically greater than about 12–15 m/h unless the water is deaerated or the design allows for efficient air separation (Section 7.18). Negative head can be overcome by providing sufficient water depth above the top of the media or by washing the filters at a head loss less than the static head down to the point of negative head development.

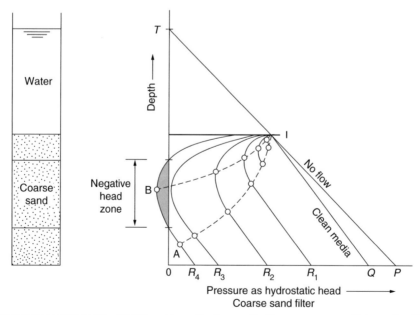

Note: Scales of axes are same so OP = OT. IQ is pressure with clean media, IR$_{1-4}$ are pressures with increasing dirt loss.

FIGURE 8.3

Pressure distribution in a rapid gravity filter.

The head loss in a clean filter is directly proportional to the viscosity of the water hence the temperature and is made up of clean media loss and underdrain loss (nozzles contribute about 0.05 mm). Clean media loss at a filtration rate of 6 m/h for sand of 0.85–1.17 mm, 1.0 m deep and voidage of 40% is 0.178 m at 10 °C and 0.115 m at 25 °C. Therefore when a filter is returned to service after washing the loss of head through the bed, underdrain system and the filter outlet should be less than about 0.35 m. The rate of head loss development is a function of solids retention capacity of the filter and is lower for coarse homogeneous sand filters than for graded sand filters; in the latter it is improved by the use of an anthracite layer (Section 8.6). The head loss allowed for retention of suspended solids (clogging head) with all the filters in service should not be less than 1.5 m on the basis of 24 hour filter run.

Solids Retention

The maximum solids retention capacity of a filter is a function of the voids which is approximately equal to 40–45%. In practice only about a quarter of this space is available for solids removal. It is considered that the solids retention capacity of a gravity filter is limited to 10 and 35 g dry solids/litre of voids for light hydroxide floc and suspended solids in river water, respectively. For example therefore, a filter of 0.9 m sand bed and filtration rate of 7.5 m³/h.m², washing every 24 hours, cannot accept more than 5.6 mg/l of light hydroxide floc or 20.0 mg/l of suspended turbidity in the influent over the run length. The respective equivalent filter loadings are 1012 and 3546 g/m² filtration area. Cleasby (1999) quotes a value less than 1580 g/m² per m of media. Some designers tend to use a more conservative loading of 350 g/m² for sand filters, whilst others use a value of 1000 g/m². Solids loading on to anthracite-sand filters would be higher and is of the order of 700 g/m². When anthracite sand filters are used for direct filtration a solids loading of about 1500 g/m² is used. The objective is to achieve a minimum filter run length of 24 hours. These values should be used for guidance only. Actual values for design purposes should be determined by pilot plant trials.

Flow Control

As the bed becomes clogged the head loss through it increases resulting in the lowering of the filtration rate if head available is constant. For best filter performance it is desirable to have a constant filtration rate and any changes in filtration rate, such as when removing a filter for washing, should be as smooth as possible. Therefore control systems which divide flow equally between filters and allow filtration without fluctuations in rate are essential for good filtration results. Equal flow is achieved by using weirs to proportion the flow equally to all filters or it may be achieved by sizing the outlet pipework and valves of each filter to limit the maximum flow hydraulically. In such cases, after a filter is backwashed, the level of water in the filter rises to such a level above the outlet head on the filter that it is sufficient to overcome the headloss through a clean filter and its underdrain system. As the sand becomes progressively clogged during a filter run, increasing the headloss through the filter, so the water level in the filter box rises to the maximum possible. The simplicity of the rising level filter design is attractive and has many advantages (Cleasby, 1969), but it suffers from floc breakage due to the free fall (about 2 m) of water into the filter and the absence of any flow control on the outlet valve at start-up after a filter wash and so is unable to incorporate 'slow start'. At the start therefore, it is desirable to have a reduced rate of flow through the bed for a while because the full output through the newly cleansed sand can result in a temporary breakthrough of turbidity (Section 8.3).

If an outlet controller is fitted, it modulates the outlet control valve so as to maintain a flow which does not exceed a permitted maximum. As the bed becomes clogged and the headloss through it increases, the controller opens the outlet valve to keep the output constant. There are also mechanical, electronic, electrical, or pneumatic linkages from this controller back to the filter, so arranged as to maintain a substantially constant water level during the course of a filter run. The principal reason for an outlet controller is to maintain a predetermined constant output from each filter under varying head-loss conditions. Equal flow division could be achieved on the inlet using weirs but the resulting free fall can break the fragile floc into small fragments which may not be effectively removed in the filters. With flow division based on outlet flow measurement this is overcome by having submerged inlets with common water level in all filters. A flowmeter in each filter outlet monitors the flow and modulates the outlet control valve to achieve the required flow which is derived by dividing the total flow by the number of filters in service. The constant level is maintained as before except that the level measurement could be made in the inlet channel because all filters have a common level. One of the drawbacks of constant rate filters is that turbidity breakthrough may occur towards the end of the filter run. This is less likely in declining rate filters where the filtration rate decreases as the headloss develops.

Declining rate filters are based on a simple design (Hudson, 1981; Cleasby, 1993a; 1993b). The system is best suited for a group of six or more filters so that the additional flow to be shared when one filter is taken out of service for washing is not excessive. It is reasonable to design on the basis of a maximum flow range through each filter of ±35% of the average filtration rate, the average being taken as the total output divided by the total sand bed area provided. The filter inlet is submerged and, to restrict the filtration rate to the maximum, it is necessary to install some form of restricting orifice or valve on the outlet. Filters are washed in a fixed sequence and individual filter instrumentation for loss of head or quality of filtrate, i.e. turbidity, is used only to detect a filter whose behaviour is out of line with the rest for some reason. In terms of hardware the system is very simple and this is its chief merit. A detailed analysis of rapid gravity filters and their hydraulics is given by Stevenson (1998).

8.3 BACKWASHING

Rapid gravity filters employing graded sand are washed by separate use of air and water through the bed by reverse flow and the used washwater is removed by a washwater collection channel. Combined air-water wash would intermix the media and result in loss of media. The first operation is to allow the filter to drain down until the water lies a few centimetres above the top of the bed. Air is then introduced through the collector system at a rate of about 6.5–7.5 mm/s. The air breaks up the surface scum and dirt is loosened from the surface of the sand grains. This is followed by an upward flow of water at a carefully selected velocity to expand and fluidise the bed. Under this condition the voids between grains of sands are increased and resulting rotation of grains and consequent attrition between grains produces a scouring action to remove attached deposits. The wash rate should be just sufficient to achieve fluidisation velocity (incipient fluidisation) with little bed expansion. Increasing the backwash rate beyond this state would be counter productive because as the distance between sand grains increases, the scouring action is reduced. High backwash rates may result in loss of sand and wastage of water and energy. The washwater collection channel cill is usually placed about 100–150 mm above the sand.

In the UK the practice is to use wash rates to produce 1–3% bed expansion. The rates are viscosity dependent and therefore are affected by water temperature, with higher rates used at warm water temperatures. Typical wash rates to give 2.0% bed expansion are given in Table 8.1.

Table 8.1 Wash rates (mm/s) for sand filters to give 2% bed expansion at varying water temperatures

Sand size range (mm)	Effective size (mm)	Water temperature (°C)					
		5	10	15	20	25	30
0.5–1.0	0.55	3.1	3.5	4	4.5	5	5.5
0.6–1.18	0.75	4.4	5.0	5.6	6.3	6.9	7.5

An empirical relationship has been developed to express bed expansion of graded media in terms of temperature (Tebbutt, 1984) which is the ratio of expansions at T °C and 20 °C, this is 1.57 $e^{-0.02255T}$:1.

Applying this equation to the UK conditions it can be shown that if summer wash rates are used throughout the year, a 40% increase in the degree of expansion is shown in the winter. This would lead to wastage of water and could result in loss of sand. Therefore, in temperate climates, means for seasonal adjustment of wash rates is advisable.

Filters comprising deep bed coarse homogeneous sand rely upon the application of air and water together in the wash phase, followed by a water rinse. In both phases the water rate is well below the fluidisation velocity and does not cause the bed to expand. This prevents hydraulic grading and maintains the homogeneity of the filter bed. The air rate is 16 mm/s of free air. The water rate in the wash phase is usually 2 mm/s and that in the rinse phase is 4 mm/s; although some designers use water rates as high as 15 mm/s in the second phase or a water rate of 4–5 mm/s in both the wash and rinse phases. In the combined air-water wash method, the influence of temperature on wash rate is less pronounced; for every 10 °C rise in water temperature wash rate increases by about 0.4–0.6 mm/s. In such filters the washwater collection channel sill is about 500 mm above the sand and the sill has a large forward chamfer to allow locally suspended sand to drop out as flow approaches the weir crest.

The duration of the wash phases depends on the method of wash and filter influent quality. For designs with air and water applied separately, air scour lasts about 3–4 minutes and the water wash lasts about 4–6 minutes; and for designs with the air and water applied concurrently, air is first introduced and after about 1.5–2 minutes to allow the air flow to become established, water is introduced and the combined air–water wash proceeds for about 6–8 minutes; the air flow is stopped while the water flow continues to rinse the bed for another 8–10 minutes. The total period a filter remains off-line for washing is about 30–45 minutes which includes about 15–30 minutes for draining the filter down depending on the headloss in the filter. The total water consumption per wash amounts to about 2.5 bed volumes.

To remove as much of the dirty backwash water as possible from the top of the filter before it is refilled and put to use, it is usual to allow the filter influent (clarified water) into the filter and to flow across the top of the bed from the side remote from the wash water collection channel as the last stage of backwashing. This is called 'surface-flush' or 'cross-wash' and increases the total washwater consumption to about 3 bed volumes. In one filter design the filter is washed as usual, but the used washwater is contained above the media by allowing the level to rise. At the end of the wash the used washwater is rapidly discharged to waste via flap gates in one side wall. An advantage of the design is that high wash rates may be applied without fear of media loss. Typical free air and water rates used in this design are 14–22 mm/s and 10–18 mm/s, respectively and can be applied either concurrently or separately. The water usage is about 1–2.5% of works throughput.

When the filter is returned to service after backwashing there is a short period, lasting about 15–60 minutes, when the filtrate turbidity is high. This is due to the displacement of residual washwater containing solids loosened from the bed during backwashing and also the lower solids removal efficiency of the freshly washed media. The options available for reducing the risk of this effect and allow time for the filter to 'ripen' are to return the first 15–60 minutes of filtrate to waste (or usually to the works inlet or stored for washing filters), to allow the filter to stand for up to about 30 minutes (so-called 'delayed start'), to start the filter at a slow rate (so-called 'slow start') or a combination of them. All of these features can be operated automatically as part of the wash sequence provided that the filters are appropriately designed.

Water used for backwashing should be filtered and preferably chlorinated particularly if there is no pre-chlorination at the works. In works using aluminium coagulants washwater is best taken upstream of any final pH correction to minimise the risk of dissolving aluminium hydroxide floc retained in the filter. The total amount of washwater used has an important bearing on the economy of a treatment works, especially in relation to the net yield of a source. The total washwater used should normally not exceed 2–2.5% of the treated water output on the basis of 24 hour filter runs.

8.4 OPERATION OF FILTERS

When the maximum permitted headloss has been reached the filter run is terminated. It is usual to design for a minimum filter run length of 24 hours. The length of the run typically varies from about 24–60 hours. Long run lengths help to make savings on backwash water but encourage bacterial growth in the filter bed, in particular if the raw water contains organic matter or is of poor bacteriological quality and is not treated with a disinfectant upstream. It is therefore desirable to restrict maximum run lengths to 48 hours in warm water temperatures and to 60 hours in cold water temperatures. Any appreciable increase in frequency above the design value would reduce output below the design level. The sequence of backwashing including the draindown usually takes about 30–45 minutes so that, for a bank of filters requiring to be washed less than every 24 hours, it could prove necessary for one filter to be draining whilst another is being backwashed. In difficult circumstances, when filter runs are as short as 6–12 hours, it is often the practice to discharge the total contents of the filter to waste instead of draining to supply in order to speed up the wash cycle: this is sometimes referred to as 'dumping' the filter.

The primary indications that a filter requires backwashing are duration of filter run, loss of head, or deterioration of filtrate quality measured by turbidity. In most installations all three parameters are monitored and can be used to initiate the washing cycle automatically. More often than not, only the first two parameters are used for automating the start of a filter washing cycle. It is useful to monitor filtered water turbidity in total and from each filter and initiate an alarm on high value. The individual filter turbidity helps to identify filter breakthrough which may be used to give early warning of *Cryptosporidium* oocysts in the treated water, although particle counters on individual filter outlets would be more appropriate. In automatic filter plants manual washing should always be possible as an alternative so that special cleansing measures can be taken when necessary.

For air scouring, compressors working in conjunction with air storage cylinders or Roots type positive displacement blowers are used. Air is usually applied at about 0.35 bar pressure at the air inlet valve. For backwashing it is usually most economic to use gravity flow from a large elevated storage tank, since the rate of flow required is large and electrical demand charges are minimised by keeping such a tank topped up by a relatively small pump drawing from the filtered water supply or from the treated water supply pumps. The tank should have two compartments with each sized for at least one filter wash. To ensure that the backwash rate does not change too much with the head in the tank, either flow rate controllers are used or the tank is arranged to have a large surface area. Pumps used to introduce water directly into filters are usually of the centrifugal type. The head required at the wash water inlet valve of the filter is about 5 m.

Flowmeters on backwash flow are important and these should have control facilities for setting the optimum rate, changing the rate between wash and rinse phases and between winter and summer periods. Butterfly valves (double fanged) should be used for all filter valves. Penstocks should be considered only for the filter inlet. On large filters or filters designed for automatic washing power assisted actuators are necessary for operating penstocks and valves. Electric power is often used for operating the actuators but pneumatically operated actuators are also successfully employed. Filters are not normally covered except in very cold climates, to prevent freezing. Once filtered, water should not be exposed to contamination and therefore all chambers and channels downstream should be closed. The water from a rapid gravity filter is not completely bacteriologically pure and, before the water passes into supply, it must be disinfected (Chapter 11). A comprehensive analysis of operation and maintenance of filters is given by Logsdon (2002).

8.5 CONSTRUCTION AND OPERATION OF PRESSURE FILTERS

Pressure filters are similar in bed construction to rapid gravity filters, except that they are contained in a steel pressure vessel. Perforated pipes or a steel plate with nozzles are used for collecting the filtered water and for distribution of the washwater and air scour. The steel pressure vessel is cylindrical, arranged horizontally (Fig. 8.4) or vertically. With a pipe lateral underdrain system the bottom of the vessel is usually filled with concrete so as to obtain a flat base. In the plenum floor design a

FIGURE 8.4

Sectional view of a pressure filter (Black & Veatch).

steel plate into which nozzles are screwed, is used. In a horizontal vessel, sometimes vertical plates are welded inside to give a rectangular shaped sand bed within the cylinder so that the bed may be washed evenly and there are no 'dead' areas beneath which air scour and water pipes cannot be placed. More commonly media is placed in the entire vessel such that the depth is equally distributed about the horizontal diameter of the filter. The whole of the cylinder is kept filled with water under pressure and at the highest point an air release valve is inserted for the release of trapped air. To avoid having to employ special transportation procedures for large loads the maximum diameter of filters is limited to 3–4 m and the length/height is limited to about 12 m.

The backwashing of such filters is very similar to that of an open rapid gravity filter. A bellmouth and pipe can be used for the removal of dirty washwater in a vertical filter; for most horizontal filters a single vertical plate located near to one of the dished ends facilitates washwater removal, but for the larger filters, a central washout channel formed by two vertical plates is necessary.

The advantage of pressure filters is that excess raw water pressure is not lost when the filtration process takes place, as is the case with an open rapid gravity filter system. About 3 m head may be lost in friction through the sand bed and the inlet and outlet fittings. This includes an allowance for dirt loss of about 1.5–1.8 m. This combined head loss is between the common inlet and outlet bus mains serving a battery of filters. Pressure filters may be interposed on a pumped or gavity pipeline without a large loss of pressure on the supply. Air binding, hence negative head, should not occur in pressure filters if the pressure in the media is always above that at the points upstream where air could have got into solution.

Pressure filters suffer from the disadvantage that the state of the bed under backwashing conditions and when the plant is working cannot be directly observed. It is of vital importance, therefore, that every pressure filter is fitted with an open box or dish in the front of it, into which the washwater is turned so that at least any washing out of the sand may be observed and the backwash rate immediately reduced.

When coagulation or other chemical treatment is required chemicals are injected and mixed under pressure and flocculation must be hydraulically carried out in pressure vessels fitted with baffles. The same applies when a contact tank is needed on the downstream side for disinfection.

Pressure filter installations are often not provided with any form of individual flow control. The result is that each operates as a declining rate filter for at least part of its filtration cycle and flow is concentrated through the clean, newly washed filters. An orifice plate may be installed in the outlet to restrict flow. However, a better option is to modulate the outlet valve using signals from a flowmeter on each filter so that flow equals the total flow divided by the number of filters in service. As the head loss through the filter increases the outlet valve opens to maintain the average flow. When employing flow controllers head loss can be monitored on individual filters; the filter run can be terminated either on headloss, length of filter run or turbidity breakthrough.

The equipment required for air scouring pressure filters and for valve control is similar to that used for rapid gravity filters. The filtered water usually has enough pressure for backwashing and this is often used so avoiding the need for backwash tanks of pumps. For a filter washed by separate air and water the required rate of application of water is about four times as great as the rate of filtration which is about 5–6 m/h. Thus in a large filter battery of about 15 filters, groups of five filters can be taken out of service at a time and the combined filtrate from four of the filters can be used to wash the fifth filter, and so on, until all filters in the group are washed. In large pressure filter plants arrangements are usually made to wash filters in groups at a specific time each day. Monitoring of individual filters for loss of head is seldom done except when outlet flow control is used. In fact, because a large battery of filters is usually supplied by a common inlet bus main and the outlets from filters also connect to a common outlet bus main, it is only meaningful to measure the headloss across the battery of filters. Individual filtrates could usefully be monitored for turbidity.

Pressure filters come under special procedures for pressure vessel design and fabrication. Condensation on the outside of the tanks is a continual nuisance as it causes corrosion of the steel shell and staining of the floor below. The pressure applied to steel pressure filters is not usually in excess of 80 m head of water. This should normally be adequate for most distribution systems. Above this pressure the thickness of plates used for the steel shell may be so great that cost rises rapidly. Due to the limitations on throughput and pre-treatment by clarification their application in preference to rapid gravity filters is restricted to small plants which are required to be installed into a system without breaking the hydraulic gradient, for iron and manganese removal primarily in ground waters, treating stored water by direct filtration or as GAC adsorbers for organics removal.

MULTILAYER AND OTHER METHODS OF FILTRATION

8.6 USE OF ANTHRACITE MEDIA

The most efficient form of media grading for a rapid gravity filter in order to obtain maximum capture of solids would be to have the sand decreasing in size in the direction of flow. Clearly this is not possible because hydraulic regrading takes place during backwash, so that the finer sand collects at the surface of the bed. This can be countered by using separate layers of different filter materials having different density and grain size, the denser materials being at the bottom of the bed and the less dense at the top. One type of multilayer filter bed in wide use is the two-layer filter using anthracite over sand. It has been found that the filtrate quality from anthracite-sand can be as good as that from conventional sand only filtration, but because of in-depth clogging of anthracite and therefore its capacity to retain solids throughout its entire depth and the less rapid development of headloss, filter runs can be 1.5–3 times longer. A layer of anthracite is sometimes incorporated in a filter to extend filter runs. They are usually operated at filtration rates similar to those used on coarse homogeneous sand filters although high filtration rates, even up to 15 m^3/h.m^2, can be achieved but, at the cost of short filter runs. The specific gravity of anthracite (1.4–1.45) is lower than that of sand (2.6–2.7), bulk density is about 0.73 g/cc and voidage is about 50%. The lighter anthracite is placed on denser sand and typical size ratio for anthracite:sand is 2:1 to 4:1. The anthracite used is usually of size 1.18–2.36 mm (effective size 1.5 mm) or 1.4–2.8 mm (effective size 1.7 mm) and uniformity coefficient <1.5. The sand is of size 0.5–1.0 mm (effective size 0.55 mm) but sometimes of 0.6–1.18 mm (effective size 0.75 mm). The depth of the anthracite bed is of the order of 0.15–0.3 m with sufficient sand to give a combined depth of about 0.75 m. The filters are washed by separate application of air and water. Backwashing with concurrent application of air and water allows deeper anthracite layers (up to 0.6 m) to be used to give more capacity but with a corresponding increase in the depth of the sand layer (up to 0.6 m). The sand should be supported on gravel (Section 8.2) depending on the type of underdrain system used. For the pipe lateral system graded sand and gravel should be used and for plenum floor design a shallow gravel layer (50 mm of 4–8 mm or 75 mm of 6.7–13.2 mm) is normally used. Sources of anthracite (carbon content at least 90%) are limited to a few countries in the world. Therefore there is a tendency to use high grade bituminous coal (carbon content at least 80%) in place of anthracite. This is generally acceptable provided it is non-friable and can meet the standards for friability and acid solubility (Section 8.2) and hardness (BW, 1996; AWWA, 2002; ASTM, 2005).

Construction of a plenum floor for a rapid gravity filter: (a) and (b) installation of pre-formed polystyrene formwork; (c) steel reinforcement, nozzle and cap; (d) placing concrete; (e) finished floor with caps removed and slotted domes; (f) satisfactory combined air/water backwash distribution.

PLATE 13

(a) Ultrafiltration membrane cartridge containing more than 15 000 fibres with total surface area about 80 m², treatment capacity 6–8 m³/mh. Inset cross section of hollow fibre membrane, internal dia. 0.8 mm. (Manufacturer: Koch Membrane Systems Inc., Wilmington, MA, USA).

(b) Ultrafiltration membrane system for treatment of lime softened water, capacity 90 000 m³/day (Plant location: Mid West USA, System manufacturer: Koch Membrane Systems Inc., Wilmington, MA, USA).

PLATE 14

Table 8.2 Wash rates (mm/s) for anthracite–sand filters at varying water temperatures

Bed expansion %	Water temperature (°C)					
	5	10	15	20	25	30
10	4.5	5.3	6.2	7.2	8.0	8.9
15	6.5	7.5	8.3	9.2	10.0	10.7
30	11.4	12.1	12.8	13.5	14.3	15.3

Beds of anthracite can be expanded or fluidised at about the same backwash rate as a sand bed when the size of the anthracite grains is about 1.5–3 times that of the sand grains. Backwash rates are used to give 10–15% expansion with an occasional wash at higher rate to give about 30% bed expansion for regrading the media. The rates are viscosity dependent and therefore affected by the water temperature. Wash rates required for a bed of anthracite (effective size 1.3 mm) and sand (effective size 0.55 or 0.75 mm) are given in Table 8.2.

Filters are cleaned by air scour followed by water backwash. The air scour rate applied is about 8–12 mm/s. The water rate is as for graded sand filters with temperature compensation. Concurrent application of air and water requires high level suspended washwater collection troughs in place of the conventional low level collection channel as otherwise lighter anthracite would be carried to waste. The sequence consists of simultaneous application of air at 16 mm/s and water at about 2 mm/s as for coarse homogeneous sand filters; air is turned off before the water level reaches the cill of the trough weir and the wash rate is then increased to about 16 mm/s for rinsing and regrading the media. The troughs are usually placed with invert level about 600 mm above the unexpanded media surface; this permits about five minutes of concurrent wash.

After rinsing at fluidisation velocities, the two-layer bed settles down again with regrading of the two layers by density and grain size. In practice the media are mixed at the interface (100–150 mm) and evaluation by Cleasby (1975) on intermixing suggests that the mixed interface is beneficial as it reduces the voidage of the anthracite layer and produces better filtrate quality but at the expense of higher headloss due to greater retention of suspended solids. To prevent intermixing a coarse sand would be required but this would defeat the objective of the dual media filter where a fine size sand must be used.

In some designs anthracite is used as the sole filter media. In one plant in the Middle East the bed depth used is about 2 m, size range 1.4–2.5 mm (effective size 1.5). Filters are washed by air at 20 mm/s followed by water (6.5–16.7 mm/s) to give a bed expansion of 5–10%.

8.7 USE OF ANTHRACITE TO UPRATE FILTERS

In recent years the output of many existing filter plants has been increased by changing their media from sand to anthracite–sand, increasing the previously conventional filtration rates of 4–6 $m^3/h.m^2$ for graded sand filters to rates of 6–12 $m^3/h.m^2$. However, these increases have frequently been associated with the improvement of floc characteristics by the addition of a polyelectrolyte as a coagulant aid. Use of polyelectrolyte may also be necessary to prevent penetration (Crowley, 1979) through the anthracite and sand by small green algae, dominated by minute species *Nannochloris* and *Ankistrodesmus* with cell diameters of 4–8 microns.

It is important to pay attention to floc size when adopting anthracite filters: too fine a floc may pass through the anthracite layer and cause too large a load to reach the sand below; too large a floc in relation to the anthracite size may place too large a load on to the anthracite, defeating the object of gaining filtration in depth by using anthracite and sand. Since anthracite is an expensive material compared to sand it is important to ensure a filter is adequately designed hydraulically before anthracite is used for uprating. This means checking that the inlet, outlet and flow control pipes and valves can accept the higher flow rates; that the distribution of water at the inlet to the filter does not cause excessive scouring of the anthracite-sand bed because of the higher flow rate; that the higher backwash rate required does not result in loss of anthracite over the washwater weir; that the washwater discharge channel is of sufficient size and gradient to accept the increased backwash without backing up; and that the filter underdrain system can accept higher upthrusts.

8.8 GRANULAR ACTIVATED CARBON ADSORBERS

Granular activated carbon (GAC) adsorbers employed for organics removal (Section 10.36) are similar to sand filters of the rapid gravity or pressure type except that gravity type adsorbers are usually enclosed in a building or fully covered. GAC is a good filter medium and therefore could be used on its own for filtration of turbidity but, for such filtration, GAC would be subjected to aggressive and frequent washing and, being more friable than sand, would be susceptible to breakdown. Consequently, losses as fines would be much greater than when used for adsorption only.

GAC should be reactivated as and when it is exhausted with respect to organic compounds. Typically the iodine number should not be allowed to fall below about 60% of the initial value (Table 8.3) as reactivation recovers only about 300 points. Filtration rates vary from about 6–7.5 $m^3/h.m^2$ for filtration and up to 15 $m^3/h.m^2$ for adsorption. Media depth is a function of the empty bed contact time which could vary between 5–30 minutes; for filtration only duties the media depth is about 1–1.2 m while depths up to 2.5 m (for gravity filters) and 3 m (for pressure filters) are used for adsorption. Effective size of GAC varies with the type and the application and usually lies in the range 0.6–1.1 mm. GAC should be tested for friability (BW, 1996). Water soluble ash should be less than 1% $^w/w$. For GAC activated by phosphoric acid the phosphate content should not exceed 1% $^w/w$.

GAC is placed in the adsorber on 50 mm of gravel or directly on appropriate filter nozzles. Since the adsorbers are washed by the sequential application of air and water most types of underdrain systems would be suitable. GAC adsorbers are normally washed using water at a rate of about 5–12 mm/s, depending on the effective size, the base material of the GAC (i.e. coal, wood, peat or coconut) and the water temperature, to give 20–30% expansion of the carbon bed. Wash rates for different grades of a coal-based GAC at varying water temperatures to give 20 and 30% bed expansion are given in Table 8.3. The use and the frequency of air scour depend on the water source and applied at the rate of 14 mm/s and 0.35 bar. Frequent air scour tends to breakdown GAC.

The frequency of backwashing of GAC adsorbers at ground water sites could vary from once every 2–8 weeks depending on the raw water quality; wash is by water only. At surface water sites, in order to maintain low bacterial counts in the filtered water, the backwashing should include air scour and the wash frequency should be 2–3 days. This can also help to control the growth of micro-animals (zooplanktons such as nematodes, chironomid midge larvae) because the wash frequency

Table 8.3 Wash rates (mm/s) for GAC to give 20% and 30% bed expansions at varying water temperature

Grade	Iodine no.[a]	BET surface area (m²/g)[b]	GAC size range (mm)	Effective size range (mm)	Water temperature (°C)					
					5	10	15	20	25	30
F200	850	900	0.425–1.70	0.6–0.7	4.5 (5.6)	5.0 (6.4)	5.8 (7.2)	6.3 (7.8)	6.7 (8.3)	6.9 (8.6)
F300	950	1000	0.600–2.36	0.8–1.0	6.4 (8.1)	6.9 (8.9)	7.8 (10.0)	8.5 (10.3)	9.0 (11.4)	9.3 (11.5)
F400	1050	1100	0.425–1.70	0.6–0.7	3.9 (5.0)	4.7 (6.0)	5.3 (6.7)	5.8 (7.2)	6.4 (8.1)	6.5 (8.2)
TL830	1050	1050	0.85–2.00	0.9–1.1	6.9 (8.9)	7.8 (9.7)	8.6 (10.3)	8.9 (10.8)	9.5 (11.7)	9.7 (11.8)

Notes:
Values for 30% bed expansion shown in brackets.
[a] Iodine number: It indicates a GAC's ability to adsorb organic compounds and be regenerated. It should be greater than 500 mg/g of carbon (AWWA, 2006)
[b] BET surface area: It indicates the surface area available for adsorbates in water. Measured by N_2-BET method (Brunauer, 1938)
Source of information: Chemviron Carbon Ltd, UK

is shorter than their reproductive cycle. The problem of micro-animals can also be overcome by chlorinating the backwash water or taking a filter out of service for a period sufficient to produce anaerobic conditions in the filter, to kill the micro-animals (Weeks, 2003). This should be carefully controlled not to loose biological activity in the filter and to prevent the formation of ammonia and nitrite in the filter. Alternatively micro-animals in the filtered water could be removed by the use of microstrainers (which would also remove any carbon fines). A combination of ultrasound and sand filtration was found to be successful in a demonstration plant (Matsumoto, 2002).

When reactivation of the carbon is required it is usually removed from the adsorber by means of a water operated eductor (5 parts of water to 1 part of GAC) or by recessed impeller centrifugal pumps (3 parts of water to 1 part of GAC) of speed less than 1000 rpm. In some designs adsorbers are provided with a sloping floor or recessed drain in the floor discharging to a collector system. All pipework, in particular bends, should be in stainless steel. Straight lengths could be in ABS or PVC-U. Bend radii should be 5–10 pipe diameters. The pipeline velocities should be maintained within 1.5–2.0 m/s. The same equipment and design parameters should be used for carbon placement in adsorbers.

Virgin GAC contains contaminants such as aluminium (0.65%), iron (0.35%), copper (0.0025%) and traces of manganese and arsenic and reactivated GAC contain chemicals adsorbed in the process and not completely removed in the reactivation. Materials that could leach into the filtrate when placed in adsorbers include sulphides, sulphites and bisulphites (causing chlorine demand and odours), alkali (resulting in high pH), phosphates (if phosphoric acid is used in the activation process) and metals such as aluminium, iron, manganese and copper (Lambert, 2002). Repeated backwashing with water followed by running to waste of the filtrate should be carried out until tests confirm that water is of acceptable quality for supply. The impact of reactivated GAC on water quality can be minimised by pre- and post-acid wash in the reactivation process. Pre-acid wash is useful for manganese, aluminium hydroxide and calcium carbonate and post acid wash is useful for manganese.

8.9 UPWARD FLOW FILTRATION

Upward flow filtration with upflow washing has been used for a few potable water treatment plants in the UK, but its use is more appropriate to industrial water applications, as roughing filters ahead of slow sand filters or to tertiary sewage filtration where a high standard of filtrate quality is not so important. The principle used in upflow filters is to have progressively finer sand in the direction of flow, which allows the filter to carry a greater load of impurity before backwashing because the larger particles tend to be held in the lower, coarser part of the filter, leaving the upper layers to deal with the smaller particles. However, unless the finer grades of sand are restrained they would be washed away at higher rates of filtration (as well as during backwashing); to prevent this a filtrate collector pipe system is buried in the top layer of fine sand, with strainers located on the side of the pipes so that filtrate water flow has to change from a vertical to a horizontal direction, thus preventing expansion of the sand. During backwashing the filtrate collector is not used and dirty washwater escapes from an elevated trough. Another method is the use of a grid, square in section, located about 0.1 m below the surface of the sand. During filtration the sand arches between individual members of the grid and prevents expansion of the sand, whilst during backwashing the arches are intentionally broken by successive applications of air and backwash water. Piped lateral floors with large orifice nozzles are used to distribute the incoming water. Screening of the raw water to remove leaves and other debris is essential to prevent blockages.

8.10 DIRECT FILTRATION

Some surface waters can be treated by coagulation, flocculation and rapid filtration (gravity or pressure), eliminating clarification. Such waters need to be carefully selected and pilot tested. In general a water source is considered to be suitable for direct filtration when average turbidity and colour values are less than 10 NTU and 25° Hazen, respectively, with peaks of 40 NTU and 40° Hazen for periods less than 24 hours. The total organic carbon value should be less than about 2 mg/l as it influences the coagulant requirement. The coagulant dose should be no more than 1 mg/l as Al or 1.5 mg/l as Fe, although higher doses for short periods are acceptable. A polyelectrolyte may be used as a coagulant aid. Algae, both the filter clogging and passing types can cause problems such as shortened filter runs, if numbers are high (Chan Kin Man, 1991); an upper limit of 2000 asu/ml for diatoms is reported (Hutchinson, 1974) (asu/ml = areal standard units/ml; 1 asu is 20×20 µm and for a medium-sized algae 1 asu/ml can be approximated to 0.1 µg/l of chlorophyll 'a'). The total flocculated solids load on to the filters should be limited to about 20–25 mg/l, with short-term peaks up to about 60 mg/l. Direct filtration operates well on micro-flocs and requires untapered high energy flocculation (Section 7.12). The filters used are either monograde sand (ES 0.9 mm) deep bed (0.9–1.0 m) or anthracite-sand containing 0.3 to 0.4 m anthracite (ES 1.3 mm) and 0.6 m graded sand (ES 0.55 mm) (Sections 8.2 and 8.6). Filtration rates should be maintained below about 7.5 $m^3/h.m^2$.

8.11 FILTER PROBLEMS

Filter problems are due to incorrect design (filtration and wash rates, media grading, flow control method, etc.) for the water to be treated, poor hydraulic design, use of unsuitable material, bad installation in particular of the underdrain system and mal-operation. These result in short filter runs, inefficient filter cleaning (dirty filters), very high starting head loss (up to 1 m), media loss, loss in capacity and shortfalls in filtered water quality and sometimes even ruptured filter underdrain systems. Installation problems include damaged or blocked or poorly fitted nozzles, incorrect levelling of the floor or pipe laterals and nozzles (outside their tolerance limits), construction debris left in the pipe laterals or the plenum, poor sealing of the floor slabs or pipe/duct joints, incorrect grout being used and differential settlement. These flaws can be checked before placing the media by a hydraulic pressure test with all the nozzles plugged and by testing for uniform air distribution with about 150 mm of water in the filter sufficient to cover the nozzles. The observation of air scour pattern during backwash is a way of identifying underdrain problems in operating filters.

Blocked nozzles (usually the result of construction debris left in the underdrains), manganese or calcium carbonate deposits, or lime or sand particles in washwater can result in high pressures in the underdrain system leading to its rupture. The risk of damage to the underdrain system can be minimised by incorporating a standpipe with free discharge in the washwater main; a pressure relief valve is not recommended. Damaged or poorly installed nozzles are a common problem with operating filters and allow sand into the underdrain system reducing its capacity; during washing they can cause channelling, sand boils, high localised velocities with sand ingress into support layers and gravel brought to the surface by jetting and loss of media. Sand boils and consequent upset of gravel layers can also be the result of sudden introduction of backwash water. Some channelling and sand leakage into the underdrain system could be attributed to incorrect sizing of sand and gravel media or use of nozzles with a slot size incompatible with the media size.

Deficient washing results in the build up of floc, iron oxide deposits and organic matter (algal and detrital) ultimately leading to the formation of mud balls and jetting (build up of columns of support gravel through the media), cracks in the filter bed and bed shrinkage with media pulling away from the walls. Sometimes polyelectrolyte when dosed in filters or carried over in the clarified water due to overuse can encourage mud ball formation. A good analysis of filter problems is given by Hudson (1981), Lombard (1994) and Baylis (1971). The problem of air binding is discussed in Section 8.2. It is important that filter washes are frequently witnessed to identify problems early to allow corrective measures to be taken. Observations to be made include: uniformity of air distribution, undisturbed areas, sand boils, mud balls and media carryover. The wash efficiency of a filter can be checked by taking core samples from several places in the bed after backwashing and analysing 250 ml of media from different depths in each sample for suspended solids by washing it thoroughly with water (up to 250 ml) and measuring the solids by volume using an Imhoff cone. For a good wash, suspended solids should be less than 2% $^v/v$. An alternative criteria defined by Bauer (1997) based on TWUL experience is that the sand after backwashing should contain particulate organic carbon less than 0.4 g and suspended solids less than 2.4 g per litre of filter medium.

SLOW SAND FILTRATION

8.12 INTRODUCTION AND HISTORY

Slow sand filters were the first effective method devised for the purification in bulk of surface waters contaminated by pathogenic bacteria. They remain equally effective today and there are a growing number of circumstances which suggest that their use in new works should be encouraged.

'Slow' sand filters are so called because the rate of filtration through them may be only one-twentieth or less of the rate of filtration through rapid gravity or pressure filters. They were first constructed in the UK in the early nineteenth century. Their capacity for purification is well illustrated by the history of the cholera outbreak in Hamburg in 1892. Both Hamburg and Altona took water from the river Elbe which became contaminated with cholera, Hamburg suffered 8600 deaths from cholera but Altona which had slow sand filters had no cases of cholera. This vividly illustrates the ability of slow sand filters to purify a water bacteriologically, as well as physically. Most of London's surface derived supplies are treated by slow sand filters although a number of less efficient works have been closed down, with filters in those remaining being uprated, with improvements to pretreatment and filtration rate; the current total filter area is about 42 ha. Slow sand filters have also been adopted for supplies to cities such as Amsterdam, Antwerp, Belfast, Paris, Stockholm and Budapest. A summary of slow sand filter plants operated by TWUL is given in Table 8.4.

Slow sand filters are an efficient method of producing water of good bacteriological, physical and organic quality which requires marginal chlorination before distribution. They achieve 2–4 log removal of coliforms, *E. coli*, pathogenic organisms, cercariae of *Schistosoma* (Pike, 1987), ova, cysts such as *Giardia* and oocysts such as *Cryptosporidium* and remove viruses (Dullemont, 2006), oxidise ammonia and biodegradable natural organic matter (<80%) and reduce turbidity to less than 1 NTU.

Table 8.4 Slow sand filter plants operated by TWUL, UK

Works	Capacity (Ml/day)	Pre-treatment	Filtration area (m²)
Ashford Common	690	Pre-ozone, anthracite—sand gravity filters, main ozone	32 × 3121
Coppermills	680	Sand gravity filters, main ozone	34 × 3400
Hampton	790	Sand gravity filters, main ozone	112 900 (25 filters)
Kempton	200	Sand gravity filters, main ozone	12 × 3640
Walton	140	Pre-ozone, coagulation (iron salt) COCO-DAFF (with anthracite—sand gravity filters), main ozone, GAC filters	10 × 3320
Fobney	75	Pre-ozone, sand upflow filters, main ozone	12 × 900

Notes:
All slow sand filters (except Walton) have a GAC sandwich of 100–150 mm in a total bed depth of 800–900 mm. In addition there is a gravel layer of 100 mm.
Filter underdrain systems are porous concrete floor type.
Filtration rate varies in the range 0.3 m/h (average) to 0.5 m/h (maximum).
For COCO-DAFF see Section 7.18.
Post-treatment is chloramination.
All works except Fobney serve London.
Source of information: TWUL, UK.

8.13 MODE OF ACTION OF SLOW SAND FILTERS

The slow sand filter acts by a combination of a complex straining process and microbiological action, the latter being the more important. The purification of the water takes place not only at the surface of the bed but also for some distance below. Van de Vloed (1955) has given clear account of the purification process. He distinguishes three zones of purification in the bed: the surface coating (the *schmutzdecke*); the 'autotrophic' zone, existing a few millimetres below the *schmutzdecke*; and the 'heterotrophic' zone, which extends some 300 mm into the bed.

When a new filter is put into commission and raw water is passed through it, during the first two weeks the upper layers of sand grains become coated with a sticky reddish brown deposit of partly decomposed organic matter together with iron, manganese, aluminium and silica. This coating tends to absorb organic matter existing in a colloidal state. After two or three weeks there exists in the uppermost layer of the sand a film of algae, bacteria and protozoa, to which are added finely divided suspended material, plankton and other organic matter deposited by the raw water. This skin is called the *schmutzdecke* and it acts as an extremely fine-meshed straining mat.

A few millimetres below this *schmutzdecke* is the autotrophic zone, where the growing plant life breaks down organic matter, decomposes plankton and uses up available nitrogen, phosphates and carbon dioxide, providing oxygen in their place. The filtrate thus becomes oxidised at this stage.

Below this a still more important action takes place in the heterotrophic zone which extends some 300 mm into the bed. Here the bacteria multiply to very large numbers so that the breakdown of organic matter is completed, resulting in the presence of only simple inorganic substances and unobjectionable salts. The bacteria act not only to break down organic matter but also to destroy each other and so tend to maintain a balance of life native to the filter so that the resulting filtrate is uniform.

The biological processes require oxygen and if it is absent anaerobic conditions would set in, resulting in the formation of hydrogen sulphide, ammonia, soluble iron and manganese and taste and odour producing substances. Therefore, to ensure satisfactory operation the feed water must contain sufficient oxygen determined by the filtrate oxygen which should not be allowed to fall below 3 mg/l (Huisman, 1974). The efficiency of the process is also temperature dependent. For example the reduction in permanganate value, a measurement of the organic content (Section 6.37) decreases by $(T + 11)/9$ where T is the water temperature in °C (Visscher, 1987). Below 6 °C ammonia oxidation ceases. At low water temperatures the rate of biological reactions and activity of bacteria consuming micro-organisms reduces rapidly and the rate of reduction in *E. coli* falls sharply, thus requiring the chlorine dose to be increased. If water temperatures less than 2 °C persist for prolonged periods, consideration should be given to covering the filters.

8.14 CONSTRUCTION AND CLEANING OF SLOW SAND FILTERS

The bed of sand in a slow sand filter is 0.6–1.25 m thick and is laid over a supporting bed of fine gravel, beneath which a filtrate collector pipe system is constructed. The water passes downwards and the whole arrangement being sited in a shallow watertight tank of large size. In some plants a porous concrete floor is used in place of pipes. It is important to note that the bed is 'drowned'. Filtration rates used are typically 0.1–0.3 m/h. Figure 8.5 shows the construction of a typical slow sand filter using the original Coppermills Works, TWUL as an example; the sand bed of 0.675 m thick, lies on 75 mm of fine gravel, which, in turn, rests on a bed of porous concrete. Each filter is about 34 m × 90 m and the filtration rate was in excess of 0.3 m/h. These filters have since been upgraded (Table 8.4). Below the concrete, collector drains feed the filtered water to the main effluent pipes. The sand is ungraded; because the filters are not backwashed, hydraulic grading of the media does not occur. Size distribution in the bed is purely random. The sand has a size range 0.21–2.36 mm (effective size of 0.30 mm) and uniformity coefficient 1.5–3.5 (less than 2 is preferable). The stated size range is generally 5 and 90 percentiles. The sand should be hard and should not be liable to break down when subjected to skimming and washing processes. The suspended solids and particulate organic content of the sand should not exceed 0.5 g and 0.1 g per litre of media, respectively.

The raw water is led gently on to the filter bed and percolates downwards. Directly after a bed has been cleaned a head of only 50–75 mm of water is required to maintain the design rate of flow through the bed. However, as suspended matter in the raw water is deposited on to the surface of the bed, the *schmutzdecke* builds up on the surface and increases the loss through the bed. To maintain the flow at a uniform rate as far as possible the headloss across the bed is gradually increased and when it reaches some predetermined value between 0.6 and 0.9 m the bed must be taken out of service and cleaned. The maximum permissible head loss should be kept to about 1 m. If fine grain sand is used then the depth should be reduced to minimise the resistance. To prevent negative head (Section 8.2) the maximum permissible headloss must be kept less than the depth of water on top of the sand. Normally the water depth above the sand bed is maintained at about 1–1.5 m with a

Source of information: TWUL.

FIGURE 8.5

Slow sand filter.

maximum of 2 m. As a further preventative measure the outlet weir (if there is one) should be set above the media surface level.

When a sand bed requires cleaning it is drained of water and the top 12–25 mm of the sand surface are carefully scraped off. The filter is then returned to service by gradually increasing the flow over 24 hours, sometimes longer. When the sand bed again requires cleaning a further 12–25 mm of sand are scraped from the surface, and this process is repeated until the bed is thinned to the minimum practical thickness for efficient filtering which should not be less than 0.5 m (Visscher, 1987). When this stage is reached the bed is then topped up with clean new sand to its original level or the old sand may be replaced if it has been adequately washed and cleaned in a sand cleaning machine.

The interval between scraping may vary from several months during the winter when pre-filtration is installed, to 10 days where no pre-filtration occurs and algal growth is at a maximum. Resanding would only be necessary every 2–5 years depending on the scraping frequency. Originally, slow sand filters were invariably scraped by manual labour (typically for scraping: 5 person-hours/100 m² and resanding: 50 person-hours/100 m²) (ASCE, 1991) but the increased cost and decreased availability of this type of labour led to the introduction of mechanical methods, in particular for large plants thereby reducing labour typically to about 2–4 person-hours/100 m² for scraping and 5–8 person-hours/100 m² for resanding. Resanding by wet slurry method is reported to reduce labour requirement further (Kors, 1996). Some of the mechanical methods are fairly simple, e.g. using small tracked skimming machines discharging to dump trucks. One of the difficulties of mechanising the cleaning of the older slow sand filters is that their sizes and geometry were often not uniform since these factors were unimportant in the days of manual cleaning. However, with new installations, provided that the filters are constructed with the correct dimensions, it is possible to span them with a standard sand lifting bridge which runs on tracks along the sides of the filters, thus greatly reducing the labour necessary for scraping (Lewin, 1961). Other cleaning methods used are wet harrowing and flushing (Collins, 1988; Joslin, 1997) which does not disturb the attached bacteria and the use of a suction dredge using laser depth control (Glendinning, 1996) which avoids the need to drain down the filter. Sand scrapings must be washed to remove dirt before being replaced in the filter. Washing must be carried out soon after removal. In small plants manual washing by high pressure water jet is used. In larger plants mechanised systems consisting of screens and hydrocyclones are used. The standards aimed for cleaned sand are: silt concentration in the range 0.5–1.0 g/l of sand and particulate organic concentration 0.1 g/l of sand or 0.1% by weight of sand (Toms, 1988; Harrison, 1997).

Each filter needs five valves: inlet, outlet, back-filling, waste and drain. The valved inlet discharges into the filter over a weir arranged to give a velocity of about 0.1 m/s for good distribution of water and not to damage the *schmutzdecke*. The weir box or the inlet pipe is provided with a valved drain to remove the top water when the filter needs cleaning. The back-filling valve is used to refill the filter (after scraping) by reverse flow to a water depth of about 250 mm to remove air entrained in the voids. The waste valve is used to discharge the filtrate for a period until the filter is 'ripened' after it is brought on line following skimming or re-sanding; filtered water is discharged to waste or returned to other filters in use until chemical and bacteriological tests prove it to be satisfactory. In the operation of filters it is important to minimise fluctuations in inflow and raw water quality. The plant should be designed to operate at constant rate with a constant level of water above the sand surface, or rising water level in the filter. With the latter the shallow water depth allows penetration of sunlight, encouraging algae growth and there is always the danger of disturbing the *schmutzdecke* by the incoming water. With the low filtration rate, manual control of filter inlet and outlet is more than adequate and cheaper but automatic control is useful for large plants. Water level in the filter can be kept constant by a level controlled inlet valve and a constant filtration rate is maintained by a flow controlled outlet valve.

8.15 USE OF PRE-TREATMENT WITH SLOW SAND FILTERS

Slow sand filters may operate as the sole form of filtration for raw water turbidities of up to about 10 NTU. The object of pre-treatment such as storage or pre-filtration is to lighten the load of suspended matter on the slow sand filters and so permit longer intervals between cleanings or a faster rate of filtration, or both. Van Dijk (1978) recommended that, if the average raw water turbidity exceeds 50 NTU for more than a few weeks or is in excess of 100 NTU for a day or so, then pre-treatment is indispensable. Pre-treatment can take the form of river bed filtration, Ranney wells, storage, plain settlement, filtration (horizontal, pebble matrix, rapid gravity or upflow) (Galvis, 1993) or combination of these depending on the turbidity of the raw water. If the filtrate has persistently low oxygen content (<3 mg/l) then aeration of the raw water would be necessary. Pre-treatment sometimes includes chemical coagulation and flocculation without clarification or with clarification (Welte, 1996) and filtration (Abrahamsson, 2006) or DAF followed by rapid gravity filtration (Walton-on Thames, London). Residual coagulant metal appears not to be detrimental to the biological activity (Dorea, 2006). If raw water *E. coli* counts are in excess of about 10 000/100 ml, the filtered water should be subjected to disinfection as marginal chlorination would be inadequate. A summary of the effects of pretreatment is given by Ridley (1967), who compared three London supply works: Hanworth Road with slow filtration only; Kempton Park with rapid gravity filters preceding slow sand filters; and Ashford Common with microstraining preceding slow sand filters. Figures for

Table 8.5 Performance of slow sand filters for London water supply (1998–99)

Station and type of pre-treatment currently adopted	Rate of filtration (m/h)	Quantity of water filtered per hectare of sand bed cleaned (Ml/ha.day)[a]
Hanworth Road[b]—no pre-treatment	(0.004–0.059)[c]	(3.15–7.05)[c]
Kempton Park; rapid gravity pre-filtration and ozone	0.3 (0.132–0.152)[c]	33–49.5 (12.6–16)[c]
Ashford Common; pre-ozone, rapid gravity pre-filtration and main ozone	0.3 (0.132–0.137)[c]	41.4–69 (10–15.7)[c]

[a] 1 Ml = 10^3 m³. 1 ha = 10^4 m². So 1 Ml/ha = 0.1 m³/m².
[b] Hanworth Road installation is now closed.
[c] Values in brackets show the information for the summer seasons in 1959–1963; since then ozone treatment and GAC sandwich have been introduced.
Source of information: TWUL, UK.

the period 1 April–30 September (when algal growths are usually most prominent) for the 5 years 1959–63 inclusive and the corresponding data for 1998-99 following upgrading of the pretreatment on operating criteria are shown in Table 8.5.

During the period 1956–63 Kempton Park rapid gravity filters used 1.15–1.62% washwater (as a percentage of the water filtered) while Ashford Common microstrainers used 1.64–2.03%. The extra washwater used and the capital and running costs of pre-filtration and ozonation are usually taken as economically justified in view of the very considerable increase in water filtered through the slow sand filters between cleanings.

8.16 LIMITATIONS AND ADVANTAGES OF SLOW SAND FILTERS

Slow sand filters tended to be ignored when new plants were being considered because of the amount of land and labour they use and their relatively large capital cost. However, there is still a place in water treatment for slow sand filters, providing their advantages and limitations are carefully weighed in each particular case. They may also play a greater role in future water treatment technologies owing to pressures for less chemical usage in treatment.

Slow sand filters do not significantly reduce the 'true colour' (Section 6.20) of a water. They are thus only suitable for treating waters of relatively low colour. Another factor is that slow sand filters are not effective in removing iron and manganese in solution and, being a biological process, there is a marked reduction in efficiency in the removal of contaminants at low temperatures (Section 8.15).

A high concentration of algae in the raw water can cause treatment difficulties due to clogging of the filters or anaerobiosis and tastes if the cells die off in the filter, due to excessive respiratory demands. The chlorophyll-a concentration of the feed should ideally be limited to 5 μg/l (ASCE, 1991; Bauer, 1995) with a peak of 15 μg/l. Thus slow sand filters are not the treatment of choice for highly eutrophic waters in which sudden peaks of algae may appear in spring, summer or autumn. Methods to prevent or control algae include pre-treatment, covering of slow sand filters and chemical treatment. Low to moderate algal concentrations in particular filamentous species on the other hand are beneficial to the process. Small algae such as diatoms increase the resistance of *schmutzdecke*.

Slow sand filters are also not very suitable for the removal of any substantial amount of finely divided inorganic suspended matter. With pre-filtration using primary filters, however, they may function successfully for years treating waters which are intermittently laden with fine silt.

Removal of organics by slow sand filters is not complete and that of pesticides is type depen-dent (Lambert, 1995). Therefore on some waters they are supplemented by additional treatment usually ozone and granular activated carbon (GAC) adsorbers used either upstream or downstream of slow sand filters. At TWUL's Ashford Common Works a 0.135 m GAC sandwich layer is incor-porated within the slow sand filters; ozonation is carried out upstream of roughing and slow sand filters(Glendinning, 1996). At Ivry and Orly plants near Paris, ozone and GAC are used downstream (Welte, 1996).

Apart from the above, the other stated limitations of slow sand filters often appear of doubtful validity on closer inspection and four examples are given to illustrate this. Firstly, it is stated that slow sand filters occupy a greater area of land than do coagulation and rapid gravity plants. This is true so long as one ignores the question of sludge disposal and chemical dosing, but if adequate sludge disposal and chemical facilities are included the land required for slow sand filters may well be no greater than that for a conventional treatment. Secondly, it is maintained that slow sand filters are expensive in capital costs. However, when the cost of chemicals is taken into account, slow sand filters may be cheaper in whole life costs than conventional plants. Thirdly, whereas it was usual to design slow sand filters for low rates shown in Table 8.5 for 1959–63, it has been found that with improvements to pre-treatment (e.g. pre-ozonation, GAC, redesigned primary fil-ters, intermediate ozone) without covering (Glendinning, 1996) or with covering (Wilson, 1996), filtration rates have been almost doubled to about 0.3 m/h. Abrahamsson (2006) reported that covering increased filter run time from 6 months to 4 years at Stockholm Works. Other techniques such as the use of replaceable non-woven synthetic fabric layer (geotextiles) on top of sand has helped to reduce the cleaning frequency (Hendricks, 1991) and a six-fold reduction in cleaning frequency is reported by Klein (1994). Fourthly, slow sand filters are said to be labour intensive, but where designed specifically for the latest methods of mechanical cleaning, they may be no more labour intensive than conventional treatment plants, taking into account the work involved in sludge disposal.

In some European countries they are used as a final polishing stage of treatment, being pre-ceded by as many as five or six stages of treatment including chemical coagulation, sedimentation (Schalekamp, 1979), ozone, GAC, softening, etc; e.g. River-Lake plant, Loenderveen and Wesperk-arspel, Netherlands (Kors, 1996).

8.17 MEMBRANE FILTRATION

Membrane filtration employs a semipermeable membrane to separate materials according to their physical and chemical properties when a pressure differential or electrical potential difference (electrodialysis—see Section 10.46) is applied. These processes can be broadly classified according to the membrane pore size and size of particles removed (Fig. 8.6). The types of membranes are microfiltration (MF), ultrafiltration (UF), nanofiltration (NF) and reverse osmosis (RO). RO and NF

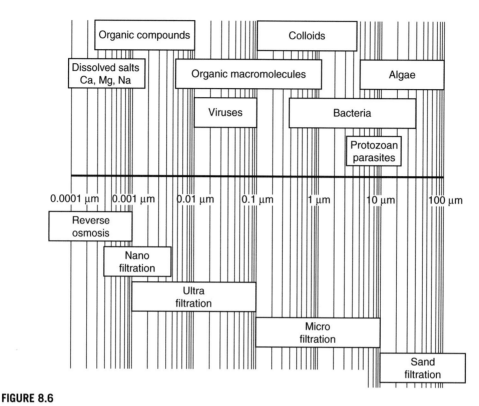

FIGURE 8.6

Comparison of treatment methods with particle size.

are used to remove materials dissolved in water (Section 10.47). MF and UF remove particles from water. MF and UF are sometimes called low pressure membrane filtration, because typical MF/UF transmembrane pressures (TMP) are 1 to 2.5 bar compared to NF operating pressures of 6 to 14 bar, 7 to 20 bar for low salinity RO, and 55 to 80 bar for seawater desalination with RO.

Filtration with MF/UF removes particles based on size exclusion applying a sieving mechanism. Unlike more traditional granular media filtration, MF/UF does not require physiochemical conditioning, such as addition of a coagulant prior to filtration to enable removal of small particles. MF membranes used for water treatment have pore sizes of 0.1 to 0.5 μm and UF of 0.01 to 0.1 μm. To define their ability to remove large organic molecules, which can be dissolved in water, some UF membranes are also rated based on molecular weight cut-off (MWCO) and nominal values of 100 000 to 200 000 daltons are typical. (Dalton is a unit of measurement defined to be 1/12 of the mass of one atom of carbon). MF provides essentially no removal of dissolved materials and UF removes very little of the dissolved materials of concern for potable water. For example, neither type removes dissolved manganese, true colour, or disinfection by-products (DBPs) such as trihalomethanes. While UF removes some large organic molecules, the precursors of DBPs generally have much lower molecular weights than the MWCO, so these solutes pass through both MF and UF. To augment removal of dissolved contaminants, MF/UF is integrated with pretreatment processes that convert the dissolved parameter into a particulate form, such as coagulant or powered activated carbon addition to adsorb dissolved organics, sources of taste-and-odour and DBPs, or use oxidants to precipitate manganese and iron.

MF/UF consistently removes small particles, thereby providing effective and well documented control of turbidity with typical filtrate values of less than 0.1 NTU. Speth (2005) summarized a literature review of 122 studies that showed that MF and UF membranes produce extremely high-quality water regardless of influent turbidity and that there is no apparent difference in turbidity removal between membrane type, manufacturer, or whether a coagulant was used. Mean filtrate values had a median of 0.06 NTU and the maximum values had a median of 0.08 NTU. Studies by Jacangelo (1991 and 1995), Coffey (1993) and others have shown that MF and UF also consistently remove *Giardia* cysts and *Cryptosporidium* oocysts exhibiting removals greater than 4 log and ranging up to 6 to 7 log. Log removal expresses the filtration removal effectiveness for a target organism, particle or surrogate and is calculated as log [feed concentration] − log [filtrate concentration]; therefore, a log removal value (LRV) of 4 = 99.99% removal and a LRV of 2.5 = 99.68%. Cysts are not detected in the filtrate of intact MF/UF membrane systems; therefore, the reported values are limited by the sensitivity of the measurements. While there is no observed difference between MF and UF regarding removal of protozoan cysts (which are 2 to 15 μm in size), UF provides greater removal of viruses (which are 1 to 2 orders of magnitude smaller). One regulatory agency that bases MF/UF ratings on carefully controlled microbial challenge tests, California Department of Health Services, routinely grants 4 log removal credit for *Giardia* and *Cryptosporidium* to accepted MF/UF membranes. For MF membranes, the virus removal credit is typically 0.5 log, while for UF it ranges from 2 to 4 log depending on performance at the specific conditions. MF with 0.2 μm pores should not reject virus-sized particles but limited virus removal is observed. It has been speculated that the mechanism for this may be rejection of larger particles on which viruses have become attached, or adsorption of viruses on the cake layer. Even given the high microbial log removal values of MF/UF, facilities for chemical disinfection of the filtrate (e.g., chlorination) are almost always incorporated into membrane filtration plants to protect the public health with a multiple barrier approach.

The use of low pressure membrane filtration has increased rapidly over the past decade—partly due to concern over microbial contaminants and the ability of MF/UF membranes to consistently provide low turbidity filtrate and high log removal values independent of chemical pre-conditioning.

There are significant differences between the various types of MF/UF systems. The most basic hydraulic difference is that some employ membranes mounted inside pressure vessels (called pressurised or encased systems) (Plate 14(b)) while others apply membranes mounted in a tank open to atmospheric pressure (called submerged, immersed or vacuum systems) (Plate 15(b)). With encased systems a positive pressure is applied to the upstream side of the membrane to transport the filtered water through the membrane while for submerged systems a negative pressure is applied to the downstream side to draw filtrate through the membrane. Either type can utilize a site's hydraulic gradient, so selection and equipment arrangement may be influenced by topography. For example, siphon-driven and mixed siphon-pump submerged UF membrane systems were selected to fit the topography and hydraulic profile and to reduce costs at the 273 000 m³/d extension of Chestnut Avenue Waterworks (Freeman, 2004) and the retrofit of membranes into existing granular media filters at the 184 000 m³/d Choa Chu Kang Waterworks (Ratnayaka, 2008).

The membrane geometry most commonly used in MF and UF is hollow fine fibre with lumen (e.g., internal bore) diameters typically ranging from 0.4 to 1 mm. (Plate 14(a)). Most systems use polymeric membranes, but ceramic materials have been developed and are sometimes employed allowing a wider range of temperature, pH and oxidant resistance, which can be beneficial during cleaning (Fujiura, 2006). Considering materials, the main polymers are polyvinylidene difluoride,

Table 8.6 Characteristics of selected membrane materials (AWWA, 2005)

Material	Type	Oxidant tolerance	pH range
Polyvinylidene difluoride (PVDF)	MF/UF	Very high	2–11
Polyacrylonitrile (PAN), rarely used	MF/UF	High	2–11
Polyethersulfone (PES)	UF	High	2–13
Polysulfone (PS)	UF	Moderate	2–13
Polypropylene (PP), rarely used	MF	Low	2–13
Cellulosic, rarely used	UF	Moderate	5–8

For example, PVDF generally has tolerance to up to 5000 mg/l chlorine residual short term and 20 mg/l continuously, chlorine dioxide to 5 mg/l, $KMnO_4$ to 5 mg/l, and even limited tolerance to ozone if special construction methods are used, while PES can generally tolerate up to 100 mg/l of chlorine.

polyethersulfone or polysulfone; but some older systems use polypropylene or cellulosic materials and one manufacturer offers a polyacrylonitrile membrane.

For most membranes the flow pattern is outside-in with the raw water applied on the outside of the fibres and the filtrate flowing inside; however, some encased systems use inside-out pattern. Another variable is flow mode. MF and UF processes are operated either in cross-flow or dead-end mode. In the former only part of the inflow is filtered through the membrane; the remainder flows tangential to the membrane surface carrying with it the particulates removed on the surface and is recirculated back to the feed water supply. If a cross-flow option is compared to dead-end, it is important to include energy usage in the cost comparison. Some older cross-flow systems exhibited high recirculation rates, but up to 10% is currently more typical. Overall, the following parameters can vary between types of MF/UF systems: hydraulic approach (encased vs. submerged), chemistry/materials of construction, chemical tolerance (i.e., tolerance to oxidants, disinfectant and pH) (Table 8.6), physical tolerance (i.e., temperature and pressure), flow pattern (outside-in vs. inside-out), flow type (dead-end vs. cross-flow), pore size, packing density, backwash type, pumping and air requirements, energy usage, arrangements (horizontal vs. vertical and single or multiple modules per vessel).

There are features which are similar for all types of MF/UF as currently practiced. These systems have five basic operating cycles: filtration; backwash, which is sometimes augmented with an air scour; chemically enhanced backwash, chemical clean-in-place (CIP); and integrity verification. Approximately 85% of the time the unit is in filtration mode, producing filtrate at a set flow or flux (filtrate flow/membrane area, generally expressed in l/m^2 h). Periodically, usually several times an hour, the particle build-up on the surface is removed by backwashing, generally with filtered water or, for some systems, the liquid backwash is assisted with an air scour or, for one type, backwashed with filtered air. Less frequently, the membrane is also chemically cleaned with a chemically enhanced backwash (CEBW) or a clean in place (CIP) operation. CEBW is a special backwash operation conducted once or twice a week with chemicals such as sodium hypochlorite, caustic soda or acid, added to the supply. CIP uses more concentrated chemical solutions to clean the membrane and are typically conducted monthly. Terminal transmembrane pressure, the condition at which cleaning must be conducted, varies in the range from 0.7 to 3 bar. Feed water recovery varies between 90 and 98%, depending on the operating mode and on raw water quality, including an allowance for backwash water which can be about 2–10% of product flow.

The most critical design parameters are determining capacity, sustainable flux and level of redundancy. Flow through MF/UF is more temperature sensitive than through granular media. Because the viscosity of water increases at lower temperature, the limiting condition typically occurs in winter. Designers need to evaluate the relationship between seasonal temperature, demand and MF/UF capacity to provide sufficient capacity year-around. Membrane capacity or flux is proportional to $1.03^{(T-20)}$ with 20 °C being the reference temperature.

The second most important design task is selecting a flux that is sustainable long term. To keep the initial costs low, the designer is tempted to select as high a flux as practical; however, if the flux is too high the facility would experience high operating costs, fouling, which leads to insufficient capacity, and stress, which leads to excessive fibre breakage. Ideally, long term piloting would be used to determine the optimum stable flux; however, it is not always feasible for a pilot program to capture the complete range of raw water quality and temperature variations and the cumulative effect of multiple filtration and CIP cycles. Therefore, full-scale results on similar applications are also considered. While it is difficult to generalize, in many cases having an ambient temperature, maximum instantaneous flux of 40–50 l/m² h is stable. For especially challenging applications, such as filtering secondary effluent for water reclamation or treating a coagulated water without an intermediate clarification process, 38–42 l/m² h is generally sustainable, while for treating low-fouling waters a flux of 60–80 l/m² h might be suitable. If a pilot is conducted, it should include any seasonal water quality and temperature variations that would be anticipated for the full-scale plant, standard full-scale modules to avoid misleading results, as well as full representation of all pretreatment and recycled streams.

The level of redundancy should also be carefully considered in the planning and design of facilities. The evaluation would consider the frequency and duration of peak-day events as well as provisions for emergencies, availability of alternative water sources, and storage in the distribution system. Operators of MF and UF systems report a significant major difference between membrane filtration and granular media. With membranes, low turbidity filtered water is easy to achieve, independent of pre-conditioning; however, it can be impossible to force more filtrate through a membrane, even to address a short-term need. Sometimes one or two spare trains are included in addition to sufficient membrane area to provide the full capacity. To avoid queuing for routine operating cycles such as backwashing, larger facilities may need to consider redundancy within sections of the plant.

One of the benefits of MF/UF membrane technology is the integrity verification test. Generally practiced once each day, this simple automated test verifies there are no flaws greater than a set resolution or size, typically 3 µm to coincide with the low end of the size range of *Giardia* cysts and *Cryptosporidium* oocysts. While there is some variation, the general description of the test is as follows. About 1 bar of air is applied to one side of the membrane, generally the filtrate side to prevent drying fouling material on the working surface. The membrane itself and the down-stream side are wet. After pressurizing the system with air and isolating it with valves, the air pressure or flow rate is monitored for 10–15 minutes. If the membrane is intact, then all of the pores should be smaller than 3 µm, in which case air leakage would be low. With intact membrane, the surface tension of the water holds the water in the pores and the low rate of leakage will be due to diffusion of air into the water. However, if the membrane is not intact, the leakage rate is higher. If the automated test indicates a leak, operators can find and repair the leak with a series of similar manual tests. It should be noted that turbidity measurement is not sensitive enough to indicate flaws in full-scale MF/UF systems (Jacangelo, 1991). It is generally not practical to demonstrate 4-log removal with particle counting, since the feed water does not exhibit 4-log concentration of target sized particles. That is why the air-based membrane integrity test, which is sufficiently sensitive, is widely practiced.

Both MF and UF systems can be applied directly on raw water with only a pre-filtration stage to reduce particle size to a range 150–500 μm depending on specific requirements. Membrane processes can be preceded by pretreatment to augment removal of dissolved materials either to meet water treatment goals (such as for colour, manganese, hardness or formation of DBPs) or to protect the membrane from fouling. Pretreatment options cover the full range of water treatment technologies, including coagulation, flocculation, sedimentation, dissolved air flotation (DAF), ballasted clarification, lime softening or activated carbon.

A typical plant consists of raw water pumps feeding a bank or several banks of membrane modules, recirculation pumps for cross-flow mode operation with capacity to give recirculation flow to raw water flow ratio in the range 3:1–6:1 depending on the raw water quality and backwash pumps. A module consists of several membrane elements in a pressure shell complete with feed inlet ports, distributors, outlets and permeate removal points. Several modules may be connected in series. In a bank several modules may be connected in parallel. Membrane banks are backwashed sequentially. Backwash pressure varies in the range 1–2.5 bar. All pumps used on a membrane plant are usually of the centrifugal type. Both MF and UF systems can be staged with used backwash water and the bleed from the recirculation water treated in a second stage. A flow schematic diagram of a typical membrane plant is shown in Plate 15(a).

The operating and maintenance (O&M) cost for a membrane plant is primarily made up of energy, membrane replacement and labour. Adham (2005) surveyed 87 full-scale MF/UF plants with capacities greater than 4000 m³/d; these reported median O&M costs of $39.8 per 1000 m³. These operating costs comprise the following categories: labour (32% of the total), energy (30%), parts (10%), chemicals (9%), chemical disposal (2%) and other (8%).

8.18 MISCELLANEOUS FILTERS

One type of filter comprises a vertical hopper bottomed cylindrical vessel filled with sand, raw water is introduced into the cylindrical section just above the hopper and flows upward through the sand bed counter-current to the sand moving continually downwards and leaves the filter at the top over an outlet weir (e.g. 'Dynasand'). The sand containing dirt is conveyed from the hopper section by an air-lift pump; the turbulent action of the pump cleans the sand with additional hydraulic or mechanical cleaning in a sand washer at the top. The cleaned sand is returned to the top of the filter. Filtration rates can vary in the range 8–15 m³/h.m². Such filters are used as primary filters ahead of slow sand filters or in direct filtration mode with coagulation. In the latter case filtrate turbidity can be less than 1 NTU. Water loss through sand cleaning is in the range 5–8% of the inflow.

Filtration of water can be achieved with diatomaceous earth filters or pre-coat filters, where the filtering medium is formed on septums or 'candles' of metal or other materials inside a pressure vessel. Excellent removal of suspended matter may be achieved. A good account of the process is given by Cleasby (1999).

Cartridge filters with elements of the woven fibre type rated at 5–10 microns are used in reverse osmosis plants preceding high pressure feed pumps. These filters remove particulate matter passing the pretreatment stage and are used as safety filters to protect high pressure feed pumps and the membrane. Once used, they are normally discarded, although backwashable elements are now available.

There are so called 'depth filters' where the filtration efficiency increases with passage as a 'dynamic filtering' medium is deposited on a filtration surface, either from particles in the water being treated or by addition of a suitable fine material such as diatomaceous earth, Fuller's earth (a fine non-cohesive clay) or powdered activated carbon. The filtration surface can be fibres wound on to a cassette or a candle with grooves, fibres arranged in a bundle around a central core or a porous cloth woven to produce a series of filtration tubes. The fibrous filters are capable of filtering particles down to about 5 μm, and achieving *Cryptosporidium* oocysts removal up to 1-log consistently (O'Neill, 1995). The porous cloth type is known to be capable of removing particles greater than 1 μm and therefore provide 2–3 log removal of oocysts.

CRYPTOSPORIDIUM OOCYSTS AND *GIARDIA* CYSTS REMOVAL

8.19 *CRYPTOSPORIDIUM*

Removal of *Cryptosporidium* oocysts can be achieved by any process that removes particles down to a size of 4 μm or smaller. However, conventional water treatment processes in current use for drinking water supply cannot guarantee complete removal. Since the effect of disinfectants on oocysts is poor (Sections 6.63 and 11.6) the removal of oocysts in the upstream solid–liquid separation processes must therefore be maximized. Well operated and maintained conventional treatment processes, which include coagulation, flocculation and clarification followed by granular media filtration, can be expected to achieve 2 to 3-log removal consistently (Standen, 1997). Criteria for good clarification performance would be turbidity less than 1 NTU and total coagulant metal ion concentration less than 1 mg/l as Fe or 0.5 mg/l as Al depending on the coagulant used (UKWIR, 1998). Filters should be operated at about 6 $m^3/h.m^2$ with all filters in service. Filtrate turbidity should be less than 0.1 NTU (Logsdon,1998). Filtered water turbidities <0.1 NTU and particle counts <50/ml are indicators of good treatment for controlling *Cryptosporidium* (Edzwald, 1998). Second stage filtration such as GAC adsorbers or manganese removal filters may provide a further barrier with removal <0.5 log but cannot be relied on for effective removal because they are not used for residual floc removal.

Slow sand filters with an active *schmutzdecke* and at filtration rates up to 0.3 $m^3/h.m^2$ are expected to achieve up to 5-log removal. US EPA (2006) gives 3-log credits for conventional treatment (coagulation, clarification and filtration) or slow sand filtration (with no pre-chlorination) and 2.5-log credits for direct filtration or slow sand filtration (as secondary filtration). Additional treatment is not required for oocysts concentration < 0.075/1. For systems supplying more than 10 000 consumers and water containing 0.075/1 ≤ oocysts <1.0/1, 1.0/1≤ oocysts < 3.0/1 or oocysts ≥ 3.0/1, (based on maximum value for 12-month running annual average or 2-year arithmetic mean if twice-monthly monitoring is conducted), plants with conventional treatment (coagulation, clarification and filtration) or slow sand filtration (with no pre-chlorination) must be provided with additional treatment sufficient to give 1-log, 2-log and 2.5-log removal respectively. However, the equivalent removal rates for additional treatment for plants with direct filtration or slow sand filtration (as secondary filtration) are 0.5-log greater. The additional treatment could be bankside filtration, cartridge filtration, membrane filtration or disinfection by chlorine dioxide, ozone or UV. Systems serving less than 10 000 consumers are not required to be monitored and do not require additional treatment.

Microfiltration and ultrafiltration can achieve consistent removal rates better than 2-log and 4-log respectively (Section 8.17); in some studies higher removals (>6-log) (Jacangelo, 1995) have been reported. *Cryptosporidium* oocysts are resistant to even high concentrations of chlorine (Section 11.6) but ozone, when used in doses typically applied at water treatment works, is reported to produce 1-log to 2-log inactivation of oocysts (Badenoch, 1990). For ozone the '*ct*' value required to achieve 2-log inactivation is about 5 mg.min/l at 7 °C and 1.7 mg.min/l at 22 °C and that to achieve 3-log inactivation is about 8 mg.min/l at 7 °C and 2.5 mg.min/l at 22 °C (Finch, 1993). Ozone if in use for other applications (e.g. oxidation, disinfection, etc.) would be beneficial for oocyst inactivation, although the installation of ozone solely for oocyst inactivation is not recommended until its effectiveness is confirmed in further studies and full scale plants. The process is used in the USA as a treatment method for oocysts (Section 11.21). Chlorine dioxide is less effective than ozone; for 3-log credits Ct values are 19 and 536 mg.min/l respectively at 15 °C. UV disinfection is widely accepted for inactivating *Cryptosporidium* (Section 11.24).

It is important to note that irrespective of the various degrees of removal achieved in different processes, the critical factor is the number of oocysts remaining in the water, which is a function of the number in the raw water and how it relates to the infectious dose.

In the UK, three reports were produced by experts (Badenoch, 1990 and 1994; Boucher, 1998) and based on the recommendation the following design and operational guide lines are suggested to minimize the risk of *Cryptosporidium* passing into water supplies:

1. The design and operation of rapid gravity filters should ensure sudden surges of flow are avoided; flow changes should be limited to 1.5–5% per minute.
2. Rapid gravity filters should not be restarted after shutdown without backwashing.
3. After cleaning, slow sand filters should not be brought back into use without an adequate 'ripening' period.
4. Bypassing part of the water treatment process should be avoided.
5. Turbidity of individual filters should be monitored to detect early turbidity breakthrough. Sudden increases in turbidity should be investigated as that could infer dislodging of oocysts.
6. Recycling supernatant water from clarifier sludge and filter backwash water treatment facilities to the treatment works inlet should not be practised unless no more than 5% return of oocysts to the works inlet can be assured. This may require membrane filtration or ozone dosing. The recycle flow should be between 5–10% of raw water flow.
7. 'Slow start' alone is inadequate. Filtrate at the start of the filter run (first filtrate to waste, typically one bed volume) should be diverted to waste, recycled to the works inlet or stored for backwash water. This could be supplemented with a 'delayed start' (standing the filter for about 15–60 minutes). Figure 8.7 illustrates the effects of these three operations.
8. Particle counters are more representative of oocyst breakthrough than turbidity monitors.

Logsdon (1998) suggested for plants treating water from low or medium risk sources the target quality for the supernatant return from the settling process being a turbidity of <5 NTU, suspended solids of <10 mg/l and total coagulant metal ion concentration of <5 mg/l as Al or Fe.

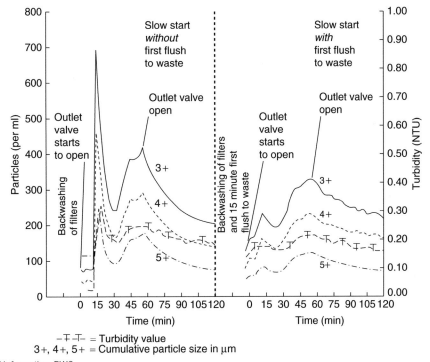

Source of information: PWC

FIGURE 8.7

Particle count and turbidity on filter start up—effect of first filtrate to waste.

8.20 *GIARDIA* CYSTS

Giardia cysts being larger organisms (8–12 µm) than *Cryptosporidium* oocysts (4–6 µm) are easier to remove by solids–liquid separation processes than *Cryptosporidium*. In most cases *Giardia* removal is about 1-log to 2-log better than the corresponding value for *Cryptosporidium* removal. Because of their larger size they can be better correlated to turbidity than *Cryptosporidium* oocysts.

 Giardia cysts are less resistant to disinfection than *Cryptosporidium* oocysts, though chlorine is not very effective, a '*ct*' value of 100 mg.min/l (Section 11.5) being required for 2-log inactivation at pH 7 and 5 °C, improving to about 10 mg.min/l at 25 °C (Clarke, 1989). Using ozone, the corresponding values for 2-log inactivation at pH 7 are 0.53 mg/l.min at 5 °C, and 0.17 mg.min/l at 25 °C (Wickremanayaka, 1984). UV disinfection is widely accepted for inactivating and *Giardia* (Section 11.24).

REFERENCES

Abrahamsson, J. and Dromberg, P. (2006). Covering of Slow Sand Filtration; Qualitative and Operational Aspects, *Recent Progress in Slow Sand and Alternative Biofiltration Processes*, Ed. Gimbel, R., Graham, N. J. D. and Collins, M. R. IWA Publishing.

Adham, S., Chiu, K., and Gramith, K. et al. (2005). *Development of a Microfiltration and Ultrafiltration Knowledge Base*. AwwaRF.

ASCE (1991). *Slow Sand Filtration*. Logsdon G. S. (Ed.). New York.

ASTM (2005). *Standard Test Method for Ball Pan Hardness*. D3802.

AWWA (2002). *Standard for Granular Filtering Material*. AWWA B100–01.

AWWA (2005). *Manual of Practice M53: Microfiltration and Ultrafiltration Membranes for Drinking Water*, 1st Edn. AWWA.

AWWA (2006). *Granular Activated Carbon*. AWWA B604-05.

Badenoch, J. (1990). *Cryptosporidium* in Water Suppliers. *Report of the Group of Experts*. London.

Badenoch, J. (1994). *Cryptosporidium* in Water Supplies. *Second Report of the Group of Experts*. London.

Bauer, M., Buchanan, B. and Colbourne, J. et al. (1995). The GAC/Slow Sand Filter Sandwich: From Concept to Commissioning. *Water supply*, **13**(3/4).

Bauer, M. J., Bayley, R. and Chipps, M. J. et al. (1997). Enhanced Rapid Gravity Filtration and Dissolved Air Pretreatment of River Thames Reservoir Water. *Proc IAWQ-IWSA Joint Specialist Conference on Reservoir Management and Water Supply*, Prague.

Baylis, J. R., Gullans, O. and Hudson Jr., H. E. (1971). Chapter 7a: Filtration, *Water Quality and Treatment*, 3rd Edn. McGraw Hill.

Boucher, I. (1998). *Cryptosporidium* in Water Supplies. *Third Report of the Group of Experts*. London.

Brunauer, S., Emmet, P. H. and Teller, E. (1938). Adsorption of Gases in Multimolecular Layers. *J Am Chem Soc*, **60.**

BW. (1996). *The Specification, Approval and Testing of Granular Filtering Material and Code of Practice for the Installation of Filtering Material*. BW:P.18.96R.

Chan Kin Man, D. and Sinclair, J. (1991). Commissioning and operation of Yau Kom Tau Water treatment works (Hong Kong) using direct filtration. *JIWEM*, 5(2).

Clark, R. (1989). Analysis of Inactivation of *Giardia Lamblia* by Chlorine. *J ASCE, Environ Eng Div*, **115**(1).

Cleasby, J. L. (1969). Filter Rate Control without Controllers. *J AWWA*, **61**(4).

Cleasby, J. L. (1975). Effect of Media Intermixing on Dual Media Filtration. *J Environ. Eng. ASCE*, 101(EE4).

Cleasby, J. L. (1993a). Status of Declining Rate Filtration Design. *Water Science Technology*, **27**(10).

Cleasby, J. L. (1993b). Declining Rate Filtration *Water Science Technology*, **27**(7–8).

Cleasby, J. L. and Logsdon, G. S. (1999). Granular Bed and Precoat Filtration, Chapter 8, *Water Quality and Treatment*. Ed. Letterman, R. D. McGraw-Hill.

Coffey, B. M., Stewart, M. H. and Wattier, K. L. (1993). Evaluation of Microfiltration for Metropolitan's Small Domestic Water Systems, *Proc. AWWA Membrane Technology Conf.*

Collins, M. R. and Eighmy, T. T. (1988). Modifications to the Slow Rate Filtration Process for Improved Trihalomethane Precursor Removal. In: *Slow Sand Filtration* (Ed. Graham, N. J. D.). Ellis Horwood.

Crowley, F. W. and Twort, A. C. (1979). Current Strategies in Water Treatment Developments. *Proc. ICE Conf Water Resources: A Changing Strategy*.

Dorea, C. C. and Clarke, B. A. (2006). Impacts of Chemical Pretreatment in Slow Sand Filtration, *Recent Progress in Slow Sand and Alternative Biofiltration Processes* (Ed. Gimbel R., Graham, N. J. D. and Collins, M. R.). IWA Publishing.

Dullemont, Y. J., Schijven, J. F., Hinjen, W. A. M. et al. (2006). Removal of Microorganisms by Slow Sand Filtration. *Recent Progress in Slow Sand and Alternative Biofiltration Processes* (Ed. Gimbel, R., Graham, N. J. D. and Collins, M. R.) IWA Publishing.

Edzwald, J. K. and Kelly, M. B. (1998). Control of *Cryptosporidium*: From Reservoirs to Clarifiers to Filters. *Water Science Technology*, **37**(2).

Farizoglu, B., Nuhoglu, A. and Yildiz, E. et al. (2003). The performance of Pumice as a Filter Bed Material Under Rapid Filtration Conditions. *Filtration & Separation*, **40**(3).

Finch, G. R. and Black, E. K. (1993). Inactivation of *Giardia* and *Cryptosporidium* Using Ozone. *Proc Eleventh Ozone World Congress*, Volume 2, San Francisco.

Freeman, S., Verapaneni, S., Presdee, J. et al. (2004). World's largest Membrane Filtration Plants, *Proc. AWWA Annual Conf.*, Orlando.

Fujiura, S., Tomita, Y. and Kanaya, S. et al. (2006). Ceramic Membrane Microfiltration with Coagulation. *Proc. AWWA Annual Conference & Exposition*. San Antonio.

Galvis, G., Fernandez, J. and Visscher, J. T. (1993). Comparative Study of Different Pre-treatment Alternatives. *J. WSRT-Aqua*, **42**(6).

Glendinning, D. J. and Mitchell, J. (1996). Uprating Water-Treatment Works Supplying the Thames Water Ring Main. *J CIWEM*, **10**(1).

Harrison, N. (1997). Dunnore Point Water Treatment Works—Phase 1, Extension. *Water & Sewage Journal*.

Hendricks, D. (1991). *Manual of Design for Slow Sand Filtration*. AwwaRF.

Hudson Jr, H. E. (1981). *Water Clarification Processes*. Van Nostrand Reinhold.

Huisman, L. and Wood, W. E. (1974). *Slow Sand Filtration*. WHO.

Hutchinson, W. and Foley, P. D. (1974). Operational and Experimental Results of Direct Filtration. *J AWWA*, **66**(2).

Ives, K. J. and Gregory, J. (1967). Basic concepts of filtration. *JSWTE*, **16**(2).

Ives, K. J. (1969). Theory of Filtration. *Proc 4th World Filtration Congress*, IWSA, Ostend.

Ives, K. J. (1971). Filtration: The Significance of Theory. *J IWE*, **25**(1).

Jacangelo, J. G., Laine, J. M. and Carns, K. E. et al. (1991). Low Pressure Membrane Filtration for Removing *Giardia* and Microbial Indicators, *J AWWA*, **83**(9).

Jacangelo, J. G., Laine, J. M. and Cummings, E. W. et al. (1995). UF with Pretreatment for Removing DBP Precursors, *J AWWA*, **87**(9).

Joslin, W. R. (1997). Slow Sand Filtration: A Case Study in the Adoption and Diffusion of a New Technology. *J New England WWA*, September.

Kawamura, S. (2000). *Integrated Design and Operation of Water Treatment Facilities*, 2nd Edn. John Wiley.

Klein, H. P. and Berger, C. (1994). Slow Sand Filters Covered By Geotextiles. *Water Supply* **12**(3/4).

Kors, L. J., Wind, A. and van der Hoek, J. P. (1996). Hydraulic and Bacteriological Performance Affected by Resanding, Filtration Rate and Pretreatment. In: *Advances in Slow Sand and Alternative Biological Filtration* (Ed. Graham, N. and Collins, R.). John Wiley.

Lambert, S. D. and Graham, N. J. D. (1995). A Comparative Evaluation of the Effectiveness of Potable Water Filtration. *J Water SRT-Aqua*, **44**(1).

Lambert, S.D., San Miguel, G. and Graham, N. J. D. (2002). Deleterious Effects of Inorganic Compounds During Thermal Regenration of GAC: A review. *J AWWA*, **94**(12).

Lewin, J. (1961). Mechanisation of Slow Sand and Secondary Filter Bed Cleaning. *J IWE*, **15**(1).

Logsdon, G. (1998). Removal of Micro-Organisms by Clarification and Filtration Processes, Special Contribution. *Water Supply*, **16**(1/2).

Logsdon, G. S., Hess, A. F. and Chipps, M. J. et al. (2002). *Filter Maintenance and Operations Manual*. AwwaRF.

Lombard, H. K. and Hanroff, J. (1994). Filter Nozzle and Underdrain Systems Used in Rapid Gravity Filtration. *Water SA*, **21**(4).

Matsumoto, N., Aizawa, T. and Ohgaki, S. et al. (2002). Removal Methods of Nematodes Contained in the Effluent of Activated carbon. *Water Science and Technology: Water Supply*, **12**(3).

O'Neill, J. G. (1995). An evaluation Of Fibrous Depth Filters for Removal of *Cryptosporidium* Oocysts from Water in Protozoan Parasites and Water. In: *RSC Special Publication 168*. Royal Society of Chemistry.

Pike, P. G. (1987). *Engineering Against Schistosomiasis Bilharzia; Guidelines Towards Control of the Disease*. MacMillan.

Ratnayaka, D. D., Lee, M. F. and Tiew, K. N. et al. (2008). Application of Membrane Technology to Retrofit Large-Scale Conventional Water Treatment Plant in Singapore. *Singapore Water Week Convention Proceedings*.

Ridley, J. E. (1967). Experiences in the Use of Slow Sand Filtration, Double Sand Filtration and Microstraining. *Proc. SWTE*, **16**.

Scardina, P. and Edwards, M. (2004). Air Binding of Granular Media Filters, *J Environ. Eng.*, **130**(10).

Schalekamp, M. (1979) The Development of the Surface Water Treatment for Drinking Water in Switzerland. *Proc IWES Symposium: The Water Treatment Scene—The Next Decade*.

Speth, T. F. and Reiss, C. R. (2005). Chapter 2, Water Quality, *Manual of Practice M53: Microfiltration and Ultrafiltration Membranes for Drinking Water*, 1st Edn. AWWA.

Standen, G., Insole, P. J. and Shek, K. J. et al. (1997). The Use of Particle Monitoring in the Performance Optimisation of Conventional Clarification and Filtration Processes. *Water Science Technology*, **36**(45).

Stevenson, D. G. (1994). The Specification of Filtering Materials for Rapid Gravity Filtration. *J IWEM*, **8**(5).

Stevenson, D. G. (1998). *Water Treatment Unit Processes*. Imperial College Press.

Tebbutt, T. H. Y. and Shackelton, R. C. (1984). Temperature Effects in Filter Backwashing. *Public Health Engineering*, **12**(3).

Toms, I. P. and Bayley, R. G. (1988). Slow Sand Filtration, Approach to Practical Issues. In: *Slow Sand Filtration* (Ed. Graham, N. J. D.). Ellis Horwood.

UKWIR (1998). *Guidance Manual Supporting Water Treatment Recommendations from the Badenoch Group of Experts on Cryptosporidium* (98/DW/06/5).

US EPA (2006). *National Primary Drinking Water Regulations: Long Term 2 Enhanced Surface Water Treatment Rule*, **71**(3).

van de Vloed, A. (1955). Comparison Between Slow Sand and Rapid Filters. *Proc. 3rd Congress of the IWSA*, London.

van Dijk, J. C. and Oomen, J. H. C. M. (1978). *Slow Sand Filtration for Community Water Supply In Developing Countries*. Tech. Paper 10. WHO.

Visscher, J. T., Paramasivam, R. and Raman, A. et al. (1987). *Slow Sand Filtration for Community Water Supply, Technical Paper No. 24*. IRC.

Weeks, M. A., Olsen, A. and Leadbeater, B. S. C. et al. (2003). Invertebrate Infestation in Granular Activated Carbon Adsorbers in Potable Water Treatment. *Proc. Developments in Activated Carbon Processes*, Society of Chemical Industry.

Welte, B. and Montiel, A. (1996). Removal of BDOC by Slow Sand Filtration: Comparison with Granular Activated Carbon and Effect of Temperature. In: *Advances in Slow Sand and Alternative Biological Filtration*. Ed. Graham, N. and Collins R. John Wiley.

Wickramanayake, G. B., Rubin, A. J. and Sproul, O. J. (1984). Inactivation of *Giardia Lamblia* Cysts With Ozone. *Applied Env Microbiol*, **48**(3).

Wilson, D. (1996). Uprating of Barmby Water Treatment Works. In: *Advances in Slow Sand and Alternative Biological Filtration*. John Wiley.

Waterworks Waste and Sludge Disposal

9.1 TYPES OF WASTE

The primary sources of waste for the conventional treatment process are the sludge from clarifiers and used washwater from rapid gravity or pressure filters and slow sand filter sand washing plants. Waste resources for membrane filtration are membrane washwater and waste-containing chemicals used in enhanced backwash (CEBW) and clean-in-place (CIP). The secondary sources are chemical wastes (from chemical delivery, storage and dosing facilities); overflows (from the main and secondary treatment processes); water quality monitoring sample flows and waste generated during commissioning (e.g. filter wash water, water used in the disinfection of water retaining structures and pipes, water used in hydraulic tests and membrane preserving chemicals). There are also other potential discharges from the site which are themselves wastes and need to be properly managed (e.g. gas leaks, chemical dust and waste arising from cleaning processes) and wastes that would arise from chemical handling operations such as wrappings, bags, drums and pallets and also workshop waste (oil and grease), laboratory wastes (hazardous and non-hazardous) and domestic wastewater. Waste generated at a treatment works and treatment prior to disposal is summarized in Table 9.1. Secondary waste generation and disposal is considered by Li (2002).

9.2 TYPES AND QUANTITIES OF SLUDGES

Sludges may be classified according to the type of water treatment process adopted.

Non-chemical sludges arise from microstrainers, pre-settlement unaided by chemicals, membrane filter backwash water, sand washings (biological) from slow sand filters and associated roughing filter backwash water. These sludges are for the most part fairly innocuous and in some countries it may still be permissible to discharge them to a watercourse or water body without treatment; exceptions are when the raw water has high populations of algae or biological sludge when treatment is required.

Coagulant sludges derived from treatment plants using coagulation. These are the most difficult to treat for disposal because of the relatively large volumes involved and the difficulties of dewatering due to their gelatinous nature. They are dealt with in detail below.

Table 9.1 Waste produced at treatment works and treatment

Waste source	Waste	Treatment/disposal
Commissioning		
Hydraulic testing water	Free of contaminants	Discharge to a water course
Treated water	Chlorinated and may not be compliant with treated water quality standards for supply during start up and process adjustment.	Discharge to water course after dechlorination if necessary. Reuse if feasible.
Water used for disinfecting water retaining structures	High chlorine residual.	Discharge to water course after dechlorination if necessary. Reuse if feasible.
Filter backwash water	Media fines.	Settlement and discharge to a water course.
Normal operation		
Clarification Sludge	Regular batch discharges; 0.1 to 3% $^w/_v$ dry solids.	Dewatered sludge to landfill.
Media filter or membrane filter washwater	Regular batch discharges; typically 300 mg/l dry solids.	Recycle with/without settlement. Settled solids to sludge treatment plant.
Slow sand filter sand washing plant washwater	Biological matter.	To sewer or settlement and disposal.
Membrane plant clean in place or chemically enhanced backwash water	Acids, chlorine and alkali.	Neutralize and dispose to sludge plant or sewer.
Process tank drainage	Contains settled sludge.	Remove clear water to site drain. Sludge to sludge treatment plant.
Tank under-drainage	May contain dosed chemicals.	
Main process overflows	Could contain low residual chlorine and/or typically pH in the range 5 to 7 depending on the location.	Discharge to watercourse (subject to discharge consents) If necessary dechlorinate and/or neutralize.
Secondary process overflows (sludge treatment plant)	Suspended solids; concentration depends on the location.	Watercourse (subject to discharge consents). Sewer or discharge to downstream process units.
Chemical wastes (dross, flushings, drainage washings, spillages)	Care necessary to prevent mixing of incompatible materials (c.f. inadvertent chemical reactions).	Neutralize and discharge to sewer or hold for tanker removal.

(continued)

Table 9.1 *(continued)*

Waste source	Waste	Treatment/disposal
Chemical tank leakage and spillage on delivery	Concentrated chemicals.	Contain in bunds and remove for off-site disposal. Use 3-way diversion valves.
Water quality sample flows	Some do not contain chemical reagents (e.g. turbidity monitors), others contain reagents.	Those with reagents to sewer or sludge plant.
Domestic water	Sewage and grey water.	Sewer or treatment on site.
Laboratory drainage	Hazardous and non-hazardous chemicals.	Collect hazardous waste for off-site disposal. Non-hazardous to site drain.
Laboratory Solid Waste	Contaminated waste including hazardous chemical and biological waste.	Off-site disposal.
Ventilation/dust handling	Dust associated with chemical handling and storage.	Hazardous: Off-site disposal.
Membrane preserving chemical	Glycerin, sodium bisulphite etc.	To sewer or off-site disposal.
Road drainage	May be contaminated with fuel oil and other spillages.	Site drain with oil traps.
Process gas leakage from storage or generation plants	Chlorine, ozone, hydrogen etc.	Scrubbers (Cl_2), destructors (O_3), dilution and venting (H_2).
Chemical packaging	Bags, drums, pallets etc.	Off-site disposal by skips.
Workshop Wastes	Oil and grease.	Collect for off-site disposal.

Softening sludges are generated from by lime or lime-soda softening and the sludges, being of granular consistency, are comparatively easy to dewater but the quantities are much larger.

Quantities

The quantities of dry solids in sludges produced in a treatment works employing coagulation, flocculation, clarification and filtration are a function of raw water quality and the chemical treatment applied. They are made up of coagulant hydroxide, suspended solids, precipitated colour, algae, iron, manganese and other chemicals contributing to solids, such as powdered activated carbon (PAC), impurities in lime and polyelectrolyte. Dry solids produced ahead of clarifiers or filters (in direct filtration) can be calculated per litre of treatment works inflow as follows:

$$Sludge\ dry\ solids\ (mg/l) = X + S + H + C + Fe + Mn + P + L + Y$$

where: X is coagulant hydroxide (mg/l) $= f \times$ coagulant dose (mg/l as Al or Fe), $f = 2.9$ for Al and 1.9 for Fe (Some include 3 molecules of waters of hydration into the hydroxide. Then $f = 4.9$ for Al and 2.9 for Fe); S is suspended solids (mg/l), when suspended solids values are not available it can be approximated to $2 \times$ turbidity (NTU); H is $0.2 \times$ colour in °Hazen; C is $0.2 \times$ chlorophyll 'a' in µg/l; Fe is $1.9 \times$ iron in water in mg/l as Fe; Mn is $1.6 \times$ manganese in water in mg/l as Mn; P is PAC dose (mg/l); and L is lime dose in mg/l as pure lime $\times (1/w - 1)$ where w is the purity of lime expressed as a fraction; and Y is polyelectrolyte dose (mg/l).

In a well-operated treatment works using clarification and filtration all but about 2–5 mg/l of the calculated dry solids will be removed in the clarifiers. When softening follows clarification, the sludge dry solids produced in the softener can be calculated as follows (Sections 10.2–10.4 for softening).

For lime softening to remove carbonate hardness:

$$Sludge\ dry\ solids\ (mg/l) = 2CaCH + 2.54MgCH + CO_2 + L = LSSDS$$

where $CaCH$ is calcium carbonate hardness removed as $CaCO_3$ (mg/l); $MgCH$ is magnesium carbonate hardness removed as $CaCO_3$ (mg/l); CO_2 is carbon dioxide removed as $CaCO_3$ (mg/l), and L is as above.

For lime-soda softening to remove both carbonate and non-carbonate hardness:

$$Sludge\ dry\ solids\ (mg/l) = LSSDS + 2CaNCH + 2.54MgNCH$$

where $CaNCH$ is calcium non-carbonate hardness removed as $CaCO_3$ (mg/l) and $MgNCH$ is magnesium non-carbonate hardness removed as $CaCO_3$ (mg/l).

For caustic soda softening to remove carbonate hardness:

$$Sludge\ dry\ solids\ (mg/l) = CaCH + 0.54MgCH$$

For caustic soda softening to remove both carbonate and non-carbonate hardness:

$$Sludge\ dry\ solids\ (mg/l) = CaCH + 0.54MgCH + CaNCH + 0.54MgNCH$$

When softening and clarification processes are carried out together then the clarifier and softener dry solids should be summated.

When coagulants are used, the water is usually filtered by rapid gravity or pressure filters. The coagulated water may pass directly to the filters (direct filtration), in which case there is only the sludge content of the filter backwash water to deal with and should be estimated as for the clarifiers assuming total removal in the filters. In direct filtration with coagulation the backwash water volume can range from 2 to 5% of plant input, with an average of 3%. With clarification, irrespective of the wash regime, a filter will use a washwater volume equivalent to 2.5 bed volumes for sand or to 3 bed volumes for dual media; this is about 1.5–2.5% of works inflow depending on the frequency of filter washing. Filter backwash water usually contains 0.01–0.05% $^w/v$ dry solids with an average of 0.03% $^w/v$ and up to 0.2% $^w/v$ for direct filtration with coagulation (x% $^w/v = x$ g of dry solids in 100 ml of liquid containing the solids either dissolved or not dissolved).

Sludge withdrawn from clarifiers may amount to about 1.5 to 2.5% of works inflow and contains 0.1–1.0% ^{w}/v solids with an average of 0.3% ^{w}/v. Dissolved air flotation tanks employing full length scraper for float removal produce sludge flows of the order of 0.5–1% of works inflow, with dry solids concentrations ranging from 2 to 4% ^{w}/v, although withdrawal at high dry solids concentration may require water to assist conveying sludge to the sludge treatment plant. Beach scrapers produce a weak sludge with flows up to 3% of works inflow with solids concentration of about 0.1% ^{w}/v. Hydraulic float removal methods increase the sludge quantity to about 1–2% of works inflow with solids concentrations of less than 0.5% ^{w}/v and usually about 0.1% ^{w}/v (Schofield, 1997). For COCO DAFF or DAFF combined losses are about 4% of inflow at a concentration of about 0.5–1% ^{w}/v (DAF) and 0.03% ^{w}/v (F). For softeners the quantity discharged is usually in the range 0.5–2% of works throughput, although in some plants it can be as high as 5%, whilst the sludge is withdrawn at concentrations between 5 and 10% ^{w}/v. About 85–95% ^{w}/w of the solids is calcium carbonate depending on whether the softener is used for combined clarification and softening or for softening only.

For pre-settlement tanks and clarifiers which have to treat waters with heavy silt loads of over 1000 mg/l, the sludge volume removed can range from 5 to 10% of works inflow and may even reach as high as 30%. The concentration of such sludges can range from 2% to 5% ^{w}/v solids with a maximum of 10% ^{w}/v. These are exceptional cases in which special removal methods are necessary, but it is as well to be aware that they can occur with some raw waters, particularly in overseas countries. At the 1365 Ml/d Al Karkh water treatment works treating river Tigris water for Baghdad, at times when the raw water suspended solids loading was 30 000 mg/l, pre-settlement tank desludging accounted for 28% of works inflow and the concentration of the sludge was 10% ^{w}/v; the downstream clarifiers accounted for a further 7.5% of inflow at 2.5% ^{w}/v concentration. Losses due to filter backwashing remained below 1.5% of inflow. Total water losses were therefore 37% of works inflow.

Membrane filtration plants have a sludge stream and a chemical waste stream. The former is made up of the blow-down or used backwash water and has a composition similar to the feed. When coagulation and other chemicals such as PAC are used quantities should be estimated as for clarifier and filters. Hypochlorite, acid or alkali, will be present in CEBW. The quantity varies between 2–10% of inflow depending on the feed water quality, chemical pre-treatment and whether a clarification stage precedes the process. The sludge concentration is of the order of 0.025% ^{w}/v, similar to that in rapid gravity filter used washwater. The quantity of chemical cleaning waste is a function of the cleaning frequency (which is usually once every few weeks). It is primarily made up of the chemicals used in cleaning and the rinse water and is neutralized before discharge.

In slow sand filter installations, the primary filter used washwater quantities and solids concentrations are similar to those of direct filtration plants although solids do not contain coagulants. The sand washing plant produces between 10 m^3 and 20 m^3 of water/m^3 sand. This wash water, which contains solids, primarily consisting of organic debris, has a volume of 100–200 litres per 100 m^3 of water treated, assuming 50 mm is scraped off every 2 months.

9.3 METHODS OF DISPOSAL

In a well operated clarification and filtration plant using a coagulant, a high proportion of the suspended matter would be removed in the clarifiers, leaving about 2–5 mg/l of suspended solids to be removed by the filters. Therefore used filter washwater is very dilute compared to clarifier sludge and it is preferable to dispose used washwater separately, or keep the two waste streams

1. Either no pre–treatment or coagulation with or without clarification, powdered activated carbon etc.
2. Siphon driven if ΔH is adequate or pumped or a combination of both
3. Pumps could be dedicated for each train
4. If there is coagulation without clarification or PAC dosing strainers are generally used upstream of pre–treatment
5. Backwash (BW) waste is recirculated to inlet if there is a clarification stage
6. CIP supply and waste returns are both represented by single line

* Media filtration (one or two stage), or membrane filtration, possibly either combined with coagulation and clarification including dissolved air floatation

** Pump can be dedicated for each train

(a) Schematic of MF/UF membrane plant—encased and submerged types.

(b) Zenon ZeeWeed® 500d immersed membrane cassette being lowered into a tank. (Photo: GE Water and Process Technologies, Canada)

PLATE 15

(a) Filter press at the Chestnut Avenue Waterworks, Singapore.

(b) A 4-step cascade aerator (each fall 0.5 m) for 2.75 m³/s flow rate at Izmit treatment works, Turkey. (Contractor: Paterson Candy Ltd. UK)

(c) Packed tower aerator of capacity 8 Ml/day.

PLATE 16

segregated until the used washwater has undergone settlement to concentrate its sludge solids. There are instances, however, where used washwater and clarifier sludge are combined and treated.

Discharging used washwater or clarifier sludge back to the river downstream of the point of abstraction of the raw water is, of course, the simplest and most economical disposal method. However, the practice is only possible where dilution is available to reduce the coagulant metal concentrations to less than their respective drinking water quality standards; and where there are no adverse environmental consequences. An environment impact assessment should be undertaken.

Used washwater could be recycled with the solids it contains to the treatment process at the works inlet, where it is mixed with raw water. The main benefit is that solids are subsequently removed in the coagulation and clarification process and can be withdrawn as clarifier sludge at a steady rate and consistent solids concentration. Recycling also helps to maximize washwater recovery. Used washwater is however generated intermittently in relatively large amounts over short periods, with the solids concentrations declining continually until it is almost clear. It contains up to about 10% of the solids removed in the treatment plant. Uncontrolled return of used washwater as shock loads to the main treatment process may adversely affect the coagulation and downstream clarification processes. It should therefore be added uniformly at a steady flow of up to 5% of the works inflow, from a flow balancing tank.

Recycling can return undesirable material (which had already been removed) to the plant. The contaminants of primary concern are micro-animals (e.g. *rotifers, cyclopoids, copepods, chironomid midge larvae and nematode worms*) and protozoan organisms such as *Cryptosporidium* oocysts and *Giardia* cysts. If the numbers of these organisms in the raw water are high, then recycling of washwater (which would contain a large number removed in the filter) would exacerbate the problems. Holding tanks used for washwater can also provide breeding grounds for some of the organisms, in which case the used washwater returns could potentially carry them in increased numbers to the treatment process. This seeding can, if it occurs, result in infestation of clarifiers and filters, increasing the risk of passing them to the treated water.

Treatment of the used washwater by settlement reduces the risk of recycling micro-organisms (Section 8.19). With 80% settlement efficiency, the increased loading of *Giardia* cysts and *Cryptosporidium* oocysts to the plant through recycling is only about 1.2 times the source loading; settlement also helps to reduce manganese, aluminium and iron present as particulates and DPB precursor concentrations (Cornwell, 1994). High *Giardia* cyst and *Cryptosporidium* oocyst risk sites should have facilities in the form of a lagoon or similar to divert recycled water in the event of an incident.

Settlement is usually preceded by a trap to remove any filter media carried over in used washwater to minimize damage to downstream pumps. The settlement process is aided by a small dose of polyelectrolyte (0.02–0.2 mg/l of polyacrylamide) and can be by batch or continuous flow sedimentation. Batch tanks take the shape of hopper bottomed clarifiers (Section 7.16) or shallow rectangular tanks (typically $L:W = 5:1$) with sloping floors and usually consist of three or more tanks each sized for at least one filter wash and arranged to operate in rotation with one filling, one settling and one being emptied. Supernatant is decanted using a floating arm draw-off arrangement, with sludge drawn-off from the bottom. Continuous flow sedimentation is in lamella plate settlers (Section 7.17) operating at about 0.75 m/h, based on the projected area. Flocculators should be provided. The settlement process concentrates the solids to 0.3–1% $^w/_v$. The supernatant recovered will have a turbidity usually less than 10 NTU and can be either discharged to a water course or returned to the works inlet at less than 10% of the works raw water flow. Dissolved air flotation has been tested for treating used washwater (Anon, 2005).

The risk of returning *Cryptosporidium* or *Giardia* to the works can be further reduced by treating the supernatant by either membrane (Section 8.17) or depth filtration (Section 8.18) or ozonation (Section 11.21). In works where pre-ozonation is practised, the high ozone dose required to inactivate *Cryptosporidium* oocysts and *Giardia* cysts can be achieved by injecting a proportion of ozone required for main stream pre-ozonation into the supernatant return. An allowance should be made in the applied dose for the ozone demand of the supernatant. The minimum '*ct*' value required is typically 15 mg.min/l with about 2–3 minutes contact time.

The membrane filter sludge stream can be treated in a manner similar to rapid gravity filter washwater treatment. The chemical cleaning wastes from membrane plants can be either treated on site or removed for off-site disposal, depending on the quantity. On-site treatment usually consists of flow balancing followed by neutralisation for acids and sodium bisulphite dosing to remove chlorine. Membrane preserving chemicals such as glycerine (high in BOD) and sodium bisulphite (oxygen scavenger) (about 1 litre per pressure vessel) can be discharged to the sewer or removed for off-site disposal.

9.4 SLUDGE THICKENING AND DISPOSAL

Clarifier sludge and settled sludge from used filter washwater settlement tanks are mixed and concentrated in a continuous flow thickener where the residence time of the supernatant and the sludge can be varied independently of each other. The thickeners must be preceded by flow balancing tanks to contain and mix intermittent sludge discharges and feed the thickeners at a consistent concentration and uniform rate (Fig. 9.1). In applications where used washwater (without settlement) is mixed with clarifier sludge, the sizing of the flow balancing tanks becomes critical because of the large surges of dilute, used washwater. The sludge concentration achieved in the thickener is independent of the feed concentration. However, the greater the feed volume, the larger the thickener will be. There are several thickener designs in use; most are of the settlement type developed for industrial applications, which uses heavy duty scrapers with a picket-fence attachment. A design developed in UK by the WRc (Warden, 1983), for waterworks sludge thickening applications, consists of a cylindrical tank of water depth 2–3.5 m with a shallow sloping floor (1 in 20). Sludge is introduced at the central feed well and the supernatant overflows a peripheral weir. The sludge is thickened by the action of a specially designed rake which also moves the thickened sludge to a central hopper, of included angle 60°, for intermittent discharge under hydrostatic head (Albertson, 1992). Lamella settlers (Section 7.17) could be combined with a thickener in a single tank about 5.5 m deep, with a settlement rate of about 0.5–0.7 m/h on the projected area. The horizontal plate separation is about 100 mm.

With the aid of polyelectrolytes, coagulant sludges can be thickened to concentrations in the range 3 to 6% $^w/_w$ solids. The polyelectrolyte dose can be in the range 0.1–1 g/kg of dry solids and the dose should be applied proportional to the feed solids concentration and flow. It should be well mixed using an in-line static mixer or similar. The supernatant overflow will have a turbidity less than about 10 NTU. The hydraulic loading should be less than 1.5 m³/h.m² and the dry solids loading is less than 4 kg/h.m². Softener sludge is rarely thickened because it concentrates well in the clarifiers. If necessary it can be further thickened to 20–30% $^w/_w$ or more at a dry solids loading rate of about 8 kg/h.m². To prevent the softener sludge becoming too thick underflow is sometimes recycled to the feed. Dissolved air flotation (Section 7.18) is sometimes used in waterworks sludge thickening (Haubry, 1983).

FIGURE 9.1

Schematic diagram of sludge treatment plant.

Each thickener should have a dedicated feed pump of the progressive cavity type. Desludging can be initiated by sludge blanket level and terminated after a pre-set time, or when the solids concentration in the discharge measured in the outlet pipe falls to a pre-set value. In a well-operated thickener the supernatant is normally of an acceptable chemical and physical quality for recycling to the treatment process, but the potential risk of micro-organism and micro-animal return in clarifier sludges is far greater than in the case of recycling used washwater. Treatment for micro-organism removal similar to that for the supernatant from used washwater settling tanks is therefore required. Chemical quality problems can also arise in thickeners when sludges containing oxides of iron and manganese and natural organic matter are allowed to age in the thickener. Anaerobic conditions can then develop and release iron, manganese, colour and organics that would impart taste and odour into the supernatant, which if recycled, could have an adverse effect on treatment plant performance. Recycling can also lead to accumulation of toxic monomer (derived from polyelectrolytes used in used washwater settlement and thickening) in the treatment process, thereby affecting the treated water. This should be calculated by mass balance based on the assumption that all the monomer will be present in water.

In some cases it is possible to discharge unthickened waterworks sludge to the public sewer, subject to the appropriate consent and provided the proportion of sludge solids is less than 10% of the sewage sludge solids and adequate velocity (>0.75 m/s) is maintained in the sewers at all times to prevent silting. Thickened sludge may be piped or tankered to a sewage works. Coagulant sludge may aid primary sedimentation of the sewage, thus reducing the solids load to the subsequent biological stages and helping in the removal of phosphate. Thickened waterworks sludge may also be mixed with digested sewage sludge for disposal. Ferric coagulant sludge can prevent hydrogen sulphide gas formation and therefore corrosion of concrete sewers (McTique, 1989).

9.5 SLUDGE DEWATERING

Dewatering of coagulant sludges is difficult because of their gelatinous nature. Materials such as lime, fly ash or diatomaceous earth give body to the sludge and ease dewatering but the quantity of dry solids is increased by 50–100% by weight and increase in pH with lime could dissolve some of aluminium hydroxide floc. The dewaterability of softener sludge depends on the Ca:Mg ratio of the sludge; those with a Ca:Mg ratio less than 2:1 are difficult to dewater while those with a Ca:Mg ratio greater than 5:1 dewater readily.

Disposal of thickened sludge by lagooning or discharging to drying beds is still widely used in some parts of the world where land is cheap and abundant. Lagoons are shallow structures (0.5–1.5 m sludge depth) excavated from, or formed by, impoundment with earth embankments of slope no steeper than 1 in 2.5, on porous ground above the water table. A complete installation should be provided with a large number of small lagoons, some being filled, some rested for drying and some being emptied of dewatered sludge. Dewatering is by percolation (although with coagulation sludges the base soon becomes impervious), settlement and decanting, and finally evaporation. The average rate of evaporation of wet sludge is about 80% of that for free water and about 60–80% of the water is removed by settlement and decantation. The structure is provided with weir boards or penstocks at several points to allow decanting of the supernatant and rainfall. In cold countries some dewatering through freezing and thawing can be accomplished, but it is unlikely that more than one application can be dewatered in this manner. In lagooning, sludge is usually placed in layers (0.25–0.5 m deep) in several lagoons in sequence and allowed to dewater before the next layer is placed. This is continued until the total depth is utilized. The coagulant sludges generally consolidate to 10–15% ^{w}/w dry solids (x% ^{w}/w dry solids = x g of dry solid per 100 g of wet sludge) and softener sludge to greater than 50% ^{w}/w dry solids. The dried sludge is removed by using draglines or front-end loaders. The use of augers to aerate the sludge and expose it to solar evaporation is reported to show a 4-fold increase in the drying rate to give dry solids concentration as high as 70% ^{w}/w.

Drying beds, unlike lagoons, have a permeable base of 150 to 250 mm of sand (effective size of 0.3–0.75 mm, uniformity coefficient less than 4), supported by about 300 mm of graded gravel laid over an underdrainage system of plastic pipes laid with open joints and covered with coarser gravel. As an alternative porous concrete floors are sometimes used. Dewatering in drying beds is by drainage and evaporation, with the former accounting for about 40–50% of the dewatering. Rainfall delays the drying process as only about 60% of the rainfall is lost by drainage whereas rainfall during the later stages of drying, when the bed is cracked, has little effect on the drying time. However, adjustable decanting facilities help discharge the supernatant and accumulated rainfall. Sludge is usually placed to a depth of about 200–300 mm and coagulant sludges are removed at about 15–25% ^{w}/w dry solids.

Drying beds are not extensively used for softener sludges because the sludge penetrates the sand bed during drainage. This could be overcome by polyelectrolyte conditioning and sludge may be dried to over 50% ^{w}/w solids. In drying beds, layering is not normally practised as it retards the dewatering by drainage. The operating and emptying of drying beds are similar to those of lagoons. For lagooning and drying beds to succeed as dewatering processes the net evaporation rate must exceed rainfall for a considerable part of the year. Meteorological data should be used for sizing lagoons and drying beds. In sizing an allowance should be made for storage of sludge during winter and wet months when drying is minimal. The ability of waterworks sludge to dewater by gravity varies a great deal and is very dependent on the characteristics of the individual sludge. Aluminium coagulant sludges usually drain more slowly than do iron coagulant sludges. The rate of draining or settlement for coagulant sludges can be increased by 50% or more by using a conditioning agent such as polyelectrolyte (Novak, 1977). Typical performance for drying beds is about 25 kg/m².annum; this could be doubled by the use of polyelectrolytes.

Dewatering of sludge by plate pressing is becoming increasingly used for polymer thickened clarification sludges. In waterworks sludge dewatering, the recessed plate design is commonly used.

A press contains a set of a horizontal stack of rectangular or square plates covered with a filter cloth to provide a series of chambers clamped between two fixed end plates. The plates are suspended either from an overhead I-beam or on two side bars (Plate 16(a)). Plates are fabricated in steel, ductile iron or polypropylene, and cloths are of nylon, polypropylene or polyester fabric. Initially the press is closed and sludge is admitted to the press by positive displacement pumps to fill the chambers. The solids are retained on the cloth and the filtrate passes through the cloth and emerges from the drainage ports. The filtrate flow rate is at its maximum at the start, remaining reasonably constant until the pressure begins to build up with the gradual formation of the cake. When the pumping pressure reaches the operation value there is a period of constant pressure filtration (which can last several hours) with a continual decline in filtration rate which eventually ceases when the chambers are full of dry cake. The pump is then stopped, the pressure is released, the press ends are unclamped and the plates are separated from one end, one at a time, to release the cake. The presses are normally operated at pressures from about 7 to 8 bar or sometimes from 14 to 16 bar. For coagulant sludges a 25 mm cake thickness is used; thicker cakes (up to 35 mm) are feasible, but at the expense of increased filtration time.

Recessed plate presses can give a cake of 20–30% ^{w}/$_w$ dry solids. The volumetric capacity of a press is determined by the chamber depth (25–35 mm), plate dimensions, (0.5 × 0.5 m up to 2 × 2 m) and the number of chambers (up to about 160). For example, the capacity of a 30 mm deep, 150 chamber, 2 × 2 m plate press is 14.25 m^3, which is equal to the volume of the cake. If the sludge is to be dewatered to 30% ^{w}/$_w$ dry solids, then the weight of dry solids per pressing is about 5.2 t. (1 t of wet sludge occupies $(1–0.6f)$ m^3 where f is the fraction of dry solids by weight in the sludge. It is assumed that dry solids in waterworks sludge have a specific gravity of about 2.5). A drawback with the recessed plate press is that the filtration time can be up to 10 hours or even longer and the cycle time which includes downtime for filling, cake drop and cloth washing, can be 12 hours or longer. This drawback is overcome in the membrane plate press where one of the recessed plates in each chamber is replaced by an inflatable rubber or polypropylene ribbed membrane moulded round a steel insert plate, over which the filter cloth is laid. Filtration is carried out as in the recessed plate design at about 7–8 bar. The feed pump is then stopped, the membrane is inflated using compressed air or water and the remaining water in the sludge is squeezed out at about 15 bar. The cake thickness is reduced by about 40%. These presses produce thinner and dryer cakes in a much shorter time. Cake-dry solids for coagulant sludges range from 25 to 45% ^{w}/$_w$ and those for softener sludges are usually greater than 60% ^{w}/$_w$. The filtration and compression time can be about 2 hours (made up of 90 and 30 minutes, respectively) and the total cycle time is about 4 hours, thus giving six pressings a day, compared to two for the recessed plate press. In the example for the recessed plate press, the volume of cake after compression to 15 mm is 7.13 m^3. If the sludge is dewatered to 30% ^{w}/$_w$ dry solids, then weight of solids per pressing is 2.6 t. The sludge processed per day by similar recessed and membrane plate presses will therefore be 10.4 and 15.6 t dry solids, respectively. The press feed pumps are usually piston ram, hydraulic diaphragm or progressive cavity type. Polyelectrolyte is not always added to the press feed; sometimes polyelectrolyte conditioning is limited to thickening only. Filter pressing is a batch process. All operations in a cycle can be fully automated. The filtrate produced from a press initially contains about 100 mg/l suspended solids reducing to less than 10 mg/l as filtration proceeds. The overall solids capture is better than 98% (99–99.5%). The filter cake is usually discharged into a hopper located underneath the press and removed by screw conveyors for disposal. The energy consumption is about 0.03–0.05 kWh/kg dry solids. The primary advantage is the high dry solids achieved. Disadvantages include mechanical complexity, limited degree of automation, high labour requirements, requirements for special support structure (a weight of a press could be up to 100 t), large floor area and building height for the equipment (press is usually located on the first floor), limitations on filter cloth life and high capital cost. Vertical travelling filter cloth presses are used in Japan. It is reported that compared to plate presses it is complex, but dewatering times are much shorter and both capital and operating costs are higher. In addition, heating sludge to 40 °C, to reduce the viscosity and improve the filtration rate by 1.3–1.5 times that at normal temperatures, has been successfully tried in Japan (Yamane, 2005). In the INOS process sludge is heated up to about 85 °C using hot water followed by applying a vacuum which reduces the boiling point of water in the cake and eases removal of water as steam through the filtrate ports. Ferric coagulant sludge is reported to have been dewatered from 3% $^w/v$ dry solids to 90% $^w/w$ dry solids over 6 hours.

Centrifuging of sludge has always had a place in the dewatering of softening sludges and, in recent years, has also been used successfully for coagulant sludges. Centrifuges of several designs are available. The solid bowl decanter type, also known as the scroll centrifuge is widely used on waterworks sludge. The bowl is a cylinder on a horizontal axis with a conveying conical section at one end, called the beach, and an inward facing flange or adjustable weir at the opposite end. A helical screw conveyor (scroll) is mounted coaxially inside the bowl with a very small radial clearance. The parameters that affect centrifuge efficiency are bowl speed, scroll differential speed, pond depth, sludge feed rate and polyelectrolyte dose. To increase cake dryness, bowl speed should be increased and feed rate, scroll differential speed, pond depth and polyelectrolyte dose should be decreased. To increase solids recovery, bowl speed, pond depth, scroll differential and polyelectrolyte dose should be increased and feed rate should be decreased. Centrifuges are typically operated at about 1800–3500 rpm to give a centrifugal force at the wall of the bowl of about 1500–2000 g. The speed differential between the bowl and the scroll is in the range 2–40 rpm but, usually about 10 rpm. Unlike filter pressing, centrifuging is a continuous process. Its performance on coagulant sludges depends to a considerable degree on polyelectrolyte conditioning of the sludge; polyelectrolyte usage is high and usually in the range 2–6 kg/t of dry solids (Piggott, 1992). The feed to the centrifuges should be maintained at <3% $^w/v$ to ensure good mixing of polyelectrolyte with the solids. Coagulant sludge can be dewatered to about 15–25% $^w/w$ depending on the nature of the raw water, softening sludge to about 40–50%, $^w/w$. Overall solids recovery is normally better than 95% (98–99%). Centrate (i.e. the effluent) water quality from a centrifuge is usually poor with a suspended solids concentration in the range 300–1000 mg/l. It may be spread on land or discharged to a sewage treatment works. The advantages of centrifuging are enclosed unit and therefore clean in appearance, fast start-up and shut-down; quick adjustment of operating variables; continuous operation if necessary, readily automated and therefore suitable for unmanned operation; low capital cost-to-capacity ratio; and high installed capacity to building area ratio. Disadvantages are low cake dry solids, high demand for energy (about 0.07 kWh/kg dry solids), high polyelectrolyte consumption and poor centrate quality.

Freezing and thawing have also provided an effective method of dewatering coagulation sludges. Natural freezing is more appropriate to countries with more severe winter conditions. There are a few plants where a mechanical freeze-thaw method is used for dewatering (Henke, 1989).

Belt presses, where pressure is applied between moving endless belts, are being used successfully for coagulant sludge dewatering in limited numbers in USA (Migneault, 1987) and France. They are very dependent on polyelectrolyte dosing and the solids content obtained varies from 15 to 25% $^w/w$. Vacuum drum filters are successfully used for dewatering softener sludges. With filter loadings in the range 300–450 kg dry solids/h.m^2, cake solids up to 65% $^w/w$ have been achieved (Anon, 1981). Research will certainly continue on sludge dewatering because of the need in many countries to comply with increasingly stringent conditions for disposal.

9.6 BENEFICIAL USES OF SLUDGE

Economic and regulatory constraints, and environmental issues are driving water treatment plant operators to examine beneficial uses, alternative disposal methods and resources recovery. The UK (USA) figures for disposal of waterworks sludge are over 50% (20%) to landfill; nearly 30% (24%) to sewage treatment works; under 10% (25%) spread on land; and only 0.5% used in construction material (brick making and aggregate production), the remainder being lagooned or discharged to a water course (Simpson, 2002). In the USA a further 24% account for stream discharge and monofill (Cornwall, 2006).

The beneficial uses of coagulant sludge include land application, the manufacture of cast iron (Henke, 1989), brick making especially for those sludges with high silt content which have properties similar to the clay used in brick making, additive to cement as it contain most of the key elements (Si, Ca, Al, Fe) used in cement making (Cornwall, 2006; Cornwell, 1990) and in the removal of phosphate in sewage treatment (Section 9.4). In agricultural use the potential benefits are nutrient control (increased plant available N and total organic C), pH adjustment, water retention, soil aeration and increased drainage capacity (Elliott, 1991). Heavy metals in sludge are a concern, but most waterworks sludge contains only trace concentrations of heavy metals. Therefore, it is not a major concern in land application, but the local regulations for heavy metal loadings on to land must be followed which may limit the size and the number of applications. Coagulant sludges when used as soil conditioners are known to affect growth of some plants due to the ability of aluminium or iron to fix phosphorous in the soil reducing its availability. Coagulant sludges can be used in eutrophied lakes to fix phosphorus to minimize algal blooms.

Softened sludge, which has a high purity calcium carbonate with little or no magnesium and is free from coagulant and suspended solids, could be used in flue gas desulphurisation and may also be sold for agricultural use or to industry for products such as cosmetics. Such a pure calcium carbonate sludge is only likely to result from the softening of well water derived from a chalk aquifer.

The recovery of aluminium sulphate from sludge has received considerable attention in the past because it appeared to offer partial reuse of the coagulant and also reduce the volume of sludge to be handled. In theory every 1 g of aluminium hydroxide in the sludge could be recovered as 2.2 g of aluminium sulphate by treating the sludge with 1.9 g sulphuric acid at about pH 2. In practice the amount of acid can be much greater if other acid-soluble material is present in the sludge. Typically recoveries greater than 75% are feasible (Anon, 1994). There are also benefits from sludge weight reduction (about 35–40%) and improved dewatering characteristics of acidified sludge. The recovered aluminium sulphate is usually mixed with the commercial product before reuse in the water treatment process. Due to concerns over the possible accumulation of metals and other impurities such as organic material, the practice lost favour and has been discontinued in many installations. It is likely to be economical only when the cost of the sulphuric acid is lower than that of purchasing aluminium sulphate.

Lime recovery from softening sludges by recalcination is economically viable since the softening process produces calcium carbonate (Section 10.3). Generally more lime is produced than is added for treatment. Recalcination is carried out in a furnace at about 900–1000 °C. The available lime in the recalcined product may be only about 60–75% depending on the inerts present, such as magnesium, iron, silica and other compounds. Carbon dioxide which is a by-product is available for use in the recarbonation and pH correction of water on the plant.

REFERENCES

Albertson, O. E. (1992). Evaluating Scraper Designs. *Water Environment and Technology*, **4**(1).

Anon (1981). Lime softening Sludge Treatment and Disposal. *JAWWA*, **73**(11).

Anon (1994). Coagulant Recovery System Wins Big Award. *Water Engineering & Management*, **14**(7).

Anon (2005). *Use of DAF for Treatment of Spent Filter Backwash Water*. Water 21, December.

Cornwell, D. A. and Koppers, H. M. M, (1990). Slib, Schlamm, Sludge. *AWWA Research Foundation Report*. AWWA.

Cornwall, D. A. (2006). *Water Treatment Residuals Engineering*. AwwaRF.

Cornwell, D. A. and Lee, R. G. (1994). Waste Stream Recycling: Its Effect on Water Quality. *J AWWA*, **86**(11).

Elliott, H. A. and Dempsey, B. A. (1991). Agronomic Effects of Land Application of Water Treatment Sludges. *J AWWA*, **83**(4).

Haubry, A. and Fayoux, C. La. (1983). Flottation des Boues: Un Avenir Assure, L'Eau, l'Industrie. *Les Nuissances*, **79**(1).

Henke, H. (1989). Application of Freeze-Thaw for Handling of Sludge from the Treatment of Great Dhunn Reservoir Water. *KIWA/AWWA Research Foundation Experts Meeting*. Nieuwegein.

Li, A., Ratnayaka, D. D. and Hieatt, M. J. (2002). Tai Po Water Treatment Works Waste Management, Management of Waste from Drinking Water Treatment. *Proc. CIWEM International Conf.* London.

McTique, N. E. Cornwell, D. (1989). Impact of Water Plant Waste Discharge on Wastewater Plants. *Proc Residuals Management Conference*, AWWA/WPCF. San Diego.

Migneault, W. (1987). Santa Clara Valley District Sludge Concentration and Recycling Program, California Nevada Section. *Proc. AWWA Fall Conf.*

Novak, J. and T. Langford, M. (1977). The Use of Polymers for Improving Chemical Sludge Dewatering on Sand Beds. *J AWWA*, **69**(2).

Piggott, G. A., Tse, Y. S. and Ratnayaka, D. D. (1992), Waterworks Sludge Treatment and Disposal Options. *Proc. Water Malaysia `92'*, Kuala Lumpur.

Schofield, T. (1997). Sludge Removal and Dewatering Processes for Dissolved Air Flotation System. *Proc International Conference on Dissolved Air Flotation*, CIWEM. London.

Simpson, A., Burgess, P. and Coleman, S. J. (2002). The Management of Potable Water Treatment Sludge: Present Situation in the UK, *Water and Environment Journal*, **16**(4).

Warden, J. H. (1983). *Sludge Treatment Plant for Waterworks*. Technical Report 189, WRc.

Yamane, Y. (2005). Water Supply Sludge Treatment in Japan, *Water 21*.

Specialized and Advanced Water Treatment Processes

10

SOFTENING OF WATER

10.1 HARDNESS COMPOUNDS

A description of hardness is given in Section 6.30. A large proportion of waters from underground sources are hard, particularly waters from chalk and limestone aquifers which often have a carbonate hardness of 200–300 mg/l as $CaCO_3$. The hardness compounds are taken into solution because the water acquires carbon dioxide from the soil formed by the oxidation of organic matter. A major source of non-carbonate hardness in surface waters is the calcium sulphate present in clays and other deposits. In contrast, many surface waters from the older geological formations are soft or very soft, e.g. 15–50 mg/l as $CaCO_3$, because the rocks are largely impermeable and insoluble.

10.2 PRINCIPAL METHODS OF SOFTENING

There are three principal methods of softening a hard water. In the first, lime (calcium hydroxide: $Ca(OH)_2$) and soda ash (sodium carbonate: Na_2CO_3) are added to precipitate hardness compounds which are then removed by clarification and filtration. In the second method the nature of the hardness compounds is changed by passing the water through a bed of 'ion-exchange' resin. In the third method, membrane processes such as reverse osmosis remove all dissolved salts from water at an efficiency of about 95–99%; nanofiltration removes bi- or tri-valent ions (e.g. Ca^{2+}, Mg^{2+}, Al^{3+}, CO_3^{2-}, SO_4^{2-}) at an efficiency of about 80–85% and monovalent ions (e.g. Na^+, K^+, Cl^-) at an efficiency of up to 40%. The differences in the three methods are important because the chemical and membrane processes reduce the total dissolved solids in a water, a feature that is often desirable for industrial applications.

Softening in drinking water treatment is usually applied to a proportion of the flow (split-treatment) to soften the water to a hardness value below the required value and then blend it with the unsoftened water.

10.3 THE LIME-SODA PROCESS OF SOFTENING

The aim of the lime-soda process is to convert calcium and magnesium compounds to the virtually insoluble forms, calcium carbonate ($CaCO_3$) and magnesium hydroxide ($Mg(OH)_2$).

Magnesium carbonate ($MgCO_3$), unlike calcium carbonate, does not precipitate in cold water. The stages of treatment involved are set out in Table 10.1. Lime is added to remove the temporary hardness and soda ash is added to remove the permanent hardness. Several applications of lime and soda ash are needed to remove magnesium hardness. In practice complete removal of hardness is undesirable because this renders a water highly aggressive. Since in most hard waters the calcium temporary hardness forms the major component it often suffices to remove only this by the addition of lime. Caustic soda (sodium hydroxide: NaOH) can be used in place of lime for carbonate and non-carbonate hardness removal; alkalinity reduction is only 50% that of lime softening. Soda ash is formed in the reactions and sometimes it may be supplemented by soda ash addition to remove non-carbonate hardness. The advantages of caustic soda are that it is easier to handle than lime, only one chemical may be required and the quantity of calcium carbonate sludge produced is less (Section 9.2). The drawbacks are its higher cost, it is a hazardous chemical, and that sodium is added to the treated water. Chemical softening, due to its high operating pH (9.5–10.5), also removes, by precipitation, many of the heavy metals (Section 10.14), arsenic, iron and manganese.

Table 10.1 Lime-soda softening processes

Reaction	Equation number
To remove carbon dioxide in water add LIME (not a softening reaction): $H_2CO_3 + \textbf{Ca(OH)}_2 = \mathit{CaCO_3} + 2H_2O$	10.1
To remove calcium temporary hardness add LIME: $Ca(HCO_3)_2 + \textbf{Ca(OH)}_2 = \mathit{2CaCO_3} + H_2O$	10.2
To remove calcium permanent hardness add SODA ASH: $CaSO_4 + \textbf{Na}_2\textbf{CO}_3 = \mathit{CaCO_3} + Na_2SO_4 \quad CaCl_2 + \textbf{Na}_2\textbf{CO}_3 = \mathit{CaCO_3} + 2NaCl$	10.3
To remove magnesium temporary hardness add LIME + more LIME, Stage 1: $Mg(HCO_3)_2 + \textbf{Ca(OH)}_2 = MgCO_3 + \mathit{CaCO_3} + 2H_2O$	10.4
The calcium carbonate precipitates but the magnesium carbonate does not, so further LIME is added, Stage 2: $MgCO_3 + \textbf{Ca(OH)}_2 = \mathit{Mg(OH)_2} + \mathit{CaCO_3}.$ The magnesium hydroxide and calcium carbonate precipitate.	10.5
To remove magnesium permanent hardness add LIME and SODA ASH: $MgCl_2 + \textbf{Ca(OH)}_2 = \mathit{Mg(OH)_2} + CaCl_2 \quad MgSO_4 + \textbf{Ca(OH)}_2 = \mathit{Mg(OH)_2} + CaSO_4.$ The addition of soda ash then converts the calcium chloride and calcium sulphate to calcium carbonate as in 10.3 above.	10.6

Notes: Compounds in **bold** are those being added and compounds in *italic* are those precipitating. H_2CO_3—carbonic acid (carbon dioxide in water); $Ca(HCO_3)_2$—calcium bicarbonate; $Ca(OH)_2$—calcium hydroxide (hydrated lime); $CaCO_3$—calcium carbonate; $CaSO_4$—calcium sulphate; $CaCl_2$—calcium chloride; H_2O—water; Na_2CO_3—sodium carbonate (soda ash); $Mg(HCO_3)_2$—magnesium bicarbonate; $Mg(OH)_2$—magnesium hydroxide; $MgCO_3$—magnesium carbonate; $MgCl_2$—magnesium chloride; $MgSO_4$—magnesium sulphate.

10.4 SOFTENING PLANT

Lime is used in the hydrated form, which is a dry powder and is dosed as a slurry (Section 7.20). The soda ash is a powder and a solution is prepared for use. The concentration of saturated solution varies with temperature (Table 7.5). Usually a solution of about 60% of the saturated concentration at the lowest anticipated water temperature is prepared. Clarification is in hopper-bottom clarifiers of the sludge-blanket type or solids recirculation clarifiers (Section 7.16).

An excess of up to 10% of both lime and soda ash should be added over the dose stoichiometrically required in order to complete the reactions in reasonable time. If magnesium is present, a portion of it may also cause precipitation but the reaction is slower than that of calcium. Softening reduces the hardness value to 35–50 mg/l as $CaCO_3$. If the water to be softened contains suspended solids and organic matter such as colour, these can be removed concurrently although coagulants and coagulant aids have to be added. The coagulants used are usually of the iron type, because of their high coagulation pH values, but aluminium sulphates can also be used; at high pH insoluble magnesium aluminate and not aluminium hydroxide is formed. Normally softeners are operated at surface loading rates in the range 3–4 m³/h.m²; when magnesium is to be removed the rates are about 2–2.5 m³/h.m² because of the gelatinous nature of the precipitate.

The dosages of lime required for softening are high, being of the order of 100–200 mg/l. The process produces a large amount of liquid sludge due to the precipitation of hardness and coagulation of suspended solids and colour (see Sections 9.4 to 9.6 for treatment and disposal).

10.5 WATER SOFTENING BY CRYSTALLISATION

Softening reactions can be accelerated by using sand grains for seeding the crystallisation of calcium carbonate. Softening takes place in a cylindrical reactor partially filled with filter sand (0.2–0.6 mm). The water is injected with the softening chemical and passed upwards at a surface loading rate of 50–120 m³/h.m² to fluidize the sand bed. Calcium carbonate deposits on the sand grains which grow to form pellets of about 1–2 mm in diameter and accumulate at the base of the reactor, from where they can be periodically removed. Make up sand is added either from the top or at the base. The reaction tanks, called pellet reactors, are typically about 6 m deep and up to 4 m in diameter (Fig. 10.1). Lime is used when the ratio of carbonate hardness to total hardness is high; in the intermediate hardness range caustic soda is used; and when carbonate hardness is very low soda ash is used. Van Dijk (1991) reported that crystal growth is adversely affected when the phosphate content of a water exceeds 0.5 mg/l as PO_4 (0.15 mg/l as P). Fluffy pellets are formed when iron is present in the water above about 1 mg/l as Fe (van der Veen, 1988).

The advantages of the pellet reactor over the conventional softening process are its high surface loading rate, pellets are easier to handle than sludge and only a small excess of softening chemical is required. The disadvantages of the process are that it does not remove magnesium, the hardness after softening is in the range 50–100 mg/l as $CaCO_3$ (van Honwelingen, 1994) and is therefore unsuitable for most industrial uses, suspended solids in the product is high (up to 30 mg/l) and the removal of turbidity and colour in raw water by coagulants cannot be performed in the same reactor. However, it is therefore ideal for softening ground waters.

Lime is used as slurry of 10–100 g/l. The use of lime water (saturated lime) is not usually practical because of its low lime content (e.g. 1.76 g/l at 10 °C); the volume to be added would be over 10% of the volume of water to be softened. Nevertheless there are plants operating with lime water. Use of

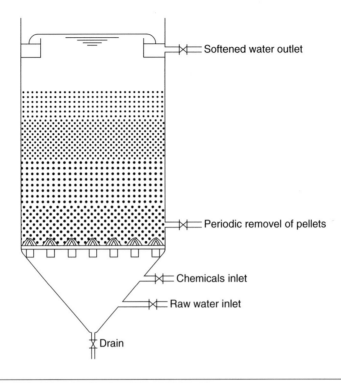

FIGURE 10.1

Pellet reactor for softening by crystallization.

lime slurry leads to carry over of calcium carbonate, un-dissolved lime and inert impurities in the lime, amounting in total to about 20–30 mg/l as suspended solids. They are removed by dosing the softened water with acid and iron coagulants followed by filtration. By improving the quality of lime and its solubility, the carry over of suspended solids can be reduced by up to 80% (van Eekeren, 1994).

An alternative to chemical crystallisation is one of a number of physical processes using magnetic, ultrasonic, electrolytic, electrostatic or electronic devices fitted to pipelines carrying hard water. The processes do not change the chemical properties of the water but modifies the crystallisation of calcium carbonate giving an increase in particles in suspension and a decrease in the formation of scale. The devices are used in waterworks on sample and lime dosing lines and on domestic hot water systems with some success.

10.6 STABILISATION AFTER SOFTENING

The softening reactions of precipitation are not usually wholly completed in the clarification tanks and therefore the water leaving is usually supersaturated with calcium carbonate and tends to form further deposits, mainly of calcium carbonate, in the later stages of the treatment plant. The water needs to be stabilized by injecting carbon dioxide (CO_2) into the water to achieve an alkalinity of about 30–50 mg/l as $CaCO_3$ (1 mg/l as $CaCO_3$ = 1.22 mg/l as HCO_3 = 0.4 mg/l as Ca) and a slightly positive Langelier Index (Section 10.41). Sulphuric acid is used in some cases but, as it converts carbonate to sulphate, care must be taken to ensure that water is not rendered corrosive due to the low calcium carbonate concentration. The objective of adding CO_2 or sulphuric acid is to prevent after-precipitation while producing a non-aggressive water. Another method of avoiding after-precipitation is to add about 0.5 to 2 mg/l as P of a sequestrant such as a polyphosphate (e.g. sodium hexametaphosphate). This method is usually preferred for industrial applications. Softening is usually followed by anthracite–sand filters.

10.7 BASE EXCHANGE SOFTENING

In the ion exchange (IX) process of softening, when water containing hardness salts is passed through a bed of strong acid cation exchange resin in the sodium form, calcium and magnesium are substituted by sodium. The hardness of the water is reduced to almost zero but the total dissolved solids concentration undergoes little change; alkalinity and pH values are unaffected. When the resin's capacity to exchange calcium and magnesium for sodium is exhausted the bed is regenerated by passing a concentrated sodium chloride solution through it. The reverse action then takes place, with the calcium and magnesium ions held in the beads being released in the effluent and the sodium from sodium chloride being substituted. The regeneration wastewater is very hard with a high concentration of dissolved salts and its disposal may present problems (Section 10.25).

The IX resins used are cross-linked polystyrene spherical particles or 'beads'. There are many different types of resins depending on the chemical functional group attached to the polystyrene matrix. Those used for base exchange softening have strong acid functionality. The total number of functional groups in the resin determines its exchange capacity. The exchange capacity of a resin for softening is most frequently stated in terms of the hardness removed by a specific volume of resin, e.g. x g eq per litre of resin, and is specific to each resin; typical values quoted vary in the range 1.5–2 g eq/l. (g eq = gram equivalent, i.e. equivalent weight in grams; 1 g eq of hardness = 50 g as $CaCO_3$). Operating capacity is less due to leakage and incomplete regeneration. A second measure of performance in base exchange softening is the amount of salt that must be used in regeneration per unit of hardness removed. The theoretical figure for regeneration is 117 g of salt per 100 g of $CaCO_3$ (or its equivalent) removed but, in practice depending on the resin used, up to 400 g of salt per 100 g of $CaCO_3$ removed would be required. This can be reduced by 50% using the more efficient counter current regeneration techniques.

10.8 PLANT FOR ION EXCHANGE SOFTENING

The plant required for IX softening is similar to that for pressure filtration. The vessels are rubber lined steel vertical type of diameter up to 4 m. The media bed would usually be about 1.2 m deep. A typical operational cycle comprises in service (i.e. softening); backwashing with softened water; regeneration; and then rinsing to remove excess regenerant. Flow during both softening and regeneration is normally downwards and operation is automated. The surface loading rate for softening varies in the range 12–20 $m^3/h.m^2$ and that for regeneration is about 2.5 $m^3/h.m^2$ to give at least 30 minute contact time. A saturated solution of sodium chloride containing about 26% $^w/w$ NaCl (33% $^w/v$) is initially prepared and diluted to about 5–10% $^w/v$ before use. Backwash water is applied at a rate to produce a bed expansion between 50 and 100%; rates vary with the water temperature and resin type. The plant is normally operated under pressure so that repumping after softening can be avoided; the loss of head is usually 4–5 m.

In general IX plants are rarely used for softening public supplies; they are not suitable for softening turbid water (e.g. suspended solids should be <1 mg/l) or water that contains significant concentrations of iron and manganese. The process also adds an equivalent concentration of sodium to the water for the hardness removed and is therefore not recommended for drinking or cooking. It is used for domestic water softeners to prevent scaling of washing machines and dish washers. In industry the tendency has been for IX softening to be superseded by demineralisation plant which can produce a water more exactly tailored to the type of process water required (Section 10.10).

10.9 REMOVAL OF HARDNESS AND ALKALINITY BY ION EXCHANGE

The process uses a weak acid hydrogen form cation exchange resin and therefore hydrogen, instead of sodium, is exchanged for calcium and magnesium equivalent to the alkalinity content of the water; sodium and potassium are not removed unless the alkalinity exceeds the hardness. Alkalinity reacts with the exchanged hydrogen ions producing carbon dioxide which is removed in a degasser, leaving a residual carbon dioxide concentration of about 5–10 mg/l in the water depending on the efficiency of the degasser. Sulphuric or hydrochloric acid is used as the regenerant instead of sodium chloride. Alternatively alkalinity alone can be removed using chloride form anion exchange resins (dealkalization). Chloride is exchanged for bicarbonates and sulphates in the water and sodium chloride solution is used as the regenerant.

Water dealkalized to give <14 mg/l of $CaCO_3$ with lime is sometimes used in the preparation of lime slurry and as carrier water for lime slurry since it minimizes the risk of scaling in tanks and pipework (Section 7.20).

10.10 DEMINERALISATION OF WATER BY ION EXCHANGE

The IX process of softening is only a particular example of ion-exchange treatment and is more specifically an example of strong cation exchange with the resin in sodium form. In demineralization a strong acid hydrogen form cation exchange resin replaces calcium, magnesium, sodium and potassium by hydrogen ions; this is followed by a second stage of treatment using a strong base anion exchange process in which chloride, sulphate and nitrate are removed. The cation exchange resin is regenerated with sulphuric or hydrochloric acid. Use of hydrochloric has advantages in that it can be applied at higher concentration than sulphuric (which is often limited to avoid calcium sulphate precipitation in the bed) resulting in higher regeneration efficiency; lower resin inventory; and smaller waste volumes. Anion exchange resin is regenerated using sodium carbonate or caustic soda solution. Carbon dioxide formed in the first stage is often removed by 'degassing' or by aeration as an intermediate stage between the cation and anion exchange vessels. The product water has a pH of 7–9 and total dissolved solids concentration is very low and has a conductivity less than 20 µS/cm. Such a treatment is therefore called demineralisation and is now most frequently adopted in industry for the production of special quality process waters.

In 'mixed' bed IX both cation and anion exchange resins as described above are mixed in one vessel. During backwashing the resins are hydraulically separated by virtue of their density

Access manway

Inlet/ backwash outlet

Air release

Withdrawable wedgewire strainer to prevent resin loss during backwash and coarse debris during service

Caustic regen inlet

Drain down

Top of resin bed

Anion resin

Regen outlet

Inert resin

Viewing window

Interphase collector laterals

Sample probe

Cation resin

Outlet/backwash inlet

Resin dump

Bottom collector laterals

FIGURE 10.2

Internal arrangement of a mixed bed demineralization unit.

difference. This allows separate regeneration of the two components (Fig. 10.2) after which the resins are remixed with an upflow of low pressure air. A mixed bed gives a water of neutral pH and conductivity of less than 0.2 μS/cm. Demineralisation can in theory be applied to brackish waters but, because of the regenerant chemical consumption, the process only finds application for waters having less than about 500 mg/l dissolved solids. For waters with higher dissolved solids the IX process would not be economic. For public supply purposes other processes should be considered (Section 10.44).

In potable water treatment, the IX process also finds application in the removal of arsenic (Section 10.14), radionuclides (Section 10.15), nitrate (Section 10.25), ammonia (Section 10.28) and organics (Section 10.36). One of the major drawbacks of IX is the difficulty in the disposal of wastewater which is highly saline and non-biodegradable.

In the production of drinking water from sea water by reverse osmosis desalination (Section 10.48), boron removal may not be complete leaving a residual >0.5 mg B/l in the permeate. IX processing can be used for further treatment of the RO permeate either from first or second pass for boron elimination (Jacob, 2007). Weak base IX resins with methyl glucamine functionality can selectively absorb boron (present as borate in solution) from practically any salt background. The total exchange capacity for boron is 0.7 g eq/l. Removal efficiency is very high with boron leakage usually <0.1 mg B/l. The treatment capacity is in the range 20 to 30 bed-volumes/h; surface loading should not exceed 40 m^3/h.m^2.

Regeneration is co-current flow and is carried out in two steps: first the absorbed borate is displaced with mineral acid, then the resin is converted into the free base form with sodium hydroxide solution. The regeneration levels should be 50 and 25 g of sulphuric acid and caustic soda per litre of resin respectively.

REMOVAL OF IRON, MANGANESE AND OTHER METALS

10.11 IRON AND MANGANESE–GENERAL

Traces of iron and manganese are found in many water sources, both surface and underground. Concentrations may occasionally range up to 20 mg/l of iron and 5 mg/l of manganese but at such high values most of the metals, in particular iron, are in particulate form so that they may be relatively easily removed by solid–liquid separation methods. It is the dissolved fractions of iron and manganese which can be troublesome and the disadvantages arising from their presence above certain concentrations are described in Sections 6.32 and 6.34. Treatment for the removal of these metals is therefore often necessary.

10.12 REMOVAL OF IRON AND MANGANESE FROM UNDERGROUND WATERS

When iron occurs in underground waters, it is usually in solution in the ferrous form in a water that is devoid of oxygen. Such waters are fairly common in aquifers that underlie an impermeable stratum for example, greensand and other sand formations underlying clays. Manganese occurs in appreciable amount in only a minority of those raw waters which contain iron. Many waters from deep boreholes in sandstone contain iron and manganese in their lowest state of oxidation as Fe(II) and Mn(II) that occur in solution. When oxidized, the ferrous Fe(II) and manganous Mn(II) are converted to the next common oxidation state ferric Fe(III) and Mn(IV) and permanganate Mn(VII); values in brackets denote the corresponding valencies. When a sample of ground water is first drawn it may appear perfectly clear, but after exposure to air for a short time it acquires a turbid appearance and after a further period a red-brown precipitate of ferric hydroxide is formed. Oxidation of Mn(II) is very slow under most conditions although it may co-precipitate with iron, if present.

Removal of iron and manganese is effected by oxidation, followed by the separation of Fe(III) and Mn(IV) as ferric hydroxide ($Fe(OH)_3$) and manganese dioxide (MnO_2) precipitates by filtration. For high concentrations of iron (>5 mg/l) filtration may be preceded by settling. Solids contact type clarifiers are suitable since oxidation is catalysed by the oxides already present in the sludge blanket or in the recirculated sludge.

Oxidation can be by oxygen in air or by the use of a strong oxidant such as chlorine, potassium permanganate, chlorine dioxide or ozone. In most cases oxidation is influenced by pH value. In oxidation reactions hydrogen ions are produced which, in turn, react with alkalinity. However, since the concentrations of iron and manganese present are generally low there is sufficient alkalinity in a water to buffer the effect of hydrogen ions and prevent a consequent reduction in pH which would otherwise reduce the reaction rates. Table 10.2 gives the stoichiometric quantities of oxidant (in mg) required to oxidize 1 mg of iron or manganese, the corresponding reduction in alkalinity and the optimum pH range for the oxidation reaction. Lower water temperature reduces the rate of oxidation.

Oxygen is added by aeration. In practice larger volumes of air are required to compensate for inefficiencies of the aeration system used. Although the oxygen requirements are small, the rate of reaction is slow and pH dependent; 90% oxidation of iron would require about 40 minutes reaction time at pH 6.9, but only 10 minutes at pH 7.2. In some waters aeration alone is adequate for complete iron oxidation because the removal of free carbon dioxide in the aeration process raises the pH above 7.5. The oxidation of manganese is much slower and also requires elevated pH for successful oxidation; at pH 9.5 about 1 hour for 90% oxidation is required. Therefore a strong oxidant such as chlorine, potassium permanganate, chlorine dioxide or ozone is usually necessary. These are also effective in oxidising iron. The application of strong oxidants to water containing iron (II) and manganese (II) results in rapid oxidation, although the rates are pH dependant. In practice excess oxidant is used to satisfy the demands due to organic matter, hydrogen sulphide and, in the case of chlorine, ammonia when present in water. The use of chlorine may be inadvisable when treating waters

Table 10.2 Oxidation of iron and manganese

Metal	Oxidant	Stoichiometric quantity of oxidant (mg/mg Fe or Mn)	Reduction in alkalinity (mg $CaCO_3$/mg Fe or Mn)	Optimum pH
Fe(II)	Oxygen	0.14	1.80	>7.5
Mn(II)	Oxygen	0.29	1.80	>10.0[a]
Fe(II)	Chlorine	0.63	2.70	>7.0
Mn(II)	Chlorine	1.29	3.64	>9.0[a]
Fe(II)	Potassium permanganate	0.94	1.49	>7.0[b]
Mn(II)	Potassium permanganate	1.92	1.21	>7.0[b]
Fe(II)	Chlorine dioxide	0.24[1]	1.96	>7.0
Mn(II)	Chlorine dioxide	2.45[1] 0.49[2]	3.64 2.18	≥7.0 ≥7.5
Fe(II)	Ozone	0.43	1.80	[A]
Mn(II)	Ozone	0.87	1.80	[A]

Notes: [a] The use of a catalytic filter medium may reduce the pH to 7.5–8.5.
[b] Reaction is known to proceed at pH > 5.5.
[A] pH value at which the reaction occurs is less dependent than for other oxidants. Low pH values are preferred as ozone performs better under acidic conditions.
Sources: [1] Knocke (1991a); [2] Faust, (1998).

containing organic substances due to the possibility of disinfection byproducts (DBPs) formation (Section 6.25).

Oxidation with potassium permanganate is more effective than chlorine and does not form DPBs in the presence of organic substances. Manganese dioxide formed in the reaction adsorbs Mn(II) and catalyses its oxidation, which brings about an improvement in Mn(II) removal and a reduction in the amount of potassium permanganate required. The potassium permanganate dose applied must be carefully controlled to minimize any excess passing into supply which could give a pink colour to the water. Products of potassium permanganate oxidation can 'muddy' the water and form 'mud balls' in filters.

Chlorine dioxide is particularly useful as an oxidant in the presence of high ammonia concentrations (Section 11.17) that would otherwise react with chlorine. A disadvantage of chlorine dioxide is the limitation on the dose that can be applied and the formation of disproportionation products.

Ozone reacts readily in the absence of organic matter to oxidize soluble Fe(II) and Mn(II) to the insoluble Fe(III) and Mn(IV) forms. When both metals are present, iron is oxidized first, followed by manganese. Excess ozone can oxidize Mn(II) to its highest oxidation state Mn(VII) which gives a pink colouration to the water. When present, organic matter is oxidized by ozone before iron and manganese and the dose would be much higher than that required for oxidation in the absence of organics. Iron and manganese are more receptive to oxidation by ozone once organics have been removed by coagulation.

In the case of iron, oxidation is followed by settling and filtration or filtration alone, depending on the concentration of iron in the water. In the presence of turbidity (and colour) and when the Fe(II) concentration is greater than about 5 mg/l, settling or flotation would be assisted by a coagulant and/or a coagulant aid. Direct filtration is used when the iron concentration is less than about 5 mg/l. Sand (effective size 0.6 mm) or anthracite–sand filters with filtration rates of 5 (Fe <5 mg/l), 7.5 (Fe <3 mg/l) and 10 (Fe <2 mg/l) $m^3/h.m^2$ are suitable for the application. Manganese is usually found in low concentrations compared to iron and, following oxidation, the water is subjected to direct filtration. In the absence of turbidity and upstream coagulation, filtration rates of 10 $m^3/h.m^2$ or higher are used; an exception being oxidation using potassium permanganate when coagulation followed by settling and filtration are used. Filtration assists in the oxidation of Fe(II) and Mn(II) through the catalytic action of either previously deposited MnO_2 on a filter medium, or a medium containing manganese oxide such as pyrolusite or a proprietary material such as 'Polarite'. This is known as contact filtration and is effective at pH >6.5. The media usually contain at least 65% by weight MnO_2. Its specific gravity is in the range 3.5–4.0 and size range is usually 0.5–1.0 mm. It is used in the ratio sand, or anthracite-sand, to oxide media of 5:1. The retention capacity in filters is about 0.2–1.2 kg of Fe and 0.1–0.7 kg of Mn/m^2 of surface area. It is reported that the effective size and density of the media are not altered by the presence of the oxide coating (Knocke, 1991b). Contact filtration can reduce manganese to values <0.01 mg/l.

An alternative filter media is manganese greensand, formed by treating greensand (glauconite), which is a sodium zeolite, with manganous sulphate followed by potassium permanganate. Mn-greensand removes soluble iron and manganese by a process of ion exchange, frequently with the release of hydrogen ions. The process is therefore pH dependent, being virtually ineffective below pH 6.0 and very rapid at pH values above 7.5. When the Mn-greensand is saturated it is regenerated by soaking the filter bed with potassium permanganate (intermittent regeneration–IR). This procedure oxidizes manganese on the surface of Mn-greensand to MnO_2 thereby reactivating the exchange sites. It is reported that the exchange capacity is 1.45 g of Fe or Mn/l of Mn-greensand and that 2.9 g of potassium permanganate (as a 1% ^{w}/$_v$ solution) per litre of Mn-greensand is required for regeneration (Benefield, 1982). Alternatively, potassium permanganate is continuously applied to the bed by dosing it at the filter inlet, which maintains Mn-greensand active (continuous regeneration–CR) and catalyses the oxidation reaction. CR also oxidizes some iron and manganese before the water reaches the filter. Mn-greensand then acts as a filter medium in addition to catalytic oxidation of any residual soluble manganese. Mn-greensand has an effective size in the range 0.30–0.35 mm and a uniformity coefficient of 1.4–1.6 and is usually capped with a layer of anthracite to achieve longer filter runs. In the CR process chlorine can be used in place of potassium permanganate. The benefits are longer filter runs, no risk of pink water and lower cost.

It is reported that manganese dioxide filter media in contact filtration behave in a similar manner to Mn-greensand with operation being carried out in either the IR or CR mode (Merkle, 1997), with chlorine being the most suitable oxidant. In the IR mode filter washwater is dosed with about 10–15 mg/l of chlorine.

Stronger oxidants such as ozone, chlorine dioxide and potassium permanganate tend to form colloidal precipitates which may not be well retained by the filters. The use of catalytic filtration media is usually limited to manganese which is otherwise difficult to remove. In most plants iron is oxidized ahead of filtration.

Organic substances such as humic, fulvic and tannic acids when found in ground waters can form soluble complexes with iron and manganese which are not easily oxidized by oxygen (aeration). These and any soluble inorganic complexes such as silicates, sulphates and phosphates are removed by oxidation using strong oxidants or sometimes by coagulation. However, when iron is strongly complexed in the presence of significant concentrations of humic and fulvic acid, strong oxidants are sometimes ineffective. The oxidation of manganese is not similarly influenced by the presence of dissolved organics because it is not strongly complexed by organic matter (Knocke, 1990 and 1991b).

Fe(II) and Mn(II) can be removed biologically by utilising the ability of certain bacteria to produce enzymes and/or polymers that, by catalytic action, promote oxidation in the presence of oxygen in the water (Sharma, 2005). Those which promote iron oxidation are generally considered to be autotrophs but the physiology of those promoting manganese oxidation is poorly defined (Rittmann, 1989). Bacteria are usually present in the ground waters which contain these metals: for example *Gallionella ferruginea* (specific to iron) *Leptothrix* sp., *Crenothrix polyspora* and *Sphaeratilus natans*. If absent, they can be introduced from a suitable source; rapid gravity or pressure sand filters being used as biological reactors. When both metals are present in water two filtration stages are usually necessary, manganese being removed in the second stage because manganese removal bacteria require a completely aerobic environment (Mouchet, 1992). For iron removal aeration should be controlled at <10% saturation (Morris, 2001), particularly at pH values greater than 7, to prevent the chemical process competing with the biological process and therefore to minimize the risk of a chemically formed precipitate breaking through the coarse media used in high rate biological filters.

Based on a number of plants operating in France it is reported that Fe(II) oxidation takes place with pH in the range 6.0–7.5 and dissolved oxygen concentration of 0.25–1.5 mg/l. Mn(II) removal needs a pH greater than 7.5 and a dissolved oxygen concentration in excess of 5 mg/l (Mouchet, 1992). The optimum water temperature is in the range 10–15 °C depending on the predominant bacteria. Toxic elements such as heavy metals (e.g. zinc), and/or compounds such as hydrogen sulphide, hydrocarbons and chlorine must be absent. Ammonia interferes with the process and, when present in excess of about 0.20 mg/l as N, the manganese removal stage must be designed for a simultaneous, slower nitrification process. When ammonia is present in excess of about 1.0 mg/l as N, a separate biological nitrification stage using biological aerated filters (BAF) should be included between the iron and manganese removal stages (Section 10.29). Due to the oxidation kinetics, filters can be operated at rates ranging from 10 to 40 $m^3/h.m^2$ using coarse sand media (0.95–1.35 mm) with a solids retention capacity of 1–4 kg Fe or Mn/m² filter surface area (Mouchet, 1992). Unlike the chemical process where a clarification stage is necessary when iron concentration exceeds 5 mg/l, the biological iron removal process allows for direct filtration even when iron concentration is as high as 25 mg/l. Back washing is by combined air–water wash, using raw or treated but unchlorinated water. Filters can be of the open gravity or enclosed pressure type; the filtered water requires aeration (in iron removal only) and disinfection. Biological filters require a seeding period before they operate at their optimum removal efficiency; this is reported to vary from one week for iron to 3 months for manganese (Bourgine, 1994). For related reasons flow variations should be minimized. Sludge produced in biological treatment is well suited to thickening (thickened sludge concentrations of 3–8% $^w\!/\!_v$) and dewatering (Mouchet, 1992).

Other merits of biological treatment, compared to conventional physical-chemical processes are: better treated water quality (Fe and Mn residuals generally not detectable and no interference with dissolved silica which otherwise forms iron-silica complexes), longer filter runs; easier operation; lower capital cost (plants are much more compact) and operation cost (no chemicals for Fe and Mn oxidation or coagulation-flocculation); less washwater losses; and reduced manpower.

10.13 REMOVAL OF IRON AND MANGANESE FROM RIVER AND RESERVOIR WATERS

Since surface waters frequently receive treatment which includes rapid sand filtration, often preceded by coagulation and sedimentation, the removal of iron and manganese is normally included, when necessary, in the same plant. Most river waters used as sources for water supplies are well oxygenated, if not saturated with oxygen. Usually significant proportions of iron and manganese in such river waters are therefore present in insoluble forms which are removed by sedimentation and filtration treatment. However, when water is drawn from the bottom of a reservoir it may be deoxygenated and the iron and manganese dissolved (Section 7.1) but in such cases the iron Fe(II) is fairly easily oxidized. On the other hand the iron and more often the manganese Mn(II) may be present as soluble organic complexes in a very stable form. A proportion of these soluble organics are removed by coagulation. The remainder requires strong oxidants to release them from complexed organics and to oxidize them to Fe(III) and Mn(IV) (Section 10.12). The lime-soda softening process, which operates at a pH value of 10.6, removes both Fe(II) and Mn(II).

With a soft reservoir water, which requires only direct filtration, manganese can be removed by raising the pH above 9.0 before filtration; the use of an oxidant and/or a catalytic filter medium would permit operation at a lower pH (7.5–8.0). More often however, when both iron and manganese are found complexed with organics in a reservoir water, a satisfactory treatment is oxidation of Fe(II) using a strong oxidant followed by coagulation using an iron salt and clarification for removal of colour, precipitated Fe(III) and turbidity, then manganese removal at pH 9.0 or at 7.5–8.0 in downstream filters, depending on the presence of a catalytic medium. If aluminium coagulant must be used, manganese should be removed in a second filtration stage after all aluminium floc has been removed in the first stage filter, otherwise aluminium floc carried over from clarifiers would dissolve at high pH. The use of chlorine at the works inlet for Fe(II) oxidation may not be desirable due to the possibility of DBP formation. In that case Fe(II) could be removed by oxidation in the first stage filter. Biological oxidation could also be used on surface waters in second stage filters.

In some cases where a trace of manganese (less than the WHO guide value of 0.05 mg/l as Mn) has passed through a filtration plant, final chlorination has assisted in precipitating the metal to form objectionable deposits in the mains. In such cases manganese value is reduced to 0.01 mg/l. Precipitation of iron and manganese in distribution systems can be controlled through sequestration (also called chelation). Sequestering agents such as polyphosphates and sodium silicates increase the solubility of the metal ion by forming a bond with it, thereby preventing precipitation.

10.14 REMOVAL OF OTHER METALS

There are several other metals that can be present in a raw water and some might be added in the distributed water due to the corrosion of water mains and plumbing systems. Metals present in dissolved form in raw water can usually be removed by precipitation as the metal hydroxide. This involves the correction of the raw water pH, usually by adding an alkali, to a value at which the metal precipitates and can be removed by coagulation and filtration. In most cases the pH value required for precipitation can be achieved during coagulation by using aluminium or iron coagulants; the latter are favoured because they have a wider coagulation pH range. Metals which show good removal during coagulation are arsenic (pH 6–8), cadmium (pH >8), chromium (pH 6–9), lead (pH 6–9) and mercury (pH 7–8) (AWWA, 1988). Other metals which require high pH values such as those experienced in lime-soda softening include barium (pH 10–11), copper (pH 10) and zinc (pH 10).

Arsenic

Arsenic (As) occurs in the soluble form as As(III) (arsenite) under anaerobic conditions predominantly in groundwaters and as As(V) (arsenate) under aerobic conditions more commonly in surface waters. Both the forms can be effectively removed by coagulation followed by solid–liquid separation processes. Arsenic As(V) is removed with equal efficiency by aluminium and iron coagulants at pH <7.5 (Cheng, 2002) but iron coagulants are more effective than aluminium coagulants in removing As(III) and As(V) at pH >7.5 (Edwards, 1994). As(III) can be oxidized to As(V) by chlorine, ozone, chlorine dioxide or potassium permanganate; the stoichiometric requirements per mg of As(III) being 0.95, 0.64, 1.8 or 1.40 respectively (Hoffman, 2006). The oxidation is rapid; 95% conversion is achieved in 5 seconds with 1 mg/l of free chlorine in the pH range 6.5–9.5 (Gottlieb, 2005). Significant removal of As(V) can also be achieved during the oxidation of Fe(II) by co-precipitation when the two coexist (Cheng, 2002). Lime softening at pH 11 to remove all magnesium also removes arsenic (McNeill, 1997).

Arsenic removal methods applied to groundwater include adsorption on to activated alumina or granulated ferric hydroxide, IX, reverse osmosis, nanofiltration and biological oxidation. The activated alumina process is more effective in removing As(V) than As(III) (Wang, 2002), but adsorption is reduced by silicate and phosphate at pH >7. The effect of silicates (at pH >5) and phosphate (at pH >6) is much more adverse on As(III) removal (Simms, 1998). Other ions which reduce the removal efficiency are sulphates and bicarbonates. The optimum pH range for adsorption is 5.5–6.5 over which As(V) is in ionic form. At lower pH activated alumina tends to dissolve (Chwirka, 2000). Because of the effect of sulphate on arsenic removal, any pH correction should be by hydrochloric acid. The process design criteria for arsenic removal using activated alumina are given in Table 10.3. Ideally a plant should consist of an equal number of duty/standby vessels (Section 10.25 gives configurations) operating in the downflow mode. Following exhaustion the bed is backwashed to give about 50% bed expansion using raw water and then subjected to co-current regeneration using caustic soda and counter current regeneration using sulphuric acid with a co-current rinsing phase in between. During operation aluminium could leach out and should be regularly measured in treated water.

The IX process removes As(V), but not As(III) unless it is oxidized to As(V). Many anions, in particular sulphate, interfere with As removal and should be applied to waters containing dissolved solids <500 mg/l and sulphate <25 mg/l as SO_4. The process is not economic when sulphate is >150 mg/l as SO_4 (Clifford, 1999). When sulphate is present in the water, a bed which has become saturated with arsenate will begin to release it on take up of sulphate which results in a higher concentration of arsenate in the treated water than in the source water. The runs must therefore be terminated early. The process uses a strong base anion exchange resin of the chloride form and is independent of pH; regeneration is by sodium chloride. Empty bed contact time is 1.5 to 3 minutes. The quoted exchange capacity of the resins is low (1.4 g eq/l) but does not exhaust rapidly because arsenic concentrations in most ground waters are very low. The presence of Fe(III) can complex arsenic and affect removal (Clifford, 1998).

As(V) is removed in biological iron removal by co-precipitation. Arsenic can be removed by adsorption on to granulated ferric hydroxide in filters (Thirunavukkarasu, 2003; Driehaus, 1998) and is available as proprietary media: e.g. Bayoxide 33® and GFH®. Bayoxide 33® media has a size range 0.5–2 mm, adsorption capacity of 4–12 g As/100 g, bulk density of 450–500 kg/m³ and a specific gravity of 3.6. Typical media depth used is 0.7–1.1 m with an empty bed contact time (EBCT) of 2.8–5 minutes. The equivalent flow rate is 14–22 BV/h. The feed pH should be in the range 5.5–8.5. The media is said to reduce As concentrations from up to 150 µg/l to 6–10 µg/l. The media is very effective against As(V); As(III) requires pre-oxidation. The bed is backwashed at 25–30 m/h. The bed is not regenerated; once exhausted it is discarded. (Source: Severn Trent Water Purification Inc., Tampa, Florida, USA).

Nanoparticle agglomerated media such as the proprietary products Adsorbsia® and MetsorbG® (titanium dioxide), ArseneXnp® (iron oxide) and Isolux® (zirconium oxide) are used in arsenic removal (Westerhoff, 2006). Arsenic can also be removed by reverse osmosis or nanofiltration (Waypa, 1997). Electrodialysis removes about 80% of arsenic.

The spent regenerants from adsorption and IX processes can be treated by aluminium or iron salts to form insoluble arsenates or by lowering the pH to 5–6.5 to form insoluble hydroxides. Waste is an issue; treatment processes are reviewed in a US EPA report (MacPhee, 2001).

Lead

Lead is rarely a contaminant of any significance in natural water. In polluted waters the total concentration of lead could be as high as 10 mg/l with the dissolved fraction usually less than 0.01 mg/l (Galvin, 1996). Particulate lead is effectively removed by settlement with the aid of coagulants and soluble lead is removed at pH >9 such as would be used in lime softening. Lead in drinking water is mainly introduced through corrosion of plumbing systems containing lead pipes and fittings, brass

Table 10.3 Process design criteria for fluoride and arsenic removal by activated alumina

Parameter	Fluoride	Arsenic[a]
Media	Alcan Granular Activated Alumina	Alcan Granular Activated Alumina
Media type 1	AA400G	AA400G
Media size	0.6–1.2 mm	0.3–0.6 mm
Media capacity		
—maximum	1.4 g/100 g	1.7 g/100 g
—practical	0.1–1.0 g/100 g	0.01–0.2 g/100 g
Bulk density	770 kg/m^3	770 kg/m^3
Specific gravity	2.6	2.6
Media type 2	AAFS50	AAFS50
Media size	0.6–1.2 mm	0.3–0.6 mm
Media capacity		
—maximum	1.4 g/100 g	2.8 g/100 g
—practical	0.1–1.0 g/100 g	0.01–0.5g/100g
Bulk density	1060 kg/m^3	1060 kg/m^3
Specific gravity	2.6	2.6
Media depth	>0.75 m	>0.75 m
Flow rate	6–12 BV/h[b]	6–12 BV/h
EBCT[c]	6 minutes	6 minutes
pH of feed	5–6	5.5–7.5
Initial raw water concentration	3–8 mg/l	10–200 µg/l
Final raw water concentration	0–3 mg/l[d]	0–10 µg/l[d]
Regeneration of media type 1. Media type 2 is not typically regenerated; once exhausted it is discarded.		
Alkali	NaOH	NaOH
Concentration	1% $^w/_w$	1–2% $^w/_w$
Regenerant volume[e]	5–10 BV	5–10 BV
Rinse water volume[e]	5–8 BV	5–8 BV
Acid	H_2SO_4	H_2SO_4
Concentration	0.25% $^w/_w$ (0.05N)	0.25% $^w/_w$ (0.05N)
Acid volume[e]	4–10 BV	4–10 BV

Source of information: Alcan Speciality Aluminas, Brockville, Ontario, Canada

Notes: [a]Removes As(V) only. Moderate removal of As(III). Pre-oxidation recommended for complete removal.

[b]BV—Bed volume: bulk volume occupied by adsorbent in a vessel.

[c]EBCT—Empty bed contact time: time required for a volume of solution equal to the bed volume of adsorbent to pass through the column.

[d]Depending on requirements.

[e]Regenerant, rinse water and acid are applied at rates moderately lower than the raw water flow rate.

fixtures and lead compounds used in pipe jointing materials. Except in very low alkalinity waters, lead enters the water not so much from direct contact with the lead pipe wall as from contact with lead-rich corrosion products formed as a scale on the pipe wall. The overall solubility of lead in drinking water, referred to as plumbosolvency, is controlled by the solubility of these corrosion products and this in turn is strongly influenced by alkalinity and pH. The waters which are most corrosive towards lead have a low alkalinity (<50 mg/l as $CaCO_3$) and pH <7. Except for very low alkalinity waters (<5 mg/l as $CaCO_3$), increasing pH above 7 dramatically reduces the lead solubility, with a theoretical minimum at a pH of approximately 9.8. At alkalinity levels above around 100 mg/l, lead solubility becomes increasingly less sensitive to changes in pH value. The least plumbosolvent water has a pH greater than 8.5 and alkalinity of between 10 and 80 mg/l as $CaCO_3$. (Shock, 1996; Sheiham, 1981).

Corrective measures are applied at the treatment works to minimize dissolution of lead in distribution. This is accomplished by appropriate control of pH and alkalinity of the water to assist in the formation of relatively insoluble lead compounds consisting of carbonates and hydroxide as a film on lead surfaces or by the dosing of orthophosphates to form a coating of sparingly soluble lead orthophosphate. For waters of low alkalinity (<50 mg/l as $CaCO_3$), lead solubility can be greatly reduced by increasing the pH into the range 9–10 (Shock, 1996) but this would contravene most national and international drinking water standards. In practice therefore elevation of pH value is restricted to about 8.5 and not less than 8.0 at the consumer's tap. There is generally no need to increase alkalinity (Sheiham, 1981) although this may be desirable if the buffering capacity of the water is inadequate to maintain the pH unaltered in the distribution system. In a well-buffered water pH variation would be limited to about 0.5 units whereas if buffering is inadequate the variation could be as much as 2.5 units. Soft and aggressive waters with alkalinity less than 50 mg/l as $CaCO_3$ could also be treated by lime and carbon dioxide. While control of pH is generally sufficient to reduce lead levels to around 50 μg/l for moderate to low alkalinity waters, any further reduction normally requires the addition of orthophosphate to form a less soluble lead phosphate film on the pipe wall. The use of pH elevation is in any case generally not suited to higher alkalinity waters where the formation of calcium carbonate scale would be a problem; orthophosphate is required in such waters.

Orthophosphate dosing provides a simple and effective treatment for achieving significantly lower lead concentrations than is possible by pH adjustment. Lead solubility in the presence of orthophosphate is comparatively insensitive to pH > about 7, especially for low alkalinity waters. For higher alkalinity waters the most effective pH range is in the region 7.2–7.8. There are suggestions that raised dissolved organic carbon levels in the water can result in increased plumbosolvency, possibly as a result of a sequestering mechanism; in such cases phosphate dosing may be less effective than elsewhere.

The formation of protective lead phosphate films is considerably quicker for new lead pipes than for older material, particularly where older pipe has extensive carbonate scale protecting the elemental lead from the phosphate. The rate of initial film formation is thus often accelerated by using a high initial dose of 1.5–2 mg/l as P (4.5–6 mg/l as PO_4), particularly in higher alkalinity waters. Thereafter a maintenance dose in the range 0.8–1.2 mg/l would normally be sufficient to achieve optimum reduction, although higher doses may continue to be used for higher alkalinity waters. Further reductions in lead concentrations often continue to be seen for two years or more after phosphate dosing commences as the system reaches a new equilibrium, particularly in areas of high water alkalinity. Lead solubility increases with increasing water temperature and some water companies operate with increased phosphate doses in summer and reduced doses in winter.

The film formation process is reversible and interruptions to treatment should be kept to a minimum (<1 week); damage or repair to lead plumbing system would also affect the protection (Colling, 1992). Older pipe may also be more susceptible to lead-rich deposits flaking off from the pipewall, giving rise to particulate lead in the water, which may on occasion exceed 200 µg/l. Such episodes may be initiated by mechanical vibration or water hammer in the supply pipework. Phosphate dosing is of little value in eliminating such flaking.

Orthophosphate is added as one of its sodium salts or as orthophosphoric acid (Table 7.5). A phosphate monitor can be used in closed loop control for phosphate dosing. Phosphate injection points should be separated from any lime dosing points. Polyphosphates have no effect in reducing plumbosolvency and are known to increase lead concentrations, on occasions, by sequestering lead into soluble compounds. Monitoring is critical to the success of plumbosolvent treatment. Parameters to be monitored should include pH, alkalinity, orthophosphate and water temperature, and should be carried out both at the treatment works and in the distribution system.

The application of plumbosolvent treatment could have adverse or beneficial effects in the following areas depending on the characteristics of the distribution system: discoloration of water due to iron release; corrosion of other metals and cement lining of pipes; biofouling; and scaling; DBP formation; effects of chloramination; and effects of mixing waters in the system. Recent studies suggest that phosphate dosing in the presence of fluoride, either naturally present or dosed artificially for dental caries control, can result in the formation of fluorapatite, a hard crystalline substance that is highly insoluble. The mechanism for its formation is not fully understood but appears to be initiated by local high temperatures, such as may be found in combination boilers or commercial calorifiers. This can result in solids deposition on heat exchange surfaces, ultimately leading to equipment failure.

Aluminium

Aluminium (both soluble and particulate) in surface waters is readily removed by coagulation using aluminium or iron coagulants in the pH range 6.5–7.2. In the presence of manganese, aluminium is first removed by coagulation and filtration (with or without clarification) followed by elevation of pH (>8), oxidation and secondary filtration for manganese removal (Section 10.12). A similar treatment regime is applied when treating raw water containing manganese with aluminium coagulants. Since aluminium hydroxide is soluble at pH greater than about 7.5, it cannot be removed together with manganese. There are exceptions to this rule as in some waters manganese can be removed at pH lower than 7.5. Aluminium is rarely found in ground waters in excess of 10 µg/l.

10.15 **REMOVAL OF RADIONUCLIDES**

An introduction to radionuclides in water is given in Section 6.43. Radon is a water soluble gas primarily found in ground waters. It has a high Henry's constant and therefore it is easily stripped-off by aeration. Packed tower aerators (Section 10.20) show removal efficiencies of about 98%, closely followed by the diffused air system (Section 10.23). Removal in a spray aerator (Section 10.21) was found to be less than 75% (Dixon, 1991). For a packed tower the packing height is about 3 m with a surface loading rate of about 75 m³/h.m². The packed tower performance was found not to be influenced by air to water flow ratio. Problems with packed towers are carbonate scaling and iron deposits if Fe(II) is present in the water (Section 10.20). Radon discharged to air from packed towers is diluted to insignificant ground level concentrations. GAC adsorbers are also known to remove radon but not as effectively as aeration. Empty bed contact time required is more than an hour and therefore suitable only for small capacities (Haberer, 1999).

All particulate radionuclides and uranium can be removed by coagulation using iron or aluminium salts followed by solid–liquid separation. Methods used for dissolved radionuclides are: lime softening at pH >10.5 for uranium and radium-226; strong base anion exchange of the chloride form for uranium, particularly in low sulphate water (Zhang, 1994); strong acid cation exchange of the sodium or calcium type for radium (Snoeyink, 1987); and mixed-bed IX for β-particles (US EPA, 2002), radium and uranium. Both the resins are regenerated with sodium chloride. Reverse osmosis could be used for uranium, radium, α- and β-particles.

In the treatment plant radionuclides become concentrated in waste streams and need careful disposal. US EPA (2005) has suggested disposal alternatives based on the concentration of radionuclides in the waste stream. These include disposal of liquid wastes to watercourses, sewer or deep injection wells and treatment by evaporation or precipitation with disposal of solid wastes (dewatered sludge) to various types of landfill facilities.

DEFLUORIDATION AND FLUORIDATION

10.16 DEFLUORIDATION

Some groundwaters contain high levels of fluoride with concentrations well in excess of 1.0 mg/l as F (Section 6.29). Levels in excess of 1.5 mg/l as F may cause dental fluorosis leading to mottling of the teeth; reduction of fluoride may therefore be necessary. Defluoridation can be achieved by chemical precipitation, adsorption or by membrane desalination processes. Lime softening at pH >10 followed by soda ash dosing removes fluoride in the presence of magnesium by adsorption on to magnesium hydroxide. The water may be dosed with magnesium sulphate or dolomitic lime; 1 mg/l of fluoride requires 50 mg/l of magnesium (Degremont, 2007). Precipitation by aluminium sulphate requires doses up to 750 mg/l and is not economical. High levels of fluoride can be reduced down to the solubility of calcium fluoride using lime, leaving about 8 mg/l fluoride in water.

Adsorption onto activated alumina is successfully used for fluoride removal. It is highly selective to fluoride in the presence of sulphate and chloride when compared to synthetic IX resins. In the presence of bicarbonates, although the fluoride concentration is reduced, the adsorption capacity shows a major decline. Silica is also known to interfere with the adsorption of fluoride. The adsorption process is best carried out under slightly acidic conditions (pH 5–6); the lower the pH the more effective the removal. Breakthrough occurs slowly and the operating cycle should be terminated before alumina is fully exhausted. During operation aluminium could leach out and should be regularly measured in treated water. Process design criteria for fluoride removal are given in Table 10.3. The plant and its operation are similar to that described for arsenic removal (Section 10.14). The spent regenerant can be treated with lime or dried in evaporation ponds.

Rejection of fluoride by reverse osmosis (Section 10.47) is pH dependent. Rejections at alkaline pH values can be greater than 99% due to fluoride being in the salt form. In the acidic pH values due to fluoride ions the rejection is less than 50%. Electrodialysis (Section 10.46) also removes fluoride.

10.17 FLUORIDATION

Many waters are deficient in natural fluoride and some regional or national health authorities consider that it should be added to reduce the incidence of dental caries. Fluoridation of water is carried out by using either hexafluorosilicic acid, which is a solution usually containing 20% $^{w}/_{w}$ H_2SiF_6 (15.8% $^{w}/_{w}$ F), or disodium hexafluorosilicate, which is supplied as a powder usually containing at least 98% Na_2SiF_6 (59.4% $^{w}/_{w}$ F) (Table 7.5). The water used for dissolving the powder should be softened to <75 mg/l as $CaCO_3$ using a base exchange softener (Section 10.7) to prevent calcium carbonate scaling. Alternatively a sequestering agent such as a polyphosphate could be used to minimize scaling.

The dosing system should be designed to ensure that the quantity of fluoride injected to supply over 24 hours does not exceed the maximum dosage allowable. Acid should be transferred from storage tank by pump to an intermediate storage tank (a day tank) which holds not more than 24 hours maximum usage. For the pre-dilution system, a day tank or a system to limit the number of batches per day is essential. For powder systems a hopper with one day's capacity should be used.

Fluoride should be accurately metered by positive displacement pump of the reciprocating diaphragm type, which should be of such capacity that it operates close to its maximum capacity most of the time. Where the works flow varies frequently by more than 5% the pump motor speed should be arranged to automatically vary with flow. Closed-loop control of the output based on a signal from a fluoride monitor could be used provided a reliable on-line fluoride analyser could be found.

The fluoride dose required varies with ambient temperature ($T°C$); higher temperatures require lower dosages, since more water is consumed in a hot climate. The dose required can be calculated by: $F = 0.34/E$, where $E = -0.38 + 0.0062(T \times 1.8 + 32)$ (Gallagan, 1957).

Fluoride should be dosed after physical treatment alongside chlorine used for final disinfection, but the injection point should be separated from lime injection point for final pH correction. Following injection it should be well mixed in the water, and a sample representative of the water to supply should be taken for monitoring of fluoride. Technical aspects on fluoridation are published in a Code of Practice by UK DWI (2005).

Both the acid and powder are highly toxic and the acid is very corrosive. Acid vapour and dust from the powder should be contained to prevent inhalation and ingestion by operators. Splashes of acid on skin should be washed with copious amounts of cold water.

AERATION

10.18 PURPOSE

Aeration has a large number of uses in water treatment; the more usual are to:

- increase the dissolved oxygen content of the water;
- reduce tastes and odours caused by dissolved gases in the water, such as hydrogen sulphide, which are then released;
- decrease the carbon dioxide content of a water and thereby reduce its corrosiveness and raise its pH value;

■ oxidize iron and manganese from their soluble to insoluble states and thereby cause them to precipitate so that they may be removed by clarification and filtration processes; and to
■ remove certain volatile organic compounds.

According to Henry's law, the equilibrium solubility of a gas in water is given by $C_s = kp$ where C_s is the saturation concentration of the gas in water (mg/l), p is the partial pressure in bar and k is the coefficient of absorption which is equal to $(55\,600 \times M) \div H$ where M is molecular weight of the gas and H is Henry's constant. The higher the H the easier the desorption.

The purpose of aeration is to speed up this process and there are two main types of aerators in general use: those in which water is allowed to fall through air, e.g. free-fall aerators, packed tower aerators and spray aerators, and those in which air is injected into water, e.g. injection aerators, diffused air-bubble aerators and surface aerators. For them to be effective it is essential they provide not only a large water–air interface (high area: volume ratio) but, at the same time, a high degree of mixing and rapid renewal of the gas–liquid interface to facilitate transfer of oxygen.

10.19 CASCADE AERATORS

Cascade aerators are of linear, rectangular or circular shape and consist of a series of steps over which water flows (Plate 16(b)). Figure 10.3 shows the design of a cascade aerator. Such aerators are widely used as water features: they take large quantities of water in a comparatively small area at low head, they are simple to keep clean, and they can be made of robust and durable materials giving

FIGURE 10.3

Cascade aerator.

a long life. The steps are usually made of reinforced concrete. The aerator should preferably be in the open air or, for protection against air-borne pollution, freezing and algal growth, in a small house which has plenty of louvred air inlets. They are efficient for raising the dissolved O_2 content but, not for CO_2 removal; reduction of CO_2 content is usually in the range of 60–70%. From work carried out by the former UK Water Pollution Control Laboratories (DoE, 1973) an empirical relationship has been developed for the ratio r of the oxygen deficit just above a weir to that just below which is:

$$r = \frac{C_s - C_u}{C_s - C_d} = 1 + 0.38abh(1 - 0.11h)(1 + 0.046T)$$

where a is 1.25 in slightly polluted water and 1.00 in moderately polluted water; b is 1.00 for a free-fall weir and 1.30 for a stepped weir; h is the height of the fall (difference in water levels) in metres, and T is the water temperature in °C, C_u and C_d are oxygen concentrations upstream and downstream, and C_s is the saturation concentration at T°C. This equation is valid only for falls up to about 0.6 m. For falls up to 3 m more representative results are obtained by replacing the term $0.38abh(1-0.11h)$ by: $ab(0.65 + 0.25h)$.

In the design of such aerators certain hydraulic criteria should also be met. These include the depth of the receiving pool y (m) defined in terms of the Drop number (D) as:

$$y/Y = 1.66\,D^{0.27} \text{ and } D = q^2/gY^3$$

where Y (m) is the height of the fall from crest to pool floor and q is the unit discharge defined as the flow Q (m³/s) per unit length L (m) of weir. The horizontal component x (m) of the trajectory of the water fall over the weir assuming no air resistance, is defined for a weir coefficient of 1.705 and where H is the head over the weir by:

$$x = [1.33H(Y - y)]^{0.5}$$

The number of steps varies from 3–10 and the fall in each step from 0.15 to 0.6 m. The rate of flow should be limited to about 125 m³/h.m weir length. To allow entrained air to mix in the water, each step should have a pool of water of depth at least 0.3 m. Weirs with serrated edges perform better (van der Kroon, 1969) as they help to break water flow into separate jets. If the water is allowed to cling to the steps especially at low discharge rates, the efficiency is reduced. This could be avoided by using a sharp edged thin plate weir or by providing a protruding lip using a steel plate. The space requirement is typically of the order of 0.5 m² per 1 m³/h water treated. The oxygen transfer efficiency for 50% saturation at 10°C and atmospheric pressure, starting at zero is calculated to be 1.56 kg O_2/kWh.

10.20 PACKED TOWER AERATORS

Packed tower aerators consist of a vertical steel, plastic or concrete cylinder filled with plastic or ceramic packing such as pall-rings, berls or saddles (Fig. 10.4 and Plate 16(c)). Water is applied at the top through a distributor and a counter current flow of air is usually blown (forced-draught) but sometimes drawn (induced-draught) using an oil-free blower or fan; air must be filtered to prevent contamination of water. The air to water volumetric flow ratio is in the range 25:1 to 75:1. The water becomes distributed in a thin film over the packing so providing the large water–air interface

FIGURE 10.4

Packed tower aerator.

required for good mass transfer. Surface loading rates are of the order of 50–100 m³/h.m². Packed towers constitute the most efficient form of aeration and are used primarily for ground waters; suspended solids in surface waters can rapidly clog the packing. They give over 90% oxygen transfer and 85% CO_2 removal efficiencies. They are also used for the removal of hydrogen sulphide, ammonia and volatile hydrocarbons and can deal with varying throughputs with no decrease in efficiency. Typically the tower needs up to about 10 m head of water.

Drawbacks of the system are that CO_2 removal increases the tendency for scale deposition, whilst in the presence of Fe(II), ferric hydroxide precipitate tends to foul the packing. A sequestering agent such as sodium hexametaphosphate can be used to eliminate scale deposition. After a number of years the packing needs replacing, depending on the deposits that have accumulated on the surfaces of the material. Capital and operating costs are high but the packed tower aerator is effective to deal with high CO_2 content, ammonia (Section 10.27) or volatile hydrocarbons (Section 10.31). The oxygen transfer efficiency for 50% saturation at 10°C and atmospheric pressure, starting at zero oxygen is calculated to be 5.0 kg O_2/kWh.

10.21 SPRAY AERATORS

Spray aerators comprise nozzles installed on a pipe grid at 1–3 m spacing. Water sprayed from the nozzles forms fine streams of small droplets which produce a large contact area with the air. They can be used on hard waters and those containing iron and manganese without risk of scale deposition or fouling problems. About 70% CO_2 removal and 80% oxygen transfer efficiencies can be obtained with the best type of spray nozzles. Nozzle diameter ranges from 25–35 mm with flow rates up to 30 m^3/h and is a compromise to prevent clogging but avoid the need for excessive pressure. Up to 20 m head of water may be required. A large collecting area (typically 0.5–1 m^2 per m^3/h water treated) is necessary to accommodate the large number of nozzles required and to minimize overlapping of jets. The receiving pool depth should be about 150 mm. Extra efficiency is obtained in some types of plant in which the spray is broken up by impinging on a plate. The sprays require protection from wind and freezing and should be enclosed in a louvred building. They are normally used on ground waters; suspended solids in surface waters can block the nozzles. The oxygen transfer efficiency for 50% saturation at 10 °C and atmospheric pressure, starting at zero oxygen is calculated to be 1.0 kg O_2/kWh.

10.22 INJECTION AERATORS

Injection aerators avoid the need to break the water pressure if this is particularly inconvenient or wasteful of energy. The water may be sprayed into a compressed air space at the top of a closed pressure vessel under pressure. The air has to be circulated by a compressor (Section 7.18). Alternatively compressed air may be mixed in water by a side stream static mixer and injected into the flowing water in a pipe upstream of an inline static mixer; or air at atmospheric pressure may be drawn into the pipe where a constriction, such as the throat of a venturi tube, reduces the water pressure below atmospheric. In the latter case the venturi tube has a much narrower throat and a much longer divergence cone downstream than a venturi designed for flow measurement. Aeration under pressure does not remove CO_2; its effect is to increase the oxygen content of the water and to saturate it with nitrogen at the operating pressure. The latter can be a disadvantage because air can appear as bubbles when the pressure is released and can lead to air binding and consequently spurious headloss in a downstream filtration process.

10.23 OTHER TYPES OF AERATORS

Mechanical surface aerators operate on the basic principle of entraining air by agitation of the water surface. There are several forms of surface aerators which are successfully used in sewage treatment works. Those where the axis of rotation of the impeller is vertical are used for aerating water in large open tanks. The oxygen transfer efficiency for a mechanical surface aerator is about 0.5 kgO_2/kWh.

   Diffused-air bubble aeration is usually carried out by passing air through some form of a diffuser placed at the bottom of a tank through which the water flows horizontally. The diffuser may be a perforated pipe or a porous ceramic plate. Aeration efficiency is a function of bubble size and tank depth; with smaller bubbles, efficiency increases with decreasing depth. Tank depth varies from 3 to 4.5 m; tank width should not exceed twice the depth. Residence time is about 10–20 minutes. Air (at atmospheric pressure) to water volumetric ratio is usually between 1:1.25 and 1:2.5. Air must be filtered to minimize contamination of water and clogging of the diffusers and the compressors used must be of the oil-free type. The oxygen transfer efficiency for 50% saturation at 10 °C and atmospheric pressure, starting at zero oxygen is calculated to be 2.5 kg O_2/kWh. This method is also widely used in sewage treatment. In water treatment, the perforated pipe arrangement finds application in the destratification of reservoirs (Davis, 1980) (Section 7.1).

NITRATE REMOVAL

10.24 GENERAL

Nitrate is found in undesirable concentrations in some water sources (Section 6.36). Blending of sources is the most basic method for achieving low nitrate supplies. For surface waters long-term storage can be used to absorb the seasonal peaks and to bring about natural biological denitrification (Section 7.1). Where such methods are not feasible, treatment of the water for nitrate reduction is necessary. The two nitrate reduction processes predominantly used are IX and biological denitrification. Membrane processes such as electrodialysis reversal and reverse osmosis can also be used for nitrate removal. All these processes bring about a nitrate removal in excess of 80%. Therefore treatment is applied only to a small part of the flow so that the desired concentration is achieved after blending with the remainder.

10.25 ION EXCHANGE PROCESS FOR NITRATE REMOVAL

The IX process is similar to that used for water softening (Section 10.8). In nitrate removal a strong base anion exchange resin in the chloride form is used. As the water passes through the bed of resin contained in a pressure vessel, nitrate and other anions in water are exchanged with chloride in the resin, thus releasing chlorides into the water. When the resin is saturated with respect to nitrate (indicated by nitrate breakthrough) the resin is regenerated with sodium chloride solution (brine) and the exchange process is reversed; anions absorbed on the resin are replaced by chloride ions. To ensure that flow is maintained during regeneration ideally 100% standby units should be provided but a total capacity of at least $3 \times 50\%$ flow for small works and $4 \times 33\%$ for large works may be considered. Either 'classical' anion exchange (Section 10.10) or nitrate selective exchange resins can be used in the process. The former has been used with success except for waters with significant sulphates. Nitrate is less well absorbed than sulphate and nitrate uptake falls off rapidly as raw water sulphate increases. If any sulphate is present in the water, a bed which has become saturated with nitrate will begin to release it on take up of sulphate; this results in a higher concentration of nitrate in the treated water than in the source water. This drawback can be overcome by terminating runs early. High levels of sulphate reduce the resin capacity for nitrate and increase regenerant (sodium chloride) consumption by reason of removal of anions other than nitrate and, as a result of sulphate removal, more chlorides are released from the resin to the water. Therefore, for waters containing mass ratio of sulphate (as SO_4): nitrate (as N) in excess of 3.43:1 or nitrate (as NO_3) in excess of 0.77:1 (1 g of nitrate as N = 4.43 g of nitrate as NO_3), it may be more economical to use nitrate selective exchange resins. These preferentially absorb nitrate and therefore do not show many of the drawbacks inherent to classical resins.

Surface loading rate for nitrate removal is about 30 $m^3/h.m^2$. Allowing about 20% of the time for regeneration, this corresponds to a mean throughput of 576 $m^3/day.m^2$. Volumetric flow rate should be less than 40 bed volumes (BV)/h; 20 BV/h is regarded as a suitable design value with all columns in service. The resin bed would be about 1.5 m deep and vessel diameters would usually be in the range 1–4 m. The length of a run (time between two consecutive regenerations) is a function of operating conditions and the resin and is usually 8–12 hours. The regeneration sequence on completion of a run comprises: backwashing of the bed to remove suspended solids; counter-current regeneration using 6–10% $^w/_w$ sodium chloride solution at 3–4 BV/h for

1.5 BVs; upflow slow rinse at 3–4 BV/h for 2 BVs; a fast rinse in the direction of service flow at 6 BV/h for 4 BV; return the bed to service. Frequency of backwashing is about every 10–25 cycles depending on the raw water quality. It is normal to use raw water for backwash and treated water for rinse. The quantity of salt required is about 160 g NaCl/l of resin for classical resins and 125 g NaCl/l for nitrate selective resins. The nitrate removal capacity for the two types of resin is about 0.25 g eq/l of resin. When treating waters containing high alkalinity, provision for acid washing of the bed using hydrochloric acid is recommended to remove calcium carbonate deposits.

Drawbacks with the use of IX for nitrate removal are that the process increases the chloride concentration and reduces alkalinity of the product water; and the need for disposal of the spent regenerant. Although the increase in chloride level of the water has no health implication, it can increase the corrosivity of the water (Section 10.42). The problem is less severe with nitrate selective resins. At the beginning of the run chloride ions increase, accompanied by a reduction in the alkalinity. The product water would therefore have a high chloride to alkalinity ratio initially; classical resins produce higher ratios than nitrate selective resins. This problem can be overcome by either or both the following methods: mixing the output of a run in a treated water reservoir before forwarding it to supply; dividing the flow between two or more parallel units and operating them out of phase. A further but more costly option is to use a bicarbonate in place of the chloride solution in the last 10–15% of the regeneration phase, to replace the chloride ion in the resin with bicarbonate ion. Thus at the start of the run bicarbonate ions and not the chloride ions are released into the water.

Waste from the plant contains high concentrations of nitrate, chloride and other anions. Depending on the nitrate removal required, the volume can amount to about 1.5–2.0% of the IX plant throughput. This may be safely discharged to a sewer or, for sites in coastal areas, to the sea. In other cases removal by tanker may be necessary as its discharge to a water course may not be acceptable. For large plants tanker removal may need to be preceded by a volume reduction process such as electrodialysis reversal or reverse osmosis; in some countries solar evaporation may be used. The Eliminate® system regenerates the resin with potassium chloride and the waste stream containing nitrate is converted to nitrogen gas by electrolysis.

The ISEP® nitrate removal process utilizes nitrate selective resin in a carousel with a multiport distribution valve arranged to feed about 30 columns of which about 20 are operating in parallel in different stages of exhaustion with the remainder in stages of rinsing (with the effluent being used to dilute the regenerant brine), counter current regeneration and displacement. Softened water is used to minimize precipitation of calcium sulphate and calcium carbonate during rinsing and displacement respectively. Recovery is about 99.7% compared to 95% for the conventional processes. Advanced Amberpack® system utilizes a fractal distribution system to ensure uniform flow across the resin bed thereby increasing the efficiency of the process and improving recovery to >99%.

10.26 BIOLOGICAL PROCESS FOR NITRATE REMOVAL

Biological denitrification of drinking water is based on the heterotrophic process that occurs in the anoxic zone in sewage treatment. In this process an organic carbon source is used to sustain bacterial growth, using oxygen bound in nitrate for respiration, reducing it to nitrogen. In water treatment, bacteria first consume any dissolved oxygen in water before the oxygen in nitrate and create an additional demand for carbonaceous matter. Organic carbon should be added to the water as most water supplies contain relatively low concentrations; the sources of organic carbon (substrate) are methyl alcohol, ethyl alcohol or acetic acid. Stoichiometric quantities of methyl alcohol, ethyl alcohol or acetic acid required for each mg of dissolved oxygen and nitrate (as N) are 2.57, 1.85 and 3.62 mg, respectively; the actual demand could be up to 1.5 times greater. Optimum pH is in the region of 7.5. The process has little effect on the alkalinity of the water. Trace concentration of phosphate of less than 0.5 mg/l as P is also needed to assist bacterial growth, and should be added if this is not present in the water. The process is sensitive to temperature and reaction rates decrease markedly below about 8 °C. The autotrophic denitrification process which uses hydrogen (Gross, 1988) or sulphur compounds (Soares, 2002) to sustain bacterial growth, is also used in some full-scale plants. The reaction with sulphur produces acidity and should be neutralized with calcium carbonate.

The biological process is usually carried out in fluidized bed or up or down flow fixed bed reactors where the biological growth is physically supported on a medium. Fine sand is commonly used in fluidized bed reactors, whilst a porous medium such as expanded clay is used in fixed bed reactors. In a fluidized bed, water mixed with the substrate flows upward at 20–30 m³/h/m² to provide 40–50% bed expansion and giving a detention time typically of 5–10 minutes. Before start up, the bed requires seeding with bacteria. This could take up to 1 month. As the biomass builds up, a proportion of the sand is periodically removed from the bed, the bacterial film is stripped in a sand cleaning plant and the sand returned to the bed. Fixed bed reactors are usually based on conventional sand filter principles, the media used being a porous medium such as expanded clay, of coarse grain size. An empty bed contact time of 10–15 minutes is usually used. Backwashing regime is similar to that used in conventional sand filters. The treated water from the biological reaction is devoid of oxygen and contains dissolved organic carbon and bacterial floc carried over from the reactor. Hence the water needs to be reoxygenated and filtered through granular activated carbon and sand filters before passing to supply.

10.27 **MEMBRANE PROCESSES FOR NITRATE REMOVAL**

Membrane processes used are electrodialysis reversal and reverse osmosis and are discussed in Sections 10.46 and 10.47. Both are desalination processes and are substantially non-selective. It can be an advantage to remove multiple contaminants in a single process step.

REMOVAL OF AMMONIA

10.28 **CHEMICAL AND PHYSICAL METHODS**

Ammonia is present in water as saline or free form (Section 6.7). The most common method used for ammonia removal is 'breakpoint' chlorination (Section 11.9) where ammonia nitrogen is completely oxidized to nitrogen leaving a residual of free chlorine. To minimize DPB formation breakpoint chlorination should be practised after DPB precursor removal. Organic nitrogen is not destroyed by chlorine. Ozone does not normally oxidize ammonia. Neither chlorine dioxide nor potassium permanganate affects ammonia.

When present as free ammonia or ammonium ion, ammonia can be removed in packed tower aerators (Section 10.20). In the latter case the pH of the water needs to be raised to 10.5–11.5 to

(a) Schematic of reverse osmosis desalination plant for sea water.

(b) Tuas SWRO plant (136 Ml/d), Public Utilities Board, Singapore. (Black & Veatch consultant to EPC contractor, Hydrochem (S) Pte. Ltd.; developer: Hyflux Ltd.; operator: SingSpring Ltd.). (Photo: courtesy of Hyflux).

PLATE 17

(a) Tuas SWRO plant, membrane units (see Plate 17(b) for credits).

(b) 88 Ml/d multi-stage seawater flash distillation plant with 320 MW output on the Mediterranean coast. (Black & Veatch).

PLATE 18

convert all ammonium ions to free ammonia. A design for a packed tower is given by Short (1973). Due to the large air:water ratio (3000:1), the height of the tower and chemicals required for pH adjustment the process is uneconomical.

Strong acid cation exchange resins used in base exchange softening could be used for ammonia removal (Section 10.7). Ammonium ion (NH_4^+) rejection by reverse osmosis is about 99.5% at pH 8.5 (Section 10.48).

10.29 **BIOLOGICAL METHODS**

Biological oxidation of saline ammonia to nitrate, known as nitrification, takes place in two steps (Richard, 1978): initially the formation of nitrite by *Nitrosomonas* bacteria, followed by oxidation to nitrate by *Nitrobacter* bacteria. Both steps require oxygen. Carbon dioxide is the carbon source; 1 mg/l ammonia (as N) consumes about 7.2 mg/l alkalinity (as $CaCO_3$). When treating some soft waters therefore alkalinity may have to be added. The process also requires up to 0.2 mg/l phosphates (as P) to allow nitrifying bacteria to develop. The seeding period is about 1-2 months. The optimum pH range for the reaction is between 7.2 and 8.2. In practice nitrification is almost 100% complete (Lytle, 2007). The water temperature should be greater than 10 °C; there is no biological activity below 4 °C. Optimum temperatures for bacterial growth are in the range 25–30 °C. The process requires oxygen at a rate of about 4.57 mg per mg of ammonia (as N) (Richard, 1978). At high ammonia concentrations simple saturation of the water with oxygen by aeration may therefore prove inadequate and oxygen must be continually added.

In biological removal of ammonia in drinking water treatment, nitrification is usually carried out in rapid gravity filters. A maximum ammonia concentration of 1.5 mg/l as N can be removed in conventional rapid gravity filters, depending on the temperature and the dissolved oxygen concentration of the influent water. Filtration rates could be in the range 5–10 m³/h.m² and to remove 1 mg/l ammonia (as N), the necessary EBCT is about 20, 10 and 5 minutes at 5, 10 and 30 °C, respectively. After nitrification the water is devoid of oxygen and needs to be reaerated. For higher ammonia concentrations biological aerated filters (BAF) have been used. These are similar in principal to the trickling and/or aerated filters used in sewage treatment where there is a continuous flow of air through the media of the filter bed. The ammonia loading of a BAF filter is typically 0.25–0.6 kg of ammonia (as N)/day.m³ (of media), depending on the type and effective size (ES) of media.

In drinking water treatment the BAF consists of a layer of coarse media (ES of between 1.5 and 3.0 mm); filtration is either downflow counter current or upflow co-current with air flow injected continuously into the bottom of the filter bed using either an independent pipe lateral system or special nozzles in a plenum floor design. The volumetric ratio of air:water is in the range 0.3–1.0 (Degremont, 2007). Upflow filters are generally 15–25% more efficient than downflow filters; to remove 1 mg/l of ammonia as NH_4^+ (0.78 mg/l as N) at pH 7.2 and water temperature of 10 °C, EBCT required for upflow and downflow filters using 2 mm ES 'Biolite' is about 3 and 4 minutes, respectively. The depth of the filter medium would be a function of EBCT and filtration rate and is dependent on the ammonia concentration in the water: a water containing 2.5 mg/l ammonia (as N) would require a depth of about 2.5 m for an upflow filter when compared to 3 m for a downflow filter. Upflow BAFs have a surface loading rate of 10–12 m³/h.m² and a coarse medium (ES 1.5–2.0 mm) while downflow BAFs operate at surface loading rates of 8–10 m³/h.m² (depending on the suspended solids loading) and use a coarser medium (ES 2.5–2.85 mm). Such high surface loading rates are feasible with expanded mineral filter media mostly of proprietary makes (e.g. 'Biolite'). Filters using naturally occurring media such as pozzolana or carbon (2–5 mm or even larger to give high specific surface per unit volume) operate at about 5 $m^3/h.m^2$ with EBCT of about 20–30 minutes (Lacamp, 1990).

BAFs are washed by concurrent application of air and water at about 16 and 7 mm/s, respectively. Washwater should be free of chlorine. When treating surface water BAFs are best used after the clarification stage and downflow filters are generally preferred. Upflow filters are generally used for ground waters particularly those containing high ammonia concentrations that require high air:water flow rates and low turbidity. BAFs are best followed by conventional rapid gravity sand or anthracite–sand filtration in order to produce a water free of suspended matter. The biological process is adversely affected by chlorine, hydrogen sulphide, heavy metals and precipitates from iron and manganese oxidation and other suspended solids in the water. When ammonia is present together with iron and manganese, the order of biological removal is iron, ammonia followed by manganese (Mouchet, 1992). Iron is usually removed separately. If the ammonia concentration is high manganese is not removed in the same filter unless adequate EBCT is provided. BAF beds also remove organic carbon effectively. Manganese dioxide coated sand filters have been successfully used to oxidize low concentrations of ammonia biologically and manganese by catalytic oxidation (Janda, 1994).

The biological reaction principle can also be applied to sedimentation tanks of the sludge blanket type. The sludge acts as the medium for the growth of nitrification bacteria. Oxygen for the nitrification reaction is limited to that which can be contained in the feed water and the ammonia removal is limited to about 0.5 mg/l as N.

REMOVAL OF VOLATILE ORGANIC COMPOUNDS FROM GROUNDWATER

10.30 GENERAL

Volatile organic compounds (VOC) (Section 6.18) are not removed by conventional water treatment processes, being stable towards most oxidants including ozone, and they are not biodegradable. The most cost effective treatment is considered to be packed towers; but adsorption onto granular activated carbon may be more economical in hard waters (Booker, 1988) (Section 10.36).

10.31 PACKED TOWER AERATORS

In packed towers (Section 10.20), also known as air stripping towers, the contaminated water flows downwards through a packing, counter-current to an air flow which strips the VOCs into the gas phase and discharges them through the top of the tower. The treated water is collected at the bottom of the tower. Since VOCs have high Henry's constants the process removes over 99.99% of VOCs (Hess, 1983).

The design parameters are air-to-water volumetric flow ratio, surface loading rate, type and size of packing and depth of packing. These are influenced by temperature, chemistry of the water and the mass transfer characteristics of the packing. Towers are usually constructed in polyethylene, glass reinforced plastic or rubber lined mild steel and should be provided with a good water distribution system and mist eliminators on the air discharge. Packing types are pall rings or Rashig rings or saddles and materials are usually of plastic or ceramic; packing height should be limited to about 6 metres. For greater packing heights two or more towers in series should be considered. Surface loading rate of the tower needs to be selected to prevent flooding of the packing and is usually about 60–75 $m^3/h.m^2$. Tower diameter should be limited to 3–4 metres. High water flow rates should be divided between two or more towers in parallel. The air-to-water volumetric ratio is usually about 25:1 to 30:1 to limit the air pressure drop across the packing to 10–40 mm of water per metre of packing.

In packed towers dissolved oxygen is increased and carbon dioxide is removed while the VOCs are stripped from the water. Therefore iron precipitation and calcium carbonate scaling are problems with packed towers (Section 10.20). The air stripping results in release of VOCs to atmosphere. Although the quantities are small, this may be of some concern for plants located in urban areas.

10.32 ADSORPTION AND CHEMICAL OXIDATION

Adsorption of organic compounds by GAC is well known (Section 10.36) and the technique can be used to remove VOCs. The adsorbers are of the fixed bed type and can be conventional rapid gravity or more commonly pressure filters. The principal design parameters are: empty bed contact time (EBCT), defined as the bulk volume of adsorbent bed (m^3) divided by the water flow rate (m^3/min); type of GAC used; bed depth and hydraulic loading. These parameters should be evaluated by pilot scale tests. For pressure filters typical values are EBCT in the range 10–30 minutes, surface loading rates between 10–20 $m^3/h.m^2$ and bed depths of 2.5–3 m. The adsorption capacity of GAC for VOCs is generally small and the media needs frequent regeneration if VOCs are present in concentrations in excess of 100 µg/l (Foster, 1991). Experience in the USA shows 12–18 month regeneration frequency (Dyksen, 1999).

VOCs can be oxidized by hydroxyl free radicals formed when ozone is used in combination with hydrogen peroxide (H_2O_2) or with UV radiation (Glaze, 1987) (Section 10.37). Ozone (O_3) is first injected into a reaction chamber consisting of two or three compartments to satisfy the ozone demand; hydrogen peroxide being injected into the second or the third compartment together with further addition of ozone. Tests must be carried out to optimize the ratio $H_2O_2:O_3$ and the contact time which depends on the concentration and the nature of the organic matter. The ratio can vary between 0.3 and 1. The oxidation stage should be followed by GAC adsorbers to remove the by-products of the oxidation and excess hydrogen peroxide. The latter dechlorinates water.

TASTE AND ODOUR REMOVAL

10.33 CAUSES OF TASTES AND ODOURS

The source of a taste or odour in a water is often difficult to identify, but the following list includes the most likely causes.

1. Decaying vegetation or algae may give rise to grassy, fishy, musty, woody, mouldy, 'pharmaceutical', fruity and cucumber type odours. Algae mostly cause offensive odours as they die off but some living algae, e.g. blue-green (*Cyanophyceae*), cause taste and odour problems.

2. Moulds and actinomycetes may give rise to earthy, musty or mouldy tastes and odours that may be wrongly attributed to algal growths. In stagnant waters and especially water in long lengths of pipeline left standing in warm surroundings, such as the plumbing system of a large building, the moulds and actinomycetes have favourable conditions for growth and the first water drawn in the morning may have an unpleasant taste or odour of the kind mentioned.

3. Sulphate reducing bacteria (*Desulphovibrio* and *Desulphotomaculum* in particular) give rise to the hydrogen sulphide (rotten egg) smell. Hydrogen sulphide is also naturally present in some ground waters in concentrations up to 10 mg/l. The taste and odour thresholds for hydrogen sulphide in water are 0.06 mg/l and 0.01–0.001 µg/l respectively. Iron bacteria (e.g. *Gallionella and Leptothrix*) often produce unpleasant tastes and odours with their death and decomposition and they also create an environment for sulphate reducing bacteria to grow.

4. Iron above a concentration of about 0.3 mg/l imparts a bitter taste to a water.

5. Excessive chloride and sulphate impart a brackish taste to a water; taste thresholds for sodium and calcium chlorides are 200 and 300 mg/l as Cl, respectively; those for sodium, calcium and magnesium sulphates are from 250, 250 and 400 mg/l as SO_4, respectively.

6. Industrial wastes are a prolific source of taste and odour of all kinds, of which those produced by phenols are the most frequently experienced. In the presence of free residual chlorine the phenols form a 'chlorophenol' medicinal taste which is quite pronounced; a quantity as small as 0.001 mg/l phenol may react with chlorine to form an objectionable taste.

7. Chlorine does not, by itself, produce a pronounced taste except in large residuals, but many taste troubles occur due to reactions between chlorine and a number of organic substances. These tastes are usually described as 'chlorinous'. Reaction of chlorine with ammonia produces tastes and odours; the limiting concentrations for monochloramine, dichloramine and nitrogen trichloride being 5, 0.8 and 0.02 mg/l respectively (White, 1999). This is a common problem where waters containing free chlorine and combined chlorine are mixed in a distribution system (Section 11.8).

10.34 METHODS OF REMOVAL OF TASTES AND ODOURS

Tastes and odours of biological origin are due to the presence of organic compounds such as geosmin and methyl isoborneol (MIB). Those of industrial origin such as phenol are also invariably of organic nature. Oxidation and adsorption onto activated carbon are considered to be the most effective methods available for reduction of tastes and odours associated with organic compounds. Of the oxidants, ozone is very effective in destroying some of the taste and odour producing compounds. In theory saturated compounds such as geosmin and MIB, which are responsible for musty and earthy tastes, and some of the chlorinated hydrocarbons are not oxidized by ozone. In practice however ozone is known to be effective in removing geosmin and MIB (Glaze, 1990; Chen, 1998). Ozone-hydrogen peroxide (Section 10.37) can be more effective in reducing tastes and odour (Ferguson, 1990), including those due to geosmin and MIB (Duquet, 1989). UV-hydrogen peroxide using medium pressure lamps is also effective against geosmin and MIB but at doses about 1000 mJ/cm^2 (Rosenfeldt, 2005).

Other oxidants such as potassium permanganate and chlorine dioxide have been successfully used for taste and odour removal. Chlorine dioxide is useful when phenols are present as it does not form chlorophenols (Section 11.17).

Activated carbon is the most effective method of removing taste and odour compounds of organic nature. It can be used in either the powdered activated carbon (PAC) or granular activated carbon (GAC) form. GAC is normally used when taste and odour removal is required continuously for long periods; for seasonal occurrence lasting several days at a time or dealing with pollution incidents it may be economical to use PAC. PAC dosages can vary between 25 and 50 mg/l with a maximum of 100 mg/l (Snoeyink, 1999) divided between raw water or inlet to clarifiers and the filters, with dosages to the filters maintained below about 5 mg/l. When applied to raw water the PAC dose may need to be increased to accommodate natural organic matter (NOM) which normally would be removed by coagulation. Addition of PAC with the coagulant can reduce the adsorption efficiency because PAC is incorporated into the floc particles and organics must therefore diffuse through the floc. When PAC is dosed after coagulation, organic removal is improved (Huang, 1995) because PAC adheres to the outer surface of the floc. Good mixing and sufficient contact time are the important design parameters. When dosed at the works inlet, a contact time of 15 minutes is sufficient. Sludge blanket and solids recirculation tanks provide adequate contact time; however, organics contained in sludge are known to utilize some of the adsorption capacity of PAC. For filter inlet dosing the filters provide the necessary contact time. PAC and oxidant addition (e.g. for manganese removal) should be separated by either the contact time required for adsorption or sufficient to ensure that there is no oxidant residual at the PAC dosing point. PAC is stored as a powder in bags or in silos or as a slurry in tanks and dosed as a slurry (Section 7.20).

GAC is used in adsorbers of the rapid gravity or pressure filter type (Sections 8.2 and 8.5). It is sometimes used in place of sand as a medium in rapid gravity filters for turbidity removal, but better overall performance is achieved when GAC adsorbers are used after rapid gravity filtration. For taste and odour removal only an EBCT of 10 minutes is appropriate. pH values in the acidic range enhances the adsorption process. The life of a GAC bed between regenerations may be 2 to 3 years or about 10 years if ozone or ozone-hydrogen peroxide oxidation is used to enhance biological activity in GAC by oxidising organic matter to biodegradable forms (Section 10.36). Biologically activated carbon filters reduce MIB to less than the threshold values (Nerenberg, 2000).

Aeration can sometimes improve the palatability of water which is made poor due to stagnation, such as the bottom waters of reservoirs. It is also effective for taste and odour caused by compounds with high Henry's constant such as chlorinated solvents and some hydrocarbons. Taste complaints resulting from iron, etc. in water that has been left to stagnate in the ends of mains can be overcome by flushing.

Hydrogen Sulphide Removal

In natural waters hydrogen sulphide (H_2S) is in equilibrium with hydrosulphide (HS^-) and sulphide (S^{2-}); H_2S is predominant up to pH 7, between pH 7 and 13 more than 50% is HS^- and above pH 13, S^{2-} predominates. Aeration is effective in removing H_2S at pH <6 when over 90% is free H_2S; a packed tower aerator (Section 10.20) removes >95%. Aeration removes CO_2 in preference to H_2S thus raising the pH. Acid dosing may therefore be necessary to maintain the pH <6.

H_2S can also be removed by chemical oxidation using chlorine, hydrogen peroxide or ozone. In the oxidation reactions elemental sulphur is formed initially and then further oxidized to sulphuric acid. The latter reaction takes place at low pH. At pH values between 6 and 9, elemental sulphur reacts with residual sulphides, even in the presence of an oxidant, to form obnoxious polysulphides and a milky blue suspension of colloidal sulphur. Treatment consists of converting colloidal sulphur and polysulphides formed in the oxidation reaction to thiosulphate by the addition sulphur dioxide (or sodium bisulphite) and then converting the thiosulphate formed to sulphate by chlorination (White, 1999; Monscwitz, 1974). Polysulphides are not formed at pH >9. Potassium permanganate oxidizes H_2S only to elemental sulphur; it is also known to introduce soluble manganese into the water. The stoichiometric quantities of oxidant required for the oxidation of 1 mg of H_2S to elemental sulphur are 2.1 mg of chlorine, 1 mg of hydrogen peroxide, 1.41 mg of ozone or 6.2 mg of potassium permanganate. Due to high cost, chemical oxidation is commonly used to remove the residual H_2S remaining in the water after aeration.

NATURAL ORGANIC MATTER AND MICROPOLLUTANTS REMOVAL

10.35 GENERAL

Natural organic matter (NOM) is mostly anionic polymeric compounds and is either hydrophobic or hydrophilic. It is measured as total organic carbon (TOC), made up of particulate (POC) and dissolved (DOC) fractions.

Micropollutants are described in Chapter 6 and include pesticides, algal toxins (Microcystin), volatile organic carbons (VOC) such as chlorinated organic solvents, taste producing geosmin and MIB, polynuclear aromatic hydrocarbons (PAH), polychlorinated biphenyls (PCB), endocrine disruptors (ECD), 1,4-Dioxane, N-Nitrosodimethylamine (NDMA) and methyl tri-butyl ether (MTBE). Disinfection by-products (DBPs) such as THMs and haloacetic acids (HAA) are rarely found in raw waters but are commonly formed during the chlorination of water containing DBP precursors, which are the natural organic substances such as humic and fulvic acids that impart colour to the water, and algal metabolic products and cell debris (Section 11.7). Micropollutants occur in concentrations of the order of micro or sub-microgram per litre and in most instances their removal is necessary to comply with standards set for drinking water quality.

Conventional chemical coagulation (and solid–liquid separation processes) removes NOM. In most instances a significant reduction in the concentration of DBP precursors is achieved by enhanced coagulation (Section 7.13). The removal of micropollutants by coagulation would be small. For effective removal of micropollutants, conventional treatment processes alone are usually insufficient and should be supplemented with advanced treatment processes including advanced oxidation processes.

It is well known that activated carbon is effective in removing organic compounds from water. PAC is normally dosed at the inlet to a treatment works and the NOM in the raw water decreases the adsorption capacity of PAC towards any organic micropollutants present. PAC is frequently used for seasonal occurrence of pesticides lasting several days at a time. The removal efficiency increases with dose and contact time. An allowance should be made in the PAC dose to satisfy competing organic matter, particularly when the concentration of NOM exceeds that of the pesticides. Complete removal of several pesticides has been reported in direct filtration plants treating ground waters using PAC doses of 5–10 mg/l supplemented by a small coagulant dose to minimize PAC breakthrough; PAC accumulated in the filter was available for adsorption throughout the duration of filter run (Haist-Gulde, 1996).

GRANULAR ACTIVATED CARBON ADSORBERS

10.36 ADVANCED TREATMENT PROCESSES

PAC is costly and is discarded after one use whereas GAC, although about two to three times the cost of PAC, can be reactivated after exhaustion and reused. Since the introduction of more rigorous standards for drinking water quality the use of GAC has become the predominant process for the removal of organic matter including micropollutants. GAC adsorbers are commonly installed downstream of rapid gravity filters used for turbidity removal (Section 8.8).

Activated carbon can be made from wood, coal, coconut shells or peat. The material is first carbonized by heating and then is 'activated' by heating to a high temperature whilst providing it with oxygen in the form of a stream of air or steam. Sometimes chemical activation by phosphoric acid is used. It is then ground to a granular or powdered form. It is a relatively pure form of carbon with a fine capillary structure which gives it a very high surface area per unit of volume. The adsorption capacity of GAC is described by various parameters including Iodine Number and BET surface area (Table 8.3). GAC is reactivated (either on or off-site) by heating to 800 °C in steam or CO_2, or chemically but in the processes up to 25% of the carbon may be lost. The loss in adsorption capacity after 1, 4 and 7 reactivations is about 5, 10 and 20% respectively (Marc, 1998). The make-up for carbon lost after each reactivation therefore restores it to almost its original capacity.

GAC adsorbers are of conventional rapid gravity or pressure filter design (Section 8.8) and the basic design parameters are the empty bed contact time (EBCT) and bed depth or hydraulic loading ($m^3/h.m^2$).

GAC characteristics vary according to the base material used. For example the adsorptive capacity for the pesticide atrazine varies in the order wood > coconut shell > peat > coal (Paillard, 1990a). However, coal based GAC finds wide use for most water treatment applications as it has a distribution of both mesopores (2–50 nm diameter) and micropores (up to 2 nm diameter), a structure suitable for medium to large (colour, taste and odour) and small organic molecules (micropollutants) respectively. Pilot plant work or laboratory accelerated column tests (ACT) should be used to optimize the GAC type and other design parameters such as adsorption capacity (by Freundlich adsorption isotherm) and to determine the life of carbon between reactivation (by ACT). Empty bed contact time varies for different micropollutants and is usually in the range 5–30 minutes; for pesticides it is 15–30 minutes and for DPBs and VOCs it is about 10 minutes.

Although GAC removes most micropollutants efficiently, the adsorption capacity towards some is low, so that frequent reactivation may be necessary which makes the adsorption process costly. For example using an EBCT of 10–30 minutes most pesticides or DBPs may show breakthrough in 6–12 months and VOCs in 12–18 months. But if only taste and odour removal is required, breakthrough normally occurs in 2–3 years when using an EBCT of about 10 minutes. In total organic carbon (TOC) removal breakthrough occurs in about 3 months. In one UK works it has been shown that the TOC removal efficiency reduced from 90% to 10% in 14 weeks, but this did not have any adverse effect on the final water THM concentration which was significantly less than that before the installation of GAC (Smith, 1996). For TOC removal a rule-of-thumb used for estimating GAC life is 50 m^3 water treated per kg of GAC (Langlais, 1991).

Biological Activated Carbon Reactors

Dissolved organic carbon (DOC) present in water is primarily in the form of refractory (i.e. poorly biodegradable) compounds with some biodegradable dissolved organic carbon (BDOC) or assimilable organic carbon (AOC). BDOC and AOC are complimentary techniques used for measurement of biodegradable organic matter but give different values with AOC normally less than BDOC for the same sample. In the GAC, all organics are adsorbed irrespective of their biodegradability but the presence of BDOC leads to some biological activity occurring after a few weeks. Although a high proportion of the organics is removed by adsorption, GAC life is ultimately determined by the refractory organic breakthrough.

Ozone, ozone-peroxide or UV-peroxide treatment of the water upstreamof GAC adsorbers can, however, oxidize a large proportion of high molecular weight refractory organics into smaller more assimilable forms (BDOCs), thus promoting biological activity in the adsorber which also helps to remove organic matter which is otherwise unaffected or only marginally affected directly by oxidation. VOCs and some pesticides are examples of such compounds. In the adsorbers the GAC acts more as a biomass substrate and less as an adsorber; hence GAC beds used in this fashion are called biological activated carbon (BAC) reactors. In the biological reaction the BDOCs are converted to carbon dioxide, thus more of the adsorption capacity of GAC is available to deal with a smaller proportion of unoxidized refractory organics. The life of the GAC between reactivations is therefore increased many-fold and breakthrough occurs when GAC is exhausted with respect to refractory organics. For example, for pesticide removal an increase in running time to 1.5 years is reported (Graveland, 1996). DBPs once formed cannot be removed with ozone or BAC treatment. BAC reactors are of similar design to GAC adsorbers (Section 8.8). Because of the biological matter present zooplankton (micro-animals) can grow in BAC reactors and filtered water could contain high heterotrophic plate count (HPC) bacteria resulting from the biological activity and zooplanktons. Therefore, reactor management should be good and washing should be efficient and frequent (Section 8.8). In some cases the filtered water may be devoid of oxygen due to the oxygen consumption during biological activity, therefore oxygenation of the filtered water may be necessary (Graveland, 1996). For most applications an EBCT of 10–15 minutes is generally considered adequate. Larger EBCT helps to increase the service life of carbon between reactivations. In BAC reactors TOC is more effectively removed than in GAC adsorbers; following a very high initial removal it stabilizes at about 30–40% thereafter (Armenter, 1996; Baur, 1996). A high proportion of the remaining TOC could be non-biodegradable refractory organic matter; therefore the use of ozone with GAC produces a biologically stable water thus reducing the risk of aftergrowth in the distribution system and minimising DBP formation in subsequent chlorination for disinfection. Biologically stable water would have an AOC value in the range 50 to 100 μg/l for water carrying a chlorine residual or <10 μg/l for water without a chlorine residual (LeChevallier, 1993). Chlorination increases the AOC concentration in water in distribution, probably by oxidation of large organic molecules. Therefore DOC should be removed by oxidation (e.g. ozone) processes followed by biological filtration (e.g. BAC).

In the ozonation stage pathogenic micro-organisms are significantly reduced. However, the water from the BAC reactors contains a higher level of non-pathogenic bacteria (HPC), thus requiring an effective disinfection stage such as chlorination for their removal.

Magnetic Ion Exchange Process

The MIEX® (magnetic ion exchange) process is used to remove DOC from water and is based on a conventional ion exchange mechanism, but uses a magnetic resin hence the name. The resin contains a quaternary amine functional group and incorporates a magnetic component. The particle size is typically

150 to 180 μm. This is significantly smaller than conventional fixed bed ion exchange resins and provides improved reaction kinetics. The surface of the resin is positively charged which allows adsorption of negatively charged organic material. A typical dose is 10–30 ml resin/l for 10–30 minutes.

The process comprises 2 or 3 continuous stirred tank reactors in series with a total retention time of about 10–30 minutes and containing the resin on which DOC is adsorbed, followed by settlers where 85–95% of the settled resin is separated and returned to the contactors. The remaining resin is regenerated in a batch process and returned to the reaction tank along with make-up resin to replace any losses (<0.1%). The magnetic properties of the resin result in heavy agglomerates, which facilitate over 99.9% recovery in the settlement tanks operating at a surface loading rate of about 10 $m^3/h.m^2$. The resin is regenerated at pH 10 with sodium chloride at a concentration of 100–120 g/l NaCl and dosed at 360 g NaCl per litre of resin (Slunjski, 2000).

MIEX removes predominantly hydrophilic organic matter (Singer, 2007) whereas coagulation removes predominantly hydrophobic organic matter. Therefore the combined operation of the MIEX process and coagulation can result in increased removal of DOC and lower coagulant doses (Budd, 2005). Test work carried out on the Wanneroo plant (capacity 225 Ml/d) showed that MIEX followed by coagulation performed better than enhanced coagulation in reducing DOC and the THM formation potential of the water (Warton, 2007).

MIEX produces a saline waste stream with a high organic content from the regeneration system. This waste stream should be disposed of either to sewer or, for coastal installations, to the sea. The waste stream volume is typically of the order 0.2 to 0.4 litres per m^3 treated.

10.37 **ADVANCED OXIDATION PROCESSES**

The effectiveness of oxidation of organic matter by ozone can be increased by using it in conjunction with hydrogen peroxide (H_2O_2). This advanced oxidation process (AOP) is sometimes called 'peroxone' and produces the so called hydroxyl free radicals which are strong oxidants with an oxidation potential of 2.80 V compared to ozone (2.07 V), and oxidizes organics including those resistant to ozone such as some pesticides and VOCs. For example this process is known to improve atrazine removal, from 15 to 40% using ozone alone, to 70–85% (Mouchet, 1991). Removal is better at low alkalinities and with pH in the range 7–7.5. The H_2O_2: ozone ratio used depends on the application and varies from less than 0.3:1 for controlling taste and odour compounds (Ferguson, 1990) to 0.5:1 for controlling pesticides and VOCs (Paillard, 1990b). In the 'peroxone' process ozone is added to the first stage of a two or three stage contact chamber in which ozone rapidly oxidizes reactive organic substances, followed by the second stage where the ozonation efficiency towards less reactive organic substances is substantially increased by the addition of H_2O_2. Alternatively the second chamber is used for disinfection with the final chamber being used for H_2O_2-ozone reaction. Again the process is best used downstream of coagulation and solid-liquid separation processes after a significant proportion of the organic load has been removed. Like ozone the 'peroxone' process converts refractory organic substances to readily biodegradable compounds which should be removed in biological activated carbon reactors. These reactors also act as safety filters in removing the by-products of oxidation and any unreacted hydrogen peroxide which would otherwise dechlorinate the water when final residual chlorination is adopted.

The high oxidative power of H_2O_2, the large volume of oxygen generated and the heat evolved from its decomposition make it a hazardous chemical to handle. Correct material selection and good facility design can eliminate these hazards. For waterworks application the commercial grade of solution strength 35% $^w/_w$ is recommended (Table 7.5).

Hydroxyl free radicals are also produced when H_2O_2 or ozone is coupled with UV-radiation (Section 11.24). In the process water is dosed with H_2O_2 followed by UV irradiation. The fluence (dose) required for this process is rather high (500–750 mJ/cm^2) compared to UV fluence for cryptosporidium inactivation (20–70 mJ/cm^2) (Ijpelaar, 2007). UV transmittance is measured by extinction coefficient (EC). EC is the absorbance of light per unit path length (cm) per unit concentration (moles/litre). The larger the EC the faster the UV intensity is extinguished as it is transmitted through the water. The EC of pure water at 254 nm is about 6×10^{-6} l mole^{-1} cm^{-1}; hydrogen peroxide has an EC of 16.6 l mole^{-1} cm^{-1} and therefore UV transmittance (UV-T) decreases as it travels through the water consequently, hydroxyl free radical production decreases unless the H_2O_2 dose is large (>10 mg/l) (Lobo, 2007). The hydrogen peroxide dose is typically in the range of 5–15 mg/l. Pilot work by Hofman (2004) showed that organic contaminants can be controlled by a UV dose of 540 mJ/cm^2 and a H_2O_2 dose of 6 mg/l. H_2O_2 is photo-reactive over the wavelength range 185–400 nm; the highest hydroxide free radical yields are over the UV range 200–280 nm. Since the output of low pressure UV lamps is set at 254 nm, medium pressure UV lamps may be more efficient in converting H_2O_2 to free radicals. Bicarbonate (HCO_3^-) and carbonate (CO_3^{2-}) are hydroxyl free radical scavengers and therefore the process is suitable for low alkalinity waters. Nitrites are formed by photolysis of nitrate (Section 11.27). The UV-H_2O_2 process oxidizes NOM to form AOC; the concentration in treated water may be as high as 140 µg/l (Kruithof, 2005). BAC reactors effectively remove AOC, nitrite and residual H_2O_2 and produce a biologically stable water (Martijn, 2006). A key advantage of the UV-H_2O_2 process is that compared to ozone based oxidation it does not produce bromate. The process is costly due to very high UV fluence and should therefore be used to target micropollutants only; other dissolved organic matter should be removed ahead of the AOP. Although AOP can be used for the micropollutants listed in Section 10.35 certain micropollutants can be treated by UV photolysis alone. These include ECD (Rosenfeldt, 2004) and NDMA (Lobo, 2007; Wong, 2000).

UV irradiation alone is reported to give 70% removal of acid pesticides such as mecoprop, MCPA and bromoxynil at an energy input of 200 Wh/m^3. However the removal of atrazine was limited to 30% (Paillard, 1987); increasing the energy input to 700 Wh/m^3 increased the removal to about 70% (Bourgine, 1992). Pilot and field trials by Richardson (2004) have demonstrated good removal of algal toxins and pesticides with a UV dose of 2000 mJ/cm^2 at a UVT of 80%.

10.38 COLOUR REMOVAL

Natural colour is primarily due to humic substances classified as humic and fulvic acids. They are colloidal, hydrophilic, anionic polymeric compounds which are refractive (non-biodegradable). Fulvic acid is more soluble than humic acid. Both can be removed by coagulation; low coagulation pH values are considered to be most effective. Iron salts are most effective at pH values of about 5–5.5 but aluminium salts are successfully used with lowland waters in the pH range 6.5–7.2 and upland waters in the pH range 5.5–6.2. Since colour is organic in nature it is susceptible to oxidation by chlorine, chlorine dioxide and ozone. Chlorine, although very effective in oxidising colour, is not recommended due to the formation of THMs (Rook, 1976) predominantly with fulvic acid. The ozone dose required to breakdown colour can be high and it is therefore beneficial to use it after chemical coagulation, sedimentation and filtration, the ozone being used to oxidize the residual colour. Pilot work has shown that up to 5.5 mg/l with 15 minutes contact reduced colour from 70 to less than 20° Hazen and the dose varied proportional to the colour value.

NF membranes of the spirally wound cellulose acetate type are used to remove colour in several plants in Scotland. It is reported that true colour removal of 100° Hazen to less than 1° Hazen and iron (mostly organically complexed) removal of 1.35 to less than 0.01 mg/l as Fe have been achieved (Merry, 1995). Membranes were operated in the cross-flow mode (Section 8.17) with recycle. Dual media filters followed by 5μ cartridge filters were used in the pre-treatment. Fouling of the membranes was caused by organic colour complexed with iron and biological film and was high. Frequent cleaning by a weak acid and detergent and periodic washing with sodium hypochlorite were necessary, with membrane replacement every three years. Fyne process which utilizes tubular membranes also of cellulose acetate was effective with minimum of pre-treatment. The use of mechanical foam-ball cleaning every 4–6 hours minimized the use of chemicals and therefore could be operated with a high fouling rate. Performance was slightly inferior with respect to iron and manganese removal. These membrane designs are suitable for small capacity plants (Irvine, 2000).

GAC with a high proportion of meso and macro pores, i.e. high Molasses Number (e.g. Lignite based GAC) adsorbs colour effectively.

CORROSION CAUSES AND PREVENTION

10.39 PHYSICAL AND ELECTROCHEMICAL CORROSION

High velocity water flow can cause deep pitting and erosion of surfaces due to the phenomenon known as cavitation. Section 13.12 describes how cavitation occurs and Sections 16.20 and 17.13 discuss its effects on valves and pumps respectively. Cavitation erosion can be avoided by hydraulic design to avoid high velocity low pressure flow impinging on surfaces where an increase in pressure can causes collapse of any vapour pockets. The use of hard corrosion resistant materials can help resist cavitation damage but care must be taken with some stainless steels whose corrosion resisting chromium oxide layer is easily removed by cavitation (and erosion).

Electrochemical corrosion occurs where two metals of differing electropotential are immersed in a common body of water. All natural waters can act as an electrolyte, but the degree to which they do so depends on the dissolved salts present. The metal which loses ions to the electrolyte and therefore corrodes is termed the 'anode'; the other metal is the 'cathode' which gains ions. Of the metals commonly used in water supply systems, zinc is anodic to iron and steel, which are themselves anodic to copper. Thus in zinc-galvanized pipes or tanks zinc provides sacrificial cathodic protection to steel, but it may corrode if fed by water which has passed through copper pipes. Corrosion of lead joints can occur in lead-soldered copper pipework, because lead is anodic to copper. Electrochemical corrosion can also occur because the pipe metal may change along its length, creating 'pockets' of differing metal which have an electropotential between them. Corrosion processes are complicated by additional chemical reactions that can occur. Thus anodic corrosion of iron piping can produce tubercles of rust, which alter the rate of corrosion and beneath which other forms of corrosion (e.g. anaerobic) can occur. Ions deposited on the cathode may reduce its cathodic effect, so the electrochemical reaction dies away; or the ions may be constantly removed by flowing water or combine with other substances in the water so the anodic corrosion continues or may even increase.

Unprotected metal pipes corrode externally when buried in wet soil which acts as an electrolyte. Corrosion can be increased if the pipes become anodic to other buried metallic structures in the soil, particularly if these structures are acting as 'earths' to electrical apparatus which introduces stray currents into the soil. Soil surveys can locate conditions likely to cause external corrosion which may include (Tiller, 1984):

- soils with a moisture content above 20%;
- acid soils with a pH of 4 or below;
- poorly aerated soils containing soluble sulphates (Section 10.40);
- soils with a resistivity less than 2000 ohm cm.

Sleeving, wrapping and cathodic protection of pipelines are measures widely adopted to prevent external pipeline corrosion as described in Sections 15.10, 15.13 and 15.16.

10.40 BACTERIAL CORROSION

Sulphate-reducing bacteria (i.e. *Desulfovibrio desulfuricans*) existing in anaerobic conditions, can cause corrosion of iron. These bacteria are capable of living on a mineral diet and, as a result of their metabolism, they produce hydrogen sulphide which attacks iron and steel to form ferrous sulphide. Thus steel becomes pitted and cast iron becomes 'graphitized' (Cox, 1964), in which state it becomes soft. Sulphate-reducing bacteria may be most numerous in waterlogged clay soils in which oxygen is absent and where sulphur in the form of calcium sulphate is likely to occur. In such clays sulphate reducing bacteria can produce one of the most virulent forms of attack on iron and steel. Backfilling pipe trenches with chalk, gravel or sand to prevent anaerobic conditions arising, in addition to cathodic protection and sleeving, are used to prevent this form of corrosion.

Iron bacteria (e.g. *Crenothrix, Leptothrix* and *Gallionella* types) may be present in a water which is deficient in oxygen (as evidenced by the presence of sulphates and hydrogen sulphide in the water) and may cause internal corrosion. These bacteria have the ability to absorb oxygen and then oxidize iron from water or iron pipes and to store it. The bacteria are aerobic and, under favourable conditions, can form large deposits of slime which are objectionable, giving rise to odours and staining. The potential for growth of such bacteria is ever-present in a water with a high iron content.

Tuberculation on the inside of an iron main is sometimes initiated by sulphate reducing bacteria, or more commonly by organic substances in the water, low pH, or high oxygen content of the water. The external surface of a tubercle or nodule consists of a hard crust of ferric hydroxide, often strengthened by calcium carbonate. Below this crust conditions tend to be anaerobic so that sulphate-reducing bacteria can flourish creating further products of corrosion. Hence two remedies for the prevention of internal tuberculation are the aeration of a sulphate or hydrogen sulphide containing water, and lining the interior of the pipeline with an epoxy resin or with cement mortar.

10.41 CORROSION CAUSED BY ADVERSE WATER QUALITY

Desalinated water which is devoid of dissolved substances is highly corrosive to metals and in the presence of dissolved carbon dioxide (CO_2) is also highly aggressive to concrete and mortar. Some waters have a tendency to corrode metals due to a high content of dissolved solids (e.g. chloride or sulphates). In some cases free residual chlorine tends to cause corrosion due to its oxidation potential. The degree of corrosion depends largely on the acidity or CO_2 content of the water and the extent to which this is countered by the presence of calcium bicarbonate alkalinity in the water and a sufficiently high pH. Where bicarbonate alkalinity is present, an excess of dissolved CO_2 is necessary to prevent the decomposition of bicarbonate ions (HCO_3^-) back to carbonate ions (CO_3^{2-}) and subsequent precipitation of calcium carbonate by the following equation:

$$CO_3^{2-} + CO_2 + H_2O \leftrightarrow 2HCO_3^-$$

The relationship between alkalinity and carbon dioxide at equilibrium is given in Table 10.4, which is based on a graph by Cox (Miles, 1948).

When the concentration of carbon dioxide exceeds the requirement to maintain the equilibrium between carbonate and bicarbonate ions the water tends to dissolve any coating containing carbonates (such as concrete or mortar) with which it is in contact. The excess of CO_2 over that required for equilibrium is termed *aggressive* carbon dioxide. For example, from Table 10.4, a water having an alkalinity of 125 mg/l as $CaCO_3$ at pH 7.5 is aggressive if the free CO_2 concentration exceeds 7.5 mg/l. Conversely, a water with less free carbon dioxide deposits some of the bicarbonate as calcium carbonate. Where the water is only slightly over-saturated this deposition occurs slowly with the calcium carbonate often forming a protective coating on the surface of metals. If however, the water is significantly over-saturated the precipitation occurs rapidly in the amorphous soft form and a protective coating is unlikely to be formed.

A useful method of determining whether or not water is in equilibrium with calcium carbonate is the 'chalk test'. This requires taking a stoppered glass bottle to which some powered chalk is added, filling the bottle carefully and completely with the water of interest, and replacing the stopper so as to exclude all air. After some 12 hours or more the final pH should be measured. This pH value, denoted as pH_s, is termed the 'saturation pH value' of the water. Alternatively, the pH_s of a water could be calculated from:

$$pH_s = (pK_2 - pK_s) - (\log_{10}[Ca^{2+}] + \log_{10}[2Alk])$$

where, terms in [ ] are expressed in gmoles/l and values for pK_2-pK_s as a function of temperature and total dissolved solids (TDS) are given in Table 10.5.

Langelier's Saturation Index (*LSI*) is then determined as the numerical difference between the pH of the water and its pH_s, i.e.:

$$LSI = pH - pH_s$$

Table 10.4 Relationship between alkalinity, equilibrium pH and dissolved carbon dioxide

Alkalinity as CaCO₃ (mg/l)	25	50	75	100	125	150	175	200	250	300
pH at equilibrium, i.e. pH$_s$	8.8	8.1	7.7	7.6	7.5	7.4	7.35	7.3	7.2	7.0
Free CO₂ at equilibrium (mg/l)	0.0	1	2.5	4.5	7.5	12	18.5	27	32	60

Table 10.5 Relationship between pK$_2$–pK$_s$, total dissolved solids and temperature

Temp (°C)	pK$_2$–pK$_s$ at TDS (mg/l)										
	20	40	80	120	160	200	240	280	320	360	400
0	2.45	2.58	2.62	2.66	2.68	2.71	2.74	2.76	2.78	2.79	2.81
10	2.23	2.36	2.40	2.44	2.46	2.49	2.52	2.54	2.56	2.57	2.59
20	2.02	2.15	2.19	2.23	2.25	2.28	2.31	2.33	2.35	2.36	2.38
30	1.86	1.99	2.03	2.07	2.09	2.12	2.15	2.17	2.19	2.20	2.22

Source of information: Water Works Engineering, Qasim, S.R. et al, Prentice-Hall, New Jersey, USA, 2000.

When *LSI* is negative the water is liable to be aggressive and if *LSI* is positive the water tends to deposit calcium carbonate. However Langelier's Index does not quantify the degree of under- or over-saturation of water. This can be measured in the 'chalk test' by determining the total alkalinity prior to the test and again, immediately afterwards. The difference between the two measurements is termed the Calcium Carbonate Precipitation Potential or 'CCPP'.

Values of CCPP between +3 and +10 mg/l suggest that the water is capable of precipitating a protective coating on the surface of metals, values much in excess of 10 mg/l indicate that protective coatings are unlikely to be formed. If the CCPP is negative the water is aggressive and will attack concrete and mortar particularly when the CCPP is lower than –10 mg/l (as CaCO$_3$). Consequently, treatment for an aggressive water consists of chemical addition, sufficient to produce a thin uniform protective coating of calcium carbonate. The treatment must be continuous and usually consists of dosing an alkali such as hydrated lime to ensure that the pH of the treated water is maintained only slightly above the pH$_s$ value (i.e. +0.2 to 0.3), the resulting deposition being in the form of a smooth hard scale. Table 10.4 is a useful practical guide for determining the saturation pH (pH$_s$) of a water.

The concentration of both chloride and sulphate ions in the water can have a significant effect on the corrosive nature of the water. Both the rate of corrosion and the dissolution of iron increase markedly as the concentration of chloride and/or sulphate increases. Work by Larson and Skold (1957) has suggested the use of the following ratio between chloride plus sulphate and the alkalinity (Larson ratio, LR) where [ ] is expressed in gmoles/l:

$$LR = ([Cl^-] + 2[SO_4^{2-}]) \div [HCO_3^-]$$

Water with a *LR* >0.2–0.3 is considered corrosive.

10.42 CORROSIVENESS OF VARIOUS WATERS

There are a wide variety of naturally occurring waters with differing characteristics which determine their aggressiveness to metals and concrete. The principal types are dealt with below.

Hard surface waters, with fairly high alkalinity. A large proportion of lowland river waters associated with sedimentary geological formations have a positive Langelier Index. Generally, therefore, such waters are satisfactory for the avoidance of corrosion of metals and are not usually plumbosolvent (Section 10.14).

Hard underground waters with substantial alkalinity. The mineral composition of many underground waters from chalk, limestone and other sedimentary formations may often be similar to that of many surface waters; but the carbon dioxide content of the water is usually much higher. The excess carbon dioxide should be removed, usually by aeration, to give a slightly positive Langelier Index. Although water with a positive Index should, theoretically, deposit calcium carbonate, in practice it will probably not do so to any appreciable extent in water mains or metal tanks. Its anti-corrosion effect, however, probably depends in part upon the rapid deposition of calcium carbonate on cathodic areas where electrochemical corrosion is initiated, the effect being to stifle this type of incipient localized corrosion. Waters with fairly high alkalinity due to calcium bicarbonate are plumbosolvent (Section 10.14).

Underground waters with low alkalinity but high free carbon dioxide. Many small sources from some of the older geological formations and some from gravel wells and springs comprise waters of this type; although many of the gravel sources can have a high non-carbonate hardness. Aeration can be adopted to reduce the carbon dioxide content and lime can be added to achieve a further reduction. If the water is already hard it may be preferable to neutralize the free carbon dioxide by adding caustic soda. This reagent is very soluble and thus can be administered as a small volume of a strong solution. For the correction of the corrosive characteristics of very soft waters, passage through a filter containing granular limestone ($CaCO_3$) may be useful.

Soft waters from surface sources. Certain lake waters are very pure, with a bacteriological quality approaching that of an underground water but not, however, containing any appreciable quantity of carbon dioxide. They are therefore almost neutral in reaction, but many may also have a very low alkalinity, e.g. 10 mg/l as $CaCO_3$. However, from the point of view of corrosion protection and of preventing plumbosolvent action, they need careful treatment (Section 10.14). In some instances an economical and satisfactory form of treatment is simply to add a small dose of lime. In cases where a greater amount of alkalinity is required, dosing with carbon dioxide followed by lime is adequate. Alternatively use of calcium chloride or sulphate with soda ash or sodium bicarbonate may be convenient.

The majority of surface waters derived from upland catchments however, are moderately to highly coloured by peaty matter, which contains organic humic and fulvic acids that render the water acidic in reaction with a pH of 6 or lower. To counter the general corrosiveness of this type of water and its plumbosolvent action, the organic acids must be removed (Turner, 1961 and 1965). This is usually achieved by chemical coagulation followed by filtration. Aluminium or iron coagulants, being acidic, destroy alkalinity; the carbonate or bicarbonate is converted to free carbon dioxide (Section 7.21). Hence lime should be added after filtration and the carbon dioxide is then largely converted back to calcium bicarbonate.

10.43 **DEZINCIFICATION**

Galvanizing on iron and steel is susceptible to corrosion by excess free carbon dioxide, which dissolves zinc, and to any water in which the pH is excessively high. However, the term 'dezincification' usually refers to the effect on brasses (alloys of copper and zinc), when the zinc is dissolved. A particular form of this type of corrosion is called 'meringue' dezincification because of the bulky

Table 10.6 Limiting values of chloride-to-alkalinity ratio (at pH values greater than 8.2) for dezincification

Chloride (mg/l as Cl)	10	15	20	30	40	60	100
Alkalinity (mg/l as CaCO₃)	10	15	35	90	120	150	180
Chloride: alkalinity ratio	1:1	1:1	1:1.75	1:3	1:3	1:1.5	1:1.8

white layer of corrosion product which appears, the effect of which is to cause failure of fittings, mainly hot water fittings when they are constructed of hot-pressed brass. The effect, investigated by Turner (1961 and 1965) was found to occur with waters having a pH >8.2 and a chloride to alkalinity ratio greater than those shown in Table 10.6.

Subsequent investigations (AwwaRF, 1996) showed that this relationship could be transferred to dezincification in general, although at pH <7.6 this type of corrosion did not occur. These findings reinforce the need for the addition of alkalinity (lime and CO_2 dosing) to very soft waters to reduce their aggressiveness to metals. If a water contains an appreciable amount of chloride, then simply raising its pH without increasing its alkalinity may increase its liability to cause dezincification and may also fail to control its corrosiveness towards other metals.

DESALINATION

10.44 INTRODUCTION

Desalination is the term that describes processes used to reduce the concentration of dissolved solids in water, which are usually referred to as total dissolved solids (TDS) and expressed in mg/l. Sometimes electrical conductivity, which is expressed as microSiemens per cm (µS/cm), is used as a surrogate measurement of total dissolved solids because it is easy to measure online or in the field with a conductivity meter. As an approximation, the TDS value of a water can be calculated by multiplying the conductivity value by 0.66. Natural waters may be classified in broad terms based on TDS value as given in Table 10.7.

Several methods have been commercially developed for the desalination of high TDS waters. Selection of the correct process requires evaluation of process efficiency and plant capital and operating costs. As a general guide, the most frequent application of the various desalination processes based on TDS concentration is given in Table 10.8.

RO is the preferred method up to at least 10 000 mg/l TDS because it is significantly more energy efficient than any of the distillation methods at these concentrations. For higher salinity applications, such as seawater desalination, the choice between RO and distillation depends on site-specific issues. In the past, larger facilities utilized distillation, but that trend may be changing. The installed capacity of desalination plants has been steadily increasing and the percentage using RO has been increasing due to improvements in pretreatment effectiveness, energy efficiency and economy of scale. Examples of these improvements include, respectively, MF/UF membrane filtration, more permeable membranes, better energy recovery devices and large diameter RO elements.

Table 10.7 Classification of waters according to TDS

Type of water	TDS value (mg/l)
Sweet water	0–1000
Brackish water	1000–5000
Moderately saline water	5000–10 000
Severely saline water	10 000–25 000
Seawater	Above 25 000

Table 10.8 Classification of desalination processes according to TDS

Process	TDS value (mg/l)
IX	up to 100
Electrodialysis	up to 1000
Nanofiltration (NF)	Up to 1000 (but depends on divalent/monovalent ratio)
Low-pressure reverse osmosis (RO)	up to 10 000
High-pressure reverse osmosis	10 000 to 40 000
Distillation	Above 30 000

10.45 ION EXCHANGE

The use of IX processes for reducing mineral or saline constituents in water is discussed in Section 10.10. IX rarely finds application for waters containing more than 1500 mg/l of dissolved solids. Even at lower concentration, it is usually more cost effective to apply RO as pretreatment to IX to produce demineralized industrial process water.

10.46 ELECTRODIALYSIS

In this process a DC electrical potential is applied between electrodes and the ionic constituents in the water are thus caused to migrate through semi-permeable membranes which are selective to cations and anions. Desalination occurs in a series of cells separated by alternate pairs of cation and anion membranes (Fig. 10.5). A number of membrane pairs can be installed in parallel between a pair of electrodes to form a stack and several stacks can be connected hydraulically together either in parallel to increase output or in series to increase salt removal. Each electrodialysis pass only removes about half of the remaining ionic material; therefore, three and four pass systems are typical

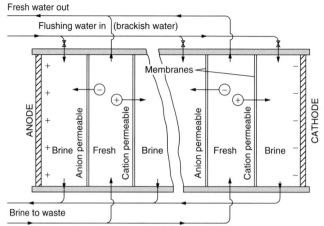

A d.c. voltage is applied across a battery of cells. The anions in the saline (or brackish) water, induced by the e.m.f. applied, pass through the anion-permeable membrane into the brine cells, but cannot pass the cation-permeable membrane. Similarly the cations are also collected in the concentrate or brine compartments which discharge to waste. The saline (or brackish) water in alternate chambers loses ions and reduces in salinity. A proportion of the saline feed is directed to the concentrate (brine) compartments to flush out the brine.

FIGURE 10.5

Principle of electrodialysis plant.

for brackish water applications. Electrodialysis is not used for seawater desalination. The reduction in dissolved solids obtained by electrodialysis is directly related to the electrical energy input and therefore in practice to the cost of electricity. Electrodialysis membranes, like reverse osmosis membranes, are subject to fouling and feed waters require pre-treatment.

In the electrodialysis process the current flow is uni-directional and the desalinated and concentrate compartments remain unchanged. Fouling and scaling material which build up on the membranes are cleaned by taking units out of service. In electrodialysis reversal (EDR), the electrical polarity is reversed periodically which results in a reversal of ion movement and provides 'electrical flushing' of scale forming ions and fouling matter. This self cleaning can allow operation at higher levels of supersaturation of sparingly soluble salts, thereby achieving higher water recoveries.

When comparing options it is important to note that, unlike the other membrane technologies, in ED/EDR the product water does not pass through the membrane; therefore, water is not filtered by this process. The ions being removed are transported through the ED/EDR membrane whilst the water which becomes product flows tangentially along the membrane surface. Only charged material is removed; uncharged material, such as dissolved silica, viruses, bacteria, protozoan cysts and other microbes are not removed by this process.

10.47 REVERSE OSMOSIS AND NANOFILTRATION

In osmosis when a salt solution is separated from pure water by a semi-permeable membrane, there is a net flow of water across the membrane until an equilibrium is achieved. This equilibrium occurs when the pressure on the pure water side is offset by the osmotic pressure of the salt solution.

In reverse osmosis (RO), the osmosis process is reversed. Applying pressure greater than the trans-membrane osmotic pressure to the higher concentration solution results in a net flow of water through the membrane into the lower concentration side. Osmotic pressure is a physical property that is a function of concentration and type of solute. As an approximation, there is about 0.7 bar of osmotic pressure for every 1000 mg/l of TDS. An operating pressure much higher than the average transmembrane osmotic pressure is required to achieve an economically feasible flow (by a factor of about 2 for seawater and more than 10 for brackish water).

Nanofiltration (NF) is essentially a type of RO. Both RO and NF are pressure driven cross flow filtration processes that utilize semi-permeable membranes to remove dissolved solutes from water. These membranes also remove particulate material but that is a misapplication, because even small (i.e., less than 1 micron) particles foul the membrane yielding resistance to flow and lowering the permeability. Therefore, RO/NF processes require high quality pre-treatment, especially for surface water applications.

On visual examination NF and RO membranes and their system components look identical. The main difference is that NF exhibits greater selectivity in rejecting dissolved material. NF provides low rejection of monovalent ions, such as sodium and chloride, (typically 10 to 40% depending on the membrane and operating conditions), whilst providing higher rejection of multivalent ions, such as calcium and sulphate (typically 80 to 85%), compared to about 95 to 99% rejection of most ionic material by RO. Both NF and RO provide greater than 98% rejection of dissolved organics with molecular weight greater than 200 Daltons and both provide some rejection of smaller organic molecules. (Dalton is a unit of measurement defined to be 1/12 of the mass of one atom of carbon). AOC removal of over 97% has been reported for NF and the efficiency depends on the pH and molecular weight cut-off of the membranes (Hofman, 2004). Neither NF nor RO remove dissolved gasses, such as radon, carbon dioxide (CO_2) or hydrogen sulphide (H_2S). For some ions, such as ammonium ion (NH_4^+), rejection is a function of pH. Ammonium, and hence ammonia, is rejected better at pH higher than 8. With pore size on the order of 1 nanometre (0.001 micron) both NF and RO reject microbial material including bacteria and viruses. NF applications include softening, colour removal, control of dissolved organic carbon (DOC) and removal of disinfection by-product (DBP) precursors. The range of RO applications include these, as well as desalination of brackish and sea-waters and removal of specific contaminants (such as nitrate, fluoride and radium). An advantage of both RO and NF is the ability to remove multiple contaminants in a single treatment step. Many communities using RO/NF benefit from softening while also receiving potable water treated for regulated parameters, such as colour, DBP precursors and nitrate.

NF should not be viewed as a weak version of RO. NF rejection is more selective than RO, which can be an advantage in cases where the goal is partial removal of organics or hardness but where the concentrations of TDS, sodium and chloride in the source water do not require treatment. NF is sometimes chosen instead of RO because of the perception that NF consumes less electricity; however, with the development of modern, ultra low pressure RO membranes, the difference in operating pressures between these membrane types has narrowed. In cases where either type could be used, a detailed capital and operating cost comparison is needed to determine the most economical option. The comparison should include operating costs for electricity and an escrow for future membrane replacement. Due to the higher rejection, RO tends to require less membrane area than NF, which in turn lowers initial capital cost and the membrane replacement budget. This may offset the lower operating pressure of NF.

When used to desalinate seawater RO feed pressure is 55 to 80 bar. On typical low salinity applications RO pressure is 7 to 20 bar while NF is 6 to 14 bar. Higher operating pressure is needed if any of the following are increased: flux, feed concentration, recovery and fouling tendency of the water; or if the temperature is decreased.

The application of devices to recover residual energy from the concentrate stream improves the energy efficiency of the process, especially on seawater desalination plants. Energy recovery devices (Veerapaneni, 2005) include rotating equipment, such as Pelton wheels and positive displacement devices, such as pressure and work exchangers (Hauge, 1999). An energy efficient seawater RO process uses about 3 kWh/m³ of desalinated water for the high pressure pumping energy. Lowering this value is an active field of research. The theoretical thermodynamic minimum is 0.8 kWh/m³. One of the lowest specific energy values reported for an entire seawater RO desalination facility with energy recovery and including pre and post treatment and pumping, is 4.346 kWh/m³ at Singapore's 136 Ml/d Tuas plant (Kiang, 2005) (Plates 17(b) and 18(a)). Brackish water RO energy usage is significantly lower, but it is difficult to generalize given the wide range of application.

The most commonly used RO/NF membrane configuration is the spiral wound type as is illustrated in Figure 10.6.

The hollow fibre configuration was used in the 1970s and 80s but is rarely used today. With spiral wound elements, layers of membrane formed on flat sheets are wrapped around a central product water tube between layers of permeate and feed-concentrate spacer material, which yields a cylindrical element that is mounted inside a pressure vessel. As feed water flows from one end of the cylinder to the other, it flows along the membrane surface and a portion passes through the membrane to the permeate side and then to the product water tube. Typically, large spiral systems have either six elements in series inside each pressure vessel if 40-inch long elements are used, or four elements in series if 60-inch long elements are used. Essentially all RO element manufacturers present

FIGURE 10.6

Spiral wound reverse osmosis membrane.

the length of spiral elements in inches, typically either 40 inches (1.016 m) or 60 inches (1.524 m). Standard diameters are 4 inch (0.102 m) and 8 inch (0.203 m). Larger diameters of 16 inch (0.406 m) and 18 inch (0.457 m) are also available.

For hollow fibre elements, the membrane was manufactured as fibres with an outside diameter of about 100 micron and a hollow lumen in the centre with a diameter about half the outside diameter. The hollow fibre module was also mounted in a pressure vessel, but with most designs there was only one module per vessel. The membrane skin was on the outside of each fibre, so the permeate flowed along the lumen on the inside.

Almost all municipal applications employ polyamide membrane chemistry, rather than the older style cellulose acetate derivatives. The main reasons are that polyamide membranes are productive at much lower operating pressures, provide higher salt rejection, do not require the same level of pre-acidification, and have longer service life. Cellulose acetate membranes can tolerate continuous exposure to residual chlorine up to about 1.0 mg/l but essentially require a chlorine residual to control biological degradation; they are sensitive to pH outside the range 5.5–6 and water temperatures >30 °C. Polyamide membranes have no tolerance towards chlorine or other oxidants but are not degraded by bacteria. Polyamide spiral wound elements have a wide pH operating range (3–10 for most types and up to 1–12 for short term cleaning exposure for some types) and temperature range (up to 45 °C); however, operating at the extremes does shorten service life.

The main controlled parameters in an RO/NF system are feed pressure and recovery, which is the ratio of permeate flow to feed flow. This is typically accomplished by an automated control system that adjusts the feed pump to yield a permeate flow set point, and a modulating concentrate control valve to yield a recovery set point. Generally the feed pump is adjusted by a variable speed drive (VSD) or variable frequency drive (VFD). Some older systems utilize a modulating control valve on the pump discharge, rather than a VSD; however, this increases energy consumption and shortens membrane life due to hydraulic stress on startup. Since the permeate and recovery controls influence each other, programming tolerances and response times should be selected to avoid cycling or 'hunting'. Lower TDS applications also have a blending control loop to bypass a portion of the feed to the finished water. The blend ratio is usually based on a flow ratio or an online conductivity measurement, which can indirectly indicate parameters such as TDS or hardness based on local experience.

The purified stream leaving an RO/NF plant, which is called permeate or product water, has passed through the membrane; therefore, it contains essentially no particulate material and only a small portion of the original dissolved material. The material rejected by the membrane is concentrated into a stream called concentrate (sometimes also referred to as reject, brine or retentate). RO/NF plants are arranged in stages and passes. While sometimes these terms are used interchangeably, with precise usage the meanings differ. Within a multi-stage train the concentrate from one stage flows via internal piping to feed the next stage. Stages are used to increase recovery. For example, a seawater plant, which would typically operate at 45–50% recovery, would employ a single stage. That stage would consist of multiple vessels in parallel which would each house four to six elements per vessel (assuming standard 40-inch long elements). For a low TDS (i.e., NF or brackish water RO) plant operating at 75% to 80% recovery, two stages with six or seven elements per vessel would typically be used; for 85%, three stages are generally used. There are fewer vessels in each subsequent stage to maintain sufficient velocity across the membrane surface; low velocity can result in localized areas of high concentration, which in turn can result in inorganic scaling of the membrane surface. In a two-stage system about 67% of the vessels are in the first stage; in a three-stage system about 57% of the vessels are in the first stage and about 28% in the second stage.

More than one pass is employed when additional RO or NF treatment is needed to meet finished water quality requirements. In a two-pass plant the permeate from the first pass is collected and re-pressurized as feed to a downstream RO/NF system (Plate 17(a)). In this case there are two RO/NF plants in series. Sometimes multiple passes are used to produce ultra-pure water for industrial facilities. Sometimes a partial second pass is applied to seawater desalination to meet boron limits. Long Beach, California is investigating two passes of NF for seawater desalination for potentially improved energy efficiency (Harrison, 2005; Leung 2005).

10.48 RO AND NF PLANT DESIGN

While detailed calculations are frequently conducted using computer modelling programs, an understanding of basic equations can assist designers as well as operators. A preliminary design can be conducted after selecting recovery and flux. Recovery and vessel arrangement are discussed in Section 10.47. RO and NF concentrate dissolved material and therefore achievable water recovery is a function of the presence of sparingly soluble salts such as calcium carbonate, calcium sulphate, barium sulphate, and silica. When the products of their molar concentrations on the feed-concentrate side exceed the related solubility product, precipitation occurs, causing scale to form on the membrane surface which lowers permeability. Solubility product is the product of the equilibrium molar concentrations of the ions (moles/l) of a sparingly soluble salt in a saturated solution in water. For $CaCO_3$ it is defined as $K_{sp} = [Ca^{2+}] \times [CO_3^{2-}] = 3.8 \times 10^{-9}$ at $25\,°C$. If identified early enough, most scales can be removed with chemical cleaning, although silica scale can be impossible to remove. However, for all types it is better to avoid scaling and the attendant higher operating costs and risk of irreversible damage. Therefore, precipitation should be prevented by limiting the recovery to the maximum allowable and for brackish RO and NF an antiscalant is applied to optimize the recovery. For seawater RO/NF, recovery is limited by the maximum allowable operating pressure, which generally allows 45% to 55% recovery with only sulphuric acid addition to control calcium carbonate scaling. For NF and brackish RO applications, commercially available antiscalants, frequently based on polyphosphates and chelating agents, are added to the feed to allow higher recovery by slowing the reactions sufficiently that the concentrated stream leaves the system before precipitation forms. In the past sodium hexametaphosphate was also used as an RO/NF antiscalant but proprietary blends are now typically used to allow higher recovery without deleterious biological decay. While detailed calculations are needed to determine maximum allowable recovery for a specific case, frequently 75% to 85% is achievable when a modern antiscalant and sulphuric acid are added to brackish RO/NF feed water.

Setting the correct maximum flux is critical for avoiding excessive fouling or decreased plant capacity. While it is impossible to be definitive in a text book, here are some suggested RO/NF fluxes that are generally achievable assuming good pre-treatment: seawater 12–14 l/m² h; brackish groundwater 20–25 l/m² h; surface water 16–20 l/m² h; water reclamation 18–20 l/m² h. Higher flux may be successfully applied in some cases but cost-benefit studies tend to favour lower flux. Even if performance is stable over the long term, higher flux consumes more electricity since the driving pressure is proportional to flux and higher flux requires more frequent chemical cleaning in almost all cases. A preliminary design based on six 8-inch diameter × 40-inch long RO/NF elements per vessel can be approximated with the following rules of thumb: 0.11 Ml/d per vessel for brackish groundwater and 0.07 Ml/d for seawater.

Three main parameters are monitored as a function of time to determine if a system is operating correctly and to plan maintenance activities: salt passage, differential pressure, and normalized permeate flow. Salt passage can be calculated for TDS or specific solutes. Conductivity passage can be monitored as a surrogate measurement. A gradual increase in salt passage (possibly up to 5% per year) may be expected and should be accounted for in the initial design calculations; however, a more rapid increase indicates damaged o-rings or membrane. With RO/NF differential pressure refers to the feed-concentrate pressure drop. If this exceeds a manufacturer specified value (~4 bar for a 6-element vessel) the force can irreparably damage the element by telescoping which causes salt leaks at the glue lines. The most complex calculation is normalized permeate flow (or the related normalized net driving force). To account for changes in operating conditions, permeate flow is normalized for feed concentration, temperature and recovery. Simplified versions of the equations are summarized in Table 10.9.

Additional major issues regarding the design of RO/NF include pre-treatment, post-treatment and concentrate disposal. Pre-treatment is needed to protect the membrane from fouling and plugging. After the feedwater has been well pre-treated, cartridge filtration with disposable elements, typically rated at 5 micron, is applied just upstream form the RO/NF as a guard filter. The only exception is that systems with MF/UF membrane filtration pre-treatment sometimes do not include cartridge filters if steps are taken to be certain that no particulate material is present in the RO/NF feedwater. For groundwater sources, pre-treatment can be as simple as cartridge filtration and addition of antiscalant and sulphuric acid to control scaling. Sometimes groundwater also requires removal of materials that can scale the membrane, such as oxidized iron, manganese and hydrogen sulphide (H_2S) gas. H_2S gas does not scale the membrane surface per se but, if oxidized, the resulting elemental sulphur sticks to the membrane, impedes permeate flow and risks damage. For some groundwaters, those with reduced and soluble forms of iron, manganese or hydrogen sulphide, it may be best to prevent oxidation of the feed upstream of RO/NF. Ionic iron and manganese will be rejected just as other ions are. The H_2S will pass through the membrane just as other dissolved gases do; the resulting H_2S in the permeate stream can be subsequently removed by aeration or chemical oxidation (Section 10.34).

In recent years one of the major issues with seawater desalination has been the boron concentration in the finished water. The boron concentration is generally in the range of 5 to 7 mg/l. WHO Guideline Value for drinking water is 0.5 mg/l although values as low as 0.3 mg/l have been used. Boron ions are weakly charged so boron rejection by RO is low compared to most ions and rejection by NF is even lower. The charge density on boron ions increases with pH. Busch et al (2003) showed that boron rejection at pH 8 is 45 to 75% using standard brackish water elements and 85 to 92% with standard seawater elements, while at pH 11 rejection is increased to 98 to 99%. Frequently, it is not practical to operate the first pass of a facility at elevated pH due to the risk of scaling by calcium carbonate; therefore, to meet stringent boron limits a second pass may be required. RO could be combined with selective IX to meet the boron standard (Section 10.10).

For surface water more extensive pre-treatment is required to remove bacteria, turbidity, colour, organic matter, oil and grease. This usually includes a combination of disinfection, coagulation, single or two-stage dual media filtration or clarification followed by dual media filtration and granular activated carbon adsorption depending on the parameters to be removed. Dissolved air flotation can be useful where oil and grease or algae are a concern. Pre-disinfection, that is disinfection upstream of the RO/NF, may be needed for control of biofouling. If needed, intermittent application of disinfectant or even intermittent pH adjustment may be more effective than continuous

Table 10.9 Family of equations that describe the RO process

Description	Equation	Equation number
Flux (F)	$F = Q_p/A$	10.7
Recovery (R)	$R = (Q_p/Q_f) \times 100$	10.8
Salt Passage (SP)	$SP = (C_p/C_f) \times 100$	10.9
Salt Rejection (SR)	$SR = 100 - SP$	10.10
Differential Pressure (ΔP_{fc})	$\Delta P_{fc} = P_f - P_c$	10.11
Normalized Permeate Flow (NPF)	$NPF = Q_p \times TCF_{25} \times \left(\dfrac{NDP_{initial}}{NDP_{today}}\right)(MC)$	
Net Driving Pressure (NDP)	$NDP = P_{fcavg} - P_p - P_{osm}$	
Temperature Correction Factor (TCF)	$TCF_{approx} = 1.03^{(25-T)}$	

Where:
F = flux, l/m² h
Q_p = permeate flow measured at ambient temperature, l/h
A = active membrane area, m²
R = recovery, percent
Q_f = feed flow measured at ambient temperature, l/h
SP = salt passage, percent
C_p = permeate concentration, mg/l
C_f = feed concentration, mg/l
SR = salt rejection, percent
ΔP_{fc} = feed-to-concentrate pressure differential
P_f = feed pressure, bar
P_c = permeate pressure, bar
NPF = normalized permeate flow, l/h
TCF = temperature correction factor, dimensionless
$NDP_{initial}$ = net driving pressure during initial operation
(i.e., typically based on readings taken during the first 24 to 48 hours of operation), bar
NDP_{today} = net driving pressure when Q_p was measured, bar
MC = membrane compaction factor or aging factor; for most modern membranes manufacturers consider $MC = 1$
NDP = average net driving pressure, bar
P_{fcavg} = average feed-concentrate pressure, bar
$= (P_f + P_c) / 2$
P_f = feed pressure, bar
P_c = concentrate pressure, bar
P_p = permeate pressure (for simplified calculation, assume ~ 0), bar
P_{osm} = osmotic pressure, bar
~approximately = $\dfrac{C_f + C_c}{2000} x(0.7)$
C_f = feed concentration of total dissolved solids, mg/l
C_c = concentrate concentration of total dissolved solids, mg/l

pre-disinfection in some cases. If pre-disinfection is used then residual disinfectant should be removed by the addition of sodium bisulphite or other reducing agent upstream of the RO/NF to protect the membrane from unwanted oxidation. The use of membrane filtration, i.e. microfiltration or ultrafiltration (MF/UF), instead of granular media is an active area of research. MF/UF provides low turbidity and low Silt Density Index (SDI) feedwater for RO/NF and may be cost-effective given the downward trend of membrane prices. However, there are currently no major seawater desalination plants in operation that employ MF/UF for filtration. In some seawater applications pre-treatment requirements can be reduced by using beach wells or bank filtration rather than open intakes (Missimer, 1994; Wright, 1997; Voutchkov, 2005).

The main pre-treatment goals are to maintain feed water turbidity less than 0.1 NTU and 15-minute SDI (ASTM, 2002) less than 5, but less than 3 is the preferred value. In addition, the concentrations of free chlorine and oil and grease should be nil, and TOC and microbial activity may need to be controlled, but there are no firm quantified values for these. As an approximation, TOC up to 3 or 4 mg/l is probably acceptable, but it depends on the attraction of the organic material to the membrane.

In general, a pilot program should be seriously considered for all surface water RO/NF applications, including seawater desalination, and even for many groundwater applications. The pilot program should include all major aspects of the full-scale design to avoid misleading results. Therefore, the pilot study should closely mimic the pre-treatment conditions, including any recycle streams within the facility. Full-scale, commercially-available RO/NF modules should be piloted since it is difficult to accurately scale-up results. Multiple RO/NF operating cycles are needed to project longer-term results. Given a stable groundwater source a pilot program may consist of three or four months of testing, while with a variable surface water a year or longer may be advised to account for the impact of seasonal changes.

Post-treatment of the permeate is very important to avoid corrosion of downstream piping and equipment. The RO product water is aggressive, because it is slightly acidic (pH 5–6) and soft with little alkalinity. Generally, NF product is less aggressive since the rejection rates are lower than with RO, but the risk of corrosion in the distribution piping should be evaluated for both RO and NF. The finished water should be remineralized to contain a minimum alkalinity of 40 mg/l as $CaCO_3$. A positive Langelier Index or Calcium Carbonate Precipitation Potential value should also be achieved (Section 10.41). The treatment methods include—hydrated lime and carbon dioxide dosing; carbon dioxide dosing followed by filtration through limestone (note: saturation pH cannot be exceeded; therefore should be followed by lime dosing); sodium bicarbonate and calcium chloride dosing (note: high chloride to alkalinity ratio is unsuitable for duplex brass fittings); or blending with a suitable brackish water source (Applegate, 1986). Seawater itself cannot be used as a means of remineralisation as it does not contain a sufficiently high ratio of alkalinity to TDS to permit this. Although RO/NF removes bacteria and viruses disinfection (chlorination) is necessary to safeguard against leakage due membrane imperfections or leaking seals.

RO/NF processes generate a concentrated waste stream which must be disposed of. While this may be more of an issue for inland RO and NF facilities, available options should also be evaluated for coastal plants. Generally the order of preference (increasing cost) is surface water discharge (the normal method for seawater RO plants); discharge to the outfall of a wastewater treatment plant (WWTP); sewer discharge to a WWTP; deep well injection; evaporation pond; or a special concentration process. In the past the main high-concentration process has been seeded slurry evaporation with mechanical vapour compression, which has very high capital and operating costs. Currently, research is being conducted for AwwaRF and California Energy Commission (Bond, 2007) on selective precipitation coupled with an additional RO step to increase the overall recovery while also reducing the waste flow rate. For coastal plants sea discharges should be designed to achieve suitable localized blending with the design ideally verified by modelling.

The flow diagram in Plate 17(a) illustrates the issues discussed in this section. This figure shows the process steps for a two-pass seawater RO facility but the example can be also be generalized to brackish water RO as well as NF. For further study readers should refer to AWWA's Manual of Practice (AWWA, 2007). Plates 17(b) and 18(a) show a large SWRO plant in Singapore.

10.49 THERMAL PROCESSES

Distillation works on the principle that the vapour produced by evaporating seawater is free from salt and the condensation of the vapour yields pure water. The majority of modern plants use the multistage flash thermal process (MSF) (Plate 18(b)) or the thermal vapour compression system multi-effect desalination (TVC-MED). Mechanical vapour compression MED (MVC-MED) units are less common; due to current limitations in compressor technology the maximum capacity of MVC-MED units is 5000 m^3/day per unit. Recent practice has tended towards installation of MSF units for distillers in the capacity range 30 000 to 75 000 m^3/day with TVC covering the 10 000 to 25 000 m^3/day range. Most large thermal distillation plants are constructed as dual-purpose stations for both desalination and the generation of power. The trend towards higher operating temperatures means that greater attention has to be paid to the reduction of corrosion and the use of cost-effective materials and chemicals to combat corrosion.

Effluents from desalination processes have a higher density than seawater from the concentration of total dissolved solids to about double that of seawater. In addition the effluents also contain corrosion products (distillation processes), e.g. copper, zinc, iron, nickel and aluminium and any additives, e.g. corrosion inhibitors (distillation process) and sequestering agents. Distillation process effluents also have elevated temperatures. Disposal of such hot, hyper saline waters may give rise to adverse ecological effects on the environment. However, the possible environmental effect of the concentrated hot effluent discharge is largely mitigated by mixing with the cooling water needed to convey the low grade rejected heat from the distiller. This cooling water is normally not concentrated and is at ambient temperature. The use of a long sea outfall provided with discharge ports designed to achieve further dilution of the effluent, followed by subsequent tidal dispersion makes it possible to meet a predetermined water quality criteria. It is usually necessary to carry out studies at each site to enable prediction of dispersal. For land installations deep-well injection, evaporation ponds and discharge to sewer are options which have been adopted.

One of the major design parameters for all distillers is the performance ratio—a measure of the efficiency of energy utilisation. The amount of energy required to desalt a given brine concentration varies according to the degree of sophistication of the plant installed, i.e. annual energy costs reduce as capital costs increase. Other factors to be taken into account include size of units, load factor, growth of demand, interest rate on capital and technical matters concerning the auxiliary services, repairs and maintenance. For detailed design some 70 design parameters have to be settled. Many of these are concerned with the safe or most economic limits for the temperatures, velocities and concentrations of the coolants, brines, brine vapour, steam, steam condensate and boiler feed water, with particular reference to the prevention of scaling, corrosion, erosion, the purity of the distillate, the efficiency of heat exchangers and the nature and cost of the auxiliary plant involved.

Distilled water approaches zero hardness with an alkalinity not likely to exceed 2 mg/l. It is aggressive to metal and asbestos cement pipes and takes up calcium from mortar-lined pipes. It is unsuitable for distribution and is unpalatable, being flat and insipid. To remedy these deficiencies the distilled water should either be remineralized (Withers, 2005) or mixed with more saline water (e.g. from an artesian source).

Major problems, which formerly occurred with seawater distillers, were scale formation on heat transfer surfaces due to the presence of carbonates and sulphates of calcium; internal corrosion due to hot sodium chloride and the presence of dissolved gases such as oxygen, ammonia and hydrogen sulphide; plant start-up problems and running at low capacity. These problems have been largely overcome and continuous unit operation in excess of 8000 hours at variable load conditions in the range 60% to 100% of full load are commonplace.

To augment this brief introduction to thermal desalination, the authors refer readers to the publications by El-Dessouky (2002) and Sommariva (2004).

10.50 THE COSTS OF DESALINATION

The subject of desalination costs can be discussed only cursorily because of its diversity and complexity. The purpose of this brief review is therefore to indicate the principal items contributing to the production costs of desalinated water and to point out how these are likely to be influenced by local conditions. The major factors to be considered in any desalting application include: type and characteristics of the saline feed; type of desalination process to be used; local infrastructure; local costs of primary energy source (e.g. oil or gas or electricity); and availability and costs of chemicals needed for feed pre-treatment, product conditioning and plant cleaning. Also important, in the case of distillation plants, is the choice of performance ratio and whether dual-purpose operation is proposed (Sections 10.47 and 10.48). Except for the smallest plants, staffing costs are not usually a major item. Table 10.10 summarizes the principal cost contributory factors of the desalination processes currently popular as applied to single purpose plants (i.e. producers of water only). In the case of brackish waters (Table 10.7) the primary energy and power demands are usually less than half those for seawater desalination, whether electrodialysis or reverse osmosis is used. Thermal processes are not usually adopted for brackish water desalting because of their unfavourable energy demands. The feed water and chemical consumption of brackish water desalination are also only about one-third of those for seawater reverse osmosis.

The costs of desalted water are very variable and site specific. Amortisation of the initial capital investment, together with energy costs, usually accounts for up to 80% of the total water cost. The energy components can be calculated from the data in Table 10.10 if local primary energy unit costs are known. The components of operation and maintenance (O&M) costs for a 40 000 m³/d brackish water RO plant are: labour 27%, chemicals 10%, energy 48%, membrane replacements 8% and miscellaneous 6%. For a 40 000 m³/d seawater RO facility the O&M breakdown is labour 8%, chemicals 4%, energy 81%, cartridge filters 1%, membrane replacements 4% and miscellaneous 2%.

Only the briefest indication of capital costs can be given in the present context for the larger sizes of the five major types of desalination plant listed in Table 10.10. Plant costs, erected and commissioned, including typical costs of civil works, local product storage, and all other site-specific costs (such as cost of intake, discharge of effluents, fuels storage and handling, etc.) in 2007 were as given in Table 10.11.

Table 10.10 Cost contributory factors for major types of seawater desalination processes major

Process	Usual maximum size of unit to date m³*/day)	Total primary fuel energy demand (MJ/m³*)[a]	Components of energy demand		Chemical consumption (g/m³*)	Feedwater consumption (m³/m³*)	Annual cost of spares and replacements as % of initial capital cost of desalter
			Power (kWh/m³)	Heat (MJ/m³)			
Multistage flash evaporation (MSF)	75 000	120–400	3–4	185–300	4–6	5–10	2%
Thermal vapour compression (TVC-MED)	20 000	120–300	2–3	170–250	5–7	5–10	2–3%
Mechanical vapour compression (MVC-MED)	1500	88–130	8–15	Nil	5–7	2.0–2.5	2–4%
Seawater reverse osmosis (SWRO)	10 000	45–65	4–6	Nil	10–20	2–2.5	4–6%
Brackish water reverse osmosis (BWRO)	15 000	22–33	2–3	Nil	6–12	1.2–1.5	4–6%

Notes:
*Output
[a] Lower value for dual purpose plant.
Data refer to stand-alone plants. Heat and power assumed generated on site.
Thermal efficiencies assumed: 80% for steam boilers for MSF plant; 33% for diesel generators for RO and MVC plant.
Maximum capacity for RO plant refer to one train within a possible multi train arrangement.
Higher energy value of diesel fuel taken as 37.85 MJ per litre (1 MJ = 0.278 kW h).

Table 10.11 Capital costs for large desalination plants

Type of plant	Plant capital cost (£ per m³/day output)
Multistage flash evaporation (MSF)	700–850
Thermal vapour compression (TVC-MED)	550–650
Mechanical vapour compression (MVC-MED)	670–990
Reverse osmosis (seawater)	565–650
Reverse osmosis (brackish water)	225–260

REFERENCES

Armenter, J. L. L. and Canto, J. (1996). Filtration by GAC and On-site Regeneration in the Treatment of Water for the City of Barcelona. Review of 15 Years in Operation. *Water Supply*, **14**(2).

Applegate, I. Y. (1986). Post Treatment of Reverse Osmosis Product waters. *J AWWA*, **78**(5).

ASTM (2002). *Standard Test Method for Silt Density Index (SDI) of Water*. D4189–95.

AWWA (2007). *Manual of Practice M46, Reverse Osmosis and Nanofiltration*. 2nd Edn. Ed. Bergman. R.

AWWA (1988). Committee Report on: A Review of Solid Solution Interactions and Implications for Controlling Trace Inorganic Materials. *J AWWA*, **80**(10).

AwwaRF (1996), *Internal Corrosion of Water Distribution Systems*, 2nd Edition.

Bauer, M., Buchanan, B., Colbourne, J. et al. (1996). The GAC/Slow Sand Filter Sandwich—From Concept to Commissioning. *Water Supply*, **14**(2).

Benefield, L. D. and Judkins, J. F. (1982). *Process Chemistry for Water and Wastewater Treatment*. Prentice-Hall.

Bond, R. G. and Veerapaneni, V. (2007). *Zero Liquid Discharge and Volume Minimization for Inland Desalination*. Sponsored jointly by AwwaRF and the California Energy Commission, Proj No. 3010.

Booker, N. A., Hart, J., Hyde, R. A. et al. (1998). *Removal of Volatile Organics in Groundwater* WRc.

Bourgine, F., Chapman, J. and Martin, S. (1992). Treatment of Pesticides by UV photolysis *Techniques, Sciences, Methodes*, **92**(7/8).

Bourgine, F. P., Gennery, M. and Chapman, J. L. (1994). Biological Processes at Saints Hill Water Treatment Plant, Kent. *J IWEM*, **8**(4).

Budd, G. C., Long, B. W., Edwards-Brandt, J. C. et al. (2005). *Evaluation of MIEX Process Impacts on Different Source Waters*. AwwaRF.

Busch, M., Mickols, W. S., Jons, S. et al. (2003). Boron Removal in Seawater Desalination, *Proc. International Desalination Assoc*, World Congress, Bahamas.

Cheng, A. S. C., Fields, K. A., Sorg, T. J. et al. (2002). Field Evaluation of As Removal by Conventional Plants. *J AWWA*, **94**(9).

Chen, T., Huddleston, J. I., Atasi, K. Z. et al. (1998). Factor Screening for Ozonating the Taste- and Odor-Causing Compounds in Source Water at Detroit, USA, Taste and Odour Preconference Seminar.

Chwirka, J. D., Thomson, B. M. Stomp III, J. M. (2000). Removing Arsenic from Groundwater. *J AWWA*, **92**(3).

Clifford, D., Ghurye, G. and Tripp, A. (1998). Arsenic Removal by IX with and without Brine Reuse. *Proceedings of AWWA Inorganic Contaminants Workshop*, San Antonio, Texas.

Clifford, D. (1999). Ion Exchange and Inorganic Adsorption, *Water Quality and Treatment: A Handbook of Community Water Supplies*, 5th Edn. McGraw-Hill.

Colling, J. H., Croll, B. T., Whincup, P. A. E. et al. (1992). Plumbosolvency Effects and Control in Hard Waters. *J IWEM*, **6**(3).

Cox, C. R. (1964). *Operation and Control of Water Treatment Processes*. WHO.

Davis, J. M. (1980). Destratification of Reservoirs—A Design Approach for Perforated-Pipe Compressed Air Systems. *Water Services*, August.

Degremont (2007). *Degremont Water Treatment Handbook*.

Dixon, K. L., Lee, R. G., Smith, J. et al. (1991). Evaluating Aeration Technology for Radon Removal. *J AWWA*, **83**(4).

DoE (1973). *Notes on Water Pollution*. DoE (UK), Note No. 61; June.

Driehaus, W., Jekel, M., Hildebrandt, U. et al. (1998). Granular Ferric Hydroxide—A New Adsorbent for the Removal of Arsenic from Natural Water. *J WSRT-Aqua*, **47**(1).

Duquet, J. P., Bruchet, A. and Mallevialle, J. (1989). New Advances in Oxidation Processes: The Use of Ozone/Hydrogen Peroxide Combination for Micropollutant Removal in Drinking Water. *Water Supply*, **7**(4).

DWI (2005). *Code of Practice on Technical Aspects of Fluoridation of Water Supplies*.

Dyksen, J., Raczko, R. and Cline, G. (1999). Operating Experience at VOC Treatment Facilities, Part 1: GAC. *Opflow; AWWA*, **25**(1).

Edwards, M. (1994). Chemistry of Arsenic: Removal during Coagulation and Fe-Mn Oxidation. *J AWWA*, **86**(9).

El-Dessouky, H. T. and Ettouney, H. M. (2002). *Fundamentals of Salt Water Desalination*. Elsevier.

Faust, S. D. and Aly, O. M. (1998). *Chemistry of Water Treatment*, 2nd Edn. Ann Arbor Press.

Ferguson, D. W., McGuire, M. J., Koch, B. et al. (1990). Comparing 'Peroxone' and Ozone for Controlling Taste and Odour Compounds. Disinfection By-products and Micro-organisms. *J AWWA*, **80**(4).

Foster, D. M., Rachwal, A. J. and White, S. L. (1991). New Treatment Process for Pesticides and Chlorinated Organics Control in Drinking Water. *J IWEM*, **5**(4).

Gallagan, D. J. and Vemillion, J. R. (1957). *Determining Optimum Fluoride Concentration*. Public Health Reports, **72**(6).

Galvin, R. M. (1996). Occurrence of Metals in Waters: An Overview. *Water SA*, **22**(1).

Glaze, W. H. (1987). Drinking Water Treatment with Ozone. *Environmental Science and Technology*, **21**(3).

Glaze, W. H., Schep, R., Chauncey, E. C. et al. (1990). Evaluating Oxidants for the Removal of Model Taste and Odor Compounds from a Municipal Water Supply. *J AWWA*, **82**(5).

Gottlieb, M. C. (2005). Ion Exchange Application in Water Treatment, Chapter 12, *Water Treatment Plant design*, AWWA/ASCE, 4th Edn. Ed. Baruth, E.E.

Graveland, A. (1996). Application of Biological Activated Carbon Filtration at Amsterdam Water Supply. *Water Supply*, 14(2).

Gross, H., Schooner, G. and Rutten, P. (1988). Biological Denitrification Process with Hydrogen Oxidizing Bacteria for Drinking Water Treatment. *Water Supply*, 6.

Haberer, C. and Raff, O. (1999). *Removal of Naturally Occurring Radionuclides from Drinking Water—An Overview*. Vom Wasser, 93, September.

Haist-Gulde, B. and Baldauf, G. (1996). Removal of Pesticides by Powdered Activated Carbon—Practical Aspects. *Water Supply*, **14**(2).

Harrison, C. J., Amy, E. C., Yann, A. et al. (2005). Bench-scale Testing of Seawater Desalination Using Nanofiltration, *AWWA Membrane Tech Conf*, Phoenix, US.

Hauge, L. J. (1999). Pressure Exchanger, *Desalination & Water Reuse*, **9**(1).

Hess, A. F., Dyksen, J. E. and Dunn, H. J. (1983). *Control Strategy—Aeration Treatment Technique—Occurrence and Removal of VOCs from Drinking Water*. AwwaRF.

Hoffman, G. L., Lyttle, D. A., Sorg, T. J. et al. (2006). *Design Manual: Removal of Arsenic from Drinking Water Supplies by Iron Removal Process*. EPA/600/R-06/030.

Hofman, J. A., Ijpellaar, G. F., Heijman, S. G. J. et al. (2004). Drinking water Treatment in Netherlands: Outstanding and Still Ambitious. *Water Science & Technology; Water Supply*, **4**(5–6).

Huang, W. J. (1995). Powdered Activated Carbon for Organic Removal from Polluted Raw Water in Southern Taiwan. *J WSRT-Aqua*, **44**(6).

Ijpelaar, G. F., Harmsen, D. J. H. and Heringa, M. (2007). *UV Disinfection and UV/H$_2$O$_2$ Oxidation: By-product Formation and Control*, Techneau, D2.4.1.1.

Irvine, E., Grose, A. B. F., Welch, D. et al. (2000). Nanofiltration for Colour Removal—7 Years Operational Experience in Scotland, *Membrane Technology in Water and Wastewater Treatment* (Ed. Hills, P.). Royal Society of Chemistry.

Jacob, C. (2007). Seawater Desalination: Boron Removal by Ion Exchange Technology. *Desalination*, **205**(1–3).

Janda, V. and Rudovsky, J. (1994). Removal of Ammonia in Drinking Water by Biological Nitrification. *J WSRT-Aqua*, **43**(3).

Kiang, F. H., Yong, W and Ratnayaka, D. D. (2005). Supply of Desalinated Water by the Private Sector Singapore's First Public-private-partnership Initiative. *Proc. International Desalination Assoc*, World Congress, Singapore.

Knocke, W. R., Van Benschoten, J. E., Kearney, M. et al. (1990). *Alternative Oxidants for the Removal of Soluble Iron and Manganese*. AwwaRF.

Knocke, W. R., Van Benschoten, J. E., Kearney, M. et al. (1991a). Kinetics of Manganese and Iron Oxidation by Potassium Permanganate and Chlorine Dioxide, *J AWWA*. **83**(6).

Knocke, W. R. (1991b). Removal of Soluble Manganese by Oxide-Coated Filter Media: Sorption Rate and Removal Mechanism Issues. *J AWWA*, **83**(8).

Kruithof, J. C., Kamp, P. C., Lute, N. W. et al. (2005). UV/H_2O_2 Treatment for Primary Disinfection and Organic Contaminant Control at PWN's Water treatment Plant Andijk, *Proceedings of IUVA World Congress*, Whistler.

Lacamp, B. and Bourbigot, M. M. (1990). Advanced Nitrogen Removal Processes for Drinking and Wastewater Treatment. *Water Science and Technology*, **22**(3).

Langlais, B., Reckhow, D. A. and Brink, D. R. (Ed.) (1991). *Ozone in Water Treatment—Application and Engineering*. AwwaRF.

Larson, T. E., Skold, R. V. (1957). Corrosion and Tuberculation of Cast Iron. *J AWWA*, **49**(10).

LeChevallier, M. W., Shaw, N. E., Kaplan, L.A. et al. (1993). *Development of a rapid assimilable organic carbon method for water. Applied and Environmental Microbiology* **59**(5).

Leung, E., Trejo, R., Rohe, D. L. et al. (2005). Prototype Testing Facility for Two-pass Nanofiltration Membrane Seawater Desalination Process, *AWWA Memb Tech Conf*, Phoenix.

Lobo, W. S. and Reid, A. (2007). Advanced Oxidation Processes (AOP) The Next 'Silver Bullet' … But Which Process is Best? A Comparison of Treatment Processes Combining Ozone, UV and Hydrogen Peroxide. *World congress on Ozone and Ultraviolet Technologies*. Los Angeles.

Lytle, D. A. (2007). Biological Nitrification in a Full-scale and Pilot-scale Iron Removal Drinking Water Treatment Plant. *J WSRT-Aqua*, **56**(2).

MacPhee, M. J., Charles, G. E. and Cornwall, D. A. (2001). *Treatment of Arsenic residuals from Drinking Water Treatment Processes*. EPA 600/R-01/033.

Marc, J-F and Pinker, B. (1998). Reactivation of Granular Activated Carbon for Drinking Water Treatment, *Journees Information Eaux 98*, Portiers.

Martijn, B. J., Kruithof, J. C. and Welling, M. (2006). UV/H_2O_2: Treatment: The Ultimate Solution for Organic Contaminant Control and Primary Disinfection, *Proceedings of the WQTC*, Denver.

McNeill, L. S. and Edwards, M. (1997). Arsenic Removal During Precipitative Softening. *J Environmental Engineering, ASCE*, **123**(5).

Merkle, R. B., Knocke, W. R., Daniel, L. et al. (1997). Dynamic Model for Soluble Mn^{2+} Removal by Oxide-Coated Filter Media. *J Environmental Engineering, ASCE*, **123**(7).

Merry, A. et al. (1995). Membrane Treatment of Coloured Water. *Proc of International Symposium of Wastewater Treatment* and *7th Workshop on Drinking Water*, Montreal.

Miles, G. D. (1948). *J Soc Chem Indust*, **67**.

Missimer, T. M. (1994). *Water Supply Development for Membrane Water Treatment Facilities*. Lewis Publishers/ CRC Press.

Monscwitz, J. T. and Ainsworth, L. D. (1974). Treatment of Hydrogen Sulphide. *J AWWA*, **66**(9).

Morris, T. and Siviter, C. L. (2001). Application of a Biological Iron Removal Process at Grove Water Treatment Works. *J CIWEM*, **15**(2).

Mouchet, P. and Capon, B. (1991). *Recent Evolution in Drinking Water Treatment Technology*. Extract Revue Travaux, Degremont, July/August.

Mouchet, P. (1992). From Conventional to Biological Removal of Iron and Manganese in France. *J AWWA*, **84**(4).

Nerenberg, R., Rittmann, B. E. and Soucie, W. J. (2000). Ozone/Biofiltration for Removing MIB and Geosmin, *J AWWA*. **92**(12).

Paillard, H., Bruent, R. and Dore, M. (1987). Application of Oxidation by a Combined Ozone/Ultraviolet Radiation System to the Treatment of Natural Water. *Ozone: Science and Engineering*, **9**(4).

Paillard, H., Legube, B., Gilbert, M. et al. (1990b). Removal of Nitrogonous Pesticides by Direct and Radical Type Ozonation. *EC Annual Conference on Micropollution*, May, Lisbon.

Paillard, H., Partington, J. and Valentis, G. (1990a). Technologies Available to Upgrade Potable Waterworks for Triazines Removal. IWEM Scientific Section, Pesticide Symposium, London, 11–12 April.

Richard, Y., Brener, L., Martin, G. et al. (1978). Study of the Nitrification of Surface Water. *Prog Wat Tech*, **10**(5/6).

Richardson, A. M. and Holden, J. B. (2004). The Evaluation of UV and UV/H_2O_2 for Pesticide and Algal Toxin Removal, *Conf. on Advanced Oxidation Processes for Water and Waste Water Treatment*, Cranfield.

Rittmann, B. E. and Huck, P. M. (1989). Biological Treatment of Public Water Supplies. *CRC Critical Reviews*, **19**(2).

Rook, J. J. (1976). Haloforms in Drinking Water. *J AWWA*, **68**(3).

Rosenfeldt, E. J. (2005). UV and UV/H_2O_2 Treatment of Methylisoborneol (MIB) and Geosmin in Water *J WSRT-Aqua*, **54**(7).

Rosenfeldt, E. and Linden, K. (2004). Degradation of Endocrine Disrupting Chemicals Bisphenol A, Ethinyl Estradiol and Estradiol during UV Photolysis and Advanced Oxidation Processes. *Environ. Sci. Technol.*, **38**(20).

Schock, M. and Oliphant, R. J. (1996). *The Corrosion and Solubility of Lead in Drinking Water, Internal Corrosion of Water Distribution Systems*, 2nd Edn. AwwaRF.

Sharma, S. K., Petrusevski, B. and Schippers, J. C. (2005). Biological Iron Removal from Groundwater. *J WSRT-Aqua*, **54**(4).

Sheiham, I. and Jackson, P. J. (1981). The Scientific Basis for Control of Lead in Drinking Water by Water Treatment. *J IWES*, **35**(6).

Short, C. S. (1973). *Removal of Ammonia from River Water*. TP101, WRc.

Simms, J. and Azizian, F. (1998). Pilot-Plant Trials on the Removal of Arsenic from Potable Water Using Activated Alumina. *Proc. AWWA Water Technology Conference*, Denver.

Singer, P. C., Schneider, M., Edwards-Brandt, J. et al. (2007). MIEX for Removal of DPB Precursors: Pilot plant Findings. *JAWWA*, **99**(4).

Slunjski, M., O'Leary, B., Tattersall, J. et al. (2000). MIEX Water Treatment Process, *Proc. Aquatech*, Amsterdam.

Smith, D. J., Pettit, P. and Schofield, T. (1996). Activated Carbon in Water Treatment. *Water Supply*, **14**(2).

Snoeyink, V. L., Cairns-Chambers, C. and Pfeffer, J. L. (1987). Strong Acid IX for Removing Barium, Radium and Hardness. *J AWWA*, **79**(8).

Snoeyink, V. L. and Summers, R. S. (1999). Adsorption of Organic Compounds, Chapter 13 in: *Water Quality and Treatment: A Handbook of Community Water Supplies*. 5th Edn. (Ed. Letterman, R. D.) McGraw-Hill.

Soares, M. I. M. (2002). Denitrification of Goundwater with Elemental Sulfur, *Proc. Water Quality Technology Conference*, San Diego.

Sommariva, C. (2004). *Desalination Management and Economics*. Faversham House Group.

Tiller, A. K. (1984). Corrosion Induced by Bacteria. *The Public Health Engineer,* **12**(3).

Thirunavukkarasu, O. S. (2003). Arsenic Removal in Drinking Water Using Granular Ferric Hydroxide. *Water SA*, **29**(2).

Turner, M. E. D. (1961). The Influence of Water Composition on Dezincification of Duplex Brass Fittings. *J SWTE*, 14.

Turner, M. E. D. (1965). The Influence of Water Composition on Dezincification of Duplex Brass Fittings. *Proc. SWTE*, **10**.

US EPA (2005). *A Regulator's Guide to the Management of Radioactive Residuals from Drinking Water Treatment Technologies*. EPA 816-R-05-004.

US EPA (2002). *National Primary Drinking Water Regulations; Radionuclides*; Office of Ground water and Drinking Water, Section I-C.10.

van der Kroon, G. M. and Schram, A. H. (1969). Weir Aeration—Part 1: Single Free Fall, *H2O*, **22**(2).

van der Veen, C. and Graveland, A. (1988). Central Softening by Crystalisation in a Fluidized-Bed Process. *J AWWA*, **80**(6).

van Dijk, J. C. and Wilms, D. A. (1991). Water Treatment without Waste Material—Fundamentals and State of the Art of Pellet Softening. *J WSRT-Aqua*, **40**(5).

van Eekeren, M. W. M., van Paassen, J. A. M. and Merks, C. W. A. M. (1994). Improved Milk-of-Lime for Softening of Drinking Water—the Answer to the Carry-over Problem. *J WSRT-Aqua*, **43**(1).

van Honwelingen, G. A. and Nooijen, W. F. J.M. (1994). Water Softening by Crystallization. *European Water Pollution Control*, **3**(4).

Veerapaneni, S. V., Jordan, B., Leitner, G. et al. (2005). Optimization of RO Desalination Process Energy Consumption, *Proc. International Desalination Assoc.* World Congress, Singapore.

Voutchkov, N. (2005). Use of Large Beach Well Intakes for Large Desalination Plants. *EDS Newsletter*, **22**.

Wang, L., Chen, A. S. C., Sorg, T. J. et al. (2002). Field Evaluation of As Removal by IX and AA. *J AWWA*, **94**(4).

Warton, B., Heitz, A., Zappla, L. R. et al. (2007). Magnetic Ion Exchange Drinking Water Treatment in a Large-scale Facility. *J AWWA*, **99**(1).

Waypa, J. J., Elimelech, M. and Hering, J. G. (1997). Arsenic Removal by RO and NF Membranes. *J AWWA*, **89**(10).

Westerhoff, P. (2006). *Arsenic Removal with Agglomerated Nanoparticle Media*. AwwaRF.

White, G.C. (1999). *Handbook of Chlorination*, 4th Edn. Van Nostrand Reinhold.

Withers, A. (2005). Options for Recarbonation, Remineralisation and Disinfection for Desalination Plants. *Desalination*, 179(1–3).

Wong, J. M. (2000). Treatment Technologies for the Removal of NDMA from Contaminated Groundwater, *Groundwater Resources Assoc of California*, **11**(2).

Wright, R. R. and Missmer, T. M. (1997). Alternative Intake Systems for Seawater Membrane Water Treatment Plants. *Proc.International Desalination Assoc.* World Congress, Madrid.

Zhang, Z. and Clifford, D. A. (1994). Exhausting and Regenerating Resins for Uranium Removal. *J AWWA*, **86**(4).

Disinfection of Water

11

11.1 DISINFECTANTS AVAILABLE

The term 'disinfection' is used to mean the destruction of infective organisms in water to such low levels that no infection of disease results when the water is used for domestic purposes including drinking. The term 'sterilisation' is not strictly applicable because it implies the destruction of all organisms within a water and this may be neither achievable nor necessary. Nevertheless the word is often loosely used, as in 'domestic water sterilizers'.

On a plant scale the following disinfectants are in common use:

- Chlorine (Cl_2);
- Chloramines (NH_2Cl, $NHCl_2$)
- Chlorine dioxide (ClO_2)
- Ozone (O_3)
- Ultra-violet (UV) radiation.

For small plants or under special circumstances the following may be used: products releasing chlorine, (e.g. calcium hypochlorite ($Ca(OCl)_2$), also called chloride of lime and sodium hypochlorite ($NaOCl$)).

The organisms in water which it may be necessary to kill by disinfection include bacteria, bacterial spores, viruses, protozoa and protozoa cysts, worms and larvae. The efficacy of disinfection depends on numerous factors: the type of disinfectant used; the amount applied and the time for which it is applied; the type and numbers of organisms present; and the physical and chemical characteristics of the water.

CHLORINE AND CHLORAMINE PROCESSES OF DISINFECTION

11.2 ACTION OF CHLORINE

The precise action by which chlorine kills bacteria in water is uncertain but it is believed that the chlorine compounds formed when chlorine is added to water rupture bacterial membranes and inhibit vital enzymic activities resulting in bacterial death. Chlorine is also a strong oxidising agent that will break up organic matter in a water; but, in so doing, because it is a highly reactive chemical, it can form a wide range of chlorinated compounds with the organic matter present. Among these are the trihalomethanes (THM) and haloacetic acids (HAA) for which limits have been set for health reasons (Sections 6.25, 11.7). Chlorine can also restrain algal growth, react with ammonia and convert iron and manganese in the water to their oxidized forms that may then precipitate. Hence there are a number of factors to be taken into consideration when using chlorine as a disinfectant.

11.3 CHLORINE COMPOUNDS PRODUCED

When chlorine is added to water, which is free from organic matter or ammonia, hypochlorous acid $HOCl$ is formed which is further dissociated to H^+ and OCl^-:

$$Cl_2 + H_2O = HOCl + HCl \tag{11.1}$$

$$HOCL \Leftrightarrow H^+ + OCl^- \tag{11.2}$$

The dissociation is favoured by high pH and temperature of the water as shown on Table 11.1. The hypochlorous acid $HOCl$ and hypochlorite ion OCl^- are together known as 'free chlorine' and

Table 11.1 Variation of HOCl as percentage of free chlorine with pH and temperature values of water

pH	Percent HOCl at water temperatures (percent OCl⁻ = 100 − percent HOCl)						
	0 °C	5 °C	10 °C	15 °C	20 °C	25 °C	30 °C
6.0	98.5	98.3	98.0	97.7	97.4	97.2	96.9
6.25	97.4	97.0	96.5	96.0	95.5	95.1	94.6
6.5	95.5	94.7	94.0	93.2	92.4	91.6	91.0
6.75	92.3	91.0	89.7	88.4	87.1	86.0	84.8
7.0	87.0	85.1	83.1	81.2	79.3	77.5	75.9
7.25	79.1	76.2	73.4	70.8	68.2	66.0	63.9
7.5	68.0	64.3	60.9	57.7	54.8	52.2	49.9
7.75	54.6	50.5	46.8	43.5	40.6	38.2	36.0
8.0	40.2	36.3	33.0	30.1	27.7	25.6	23.9
8.25	27.4	24.3	21.7	19.5	17.6	16.2	15.0
8.5	17.5	15.3	13.5	12.0	10.8	9.8	9.1
8.75	10.7	9.2	8.0	7.1	6.3	5.8	5.3
9.0	6.3	5.4	4.7	4.1	3.7	3.3	3.0

are the most effective forms of chlorine for achieving disinfection. Of the free chlorine, hypochlorous acid is a far more powerful bactericide than the hypochlorite ion. Thus free chlorine acts more rapidly in an acid or neutral water. Therefore when final pH correction is practised, the alkali should be added after the disinfection process has been completed.

If ammonia is present in the water or if ammonia is added to the water, chloramines will be formed in a stepwise manner: monochloramine NH_2Cl; dichloramine $NHCl_2$; and trichloramine NCl_3 (nitrogen chloride) (Section 11.8). Of these compounds, the monochloramine and dichloramine together in total are known as the 'combined chlorine'; total chlorine is the sum of combined chlorine and free chlorine.

The free chlorine is many times more powerful as a bactericide than combined chlorine (i.e. mono and dichloramines). Of the chloramines, the dichloramine is more powerful than monochloramine requiring only about 15% of the monochloroamine dose for inactivation of *E. coli*. Butterfield (1943–46) estimated that 25 times as much combined chlorine is needed to achieve the same degree of kill of bacteria as free chlorine in the same time. Since ammonia is often naturally present in a water, it is usual to add sufficient chlorine to react with all the ammonia present and produce an excess of free chlorine sufficient to achieve speedy disinfection. As a consequence the efficacy of chlorine as a disinfectant is influenced by a number of conditions.

11.4 FACTORS RELATING TO THE DISINFECTION EFFICIENCY OF CHLORINE

The following factors have to be taken into account when treating water with chlorine.

The stage at which chlorine is applied. Chlorine is often applied at more than one stage in the treatment of a water. 'Pre-chlorination' comprises the application of chlorine to a water (often a raw water) before it is processed through treatment works, e.g. before clarification and filtration. 'Intermediate-stage chlorination' is chlorine added between stages of treatment. 'Final chlorination' means the final disinfection of a water before it is put into supply. The purposes of pre-chlorination and intermediate-stage chlorination are often partly biological so as to reduce bacterial content, prevent bacterial multiplication and restrain algal growth; and partly chemical, so as to assist in the precipitation of iron and manganese and achieve other oxidation benefits. Final chlorination is always for the purpose of disinfecting the water and to maintain a residual in the distribution system so that it is safe for drinking.

Effect of turbidity. The effect of turbidity in a water is to make it difficult for the penetration of chlorine and therefore the destruction of bacteria taking shelter in particles of suspended matter. It is always necessary therefore, that final disinfection by chlorine is applied as a final stage of treatment in water which contains low turbidity. For effective disinfection the WHO (1993) suggests a guide level value for turbidity of less than 1 NTU.

Consumption of chlorine by metallic compounds. A substantial amount of chlorine may be used to convert iron and manganese in solution in the water into products which are insoluble in water (Section 10.12). Reduction of these parameters by upstream processes is therefore essential. Typically iron and manganese should be less than 0.1 mg/l as Fe and 0.05 mg/l Mn, respectively. If at the point of chlorine application their levels are too low to justify removal, the dose must take their demand into account.

Reaction of chlorine with ammonia compounds and organic matter. The ammonia compounds may exist in organic matter or, alternatively, ammonia may exist separately from organic matter

(Section 6.7); in either case they will form combined chlorine which is not as effective a bactericide as free chlorine. Chlorine may be used in the oxidation of some organic matter, but at the risk of forming disinfection byproducts (Section 11.7). Ammonia should not exceed 0.01 mg/l as N. When this value is exceeded or when organic matter is present, an allowance should be made both in the chlorine dose and contact time to satisfy the chlorine demand prior to disinfection. Therefore the substances that are causing a chlorine demand must be removed prior to disinfection by upstream treatment or an allowance for them must be made in the chlorine dose, otherwise disinfection could be compromised.

Low temperature causes delay in disinfection. A very substantial decrease in killing power takes place with lowering of temperature. The difference in kill rate of bacteria between the temperatures of 20 and 2 °C is noticeable both with free and combined chlorine. This must be borne in mind when fixing the contact period.

Increasing pH reduces effectiveness of chlorine. In free chlorine and in combined chlorine the more effective disinfectants in each case, i.e. hypochlorous acid and dichloramine respectively, are formed in greater quantities at low pH values than at high values. Thus disinfection is more effective at low pH values; the guide level value suggested by WHO (1993) is less than 8.

The number of coliforms presented for disinfection. This has an influence on the disinfection efficiency. To be confident of achieving 100% compliance with the requirement of zero coliforms after the disinfection stage, the water subjected to disinfection ideally should not contain more than 100 coliforms/100 ml. Most ground waters satisfy this criterion. In surface waters, coagulation followed by solid–liquid separation processes, achieves up to 99.9% bacteria removal (Section 7.13). Consequently pre-disinfection in addition to conventional treatment is only required for heavily polluted surface waters.

Time of contact is important. The disinfecting effect of chlorine is not instantaneous. Time must be allowed for the chlorine to kill organisms. This important factor is dealt with in the next section.

11.5 CHLORINE RESIDUAL CONCENTRATION AND CONTACT TIME

The principal factors influencing the disinfection efficiency of chlorine are free residual concentration, contact time, pH and water temperature. The term 'free residual' means the amount of free chlorine remaining after the disinfection process has taken place. Given adequate chlorine concentration and contact time, all bacterial organisms and most viruses can be inactivated. Thus a useful design criterion for the disinfection process is the product of contact time (t in minutes) and the chlorine-free residual concentration (C in mg/l) at the end of that contact time. This is known as the 'Ct value' or 'exposure value' (WHO, 1989). On this basis the WHO guide level of 0.5 mg/l free residual concentration after 30 minutes contact would have a Ct value of 15 mg.min/l. This is shown to provide a 12.5-fold factor of safety so that a degree of inefficiency in the contact tank performance can be tolerated (Stevenson, 1998).

The WHO Ct criterion is for faecally polluted water and therefore may be varied according to the bacteriological quality of the source water. For example a groundwater free of *E. coli* but containing no more than 10 coliforms/100 ml could have a Ct value of 10 mg.min/l with t not less than 15 minutes and for groundwaters where coliforms and *E. coli* are absent, chlorination sufficient to maintain a residual in the distribution system with no contact at the treatment works could be acceptable. On the other hand for surface waters a higher Ct value would be used, e.g. 30 mg min/l, with t being not less than 30 minutes, and C being 0.5–1.0 mg/l depending on the degree of bacterial pollution. If chlorine demand is to be satisfied in the disinfection stage C and t should be increased.

In the USA the Surface Water Treatment Rule (SWTR) (US EPA, 1991) for disinfection specifies the water treatment required on raw water that averages 1 *Giardia* cyst per 100 l to achieve removal by solid–liquid separation processes and inactivation by disinfection processes, of 3-log (99.9%) for *Giardia lamblia* cysts and 4-log (99.99%) for enteric viruses, these two being chosen for their resistance to disinfection when compared to *Legionella*, heterotrophic bacteria and coliforms. According to the SWTR direct filtration treatment is given 2-log credit for *Giardia* removal and 1-log for virus removal, whilst conventional treatment of clarification and filtration is given 2.5-log credit for *Giardia* and 2-log for viruses. These requirements increase for higher raw water *Giardia* concentrations. This leaves 1 to 0.5-log inactivation of *Giardia* and 2- to 3-log inactivation of viruses to be achieved by disinfection. Tables 11.2 and 11.3 give the Ct values stated in the SWTR to achieve 1-log inactivation of *Giardia* and 2 and 3-log inactivation of viruses.

The WHO Ct criterion is applicable to the inactivation of bacteria and most viruses and therefore cannot be compared with the Ct values for inactivation of cysts given in Table 11.2. Comparison with Ct values in Table 11.3 confirms that the WHO criterion has a high factor of safety which is desirable to ensure complete inactivation.

In the expression Ct, the contact time t is the time the water remains in the contact tank. Due to eddies and short-circuiting, the total volume of the tank may not be available for contact, some water may pass through the tank in less time than the theoretical residence time (t_T) which is the volume of water in the tank divided by the rate of flow. The inefficiencies are primarily due to jetting at the inlet, turning flow at the end of baffles and the weir outlet. The contact tanks are designed so that at least 90% of water passing through the tank remains in the tank for more than the required contact time at the design flow rate. This is also called the t_{10} time, i.e. the time it takes 10% of water to pass through the tank. The ratio of t_{10}/t_T is a measure of short-circuiting in the contact tank and varies in the range 0–1. A value of one is an indication of total plug flow but in practice it varies between 0.6 to 0.8 depending on the design.

Table 11.2 *Ct* values for achieving 1-log inactivation of *Giardia Lamblia*

| Disinfectant | pH | Ct values at water temperature | | | | | |
		0.5 °C	5 °C	10 °C	15 °C	20 °C	25 °C
Free residual chlorine of 2 mg/l	6	49	39	29	19	15	10
	7	70	55	41	28	21	14
	8	101	81	61	41	30	20
	9	146	118	88	59	44	29
Ozone	6–9	0.97	0.63	0.48	0.32	0.24	0.16
Chlorine dioxide	6–9	21	8.7	7.7	6.3	5	3.7
Chloramines	6–9	1270	735	615	500	370	250

Source of information: Surface Water Treatment Rule (SWTR) (US EPA, 1991)

Table 11.3 Ct values for achieving 2- and 3-log inactivation of enteric viruses at pH values 6–9

Disinfectant	Log inactivation	Ct values at water temperature					
		0.5 °C	5 °C	10 °C	15 °C	20 °C	25 °C
Free residual chlorine	2	6	4	3	2	1	1
	3	9	6	4	3	2	1
Ozone	2	0.9	0.6	0.5	0.3	0.25	0.15
	3	1.4	0.9	0.8	0.5	0.4	0.25
Chlorine dioxide	2	8.4	5.6	4.2	2.8	2.1	1.4
	3	25.6	17.1	12.8	8.6	6.4	4.3
Chloramines	2	1243	857	643	428	321	214
	3	2063	1423	1067	712	534	356

Source of information: Surface Water Treatment Rule (SWTR) (US EPA, 1991)

An optimum design with a t_{10}/t_T of 0.8 can be achieved either by physical modelling or computational fluid dynamics (CFD) models (Appendix to Chapter 12). In the absence of these tools the following basic design parameters are suggested: the inlet jet should be baffled; the tank should be divided into long straight flow and return channels of length:width ratio greater than 10 and depth:width ratio less than 1.5; and a weir outlet should be provided to maintain the required volume of water under all flow conditions. With these provisions, a value of t_{10}/t_T of 0.5–0.7 may be achieved. The actual time of contact provided by newly designed or existing tanks can be checked by timing the passage of a slug dose of a tracer chemical such as lithium chloride at a concentration of <0.1 mg/l through the tank (Falconer, 1986) or by physical or CFD models.

Pipelines are ideal for contact as they provide good plug flow and the ratio of t_{10}/t_T could be almost 1.0. However, in practice it should be reduced to about 0.9 to allow for bends and exit and entry conditions. The pipeline should be within the treatment plant site boundary so that control and monitoring of the residual chlorine at its downstream end is possible and convenient.

The free residual chlorine concentration of the water leaving the contact tank, after the requisite time of contact, can be reduced if desired by partial dechlorination (Section 11.13) to suit the needs of the distribution system.

11.6 EFFICIENCY OF CHLORINE IN RELATION TO BACTERIA, ENTERIC VIRUSES AND PROTOZOA

Bacterial kill. The work of Butterfield (1943–46) has shown that under nearly all conditions the typhoid bacillus and other enteric pathogenic bacteria are at least as susceptible to chlorination as *E. coli*. Due to the far greater concentrations of *E. coli* present in pollution of human or animal origin (Section 6.66) it is therefore practicable to assume that, if *E. coli* are absent in a 100 ml sample of disinfected water, then the water is also free of pathogenic bacteria. The spores of bacteria are, however, more resistant to the action of chlorine than are the bacteria; fortunately the bacteria causing most waterborne diseases are not spore formers. The spore-forming *Clostridium perfringens* (*Cl. welchii*) used as an indicator of pollution (Section 6.69) is not considered significant for health.

Virus kill. The pathogenic enteric viruses, described in Section 6.64, occur in far less numbers than *E. coli* in a polluted water. However they can survive for long periods in water and the minimum dose causing human infection is believed to be very low. Test methods available for detecting the presence of viruses (Section 6.69) cannot be used for routine monitoring. The enteric viruses have also been shown to be more resistant to chlorine than *E. coli*. Poynter (1973) reported Russian experiments had indicated that higher levels of residual chlorine and longer periods of contact were required to free water from viruses than were required to destroy *E. coli*; similar results were obtained by Scarpino (1972).

The consequence of these difficulties is that tests showing the absence of *E. coli* in 100 ml samples of disinfected water do not give the same level of confidence that viruses are absent as they do for the absence of pathogenic bacteria. However the *E. coli* test remains the only practicable one at present for routine monitoring and the WHO (1993) considers: 'it has been demonstrated that a virus-free water can be obtained from faecally polluted source waters' if the following chlorine disinfection conditions are met:

- the water has a turbidity of 1 NTU or less;
- its pH is below 8.0;
- a contact period of at least 30 minutes is given; and
- the chlorine dose applied is sufficient to achieve at least 0.5 mg/l free residual chlorine during the whole contact period.

Protozoa resistance to chlorine. The principal protozoa pathogenic to humans—*Entamoeba histolytica*, *Giardia lamblia*, and *Cryptosporidium parvum*—were described in Section 6.63 where it was mentioned that human infection takes place through ingestion of them in their cyst stage. Like viruses the cysts can remain viable in the environment for long periods; the ingestion of a very few may be sufficient to cause human infection. Detection of their presence by routine testing is impracticable.

The cysts are more resistant to chlorine than viruses. WHO describes the cysts of *E. histolytica* as 'among the most chlorine-resistant pathogens known', and the resistance of *Giardia lamblia* to chlorine as falling 'somewhere between *E. histolytica* and enteric viruses' (WHO, 1996). The cysts of *Cryptosporidium* (called 'oocysts') were also found to be very resistant to chlorine, concentrations of 16 000 mg/l chlorine or more having been found necessary to reduce the viability of the oocysts to zero (Smith, 1989). The small size of cysts—*E. histolytica* 10–20 μm; *Giardia* spp. 8–12 μm; *Cryptosporidium* 4–6 μm—means their complete removal by conventional processes of coagulation, clarification and filtration cannot be relied on although these processes can be designed and operated to give between 2- and 3-log removal for *Cryptosporidium* (Section 8.19) and higher log removal for others.

No completely reliable disinfection procedure to eliminate these protozoan cysts and oocysts has yet been found. Irrespective of the treatment methods available, it is paramount that the sources of pollution likely to give rise to the presence of such pathogenic protozoa in a water are eradicated. If their presence is detected in a water an intensive search for the source and its elimination is necessary. When they are detected or suspected in the treated water, it is necessary to advise consumers to boil water used for drinking until evidence of their elimination from the supply is obtained (Section 11.29).

11.7 CHLORINATION AND THE PRODUCTION OF DISINFECTION BY PRODUCTS

When chlorine is applied to water containing precursors, which arise from the presence of natural colour and algal metabolic products, disinfection byproducts (DBP) e.g. THMs and HAAs (haloacetic acids) are formed (Section 6.25). The reaction rate is favoured by elevated free chlorine residual and precursor concentrations: at TOC > 4 mg/l, THM will exceed 100 µg/l for a travel time of 2–3 days in distribution if a free residual is to be maintained at the tap. It is also favoured by alkaline pH: at pH 9 about 10–20% more THM will form than at pH 7; increase in temperature: below 10°C there is no appreciable increase; and extended contact time (UKWIR, 2000). On the contrary HAA formation decreases with increase in pH but dicholroacetic acid formation is independent of pH. Limits to DPB concentrations in drinking water have been set because of their possible health effects. Emphasis is now placed on avoiding or limiting the pre-chlorination of a raw water containing organic matter in order to minimize the formation of DBPs. If a pre-disinfection stage is necessary, chlorine dioxide or ozone may be used instead of chlorine (Sections 11.17 and 11.22). Alternatively, since chloramine does not react with organic matter to produce DBPs to the same extent as free chlorine, the chlorine dose may be kept low enough to produce only chloramines by making use of ammonia when present in the water, or by using chloramination of raw water (Section 11.8). If relatively high levels of DBPs are expected at the treatment works outlet, the degree of removal of organic substances prior to chlorination must be maximized. In some instances, it has been sufficient to move the pre-chlorination dosing point further downstream in the treatment process, e.g. after coagulation and clarification has removed a high proportion of DBP precursors. If that is insufficient, advanced treatment processes, described in Section 10.36, can be used to prevent DBP formation by the removal of precursors before final chlorination is applied.

The maintenance of residual chlorine in the treated water may cause DBP formation to continue within the distribution system by the reaction with residual precursors, depending on the pH. There is some evidence that HAAs undergo some biodgredation in distribution. DBP formation reaction is initially fast with up to 50% formation in the first hour or so, but takes several hours or even days to complete and therefore its concentration at the consumer's tap could be much higher than in the treated water leaving the works. The position can be exacerbated in nutrient-rich waters when 'booster chlorination', i.e. addition of further chlorine at some key point(s) in the distribution system, has to be adopted to limit biological aftergrowths within the system. The formation of THMs in the distribution system can be minimized by controlled dosing of ammonia to convert free chlorine residual to combined chlorine residual. However, monochloramine increases the formation of HAA (Parsons, 2006; Hong, 2007).

Whatever control measures are adopted, the WHO Guidelines emphasize that the disinfection process must not be compromised and that 'inadequate disinfection in order not to elevate the DBP level is not acceptable'.

11.8 THE AMMONIA–CHLORINE OR CHLORAMINATION PROCESS

Combined chlorine, i.e. chloramines (mono and dichloramines) resulting from the reaction of chlorine with ammonia in water, is not commonly used as a primary disinfectant because of the acknowledged greater efficiency of free chlorine. In some instances a theoretical contact period of several hours would be required for chloramine to achieve adequate disinfection of certain difficult waters (Smith, 1990). However, ammonia is sometimes deliberately added after final chlorination to produce a chloramine residual in the final water passing into the distribution system. The primary reasons for using chloramines rather than chlorine are the residual is longer lasting than chlorine; reduction in THM (but not HAA) formation; superior control of biofilm growth (aftergrowth of bacteria or low forms of animal life) in the distribution system; and high doses (about 2 mg/l) can be applied with less risk of producing chlorinous taste. The weight ratio of chlorine to ammonia (as N) is usually in the range 3:1 to 4:1; when the ratio exceeds 5:1 monochloramine is destroyed. Ammonia is added after final chlorination when the free chlorine has acted for the requisite contact time. The reaction is fast, taking only a few seconds to form a combined residual. The most effective pH range for chloramination is 7.5–8.5 when monochloramine is predominant. Also excess ammonia can be used to ensure monochloramine is predominant. Dichloramine is considered to be a stronger disinfectant than monochloramine but it decays faster. Trichloramine has no disinfection properties and also it is volatile. Di and trichloramines have very low taste and odour threshold concentrations. Monochloramine can be preformed as a solution of 1.5 g Cl_2/l by reacting ammonium sulphate with sodium hypochlorite in the presence of caustic soda at pH of about 10. About 5% excess ammonia should be maintained to ensure monochloramine is predominant. The product should not be stored for more than about 12 to 18 hours depending on the ambient temperature.

An advantage of maintaining residual chloramine in water passing into the distribution system is that, if routine examinations show chloramine is present at the ends of the distribution system, this should indicate that no serious pollution has entered the pipes en route. The residual cannot be large enough to disinfect pollution entering the system, but it is useful for monitoring the state of the distribution system. The principal disadvantages of chloramination are the formation of HAAs and the nitrification of any excess ammonia or free ammonia released by the decay of chloramines, to form nitrite in the distribution system by nitrifying bacteria. Nitrites are toxic (Section 6.36). Control methods include: optimising the chlorine-to-ammonia ratio (Lieu, 1993); using chlorine to ammonia ratio closer to 5:1 (the stoichiometric ratio); flushing out the affected sections; decreasing residence time in service reservoirs; removing excess ammonia locally by break-point chlorination (Section 11.9); chlorite addition; reducing natural organic matter; periodic changes to free chlorine; pH control; and re-chloramination of the affected sections to eliminate bacterial growth (AwwaRF, 2003). Chloramine also forms N-nitrosodimethylamine (NDMA) (Section 6.80).

Blending of chloraminated water with water containing free residual chlorine in distribution systems could result in breakpoint chlorination or in the formation of dichloramine and nitrogen trichloride, which are well known for causing taste problems. If blending is necessary it should be carried out upstream of a service reservoir. Barrett (1985) developed a blending model which could be used to predict acceptable blends.

11.9 BREAKPOINT CHLORINATION

Where a water already contains natural ammonia the production of chloramine is unavoidable when chlorine is added. To ensure the production of free chlorine to enhance bacterial kill, substantially more chlorine may have to be added because the additional chlorine at first only causes a reduction of the chloramines by oxidation. Only when this reaction is completed does the addition of further chlorine produce free chlorine. Stoichiometrically the breakdown of ammonia to nitrogen commences at a chlorine:ammonia as N ratio of 5:1 and completes at a ratio of 7.6:1. (Chlorine: ammonia as NH_3 or NH_4^+ is 6.26:1 or 5.92:1 respectively). In practice the ratio for complete breakdown could be as much as 10:1 and is pH dependent. The point at which the free chlorine begins to form is

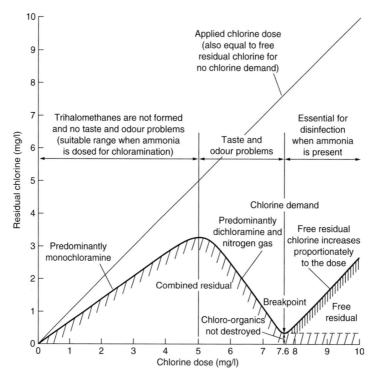

FIGURE 11.1

Theoretical breakpoint curve for a water containing 1.0 mg/l ammonia (as N).

called the 'breakpoint' for the water, and adding enough chlorine to exceed this is called 'breakpoint chlorination'. This is illustrated in Figure 11.1. In laboratory experiments Palin (1950) observed that for neutral to slightly alkaline pH when the ratio of chlorine to ammonia (as N) is less than 5:1 (by weight) the residual was mainly NH_2Cl; breakpoint occurred at 9.5:1 for pH 6; between 8.2:1 and 8.4:1 for pH 7 to 8; and 8.5:1 for pH 9. As the ratio increased to 10:1 and above there was a decrease in combined chlorine accompanied by increases in NCl_3 and free Cl_2. Apart from the advantage of producing free chlorine, breakpoint chlorination can sometimes reduce taste and odour problems resulting from di and trichloramines. The foregoing reactions are complex, being dependent on numerous factors such as temperature, pH and contact time. The breakpoint reaction could take about 20 minutes to complete and depends on the water quality. In some waters the ammonia content may be so high (0.5–1.0 mg/l) that the amount of chlorine to be applied to achieve breakpoint is uneconomic and other means to reduce the ammonia first should be adopted (Section 10.28).

11.10 SUPERCHLORINATION

'Superchlorination' means the dosing of a water with a high dose of chlorine, often much larger than the usual condition of the water demands. The method is most often used on borehole or well waters which, though normally free of pollution, may be subject to the onset of pollution to an unknown degree following heavy rainfall or some other circumstances. The normally unpolluted water may only require a small protective dose of chlorine of the order of 0.2 mg/l. To wait for the pollution to occur, detect it, and then increase the dose is impracticable since action could not take place in time to prevent some of the pollution passing into supply. It is also used on heavily polluted river waters, where the pre-treatment process does not include a pre-disinfection stage. Therefore a continuous high dose and adequate contact time is given sufficient to counter the worst conditions likely. After the contact period the water can be partially de-chlorinated by the injection of sulphur dioxide or sodium bisulphite (Section 11.13), and the dose may be so controlled that only a part is removed, leaving a residual to go into supply.

11.11 TYPICAL CHLORINE DOSE; TASTE AND ODOUR PROBLEMS

Typical chlorine dosages to final treated waters are frequently in the range 0.2–2.0 mg/l of free chlorine to give a residual of about 0.02–0.3 mg/l at the consumer's tap. The lower doses (0.2–0.5 mg/l) tend to be those used on clear well waters not subject to pollution; the higher doses relating to treated surface waters or to well or borehole supplies which are liable to experience sudden pollution, where super-chlorination followed by partial dechlorination after adequate contact time may be advised. Some raw waters, particularly surface waters, can have a high chlorine demand of 6–8 mg/l (Smith, 1990).

Free chlorine has a taste threshold concentration of 0.6 to 1.0 mg/l. Therefore, application of chlorine causes taste and odours, principally by the reaction of chlorine with some of the many trace compounds in the water (Section 10.33).

11.12 USE OF CHLORINE GAS

Chlorine is contained as a liquid under pressure in drums or cylinders. The liquid occupies about 95% of the volume; the remaining 5% is occupied by the gas. Most cylinders are designed to draw gas only, but the drum design allows either gas or liquid to be withdrawn. Typically cylinders are available in 33, 50, 65 and 71 kg capacities whilst drums are available in 864, 966 and 1000 kg capacities. Gas withdrawal rate at 15°C and 2 bar back pressure is 1.0 kg/h for 33 and 50 kg and 1.5 kg/h (or 2.3 kg/h at 25°C) for 71 kg cylinders; that for a drum is 7 kg/h. Higher rates can be achieved by connecting more than one container in parallel. A practical limit to the number of containers connected to a header is about six for cylinders and four for drums. The larger the number of containers connected the greater the number of connecting joints and hence the risk of a leak and also greater the risk of liquefaction in the pipe. When higher rates are required the contents should be drawn as liquid and vaporized in evaporators before use. The liquid should not be drawn from more than one container and the withdrawal rate should not exceed 180 kg/h. In a few large works where chlorine usage is high, chlorine is stored as a liquid between 4 and 8 bar pressure at 5–30°C in bulk storage vessels which are filled from road tankers. Chlorine storage facilities must be designed to very high standards of safety (HSE, 1990). Plate 19(a) shows a drum store with evaporator.

Chlorine gas is drawn via a vacuum regulating valve and made into a chlorine–water solution containing up to 3500 mg/l Cl_2 for application to the water supply by means of an auxiliary water flow passing through an ejector which, at the same time, creates the operating vacuum. The valve reduces the gas supply pressure from up to 10 bar down to a vacuum of about 170 mb The interruption

(a) Chlorine drum store with evaporator in the background. (Manufacturer: Portacel, UK; supplier: Bisan Inc. Canada.)

(b) Chlorine gas system with catchpot, changeover device and vacuum regulator. (Manufacturer: Portacel, UK)

(c) Floor mounted chlorinators. (Manufacturer: Portacel, UK)

(d) 580 kg Cl$_2$/d on-site hypochlorite generating plant at Mammashia WTW, Botswana. (Engineer: Burrow Binnie)

PLATE 19

(a) Earl Schmidt Filtration Plant ozone generators, Castaic Lake Water Agency, Santa Clara, CA, USA; capacity 2×17 kg/h @ 10% $^w/_w$ ozone. (Engineer: Black & Veatch. Manufacturer: Ozonia)

(b) 3×40 kW UV plant treating 60 Ml/d at Elmer water treatment works, Sutton & East Surrey Water Plc. (Consultant: Black & Veatch and MWH; Manufacturer; Trojan Technologies).

PLATE 20

or failure of the operating water supply will immediately shut down the flow of gas. In the unlikely event that positive pressure should reach beyond the vacuum regulating valve a pressure relief valve would vent a small quantity of gas to outside and operate an internal check valve until such time as normal vacuum conditions are restored. Plate 19(b) shows a vacuum regulator valve.

The flow of gas is metered by maintaining a constant differential pressure drop across a manually or automatically adjusted variable area orifice through which the gas passes, the rate of flow being indicated by a glass tube flowmeter. Plate 19(c) shows a bank of chlorinators. Should an excess vacuum condition occur within the system, as when the gas supply is exhausted, a relief valve operates downstream of the metering section and a no-flow condition is indicated on the flowmeter. Depending upon the type and capacity of the equipment the ejector may form an integral part of the unit, or be separately mounted either adjacent to the chlorinator or near the point of application. The chlorine dose rate may be automatically adjusted to accommodate variations in water flow or quality, or both, by means of an electric positioner fitted to the variable area orifice (Fig.11.2) to accept a control signal from a chlorine residual monitor to maintain a pre-set level of residual chlorine.

Chlorine is known to contain up to 0.1% $^{w}/_{w}$ of bromine. This, if allowed to react with ozone could form up to 1.6 µg/l of bromate for each 1 mg/l of chlorine dosed upstream of ozonation.

Safety precautions. Chlorine gas reacts explosively with many organic compounds. When mixed with water or moisture it is extremely corrosive. Pipework carrying liquid or gaseous chlorine should be of carbon steel and must be designed to exclude water or moist air; PVC-U pipework can be used for chlorine gas under vacuum and for chlorine solution applications.

Chlorine is very toxic with a 15-minute exposure limit of 0.5 ml/m³ (HSE, 2005). There is no 8-hr exposure limit. Liquid chlorine leaks are far more dangerous than chlorine gas leaks since, on

Notes: Distance B to C should be about 10 hydraulic pipe diameters (if no mixer). Process time A (BC) to D should be <4 minutes at average flow.

FIGURE 11.2

Vacuum-operated gas metering unit for automatic dosing with flow and pre-set chlorine residual control. Suitable for chlorination, partial dechlorination or chloramination.

evaporation, 1 kg of liquid chlorine yields about 335 litres of gas at 15 °C and 345 litres at 25 °C. Chlorine gas is heavier than air. A chlorine solution leak gives chlorine fumes laden with moisture and is very dangerous as it seems more tolerable to the respiratory tract than inhalation of dry chlorine (White, 1992).

Safety precautions must be taken in the design and layout of chlorine installations to safeguard the operators and the public (HSE, 1990; 1999). For sites in the European Union countries the Control of Major Accident Hazards Regulations (COMAH, 1999) must also be complied with.

Chlorine facilities should be designed to minimize leaks and to contain them if they should occur. A chlorine building should house storage and dosing equipment and should be separated from any ventilation intakes of other buildings by at least 25 m and from the site boundary by at least 20 m (for cylinder installations), 40 m (for drum installation using gas) and 60 m (for drum installations using liquid). Chlorine containers should always be installed in a separate 'gas tight' store constructed in substantially fire-resistant material and should be provided with access from open air. All doors should open outwards and emergency exit doors should have a push bar operated panic bolt. External windows should be avoided, artificial illumination being used throughout. A useful safety measure at chlorine installations is to fit the containers connected to headers with a remote shut-off pneumatically operated device (e.g. ChlorGuard) with a motive air system, for effective isolation of the chlorine supply at the container in the event of a downstream leak. Risk of chlorine leaks should be restricted to the container store by locating the vacuum regulating valves in the store and ejectors should be located remote from the chlorinators, local to the point of application so that the risk of a chlorine solution leak is confined to a short length of pipe at the point of application. Heaters (non-radiant type) may need to be installed in the container store to maintain a temperature greater than 10 °C. The ventilation system in the container store should have high level fresh air inlets and low level (500 mm above the floor) extraction of foul air discharging to outside at high level. Ventilation systems should be designed to give not less than 10 air changes per hour and should be arranged to start at a low leak level and to shut down at a high leak level to contain major leaks. Leak detectors with a 0.5 ml/m^3 detection limit should be installed in enclosed areas where chlorine is handled.

At chlorine sites in urban areas or near housing developments or with inventories of 10 or more tonnes, the use of a chlorine absorption plant for neutralising a leak should be considered. The absorption plant should be designed to treat the contents of one container (e.g. 1000 kg for a drum installation). The leakage rate could vary from 1.5 kg/min to about 35 kg/min. A typical system consists of a packed tower where the foul air flows counter-current to a flow of dilute caustic soda (10–20% ^{w}/$_w$ NaOH) which is continually recycled back to the top of the tower. Stoichiometrically 1000 kg of chlorine (about 340 m^3 as gas) requires 1127 kg of caustic soda (100% ^{w}/$_w$ NaOH); in practice about 20% more caustic soda would be required. An alternative absorbent is ferrous chloride which could be regenerated by adding iron fillings to the tank.

11.13 DECHLORINATION

Dechlorination can be achieved with sulphur dioxide or sodium bisulphite ($NaHSO_3$) or sodium metabisulphite ($Na_2S_2O_5$) and should be carried out after chlorine has had adequate contact time for disinfection. In disinfection treatment (superchlorination, Section 11.10) only partial dechlorination is required. In pre-treatment for reverse osmosis, complete dechlorination is the objective (Section 10.47). Chloramines can be similarly dechlorinated. Stoichiometrically 0.9 mg of sulphur dioxide or 3.6 mg of sodium bisulphite (25% ^{w}/w SO$_2$) or 1.38 mg of sodium metabisulphite (65% ^{w}/w SO$_2$) removes 1.0 mg of chlorine or 0.742 mg of monochloramine; in practice at least 15% more sulphur dioxide or twice more sulphites would be required; the dechlorination reaction is fast taking less than 1 minute to complete.

Sulphur dioxide, similar to chlorine, is contained in cylinders (30 and 65 kg) or drums (865 and 1016 kg) and is drawn for use as a gas or liquid. The gas withdrawal rates, at 15 °C and 2 bar back pressure, are 0.3 and 0.45 kg/h from 30 and 65 kg cylinders, respectively, and 2.3 kg/h from a drum. Higher withdrawal rates can be achieved by connecting several containers in parallel or withdrawing sulphur dioxide as a liquid and vaporising in evaporators. The liquid should not be drawn from more than one container and the withdrawal rate should not exceed 140 kg/h. The equipment used for sulphur dioxide is very similar to that used for chlorine. The dose rate may be automatically controlled proportional to water flow to maintain a pre-set level of chlorine residual, or both.

Safety precautions. Sulphur dioxide is corrosive; materials of construction used for chlorine can be used for sulphur dioxide. Sulphur dioxide is a toxic gas with an 8-hour exposure limit of 2 ml/m^3 and a 15-minute exposure limit of 5 ml/m^3 (ISCS, 2004). Liquid sulphur dioxide leaks are far more dangerous than a corresponding gas leak since, on evaporation, 1 kg of liquid sulphur dioxide yields about 370 litres of gas at 15 °C and 380 litres at 25 °C. It is heavier than air and is compatible with chlorine and therefore can be stored in the same room as chlorine.

Sodium bisulphite is available as a solution (Table 7.5). If the powder is required sodium metabisulphite should be used. Any spillages should be neutralized with soda ash, to prevent sulphur dioxide emission, and then be oxidized to neutral sulphate with sodium hypochlorite.

Other dechlorination methods include filtration through granular activated carbon (GAC) or use of ammonia (Section 11.9) or hydrogen peroxide. GAC is sometimes used to protect chlorine sensitive reverse osmosis membranes; flow rates of 10–20 m^3/h.m^2 are typical with EBCT between 10 to 15 minutes. GAC catalyses the reaction; therefore in theory GAC is not consumed by chlorine. However, it should be replaced on a regular basis due to adsorption of other material; it is not suitable for reactivation for dechlorination use because dechlorination weakens the GAC and it breaks down during reactivation.

Sulphur dioxide leaks are treated in absorbers of a similar design to those described for chlorine, using caustic soda as the absorbent.

11.14 AMMONIATION

Ammonia (anhydrous or aqueous) or ammonium sulphate can be used for ammoniation and should be added after disinfection and partial dechlorination (if applicable). Final pH correction should follow ammoniation. Ammonia is available as a liquid in cylinders (49 and 65 kg) or drums (500 kg). It is withdrawn as a gas at 0.5 kg/h from a cylinder or 2 kg/h from a drum at 15 °C, or as a liquid to evaporators. Apparatus used is very similar to that used for chlorine. Ammonia dosing may be automatically controlled proportional to water flow or to a pre-set ratio in the range 3:1–5:1 of chlorine:ammonia (as N), or both. Ammonia is very soluble in water and is corrosive. Steel piping is suitable for conveyance of ammonia liquid and dry gas. Iron, copper and zinc are attacked by ammonia solution, but PVC-U is suitable. Motive water for ammonia dosing units should be softened to a hardness value of less than 25 mg/l as CaCO$_3$ to prevent calcium carbonate scaling of fittings.

Ammonia is toxic with an 8-hour exposure limit of 25 ml/m^3 and a 15-minute exposure limit of 35 ml/m^3 (HSE, 2005) and is lighter than air. It is flammable and limits are 15.5 and 27% by volume.

Electrical apparatus in areas where ammonia is handled should therefore be suitably protected. Ammonia gas forms explosive mixtures with chlorine and sulphur dioxide gases and should therefore be stored away from them. Storage room design requirements are similar to those for chlorine, except for the ventilation system which requires low level fresh air inlets and air extraction and disposal at high level.

Absorbers for treating ammonia leaks are of a similar design to those used for chlorine. Absorbent used for ammonia is sulphuric acid (10% ^{w}/w H_2SO_4). Since ammonia has a high affinity for water the absorbent could be water as well.

Ammonia is also available as aqueous ammonia of concentration up to 40% ^{w}/w NH_3, which is called aqua ammonia. Ammonia vapour may evolve from the solution and so should be treated in the ame way as anhydrous ammonia.

Ammonium sulphate is a crystalline powder (Table 7.5) although it may be available as a 40% solution and is safer than anhydrous or aqueous ammonia. A solution is prepared from the powder for dosing. Dose control is similar to gaseous ammonia.

11.15 HYPOCHLORITE PRODUCTION ON SITE BY ELECTROLYSIS

Sodium hypochlorite can be produced by the electrolysis of brine (a solution of sodium chloride) in a fully mixed cell. A direct current passed through a solution of sodium chloride (common salt) containing Na^+ and Cl^-, produces chlorine at the anode, and hydrogen at the cathode. With mixing of the catholyte, anolyte and sodium ions in the solution, sodium hypochlorite (Na^+OCl^-) is produced. The principal reactions are as follows:

At the anode	$2Cl^- - 2e = Cl_2$
At the cathode	$2H_2O + 2e = 2OH^- + H_2$
On mixing	$2Na^+ + 2OH^- + Cl_2 = Na^+OCl^- + Na^+Cl + H_2O$

The process has the advantage that the hypochlorite solution can be manufactured on site, thus avoiding the risks of transporting, storage and handling liquid and gas chlorine and the difficulty of meeting all the associated safety measures required. The on-site generation produces a hypochlorite solution easy to handle, with favourable running costs as compared with the use of purchased liquified chlorine. The process can be used on large or small supplies.

A diagram of the process used is shown in Figure 11.3; a small plant is illustrated in Plate 19(d). The brine for electrolysis is prepared by withdrawing saturated brine from a salt saturator at a concentration 358 g/l (10 °C), 360 g/l (20 °C) or 363 g/l (30 °C) and diluting this to 20–30 g/l. About 3.9 kg of sodium chloride is required to form 1 kg of chlorine; approximately 50% of sodium chloride is converted to hypochlorite. The salt should be of high purity grade, free from calcium, magnesium and heavy metals in particular manganese (<10 µg/l). Pure dried vacuum salt containing at least 99.7% ^{w}/w NaCl is most suitable (Table 7.5). Lower quality salt should be treated to remove hardness, turbidity, iron and manganese. Water for saturation and dilution of salt must have a hardness less than 15 mg/l as $CaCO_3$ and may therefore need to be softened; base exchange softening being used. The process water requirement is about 125–150 litres per kg of chlorine. The feed water temperature should be maintained above 5–10 °C to maximize anode coating life; the

FIGURE 11.3

On-site hypochlorite production by electrolysis of brine.

product solution temperature should be below 40 °C to inhibit chlorate formation. The temperature rise in the electrolyser is limited to about 20 °C. In cold climates it may be necessary to warm the feed water usually using the product in the secondary circuit, and in hot climates to cool the dilution water using chillers.

The chlorine content of the hypochlorite produced is in the range 6–9 g Cl_2/l. The electrical consumption is 4–5 kWh per kg of chlorine produced; a low voltage direct current (e.g. 40 V) is used. Hydrogen gas is produced at the cathode at the rate of 330 litres at 15 °C and 1 bar pressure, per kg of chlorine produced. The sodium hypochlorite solution with the hydrogen is fed to an enclosed hydrogen disentrainment tank which can also act as a solution storage tank. The hydrogen degasses very rapidly to an airspace at the top of the tank, which is force ventilated with air at a rate 100 times the maximum hydrogen production rate. This reduces the hydrogen concentration to one-quarter of the 'lower explosion limit' of 4% of hydrogen in air by volume. The diluted gas is then vented to the atmosphere. The hypochlorite solution, having checked for chlorine content, is injected by a positive displacement metering pump into the water to be treated. Hypochlorite solution tanks can be constructed in high density polyethylene or GRP with a PVC or polypropylene lining.

At coastal installations hypochlorite can be produced from seawater. Due to the hardness of the seawater, heavy metals and suspended matter, a special cell design with appropriate anodes and cathodes is used, and the chlorine content of the hypochlorite solution is maintained at about 1 g/l. The feed water should be free of pollution and have a relatively constant salinity and must be passed through a strainer to remove coarse suspended particles. Hypochlorite produced from seawater is used primarily for biological growth control in raw water, or cooling water treatment at desalination, or coastal power generation plants. The hypochlorite solution is unstable with decay rate of 2–3% per hour due to the presence of heavy metals (White, 1999) and is not suitable for storage.

Safety measures. Electrolytic plant presents a potential fire and explosion hazard due to the hydrogen produced. Precautions should therefore be taken in the siting, layout, and design of the plant (HSE, 1987). The design of electrical apparatus needs particular attention: as much as possible of such apparatus should be sited in a non-hazardous area or else specially protected.

By-products. Several side reactions take place during the electrolysis of brine. One such produces chlorate (ClO_3^-) (Section 6.25) which has a WHO Guideline value of 0.7 mg/l. Commercial units are known to produce less than 50 μg of chlorate as ClO_3^- per 1.0 mg of chlorine and this acts as a limiting factor to the amount of hypochlorite that can be applied. The process also produces bromate from bromides present as an impurity in the salt. The use of high purity salt containing 0.015% $^w/_w$ of bromide restricts the bromate concentration to less than 1.0 μg per 1.0 mg of chlorine.

All the sodium chloride used in the feed to the electrolyser is injected into the water being treated as sodium and chloride. Account therefore has to be taken of the limits set for the sodium content of drinking water (Section 6.47) and the design of the disinfection process has to take this factor into account. In addition, chloride coupled with low alkalinity in a water can contribute to corrosion of metals (Section 10.41). The quantities of sodium and chloride produced are approximately 1.4 and 2.1 mg respectively per mg of Chlorine produced.

11.16 TESTING FOR CHLORINE

On-line measurement of the chlorine content of a water is achieved by passing a continuous sample of the chlorinated water through a cell containing electrodes of dissimilar metals, e.g. platinum and copper or silver. The galvanic current induced between the electrodes reduces because of polarisation of the electrodes but chlorine acts as a de-polarising agent, therefore the current produced by the cell varies in proportion to the concentration of chlorine present in the cell water. The resulting variations of the current, measured in amperes, are used to measure the chlorine concentration. The electrodes have to be constantly cleaned and the pH of the sample water may have to be automatically adjusted if it is outside the range for which the apparatus is designed.

Manual testing for chlorine comprises the addition of reagents, in the form of tablets, powders or liquid to a prescribed quantity of water sample. The most commonly used reagents for measuring chlorine content comprise diethyl-p-phenylenediamine (DPD) to produce a pink colouration, the intensity of which is proportional to the concentration of free chlorine present. The subsequent addition of potassium iodide to the same sample produces a colour intensity proportional to the total chlorine concentration (i.e. free plus combined chlorine), the difference between the two readings (i.e. total free chlorine) gives the combined chlorine concentration.

The DPD-chlorine colour may be measured on a bench-top spectrophotometer at a wavelength of 515 nm, a hand-held photometer or, less commonly nowadays, by comparison against a standard colour disc. Photometers may be either single parameter at a fixed wavelength or multi-parameter with a series of preset wavelengths. A typical measuring range for such instruments would be 0 to 5.00 mg/l Cl_2 with a resolution of 0.01 mg/l. The sample cell path-length should be at least 10 mm. Standard discs containing a series of pre-calibrated glass filters are used in a handheld comparator in which the discs available most commonly cover the range 0 to 1.0 mg/l or 0 to 4.0 mg/l Cl_2.

11.17 USE OF CHLORINE DIOXIDE

Chlorine dioxide is produced on-site due to its relatively short half-life, commonly by the reaction between a solution of chlorine (in water) and sodium chlorite ($NaClO_2$) in a glass reaction chamber packed with porcelain Rashig rings. Sodium chlorite is supplied as a liquid containing 26% $^w/_w$ of $NaClO_2$ in small containers or in bulk or as a powder containing 80% $^w/_w$ $NaClO_2$ from which a solution of up to 31% $^w/_w$ is made for use (Table 7.5). Spillages of sodium chlorite should be washed quickly as evaporation leads to deposits of highly flammable sodium chlorite powder.

The proportion of chlorine to sodium chlorite (100% $^w/_w$ $NaClO_2$) will vary from the stoichiometric ratio of 0.39:1 to as much as 1:1, depending on the alkalinity of the water. Excess chlorine, above the stoichiometric requirement, should be limited to that required to neutralize the alkalinity, otherwise any further excess chlorine will promote chlorate (ClO_3^-) production and cause DBPs to form if precursors are present. Some of the alkalinity also reacts with the hydrochloric acid produced by the action of chlorine on water which otherwise would have reduced the chlorine dioxide yield by about 20%. The pH value should be about 4 and therefore for most waters the chlorine concentration needs to be over 500 mg/l. In practice chlorine dioxide solution concentrations are maintained at less than 1 g/l in open systems, and 10 g/l in fully enclosed pressurized systems.

The other processes include (a) reaction between $NaClO_2$ and hydrochloric acid; (b) reaction between $NaClO_2$, sodium hypochlorite and hydrochloric acid; (c) reaction between Purate® (a mixture of sodium chlorate and hydrogen peroxide) and sulphuric acid; and (d) electrolysis of a $NaClO_2$ solution (e.g. Pureline®).

The acid process uses about 1.25 times more sodium chlorite than the chlorine process to produce the same amount of chlorine dioxide. Stoichiometrically 1.67 g of sodium chlorite (100% $^w/_w$ $NaClO_2$) and 0.54 g of HCl are required to produce 1 g of chlorine dioxide. In practice about 50% more sodium chlorite is required. Furthermore between 300–350% of the stoichiometric quantity of acid is required to lower the pH (to ≤0.5) and neutralize alkalinity and maximize the yield. The electrolysis process is claimed to produce 99.5% ClO_2.

Chlorine dioxide (ClO_2) is most commonly used as a disinfectant in cases where problems of taste and odour arise with chlorine, particularly those due to the presence of phenols (Walker, 1986). It is a powerful oxidant but at the limited dose levels it can be used (because of by-products formed), its oxidation potential is not fully utilized. It is known to oxidize iron (II), manganese (II) (Knocke, 1991) (Section 10.12), colour (Aieta, 1986) and certain types of tastes and odours. It does not produce THMs, although known to increase HAA (Parsons, 2006), nor oxidize DBP precursors; nor does it react with ammonia or phenols in water. In the USA it is primarily used as a substitute pre-oxidant for chlorine at the inlet to the works for taste and odour control, colour removal, pre-disinfection and iron and manganese oxidation because unlike chlorine it does not produce DBPs. Its bactericidal efficiency is comparable with that of free chlorine in the neutral pH range (Bernade, 1965) but, unlike chlorine, its efficiency increases with increasing pH (Aieta, 1986). Chlorine dioxide therefore has particular advantages for disinfecting waters liable to produce chlorophenol tastes, or which have a high pH or which contain substantial concentrations of ammonia.

The principal drawback with chlorine dioxide use is the formation of chlorate (ClO_3^-) and chlorite (ClO_2^-) in the generation process and in the water. Chlorates are produced when generating chlorine dioxide at too low pH and with high excess chlorine. At concentrations less than 10 g/l chlorine dioxide disproportionates under alkaline and acidic conditions to form chlorate and chlorite. Disproportionation is the transformation of a compound into two dissimilar compounds by a process involving simultaneous oxidation and reduction. In practice the disproportionate products are kept to a minimum by maintaining the pH of the solution in the range 3.5–7.5. Chlorates are also known to form by the exposure of water dosed with chlorine dioxide to sunlight, by increased pH (such as in softening) or by the action of chlorite ion with free residual chlorine in the contact tank or distribution system. Chlorite originates from the reactants in the generation process and from the disproportionation of chlorine dioxide. Typically about 65–75% of the chlorine dioxide dose will disproportionate to chlorite. The chlorite so formed is effective in minimising nitrite formation in the distribution system where chloramination is practised.

In the UK, a limit has been set for the combined residual levels of chlorine dioxide, chlorite and chlorate of 0.5 mg/l as ClO_2 in the water entering supply (UK, 2000), which restricts the chlorine dioxide dosage to about 0.75 mg/l. US EPA has a maximum residual disinfectant level of 1.0 mg/l as ClO_2 which would allow a dosage of about 1.5 mg/l.

When used as a pre-oxidant it is seldom applied at a dose greater than 1.0 mg/l. At a dose of 1.0 mg/l, chlorate ion in the treated water would be in the range 0.2–0.4 mg/l. Chlorates can be removed by ferrous chloride and it is reported that 6–7 mg/l ferrous chlorides as Fe per 2.5 mg/l of chlorine dioxide dose was effective in reducing the combined species concentration to 0.2 mg/l (Hurst, 1997). Sodium bisulphite or sulphur dioxide is effective in removing chlorite ions; 95% removal efficiency in the pH range 5–6.5. Due to restrictions on its by-products in the finished water chlorine dioxide is rarely used as the primary disinfectant.

11.18 CALCIUM HYPOCHLORITE POWDER

Calcium hypochlorite powder, commonly known as 'bleaching powder' or chloride of lime, is widely used in the less developed countries of the world for disinfection of water supplies. The bleaching powder contains 30–35% $^w/w$ of releasable chlorine and excess lime. The powder has a bulk density of about 400 kg/m^3 and has the advantage that sealed drums of it can be held in store for long periods without serious loss of chlorine. It is best to make up a solution in batches. Assuming the powder to contain 33% $^w/w$ of chlorine, a 1% $^w/v$ Cl_2 (10 g Cl_2/l) solution would be made up by mixing 30 kg of powder in 1000 litres of water. A 100 litres of this 'batch' would be sufficient to give 1.0 mg/l in 1000 m^3 of water. It should be allowed to settle out excess lime. The supernatant containing chlorine is drawn off and diluted in a storage tank to the dosing concentration, which is injected into the water to be treated by means of a positive displacement reciprocating pump of the diaphragm type or by constant feeders. In large plants, saturators may be used (Section 7.20). The resulting chlorine solution rapidly loses its chlorine content if exposed to air or sunlight; hence it needs to be made up daily or every second day. Likewise a drum of bleaching powder begins to lose its chlorine content once opened.

11.19 CALCIUM HYPOCHLORITE GRANULES

Calcium hypochlorite granules contain 65–70% $^w/w$ chlorine and can be supplied in 45/50 kg drums with plastic liners. It has a bulk density of about 800 kg/m^3. The granules are readily soluble and solutions of concentration up to 10% $^w/v$ (100 g/l) can be made up for dosing. A 1% $^w/v$ Cl_2 (10 g Cl_2/l) solution is prepared by mixing 15 kg of granules containing 65% $^w/w$ Cl_2 in 1000 litres of water. As with solutions made from chloride of lime, the chlorine content reduces substantially in a few days if left exposed to the air.

Table 11.4 Decomposition of chlorine in sodium hypochlorite of 14.5% $^w/_w$ Cl_2 due to age and temperature

Time (weeks)	Chlorine content % $^w/_w$ at solution temperatures				
	10°C	15°C	20°C	25°C	30°C
1	14.43	14.39	14.08	13.57	12.41
2	14.39	14.08	13.57	12.41	10.88
4	14.17	13.58	12.41	10.52	8.70
26	11.72	9.54	7.00	4.5	2.59

11.20 SODIUM HYPOCHLORITE SOLUTION

Sodium hypochlorite is available for waterworks purposes as a clear solution containing 14–15% $^w/_w$ of available chlorine. It is frequently used in place of chlorine gas for safety reasons. It can be supplied in small containers or in bulk, but loses its chlorine strength when exposed to atmosphere or sunlight (for properties and storage see Table 7.5). The rate of decomposition increases with increased temperature as shown in Table 11.4. The results are based on tests carried out in the dark, in the absence of metallic contaminants except traces of iron, copper and nickel present in the product. Therefore, for stability under high ambient temperature and long storage the solution should be chilled to about 10°C using an external cooling circuit.

Dosing of hypochlorite can be by positive displacement reciprocating diaphragm pumps. Sodium hypochlorite dissociates to form oxygen. The features to minimize the problems associated with trapping of oxygen are keeping the pump suction pipework short and simple with minimum of valves, providing vents or drains where entrapment of liquid could occur with vents on pump suction to above tank liquid level, drilling ball plugs of ball valve in the upstream direction, arranging pipework with flow in an upward direction. Attention has to be paid to limiting the introduction of chlorate and bromate to the treated water. Commercial hypochlorite of 15% $^w/_w$ chlorine, produced using the membrane process, contains chlorate (as ClO_3^-) and bromate (as BrO_3^-) of about 0.25 and 0.035% $^w/_w$, respectively. This limits the dose that may be applied to about 4.3 mg/l as Cl_2, provided hypochlorite is the sole contributor to bromate in the water.

OZONE PROCESS OF DISINFECTION

11.21 ACTION OF OZONE

Ozone (O_3) gas is a powerful oxidising agent widely used for primary disinfection and oxidation. The bactericidal effect of ozone is rapid, the usual contact time being between 4 and 10 minutes with dosages of the order of 1–3 mg/l. It is also known to be more effective than chlorine in killing viruses, cysts and oocysts (Tables 11.2 and 11.3). The WHO Guidelines (1993) state: 'Ozone has been shown to be an effective viral disinfectant, preferably for clean water, if residuals of 0.2–0.4 mg/litre are maintained for 4 minutes'. Therefore the criterion adopted by European designers for disinfection is to maintain a

Table 11.5 *Ct* values for cryptosporidium inactivation by ozone

Log Inactivation	*Ct* values at water temperature					
	5°C	**10°C**	**15°C**	**20°C**	**25°C**	**30°C**
1	16	9.9	6.2	3.9	2.5	1.6
2	32	20	12	7.8	4.9	3.1
3	47	30	19	12	7.4	4.7

Source of information: Surface Water Treatment Rule (SWTR) US EPA

free ozone residual of 0.4 mg/l for 4 minutes (i.e. $Ct = 1.6$ mg.min/l); this residual being the required level after the initial ozone demand is satisfied. USA practice is to design for the *Ct* values based on the inactivation of cysts and viruses (Tables 11.2 and 11.3) or oocysts (Table 11.5) (US EPA 2006).

Ozone decay can be modelled using first order kinetics and has a half life in water between 0.5 and 5 minutes. The half-life is a function of water quality, temperature and pH. As the concentration of organics, temperature and pH decreases, the half-life increases. The half-life of ozone is too short for it to be effective as a residual disinfectant in the distribution system. Consequently to provide a residual, chlorination, chloramination or chlorine dioxide dosing has to be used as a final stage of treatment.

Ozone oxidizes many organic substances to less complex compounds (Sections 10.36, 10.37). It oxidizes colour, some organic substances responsible for taste and odour, hydrogen sulphide and reduced metals, such as iron(II) and manganese(II) to form precipitates. Ozone does not react with ammonia; therefore it is useful for the disinfection of waters containing a high amount of ammonia which would otherwise require a large dose of chlorine. Ozone in conjunction with GAC filtration has been found useful for the removal of some pesticides in water (Foster, 1991). Pre-ozonation results in improved coagulation of particles and turbidity (Tobiason, 1995) and in sand and GAC filtration (Bourgine, 1998). To enhance coagulation, the ozone dose required is about 0.4 mg/mg total organic carbon (Langlais, 1991).

Experiments have shown that ozone does not remove organic matter but breaks it down into smaller, more bio-degradable organic matter (BOM) (Huck, 2000). These organic by-products, being more biodegradable than their precursors, can provide nutrients for biological growth and thus promote aftergrowth in the distribution system (Langlais, 1991). Therefore BOMs need to be reduced to prevent aftergrowth (Section 10.36).

The use of ozone has been found to limit the formation of chlorination disinfection byproducts, specifically trihalomethanes and haloacetic acids. The mechanism for lower concentrations of chlorinated disinfection byproducts is through oxidation of organic precursors and reduction in chlorine demand (Langlais, 1991). However, some of these ozonated precursors react more readily with chlorine, leading to an increase, rather than a decrease, in the formation of disinfection byproducts such as chloroacetic acid and chloro aldehydes and ketones during final chlorination (Foster, 1991; Hyde, 1984). The potential health significance of the many by-products of ozone is still not well understood. Among the identified by-products are formaldehyde, organic peroxides, unsaturated aldehydes, epoxides, haloacetic acid and the inorganic by-product bromate. However, the addition of biological activated carbon reactors (Section 10.36) after ozonation can remove a high proportion of the organic by-products, but not the inorganic by-product bromate.

Bromate (BrO_3^-) is formed during ozonation of waters containing bromides. The current drinking water standard for bromate set by the EU, USA and WHO is 10 μg/L. Bromate formation can be minimized through several control strategies. Bromide is present in water and is first oxidized to the weak acid base pair of hypobromous acid (HOBr) and hypobromite (OBr^-) which has a pKa of 8.68. Bromate is then formed through the reaction of ozone with hypobromite. Hypobromous acid does not react with ozone; therefore one control strategy is to limit the concentration of hypobromite by reducing the pH. Depression of pH to less than 7 minimizes the bromate formation but may not be economical for high alkalinity waters (Siddigui, 1995). A second control strategy is to add ammonia before ozonation, which reduces bromate concentration through monobromoamine (NH_2Br) formation (Ozekin, 1998). A third strategy is addition of chlorine to oxidize the bromide to hypobromous acid prior to ozone addition and then add ammonia to form monobromoamine (Buffle, 2004). A fourth strategy is addition of chlorine dioxide (ClO_2) prior the addition of ozone although, following ozonation, any unreacted chlorine dioxide will form another disinfection byproduct, chlorate (ClO_3^-) (Buffle, 2006). Hydrogen peroxide (H_2O_2) dosed after ozonation decreases the bromate formation with the increase of its dose. For maximum reduction, the weight ratio of $H_2O_2:O_3$ would be greater than 2:1 (Kruithof, 1995). Hydrogen peroxide is usually added immediately after the oxidation stage in the post-ozone contactor and would therefore prevent disinfection because it reacts with the residual ozone. Bromides are usually found in very small concentrations (<0.2 mg/l) in water and their removal (by ion exchange or reverse osmosis), is not cost effective. There are no proven methods in full scale operation for the removal of bromate after its formation.

11.22 PRODUCTION OF OZONE

Ozone is produced on a commercial scale by passing air or oxygen through a silent electrical discharge. A high voltage AC current is applied between two electrodes separated by a dielectric and a narrow gap through which the air or oxygen to be ozonized is passed. The gap width is a function of the feed gas, the dielectric material, and the power supply's characteristics. Typical gap widths are between 0.5 and 3 mm. Electrical power options commonly available for electrolysis are low frequency (50–60 Hz at 14–19 kV); medium frequency (60–1000 Hz at 9–14 kV); and high frequency (greater than 1000 Hz typically 5000–7000 Hz at 10 kV). Medium frequency generators are used primarily for oxygen feed systems; they have a smaller discharge gap and allow higher ozone production per tube. Low frequency systems have the simplest power supply. High frequency systems have the advantage of lower voltage and therefore few dielectric failures.

In practice, the two electrodes are concentric tubes but may also be arranged as plates. In concentric tube designs, the outer electrode is a stainless steel tube. The dielectric is a glass or ceramic material, which can be plated onto the outside tube or the inner electrode. Historically a glass tube with an inner metallized coating forms the inner electrode. A typical ozone generator consists of several hundred to a thousand of such tubes assembled in a large vessel (Plate 20(a)). The high voltage is applied to the inner tube; the low voltage is connected to the stainless steel outer tube.

Approximately 90–95% of the energy input appears as heat and must be removed by applying cooling water. Manufacturers recommend that cooling water has a chloride content <30 mg/l to minimize corrosion of stainless steel. Depending on the temperature and the chloride content, the cooling water system could be once-through type using filtered plant water or closed-circuit type that uses a water-cooled heat exchanger or refrigerated water chillers. The system capacity should be adequate to limit the increase in gas stream temperature to less than 5 °C. The feed (air or oxygen) to the ozonizer must be completely oil-free to prevent detonation (hydrocarbon content <10 ml/m³), dust free to a level better than 99% at 0.5 micron to reduce electrode fouling and very dry; the normal requirement is that the feed should have a dew point below −80 °C to prevent sparking and formation of nitric acid. To achieve this dryness, it is usually necessary to refrigerate the feed gas and pass it through a desiccant. The feed gas should be cool with a temperature below 25 °C and any compressors or blowers that deliver gas to the generators must be of the oil-free type. A schematic diagram of an ozone plant is shown in Figure 11.4.

With air as the feed, typical production is up to 4% ʷ/w ozone in air (52.3 g O₃/m³ of air at 0 °C, 1.013 bar). Air feed systems can be either low pressure (<2 bar g) or high pressure (>4 bar g). In a high pressure system, most of the moisture can be removed in the aftercooler, which eliminates the need for a refrigerant dryer. High pressure systems are commonly used in packaged ozone generators.

With oxygen as the feed gas, typical production is about 10–12% ʷ/w ozone in oxygen (119–179 g O₃/m³ of oxygen at 0 °C, 1.013 bar) although ozone generators capable of 16% ʷ/w (190 g O₃/m³) are entering the marketplace. Methods of supplying oxygen include use of liquid oxygen (LOX), delivered in bulk to site by tanker or pipeline, on-site production of LOX, or on-site generation of gaseous oxygen (GOX). LOX delivered in bulk is almost 100% pure oxygen with a dew point lower than −80 °C and therefore does not normally require further treatment. It is stored on site in a vacuum insulated storage tank. Vaporizers, which use ambient air, warm water, steam or electrical energy, convert

FIGURE 11.4

Ozone plant.

LOX to GOX for use in the ozone generators. Ozone generators have been found to be as much as 10 to 20 percent more efficient when a small amount (0.1 to 4%) of the oxygen flow is composed of nitrogen (Rackness, 2005). It is now common to install a supplemental air system that removes the moisture and hydrocarbons to the required limits and blends air (composed of nearly 78 percent nitrogen) with the vaporized LOX.

LOX may be produced on-site in pre-engineered packaged plants with capacities >20 t/day by cryogenic air separation where liquefaction of air is followed by fractional distillation to separate oxygen and nitrogen. This method is commonly used in commercial gas production plants where both oxygen and nitrogen are required as products. Oxygen concentration in the product gas is usually greater than 95%.

More commonly, GOX is produced on site by pressure swing adsorption (PSA) or vacuum swing or vacuum assisted pressure swing adsorption (VSA/VPSA) processes. These comprise of one to three vessels containing a synthetic zeolite material which selectively adsorbs nitrogen from the air at elevated pressures allowing oxygen to pass through. A second adsorbent also removes moisture, CO_2 and hydrocarbons. Once the bed is saturated, it is regenerated by subjecting the bed to a lower pressure to release the byproducts: nitrogen, moisture, CO_2 and hydrocarbons. A continuous stream of oxygen is maintained by switching the beds periodically at 30 second to 2 minute intervals. In PSA, adsorption is at 1–3 bar and regeneration is at atmospheric pressure whilst in VSA/VPSA adsorption is at 0.2–1 bar and regeneration is under vacuum. The processes produce oxygen of purity 90–94% with low hydrocarbon content (<10 ml/m³) and dew point lower than −80 °C. If the system has less than three adsorbers, an oxygen buffer storage vessel is usually provided. For VSA/VPSA systems, an oxygen booster compressor is required for feeding the ozone generator.

In general, bulk delivered LOX is the most economic choice for oxygen capacities up to 20 t/day. Absorptive processes are suitable for small to intermediate capacities (PSA: 0.1–10 t/day and VSA/VPSA 10–60 t/day) and cryogenic plants are best for capacities of more than 20 t/day. The source of oxygen should be determined through an economic evaluation that compares the cost and availability of delivered liquefied oxygen with the electrical and maintenance requirements of onsite generation alternatives.

The output of ozone generators increases with the increase in oxygen concentration, improved quality of the feed gas (i.e. dryness, hydrocarbon content and dust content), decrease in the gas discharge gap between the electrodes and increase in the frequency of the current applied to the dielectric.

The specific energy consumption for ozone generation is dependent on a number of factors and is in the range 20–27 kWh per kg O_3 produced for low pressure air feed plant, and about 15 kWh per kg O_3 for oxygen feed plants. Specific energy consumption for on-site oxygen production processes are about 0.3–0.35, 0.3–0.4 and 0.4–0.45 kWh/kg O_2 for cryogenic, VSA/VPSA and PSA, respectively. Ozone production plants require cooling water between 1.5 and 3 m³/kg O_3 produced; the amount being a function of maximum cooling water temperature and heat transfer within the generator shell.

11.23 OZONE DISSOLUTION AND CONTACT

Ozone must be transferred from the gas phase to water efficiently to minimize losses. Therefore the design of the ozone-water contacting units is critical to the transfer process. In ozone contactors for disinfection, fine bubble diffusers are commonly used to transfer ozone into water. They are either porous rod, dome or disc type and require submergence depths between 5.5–7.5 m. They are operated under the ozone generator discharge pressure. Transfer efficiencies achieved are usually greater than 93%. Transfer efficiency is a function of the diffuser pore size and submergence. The number of diffusers may be calculated based on the rated gas flow for each diffuser. Usually an additional 10–30% may be installed to account for fouling and blow off diffusers are required to release water from the submerged section of the ozone gas conveyance pipeline. Systems that produce low concentrations of ozone require more diffusers than systems capable of producing ozone at 8–12% $^w/_w$. The number of diffusers controls the dimensions of the ozone contractor transfer cell. Clogging of the diffuser pores due to the precipitation of iron and manganese is a disadvantage and tendency to foul precludes their use in pre-ozonation applications if raw water suspended solids concentrations are high. The gaskets and diffusers require annual inspection and periodic replacement based on the deterioration of the gaskets and the fouling.

Alternative, transfer methods include turbine mixers, static radial diffusers, venturis, or in-line static mixers. The turbine mixers aspirate ozone-feed gas into the water and eject the resulting mixture into the contactor in a manner which encourages mixing with the bulk liquid (Langlais, 1991). In some designs the mixer motor is submerged. Turbine mixers require up to ten times the energy required for bubble diffusers. In static radial diffusers about 10% of the water flow is pumped to a submerged radial diffuser where it meets the ozone gas stream and divides this up into very fine bubbles when injected into the contactor through the diffuser head (Anon, 1994). If a pressurized water flow is available, a vacuum can be created through the throat of a venturi to draw ozone gas into solution, resulting in excellent ozone transfer, but at the cost of energy required for pressurizing the sidestream. Sidestream injection systems commonly achieve 93–98% transfer efficiency but the sidestream must be pressurized to 2 to 4 bar and the gas to liquid ratio limited to less than 0.3–0.5 volume gas/volume liquid (Rakness, 2005). Pipeline static mixers are similar to those described in Section 7.10. If the flow in the pipe has a turn down of less than 2:1, a single in-line static mixer in the main pipe is adequate. If the turn down in flow is greater than 2:1, the main line static mixer needs to be supported by a sidestream static mixer in a pumped recirculation system with ozone injected into the side stream (Boisdon, 1995). The side stream flow is usually less than 10% of the main stream flow. The main line static mixer may require about 0.5 m headloss. The static mixer should be followed by a contactor for oxidation and disinfection reactions to complete.

Ozone contactors with diffusers are commonly designed to incorporate at least two transfer stages in series. In the first transfer stage, the initial ozone demand is satisfied, and in the second, a sufficient quantity of ozone is added for disinfection. Retention time provided for disinfection should ensure a Ct value of 1.6 mg.min/l where the contact time t is equal to 4 minutes and the ozone residual C is equal to 0.4 mg/l or values to suit an alternative criterion, such as US EPA (2006) guidance. A further stage is usually included to provide retention for ozone to decay. At the outlet of the ozone contactor, excess ozone may be removed from the water by physical or chemical methods. One such method involves cascading the water over weirs. Ozone is a highly volatile gas and is easily stripped from water (Henry's Constant = 100 000 atm.m^3/mol (AwwaRF, 1991)). A second more widely accepted method for removing ozone is by dosing an ozone scavenger such as sodium bisulphite, hydrogen peroxide or calcium thiosulphate.

According to US EPA (1990), counter-current ozone contactors with diffusers may receive flat inactivation credits of 1-log for viruses and 0.5-log for *Giardia*, if ozone residuals of 0.1 mg/l and 0.3 mg/l, respectively, are maintained at the outlet of the first cell of the contactor. Therefore, maintaining an ozone residual of at least 0.3 mg/l at the outlet of the first cell may reduce the Ct required in subsequent stages. The procedure for calculation of Ct values in the subsequent cells is described in the US EPA design criteria. (US EPA, 1990; Langlais, 1991). WHO Guidelines do not define the t value, whereas the European practice has been to define t as the 'hydraulic' residence time (volume divided by mean flow). Ideally t should be the t_{10} value which for ozone contactors could vary in the range of 0.5 to 0.65 of the t_T (Section 11.5).

The number of stages in a baffled contactor is determined by the need to achieve uniform flow and to minimize short-circuiting and each stage may be provided with 2 or 3 cells formed by over and under baffles. The greater the number of baffles, the closer is the flow regime to plug flow. The preferred flow configuration in the transfer stage is one in which the ozone gas injected by diffusers in one cell flows upwards counter-current to the water flowing downwards. In the second cell the flow is reversed using underflow baffles to enable counter-current flow to be re-established in the second transfer cell (Figure 11.4). This provides good liquid–gas contact and helps to minimize short-circuiting. To minimize capital costs, transfer may occur in adjacent counter-current and co-current cells. Where large flows are treated, the flow is split into several equal parallel streams. The contactors should be designed using CFD modelling.

With the advent of sidestream injection, ozone may be injected into a new or existing pipeline, provided it is constructed or coated with ozone compatible materials. Pipeline contactors better approximate plug flow than rectangular baffled contactors and may serve a necessary purpose of conveying water between processes. Sidestream injection systems typically consist of only one location through which ozone can be delivered to the process flow.

In the peroxone process (Section 10.37) hydrogen peroxide may be added upstream of ozone. However, hydrogen peroxide reacts with ozone therefore when disinfection is also necessary, the hydrogen peroxide must be added after disinfection.

About 5–10% of the ozone introduced into the water remains in the spent gas at the top of the contactors. A destructor is required to convert unused ozone to oxygen for safe discharge. This conversion can be achieved by heating the gas to 350°C at which temperature decomposition takes place in 5 seconds, or by heating the gas 5–10°C above the inlet temperature (to decrease relative humidity) and passing it through a catalyst to accelerate decomposition. The higher temperature thermal system usually incorporates a heat recovery system; its main drawbacks are a longer start-up time and increased energy use. The catalytic system on the other hand can be prone to fouling by moisture. Destructors are usually of stainless steel construction and any chlorine from pre-chlorination could cause severe corrosion at high operating temperatures.

Ozone is toxic and a dangerous gas to handle; its odour perception threshold is less than 0.02 ml/m³. The 15-minute exposure limit is 0.2 ml/m³ (HSE, 2005). Ozone leak detectors must be installed with facilities to shut down generators in the event of a leak. Other precautions must be taken in the design and layout to minimize hazards to health (HSE, 1996). In addition, ozonized air is highly corrosive in the presence of moisture; hence piping and other equipment must be of special materials, mostly stainless steel grade 316L. Concrete surfaces in ozone reaction tanks do not require any special protection; rich concrete mix with a minimum cover of 50 mm over reinforcement bars and high standard of compaction are paramount. Stainless steel reinforcement bars could be used with less cover. Electrical plant and insulations may also need special protection against the possibility of ozone leakage.

ULTRAVIOLET RADIATION

11.24 **UV DISINFECTION**

UV radiation is a form of electromagnetic radiation that extends from 100 to 400 nanometers (nm) in the electromagnetic spectrum between X-rays and visible light. There are four classes of UV radiation: UV-A (315 to 400 nm), UV-B (280 to 315 nm), UV-C (200 to 280 nm) and vacuum UV (100 to 200 nm) (Meulemans, 1986). UV disinfection occurs at the germicidal wavelengths of 200 to 300 nm, in the UV-B and UV-C ranges of the electromagnetic spectrum.

The main mechanism of UV disinfection is the absorption of UV radiation and the subsequent photochemical breakdown of an organisms' deoxyribonucleic acid (DNA) or ribonucleic acid (RNA). In order for UV disinfection to be effective, the target organism must absorb energy from the appropriate range of the electromagnetic spectrum. DNA and RNA, and thus all living cells containing either as its genome, absorb radiation in the range of 200 to 300 nm, with a peak of approximately 260 nm.

Of the pathogens of concern in drinking water sources, the most resistant to inactivation by UV irradiation are viruses, followed by bacteria; *Cryptosporidium* oocysts and *Giardia* cysts are the most susceptible to UV disinfection. The most UV resistant virus strains identified to date are Adenovirus Types 40 and 41, which require UV doses three to four times higher than needed to inactivate other water-borne viral pathogens.

Micro-organisms exposed to UV radiation may be repaired by photoreactivation (called photorepair). This occurs as a result of exposure to radiation between wavelengths of 310 and 490 nm and/or 'dark repair', during which, an entirely new DNA or RNA segment is constructed. Photorepair is relatively limited and slow, requiring much higher doses of relevant wavelengths than does the 4-log inactivation at the DNA/RNA peak absorption wavelength of 260 nm. Bacteria are capable of repairing the damage from UV exposure through photorepair; however, not all bacteria contain the enzyme required for dark repair. Although viruses lack the enzymes required for photorepair or dark repair, some of them can utilize the enzymes of their host cell for DNA repair. Recent research has demonstrated lack of photorepair or dark repair in *Giardia* exposed to UV doses typically used for drinking water disinfection (Linden, 2002). Research on *Cryptosporidium* concluded that this organism is capable of both photorepair and dark repair to its DNA; however, it can not regain its infectivity (Shin, 2001; Oguma, 2001). In order to avoid reactivation and growth of inactivated bacteria and viruses, the general strategy for UV disinfection has been to apply a dose high enough to cause sufficient damage to the microbial genome to make its reactivation mechanisms permanently ineffective. The doses required by regulations usually take into account the possibility of photorepair. In practice, there is no reactivation or repair after exposure to UV radiation from either LPHO or MP lamps and they are both equal with respect to inactivation of protozoa and hence safeguarding health.

11.25 **GENERATION OF UV RADIATION**

UV radiation is generated by passing an electrical arc through mercury vapour inside a hermetically sealed tube of UV-transmitting material, typically constructed of high-purity quartz. The mercury vapour is excited by the discharge, which then returns to a lower energy state, resulting in the emission of UV radiation.

Historically, the most common method of generating UV radiation involves the use of a Low Pressure (LP) mercury vapour continuous wave lamp with a monochromatic output of 253.7 nm. The 'low pressure' refers to the pressure of the mercury vapour inside the lamp, which is approximately 0.93 Pa for conventional LP lamps (US EPA, 2006a). Advances in LPUV technology have led to the development of Low Pressure High Output (LPHO) UV lamps, which contain additional elements (bismuth/indium/gallium) rather than pure mercury (Heering, 2004; Bolton, 2001). Referred to as 'amalgam' lamps, the LPHO lamps operate at vapour pressures ranging from 0.18 to 1.9 Pa and a monochromatic output at 253.7 nm, which is identical to that of conventional LP lamps; however, their input per unit length is several times higher.

Medium Pressure (MP) mercury vapour continuous wave lamps operate at higher pressures, between 40 000 and 4 000 000 Pa (US EPA, 2006a). The higher vapour pressure enables the lamps to carry much more power as well as broadening their emission spectra, resulting in polychromatic output. MP lamps also have a much higher UV output per unit length than LP and LPHO lamps; however, they are less efficient, since not all of the radiation they emit is within the UV-B and UV-C germicidal range. A comparison of the operating characteristics of LP, LPHO, and MP lamps is presented in Table 11.6.

Generation of UV light via LP, LPHO and MP lamps represents those commercially available technologies that are currently the most commonly used for the disinfection of drinking water. Besides the lamps, UV disinfection systems include reactors, sleeves, sensors, ballasts and cleaning systems and possibly on-line UV transmittance (UVT) monitors. UV reactors used for disinfection of drinking water are typically closed in-line vessels, with lamps arranged parallel, perpendicular or at an angle to the direction of flow. Open-channel UV reactors are available, but are usually used for disinfection of wastewater.

Table 11.6 Operating characteristics of mercury vapour lamps

Parameter	LP	LPHO	MP
Germicidal UV radiation	Monochromatic (254 nm)	Monochromatic (254 nm)	Polychromatic, (inc. 200–300 nm germicidal range)
Mercury vapour pressure (Pa)	Approx. 0.93	0.18–1.6	40 000–40 000 000
Operating temperature (°C)	Approx. 40	60–100	600–900
Electrical input (W/cm)	0.5	1.5–10	50–250
Germicidal UV output (W/cm)	0.2	0.5–3.5	5–30
Electrical to germicidal UV conversion efficiency (%)	35–38	30–35	10–20
Arc length (cm)	10–150	10–150	5–120
Relative number of lamps needed for a given dose	High	Intermediate	Low
Lifetime (hrs)	8000–10 000	8000–12 000	4000–8000

Source of information: US EPA (2006a).

Based on information from: Hanovia Ltd.
FIGURE 11.5

UV treatment chamber.

The lamps inside the reactor are encased in sleeves, usually of high purity quartz, which isolate the electrical components from the water, help maintain optimal operating temperatures, and serve to protect the lamps from liquid-generated forces and thermal shock. The intensity of UV light is measured at a fixed point within the reactor by UV sensors. Figure 11.5 illustrates a simplified UV reactor; Plate 20(b) shows a UV installation.

The two general types of sensors are wet sensors, which are in direct contact with the wetted interior of the UV reactor; and dry sensors, which include a barrier against the wetted interior of the reactor but allow passage of UV radiation. The advantage of a dry sensor is that it can be removed for replacement, maintenance, or calibration without shutting down or draining the reactor.

Ballasts are used to control the incoming power to the lamps for ignition and operation. Two general types of ballasts are used in UV applications: electronic and magnetic. Electronic ballasts and inductor-based magnetic ballasts allow continuous adjustment of lamp output (i.e. intensity); transformer-based magnetic ballasts allow only step adjustment or fixed lamp output.

Three general methods are used to clean lamp sleeves and UV sensors to remove fouling: (a) off-line chemical cleaning, which consists of flushing or spraying the interior of a drained UV reactor with a cleaning solution; (b) on-line mechanical cleaning, using wipers (Teflon rings, stainless steel brush) that physically remove fouling as they are driven up and down the quartz sleeves by electric motors or pneumatic piston drives; and (c) on-line chemical/mechanical cleaning, using a collar filled with cleaning solution to remove fouling by physical contact with the sleeve in combination with chemical cleaning. The chemical cleaning solutions usually consist of phosphoric or citric acid.

Some UV systems may be fitted with on-line UVT monitoring systems, which allow continuous, real-time measurement of the UVT. The monitors are typically either flow-through spectrophoto-meters that measure the absorbance of UV light at 254 nm through a defined path length and convert it to a UVT value, or a flow-through device that measures the intensity of UV light at various distances from a LP lamp. The difference in the sensor readings coupled with the individual path lengths is used to calculate the UVT.

Disinfection of water by UV radiation has enjoyed a spike in popularity as a result of its many advantages over conventional disinfection methods by chemicals such as chlorine. However, UV, like any other disinfection technology, also has its share of disadvantages. A comparison of UV disinfection vs. chemical disinfection is presented in the Table 11.7.

Table 11.7 Comparison of UV disinfection vs. chemical disinfection

Advantages	Disadvantages
• Short contact time	• Higher power consumption
• Minimal space requirements	• May require uninterruptible power supply (UPS) and standby power source
• No transportation, storage, or handling of hazardous chemicals	• Effectiveness reduced by suspended solids, colour, soluble organic matter, and turbidity
• No significant formation of toxic disinfection byproducts (Section 11.27)	• Microbial reactivation by light and dark repair mechanisms
• Simple equipment, easy to operate and maintain	• No disinfectant residual
• Good reliability	• Need for periodic replacement of lamps
• Effective against *Giardia* and *Cryptosporidium*	• High UV doses required for inactivation of Adenovirus Types 40 and 41.
• Performance not affected by pH of water.	

11.26 CONCEPT OF UV DOSE AND FACTORS INFLUENCING DOSE DELIVERY IN UV REACTORS

Mathematically, UV dose (or fluence) is the product of the intensity I of the UV light that an organism is exposed to and the time of exposure T of that organism to the radiation. Intensity is typically measured in mW/cm^2 or W/m^2 and time is measured in seconds (s). The resulting units of UV dose are mW-s/cm^2 or W-s/m^2, which are equivalent to mJ/cm^2 and J/m^2, respectively.

However, the application of this equation for the determination of the UV dose delivered to an organism within water flowing through a UV reactor is problematic. The $D = IT$ equation assumes that the intensity field is uniformly distributed and that the flow through the reactor is plug flow, having a homogenous velocity throughout the cross section of the irradiation zone. However, intensity fields and residence times are not uniform as a result of Light Irradiance Distribution (LID) and Residence Time Distribution (RTD).

Light irradiance distribution throughout a UV reactor is a function of both direct and indirect water matrix effects. Direct water matrix effects are water quality parameters that directly impact the ability of UV to inactivate an organism once it is emitted from a light source. UVT is a measurement of the ability of UV radiation to penetrate water and is reported as a percentage for a sample of a defined path length (typically 1 or 10 cm). UVT is the primary water quality parameter used as the basis of UV system design and is a function of the physical (suspended solids, turbidity), chemical (humic/fulvic acids), dissolved organics (phenols) and dissolved inorganics (iron and manganese) characteristics of the water through which the UV radiation must pass. Typical UV transmittance values for drinking water range between 75 and 99 percent.

The attenuation of UV light through water can also be expressed in terms of UV Absorbance (UVA). The mathematical relationship between *UVA* and *UVT* is demonstrated in the following equation:

$$UVT = 100 \times 10^{-UVA} \qquad \textbf{(11.3)}$$

Taking into account attenuation, the intensity of UV radiation decreases with increasing distance from the lamp; and directly related to this, as the *UVT* of a water decreases (the *UVA* increases), the intensity of UV radiation at a fixed point from a lamp in a UV reactor would decrease.

Suspended solids in the water affect the efficiency of UV disinfection due to their ability to scatter the radiation and shade or shield the organisms. Research by Qualls (1983) concluded that particles larger than 8 to 10 μm have an adverse effect on UV disinfection efficiency.

Indirect water matrix effects are water quality characteristics that impair the ability of the UV radiation to penetrate the water stream, which leads to fouling of wetted components of the system, such as the quartz sleeves and UV sensors. Fouling of quartz sleeves reduces their ability to transmit UV and thus decreases the UV efficacy. Fouling can be inorganic, typically consisting of a build up of iron, manganese, and/or calcium and magnesium (hardness) deposits, or it can be organic, caused by the growth of algae or biofilms on while reactors are out of service but remain flooded.

RTD in the UV reactor is the result of the velocity distribution of the water flowing through the vessel and its components rather than a single, uniform velocity field such as would be found in a hollow pipe. The velocity distribution is a function of the reactor's inlet and outlet piping, reactor geometry, lamp arrangement and the flow rate through the vessel.

The result of the variations in LID and RTD is the distribution of delivered UV dose that is not possible to calculate with a simple $D = IT$ equation. Although these factors can be used to derive more complex equations that take into account the distribution of intensity and velocity profiles throughout the UV reactor, the most practical methodology of determining the dose delivery of a reactor is validation by biodosimetry.

US EPA (2006a) gives 3-log credits for *Giardia* and *Cryptosporidium* inactivation at a UV dose of 11 and 12 mJ/cm^2 respectively and 4-log credits for inactivation of both at a UV dose of 22 mJ/cm^2. These regulatory doses are theoretical values based on collimated beam apparatus. In practice, however, the dose applied should account for uncertainties in the full scale plant such as hydraulic effects, reactor equipment and monitoring. The reactors should therefore undergo validation testing to determine the operating conditions under which the reactor delivers the required dose. The dose applied (termed validated dose) in practice to achieve 3-log inactivation is approximately 40 mJ/cm^2.

11.27 BY-PRODUCT FORMATION

UV disinfection does not form any of the halogenated byproducts such as trihalomethanes (THMs) and haloacetic acids (HAAs) because it is not a chemical disinfection process utilizing halogens such as chlorine. Nitrate absorbs UV radiation predominantly in the 200–240 nm range, which leads to formation of nitrite by photolysis. Therefore nitrite formation with MP UV would be significantly higher than with LP or LPHO UV lamps. The nitrite formation could be significantly reduced in MP lamps by incorporating a filtering system on the quartz to block the wavelengths below 240 nm when high levels of nitrate are present in the water. The formation of assimilable organic carbon (AOC) during UV disinfection is not significant at UV doses applied in disinfection irrespective of the lamp type (Ijpelaar, 2007).

11.28 VALIDATION OF UV REACTORS

Biodosimetry is the technique used for validating UV disinfection systems; it uses a microorganism to measure the average UV dose delivery in the reactor. The testing takes place under a variety of conditions and the dose delivery is measured as a function of parameters such as UVT, flow rate, number of operating banks of lamps, UV intensity, etc. An indicator organism and a UV absorber to adjust the UVT are introduced to the influent water and samples are collected before and after exposure to UV. Additional samples are collected from the system influent for bench-scale testing using a collimated beam device.

The degree of inactivation of pathogens under each field test condition is established by comparing the viable post-UV microbial concentration with the pre-UV viable microbial concentration as determined by laboratory analysis. The dose-response curve from the collimated beam analysis is then correlated to the field test inactivation results and an equivalent inactivation dose between the collimated beam analysis and field tests is determined. The correlated dose is commonly referred to as the Reduction Equivalent Dose (RED).

UV systems can be validated on-site or off-site at a validation test centre. There are four UV validation test centres, one in Austria, one in Germany, and two in the USA, each of which employs validation protocols dictated by the country's regulatory agency: the Austrian Standards Institute (Österreichisches Normungsinstitut ÖNORM); the German Technical and Scientific Association for Gas and Water (Deutsche Vereinigung des Gas- und Wasserfaches e.V.—Technisch-wissenschaftlicher Verein: DVGW); and US EPA.

Although the European and USA validation centres use the same basic biodosimetric techniques, they vary in several ways. ÖNORM and DVGW utilize bacterial (*Bacillus subtilis*) spores as the microbial indicator organism, with standardized safety factors applied to the validation results. This approach requires a RED of 40 mJ/cm^2 for the disinfection (99.99 percent – 4-log inactivation) of human pathogens including: bacteria, *Cryptosporidium* and a majority of viruses.

The UV Disinfection Guidance Manual (UVDGM) of US EPA (2006a) allows the use of a variety of microbial indicator organisms, including *Bacillus subtilis* spores and bacterial-specific viruses, known as bacteriophage, which include (in order of increasing sensitivity to UV inactivation) MS-2, QB, T1 and T7. The safety factor is not a fixed value but depends on the experimental uncertainty and the RED bias factor of the individual validation. The target RED is a not a fixed general value but can vary as a function of the target pathogen (*Cryptosporidium*, *Giardia*, or virus) and its required level of inactivation (1, 2, 3 or 4-log).

11.29 BOILING WATER

Boiling water is an extremely useful process for disinfecting water because boiling kills bacteria, viruses, ova and cysts present in polluted water. It is reported to be equally effective whether the water is clear or cloudy, relatively pure or highly contaminated (though obviously contaminated or cloudy water should be used only as a 'last resort'). Turbid water should preferably be filtered through a clean cloth before boiling. Alternatively the water should be boiled for up to 5 minutes. In an emergency because a normally safe supply is contaminated with micro-organisms, it may be necessary for a water utility to advise consumers to boil all water used for drinking or cooking or brushing teeth. Then it is sufficient to bring the water to boil and the use of electric kettles with automatic switch-off is acceptable (Boucher, 1990). In the UK, immuno-compromised individuals are advised to boil all drinking water from any source (Boucher, 1990). For the complete sterilisation of a polluted water the WHO recommends bringing the water to 'rolling boil' (large bubbles continuously coming to the surface) and then maintaining this for at least 1 minute for a clear water. At high altitude 1 minute extra time should be given for every 1000 m above sea level because of the lower temperature at which boiling takes place (WHO, 1997). Boiled water can become recontaminated once it has cooled as it has no residual disinfectant and therefore should be stored in a clean closed container.

DISINFECTION OF WATERWORKS FACILITIES

11.30 DISINFECTION OF WATER MAINS AND TANKS

The UK recommended practices for disinfecting mains have been set out in Technical Guidance Notes published by UK Water (1998).

Before a new or renovated pipeline is put into service, it should first be swabbed clear of dirt and debris with a foam swab and flushed with water. It should then be filled with water containing about 20 mg/l of free chlorine and allowed to stand for 16 hours, after which it should be flushed and recharged with mains water and allowed to stand for a further 24 hours. Samples should then be taken from a number of points along the main and at its extremities and, if samples are found to be free of coliform organisms and give satisfactory results for residual chlorine, taste, odour and appearance, the main can be brought into service. All service pipes connected to the main should be flushed out. Alternatively a renovated main may be disinfected for a minimum period of 30 minutes with 50 mg/l of free chlorine followed by flushing, filling with mains water and sampling. A similar treatment is advised for an operation which involves cutting the main, and where there is a risk of contamination from water in the trench or other foul water (e.g. from sewer). For repairs to a live main which has to be cut with minor soiling at opening, all surfaces which come in contact with the drinking water must be cleaned down with water containing 1000 mg/l free chlorine. In all cut main repairs, following disinfection and flushing the main must only be returned to service when bacteriological and other qualitative tests have proved satisfactory.

Disinfection of reservoirs is set out in AWWA standard C652-02 (2002). Alternatively the following variation could be used. Tanks and service reservoirs are usually hosed down with strong jets of clean water and the walls and floor are then brushed down with a chlorine solution containing not less than 20 mg/l chlorine as against 200 mg/l recommended in the AWWA standard. The reservoir is then half filled with water containing at least 0.5 mg/l chlorine and, after standing for 24 hours, a sample of the water is tested for coliform organisms. If the sample fails, the chlorine dose of the inlet water is increased to 1.0 mg/l and the reservoir is filled, the water being retested after a further 24 hours. If coliforms are absent, the reservoir can be filled with water having 0.5 mg/l chlorine, which is resampled and tested when the reservoir is full. Personnel entering a service reservoir which has been emptied for inspection, etc. should scrub their footware in a tray containing 1000 mg/l chlorine immediately before entering the reservoir. Normally chlorine tablets, bleaching powder, calcium hypochlorite granules or sodium hypochlorite are used as chlorine sources, but for large new transmission mains and reservoirs chlorine water from a gas chlorinator might be injected into the filling water. All test water containing chlorine should be dechlorinated before disposal; chemicals available include sodium bisulphite, sulphur dioxide, sodium metabisulphite and sodium thiosulphate.

11.31 CONTROL OF AFTERGROWTH IN DISTRIBUTION MAINS

Considerable aftergrowth may occur in distribution mains especially with nutrient rich waters. Its cause and control is discussed in detail in a WHO publication (2004). Organic matter, iron and manganese deposits, algae, corrosion products may foster growth in the mains of bacteria and other forms of life such as biofilms, which are resistant to control measures (Flemming, 1998). The problems caused by biofilms include taste and odour, brown water, corrosion of mains, bacterial contamination, increased disinfectant demand, greater head loss in mains, habitats for bacteria, viruses, protozoa and fungi and slough of biomass. In systems with biofilms it is often difficult to retain any residual chlorine in the water. No practicable level of residual chlorine prevents problems of aftergrowth if algae, residual iron or aluminium floc, suspended solids or substantial amounts of biodegradable organic matter (Section 10.35) are allowed to remain in a treated water. Phosphates used in plumbosolvency control can also contribute to aftergrowth. Swabbing followed by slug dosing with a heavy dose of chlorine passed slowly along the mains affected is a short-term remedy, but the permanent cure lies in optimising the treatment process and thus improving the quality of the water distributed (Mouchet, 1992). In extensive distribution systems carrying nutrient-rich waters it may be necessary to adopt 'booster chlorination', i.e. the addition of further chlorine or chloramination of the water at some key distribution point or points in the system, usually at the inlet or outlet of a service reservoir. Chloramination has also become more common, where there is a need to control biofilms and avoid exceeding chlorination by-product concentrations (Section 11.7). This has become essential in some utilities (Norton, 1997). As noted in Section 6.68 bacteriological testing of water in service reservoirs is required under the UK Water Regulations.

11.32 DISINFESTATION OF DISTRIBUTION MAINS, WELLS AND BOREHOLES

Some distribution systems can be become infested with small aquatic animals, especially systems with old mains carrying treated lowland surface waters which contain considerable organic matter. The most commonly experienced animals are—*Asellus* the 'water louse'; *Gammarus* the 'freshwater shrimp'; nais worms and nematode worms. Occasionally the larvae of midges and flies may be found, having passed through filter beds or gained entry to a service reservoir. Many other small aquatic organisms can occasionally be present. The flushing of mains and chlorination is largely ineffective where animal growth is prevalent. The ability of the animals to leave reproductive spores or to reproduce from fragments means that reinfestation tends to be rapid: hence disinfestation is necessary and is effected by Permasect WT which contains pyrethin and permethrin. The application has to be controlled; the average and maximum concentrations of pyrethin and permethrin must not exceed 10 and 20 µg/l, respectively, under Regulation 31 of the UK Water Regulations (DWI, 2007).

In cases where corrosion of the main is the cause of high turbidity or coloured water flushing and pigging with polyurethane foam swabs have been shown to be effective (Smith, 1986).

In mains iron bacteria may develop and occasionally other types of bacteria may multiply into substantial colonies. One of the aeromonas strains, *Aeromonas hydrophila*, is believed capable of causing gastroenteritis (Edge, 1987). The bacterial infestation of mains can usually be dealt with by flushing, swabbing and then chlorinating. This has usually to be done in sections when, as is likely, a whole system is affected. Maintenance of an enhanced chlorine residual can assist in preventing regrowth, but this practice may be inhibited by the need to keep DBPs below the permitted maximum (Section 6.25). In the case of iron bacteria in mains the proper remedy is to revise the treatment of the water to remove the iron or, if the bacteria grow because of deterioration of old iron mains, relining them is effective.

Iron bacteria are particularly prone to develop in wells and boreholes drawing water from ferruginous formations. The bacteria are sessile i.e. attach themselves to a surface, the majority produce large masses of extra-cellular covering material in the form of slime which clings to well screens and borehole linings. Often the presence of large iron bacterial growths remains unevidenced until part of the slime detaches and is discovered in the water, or the yield of the well or borehole falls off because of clogging of screens. The slimes protect the bacteria against any biocide; hence physical removal of the slimes is necessary. Pumps and rising mains in wells and boreholes can be withdrawn and cleaned. However, to clean well screens, surging, jetting, chemical applications or steam injection may have to be adopted. Prevention of regrowth in well screens and borehole linings after physical cleaning may have to comprise slug-dosing with chlorine and then flushing. Chlorine should never be applied direct to a well or borehole; this prevents knowledge of the degree of pollution of the water and may encourage corrosion of the well or borehole lining and of the pump inserted. The continuous application of chlorine may also be inadvisable since it tends to precipitate the iron and manganese in solution.

REFERENCES

Aieta, E. C. and Berg, J. A (1986). Review of Chlorine Dioxide in Water Treatment. *J AWWA*, **78**(6).

Anon. (1994). Ozonia Radial Diffuser. *Ozonia Product Information*. Zurich.

AWWA (2002). *Disinfection of Water Storage Facilities*, C652-02, AWWA.

AwwaRF (1991). *Ozone in Water Treatment: Application and Engineering*. Lewis Publishers Inc.

AwwaRF (2003). *Ammonia From Chloramine Decay: Effects on Distribution System Nitrification*, Report 90949.

Barrett, S. E. Davis, M. K. and McGuire, M. J. (1985). Blending Chloraminated and Chlorinated Waters. *J AWWA*, **77**(1).

Bernarde, M. A., Israel, B. M. and Olivieri (1965). Efficiency of Chlorine Dioxide as a Bactericide. *Applied Microbiology*, **13**(5).

Boucher, I. (1990). *Cryptosporidium in Water Supplies*, Third report of the Group of Experts, London.

Boisdon, V., Martin, N. and Galey, G. (1995). The Effect of Static Mixer on Mass Transfer in a New Design of Ozone Reactor. *12th World Congress*, International Ozone Association, 15–18 May, Lille, France.

Bolton, J. R. (2001). *Ultraviolet Applications Handbook*, 2nd Ed. Bolton Photosciences Inc.

Bourgine, F. P., Chapman, J. I., Bastment, R. J. et al. (1998). The Effect of Ozonation on Particle Removal in Drinking Water. *J CIWEM*, **12**(3).

Buffle, M. O., Gailli, S., von Gunten, U. (2004). Enhanced Bromate Control during Ozonation: The Chlorine-Ammonia Process. *Environ. Sci. Technol.*, **38**.

Buffle, M. O. (2006). Enhanced Bromate Control during Ozonation: Pre-oxidation with ClO_2 (on line) http://e-collection.ethbib.ethz.ch/ecol-pool/diss/fulltext/eth16266.pdf. (January 2008).

Butterfield, C. H., Wattie, E., and Megregian, S. (1943–46). Influence of pH and Temperature on the Survival of Coliform and Enteric Pathogens when Exposed to Free Chlorine, *US Public Health Reports 58, 59, 61*; also *J AWWA*, (1948), **40**.

COMAH (1999) Control of Major Accident Regulations and amendments of 2005, HSE.

DWI (2007), *List of Approved Products*, (updated annually), London. http://www.dwi.gov.uk/31/pdf/soslist06.pdf.

Edge, J. C. and Finch, P. E. (1987). Observations on Bacterial Aftergrowth in Water Supply Distribution Systems. *J IWEM*, Aug.

Falconer, R. A. and Tebbutt, T. H. Y. A. (1986). Theoretical and Hydraulic Model Study of a Chlorine Contact Tank. *Proc ICE*, Part 2, **81**(6).

Flemming, H. C. (1998). Biofilms in Drinking Water System *GWF Wasser/Abwasser*, **139**(13).

Foster, D. M., Rachwal, A. J. and White, S. L. (1991). New Treatment Processes for Pesticides and Chlorinated Organics Control in Drinking Water. *J IWEM*, **5**(4).

Hong, T., Liu, S., Song, H. et al (2007). HAA Formation During Chloramination—Significance of Monochloramine's Direct Reaction with DOM. *J AWWA*, **99**(8).

HSE (1987). *Fire and Explosion Hazards at Electrochlorination Plant*. Document No. Nov. HSE 490/11.

HSE (1990). *Safety Advice for Bulk Chlorine Installations*. HS/G28.

HSE (1996). *Ozone Health Hazards and Precautionary Measures*. EH 38.

HSE (1999). *Safe Handling of Chlorine from Drums and Cylinders*. HS/G40.

HSE (2005). *Occupational Exposure Limits*. Guidance Note EH40/2005, March (updated regularly).

Heering, W. (2004). UV sources—Basics, properties and applications. *IUVA News*, **6**(4).

Huck, P. M., Coffey, B., Amirtharajah, A. et al. (2000). *Optimizing Filtration in Biological Filters*. AwwaRF.

Hurst, G. H. and Knocke, W. R. (1997). Evaluating Ferrous Ion for Chlorite Removal. *J AWWA*, **89**(8).

Hyde, R., Zabel, T. and Green, L. (1984). *Proceedings, Seminar on Ozone in UK Water Treatment Practice*, IWES, Sept.

Ijpelaar, G. F., Harmsen, D. J. H. and Heringa, M. (2007). *UV Disinfection and UV/H_2O_2 Oxidation: By-product Formation and Control*, Techneau, D2.4.1.1.

ISCS (2004). International Programme on Chemical Safety, ICSC-0074, Sulphur Dioxide.

Knocke, W. R., Van Benschoten, J. E., Kearney, M. J. et al. (1991). Kinetics of Manganese(II) and Iron(II) Oxidation by Potassium Permanganate and Chlorine Dioxide. *J AWWA*, **83**(6).

Kruithof, J. C. and Meijers, R. T. (1995). Bromate Formation by Ozonation and Advanced Oxidation and Potential Options in Drinking Water Treatment. *Water Supply*, **13**(2).

Langlais, B., Reckhow, D. A. and Brink, D. R. (Eds). (1991). *Ozone in Water Treatment, Application and Engineering*. AwwaRF. Lewis Publication.

Lieu, N. I., Wolfe, R. L. and Means, E. G. (1993). Optimising Chloramine Disinfection for the Control of Nitrification. *J AWWA*, **85**(2).

Linden, K. G., Shin, G. A., Faubert, G. et al (2002). UV Disinfection of *Giardia lamblia* Cysts in Water. *Environmental Science and Technology*, **36**.

Meulemans, C. C. E. (1986). The Basic Principles of UV-sterilization of Water. *Ozone + Ultraviolet Water Treatment*, Aquatec Amsterdam, Paris: International Ozone Association.

Mouchet, P. (1992), Physico-Chemical Degradative Changes in Water While in the Distribution System. *Techniques, Sciences, Methods*, **6**.

Norton, C. D. and Le Chevallier, M.W. (1997). Chloramination, Its Effect on Distribution Water Quality. *J AWWA*, **89**(7).

Oguma, K., Katayama, H., Mitani, H. et al. (2001). Determination of Pyrimidine Dimers in *Escherichia coli* and *Cryptosporidium parvum* During UV Light Inactivation, Photoreactivation and Dark Repair. *Applied and Environmental Microbiology*, 67.

Ozekin, K. Westerhoff, P., Amy, G. et al. (1998). Molecular Ozone and Radical Pathways of Bromate Formation During Ozonation. *J Environmental Eng ASCE*, **124**(5).

Palin, A. T. (1950). Chemical Aspects of Chlorine. *J IWE*, **IV**(7).

Parsons, S. (2006). The Impact of Disinfectant Choice on DBP Formation, *Disinfection for Water Supply—UV or not UV*, Proceedings of CIWEM Seminar.

Poynter, S. F. B., Slade, J. S. and Jones, H. H. (1973). The Disinfection of Water with Special Reference to Viruses. *J SWTE*, **22**.

Qualls, R. G., Flynn, M. P. and Johnson, J. D. (1983). The role of suspended particles in ultraviolet disinfection. *Journal of the Water Pollution Control Federation*, **55**.

Rackness, K. (2005). *Ozone in Drinking Water Treatment: Process Design, Operation, and Optimization*. AWWA.

Scarpino, P. V., Berg, G., Chang, S. L. et al. (1972). A Comparative Study of the Inactivation of Viruses in Water by Chlorine. *Water Research*.

Shin, G. A., Linden, K. G., Arrowood, M. G. et al. (2001). DNA repair of UV-irradiated *Cryptosporidium parvum* oocysts and *Giardia lamblia* cysts. *Proceedings of the First IUVA Congress*, Washington, DC, June 14–16.

Siddiqui, M. S, Amy, G. L. and Rice, R. G. (1995). Bromate Ion Formation: A Critical Review. *J AWWA*, **87**(10).

Smith, D. J. (1990). The Evolution of an Ozone Process at Littleton Water Treatment Works. *JIWEM*, **4**(4).

Smith, G. A. (1986). A Case Study in Pipeline Maintenance. *Water & Wastewater Int.*, **1**(3).

Smith, H. V. Smith, A. L. and Girdwood, R. W. A. (1989). The Effect of Free Chlorine on the Viability of *Cryptosporidium* Oocysts. WRc report PRU 2023-M.

Stevenson, D. G. (1998). *Water Treatment Unit Processes*. Imperial College Press.

Tobiason, J. E. Edzwald, J. E. and Edzwald, J. K. (1995). Effects of Ozonation on Optimal Coagulant Dosing in Drinking Water Treatment. *J Water SRT-Aqua*, **44**(3).

UK (2000) Water Supply (Water Quality) Regulations.

UK Water (1998), *Technical Guidance Notes 1 to 7*.

UKWIR (2000). *Toolboxes for maintaining and improving drinking water quality*. Report 00/DW/03/11, Section 3. Toolbox for Trihalomethane Formation.

US EPA (1990). *Guidance Manual for Compliance with Filtration and Disinfection Requirements for Public Water Systems Using Surface Water Sources*, Appendix O. Guidance to predict performance of ozone disinfection system.

US EPA (1991). *Guidance Manual for Compliance with the Filtration and Disinfection Requirements for Public Water Systems using Surface Water Sources*.

US EPA (2006). *National Primary Drinking Water Regulations: Long Term 2 Enhanced Surface Water Treatment Rule*, 71(3).

US EPA (2006a). *Ultraviolet Disinfection Guidance Manual for the Final Long Term 2 Enhanced Surface Water Treatment Rule*. EPA 815-R-06-007.

Walker, G. S., Lee, F. P. and Aieta, E. M. (1986). Chlorine Dioxide for Taste and Odour Control. *J AWWA*, **78**(3).

White, G. C. (1999). *Handbook of Chlorination*, 4th Edn. Van Nostrand Reinhold.

WHO (1989). *Disinfection of Rural and Small-Community Water Supplies*. WRc.

WHO (1993). *Guidelines for Drinking-Water Quality*, Vol. 1, Recommendations. 2nd Edn.

WHO (1996). *Guidelines for Drinking Water Quality*, Vol. 2, Health Criteria and Other Supporting Information, 2nd Edn.

WHO (1997). *Guidelines for Drinking Water Quality*, Vol. 3. *Surveillance and Control of Community Supplies*. WHO.

WHO (2004). *Safe Piped Water: Managing Microbial Water Quality in Piped Distribution Systems*. Ed. Ainsworth, R. IWA Publishing.

Hydraulics

12.1 THE ENERGY EQUATION OF FLUID FLOW

A fluid moves, in accordance with Newton's laws, under the action of external forces. If there is a net force acting on an element of the fluid, then that element will either accelerate or decelerate depending on the direction of that force; or, if the forces are in balance, then the element will remain at rest or at the same velocity. There is a resistance to motion, however, in the form of drag on that element of fluid and, in moving, energy is expended in overcoming that drag. This expenditure of energy appears in the form of turbulence in the water created by the drag of the surfaces of the conduit and by any obstructions or changes to the shape and direction of the conduit. As the eddies of turbulence decay, their kinetic energy is transmitted to the motion of individual water molecules, so the temperature of the water increases slightly; except for the turbulence found in some pumps (Section 17.12) the temperature change is scarcely detectable and too small to be of any practical use. The balance of energy remaining is important because it determines the subsequent level, pressure and kinetic energy of flow.

The energy of a unit mass ρ of water can be expressed as:

$$E = \qquad (\rho u^2/2) \qquad + \qquad (p) \qquad + \qquad (\rho g z)$$

$$\text{Kinetic energy} \qquad \text{Pressure energy} \qquad \text{Potential energy}$$

where u is the velocity of the water, p its pressure and z its height above some given datum. For convenience this expression is divided by ρg so that the energy and pressure are expressed in terms of a height, or 'head' of water to give the expression in the form:

$$\frac{E}{\rho g} = H = \frac{u^2}{2g} + h + z \tag{12.1}$$

For flow between two points, A and B, as shown in Figure 12.1:

$$H_A = H_B + H_L$$

where HL is the energy head lost by the water flowing from A to B. Hence:

$$\frac{u_A^2}{2g} + h_A + z_A = \frac{u_B^2}{2g} + h_B + z_B + H_L \qquad (12.2)$$

This is the general form of the energy equation; it is fundamental to almost all hydraulic calculations and is referred to several times below. This general energy equation is also referred to as the modified Bernoulli equation. Bernoulli's equation itself, which is one of the most widely quoted equations in fluid dynamics, is a particular case of this general energy equation in which there is no energy loss between points 1 and 2. Bernoulli's equation is:

$$\frac{u^2}{2g} + h + z = \text{constant}$$

The limitations of Bernoulli's equation must be carefully noted. It applies only to steady flow within a streamline and to flow where no energy is lost through turbulence. Therefore, it can only be applied as an approximation over short distances. In all civil engineering applications energy is lost as flow moves from point A to point B and the challenge is the determination of this loss.

FIGURE 12.1

Variation in energy as unit mass of water moves from position A to B.

12.2 BOUNDARY LAYERS

The concept of streamlines is most relevant to the idealized condition of flow moving uniformly in a large body of water. Away from the influence of any solid boundary, a particle of water follows its streamline, a condition known as potential flow. However, when a boundary is introduced into the flow the water in immediate contact with the solid surface must be stationary. Away from the surface the velocity increases up to a point where the flow is unaffected by the surface boundary. In this region adjacent to the surface there is a varying velocity gradient, with adjacent streamlines at different velocities. This region is known as the boundary layer. The thickness of the boundary layer depends on a number of factors but the boundary layer can be envisaged as growing from the point of contact of the flow with the solid boundary. In a conduit of finite size the boundary layer increases in the direction of flow until it fills the full depth of flow in an open channel or the full cross-section of a pipe.

The velocity profile within a boundary layer is primarily a function of the size of the conduit, the velocity of flow and the density and viscosity of the fluid. These parameters are combined in a single dimensionless grouping known as the Reynold's number Re where:

$$Re = \frac{VD}{v}$$

V being the average velocity of flow, D some representative dimension of the flow (for example, the depth of flow in an open channel or the diameter of a pipe) and v is the kinematic viscosity of the fluid, defined as μ/ρ where μ is the viscosity and ρ the density.

At low Reynold's numbers the flow is described as *laminar*. In laminar flow viscosity is dominant and the momentum and inertia of the flow have little effect. The boundary layer is small and the velocity gradient within it may be high. A good example is treacle flowing over an object clinging to the surface without separation or turbulence. Laminar flow occurs rarely in water supply systems as it occurs when the Reynold's number is below about 2000. Since the value of the kinematic viscosity of water is about 1.1×10^{-6} m²/s at 15 °C, this requires that $V \times D$ be less than about 0.002 m²/s. Thus, for example, it would apply to a case where the flow velocity was less than 0.002 m/s in a pipe of 100 mm diameter or 0.2 m/s in a pipe of 10 mm diameter.

As the Reynold's number increases so turbulence becomes increasingly important; above about 2000 there is a sudden transition in the flow as turbulence begins to affect the motion and laminar conditions no longer apply. Resistance to motion then increases markedly. Viscosity remains a major influence at Reynold's numbers beyond this transition and the flow is described as 'smooth turbulent'. As the Reynold's number increases further the effect of viscosity gradually reduces and the influence of turbulence increases. At some higher Reynold's number, depending on the roughness of the boundary, the effect of viscosity becomes negligible and the flow is said to be '*rough turbulent*'. The region between smooth and rough turbulent flow is described as the '*intermediate zone*' where both viscosity and turbulence have an influence.

This relationship between resistance to flow and Reynold's number is best illustrated by the diagram developed by Moody (1944) (Figure 12.2) for pipe flow. The flow resistance is characterized by the friction factor f, which is explained below. In the laminar flow region f is a function solely of the Reynold's number but above the transition region, f is a function of both Reynold's number and the roughness of the pipe surface, which is expressed in terms of the relative roughness k_s/d where k_s is a linear measurement of the surface roughness and d the pipe diameter.

FIGURE 12.2

Moody diagram illustrating the variation of hydraulic resistance with Reynold's number. Note: i is the gradient of the energy line, the 'hydraulic gradient'.

Turbulence is generated by water moving at a different velocity from that in a streamline adjacent to it. This generates a shear force between the streamlines tending to retard the faster flow and speed up the slower flow. Thus, in a boundary layer, the faster flow is dragged down towards the boundary and the slower flow pulled away from the boundary, creating a lateral velocity component and the formation of turbulent eddies. There is a transfer of momentum across the streamlines and the velocity profile is no longer linear but approximately logarithmic with zero value at the boundary.

12.3 PIPE FLOW

Two types of water flow are covered in this chapter: open-surface channel flow and closed conduit flow. In the former the depth of flow can vary; in the latter the area of flow is fixed and for a known flow in a given size of conduit the velocity can be calculated directly. Pipe flow is considered first as it is more straightforward.

At the entry of water into a pipe from a large tank or reservoir, the flow, as it accelerates into the inlet, approximates to the idealized condition of potential flow. However, a boundary layer is generated from the lip of the inlet and, within a relatively short distance downstream of the entry, expandeds to fill the whole pipe. At Reynold's numbers just above the laminar transition the velocity profile is logarithmic from the boundary wall to the centreline of the pipe. At higher Reynold's numbers the turbulence is such that momentum is transferred across the streamlines. Across much of the section the time-averaged velocity is constant but there is a steep velocity gradient near the pipe wall.

(a) Flow in smooth-turbulent zone (b) Flow past an obstacle (c) Flow in rough-turbulent zone

FIGURE 12.3

Velocity profiles in a pipe flowing full.

Figure 12.3(a) illustrates the velocity profile across a pipe operating in the smooth turbulent zone. At the centre of the pipe the velocity is greatest and is about 1.2 times the average value, although the ratio of the maximum speed of the flow to the average varies as a function of the Reynold's number and the roughness of the walls. Of course, it may vary very greatly if an obstacle in the pipe affects the flow as illustrated in Figure 12.3(b). Figure 12.3(c) shows the velocity profile across a pipe operating in the rough-turbulent zone, with a much more uniform velocity across much of the cross-section, but with steep velocity gradients close to the walls.

The average velocity in a pipe is simply the rate of discharge divided by the cross-sectional area of the pipe, i.e. Q/a, where Q is the discharge and a is the cross-sectional area. This average velocity is denoted as V to distinguish it from the velocities in individual streamlines, u_1, u_2, u_3, etc. The energy equation applies to any streamline flow but to apply the same equation to the whole flow within a pipe the energy equations for the separate streamlines must be summed. Thus the total kinetic energy of the flow is:

$$\Sigma \left(\frac{u_1^2}{2g} + \frac{u_2^2}{2g} + \frac{u_3^2}{2g} + \ldots + \frac{u_n^2}{2g} \right)$$

However this sum does not equal $V^2/2g$. The ratio of $(\Sigma u^2/2g)/(V^2/2g)$ is therefore given the symbol of α and the total kinetic energy of the flow is $\alpha V^2/2g$. In a similar fashion the summation of the pressure energies is βP, where P is the average pressure across the cross-section of the pipe. As it is a linear function, the average potential energy in a circular pipe is given by the level of the centre of the pipe, provided the pipe is flowing full. For flow in a straight pipe, away from the influence of bends and obstructions, the pressure distribution is hydrostatic and $\beta = 1.0$. Thus for the whole pipe flow the energy equation is correctly:

$$H = (\alpha V^2/2g) + \beta P + z$$

where V and P are the average velocities and pressures across the section, respectively, and z is the level of the centre of the pipe above some datum. In practice, the values of α and β are close to unity and in most practical applications the errors involved in ignoring the differences are so small as to be insignificant. (The one case where α may be significant is in the high-velocity flow down a spillway chute where the velocity head is the major part of the total energy head). Hence, this assumption permits the flow at any section of a pipeline to be related to that at any other point of the same line.

Units Used

As indicated in Section 12.1, for water supply systems the common practice is to express the terms in the energy equation in '*head of water*'. In metric units this is normally expressed in metres. This is convenient as levels are expressed in the same form. Thus, for example, the difference in level between two reservoirs connected by a pipeline can be considered as a direct measure of the available potential energy between the two locations. In metric units, if the energy, H is expressed in 'metres head of water' then z must also be expressed in metres and the kinetic energy in the same units. This requires that the velocity is in m/s and the gravitational acceleration, g, is 9.81 m/s^2.

The term 'head' is often used loosely applied; for example 'pump head' is used to denote the pump lift, 'head' is also sometimes used in reference to a pressure level or elevation, relative to a datum. In this text 'head' is used simply to mean an equivalent depth of water.

12.4 HEADLOSSES IN PIPES (1)—THE COLEBROOK–WHITE FORMULA

The energy equation (Equation 12.2) for the total flow in a pipe can now be expressed as:

$$(V_1^2/2g) + h_1 + z_1 = (V_2^2/2g) + h_2 + z_2 + H_L$$

where H_L is the energy loss in the pipeline between the two sections 1 and 2. Knowing the total energy level at location 1 (for example, the level of a reservoir) the energy level at any other point can be calculated if the energy loss H_L can be estimated.

The energy lost through turbulence is caused by two mechanisms: (a) the drag of the pipe walls on the flow, and (b) turbulence generated whenever there is a change to the direction or area of the flow. A flowing fluid has momentum and does not want to change direction; any change to the angle of the boundary walls, particularly if the boundary turns away from the direction of flow, may lead to the flow 'breaking away' from the surface leaving an area of turbulence, the 'wake'. The more abrupt the boundary change the greater is the potential for energy loss. The former mechanism for energy loss is known as the hydraulic resistance or 'friction' loss; the latter is the 'form' loss due to the geometry of the change of cross-section or obstruction in the flow. The friction losses are continuous over the length of a pipeline; the form losses are localized in the immediate vicinity of the element causing the energy loss and are also referred to as 'local' or 'fitting' losses.

The loss of energy due to the hydraulic resistance of a pipe is a function of the velocity of the flow, V, the internal diameter of the pipe, d, the length of the pipe, L, the roughness of the surface of the pipe and the characteristics of the flowing fluid, expressed in terms of the kinematic viscosity, ν.

Darcy–Weisbach Formula

A dimensionally correct formula for the head loss is the Darcy–Weisbach equation, which gives the head loss in a length of pipe as:

$$H_L = fLV^2/(2gd) \qquad\qquad (12.3)$$

where f is a non-dimensional coefficient, known as the *friction factor*, which includes the effects of pipe wall roughness and the fluid viscosity. (Note that the friction factor f is also commonly designated by λ.) Unfortunately f is not constant but varies with the size of pipe and the degree of turbulence of the flow.

Colebrook–White formula

Colebrook and White showed that f in the Darcy–Wiesbach formula is a function of the relative roughness of the pipe surface, the viscosity of the flow, and the Reynold's number, R_e. From a combination of theoretical analysis and empirical data they showed that:

$$\sqrt{\left(\frac{1}{f}\right)} = -2\log_{10}\left[\frac{k_s}{(3.71d)} + \frac{2.51}{\left(Re\sqrt{f}\right)}\right]$$

which can also be written as:

$$\sqrt{\left(\frac{1}{f}\right)} = -2\log_{10}\left[\frac{k_s}{(3.71d)} + \frac{2.51v}{\left(d\sqrt{2gdi}\right)}\right]$$

where i is the *hydraulic gradient*, H/L; k_s is the roughness of the internal surface of the pipe; v is the kinematic viscosity of the water.

The experimental data showing variation of f with Reynold's number and relative roughness was plotted by Moody and forms the basis of the diagram illustrated in Figure 12.2. Thus the Colebrook–White equation is, in effect, a mathematical representation of the Moody diagram. The Colebrook–White equation is a formulation of the Darcy–Weisbach equation with f replaced and is usually written as:

$$V = -2\sqrt{(2gdi)} \cdot \log_{10}\left[\frac{k_s}{(3.71d)} + \frac{2.51v}{\left(d\sqrt{2gdi}\right)}\right] \qquad \textbf{(12.4)}$$

Equation (12.4) can be solved directly only for V (and hence Q), knowing d and i. More commonly it is required to find i, knowing Q and d. The equation is then not explicit but must be solved via an iterative technique of successive approximations. Initially this was a major objection to its widespread use but is no longer a significant drawback. The Colebrook–White equation, or one of the approximations given below allowing a direct solution, is now widely used and is recommended as the formula that should be used to estimate pipeline head losses. One of the advantages of this equation over the empirical formulae discussed below is that the roughness coefficient, k_s is a function only of the surface roughness of the pipe and does not change with the size of the pipe or velocity of the flow. The factor k_s is sometimes referred to as the *equivalent sand roughness* because the original experiments carried out by Nikuradse (the data from which Colebrook and White used in the development of their formula) utilized sand grains stuck to the inside of the pipes. The value of k_s is meant to represent the equivalent diameter of the sand particles giving that degree of roughness. Although this is only a notional concept, it does provide a physical meaning to the roughness measurement, which does not apply to the coefficients in any of the empirical formulae given in Section 12.5.

A second advantage of the Colebrook–White formula is that it applies over the full range of turbulent flow from the smooth turbulent condition at Reynold's numbers as low as 3×10^3 to the rough turbulent flow condition at Reynold's numbers in excess of 1×10^7. The first term in the logarithmic function, $k_s/3.71d$, represents the effect of the pipe roughness and dominates at high Reynold's numbers, whilst the second term includes the dynamic viscosity and dominates at low Reynold's numbers.

There are approximations to the Colebrook–White formula that allow direct calculation of d or i knowing the flow, Q, and the other of the two parameters. These provide agreement with the Colebrook–White formula within 0.5%—well within the accuracy with which the roughness is known. Two such equations which can be recommended are those of Barr, to solve for i, and Pham, to solve for D. Thus the three explicit equations can be written in the following similar format.

The Colebrook–White equation, to solve for Q:

$$\frac{V}{\sqrt{(2gdi)}} = \frac{0.9003Q}{d^2\sqrt{(gdi)}} = -2\log_{10}\left[\frac{k_s}{3.71d} + \frac{2.51v}{d\sqrt{(2gdi)}}\right] \tag{12.5}$$

The Barr approximation, to solve for i:

$$\frac{0.9003}{d^2\sqrt{(gdi)}} = -1.9\log_{10}\left[\frac{k_s}{(3.71d)^{1.053}} + \left(\frac{4.932vd}{Q}\right)^{0.937}\right] \tag{12.6}$$

The Pham approximation (HR Wallingford, 2006), to solve for d:

$$\frac{0.9003Q}{d^2\sqrt{(gdi)}} = -1.8841\log_{10}\left[0.365(gi)^{0.2}\left(\frac{k_s}{Q^{0.4}}\right) + 3.55\left(\frac{v}{Q^{0.6}}\right)(gi)^{0.2}\right] \tag{12.7}$$

Units must be consistent, e.g. d (m); Q (m³/s); g (9.81 m/s²) and k_s must be in m (although quoted below in mm). The kinematic viscosity, v, of clean water is 1.310×10^{-6} at 10°C, and 1.011×10^{-6} at 20°C.

For water mains there is considerable guidance available on the choice of k_s values to adopt for design (HR Wallingford, 2006). Typical values for new clean pipes and indicative values for design purposes that allow for deterioration of interior condition are shown in Table 12.1.

In raw water mains there may be a tendency for organic slimes to develop on the walls. This slime tends to flatten as the velocity increases and there is evidence that the roughness reduces at higher flows (HR Wallingford, 2006). The designer of a new pipeline must make a judgement as to whether this effect can allow a reduction in the design head losses. It may also be necessary in a raw water pipeline to allow for some increase in roughness due to presence of sediment in the invert and possibly fresh water organic growth such as mussels. The latter may greatly increase the roughness. Also if the water quality is such that tuberculation is likely over the life of the pipeline, then much higher values may be appropriate.

The development of computer programs for the analysis of water distribution system flows allows a much more detailed approach (Section 13.15). Such programs allow calibration of network models by the adjustment of roughnesses in each pipe. Once a model using the Colebrook–White formula is calibrated then the prediction of the system's performance under other flow regimes or the design of improvements can be undertaken with some confidence, without the need for adjustment of the roughness values. However it is important to remember that measurements are rarely more accurate than ±5% in practice, with errors often greater. Some network programs still cater

Table 12.1 Values of k_s for design purposes

New clean pipes	k_s mm
Steel or ductile iron pipes:	
with spun bitumen or enamel finish	0.025–0.05
with cement mortar lining	0.03–0.1
Concrete pipes	0.03–0.3
Plastic pipes	0.003–0.06
For design with allowance for deterioration:	
Raw water mains	1.5–3.0
Treated water trunk mains	0.3–1.0
Distribution systems	0.5–1.5
For new clean service pipes:	
Galvanized steel	0.06–0.30
Copper	0.002–0.005
MDPE	0.003–0.006
PVC-U	0.003–0.06

Notes: [a]k_s values below 0.01 mm show no significant change in V or i from that at 0.01 mm.
[b]Loss at joints, elbows, tees, etc. on the line to the consumer's tap may add 50–70% to pipe losses (see also Table 14.5).
[c]It is not possible to quote typical k_s values for old service pipes due to the wide range of interior conditions that can apply.

for the use of the Hazen–Williams equation, which is discussed below. Given the uncertainties involved in network modelling this does not necessarily lead to greater inaccuracies but it does mean that a model calibrated for one set of flow conditions must be used with greater caution for other flow conditions.

12.5 HEADLOSSES IN PIPES (2)—EMPIRICAL FORMULAE

There are several other formulae for the calculation of headloss in pipes, which have been and are still used by water supply engineers. They have to be used with caution as each applies over a limited range of Reynold's numbers and in different areas of the Moody diagram (Fig. 12.2). Thus, for example, the Blasius formula applies at low Reynold's numbers in the smooth-turbulent zone, where viscosity dominates; the Hazen–Williams formula applies in the intermediate zone and Manning's formula applies in the rough-turbulent zone where the pipe roughness dominates. The Blasius equation has relatively limited applicability in the civil engineering context but the latter two are discussed in more detail below.

The *Hazen–Williams formula* has been used for many years in water supply and is still used widely in the USA. It is well documented and, until the advent of programmable calculators and computers, was considerably easier to use than the Colebrook–White equation. The equation can be expressed in metric units as:

$$H(m) = \frac{6.78L}{d^{1.165}}(V/C)^{1.85}$$

or

$$V(m/s) = 0.355Cd^{0.63}i^{0.54} \qquad (12.8)$$

where C is a coefficient and i, the hydraulic gradient ($= H/L$), d, V, H and L are as defined in Section 12.4.

The coefficient, C, is not dimensionless; it has units and is therefore a function of the other parameters. As noted above, the Hazen–Williams equation is most accurate for the pipe sizes and velocities typically found in water supply practice. The flow in a pipe of 0.6 m diameter with a velocity of 1.0 m/s has a Reynold's number of about 5×10^5. This is in the intermediate zone in the Moody diagram and the Hazen–Williams formula can be applied with reasonable accuracy provided that velocity and pipe size do not vary greatly from these values. The formula becomes increasingly inaccurate as the Reynold's number varies further away from this mean value. Thus different values of C apply for different pipe sizes and even for the same pipe at different flows.

The value of C can be adjusted to provide a more accurate answer if the parameters do vary significantly. Figure 12.4 shows how the coefficient varies with pipe diameter for a range of pipe roughnesses, and indicates the approximate adjustment needed for velocities varying from 1.0 m/s. Attempts have been made to define the variation of 'C' in terms of the other parameters, d and V, to make the Hazen–Williams formula directly applicable over a greater range. Since the Colebrooke–White formula already provides an accurate estimate over the full range of turbulent flow conditions, this seems a pointless exercise and the use of the Hazen–Williams formula should be abandoned for detailed design.

Manning's equation is appropriate for use when the flow is in the fully-turbulent range, at either high Reynold's numbers or when the conduit is particularly rough. It is widely used in open channel flow, for which there are extensive data, and is referred to below in that context. It is not generally recommended for pipeline systems, except possibly in large, rough conduits such as unlined tunnels. Manning's equation, in metric units, is normally written in the form:

$$V = \frac{R^{2/3}i^{1/2}}{n} \qquad (12.9)$$

where R is the hydraulic mean depth (also known as the 'hydraulic radius'): the area of flow divided by the wetted perimeter, i is the hydraulic gradient ($= H/L$), and n is a roughness coefficient known as the Manning coefficient. For a circular pipe, $R = \pi d^2/4\pi d = d/4$, so for a pipe the equation can be written:

$$V(m/s) = \frac{0.397d^{2/3}i^{1/2}}{n} \qquad (12.10)$$

Notes:
1. For 0.5m/s *add* to C value. For 2.0 m/s deduct 5 from C value shown above.
2. For asbestos cement pipes use k = 0.50 mm
3. For concrete line pipes the following are Scobey classes
   Class 4(K = 0.25 mm) first class finish
       3(K = 0.50 mm): good finish, all joints filled
       2(K = 1.25 mm): imperfect finish
       1(K = 5.0 mm): old concrete pipes
4. For tuberculated cast iron pipes carrying raw water K may be 10→30 mm and Mannings formula should be used in preference to Hazen Williams

FIGURE 12.4

C values in the Hazen–Williams formula as a function of pipe size and pipe roughness for a velocity of 1.0 m/s.

Again, it should be noted that n has dimensions (in US units a factor of 1.49 must be introduced into equation (12.9)) and is a function of the size of the conduit. However it is relatively insensitive to the diameter of a pipe and for many calculations can be assumed to be constant. Appropriate values for n are given in Table 12.3.

While both Hazen–Williams equation and Manning's equation have been used in the past for pipe flow and, in the case of the former equation, quite widely used, the simplicity of their formulations is no longer a significant advantage over the Colebrook–White equation, which is applicable over the full range of conditions likely to be found in water supply. The use of the Colebrook–White equation, or the direct solution approximations of it, is therefore strongly recommended for all pipeline calculations.

12.6 LOCAL HEAD LOSSES AT FITTINGS

As discussed earlier, headlosses occur at every location where there is a geometric change to the conduit, such as at a bend in the pipeline, a change of section, an obstruction to flow such as a valve, or simply the entry to a pipe or exit from a pipe into a tank. In such cases the headloss is usually expressed as a proportion of the velocity head, or kinetic energy, of the flow, i.e. ΔH (or H_L) = $KV^2/2g$, where K is a coefficient depending primarily on the type of fitting or fixture in the pipeline. V is normally taken as the velocity in the upstream pipeline: not, in the case of valves or other obstructions, the velocity through the opening in the fitting itself.

Values of K are almost entirely empirical but there have been extensive experimental measurements on standard fittings on which estimates can be based. Table 12.2 gives values of the headloss coefficients for some standard fittings and suggested values for design. Some standard data is available for valves but, in most cases, it is advisable to obtain values from manufacturers. The suggested design values in Table 12.2 are generally conservative, suitable for estimating losses in short conduits containing several fittings or changes of direction, etc. Whereas in a long pipeline the fittings losses may be a small proportion of the total losses, for example 5%, in a short line with an inlet, several bends and an outlet loss, the fittings losses form a high percentage of the total. Short conduits conveying water from one tank to another often occur in water treatment works, and it is important not to under-estimate losses when designing weir and overflow levels. For some calculations such as surge and transient flow analyses and assessing the range of duties on a pump, it may be necessary to take minimum losses possible. In addition, allowance for imperfection in lining up fittings and for increase of roughness with age may need to be made. If accurate analysis is required, for example on an existing installation or where fittings are in close proximity, then reference to more accurate data may be required (Miller, 1990). It should be noted that the net head loss through a series of adjacent fittings is less than if they were more widely spaced due to way the high energy jet created by one is transmitted through the next. It should also be noted that the losses at some fittings, particularly bends, are a function of the pipeline roughness as well as the geometric shape. For complicated arrangements a hydraulic model may be advisable (Section 13.15).

12.7 OPEN CHANNEL FLOW

The flow in an open channel follows the same principles that have been developed for pipe flow. Thus the energy equation, Equation (12.1), is still valid and energy is lost in the same way through the resistance of the channel surfaces and locally where the geometry of the channel changes. Whereas, in a given pipe, the area of flow is fixed and hence the velocity is a function only of the flow, in a channel of known dimensions, the velocity is not only a function of the flow but also of the channel width and depth of flow, which can vary.

Returning to the energy equation (12.1):

$$H = \frac{V^2}{2g} + h + z$$

the pressure head, h, is now the depth of water, y; and z is the level of the channel invert above some datum. For a case where the energy is constant any increase in velocity, and hence the kinetic energy must be accompanied by a fall in the potential energy and hence a drop in water surface. Similarly a retardation of the flow and reduction of the kinetic energy must be accompanied by a corresponding rise in the water surface if there is no energy loss. In practice, whilst there may be very little energy loss in a case where the velocity is increased and potential energy is converted to kinetic energy, it is always the case that there is some loss of energy in the reverse process. The latter process, known as *recovery of head*, is never 100% efficient.

Table 12.2 Loss coefficients through pipeline fittings

	K values in $KV^2/2g$	
	Laboratory test	**Suggested field[a]**
Entrances: V = velocity through pipe or gate		
Standard bellmouth pipe	0.05	0.12
Pipe flush with entrance	0.50	1.00
Pipe protruding	0.80	1.50
Sluice-gated or square entrance	—	1.50
Bends – 90°: (45° half values given)		
Medium radius (R/D = 2 or 3)	0.40	0.50
Medium radius—mitred	0.50	0.80
Elbow or sharp angled	1.25	1.50
Tees – 90°: Assumes equal diameters		
In-line flow	0.35	0.40
Branch to line, or reverse	1.20	1.50
Exits:		
Sudden enlargement: ratio 1:2	0.60	1.00
Gradual (well tapered) exit	0.20	0.50
Sudden contractions: Loss on contraction and subsequent expansion; V = velocity through contraction		
Contraction area ratio:		
1:2	1.00	1.50
2:3	0.65	1.00
3:4	0.40	1.00
Expansion only	—	1.00
Gate valve fully open	0.12	0.25
Butterfly valve fully open	0.25	0.5

Note: [a]The k values in the second column are recommended for assessing loss through short conduits containing several fittings, bends, etc. in close proximity.

The *specific energy head*, H_s, is defined as the energy head above the channel invert and equation (12.1) can be written as

$$H_s = \frac{V^2}{2g} + y$$

where y is the depth of flow.

Since $V = Q/A$ where Q is the flow and A the area of the flow,

$$H_s = \frac{Q^2}{(A^2 2g)} + y \qquad (12.11)$$

Since A is a function of the depth y; equation (12.11) relates the specific energy of flow H_s to a cubic function of y. This is made clearer if a rectangular channel is considered, for which $A = y \cdot b$ where b is the width of the channel. Substituting for A gives:

$$H_s = \frac{Q^2}{(y^2 b^2 2g)} + y$$

Substituting q for Q/b, where q is known as the unit discharge, i.e. the flow per unit width in a rectangular channel, gives

$$H_s = \frac{q^2}{(y^2 2g)} + y \qquad (12.12)$$

For any value of H_s (above a certain minimum value, to which reference is made later) there are two possible depths y which satisfy equation (12.12) for the same flow q, i.e. H_s can be made up of two different combinations of velocity energy and depth energy. There are a number of common examples that demonstrate this, two of which are shown in Figure 12.5.

The first is the flow over a dam spillway or weir. Upstream the flow is deep and the velocity very low. Downstream the same flow has a much higher velocity and the depth is much reduced. The second case illustrates the flow under a freely-discharging sluice gate. Again the upstream flow is deep and slow; the downstream flow is shallow and fast. In both cases it is assumed that there is negligible energy loss across the structure. For reasons which are explained below, the slow deep flow is known as sub-critical flow, whilst the fast shallow flow is known as super-critical flow.

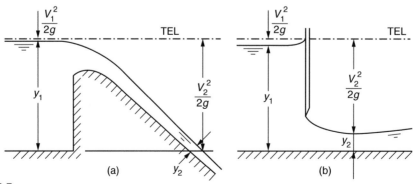

FIGURE 12.5

Examples of alternative flow depth.

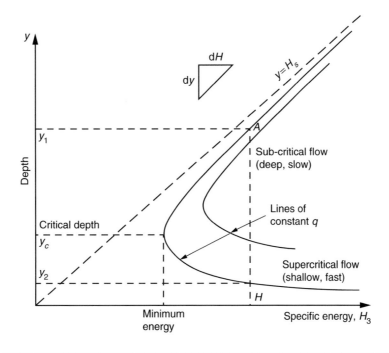

FIGURE 12.6

Relationship between specific energy and depth for a given unit discharge.

Equation (12.12) can be plotted for y as a function of the specific energy head, H_s, with a constant unit discharge, as illustrated in Figure 12.6. This shows how the equation provides two answers for a particular value of the specific energy. It also shows that there is a minimum value of the specific energy at which there is only a single value of depth which solves the equation; this is known as the *critical depth*. The critical depth is an important concept in open-channel flow; it marks the boundary between sub- and super-critical flow. If the depth is less than the critical depth then the flow is supercritical; if the depth is greater than the critical depth the flow is sub-critical.

12.8 CRITICAL DEPTH OF FLOW

The graph in Figure 12.6 shows that, for unit discharge, the minimum energy use and critical depth of flow occur when dH/dy equals zero. Differentiating equation (12.12) and putting the differential equal to zero for the minimum value gives

$$\frac{dH}{dy} = \frac{-2q^2}{(y^3 2g)} + 1 = 0$$

i.e. $y^3 = q^2/g$, from which the critical depth,

$$y_c = \sqrt[3]{(q^2/g)} \qquad\qquad (12.13)$$

This equation is important because it demonstrates that the critical depth in a rectangular channel is a function only of the discharge per unit width. Furthermore, substituting y_c^3 for q^2/g in equation (12.12), and simplifying, the following simple relationship between critical depth and the minimum specific energy is obtained.

$$H_{s(min)} = 1.5 y_c = 1.5(q^2/g)^{1/3} \qquad \textbf{(12.14)}$$

Thus the minimum energy is also a function of the unit discharge alone. Hence, given the unit discharge, the critical depth and minimum energy head at any location can be calculated.

Consider the case of flow in a rectangular channel with unit discharge, q, and assume the flow is sub-critical. As shown in Figure 12.7, the specific energy level (the energy head above the invert) is denoted by the chain dotted line and comprises the depth plus the kinetic energy of the velocity. If part of the bed is raised, as indicated in the diagram, then the specific energy reduces. The energy level over the section of raised bed cannot increase so the local depth must decrease and the velocity must increase. The velocity head therefore increases and there must be a drop in water level over the raised section. If the bed is raised further, eventually the specific energy reaches the minimum value and the depth of flow over the raised section drops to the critical depth. Any further raising of the bed would reduce the energy below the required minimum value for that unit flow, and the only possible result is a reduction in flow over the raised sill. If the flow is to be maintained then there must be a rise in upstream water level to provide more energy to enable that flow to pass over the sill.

Critical depth of flow thus occurs when there is free discharge over a weir or gate. The flow utilizes minimum energy to pass the maximum discharge possible for the available energy head. The expressions *free* discharge, or *modular* discharge, are used to denote that the downstream conditions do not affect the flow. Clearly, if there were some control downstream, such as a gate, which was progressively closed then eventually the downstream level would back up to the extent that the free discharge over the weir would be drowned out. Critical depth would not occur and the upstream water level would rise accordingly.

Critical depth conditions can also occur because of an increase in the unit discharge. If instead of raising the bed, the sides of the channel are brought in, squeezing the flow, the specific energy

(a) Low sill   (b) High sill

FIGURE 12.7

Flow over a raised sill: (a) low sill, (b) high sill.

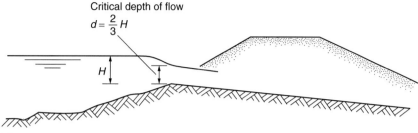

Critical depth of flow

$$d = \frac{2}{3} H$$

FIGURE 12.8

Critical depth of flow at a channel entrance with the condition that the slope is greater than the critical slope.

remains constant but the discharge per unit width must increase. Again, this can continue until the unit discharge reaches the value defined in equation (12.12). If the channel is narrowed beyond this point then the specific energy must increase and the upstream water level must rise to provide this additional energy head. This effect is utilized in measurement flumes, which are discussed in Section 12.15. The entrance to a steep culvert is another example where critical depth may occur because the flow is squeezed through a narrower conduit. There are many instances where the flow passes through critical depth due to both a narrowing of the channel and the raising of the bed. The spillway of a dam is one such example. Another is illustrated in Figure 12.8 which illustrates the flow from a reservoir into a channel. The channel is narrower than the reservoir and clearly the bed is raised, so critical depth would be expected at the channel entrance, provided the channel is steep and that the downstream water level is not sufficient to drown out the entrance to the channel. The water surface would also be seen to drop as the flow accelerates into the channel.

The proviso above that the water level downstream of the entrance must not be so high as to drown out the entrance is important. This depends on the slope of the downstream channel. If the slope on the channel is too flat to maintain the flow, the water level rises and the entrance becomes drowned out with sub-critical flow throughout. At one particular slope the depth remains at critical depth and the control at the entrance remains. At steeper slopes than this *critical slope*, the flow accelerates away from the inlet with super-critical flow conditions.

The occurrence of critical depth in a system is described as a hydraulic control. This is because flow conditions in the supercritical reach downstream can not affect upstream conditions. At the point of critical depth the *rating curve*, i.e. the relationship between depth (or level) and flow, is fixed and a function only of the local geometry. If such locations can be identified in a system they form a starting point for assessing water levels upstream and downstream of the control.

12.9 WEIRS, FLUMES AND GATES

The above discussion considered the case of flow passing from sub-critical to super-critical through the critical, minimum energy condition. This can be achieved smoothly and without significant energy loss over a weir or through a flume or gate. In each of these cases the flow is accelerating and this tends to dampen down turbulence. The reverse, passing from super-critical flow to sub-critical flow, involves the deceleration of the flow. The flow is less stable in such conditions and it is very difficult to achieve the transition smoothly without major energy loss. Generally, a region of high turbulence, known as a *hydraulic jump*, forms (Section 12.12).

Weirs

Equation (12.14) can be re-arranged in the form $(2/3H_s)^3 = q^2/g$

Hence, flow per unit width $q = \dfrac{2}{3}\sqrt{\left(\dfrac{2g}{3}\right)} \cdot H^{1.5}$

and, for a rectangular channel of width, b,

$$Q = \frac{2}{3}\sqrt{\left(\frac{2g}{3}\right)}bH^{1.5} \tag{12.15}$$

This is now in the form of a *weir* equation, which can be generalized as

$$Q = C_d bH^{1.5} \tag{12.16}$$

where C_d is a discharge coefficient.

C_d as defined by equation (12.16) has units as it involves $\sqrt{g}$. There are a number of other forms for this equation (Section 12.14), with the simplest involving a non-dimensional coefficient being:

$$Q = C_d^1\sqrt{g} \cdot bH^{1.5} \tag{12.17}$$

In equation (12.15) the discharge coefficient $C_d = (2/3)\sqrt{(2g/3)} = 1.705$ in metric units and $C_d^1 = (2/3)\sqrt{2/3} = 0.544$ in non-dimensional units. If the flow passes through critical depth over a weir crest then it might appear that C_d would always take that value. However, it is important to appreciate the inherent assumptions lying behind the theory that led to equation (12.1) and all the subsequent equations derived from it:

- there is a uniform velocity distribution across the section so $\alpha = 1.0$ (Section 12.3);
- the streamlines are straight and parallel (i.e. there is no lateral pressure set up by curvature of the streamlines);
- the vertical pressure distribution is hydrostatic (i.e. the pressure is a linear function of depth);
- the effect of the longitudinal slope is negligible.

Provided these conditions are met, as is very nearly the case with a broad-crested weir such as illustrated in Figure 12.7(b), then the discharge coefficient, C_d, is indeed about 1.705. However, many other crest profiles are used, ranging from simple plate *or sharp-edged* crests, to rounded tops of walls and triangular or ogee-profile crests. Their use depends on whether accurate flow measurement is required or whether they act merely as simple overflow hydraulic controls. Weir shapes used for flow measurement are discussed further in Section 12.14. The discharge coefficient can vary significantly from the basic broad-crested value of 1.705 and depends largely on the geometry of the crest, but it is also a function of the depth and velocity of the approach flow. The subject is wide-ranging and cannot be fully covered here. As a rule of thumb, however, the discharge coefficient is likely to be greater than 1.705 if there is strong curvature to the flow, for example over a half-round crest or an ogee crest. In the latter case, the profile of the crest is that of the underside of a free-falling jet of water, so, in theory, there should be little if any pressure on the solid surface. The pressure distribution through the depth of flow cannot, therefore, be hydrostatic and one way of considering the problem is that the back pressure on the flow over the crest is reduced, thus allowing an increased discharge and a corresponding increase in the discharge coefficient. Increases in C_d of 30–40% above the broad-crested weir value are possible.

Similarly, the discharge coefficient may be reduced if the weir crest is long, (in the direction of flow), or very rough—as might be case of flow over a grassed embankment. A value of 1.71 is, however, a good starting point for initial design or in the absence of more details of the weir shape.

It must be emphasised that a weir only acts as a hydraulic control if it has free or 'modular' discharge, i.e. the downstream water level is low enough to allow the flow to pass through critical depth. If the tailwater depth is high enough to affect the flow over the weir, the weir is said to be drowned. This condition is usually catered for in the weir equation by introducing a drowning factor f_d, which is a function of the height of the tailwater level above the crest. Referring to Figure 12.7(b) it would appear that, provided the tailwater level is not more than the critical depth, i.e. two-thirds of the upstream head above the crest, then critical depth flow will occur. Strictly, the tailwater level has a small effect even when it is at the level of the crest, particularly if lowered pressures can be generated below the nappe of the falling water, but this two-thirds criterion is a useful rule of thumb. Although the crest shape does affect the drowning factor, in many instances where a weir is not being used for measurement it is reasonable to assume that the hydraulic control remains at the weir crest until the tailwater level rises to a depth greater than two-thirds of the upstream head above the crest.

Flumes

The foregoing comments also apply to flumes. Figure 12.15 shows a typical flume with a narrow throat forcing the flow through critical depth. The weir equation can be applied to the throat and, provided the underlying assumptions as mentioned for weirs apply, the discharge coefficient will again be 1.705. Thus a flume, with parallel sides and level invert to the throat, has a discharge coefficient close to this base value, i.e., in metric units:

$$Q = 1.705bH^{1.5}$$

If however the calculation of flow is related to the more easily measured water level and the water depth above the throat invert level, h, rather than the total energy head H, then,

$$Q = 1.705C_v bh^{1.5}$$

where C_v is a coefficient related to the approach velocity in the channel, b is the width of the throat of the flume. As a general guideline, a flume will also be drowned out when the downstream water level rises to a level greater than two-thirds of the upstream head. In practice it is often possible with a well designed exit transition from the throat to recover some of the velocity head; it is also possible to design a flume so that the downstream water level can rise above this critical level and so that overall head loss is less than $H/3$.

Gates

Water control gates, as distinct from other types of gates as used in navigation locks, etc., may be either undershot or overshot. The latter case is akin to the flow over a weir and only the former type is discussed here. There are many designs of gates within these broad categories. The hydraulic principles are similar and are illustrated here for vertical-lift gates, also called *sluice gates* (and, in the UK, *penstocks*), which are by far the most common in the water supply systems. This is another case of flow passing from sub- to super-critical flow, but with a gate the flow does not pass through the critical depth at any meaningful location.

Figure 12.9(a) illustrates free discharge through a vertical penstock gate in the wall of a tank. The gate opening forms an orifice. The flow passing under the gate has a vertical contraction and, in the plane of the gate itself, there are vertical components of flow. Thus the minimum depth occurs downstream of the gate at a point referred to as the *vena contracta*, where the streamlines are parallel. Assuming negligible energy loss between the upstream section and the vena contracta, Bernoulli's equation can be applied along a streamline flow between points 1 and 2.

$$H_1 + z = V_1^2/2g + y_1 + z = V_2^2/2g + y_2 + z$$

where h_1 and h_2 are the upstream and downstream depths from the water surface to the centre of the opening. Assuming that the pressure throughout the free-falling jet at the vena contracta is atmospheric, then:

$$V_2^2/2g = H_1$$

and, since $Q = VA$

$$Q = A_2 \sqrt{(2gH_1)}$$

If the area of the gate opening is A_0 and the area of the vena contracta A_2 is $C_C A_0$, where C_C is a contraction coefficient, then

FIGURE 12.9

Orifice and undershot gate flow: (a) orifice flow, (b) undershot gate.

$$Q = C_C A_0 \sqrt{(2gH_1)}$$

In practice, as indicated for flumes, it is easier to measure the upstream water surface and depth and to relate the flow to the upstream depth using the more general equation

$$Q = C_D A_0 \sqrt{(2gy_1)}$$

where C_D is a discharge coefficient, which takes into account the approach velocity and the flow, is related to the upstream depth rather than the energy head. Typically, for a sharp-edged opening, C_D is about 0.6. If the edges of the opening are more rounded, then the value approaches closer to 1.0.

For the case where the jet is not freely discharging but the opening is drowned, the assumption that the pressure in the vena contracta is atmospheric is not true and, referring back to Bernoulli's equation above and going through the same procedure,

$$Q = C_D A_0 \sqrt{2g(y_1 - y_2)} = C_D A_0 \sqrt{(2g\Delta y)}$$

where Δy is the difference in water levels across the gate.

The same equations apply to the case of a sluice gate in a channel, as illustrated in Figure 12.9(b), but now the depth is the total water depth, y, to the channel invert rather than to the centre of the opening. For a full-width, freely-discharging gate with sharp-edged lip, the discharge coefficient varies between about 0.5 for low depths of submergence ($y_1/w = 2.0$) up to about 0.58 for submergence ratios of 10 and more, where w is the height of the gate opening as shown in Figure 12.9(b). For a submerged gate opening the discharge coefficient is again about 0.6.

12.10 FROUDE NUMBERS

The state of flow can be characterized by its Froude number. This is a non-dimensional grouping of the flow parameters, which can be shown to represent the relative magnitudes of inertial and gravitational forces acting on the fluid. The Froude number, F, is defined as

$$F = \frac{V}{\sqrt{(gy)}} \qquad \text{(12.18)}$$

where y is depth of flow, defined for non-rectangular sections as A/B where A is the flow area and B is the surface width. Its usefulness lies in the fact that at the minimum energy condition, the critical depth:

$$y_c = \frac{2H_s}{3}$$

as shown in equation (12.14). Hence the kinetic energy of the critical velocity, V_C, equals the balance of H_S available, i.e.

$$\frac{V_c^2}{2g} = \frac{H_c}{3} = \frac{y_c}{2}$$

Thus

$$V_c = \sqrt{(gy_c)}$$

Therefore, as flow passes through critical conditions its Froude number, $V\sqrt{(gy)}$, has a value of 1.0. For sub-critical flow the depth is greater and the velocity lower, therefore the Froude number is always less than 1.0; for supercritical flow the opposite is true and the Froude number is always greater than 1.0. Calculation of the Froude number thus provides an immediate check on the type of flow and how near the flow conditions are to those at critical depth. The closer the Froude number is to 1.0 the more unstable the water surface becomes since small disturbances can cause flow to flip locally between two possible energy states (Section 12.7). This can lead to waves and surface disturbances. As a general rule it is preferable to design channels with Froude numbers outside the range of 0.6 to about 1.5.

It should be noted, however, that the control of super-critical flow is more difficult than sub-critical flow. Any changes to the alignment of a channel with super-critical flow result in the development of standing surface waves. Moreover, the high momentum of super-critical flow results in significant super-elevation effects at any bend. Whenever possible channels should be designed for sub-critical conditions. If super-critical conditions must exist, such as in a spillway chute, then the channel should be straight and parallel sided and, if possible, any changes of direction effected in sub-critical flow sections.

12.11 HEAD LOSSES IN CHANNELS

The concepts of head losses in channels are exactly the same as those in pipes. There are both 'friction' losses and local losses and the same equations can be applied. However, as noted in Section 12.7, water depth can vary so that the velocity is not a function only of the flow.

Normal depth. In a long straight channel of constant cross-section and bed gradient the flow reaches an equilibrium depth when the rate of loss of energy through the turbulence generated by the drag of the boundaries equals the rate at which potential energy is given up by the fall in elevation. When the flow reaches this condition, the water surface and the energy line are both parallel to the channel bed. This equilibrium depth of flow is known as *the normal depth* and applies to both sub- and super-critical flow. Strictly, the resistance equations for pipe flow apply to this condition of normal depth flow in channels but they need to be modified for the different cross-sectional parameters. Instead of the pipe diameter, the relevant dimension is the hydraulic mean depth, R, also commonly referred to as the *hydraulic radius*. This is defined as:

$$R = A/P$$

where A is the area of the cross-section of the flow and P is the wetted perimeter, i.e. the portion of the perimeter of the channel surface in contact with the flow.

Manning's formula. Although it is possible to use the full Colebrook–White equations for open-channel flow, by far the most widely used equation for calculation of normal depth in open-channel flow is Manning's equation (Equation (12.9) in metric units). In some countries, particularly in Europe, it is also known as Strickler's equation. It should be noted that this equation is not dimensionally balanced and hence n has dimensions of length to the power of one-sixth. Some typical

Table 12.3 Values of Manning's roughness coefficient, n

Surface	Manning's coefficient n
Smooth metallic	0.012
Large welded steel pipes with coal-tar lining	0.011
Smooth concrete or small steel pipes	0.012
Riveted steel or flush-jointed brickwork	0.015–0.017
Rough concrete	0.017
Rubble (fairly regular)	0.020
Old rough or tuberculated pipes[a]	0.025–0.035
Cut earth (gravelly bottom)[a]	0.025–0.030
Natural watercourse in earth	0.030–0.040
Natural watercourse in earth but with bank growths	0.050–0.070

Note: [a]Values are for half-bankside depth of flow; at bankfull, the discharge may be 20% less.

values of Manning's coefficient are given in Table 12.3. Despite the dimensional nature of the Manning's equation, it is almost universally used for open-channel head loss calculations and there is a great deal of guidance on suitable values for the coefficient. Its use in this context is strongly recommended. For more detailed information on channel roughness, see Ven te Chow (1959).

Chezy formula. The equivalent equation to the Darcy–Weisbach equation is the Chezy equation, which is written as:

$$V = C(Ri)^{0.5} \qquad\qquad (12.19)$$

where C is a coefficient related to the roughness of the channel and i is the hydraulic gradient $= H/L$ (the slope of the energy line). Like the Darcy equation this equation is dimensionally correct so that C has no units.

Comparing equation (12.19) to the Darcy–Weisbach formula, equation (12.3), and noting that for a pipe running full $R = \pi d^2/4\pi d = d/4$

$$C = \sqrt{(8g/f)}$$

Therefore, it can be seen that, strictly, C varies, in a similar manner to the friction factor, as a function of relative roughness and Reynold's number. In practice the Reynold's numbers related to most open-channel conditions are such as to put the flow regime into the rough turbulent region of the Moody diagram, where the variation of friction factor and hence of C is small. Thus the assumption that C is constant is reasonable in most cases.

The value of C can be derived from values of Manning's n because, for given values of i and V:

$$R^{2/3}/n = CR^{1/2}$$

i.e.
$$C = R^{1/6}n.$$

Local head losses. As with pipeline fittings, head losses occur in open channels at any change of channel geometry. The calculation of losses is made in a similar manner, i. e. as a function of the velocity head. The complication with open channel losses is that it is often not possible to use a single, standard velocity as in the case of a pipe of constant diameter. It is common, therefore, to express the headloss at a change of section in terms of the difference in velocity heads, thus:

$$\Delta H = K \left| \frac{V_1^2}{2g} - \frac{V_2^2}{2g} \right| \quad \text{where } |....| \text{ denotes positive value.}$$

where V_1 and V_2 are the upstream and downstream velocities respectively and it is the positive value of the difference to which the empirical coefficient is applied. Care must be taken in interpreting data on the coefficients, however, as this form of the expression is not universal. Some data may be presented in terms of a single reference velocity as is done for pipes. For example, losses at a channel bend of constant cross-section can only relate to a single velocity. Table 12.4 provides some values of K.

Table 12.4 Local channel headloss coefficients

Feature		Loss coefficient[a]
Bends[b]		
$r/B > 3$ (B is surface width)		0.15
$r/B < 3$ but > 1		0.15–0.6
90° single mitre		1.2
45° single mitre		0.3
Transitions		
square ended	– inlet	0.5
	– outlet	1.0
cylinder quadrant	– inlet	0.4
	– outlet	0.6
smooth tapered	– inlet	0.1
	– outlet	0.2

Notes: [a]Loss coefficients for other elements can be taken as the same as for pipe elements provided the Froude Number of the flow is less than about 0.3.
[b]Based on average channel velocity.

12.12 HYDRAULIC JUMP

In the discussion of critical depth and its development over a weir crest or in a flume, it was assumed that the flow could smoothly accelerate from sub-critical conditions through critical depth to super-critical flow. Practical observation confirms this; accelerating flow tends to damp down any turbulence. The reverse is not true. To pass from super-critical to sub-critical flow is much more difficult without a significant energy loss. The flow is decelerating and expanding and almost always this leads to a region of high turbulence and energy loss. The result is *a hydraulic jump* in which the water surface rises abruptly from the fast shallow flow to the deeper sub-critical flow downstream. Figure 12.10 illustrates a hydraulic jump downstream of a sluice gate.

A hydraulic jump is a good way of 'destroying' surplus energy (in reality, converting kinetic energy to heat) and is often deliberately introduced for that purpose, such as at the foot of a dam spillway or downstream of a control gate. It enables the potentially erosive power of the high velocity flow to be reduced in a controlled fashion before the flow is released into the river downstream. The region of high turbulence is contained locally within the structure.

Because there is a high energy loss across a hydraulic jump, it is not possible to use the energy equation to analyse the phenomenon. Instead, the theory of a hydraulic jump is based on the theory of momentum. With reference to Figure 12.10 it can be seen that the only external force on the flow is the friction on the floor of the channel. Within the jump itself this is relatively small and can be ignored, so that the assumption made is that the net pressure force acting on the flow—the difference in the downstream and upstream hydrostatic pressures—is equal to the net momentum loss of the flow. This is treated more fully in other hydraulic texts (Ven te Chow, 1959 and Henderson, 1966) but it can be shown that for a jump to form, the downstream or *tailwater* depth, y_2, must be equal to or greater than that given by:

$$y_2 = \left(\frac{y_1}{2}\right)\left[\left(1+8F_1^2\right)^{0.5}-1\right] \tag{12.20}$$

where F_1 is the Froude number of the upstream, super-critical flow of depth y_1. The value of y_2 given by this equation is known as the *sequent depth*.

If the tailwater level is less than the sequent depth the super-critical flow continues further downstream until it has lost enough energy or until there is sufficient tailwater depth for the jump to form. If, on the other hand, the tailwater depth is greater than the sequent depth the jump is forced

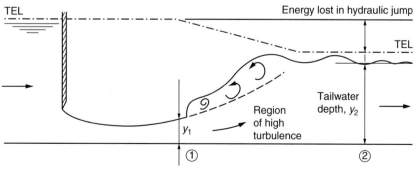

FIGURE 12.10

Hydraulic jump downstream of a gate.

upstream until the balance is re-established. Thus, for example, considering the jump downstream of the sluice gate shown in Figure 12.10, if there were another gate downstream controlling the tailwater level, then closing this second gate would increase the tailwater depth and force the hydraulic jump to move upstream. Ultimately it would drown out the first gate, which would then operate with a submerged, drowned discharge as discussed earlier.

12.13 NON-UNIFORM, GRADUALLY VARIED FLOW

So far only flow in a long straight channel of uniform gradient where the equilibrium of normal depth can be reached, has been considered. In practice, long straight channels are something of a rarity; in many instances the channel shape changes or the slope varies. Even if the channel is straight and uniform, it may not be long enough for the flow to reach normal depth. Thus, although normal depth and uniform flow are useful concepts, in many cases this condition will exist and neither the water surface nor the hydraulic gradient will be at a uniform slope.

One obvious example of conditions where the longitudinal profile of the water surface is curved, is the local drawdown of the flow as it speeds up to pass over a weir. This is a local effect and, as discussed earlier, the simplifying assumptions underlying the theory no longer hold true. It is also the case that the depth of flow upstream of a weir, dictated by the hydraulic control of the weir, is unlikely to be the normal depth in the channel, but moving further upstream the depth gradually changes towards normal depth.

This can be more clearly visualized by considering the flow in a river entering a reservoir. The flow slows down and gets deeper, with the water surface gradient changing from a slope approximately parallel to the average river bed slope, to horizontal at the reservoir, as illustrated in Figure 12.11. On the other hand, downstream of the crest of the dam spillway, the flow accelerates away from the critical depth and gets shallower and faster, as shown in Figure 12.11. Another good example of the curvature of the water surface is that of flow under a sluice gate into a horizontal channel downstream as illustrated in Figure 12.10. Energy must be lost through the resistance of the channel bed and the flow must slow down. This means that the depth must increase and the water surface curves upwards.

It is necessary to calculate the water surface profile in any of these cases or other situations, such as that of a river with varying cross-section and with occasional obstructions such as bridges or weirs. In the following explanation, steady flow is assumed; the depth and velocity of flow may vary with position, but the total flow at any point remains constant. The introduction of changes to the flow with time adds another level of complexity to the problem; such conditions are then described

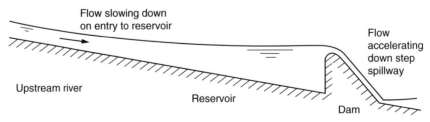

FIGURE 12.11

Examples of gradually-varied flow profiles.

as *dynamic* or *transient* and are almost always analysed using computer programs. They are beyond the scope of this book and reference should be made to any of the several open-channel flow text books available, for example, Ven te Chow (1959) and Henderson (1966).

Returning to the energy equation, consider the flow at two points, distance L apart, in a non-uniform channel, as shown in Figure 12.12.

Between the two points, ΔL apart, the head loss, ΔH is given by

$$\Delta H = H_1 + \Delta z - H_2 = \left(\frac{V_1^2}{2g} + y_1 \right) + \Delta z - \left(\frac{V_2^2}{2g} + y_2 \right)$$

Substituting Q/A for V

$$\Delta H = \left(\frac{Q^2}{A_1^2 2g} \right) + y_1 + \Delta z - \left(\frac{Q^2}{A_2^2 2g} \right) - y_2$$

If the length ΔL is reasonably short, the average slope, i, of the energy line can be taken as approximating to the mean of i_1 and i_2 so that

$$\Delta H = \Delta L (i_1 + i_2)/2$$

where i_1 and i_2 are the slopes of the energy line at Sections 1 and 2.

There are thus two equations for ΔH. From the known geometry of the channel and starting from a known depth of flow y_1 at Section 1 for flow Q, the hydraulic gradient i_1 can be calculated from the Manning or Chezy formula. Using $i_1 \Delta L$ as the loss of energy ΔH to Section 2, a first approximation of the depth of flow y_2 can be obtained (because $H_1 - \Delta H = Q^2/(A_2^2/2g)+ y_2$). From this i_2 can be calculated, so that a revised estimate of the hydraulic energy loss $\Delta H' = \Delta L(i_1 + i_2)/2$ can be used to calculate a more accurate value of y_2. By further iterations, y_2 can be calculated to the desired degree of accuracy. Thus the computation can be progressively carried out for further points downstream

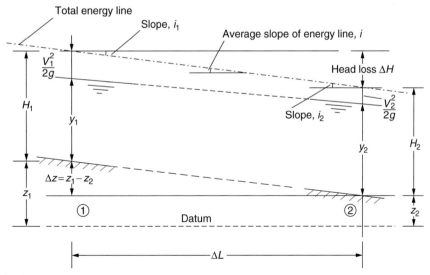

FIGURE 12.12

Energy loss in non-uniform flow.

giving the profile of the water level. For reasonable accuracy, the distance steps ΔL should be reasonably short, having regard to the change of slope of the hydraulic gradient.

Such calculations can be carried out by hand (though for non-rectangular channels they are very tedious), but are now normally undertaken with a spreadsheet or more probably with a computer program developed for the purpose. Many organisations have developed their own programs and there are a number commercially available, the most widely known probably being HEC 2 (now marketed as HEC2-RAS), developed originally by the Hydrological Engineering Centre in the USA.

There are three further points to make about this type of calculation.

1. The basic theory includes the inherent assumptions as listed in Section 12.9 for weirs. Thus, the approach cannot be used accurately where there are rapid changes to the water surface, for example close to a weir crest or through a flume.
2. The starting point for any calculation must be a known flow and depth. In many cases this is a hydraulic control, such as flow over a weir crest or through a gate where the water level can be directly calculated. Such a control defines the water profiles both upstream and downstream (Section 12.8). Thus for the sub-critical flow, the calculation proceeds upstream from the control, whilst for super-critical flow the calculation can proceed in the downstream direction, i e. the profile in sub-critical conditions is affected by what happens downstream, and the super-critical flow profile is affected by what happens upstream.
3. In some cases there may be no hydraulic control that defines a specific depth at any point; for example, a long river reach with no control structures. In such a case it may be adequate to assume a normal depth control in the reach downstream of the length of interest (assuming sub-critical flow) and to check the sensitivity of the water levels to that assumption.

The most common calculation relates to finding the water profile of sub-critical flow approaching some hydraulic control. This type of calculation is known as a *backwater calculation* and the water–surface profile is known as a *backwater profile* as the calculations proceed upstream from the control point.

12.14 MEASUREMENT WEIRS

In Section 12.9 it is shown that, provided there is free, undrowned discharge over the crest of a weir, the flow passes through critical depth and the upstream water level is uniquely controlled by the flow rate and the geometry of the crest. In theory, therefore, any weir can be used for measurement but, in practice, a limited number of standard crest shapes tend to be used, partly because experience has shown that these are the most accurate and practical structures for a particular application and partly because the discharge coefficients and the limiting conditions for their accuracy are well known. British and other international standards are available for such weirs, covering the standard geometries, measurement requirements and other features. Ackers (1978) provides a definitive discussion of the weirs and of flumes used for open-channel flow measurement. For non-standard shapes, either detailed model tests or site calibration must be carried out, but the resulting rating curve must have a weir-type relationship taking into account the overall geometry of the channel section at the weir.

For conditions which meet the underlying assumptions stated in Section 12.9 the upstream total head H is given by equation (12.15), i.e.

$$Q = (2/3)\sqrt{(2g/3)}bH^{1.5} = 1.705bH^{1.5} \qquad (12.21)$$

with $g = 9.81$ m/s^2 in metric units, and the general equation is therefore

$$Q = C_d b H^{1.5} \tag{12.22}$$

For *measurement weirs*, the general weir equation is often written as

$$Q = C_d'' (2/3)\sqrt{(2g)} b H^{1.5} \tag{12.23}$$

where, for equation (12.23) to be equivalent to equation (12.15), C_d'' has the value $1/\sqrt{3}$

However, as noted in Section 12.9, there are other formulations of the weir equation. In particular, equation (12.17) has a C_d'' varies from the value $1/\sqrt{3}$ according to the shape and type of weir crest.

Broad-Crested Weir

For the broad-crested weir, illustrated in Figure 12.7(b), the discharge coefficient, C_d'' is indeed about $1/\sqrt{3} = 0.577$. However, as indicated earlier it is easier to measure the actual water depth, h, above the weir crest, than the total energy head H (including the velocity of approach head), hence the equation is normally written as:

$$Q = C_v 0.577(2/3)\sqrt{(2g)} b h^{1.5} = C_v 1.705 b h^{1.5}$$

where h is measured away from the local draw-down of the water surface as it passes over the crest, and C_v is a *velocity coefficient* to account for the approach velocity head. If the approach velocity is small then C_v can be taken as 1.0 and for other cases the effect of approach velocity head can be calculated and included in C_v.

The generalized weir equation can be written as $Q = C_D C_V C_S C_P C_{bl} \sqrt{g} b h^{1.5}$ where C_D is a discharge coefficient, C_v is the velocity coefficient, both as defined before, C_s is a shape coefficient depending on the lateral shape of the crest (e.g. shallow 'Vee'), C_p is a coefficient to take into account the height of the weir and C_{bl} is a boundary layer coefficient to take into account the growth of the boundary layer across the length of the weir crest in the direction of flow. For accurate measurement, all these factors need to be taken into account and are discussed in detail in Ackers (1978).

Sharp-Crested, or Thin-Plate Weirs

Sharp-crested weirs, usually formed from a metal plate, are used in a variety of situations. They are simple, and because the plate can be machined to a high degree of accuracy, they can be very accurate flow measurement devices. However, because the plate becomes heavy and unwieldy for large flows, they are used particularly for small and medium flows, and not generally for river flow measurement. There are a number of standard shapes adopted, which are described by the shape cut out of the plate—rectangular of full channel width; rectangular with side contractions as illustrated in Figure 12.13; V-notch with a central angle, θ; or, less commonly, trapezoidal.

For a rectangular weir with no end contractions, i.e. across the full width of the channel, the discharge coefficient, for use in equation (12.17), can be taken from the empirically-derived Rehbock formula:

$$C_d' = 0.602 + 0.083 h/P$$

where h is the water depth over the weir crest, and P is the height of the weir crest above the floor of the channel.

FIGURE 12.13

Rectangular weir with fully developed end contractions.

For a rectangular weir with fully developed end contractions as illustrated in Figure 12.13

$$C_d' = 0.616(1 - 0.1h/P) \qquad (12.24)$$

For a 90° V-notch weir, i.e. with the central angle = 90°, C_d'' is given by the following table:

h(m)	0.050	0.075	0.120	0.125	0.150	0.200	0.300
C_d'	0.608	0.598	0.592	0.588	0.586	0.585	0.585

In this latter case correction factors must be applied if the base of the channel is less than $2.5h$ below the base of the notch, or if the width of the approach channel is less than $5h$, where h is the depth of water measured above the vertex of the notch.

In all cases for thin plate weirs there are a number of requirements for accurate measurements:

- the water depth should be measured at a distance between $3h$ and $4h$ upstream;
- if the waterway of the approach channel is less than 12 times the area of the waterway over the weir crest then a correction factor must be included to take into account the velocity of approach;
- the downstream water level must be below the level of the crest, i.e. no submergence can be allowed; and
- most importantly, the nappe (or underside) of the falling water must be well aerated to ensure atmospheric pressure there. This means that, for a full width rectangular weir in a channel whose walls extend downstream, an air pipe must be provided to allow aeration of the pocket beneath the nappe.

A limitation of sharp-edged weirs is that they can be damaged by debris brought down by the flow, and the shape of the machined edge to the plate is important for accurate measurement.

Crump Weirs

Where new measurement weirs are constructed on rivers and stream in the UK by far the most widely used weir is the Crump weir, so-called after the engineer who developed it as a measuring device. The profile is triangular with an upstream slope of 1:2 and a downstream slope of 1:5 as shown in Figure 12.14.

There are several advantages to the Crump weir, not least of which is that the discharge coefficient is very nearly constant over a wide range of discharges. C_d for use in equation (12.16) is almost exactly 2.0 (C_d'' in Equation (12.23) thus equals 0.677). Other advantages are: (a) the weir allows movement of sediment past the weir (deposition of silt against the vertical upstream face of most other types of weirs being a problem); (b) the weir is easy to build accurately, particularly with a pre-formed bronze crest angle built into the concrete structure; and (c) the high modular limit which the weir can tolerate before the discharge is significantly affected—the limit for the ratio of tailwater depth above the crest to the upstream depth being about 0.8.

Furthermore, the Crump weir can provide reasonably accurate flow measurement even if it is operating beyond the modular limit. The flow over a weir, once it is drowned, becomes a function of both the upstream and downstream water levels. For most weirs, this requires measuring the water level downstream in the area of high turbulence where the jet expands. This can be difficult and inaccurate, particularly as the critical difference between the tailwater and upstream depth is small. With a Crump profile weir, instead of the direct measurement of the tailwater level, the pressure in the separation pocket immediately downstream of the weir crest is measured (Fig. 12.14). The pressure in this pocket reflects the downstream tailwater level when the weir is drowned. It is a much more stable parameter than the turbulent water surface level downstream and can be measured more accurately. It thus enables the Crump weir to be used relatively accurately for measuring flow even when drowned. A rating curve can be developed for drowned flow using this parameter.

The one disadvantage of a Crump weir is that it can be a fairly long structure with consequent extra costs but it has been extensively model tested and criteria have been developed for truncation of the shape, i.e. neither the upstream nor downstream slopes need necessarily extend to the floor of the channel. Also investigated thoroughly has been the calibration of Crump weirs constructed with the crest sloping towards the centre so that the crest, in elevation, appears as a shallow Vee. This has the advantage of concentrating low flows towards the centre of the weir and giving greater accuracy of measurement over a wide range of flows.

FIGURE 12.14

Crump weir profile.

12.15 MEASUREMENT FLUMES

The principles behind the use of a flume as a measurement device are identical to those for a weir. In essence a flume is a constriction in the channel such that the width is reduced, forcing the flow through critical depth (Section 12.9). Provided the requirements behind the simple theory, set out in Section 12.9 are met, equation (12.15) applies to undrowned flow through a flume, i.e.

$$Q = 1.705bH^{1.5}$$

Rectangular-Throated Flume

The flume equivalent to the broad-crested weir is the rectangular-throated flume as illustrated in Figure 12.15. A flume can also have a raised floor as well as a narrowed throat but since one of the advantages of a flume over weir is that it can more readily pass sediment, any raising of the floor probably needs to be limited and with as smooth an inlet transition as practical. Downstream of the throat the flow is super-critical and a hydraulic jump forms in the downstream channel. It is important that the tailwater level downstream of the jump does not drown out the critical depth in the flume and, if necessary, a fall must be introduced into the downstream channel to ensure that the water flowing away is at a low enough level.

For a rectangular-throated flume as in Figure 12.15, the length of the throat must be not less than 1.5 times the maximum total head upstream. It is also necessary that the surfaces of the flume are smooth, whether of concrete or constructed with a pre-formed steel or fibre-glass insert, and that the divergence downstream must not exceed 1:6 and that the approach flow must be sub-critical. The upstream level measurement should be between $3h$ and $4h$ upstream of the flume inlet, where h is the upstream depth of water above the invert of the throat.

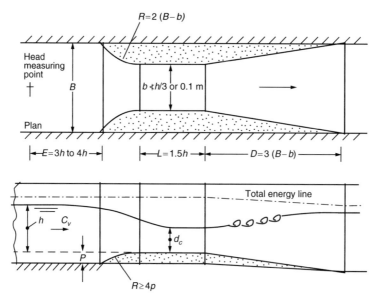

FIGURE 12.15

Rectangular-throated flume.

The discharge equation is usually modified to relate to the upstream depth of flow in the channel above the invert of the throat of the flume, h:

$$Q = C_v 1.705 bh^{1.5} (or \ Q = C_d bh^{1.5})$$

where C_v is a coefficient to account for the approach velocity. More generally the equation is written as:

$$Q = C_v C_s 1.705 bh^{1.5} \qquad (12.25)$$

where C_s is a shape coefficient to take into the particular geometry of the flume. For the smooth-entry flume illustrated in Figure 12.15, the values for C_s depend on the ratios, L/b and h/L as shown in the table below, where L is length and b the width of the throat.

h/L	L/b				
	0.4–1.0	2.0	3.0	4.0	5.0
			C_s		
0.1	0.95	0.94	0.94	0.93	0.93
0.2	0.97	0.97	0.96	0.95	0.95
0.4	0.98	0.98	0.97	0.97	0.96
0.6	0.99	0.98	0.97	0.97	0.96

Other Standard Flumes

There are, of course, many other designs of flumes and the geometry determines the particular discharge coefficient C_d'' for use in equation (12.24), or the shape coefficient for use in equation (12.25). In the USA, the Parshall flume is widely used but it has a relatively complicated geometry which varies depending on the range of flows to be measured and is rarely used in the UK. One dis-advantage of the rectangular-throated flume is that it is relatively long and the cut-throat flume has been developed as a shorter alternative. This has tapering inlet and outlet sections which meet at an angle at the throat. The shape coefficient is reduced from that for a rectangular-throated flume.

For free or modular discharge for flumes and weirs, the tailwater level should, as a general approximation, be no higher than the level of the critical depth flow at the throat or crest. (For thin-plate weirs a much lower limit must be set.) Generally, therefore, at least one third of the upstream specific energy head must be lost at the structure. Since flumes force critical depth by increasing the unit discharge rather than reducing the specific energy as do weirs, a flume has a greater critical depth in the throat than a weir in the same location. Thus a flume causes a greater head loss than a weir. (For a flume the drop in water level is one-third the depth of flow in the approach channel; for a weir it is one-third the height of water above the weir cill—which is less—see Figures 12.7b and 12.15.) Hence, if head loss is a critical issue, then this must be a factor in the choice of measurement structure.

12.16 VENTURI AND ORIFICE FLOW METERS

These types of flow meter apply to flow in closed conduits and work on the principle demonstrated by the Bernoulli equation: if the velocity of flow is increased then the pressure must drop. In both Venturi and orifice meters the flow is passed through a constriction in the pipeline causing the velocity to increase. The two types of meter are shown in Figure 12.16 and the measurement principle is illustrated in Figure 12.17.

From the energy equation (12.2) and assuming the pipeline is level and that head losses between sections A and B are negligible, the head difference between the two sections, A and B, is given by:

$$\Delta h = h_A - h_B = \frac{(V_B^2 - V_A^2)}{2g} \tag{12.26}$$

Since:

$$Q = \pi D^2/4 V_A = \pi d^2/4 V_B$$

FIGURE 12.16

(a) Orifice meter, (b) Dall-type Venturi meter.

FIGURE 12.17

Venturi and orifice flow meter measurement principles.

where D and d and A and a are the pipeline and throat diameters and areas respectively, equation (12.26) can be re-written as

$$Q = \frac{\pi d^2 \sqrt{2g\Delta h}}{4\sqrt{1-(d/D)^4}}$$

or:

$$Q = ak\sqrt{\Delta h}$$

where:

$$k = \frac{\sqrt{(2g)}}{\sqrt{1-(d/D)^4}}$$

In practice there are head losses between A and B and it is not necessarily the case that the stream lines at the constriction are parallel even for the venturi. Thus a further coefficient C must be introduced, so that

$$Q = Cak\sqrt{\Delta h} \qquad\qquad (12.27)$$

For a venturi meter C varies from about 0.95 to 0.99 depending on the exact geometry. Typically a well-designed meter with a parallel-sided throat has a value of about 0.98. The Dall tube illustrated in Figure 12.16 is a shortened version of the venturi meter and the value of C is less.

For an orifice meter Equation (12.27) equally applies. However, C varies much more, depending on the velocity and the ratio d/D. If the latter is in the typical range of 0.4–0.6 and for pipe diameters greater than 200 mm, the value of C will be in the range 0.60 to 0.61 for the usual velocities experienced in a pipeline.

The accuracy of measurement of both venturi and orifice meters depends on the lateral flow distribution through the device and can be severely affected by flow disturbances created by fittings in a pipe system. Detailed conditions for accurate measurement are laid down in BS 1242 Part 1. These can generally be met by ensuring that for d/D ratios not exceeding 0.6, there are at least 20 diameters of straight pipe without a fitting upstream of the meter and seven diameters of straight pipe downstream. The venturi is designed to minimize the head loss and this can be made very small with a well-designed expansion downstream of the throat providing good recovery of pressure head. The head loss through an orifice is substantially higher because of the sudden expansion of the diameter downstream. If head loss is an important consideration a venturi meter, such as the illustrated Dall tube, should be considered but venturis are more expensive and require greater space than an orifice meter.

12.17 OTHER FLOW METERS

A number of other types of flow meter, such as turbine, electromagnetic and ultrasonic are widely used in the water industry. The principles on which they operate are not the same as those outlined in this chapter but they are described in Chapter 16.

0.0   0.5   1.0   1.5   2.0   2.5
Velocity (m/s)

(a)

(b)

Alum (mg per L Al)
(YZ Plane 2)
— 3000

— 2000

— 1000

— 0

0.00   0.13   0.25   0.38   0.50
Velocity (Streamline 1)
(ms^ − 1)

(c)

(d)

CFD modelling (Block & Veatch):

(a) Free surface simulation.

(b) Pump sump.

(c) Contact tank flow—mixing and streamlines.

(d) Contact tank pollution.

PLATE 21

55 57 59 60 62
Age (hours)

(a)

54 55 56 57 58
Age (hours)

(b)

45 49 53 56 60
Age (hours)

(c)

0.0 0.2 0.4 0.6 0.8 1.0
Fraction of Fresh Water

10.0 12.0 14.0 16.0 18.0
Temperature (C)

(d)

CFD modelling (Black & Veatch):

(a) Rectangular tank with high level inlet and low level outlet at same location—age of water at mid depth.

(b) Circular tank with high level central inlet and low level outlet—age of water at mid depth.

(c) Rectangular tank with high level inlet and low level outlet on opposite sides—age of water at mid depth.

(d) Circular tank with low level inlet and outlet—isotherm at tank base and cross sections of fresh water fraction and temperature.

PLATE 22

APPENDIX—COMPUTATIONAL FLUID DYNAMICS (CFD)

CFD modelling requires expert modelling skills and high specification computing facilities. The models, which are relatively quick to set up, enable the simulation and visualisation of 3-dimensional hydraulic behaviour. They are used to understand the cause and solution of complex hydraulic problems which would be difficult or expensive to analyse by alternative methods such as physical models. A mesh is defined that splits the fluid into a large number of small elements. The software predicts flow patterns by solving a series of simultaneous equations for conservation of mass, momentum and energy. Basic fluid flow, chemical reactions, heat transfer and multi-phase flow (liquids, gases and particles) can all be simulated. Simple models may be completed within a few hours; more complex models may require longer running times to solve. CFD is used for a wide range of water engineering applications including:

- Free surface flow (Plate 21(a)).
- Simulated tracer tests for assessing contact time or mixing. This is probably the most common application for CFD. Simulated tracer tests enable the modeller to predict the contact time and analyse options to improve hydraulic mixing and tank efficiency. The analysis of existing tanks often demonstrates that installed baffles do not control the flow as intended (Plate 21(c)). The technique can also be used to analyse the dispersion of contaminants through contact tanks and service reservoirs (Plate 21(d))
- Head losses for varying flow through complex structures and pipe details.
- Pump sumps to simulate approach conditions for different operating scenarios and to model improvements to the design such as installing baffles (Plate 21(b)).
- Reservoir turnover can be assessed by conducting simulated tracer tests and evaluating the coefficient of variation for the tracer concentration. The effect of temperature can also be simulated to assess stratification problems and to develop designs for the inlet to improve mixing within the tank. (Plates 22(a) to 20(d))
- Dynamic forces acting on hydraulic structures.
- Particle transport.
- Air in pipelines.

REFERENCES

Ackers, P., White, W. R., Perkins, J. A. and Harrison, A. J. M. (1978). *Weirs and Flumes for Flow Measurement*. John Wiley.

Henderson, F. M. (1966). *Open Channel Flow*. Macmillan.

HR Wallingford (2006). *Tables for the Hydraulic Design of Pipes, Sewers and Channels*, 8th Edn. Thomas Telford.

Miller, D. S. (1990). *Internal Flow Systems*. 2nd Edn. BHRA.

Moody, L. F. (1944). Friction factors for pipes. *Trans ASME*, **66**.

Ven te Chow. (1959). *Open Channel Hydraulics*. McGraw-Hill.

System Design and Analysis

13

13.1 INTRODUCTION

Pipelines in a water distribution system can be divided into two functional categories:

Trunk mains convey water in bulk between facilities in the system, such as raw water intakes, treatment works, pumping stations and service reservoirs and the centres of demand. Their operational regime is a function of the demand characteristics of the area they supply, the availability of local storage and their transfer capacity. Individual consumers are not normally supplied directly from a trunk main.

Distribution pipes deliver water to the consumer from local storage and from connections off a trunk main through an integrated network. The pipes are sized to meet the hourly variation of consumers' demand and provide fire flows to any location in the network.

Within these two categories there are *'critical mains'*. A critical main is one where the consequence of failure poses an unacceptable risk of:

- interruption of supply to consumers;
- contamination of the water supply; or
- damage and disruption to third parties

Most trunk mains are critical mains and some key feeder pipes within distribution are also critical mains where failure of the pipe may affect the supply to sensitive consumers, consumers with a special need for an uninterrupted supply, hospitals, large industrial and commercial consumers and communities with a single source of supply.

Trunk mains generally operate at relatively constant flow rates up to the service reservoir; the storage being the buffer to even out the diurnal demand changes over 24 hours. There are day-to-day and seasonal variations of demand and trunk pipelines must be designed to carry the maximum day demand, typically in the range 120 to 140% of the average daily demand (Section 1.12). Design should also take account of the possible use of the pipe for short term emergency water transfers (for example re-routing supplies following a burst main) and for transfers between zones in order to maintain a balanced storage in service reservoirs on a weekly or seasonal cycle of demand.

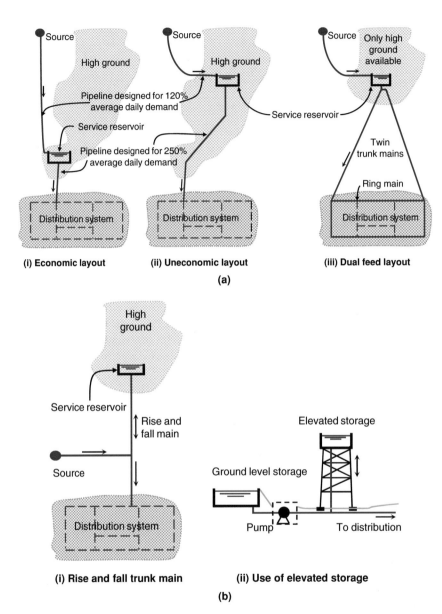

Distribution are supply pipeline options:

(a) Single trunk main and ring main feeders to distribution system.

(b) Rise and fall trunk main and elevated storage.

PLATE 23

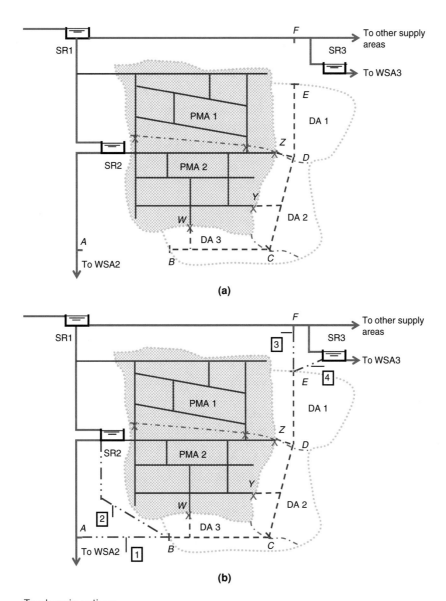

Trunk main options:

(a) Existing system with potential development areas.

(b) Four trunk main options for supplying new development areas.

PLATE 24

Trunk mains supplying a zone without local storage should be designed to transfer the peak hour demand to the distribution network. However, since these pipes would normally convey water to large areas, the peak hour factor applied to them tends to be less than that for individual distribution mains because a *'diversity factor'* applies, related to the size of the area served (Section 1.12).

The diurnal demand in a distribution system can vary by a factor of 2.5 to 3.0 or more of the annual average annual daily demand. A typical hourly variation of demand over 24 hours is shown in Figure 18.1. Urban, industrial and rural distribution systems exhibit different diurnal demand patterns between weekdays and weekends. They can also vary from one year to the next.

13.2 SYSTEM LAYOUTS

The geographic and physical characteristics of an area influence the layout of a system. An interconnected looped layout which enables water to flow in multiple paths to any part of a network provides maximum flexibility. Urban systems tend to comprise looped networks, albeit that the network is subdivided into hydraulically discrete areas for leakage and demand management (Chapter 14). Dendritic (tree like) layouts are more common for trunk mains and local distribution in rural areas.

The most economic layout for a system is one that is gravity fed from a local service reservoir located as near as technically feasible to the distribution area it serves, as shown in Plate 23(a)(i). Because the storage is used to meet the peak demands for water, the greater the distance from the service reservoir to the distribution area, the longer the pipeline that is designed for peak hourly flow rates (Plate 23(a)(ii)), and hence the more costly is the system. Storage close to the demand centre can better maintain supplies under emergency conditions and for fire-fighting, help reduce pressure fluctuations in the distribution system and aid economic development of the system.

If the service reservoir cannot be sited close to the distribution area the preferred layout is to have at least two major supply pipes from the reservoir, connected together at their extremities to form a ring main through the distribution area (Plate 23(a)(iii)). This 'redundancy' enables the network to operate more efficiently under peak flow conditions and allows supplies to be maintained while one of the pipes is being rehabilitated.

Boosting is a pumping arrangement which augments the pressure or quantity of water delivered through a system. The term is sometimes wrongly used to mean simple pumping. Of the many possible arrangements, the three most important purposes are to provide a fixed extra flow, to provide a fixed extra pressure and to maintain a given pressure, irrespective of the flow.

One of the most frequent uses of a booster is to increase the pressure of water in a distribution system at times of high demand. At low demand, the pressure may be adequate, but when demand is high system pressures may be too low. Instead of laying additional feeder mains into the distribution area, it may be more economical to boost the pressure at times of high flow. Figure 13.1 shows the hydraulic gradients that might apply before and after boosting.

A consequence of raising pressure is that flow rates increase. However, prevailing low pressures may have restricted flow taken by consumers so that flow records will not indicate the true demand when the pressure is raised. Some assessment of increased demand should be made so that an appropriate design duty for the booster pumps can be established. This can also include a margin for future increases in demand.

Much of the complication of controlling a booster pumping station can be avoided if *balancing storage* can be built into the system at an appropriate location and elevation. Balancing storage

FIGURE 13.1

Effect of boosting pressure into a distribution system.

generally improves security of supply and can be used to optimize operating costs. With balancing storage connected to a system, pumps can be operated by the simple use of level switches in the tank. When peak drawoffs occur, the water level in the storage falls thereby initiating a pump to start. With progressive lowering of the water level, further level switches may bring additional pumps into operation or increase the speed of the first pump. This arrangement has many valuable characteristics:

- sudden large but short duration increases of drawoff can be handled by the balancing reservoir without causing the booster pumps to start;
- the maximum capacity required from the booster pumps is reduced;
- the head range against which the pumps have to work is diminished;
- pumping, once started, can continue until the balancing reservoir is refilled thereby increasing the load factor and efficiency of the pumps;
- control of the pumps is simple and positive and repeated stopping and starting of the pumps is much reduced;
- balancing storage in an existing boosted system allows a range of options for extending the system for future network development.

The size of balancing tank or water tower required is often small because the periods of peak drawoff on a water distribution system are quite short. If a 5 Ml/day supply suffers from a peak drawoff 75% above normal for a period of 4 hours, the theoretical size of tank required to deliver an uninterrupted supply would be about 625 m³, but a tank of one-half or even one-third of this size would greatly reduce the maximum duty required from the pumps.

A *'rise-and-fall' main* as illustrated in Plate 23(b)(i) can be adopted where the only possible location of a service reservoir is further away from the distribution system than is the source of water. If the rate of output from the source is equal to the average daily supply, water flows out of the service reservoir whenever the demand rate exceeds the average and flows into the reservoir when demand is less than the average. Alternatively, if the source output is made sufficiently large to fill the reservoir during part of a day—for example during a day shift or to optimize the use of electricity tariffs by pumping only at night—when pumping ceases the distribution area is fed only from the service reservoir. Lower distribution pressure at times of low demand also helps to minimize night-time leakage. However, the design needs to ensure adequate turnover of water in storage to prevent water quality problems arising, particularly during periods of low demand. Hydraulic models can be used to analyse this type of system using *'extended period simulation'* (quasi-dynamic) modelling over 24 hours (Section 13.15) to optimize the pumping and storage requirements and to ensure the age of the water, particularly in the reservoir, does not become excessive and hence require additional local disinfection facilities.

Elevated storage is often necessary on flat ground; Plate 23(b)(ii) shows a typical arrangement with a rise-and-fall main to the elevated storage. If, however, the pump delivers water to the top of the tank the distribution system is at all times fed by gravity from the storage. In either case a float-operated valve can prevent the elevated tank from overflowing and water level sensors in it can control the number of pumps running or their output, to keep the supply roughly in step with demand. The advantage of the rise-and-fall main is that, during pumping, a higher pressure can be maintained in the distribution system than would be possible by gravity alone from the elevated tank which is costly to construct more than about 30 m high.

The same effect could be achieved, without elevated storage by using pressure or flow measurements in the mains feeding the distribution system to control pump output. This is considerably more difficult to set up, operate efficiently and maintain and requires a thorough understanding of system demands and how they vary. Even a small tank is preferable because it permits simpler water level control of the pumps.

Elevated storage is more expensive than ground storage to construct. Therefore it is seldom possible to provide the same amount of elevated storage as could be provided at ground-level. Consequently both ground-level and elevated storage may be provided for systems with flat topography, the ground-level storage providing the larger capacity. To ensure that the reserve supply in the ground-level tank can be used at all times, the transfer pumps filling the elevated storage usually have 50% standby capacity or more; an independent standby source of power, such as a diesel generator, may also be provided.

Many water companies prefer to pump directly into distribution. Although there are significant disadvantages with pumped distribution systems, as outlined above, the need to limit capital investment often outweighs the control advantages of elevated storage. It is true that control systems for direct pumping into a distribution system are becoming increasingly sophisticated, but they do require an appropriate level of technical maintenance which may not be available to water utilities with limited resources or in remote places where simple fail-safe systems are more reliable.

13.3 PIPELINE PLANNING

New pipelines are required either to meet forecast increased demand from within an existing supply area and from proposed developments adjacent to the area or to replace an existing pipe for structural, capacity or water quality reasons. The route of a new pipeline is often dictated by alignment and planning constraints. The planning and design processes will size the pipe and determine its horizontal and vertical alignment within the imposed physical constraints and operational conditions. However, a utility often has to plan a new pipeline not knowing precisely either where the forecast demand will be or how the future developments will be phased. Therefore pipeline planning comprises finding a strategy that best covers a range of feasible medium and long term development options while maintaining flexibility in case the anticipated development sequence changes.

The design process must include decisions as to which source should supply each area, which route is practical for new mains and how peak day demands are to be met when some sources have spare capacity and others have not. Oversized mains are advisable in some areas to cater for possible additional future demand and ring mains may be needed to transfer water across the distribution system under different phases of development. Factors to be considered include:

- proposals and planning forecasts for new housing and industry;
- where uncontrolled developments might occur;
- laying sections of the pipe in stages to match the development programme;
- delaying construction into those areas with the greatest uncertainty (confidence in demand forecasts and development programmes);
- peak outputs from existing sources and differences in their production costs;
- possible locations for future sources;
- service reservoir and booster pumping capacities;
- alternative supply regimes and operating modes for both before demand has built up and for the ultimate forecast demand;
- availability of range of sources versus seasonal and diurnal variation;
- fire risks, critical mains and system failure scenarios;
- minimum supply pressures, pressure management and district meter areas to manage and monitor water losses;
- managing water quality by establishing 'discrete' zones or areas, fed only by one quality of water;
- developing ring mains or integrated networks of pipes to maintain supply flexibility;
- avoiding the need to duplicate mains back to a source at a later date, unless the current scheme includes provision for additional land, cross connections and the infrastructure for its future control;
- the optimum engineered or most physically practical route; and
- capital versus operating costs (whole life costs including environmental and social costs).

The analysis involves a logical assessment of all viable options for which there is either a clearly identified preferred solution that satisfies known current constraints or a compromise of the options that best fits with short and medium term uncertainties. The design should achieve the optimum combination of pipe sizes and alignments to meet satisfactorily the various demand conditions. The first step is to identify route options using desk studies of maps, GIS, records of physical obstructions including the locations of other utilities and site visits to assess route viability and construction constraints. The studies should include reviewing archaeological, historic and environmental factors to minimize the disturbance of areas of special interest.

The design of a critical main must take account of how supplies are maintained if it were to fail. Options can include duplicating the whole pipe or sections of it, strategic links into different parts of the system for supply diversion and additional valves and cross connections to allow local isolation and rerouting of supplies.

Plate 24(a) illustrates part of a network which includes a supply area (WSA1) comprising two pressure areas PMA1 and PMA2, and supply areas to the south and east (WSA2 and 3). Service reservoir SR1 supplies WSA 2, PMA1 and other areas to the east. The supply to WSA2 from SR1 passes through service reservoir SR2 which separately supplies PMA2. PMA1 operates at a higher supply pressure to PMA2. In this example it is assumed that there is a shortage of storage for the local WSAs and that no future storage location has been identified, but there is spare ground adjacent to reservoir SR3 which supplies WSA3.

Plate 24(b) illustrates options for supplying proposed future developments DA1 and DA2 to the east of the urban area and DA3 to the south. The problem confronting the designer is that although planned to be built within the next 25 years the order and phasing of the developments is not known.

The options for extending the network include:

(a) *Using the existing network* (not illustrated): Supply DA1 from PMA1 and DA2 and DA3 from PMA2 by extending the existing networks, provided there is spare capacity. This would require minimal additional pipework, but increases the number of consumers at risk of interruption to supply due to pipe failure.

(b) *Discrete supply areas for the new development:* DA1 can be supplied direct from either SR1 (SR1 – E), a trunk main (F – E) [3] or a new reservoir located next to SR3 (SR3A – E) [4]. DA3 can be supplied direct from either SR2 (SR2 – B) [2] or the trunk main supplying WSA2 (A – B) [1].

(c) Supplying DA2 direct (not illustrated) would require long lengths of addition pipework into the area or laying duplicate pipes. It would be more practical to either extend the pipe supplying DA1 (E – D – C) and DA3 (B – C – D) in order to supply DA2 or to add area to PMA2 by opening the boundary valve (Y).

(d) *Laying a new main through the three development areas* [1], [2] and [3] to create a ring thereby improving hydraulic capacity, operational flexibility and reducing failure risk, but potentially incurring capital expenditure too early.

(e) *Phasing* the laying of the pipework to the development areas to suit the building programme.

The location of the additional storage depends on how each new area is supplied and where there are operational storage shortfalls. However if the ring main is developed, its location is less critical because the integrated network provides the hydraulic transfer capacity to manage risks. Alternative locations for the additional storage may include using SR2 only to supply PMA2, DA2 and DA3, and building additional storage closer to the centre of the demand for WSA2.

13.4 DISTRIBUTION SYSTEM CHARACTERISTICS

The performance requirements of a distribution system can be summarized as:

1. *Minimum pressures at peak hour demand* sufficient to serve the highest supply point in the network. Typically a mains pressure of not less than 15 to 20 m would be required to serve buildings up to three storeys high. Higher pressures may be necessary in some areas where there are significant numbers of dwellings exceeding three-storey height; but high rise buildings are normally required to have their own boosted supply.

2. *Maximum static pressures during low demand periods,* typically at night, should be as low as practicable to minimize leakage. For flat areas a maximum static pressure in the range 30 to 45 m is desirable. Pressure reducing valves can be installed to reduce high operating pressures but the valves must be regularly maintained if they are to operate reliably. Nevertheless, the pipework in the reduced pressure zone must be rated to resist the maximum possible static pressure if the valve fails or has to be by-passed for operational reasons. Higher static pressures may be unavoidable where undulating topography limits the opportunities for sub-dividing the network into discrete pressure zones, typically where ground levels vary significantly over short distances.

3. *Pressure allowance for transients* should be not less than 5 m head.

4. *Fire demand.* In the UK, fire demand requirements range from 8 l/s from a single hydrant in a one- or two-storey housing development to up to 75 l/s from one or more hydrants serving an industrial estate (Table 14.1). Owners of properties requiring supplies for sprinklers or hydrants on their premises may need to enter special arrangements with the water supplier. In the USA higher fire flows are required (Table 14.2).

5. *Consistent water quality* must be maintained throughout the system by establishing discrete hydraulic and source water quality areas. Wherever possible dead ends, long retention times, mixing of different waters within the distribution system, diurnal reversals of flow direction in mains and exceptionally high flow rates should all be avoided as they can cause water quality deterioration that may result in "supplying water unfit for consumption"; a non-compliance of the water quality regulations in the UK.

6. *Spare flow capacity* must exist in the system sufficient to meet foreseeable rises of demand over the next few years and to provide flexibility to minimize the number of consumers affected by interruptions to supply during network maintenance activities.

Lower supply pressures of about 10 m head are desirable where only standpipes are to be supplied; making it possible to install low pressure, easily operable standpipe taps which give less maintenance problems and less wastage. However, standpipe-only systems are rare and it is not practicable to supply water over large areas at very low pressure. Further design considerations determined by the level of service standards provided by a water utility to its customers are discussed in Chapter 14.

Table 13.1 illustrates the proportions of pipes laid in different sizes in urban and suburban environments in regions of the world. The proportion of pipes over 450 mm size reflects the distance between sources or reservoirs and the centres of demand of the area served and hence the proportion of trunk mains in the system. The table indicates that pipes used for distribution are typically in the range 100 to 250 mm diameter and that these comprise the majority of pipes from which properties are served. Older systems often have a significant proportion of small mains, 80 mm and below. Smaller pipes can also be used as spur or rider mains feeding small groups of properties. Often

Table 13.1 Representative percentages of mains by diameter

Diameter (mm)	50–80	100–150	200–225	250–350	375–450	500–600	>600
Diameter (inches)	2–3	4–6	8–9	10–14	15–18	20–24	>24
UK: city/urban	4.5	73.0	9.9	6.1	1.9	0.8	3.8
UK: urban/rural	26.5	49.7	7.9	6.0	3.8	3.3	2.8
SEA: city/urban	0.5	50.1	21.3	22.5	0.4	3.0	2.3
SEA: urban/rural	8.0	62.5	7.5		14.0		4.0
USA: city/urban	1.0	42.0	24.0	16.0	4.0	4.0	9.0
USA: small urban		44.0	30.0		26.0		

Notes: [1]225 mm (9 in) and 375 mm (13 in) diameter pipes are no longer in production, but many were laid and still exist.
[2]175 mm (7 in) mains are also found in some older systems (6.9% for this network).

there are also larger diameter critical distribution pipes that are used to convey water to an area for further distribution. Critical mains vary in diameter according to the size of area served and some pipes may supply properties en route; for example a critical main in a rural area may be as small as 100 or 150 mm diameter, whereas in most urban areas the minimum size of these feeder pipes may be 200 mm or larger.

The cost and difficulty of making a service connection increases with the size of the pipe and in general service connections are not installed on 300 mm diameter pipes and larger. A 300 mm main is also of such large capacity that it will normally be used to convey water in bulk to an area for further distribution. Ideally service connections should therefore be restricted to the smaller diameter pipes. In areas where there are large industrial consumers, or where the per capita consumption for domestic purposes is exceptionally high, some distribution pipework may be 400–450 mm diameter, however connections off larger mains are more typically created using pipe fittings rather that tapping the main.

13.5 DESIGNING TRUNK MAINS

There are few occasions when the design of a completely new distribution system is required, hence the majority of hydraulic studies and designs involve analysing the performance of an existing system and, where necessary, improving its performance to meet demands from new developments and increased consumption by existing consumers.

The layout must form a sensible hierarchy, in which the larger mains feed the smaller ones. The designer may need to take into account a water utility's preferred range of pipe diameters and materials, usually adopted to limit the quantity of spares that need to be held in store for emergencies. Care needs to be taken also to ensure the chosen network of preferred pipe diameters does not restrict potential growth or create unacceptable velocities, high or low. Pipe velocities at high demands should be limited to keep friction head losses within acceptable limits. However pipe sizes should also be chosen to achieve minimum velocities at least once in each diurnal cycle so that the age of the water in the pipes does not become excessive or that loose deposits do not build up locally to the extent that they could cause a dirty water incident if disturbed.

Where a network is open and highly interconnected, there are opportunities to optimize the layout to maintain acceptable operating conditions under emergency flows while reducing retention times and minimising pressure variations. However, the layout should also recognize that network operators need to manage the network efficiently by being able to monitor and control leakage and pressures and by operating stable hydraulic supply areas to support consistent water quality.

The factors to be considered when designing trunk mains include:

- the optimum engineering or most physically practical route;
- proposed mode of operating the system; available resources versus seasonal and diurnal variations;
- location and capacity of sources and storage;
- provision of system flexibility without creating diurnal reversal of flow in mains;
- capital versus operating cost;
- confidence in demand projections.

The flow in a pipe is determined from the summation of demands the pipe feeds. The network demand is summed from its individual components working upstream from the extremities of the system towards the sources. The demand assessment process is similar for both trunk mains and distribution pipework except that the demands on trunk mains are the summations of the demands from the individual pipes they supply.

13.6 DESIGNING DISTRIBUTION PIPEWORK

The design of new local distribution pipework is largely empirical. A pipe must be laid in every street along which there are properties requiring a supply. Pipes most frequently used for local distribution are 100 or 150 mm diameters (Section 13.4). However distribution pipes tend to be oversized having been designed for fire flows, future growth, to provide operational supply flexibility and where a water company operates a policy to standardize pipe diameters. When replacing or rehabilitating pipes, there is therefore often an opportunity to reduce the size of the pipes while still achieving the required hydraulic capacity. In the Netherlands, following discussions with, and the agreement of, relevant stakeholders to reduce fire flow capacity, pipe design criteria were revised. Pipes are sized to achieve a daily "self cleansing" peak velocity of at least 0.4 m/s (van den Boomen, 1999) in order to prevent sediment accumulating. This approach produces new networks comprising of 40 mm and 63 mm diameter pipes feeding domestic consumers in local networks and dead ends where previously the minimum size pipe would have been 100 or 150 mm diameter to cater for fire flows.

The diameter of local distribution pipes depends primarily on the population density of the area to be served and how the pipe is supplied. Although guidelines can be developed for sizing new pipes when extending an existing network, the designer needs an understanding of the site specific conditions and performance of the system. At any time, t, of the day, the demand from an area comprises:

$$\text{Average day domestic demand} \times D_{t1} + \text{Average day non-domestic}$$
$$\text{demands} \times D_{t2} + \text{Allowance for leakage at time } t$$

where D_{t1} and D_{t2} are the diurnal factors applying at time t to the domestic and non-domestic demands and the leakage allowance varies with the system pressure at time t.

The peak hour demand occurs at different times for different types of demand in different parts of the system. Therefore the demand characteristics need to be developed from field measurements of flow and pressure. Alternatively a system demand profile can be derived from its constituent parts using available records of demand, patterns of usage and recorded leakage, making assumptions only where better information is not available. A similar approach can be applied for deriving seasonal variations and to assess demands in trunk main systems.

Where dwelling numbers and occupancy ratios cannot be readily assessed from utility records and census data, figures for the average number of people per hectare or per kilometre of mains may be used, but with caution. Table 13.2 gives guidance on indicative population density and typical average numbers of connections per km of main for large utilities in regions of the world. They include both domestic and trade connections. The number of people served by each pipe in a street is influenced not only by the type and arrangement of the dwelling units and their occupancy ratio, but also by the kind of water supply given. In densely populated low income areas typically found in the developing world a large proportion of the population may be served by street standpipes. High-rise buildings result in high population densities, but each building may be fed by one or two large connections.

Table 13.2 Population served per kilometre of main laid

		People	Connections
	People per household	Average number per km of main	
Densely populated low income areas	7–9	3000–4000	300–550
Planned high occupancy dwellings	5–8	1000–2000	100–250
Residential urban areas:			
• Asia	4–7	400–800	70–100
• Europe & Australia	2–4	250–350	100–130
• North America	2–4	200–400	80–120
Medium to low density housing:			
• Asia	3–8	200	20–75
• Europe & Australia	2–5	100–200	50–80
• North America	3–5	20–200	5–65
Suburban areas with gardens	3–5	130–200	30–50
Populated rural areas with villages	1–4	110–160	10–50
Dormitory accommodation	n/a	750	n/a

Sizing distribution mains should take account of the fire risk along the pipe alignment and hence the minimum fire demand associated with the risk. Fire flows are discussed further in Section 14.4. The full available head at a hydrant can be utilized to meet fire demands. A single fire hydrant flow of 2 m³/min can be relatively easily delivered by a 150 mm main fed from both ends; the velocity in the main being 0.9 m/s with a rate of loss in the main of about 12 m/km. If only fed from one end, the velocity increases to 1.9 m/s and the head loss becomes 46 m/km which may be acceptable in urban areas if the lengths of pipe between cross connections are relatively short. Further away from the hydrant being used, the number of mains contributing to the flow increases so that the headloss in them is much less. The same flow in a 100 mm pipe fed from both ends would produce both high velocity (2.1 m/s) and head loss (100 m/km).

13.7 HYDRAULIC DESIGN OF PIPELINES

Figure 13.2 illustrates the elements of a simple pumping system. It comprises a source tank or sump, a length of pipeline to the pumping station, a pump and non-return valve (essential to prevent reverse flow when the pump is turned off) and a delivery pipeline discharging into a tank.

The *static lift* is the height, or equivalent head of water, between the water level in the source tank and that in the delivery tank or the level of the tank inlet if higher. This is the head across the

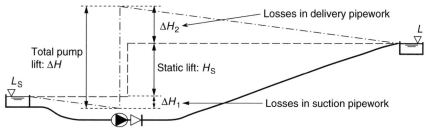

FIGURE 13.2

Definition of terms in pumping systems.

pump (or, more correctly, the non-return valve) when there is no flow. The static lift does not depend on the level of the pump. To deliver any flow the pump must overcome the static lift and the *pipe friction and other losses* in both the suction and delivery pipelines (Chapter 12). 'Station losses', which would normally be the responsibility of the supplier of the pumps and station pipework, are assumed here to be included in losses in the suction and delivery pipelines.

In a pipeline of constant diameter and roughness the rate of energy, or head, loss along the pipeline is constant. Thus, ignoring local losses, the hydraulic grade line can be represented by a straight line at constant gradient—the hydraulic gradient. This line represents the change in total energy level of the flow. Strictly, local losses should be included as abrupt steps in this line where they occur but they are often lumped in with the friction losses and spread over the length of the line. In systems where local losses dominate—e.g. in a short length of convoluted pipework in a treatment works—it may be necessary to identify each local loss and draw the hydraulic grade line correctly.

For a particular flow the hydraulic grade line (or total energy line) is below the static level on the suction side of the pump but above the static level on the delivery side. The total **pump lift** (or 'pumping head') is thus the sum of the static lift (H_S) plus the head losses in both the delivery system and the suction pipework (ΔH_1 and ΔH_2 respectively) including the 'station losses':

$$\text{Total pump lift (Pumping head)} = \Delta H = H_s + \Delta H_1 + \Delta H_2$$

The *pressure* at any point along the pipeline is the height of the hydraulic grade line above the pipe centre line. Strictly, the velocity head should be deducted from this value but in most pipeline systems this is small in comparison with the pressure and can be ignored. Only if the velocity is high or the pressure low would the velocity head component of the pressure become significant. It should be noted that the pressure is not necessarily greatest at the upstream end of a pipeline. The total energy level must decrease in the downstream direction as energy is lost, unless energy is put into the system, for example by a pump.

13.8 SYSTEM CURVES

A *system curve* is the relationship between the flow delivered into a system and the head required to deliver that flow. Figure 13.3 illustrates such a curve. It is made up of the static lift and the losses in the suction and delivery pipework. As shown in Chapter 12, both the friction and fittings head losses are approximately proportional to the square of the velocity and hence, in a pipeline of known size, to the square of the flow rate. Thus the system curve is approximately parabolic.

FIGURE 13.3

Simple system curve.

For a design flow, Q_D, the required pump lift is H_D as illustrated. This is known as the pump duty point. However, a single system curve is rarely enough to define the range of the possible pump operation. The static lift varies with the levels in the source and discharge tanks. Furthermore the friction losses may change with time so it is necessary to consider the conditions when the pipeline is new and when it has been in service for many years. In addition, the fittings losses quoted for design (Table 12.1) include a safety factor. If the fittings represent a significant proportion of the total losses then it may be necessary to consider the possible range of these losses with reference to more accurate, minimum values for these losses (Miller, 1990).

Thus, when considering a real system a range of system curves can be drawn. For the simple pipeline system considered in Figure 13.2, the envelope of possible pumping conditions can be drawn with two curves with the following parameters:

- maximum static lift (low sump level/high delivery tank level), maximum roughness and maximum fittings losses;
- minimum static lift (high sump level, low delivery tank level), minimum roughness and fittings losses.

Figure 13.4 shows this range of system curves with indicative pump characteristic curves added. For any flow, there is a range of pump lifts required: at the design flow, Q_D, the maximum pump lift under the most severe conditions is given by Point A. A fixed-speed pump selected for this duty would be delivering a flow greater than Q_D under all but the most severe conditions. Depending on the static lift and pipeline roughness the pump would be operating along its characteristic curve somewhere between Points A and A'. Conversely, a pump selected for duty on Point B would in all but the most optimistic conditions provide less than the design flow, operating between points B' and B.

A more appropriate duty point might be at Point C. This represents the pump lift needed for the design flow at average static lift and maximum pipeline roughness. If the pump was sized for this duty point, it would operate between points C" and C'. Over part of this range the pump would deliver less than the design flow and over part of the range more than the design flow. Provided the average flow rate is equal to or more than the design requirement then the pump should be adequately sized. By considering the average static lift and the possible long term deterioration of the pipe roughness adequate performance should be ensured. However, if it is necessary to ensure the instantaneous design flow under all conditions then point A would be the appropriate pump duty.

The need to consider the full range of possible pump operation is further illustrated in Figure 13.5. This shows the same system curves as in Figure 13.4 with a duty based on Point C but met by two pumps working in parallel. It is common for two or more pumps to be provided to meet a design duty.

With reference to Figure 13.5, the duty point for operation of the station has been chosen as Point C2. The duty point for sizing the individual pumps is given by point C1: the same head requirement but with half the total flow through each of the parallel pumps. With two pumps

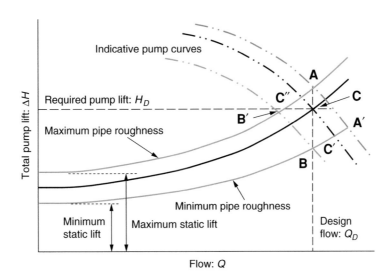

FIGURE 13.4

Maximum and minimum system curves and range of pump operation.

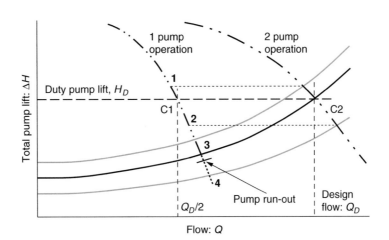

FIGURE 13.5

Multiple pump operation.

operating, each pump may operate over the range between points 1 and 2 but, with a single pump operating, the range of duties is between points 3 and 4. It is essential to plot the range of system curves and proposed pump characteristics to check that the pump can be allowed to operate over the whole range with respect to NPSH requirements (Section 17.4). As drawn in Figure 13.5 point 4 is beyond the run-out point of the pump and this would be unacceptable. It would require either consideration of different pump units or the inclusion of throttling to raise the system losses when one pump is operating. Although this illustration is for two fixed-speed pumps, the same concerns apply to any multiple pump station whether with fixed- or variable-speed drives. The full range of potential operating conditions must be checked against the pump characteristics.

A pump in a single pump installation operates at all times near to its duty point so it is logical to choose a pump with its best efficiency point close to that duty. In a multi-pump station there is a much wider range of operating conditions for any one pump. If, for example, the more common mode of operation is actually with only one pump running then it may be sensible to choose a pump with its best efficiency at such a duty. In the two-pump case illustrated in Figure 13.5 this might be a duty between points 3 and 4.

13.9 LONGITUDINAL PROFILE

The starting point for any pipeline design is the longitudinal profile of the system. It is essential that this is drawn out—more commonly from the spreadsheet or hydraulic analysis program but if necessary by hand—with the hydraulic grade lines imposed. Three example profiles of a pipeline with the same diameter, horizontal length and suction and delivery levels, L_S and L_D, are shown in Figure 13.6. Figure 13.6(a) illustrates the case where the hydraulic grade line is above the pipe profile and the static lift is the same for all flows up to the design discharge. The system curve is shown as the chain-dot line in Figure 13.7.

The profile for the system illustrated in Figure 13.6(b) however, has a summit on route, higher than the delivery point. At low flows this becomes the delivery point as 'seen' by the pumps. The static lift is higher but the effective length of the pipeline is reduced and the friction losses are smaller. Downstream of the summit, assuming that there is an air valve at that high point, the pipeline runs part full to the point at which the head is just sufficient to drive the flow through the full pipe. Note that a large-orifice air valve would almost certainly be located at the summit (Section 16.24). Above some flow, when the pipe full losses downstream of the summit equal the available head, $L_H - L_D$, the pipeline runs full over the full length and the system curve is the same as for Case (a). The full system curve for Case (b) is shown as the solid line in Figure 13.7.

The profile shown in Figure 13.6(c) has an even higher summit upstream of the delivery tank and even at maximum flow the controlling delivery point is at that summit with part full flow downstream. The system curve must be based on the higher static lift and shorter pipeline and is shown as the chain-double dotted line in Figure 13.7. The pump requires a higher lift to meet the required duty.

The profile of the pipeline may thus affect the hydraulics and choice of pump significantly and it is essential that the long section and hydraulic grade lines are drawn. Similar issues can arise on a gravity system such as illustrated in Figure 13.8.

In a raw-water system the conditions illustrated in Figure 13.6(c) might be allowed but in a potable water system, part-full flow and open air valves are normally not acceptable as the latter could

(a) Hydraulic grade lines (HGL) above pipe profile

(b) Summit higher than delivery reservoir

(c) Summit height controlling delivery head

FIGURE 13.6

Effect of high point on pumped system.

form entry points for contamination. Thus, even Figure 13.6(b) would be an unacceptable design as at low flows or when the pumps are turned off contamination could enter the system. If there is a risk of part-full flow in a potable-water pipeline then means of maintaining pressurized flow and preventing air valve opening must be considered:

- re-routing the line to avoid the high point;
- including a pressure-sustaining valve at the discharge end of the line;
- providing a break tank at the summit with a control valve at the discharge operating in response to the water level in this tank.

FIGURE 13.7

System curves for pipeline illustrated in Figure 13.6.

FIGURE 13.8

Effect of high point on gravity system.

13.10 AIR IN PIPES

Air can be introduced into a pipeline during operational activities and as air dissolved in the water is released. Sources include air left after pipeline filling, poorly designed pipe and pump inlets with insufficient submergence that allow large amounts of air to be sucked in via vortices, air valves installed near the hydraulic profile that allow air to enter the pipe under certain conditions or locations where dissolved air comes out of solution, typically as internal pressure reduces along a pumped system.

The amount of air absorbed in water varies with water temperature and is directly proportional to pressure at a given temperature (Henry's Law). Water saturated with air at a given pressure and temperature releases some in order to reach equilibrium if the pressure reduces or the temperature rises so forming pockets. Table 13.3 shows the relationship between temperature and maximum air content in water at Standard Pressure. However, the amount of dissolved air can not normally be greater than the amount introduced when the water was last exposed to air and may be less if the water has

Table 13.3 Temperature effect on maximum air content in water (at standard pressure)

Temperature °C	0	5	10	15	20	25	30	40	50
Maximum content (volume) %	2.78	2.53	2.53	2.12	1.95	1.80	1.66	1.43	1.25

previously been subjected to higher temperature or lower pressure. In normal operating conditions the pressure in a pipeline is above atmospheric. Therefore, as a general rule, provided temperature remains constant, air does not come out of solution unless pressure falls below atmospheric. There are three exceptions: where air is deliberately introduced at high pressure in a treatment process, for example dissolved air flotation (Section 7.18); where surge vessels with air-water contact are used for surge suppression (Section 13.11) and where air trapped at joints or in the pipe lining gets absorbed as the pressure rises in test or operation. The first case is only relevant to pipework in a treatment works; the third may occur in any pipeline but should not represent large volumes of air. Likewise the effect of temperature rise is usually small and would only lead to air coming out of solution if pressure is near atmospheric.

A well designed system should exclude all air on filling, prevent entry of air in operation and provide air release points. It is usual to maintain pressures at least five metres above atmospheric at high points to ensure that air valves do not leak and, in potable water systems, under all operational conditions including surge and transient events, to eliminate the risk of contaminated water being drawn in. However, air must be allowed to enter a pipeline in a controlled way and at predetermined locations to enable the pipe to be emptied and to limit sub-atmospheric pressures and prevent pipe buckling in the event of a burst.

On pipeline filling air is exhausted at large orifice air valves (Section 16.24) where the pipe is not yet full. Figure 13.9 illustrates where air can collect at high points and at changes of gradient in a pipeline in operation. Trapped air may get carried forward and accumulate at high points or may get absorbed into solution as the pressure rises. Air trapped at high points in a pipeline increases hydraulic losses with the accumulative effect of a series of high points along a pipeline profile seriously restricting flow. An air lock in domestic plumbing is an extreme example where water flow capacity is reduced to zero. Air expelled at the pipe discharge may also cause problems.

Air bubbles tend to rise to the soffit of the pipe where they may accumulate into larger pockets. Evidence from tests carried out at HR Wallingford (Escarameia, 2005a) confirms that air is transported along a pipeline mostly in the form of air pockets moving along the soffit of the pipe rather than as individual small bubbles. The air is moved forward in the direction of flow in an upward sloping pipeline but can also move forward along level pipes and along downward sloping pipes if the fluid velocity is high enough. However if air is transported forwards in a pipe at shallow downward gradient it may not be transported forwards if the pipe gradient increases. Thus air can accumulate both at high points in a system and at 'knees' in downward sloping pipes. Air may move backwards against the flow in downward sloping pipelines operating at low velocity or at steep gradient, but there is a range of flow velocities below the critical velocity (V_C) where air bubbles and pockets may 'hover' with no clear movement in either direction. Even if the air does move in one or other direction it may be very slow, for example, when filling a 2 km length of main, although the flow from a downstream hydrant may soon be steady, it can be interrupted by occasional bursts of air for perhaps half an hour or more.

Arrows show direction of possible air accumulation

Flow of water

Large
AV

AV

AV

AV

"Knee"-lessening of HG and
steeper downgrade in direction of
flow; air can accumulate

Bridge

Upgrade reducing in
direction of flow; air can
accumulate

Summit-large air
valve required for
filling main

FIGURE 13.9

Locations of air accumulation on pipeline.

The HR Wallingford design manual for air problems in pipelines (Escarameia, 2005b) indicates that the critical velocity, V_C, needed to move air pockets forward in a downward sloping pipe is given by:

$$V_C / (gD)^{0.5} = a + 0.56(\sin \theta)^{0.5} \qquad (13.1)$$

where a is a coefficient depending on the air pocket size represented by the parameter n = volume of air pocket/$(\pi D^3/4)$; a ranging from 0.45 for $n < 0.06$ to 0.61 for n between 0.30 and 2.0; θ is the slope of the pipe from the horizontal. Tests carried out to define the relationship in equation (13.1) were done in a 200 mm diameter pipe but guidance in the HR Wallingford manual suggests that the relationship is valid for pipe diameters up to about 1000 mm and slopes between 0 and 22.5 degrees. CFD modelling (Little, 2008) confirms this and suggests that the relationship generally remains valid for large diameter pipes although it is probable that a small bead of air remains in the soffit of larger pipes, particularly at undulations in pipe profile, even though pockets of air are moved forward. The tests at Wallingford show that in pipes laid horizontal or at near horizontal downward slopes the air pockets are long and thin and do not occupy a large proportion of the pipe, typically less than about 5%. This can be significant for behaviour of pipe bursts which may be more explosive than expected; may affect surge pressures under certain conditions and needs to be taken into account when designing submarine pipelines.

The HR Wallingford tests also showed that the *'hovering velocity'* was between about 0.85 and $0.95V_C$ with a recommended mean value of 0.9. Designers should try and avoid designing sections of pipeline where the mean velocity in normal operation is between say 0.85 and $1.0V_C$.

13.11 TRANSIENT PRESSURES: WATER HAMMER AND SURGE

If a valve in a pipeline is suddenly closed the water immediately upstream is brought to an abrupt stop and compressed by the momentum of the upstream water column. This results in a sudden large increase in pressure which propagates as a positive pressure wave at the speed of sound in water; noise made on reflection of this wave is known as water hammer. Similarly, downstream of the valve or downstream of a pump which has suddenly stopped, a rapid drop in pressure occurs as the momentum of the water column downstream moves it away; this is transmitted down the line as a negative wave, that is one in which the pressure drops.

Secondary pressure waves are generated as the initial wave passes any fitting or change to the pipeline, such as an enlargement or tee, but when a pressure wave reaches the closed end of a pipeline (such as a shut control valve at entry into a reservoir) it is reflected as a wave of the same type, i.e. a positive wave is reflected as a positive wave, a negative wave as a negative one. Conversely, a pressure wave reaching an open end to a pipeline discharging to a reservoir or tank is reflected as a wave of the opposite type. Even in a relatively simple system, therefore, the pattern of secondary and reflected waves can become complicated. Wave amplitude is damped by friction and pipe wall elastic losses so it is often the passage of the initial pressure wave which gives rise to the most critical pressures. However, this is not always the case as secondary and reflected waves can interact positively and may cause air or check valves to shut suddenly, so generating further high pressures. Similarly, if the pressure falls low enough, vacuum cavities may form, which may also cause high shock pressures when they subsequently collapse on a rising pressure. The analysis of transient conditions can thus be very complicated and the results depend not only on the elements of the system, but also on the profile of the pipelines. Although methods have been developed in the past for hand or graphical computations, computer programs are now used almost universally to analyse transient conditions.

The main concern with transient pressures is that they should not be high enough to cause bursting of the pipes or fittings. To put the potential for damage into perspective, it is worth noting that the rise in pressure head, on a sudden change of velocity ΔV in a pipeline, is given by $a.\Delta V/g$, where a is the velocity of wave propagation and g is the acceleration due to gravity. In a ductile iron pipeline the wave speed may be as much as 1200 m/s although it is normally between about 900–600 m/s as a result of small quantities of air present in the water. With this high value for the wave speed, the slamming of a valve in a pipeline operating at a velocity of 1.5 m/s could lead to a pressure rise of 180 m. Even with a wave speed of 600 m/s the pressure rise would be 90 m. Such surge pressures undoubtebly occur, but are often not recorded to the full extent by the ordinary Bourdon pressure gauge which is too 'sluggish' in operation to record the peak transient pressure.

Codes of practice for most pipe materials allow some transient overstressing above the allowable operating pressure (defined as the internal pressure, exclusive of surge, that a component can safely withstand in permanent service). However, other elements of the system including valves and jointing systems and the resistance of thrust blocks should be considered, particularly if an existing system is being uprated.

In potable-water systems it is normal practice to avoid any negative pressures and consequent risk of contamination being drawn in through open-air valves (particularly if located in chambers below ground in areas of high water table) or through joints designed primarily to prevent leakage from high internal pressure. In addition, large diameter, thin-walled, pipes may collapse if negative pressures fall enough to induce buckling (Section 15.5) and certain plastic materials, particularly PVC-U, may suffer from fatigue failure if there are repeated excessive transient pressure fluctuations above a certain magnitude over the life of the system.

In the majority of systems the most likely causes of transient pressures are valve closure and pump stoppage. Valve closure can be controlled and it is always advisable to ensure that valves cannot be slammed shut and have a closure time long enough to limit the pressure rise to acceptable values. The minimum time of closure should, at least, be greater than the time it takes for the pressure wave to be reflected back to the valve from the far end of the line, $2L/a$, where L is the length of the pipeline and a is the wave speed as defined above. A longer closure time, or closure that allows the last 10–20% of the movement to be much slower than over the first 80%, may be required (Fig. 16.1). However, the most critical case is usually a power failure causing simultaneous stopping of all the pumps. Most modern low-inertia pumps stop producing forward flow of water in a few seconds when turned off.

Once the potential problems are identified, it is necessary to consider their alleviation. It is rarely economic to increase the strength of a pipeline solely to cope with surge pressures so it is usually necessary to provide some other form of protection for the system. There are a number of protective measures that can be adopted as listed below (Thorley, 2004). An indication of suitable locations for installation of some of the devices mentioned is shown in Figure 13.10.

1. *Slower valve closure* by various mechanical means.
2. *Increased pump inertia.* Fly-wheels fitted to the pumps reduce the rate of deceleration of the pump and the corresponding rate of change of flow.
3. *Air vessels* (also referred to as 'surge vessels') which comprise pressure vessels connected directly to the pipeline, part of their volume being occupied by compressed air. They are commonly used to feed water into the pipeline when the pumps stop but they also provide a cushion to absorb high pressures on the returning wave and on pump start up. Air is gradually absorbed into the water and compressor facilities are required to provide occasional topping up of air in the vessel. For this reason they are generally installed only at pumping stations. The absorbed air can find its way into the pipeline and come out of solution as pressure falls (Section 13.10).
4. *Accumulators* are similar to air vessels except that the air is separated from the water in the vessel by a flexible rubber membrane thus greatly reducing the loss of air by absorption. This eliminates the need for compressor facilities and allows the use of a gas such as nitrogen in place of air, topped up periodically from a portable cylinder.
5. *Surge shafts* can be constructed, if the topography permits, but they must extend above the hydraulic grade line.
6. *Feed tanks* operate by feeding water into the line to relieve low pressures. They can be located at high points below the hydraulic grade line as they are isolated by non-return valves which only allow flow into the pipeline.
7. *Air valves* of the large orifice type may be used to prevent low pressures in the line by opening to admit air when the pressure falls below atmospheric. Their use for this purpose is not generally permitted on potable water schemes because of the risk of contamination as mentioned above. There are potential drawbacks including the generation of high shock pressure on slamming, unless special non-slam valves are used.
8. *Pressure relief valves* can be set to open at a given pressure or operate in response to an initiating event thus limiting the maximum pressures at their location.

FIGURE 13.10

Pipeline profile with locations for installation of surge protection devices.

9. *By-pass pipework* can be fitted around the pumps to allow water to be drawn from the sump provided the pressure on the delivery side falls low enough. However, this may be insufficient to prevent other low pressure problems occurring down the line on a simple system, but they may be effective at booster stations where the pressure on the suction side of the pumps rises significantly on pump stoppage.

10. *Non-return check valves* can be used along a pipeline to reduce the effect of the returning positive pressure or water column, but they may give rise to adverse effects themselves so must be analysed carefully and used with care.

With the emphasis on producing lighter pumps and motors, increasing pump inertia is out of fashion but, in the right circumstances, it is the most reliable and effective form of protection. More commonly on pumping systems a surge vessel or accumulator is used. Feed tanks may be necessary where the pipeline passes over a high point near the elevation of the discharge point, but must be designed to prevent the possibility of contamination entering the chamber and need a small sweetening flow (almost certainly having to be discharged to waste) to prevent stagnation of the water. Air valves, particularly of the non-slam type, can be helpful but should not normally be considered as the primary form of protection; regular maintenance is necessary but can often be neglected, particularly in remote locations.

Pump delivery non-return valves need to be suitable for the system and its transient response, especially if a surge vessel is also provided because the flow in the connecting pipe to the air vessel may reverse very quickly. Ideally, the non-return valve should shut at the moment of flow reversal but if it reacts more slowly the reversed flow may slam the valve shut with the generation of a high shock pressure (Section 16.18). The dynamic response of the non-return valves should thus be matched to the transient characteristics of the pipeline system.

Surge protection system design must also take account of long term performance; operations staff may be aware of the need to maintain surge equipment when the station is new but after 20 years, and after several changes of staff, the importance of components such as air valves may be forgotten. The approach should be for robust solutions in most cases but particularly for systems in remote locations.

13.12 **CAVITATION**

Cavitation occurs when the absolute pressure falls to the vapour pressure of the fluid. In water systems at normal temperatures this vapour pressure is close to absolute zero pressure (0.2m absolute at 15 °C but rising exponentially to atmospheric pressure at 100 °C). With cavitiation, in effect, the water tears apart forming cavities filled with water vapour. Collapse of vapour pockets against a surface can quickly erode material and may destroy critical components but collapse well clear of surfaces should cause no damage. The mechanism of surface damage is thought to be either local material (metal) fatigue under repeated intense pressure shocks or intense very high velocity jets towards the surface as a result of the surface inhibiting pocket collapse on one side (Borden, 1998).

Cavitation may be bulk or local. It is a common risk where the velocity is high and pressure low but can be manifest at higher pressures if the velocity is high enough. Bulk cavitation occurs across the full flow cross section wherever pressure falls to below the vapour pressure of water for some reason. Local cavitation occurs where a jet of water is flowing fast past an abrupt or smooth irregularity and then separates from the boundary; pressure falls in the space left by the jet and cavitation pockets may be generated.

Cavitation may be a problem in a number of situations including:

- during a transient event as discussed Section 13.11;
- in a pump due to low pressure on the suction side;
- in a valve used for throttling or energy dissipation;
- at dam spillway or outlet works due to high velocities and low pressures.

In the case of a transient or surge event the low pressure occurs as a result of the passage of negative pressure waves through the system. If the low pressure occurs at a defined 'knee' or high point in the system a vacuum cavity may form, separating the upstream and downstream water columns. After the separate columns come to rest they are accelerated back towards each other under high differential head and on slamming into each other may generate very high shock pressures. This must be avoided and most analysts and designers of surge protection recommend that minimum pressures in any system are limited to no lower than about 0.5 bar absolute (i.e. −0.5 barg) even if the low pressure occurs for only a very short period and the risk of vacuum cavity formation is small. This applies even if the vacuum pressure is more widespread and there is no defined point for a large cavity to form preferentially. In a potable water system no sub-atmospheric pressures at all should be allowed (Section 13.10).

In a pump cavitation can occur if the velocities in the rotating impeller are high enough such that the velocity head is greater than the absolute ambient pressure upstream. By Bernouille's theorem (Section 12.1) if this occurs the local pressure drops to zero and as a result cavitation bubbles are formed. Air may also come out of solution at this point (Section 17.4). The subsequent collapse of the cavitation bubbles as the pressure increases may cause erosion damage to the impeller and with a stream of cavitation bubbles generated in the low pressure region the blades of the impeller may be worn away rapidly.

The formation of cavitation occurs when the pressure on the suction side of a pump is insufficient. The absolute pressure on the suction side of a pump is referred to as the Net Positive Suction Head (NPSH) (Section 17.4) and every pump has a minimum requirement related to the flow it is producing. This is data supplied by the pump manufacturer. It is important that this requirement is met over the full range of pump operation. For most small typical installations with centrifugal pumps the NPSH requirement is likely to be of the order of 6 to 8 m. This is the absolute pressure and represents a pressure of 4 to 2 m below atmospheric. Thus provided the pump casing is below minimum water level the pressure should be adequate. If not then it may be necessary to lower the pump relative to the sump or suction level. In larger systems with high head pumps it is more common to provide second low-lift pumps in series to boost the pressure on the suction side of the pumps. It is essential when choosing pumps to consider the NPSH requirements and to ensure that good design of the suction system enables those requirements to be met.

In the same way that high velocities in pumps can lead to cavitation, the generation of high velocities in valves may also lead to low pressures and cavitation (Section 16.20).

13.13 EXAMPLE OF SIZING A PUMPING MAIN

Table 13.4 illustrates a calculation to determine the minimum overall cost of a pumping main in terms of capital cost of the pipe and pumping plant and annual operating costs including power consumption. The system is to convey 20 Ml/day (average) and 25 Ml/day (maximum) against a static lift of 50 m through 30 km of ductile iron main (Fig. 13.6(a)). The data used in the calculation are:

- capital costs of pipeline in rural conditions = £0.45 per mm diameter per m laid;
- Colebrook-White pipe friction (k_s) at 15 °C = 0.5 mm

- capital cost of pumping plant = £2,250 per kW installed;
- combined motor and pump efficiency = 73%
- total installed power = 133% (33% standby)
- unit cost of power = £0.10 per kWh.
- capital repayment charges = 6% per annum.

The calculation compares the total annual cost for three pipe diameters using the true internal pipe diameter for ductile iron with internal lining.

Table 13.4 Overall costs of pumping main options

		Diameter of main			Notes
		500	600	700	
	Nominal internal diameter (mm)	500	600	700	
	Internal diameter with lining (mm)	504	605	704	
a	Static lift (m)	50	50	50	Hs (Figure 13.4)
b	Friction loss @ 20 Ml/d (m)	82	32	15	$\Delta H_1 + \Delta H_2$
c	Friction loss @ 25 Ml/d (m)	128	50	23	
d	Maximum water power (kW)	506	283	206	Note 1, $\Delta H = a + c$
e	Installed motor power (kW)	921	515	376	Allow 33% standby
	Capital costs	£ × 10³	£ × 10³	£ × 10³	
f	Pipeline	6,750	8,100	9,450	30 km × £/mm dia
g	Pumps	2,072	1,160	846	e × £/installed kW
h	Total capital costs	8,822	9,260	10,296	
	Method 1 – Annual costs	£ × 10³	£ × 10³	£ × 10³	
	Cost of capital	529	556	618	
	Energy for 20 Ml/d average	361	224	176	
	Total annual costs	890	779	794	
	Method 2 – NPV calculation	£ × 10³	£ × 10³	£ × 10³	Chapter 2
	Initial capital	8,822	9,260	10,296	
	Pump renewal at 20 years	445	249	181	Note 2
	Energy costs over 25 years	3,853	2,386	1881	Note 3
	Total present value	13,120	11,894	12,358	

Notes: [1] Power (kW) = 0.1135 × Q × ΔH; Q in Ml/d and ΔH in metres.
[2] $PV = Cost \times 1/(1 + r)^{20}$ where $r = 0.08$ (Section 2.16).
[3] $PV = annual\ power\ cost \times [(1 + r)^{25} - 1]/r(1 + r)^{25}$.

Method 1 shows that the 600 mm main is the least cost option, although the difference is not large. Alternatively options can be compared using Net Present Value (NPV) (Chapter 2) illustrated by Method 2. In this example, the two methods conclude the same optimum pipe size.

In practice scheme options would be analysed in more detail and include alternative supply, demand and physical layout scenarios. The factors include:

- *change in annual energy requirements*: growth of flow over time resulting in increasing pumping costs until the pipe and pumps are operating at their capacities; pipes are seldom designed to operate at their maximum capacity from commissioning;
- *alternative demand scenarios*: phasing construction and capital expenditure, managing longer term demand forecasting risk;
- *change in Δh*: probable increase of pipe friction characteristics with time;
- *more accurate cost information*: pipe pressure/class related costs, pumps, valves, electrical and surge protection equipment, structures, etc. including their annual maintenance costs;
- *pumping equipment and operating regimes*: variable, fixed speed, manual control or full automation;
- *energy costs*: the effect of future increases in the price of energy;
- *financing costs*: the scheme owner's financing and capital repayment terms.

If the calculation is set up using a spreadsheet, it can be extended to evaluate the different parameters and to determine the optimum year for possible phasing. However the preferred solution may only emerge after considering also environmental and social costs or 'carbon footprint' differences between the options and after consultation with the client and specialists to ensure that there are no hidden engineering or operational problems, such as transient pressures or risks to water quality. The proposed pipeline route should be walked to see that it is practicable and some preliminary layout drawings may be necessary to ensure the costing is realistic.

13.14 DESIGN OF A GRAVITY MAIN

A gravity main can be designed using the principles described in the previous section. However as discussed in Section 13.9 and illustrated in Figure 13.8 the longitudinal profile of a gravity pipe can be more critical because:

- *source supply and discharge delivery heads*: pipe diameter options are determined by the available head including exit and entry (minor) losses (hydraulic gradient);
- *maximum available head*: the pipe must be designed to withstand the maximum available head which occurs under no flow conditions; the critical point is generally the lowest point in the pipeline;
- *hydraulic gradient and the pipe profile*: there is a set of hydraulic gradient lines for a range of flows determined by the downstream control conditions and pipe capacity; if the hydraulic gradient intersects the vertical profile (Fig. 13.8) there is a risk of negative pressures and air introduction at the high points if an air relief valve is fitted;
- *maximum working pressure*: pressures need to be managed along the length of the pipe to ensure that maximum working pressures are not exceeded; except for major transmission pipelines, particularly of steel, use of pipe with different pressure ratings to match working pressures along the pipeline is undesirable since it complicates maintenance and spares

stocking; it may be preferable to introduce pressure reduction, break pressure tanks or valves in order to limit operating pressures;

- *excess head to be dissipated*: where the static head is too great for the system operating conditions or where the headloss required can not be achieved by losses through the pipe or at discharge control valves, head may have to be dissipated; additionally a break on the pipeline could result in a high velocity discharge that would represent a significant risk to public safety and property; controlling such events by use of valves may be expensive because of the high pressures and back-up control valves may be needed for operational safety;

- *locating the downstream service reservoir*: where the downstream discharge is into a service reservoir, the top and bottom water levels are determined by the pipeline hydraulic gradient; ideally the top water level should be set as high as possible because once head is lost, it is expensive to recreate and it is always prudent to provide spare head on the distribution system to allow for unexpected growth in demand, pressure reduction being introduced for short term operational efficiency until demand has built up;

- *break pressure tanks*: where pressures can not be managed without using a break pressure tank, it should preferably be located so that the maximum static heads on the upstream and downstream of the tank are equal so that pipes and valves of the same class strength can be used in both sections of main; break-pressure tanks should not be provided with a by-pass, to ensure that the high pressure can not be transferred onto the downstream length of pipeline; instead the tank should have duplicate compartments to permit maintenance. An example of a break-pressure tank is shown in Figure 13.11.

FIGURE 13.11

Break pressure tank.

The hydraulic analysis and cost estimates may need to be repeated for a range of sizes of pipe and alternative schemes. Cost estimates tend to favour the smallest practicable pipe diameter. However high flow velocity may reduce opportunities for increasing capacity in the future, in which case a larger diameter main may be preferred. Other options can be considered, such as relocating the treatment works closer to the service reservoir. However, the consequences for the performance of the raw water main and the treatment works would also need to be taken into account in the comparative analysis. Raw water pipes should be sized for the additional usage at the treatment works for process and cleaning water and to allow for higher pipe friction coefficients over the life of the pipe due to the quality of the raw water being transported.

13.15 PIPELINE DESIGN TECHNIQUES

Manual calculations can be used to analyse flows in simple pipe layouts and simple small networks. Hydraulic analysis software is appropriate for analysing more complex pipe arrangements and interconnected networks, for resolving operational management issues such as blending different sources in order to maintain water quality, for optimising source and storage capacity, for testing security of supply, critical mains failure and fire demand scenarios, or where the design is integrated with base mapping (GIS) or computer aided drafting software (CAD).

Manual and Spreadsheet Calculations of Network Flows

Manual analyses can be a reliable basis for planning distribution improvements and extensions. Where they take only a day or two to complete, manual and spreadsheet calculations are cheap and give the engineer an invaluable understanding of the way a system operates. The analysis can show which principal mains are overloaded, with high flow velocities and large friction losses, or have spare capacity; why areas of low pressure occur; and where new mains are required to meet additional demands or improve pressures.

Manual and spreadsheet calculations are particularly suited for analysing dendritic layouts. They can also be used for looped networks provided the network can be reduced to a 'skeleton layout' of the more significant and larger pipes. The total average daily demand is divided out over areas and allocated to 'nodes'. The simplest way of analysing the system is to divide the network into ring mains and 'tree branches'. Flows in the ring mains are analysed by assuming a division of flows at the input point and calculating the headloss along each branch of a loop. The assumed flow split is adjusted and headlosses are recalculated until a 'point of balance' is found in the ring main where friction losses in either leg are equal. Equilibrium can generally be reached within three or four iterations. This process can also be used to analyse multiple looped networks but may require more iterations to achieve a balanced solution. The analysis can be repeated using a peaking factor applied to the average daily nodal demands.

It is now common operational practice to monitor distribution system pressures either continuously or periodically both for operational management and for level of service compliance reporting. The field pressure records for average and peak hourly conditions can be used to draw pressure contour maps along a pipeline route or across a network. Provided inline valves are open, the contours can be used to estimate the headloss and hence friction coefficients in sections of pipe. The pressures can be used to check and, if necessary, adjust the friction coefficients used in the analysis.

Where pressures records are not available, it is necessary to measure pressures and flows along the length of the pipe being analysed under a range of flow conditions.

A distribution system must have spare capacity to meet future increases of demand. However, the magnitude and locations of future demands are often not accurately known. Hence planning the development of a system is more often the choice of a strategy that best meets a range of viable future options than using precise calculation to satisfy some fixed theoretical future condition. Small mains of 200, 150 and 100 mm diameter are mostly ignored in a manual analysis, because they are normally designed according to local demand and fire flow conditions.

Analysing Existing Systems using Modelling Software

Mathematical models can be used to analyse the hydraulic performance of existing trunk mains and distribution pipework, to design new networks and to assess system operational performance under a variety of supply and operating conditions. A detailed hydraulic analysis can define both the physical pipe performance and design parameters, control regimes for pumps, control valves and storage, establish and determine the behaviour of reservoirs, their inlets and outlets and the effects of a burst main or major fire demand. However, the modeller should always remember that the accuracy of a model is no better than the quality and availability of the data from which it has been developed. Confidence in the results requires a corresponding confidence in the input data, especially for estimated nodal demands.

Most modelling software use similar data formats for physical system attributes and base demand information. Variations are usually associated with how network apparatus (pumps, valves, reservoirs) are represented mathematically. Each package has its strengths and weaknesses and the choice of software depends on client or user preference, the intended application, the availability and quality of source data, the type of output required, and whether the software is a component of the utility's integrated management information system. In practice, experienced modellers now regularly convert a model from one format to another. Software and hardware improvements have enabled larger models to be developed and to extend the models to analyse water quality parameters. It is now current practice to develop 'all mains models' that include all the pipework between sources and the service connections. Some software packages do impose physical size restrictions, but the overriding constraint is the ability of the analyst to assimilate and interpret the results from a large complex model.

With increasing need to achieve operational efficiency, maintain service levels and produce the most economic rehabilitation and reinforcement proposals, models need to be constructed and validated to more exacting criteria. A thorough systematic methodology must be used to create both hydraulic and water quality models to ensure that they are to a consistent standard and are credible and so that they can be updated in the future. Where model construction procedures have been carefully developed from the outset, future updating becomes a relatively straightforward and mechanical process.

Model data are derived from water utility corporate databases. It is increasingly common to find the hydraulic modelling software integrated with a utility's GIS system. This can simplify model updating provided that GIS errors identified and resolved through modelling are corrected in the GIS before the next model update. Data abstracted from telemetry systems, such as from SCADA and levels of service permanent monitoring instruments, and bespoke field tests designed to derive specific localized information are used for model calibration and operational control of the network. Telemetry data is now being used for 'real time' modelling using current flows and pressures to analyse an event as it is occurring. Real time modelling is a useful tool for improving the quality of short term operational forward planning, for example: resource optimization, pump scheduling, and when managing an emergency. Database links can also be used to abstract demand data from customer information systems (CIS) for updating model demand.

The physical characteristics of the system are represented in the model by nodes and pipes (or 'elements'). The nodes, joined together by pipes, represent: pipe junctions, changes in pipe diameter and the locations of system attributes such as valves and of large demands. The node and pipe data sets contain geographic co-ordinates, ground levels, basic demand information, internal diameter and friction coefficients, pump curves, service reservoir geometry and valve performance characteristics.

Water demand is allocated to the node nearest to its draw off point. Demands are distributed by the area the nodes serve and by categories of demand including metered and unmetered domestic and non-domestic, industrial, commercial and institutional, leakage; demand data also includes patterns of usage. Where all consumers are metered, the demand is the measured consumption by user category. Demand is typically related to the postal address or postcode using a geographic reference system within the billing system and GIS. Where customers are not metered, demands are assessed based on unit consumption estimates applied to property counts or population estimates served by the node.

Each demand category exhibits its own diurnal demand profile; for example, industrial consumers may operate shift systems based on 24, 16 or 8 hour working days. Institution, commercial and office consumers have different usage patterns and some consumers, such as food processing industries, schools and colleges and sports facilities can display both diurnal and seasonal variations. The incidence of on-site consumer storage may also have a significant impact on the profile. The number of modelled demand types depends on the capacity of the software and the accuracy of the base data used to develop each demand type.

The analysis applies operational conditions to the network data such as diurnal demand patterns, times at which pumps start and stop or when valves are opened and closed. The analysis may be for a single 'snapshot' in time or a sequence of time steps, know as extended period (EPS) or quasi-dynamic simulation; each step representing a unique set of demand and operational conditions. An EPS analysis uses the initial set of demand profiles, reservoir levels and network operational conditions to calculate demands, pressures and flows in the network over the first time period to determine the operational status of automated pumps and control valves and the net reservoir inflows/outflows and thereby, using the reservoir geometry, changes in reservoir levels. The new reservoir levels together with the diurnal demands and operational changes become the starting values for the second time step. The analysis is repeated for each subsequent time step. The results can be displayed both graphically and in tabular form either for a single time step or a sequence of steps to illustrate the changing performance of the network and individual elements of the system over the period of the analysis.

The analysis is an iterative solution of a set of algorithms that simulate the hydraulic behaviour of the flow of water through the piped network, solving the equations to specified tolerances by successive approximations subject to the following rules:

- the algebraic sum of the flows entering and leaving a node must be zero;
- in any closed loop in the system the algebraic sum of the pressure losses must be zero;
- the combined inputs to the system must equal the total of the nodal demands.

For hydraulic design, simulation time steps are typically 60, 30 or 15 minute intervals over a 24 hour period. Operational management and predictive planning analyses may can be used to simulate a 7 day period, seasonal variations or longer-term planning and resource scenarios, for example for a 25 year planning study.

Complex design and operational simulations should only be undertaken where the model has been calibrated either in steady state or preferably in extended period mode. Model calibration involves collecting and analysing contemporary flow and pressure measurements from the system being modelled over a period of time, typically 7 days, and comparing the measurements with the calculated flows and pressures derived from the model which has been set up to represent the field supply and demand conditions. As is true for the physical data used to develop the model, the quality of the data obtained from the field test has an impact on the credibility of the calibrated model. Field test activities include:

- measuring flows and pressure at temporary monitoring points
- acquiring telemetry and compliance monitoring flow and pressure data, including preliminary data review to identify malfunctioning instruments;
- establishing pump head/discharge and efficiency characteristics;
- checking valve status to confirm that the system is operating as stated by the operations staff and as set up in the model;
- carrying out reservoir drop tests to quantify reservoir leakage;
- surveying all pressure monitoring points to ordnance datum;
- measuring internal diameter and velocity profiles at insertion probe flow locations;
- measuring metered consumer demands and establishing diurnal demand patterns;
- estimating pipe friction characteristics by internal visual inspection (if feasible);
- resolving anomalies identified during the field test and calibration processes.

Flows should be monitored at all significant meters including: source meters; internal system flow meters used to determine area or sub-area flows; inlet and outlet flow meters at storage facilities; and large usage non-domestic meters, typically top ten consumers by usage or those that represent 10 to 20% of the model demand. Pressures are generally measured at fire hydrants, the coverage being equivalent to between 20 and 35% of the demand nodes in the model. Pressures are also measured at control valves and pump suction and discharges. The quality of calibration improves and the time required to complete the process reduces with increased numbers of reference points. However a field test can be time consuming to set up and expensive to run, especially in terms of the temporary equipment to measure flows and pressure. There is always a trade off between maximizing data collection and minimizing equipment numbers. It is also prudent to initiate extensive preliminary field investigations to investigate and minimize the number of unknowns before a field test takes place and thereby avoid an abortive test. The preliminary investigations can identify equipment failures and lack of flow meters and where equipment can and can not be installed, enabling work needed to install equipment or repair network facilities, all of which need to be resolved before the test commences. A preliminary water balance should also be calculated to satisfy the modeller that the system and sub-system demands and diurnal profile appear reasonable and that there are no unrecorded inflows, exports or unmeasured large consumers.

Model calibration involves setting up the model to reflect the demands for a typical day during the field test and fixing initial reservoir, pump and control valve settings. The model analysis results are compared with the measured flow and pressure data. Adjustments are made to pipe characteristics using engineering judgement until the two sets of data agree within specified tolerances. Where pipe roughness characteristics have to be adjusted beyond reasonable values for the type, age and condition of the pipe, the anomaly should be recorded, and data searches and field investigations initiated to justify the use of the anomalous value. Where pipe flows or demands appear incorrect, demand data should never be arbitrarily reallocated to 'fit the field measurements'. All adjustments must be justified or identified as an unresolved anomaly to be investigated later or taken into account when using the model for predictive work. A model report should include the modelling techniques used, model characteristics, how to use the model, assumptions made and outstanding anomalies. The model can then be considered calibrated.

13.16 WATER QUALITY MODELLING

Water quality modelling software is now commonly available as a module of the hydraulic analysis package. Although there are differences in the modelling features and ways of presented the results, they can generally analyse:

- age of water and thereby the retention time;
- the proportion of water from different sources, where there are multiple sources;
- concentrations of substances altered by mixing but not subject to growth or decay;
- decay or growth of substance concentration that is time dependent.

The calculated ages and relative proportion of source waters at model nodes are used to correlate with analysis results from water quality sampling programmes, or from a sample taken following a consumer complaint. Source contribution analyses are used to track and predict the mix of water from different sources and the effects and extent of pollution incidents. Chlorine residual models, calibrated against field chlorine measurements, can be used in conjunction with chemical and bacteriological sampling data and statistical predictive modelling techniques to improve the efficiency of disinfection dosing and improve network retention times. Complex parameters, such as disinfection by-products, can also be modelled either directly or by modelling a surrogate parameter.

Water quality modelling must be based on a validated hydraulic model that accurately replicates velocities in the pipes and thereby the time of travel of water throughout the network; essentially a model is required that includes all pipes down to those in which the water quality parameters are being analysed. A 'skeletal' model, a simplified hydraulic model that does not include some smaller and local distribution pipes, is generally not suitable because it over estimates velocities in the modelled pipes and thereby underestimates the time of travel and hence age of water at the modelled nodes. Age of water models are frequently included in model development specifications.

The key to all mathematical modelling is verification. The field instruments used to produce calibrated hydraulic models have developed to meet the need for increasingly accurate hydraulic models. However equivalent field deployable instruments for providing data for calibrating water quality models are still relatively expensive. They are therefore currently not available in sufficient numbers to achieve the coverage necessary for comprehensive calibration of all but relatively small models. There is no question that water quality modelling represents a potentially powerful tool for the future, but until field sampling equipment is sufficiently available, validation is constrained by the quality of data obtainable using existing equipment, sampling techniques and surrogate parameters.

13.17 UPDATING OF NETWORK MODELS

Calibrated models should be validated periodically to confirm that they are still 'fit for purpose' within acceptable tolerances. Modellers have conflicting views on the frequency of re-validating a model, whether to undertake a full or partial model rebuild or calibration field test. Unless there is strong evidence that the model is still performing within its specified tolerances of flows and pressures over the diurnal period, the model should be reviewed. Its performance should be validated against field measurements, using flow and pressure data at key locations derived from telemetry or a small field test. Where the model is assessed to be outside specified or acceptable tolerance, the decision is then whether to update the physical, demand and operational data in the existing model or to rebuild the model from scratch. In both cases the new model needs to be calibrated against a full field test. The extent of a review and revisions should also reflect the purpose for which the model will be used. Possible changes include:

- physical attributes: existing pipes and assets rehabilitated or replaced; new mains and extensions; valve status changes (hydraulic boundaries);
- base demand changes: changes in existing consumer usage and new consumers;
- operational changes; control rules, settings and profiles.

Small changes can be accommodated by revising the model, provided there are comprehensive records of the physical and operational changes since the model was developed. Changes in mains layouts or mains condition, for example scraping and relining a pipe to improve its internal roughness or hydraulic boundary changes should be recorded on the GIS. Demands should be reviewed using billing records and telemetry flows to assess changes in average day demands, seasonal variations, reductions in leakage levels, increased per capita consumption or changes in consumption volumes recorded in metering records. Trade, industrial and institutional consumers are generally metered and any changes can be derived from billing records. Operational control rules are documented and can be validated from telemetry records.

Most hydraulic analysis software can be integrated with the range of databases used by water utilities to manage their businesses. Although some software is designed to deal with particular construction issues, none has yet resolved the problem of repeatability or updating only where there are changes in the datasets. The question is therefore whether to rebuild completely each time a new generation of a model is required, or whether to update by exceptions; that is only revise sections of the model where there are known changes to the data. Either way, the decision must reflect the purpose for which the model is required.

13.18 SOFTWARE DEVELOPMENTS

Using information from mains record, customer billing, telemetry, levels of service, and water quality data accessed through an integrated management information system, combined with on line 'real-time' hydraulic analysis will provide future planners and operations engineers with a powerful suite of engineering analysis tools. The tools are being developed by a range of software developers, some as integrated suites of modules others as stand alone purpose made operational support tools.

The increasing use of GIS, CIS and associated system performance databases provides opportunities for building and updating hydraulic models electronically, thereby eliminating the costly manual process which has inhibited model revision in the past. Genetic algorithms add another dimension to model calibration, network optimisation and asset management, all potentially deriving cost savings. Real time modelling will also have delivered demonstrable efficiency savings within the next five years. However, whatever tools are available to the modeller, the quality of a model and credibility of the outputs are only as good as the data from which the model was developed and the technical ability and operational management understanding of the modeller. Where data standards are not maintained and critical business decision making relies on inexperienced modellers, the water utility is exposing its business to a risk of performance failure.

REFERENCES

Borden, G. and Friedmann, P. G. (1998). *Control valves*. 3rd Edn. Instrument Society of America.

Escarameia, M., Dabrowski, C., Gahan, C. and Lauchlan, C. (2005a). *Experimental and numerical studies on movement of air in water pipes*. Report SR661. HR Wallingford Ltd.

Escarameia, M. (2005b). *Air problems in pipelines. A design manual*. HR Wallingford.

Little, M. J., Powell, J. C. and Clark, P. B. (2008). *Air movement in water pipelines: some new developments*. 10th International Conference on Pressure Surges, BHR Group, May, Edinburgh.

Miller, D. S. (1990). *Internal Flow Systems*, 2nd Edn. BHr Group Ltd.

Thorley, A. R. D. (2004). *Guide to the Control and Suppression of Fluid Transients in Liquids in Closed Conduits*, 2nd Edn. Professional Engineering Publishing Ltd.

van den Boomen, M. and Vreeburg, J. H. G. (1999): *Report No. SWE 99.011*. Nieuwe Ontwerprichtlijen Yoor Distributientten. KIWA.

Distribution Practice

14.1 MANAGEMENT OF NETWORK ASSETS

The quantity and quality of water delivered to the customer is directly related to the physical condition and operational management of all the assets that comprise a system from source to the consumer's tap. In the past water utilities have tended to focus on increasing capacity by developing resources and refurbishing and replacing treatment plants. However, with increased financial and environmental pressures on utilities to improve operational and financial efficiency and reduce leakage and wastage, the focus has shifted to the better management of the underground assets.

The pipelines of a distribution system normally represent the largest capital asset that utilities possess. The cost of maintaining these assets in good condition has become a major financial consideration for utilities which are under increasing competition pressures. This is because historically investment has been focussed on meeting demand for higher treatment quality standards and network condition has been given a low priority due to the now usually accepted long asset life (60 to over 100 years) for pipelines. However a large proportion of distribution pipework is reaching the end of its serviceable life. In addition, stricter requirements for levels of service to consumers have been imposed at the tap including stricter water quality standards. These now oblige utilities to ensure that the condition of the distribution system does not cause the quality of the water to deteriorate before it reaches consumers.

Asset condition assessments are used to forecast the cost liability such assets pose as their performance or condition deteriorates, or as their current hydraulic capacity becomes insufficient to meet foreseeable rises of demand. The assessment process takes into account the type, age, condition, leakage and burst frequency of mains; their flow, pressure and water quality performance; and hence the potential length of asset life remaining and when rehabilitation or replacement is likely to be required. The output from the analysis is an *Asset Management Plan* for maintaining the underground assets in a condition sufficient for the company to meet its obligations and capital expenditure programme. The plan defines the short and medium term rehabilitation and replacement requirements of the assets in a system and the longer term asset needs by broad categories of assets.

For the distribution system alone, managing the assets is a major task due to the lengths of pipe involved and the coverage of associated assets such as control valves, pumping and storage. The analysis uses data derived from the utility's management information systems (MIS), typically starting with comprehensive mains record drawings which are increasingly held in electronic geographic mapping systems. Pipework data and system performance information may also be held electronically in one of these databases. Data management and processing activities can be simplified and conclusions more easily derived and visually represented if the mapping and databases are integrated within a single geographic information system (GIS). Further comments on the use of computer software for analysing the behaviour of a distribution system are given in Sections 13.15 to 13.18.

Management of distribution system assets necessitates study of many options. Water mains can be refurbished in a number of ways (Section 14.9 *et seq.*); but it may be more economical in the long term to replace an old main with a larger one which can meet forecast rising demand and thereby last longer. Re-routing flow is an alternative. New mains can bring benefits in reducing leakage, improving pressures and reducing bursts. Economic studies can also lead to other solutions, such as upgrading or automating pumping plant or adopting some re-zoning of supplies. The latter can promote better pressure regimes, aid leak detection activities and reduce problems of water quality deterioration through the distribution mains. All this work has to be repeated at intervals to keep the asset renewal plans and costs up-to-date.

In the past, asset management plans have been based on predicting future asset performance derived from historic condition and performance records. This approach can lead to planning asset replacement or refurbishment before the end of the life of the asset and thereby overestimating the financial commitment. Risk based asset management (RBAM), a more analytical approach, is now being used to decide the most cost effective capital investment strategy. The risk of failure of each asset is assessed for its likelihood based on the history of failure of the asset and of comparable assets, together with the financial, social and environmental consequences of its failure. This approach provides a more precise assessment of the condition each asset and its performance, and gives a statistically based estimate of its life. It also provides a plan for maintenance, replacement or refurbishment that is linked directly to the risk its failure represents. In the UK, utilities are developing suites of 'deterioration models' and tools to estimate future deterioration of network assets for ranges of condition and failure mechanisms. Generally each approach is bespoke to that utility's operational circumstances, business model, quality of data and data sets used. Intervention options are generally common to all utilities but the reasons for, and expectations of, the outcome of an intervention may be specific to a utility company, water quality zone or a length of main in a network. UK research in 2008 is reviewing the different modelling approaches, the opportunities and constraints related to the quality and availability of data with the intention of establishing industry best practice models. However, for many utilities there is still a dearth of accurate and comprehensive failure data for network assets in different physical and operational conditions. Consequently, asset management plans for pipes and ancillary equipment are likely to be a combination of both approaches until deterioration models and failure mode data become more openly available.

A large quantity of data is collected and processed in a variety of ways by a utility in the course of managing its business. The following table illustrates the types of data and some of the underground assets processes and systems needed to manage a network.

Examples of raw data held in company information systems	Systems and processes used to collate and interpolate data
Customer demand (billing information)	Asset data support
Customer complaints	Burst records/deterioration models
Customer meter locations	District Meter Area monitoring
Customer surveys	Distribution Operation Maintenance Strategies
Demand surveys	Emergency plans
Facilities failures	Geographic Information Systems
Flow and pressure measurements	Health & Safety records
Inspectors' site reports	Hydraulic, surge and water quality modelling
Internal pipe diameter	Information and models from other utilities
Mains and facilities record drawings	Job scheduling
Measured leakage and other Non Revenue Water uses and wastage	Leak management
Meter maintenance/audits	Meter maintenance/audits
Number of shut valves and operations	Number of shut valves and operations
Network maintenance reports	Network maintenance reports
Network performance by location	Network performance by location
Records of repairs and rehabilitation	Records of repairs and rehabilitation
Pipe age and material	Pipe age and material
Pipe failures by location	Pipe failures by location
Pipe sampling (e.g. cut outs)	Pipe sampling (e.g. cut outs)
Sediment size, SG, etc	Sediment size, SG, etc
Staff training records	Staff training records
Water quality samples by location	Water quality samples by location

14.2 SERVICE LEVELS

Service levels are the standards of supply which a water utility affords its customers. The standards can be targets for achievement set by a utility for itself, or set by some outside authority, such as the service targets set for the privatized water companies in England and Wales by the WSRA, also known as Ofwat. Levels of service are often set by, or agreed with, international funding agencies such as the World Bank, to define what a programme of rehabilitation or improvement of a utility should achieve. Records of how far such levels of service have been achieved can be used as performance indicators by regulators and customers to gauge the utility's annual service delivery performance. The three principal levels of service relating to distribution systems are hydraulic performance, continuity of supply and water quality.

Hydraulic performance defines the minimum pressure and flow domestic consumers should experience. The reference level of service required by the Ofwat when demand is not abnormal is a flow of 9 l/min at a pressure of 10 m head on the customer's side of the main stop tap at the property boundary (Ofwat, 2007). In practice the pressure is difficult to measure at this point and therefore a 'surrogate' pressure of typically 15 m in the main supplying the property may be used. The use of lower surrogate pressures is acceptable only where justified with supporting evidence. Where two properties are supplied through a common service pipe the pressure reference level is the same for twice the flow. For common services feeding more than two properties, the increase in minimum mains pressure is linked to the size of the common service, number of properties supplied and the concept of '*loading units*' defined in BS EN 806-3 (2005).

Continuity of supply is measured by the number, duration and circumstances relating to interruptions or deficiencies of supply. In the UK Ofwat uses a scoring system for assessing a water company's performance. The utility is required to report the number of properties affected by interruptions to supply of more than 6, 12 and 24 hours by the four categories of:

- unplanned interruptions due to bursts, etc.;
- planned and warned interruptions due to planned maintenance, new connections, etc.;
- unplanned interruptions caused by a third party, for example another utility damaging a water pipe while excavating for their own service; and
- unplanned interruption due to overrun of a planned and warned interruption.

Under certain circumstances customers affected are entitled to financial compensation, typically £10 for each event plus a further £10 for each 24-hour period the supply remains interrupted. Ofwat has also stated the maximum frequencies it considers acceptable for restrictions of supply to consumers by water companies in England and Wales but these targets are not mandatory and have been exceeded.

The requirements quoted above apply to privatized water companies in England and Wales. Similar targets can be appropriate where operation and maintenance of a public water supply system has been contracted out to a company or private firm for a term of years. In poorer countries of the world, particularly where utilities are publicly owned, similar target service levels may be the aim but achieving them is often not possible. For some a more realistic target is to achieve 4-hour supply morning and evening to connected consumers and with standpipe coverage for all other householders.

Water quality standards for water delivered to the consumer, together with sampling and reporting requirements are set by legislation in most developed countries (Section 6.53). They apply universally to the water reaching consumers' taps. Hence, though the quality of water leaving a treatment works may be acceptable, any decline of quality as it passes through the distribution system must be taken into account. All waters contain nutrients and long retention in mains can promote '*aftergrowth*', i.e. increase of bacteria and small organisms in the mains. Taste, odours and suspended matter in the water can occur. Thus, the design of the distribution system has to minimize these effects by ensuring good circulation of water, flow velocities kept within an acceptable range, short retention times and cleanliness of the interior of mains. Sometimes it is found necessary to introduce additional disinfectant (usually chlorine) to the water at service reservoirs in order to inhibit aftergrowth in mains.

14.3 DISTRIBUTION ORGANISATION

The general aspects of a water utility organisation are described in Chapter 2. However the management of a distribution system requires both central and local activities. Utilities in the more industrialized countries are increasingly making use of telemetry, remote sensing and control and manning of central or regional Operational Control Centres manned 24 hours a day to manage their networks. From the control centre remote sources, pumping stations and control valves can be monitored and controlled in order to maintain supplies to consumers under varying supply and demand conditions. For utilities with no central control facility, it is usual for local distribution managers to keep in touch with the source output personnel, so that source outputs are altered as necessary to meet demand variations. Regular reporting times are arranged; for example at 09.00 h to consider what source output is required according to the previous day's consumption and the water levels in service reservoirs and at 16.00 h to agree further adjustments in the light of the current day's demand. Further communication would also be needed to respond to an emergency, for example fire demand or a burst main.

Activities normally managed centrally include:

- meter reading, customer billing and income collection: Billing and Finance;
- negotiating agreements for extension of mains and new services, often called 'Developer Services';
- response to customer complaints: Customer Services and Call Centre often linked to an Operational Control Centre;
- engineering design and procurement: Engineering and Procurement;
- strategic planning and regulation: Asset Management, Strategic/Forward Planning, and Regulation;
- training: Human Resources;
- legal, budgetary and financial control measures: Finance and Corporate Control.

The activities managed locally, usually within a network operations team, comprise:

- maintenance of supplies: Zone Engineer, Supply Engineer;
- customer contact, investigation of customer difficulties and network problems: District Inspectors, Network Technicians, also known as 'Turnkeys';
- repair and maintenance of mains: Direct Labour Gangs or Contractors;
- monitoring of levels of service, flows and pressures: Network Technicians and Analysts;
- leak detection and repair, and reduction of waste: Leakage Technicians, Direct Labour Gangs and Contractors;
- inspection of plumbing systems and enforcement of water byelaws: Byelaws Inspectors/ Technicians.

The key operative at the local level is usually designated the 'District Inspector' or 'Network Technician'. The inspector liaises with the public and investigates customer complaints, monitors levels of service, inspects properties for waste, and keeps the area under surveillance for signs of visible leakage. The inspector can also be involved in leak and waste detection activities. Using their intimate local knowledge of consumers and operational anomalies, often derived over many years working on the same part of the system, they are pivotal to the efficient and effective management of the network, diagnosing the causes of operational problems and directing remedial measures where necessary. The inspector would also be responsible for operating ('turning') valves which are likely to impact flow routings and hence water quality or pressure at the customers tap. Prior to turning key supply valves, the inspector and zone engineer would complete a risk assessment in accordance with the utility's operational procedures and take appropriate mitigating actions to minimize the impact of the change on consumers.

Local networks tend to be managed from a district or zone depot office. Initial customer contact is usually received centrally, typically through a call centre. The contact details are logged and passed to the control centre for action. The district is advised of the location and nature of the contact and the zone engineer initiates investigation procedures and remedial actions, initially by the inspector and then, if necessary using a repair gang or contractor. Where the utility retains direct labour gangs to carry out pipe repairs and maintenance, lay short lengths of main and install service connections, valves and meters, the local depot will include a stores, plant yard and the facilities to support the gang. With the increasing use of contractors to detect and locate leaks and repair and maintain pipes and fittings, some utilities have adopted information technology to give their inspectors a greater degree of autonomy. They may now operate out of fully equipped vans and, using remote communication devices and computers, are sent details of customer contacts and network jobs directly from the control centre. Their communications equipment can also be linked to corporate databases so that they have access the latest mains records and asset data. Adoption by many utilities of the latest communication technology has enabled inspectors to better manage their areas independently and to be able to respond swiftly to operational requirements.

Water utilities in the UK are increasingly reorganising their distribution networks into hydraulically discrete 'stable' supply areas in order to control the network, maintain service levels and manage leakage. A supply area is typically divided into separate areas termed '*zones*' or '*districts*', which are further subdivided into '*district meter areas*' (DMAs) and '*pressure management areas*' (PMAs). Preferably zones are configured to be supplied from only one source; but if two sources of different quality waters have to be used, the utility attempts to blend the separate sources in a predefined ratio to ensure a consistent water quality is delivered to the consumers.

DMAs, used for monitoring consumption and evidence of leakage, are configured to achieve stable flows under normal conditions to avoid flow reversals or excessive velocities, which could cause water quality deterioration (Section 14.2). PMAs are used to reduce and manage pressures in parts of a network that would otherwise be subjected to high or excessively variable diurnal pressures. PMAs can comprise whole zones, parts of a zone containing a number of DMAs or be a subdivision of a single DMA.

Measurements of flows into and pressures within a zone or DMA provide the data to monitor and control each part of a system to ensure adequate supplies to consumers and to support leakage and waste detection activities (Section 14.12). The operation of key valves including boundary valves between adjacent hydraulic areas needs to be managed and controlled, in order to prevent unauthorized changes of valve setting that could affect the quality of the supply and the integrity of the flow monitoring. Data is derived from key monitoring points throughout the distribution system including:

- flow meters at sources of supply, on large consumer connections and for district metering;
- pressure monitors for system management, valve control and for level of service management and reporting.

The data can be transmitted by telemetry to the Operational Control Centre where it can be monitored, stored, processed and displayed graphically on schematics, as trend graphs and to trigger alarms; all to provide information to the operations staff to support their actions for maintaining supplies. This real time decision support can also be used to identify deficiencies of supply, burst mains, fire demands, to ensure that the necessary staff are called out to deal with the emergency and to provide information to staff responsible for dealing with consumer complaints.

Differences Under Intermittent Supply Conditions

In many countries, utilities are only able to provide intermittent supplies due to lack of source capacity and limited financial resources. Water rationing has then to be practised by operating distribution valves daily to apportion water in rotation to different areas, or the whole supply to an area has to be turned on and off periodically, for example twice a day, to provide water to consumers morning and evening. Under these circumstances other operational difficulties arise and maintaining water quality may be problematic. The operation of valves needs to minimize the risk of subjecting parts of the network to vacuum pressures which could cause groundwater infiltration and contamination of the network. This is a specific problem for branch mains off principal feed mains where valves are operated to shut-off or re-open supplies and where a service reservoir is allowed to drain down and empty during each supply cycle resulting in the network also emptying by draining to the lowest point. The principles for designing DMAs are applicable also for designing areas for controlling rationing.

Most consumers would use times of supply to fill storage containers for use during periods of non-supply and may leave their taps on causing the containers to overflow and waste water. If the quality of water is suspect, it is quite common for consumers with intermittent supplies to boil water for drinking; they may even be advised to do so by the water utility. Drinking water may also be purchased from vendors though, unless the utility controls the vendors, the quality of water they supply can be very variable. For systems subjected to severe supply constraints and water quality problems associated with infiltration and poor asset condition, it may be prudent to consider alternative methods of supply including tankering water to local storage connected directly to standpipes or to provide mobile storage at strategic locations in the supply area.

Distribution Network Extensions

The design and supervision of construction of network extensions and reinforcements, or the rehabilitation of mains (Sections 14.9 to 14.11) is normally managed by the central engineering department of a water utility. However, in countries where local water utilities have limited technical resources, a separate regional organisation may be responsible for all the design and construction of network extensions in its region. Where water companies come under regulatory control, as in England and Wales, a company's in-house design team may be required to be competitive against outside designers, and its direct labour section competitive against construction contractors.

14.4 FIRE-FIGHTING REQUIREMENTS

Under current UK legislation (FSA, 2003) a water utility must install fire hydrants on mains where required by the fire authority. The fire authority must pay for the installation and maintenance of the hydrant. The only exception is that the utility is not required to install a fire hydrant on a trunk main. There are no requirements on the utility to provide any given pressure or flow at a hydrant, except that when the fire authority is dealing with a fire it can call upon the utility to provide a greater supply or pressure by shutting off water from mains or pipes to any area. The utility has, however, a duty to maintain a constant supply of water in its mains at a pressure sufficient to reach the topmost storey of every building in its area—but not to a greater height "*than that to which it will flow by gravitation through its water mains from the service reservoir or tank from which that supply is taken*".

Table 14.1 summarizes the UK guidance on minimum flow requirements recommended for firefighting (WaterUK, 2007). The minimum requirement is 0.5 m³/min for two-storied housing, with up to 2.1 m³/min for larger dwellings. This compares with the specified minimum capacity of fire hydrants of 2 m³/min at a pressure of 1.7 bar at the inlet (BS 750:2006). When larger fires occur, suction hose lines are put out to additional hydrants further afield, thus spreading the water demand over more mains. In the USA the Uniform Fire Code (NFPA, 2007) stipulates minimum fire flow capacity of mains according to size of area and nature of property.

Table 14.1 UK Guidelines on flow requirements for fire fighting (Water UK, 2007)

Category	Description	'Minimum' requirements
Housing	Detached or semi-detached houses of not more than two floors	8 l/s through a single hydrant
	Multi-occupied housing with units of more than two floors	20–35 l/s through a single hydrant
Transportation	Lorry/coach parks—multi-storey car parks—service stations	25 l/s from any single hydrant on site or within 90 m
Industry	Recommended for industrial estates:	Site mains normally at least 150 mm diameter.
	• up to 1 ha • between 1 and 2 ha • between 2 and 3 ha • over 3 ha	• 20 l/s • 35 l/s • 50 l/s • 75 l/s
Commercial	Shopping, offices, recreation and tourism development	20 to 75 l/s
Institutional	Village halls	15 l/s through any single hydrant on site or within 100 m of the complex
	Primary schools and single storey health centres	20 l/s through any single hydrant on site or within 70 m of the complex
	Secondary schools, colleges, large health and community	35 l/s through any single hydrant on site or within 70 m of the complex

Table 14.2 Fire hydrant and fire flow requirements

	England & Wales	USA[a]	Continent
Hydrant characteristics	2 m³/min at 1.7 bar	2 m³/min	1.0–1.5 m³/min minimum
Initial flow in close proximity to fire	1.4 m³/min upwards	3.8 m³/min for 1 hour	Generally 1.0 m³/min
Subsequent rates of flow (as required) for a single fire	9 m³/min or more, up to 25 m³/min for major fires	Minimum flow 6 to 30 m³/min related to fire area and construction type	3.6–6.0 m³/min in high risk areas, or more
Minimum residual pressure in mains	Preferably 0.7 bar but not less than zero	Not less than 1.4 bar at the required flow rate	Not stated
Possible quantity of water used in a fire	Maximum flow times 'several hours' say 6 hours	Minimum flow duration 2 to 4 hours	Maximum not stated; average incidents last less than 2 hours (France)
Spacing of hydrants	Generally 100–150 m but 30 m in high risk areas	Generally 60–100 m, Maximum 90 m on dead ends, and 200 m for looped systems	80–100 m in urban areas; 120–140 m in residential areas; wider in rural areas

[a] Reduced flow requirements of up to 75% are permitted where sprinklers are installed in buildings
Source: Bernis 1976; BS 750:2006; NFPA1 2006

These flow requirements are *"over and above domestic consumption"* and therefore dominate the design of distribution mains. Some engineers question whether such high rates are necessary and, although not yet formalized, discussions on guidelines are taking place in the industry in the USA with the objective of limiting supply pressures; thereby also reducing non revenue water (Section 14.12 *et seq.*).

Fire appliances in common use have built-in pumps of capacities 2.3 and 4.5 m³/min. Nozzle sizes are commonly 13 and 19 mm, discharging flows of 0.16 and 0.45 m³/min, respectively. Larger nozzles, of 25 mm diameter with capacities up to 1.1 m³/min, represent the largest practicable size for hand-held branches. Table 14.2 summarizes hydrant spacing and flow fire requirements in Europe and the USA.

14.5 SERVICE PIPES

Service pipe connections from a main to a property are usually laid as shown in Figure 14.1. In the UK the length of service pipe from the main to the company stop tap (or consumer's meter if fitted) is termed the '*communication pipe*'; the balance of service pipe to the consumer's internal stop tap being termed the '*supply pipe*'. The communication pipe is maintained by the company and the supply pipe is the responsibility of the customer.

FIGURE 14.1

Typical service pipe connection.

Connections to the distribution main are usually made under pressure, the size of the tapping being 6 mm less in diameter than the service pipe diameter, except in the case of a 13 mm service pipe diameter (the minimum size) which has a tapping of the same diameter as itself. There are usually two stopcocks on the service pipeline: one at the boundary to the consumer's property, which may be operated only by the water utility, and one just inside the consumer's property for his own operation. The ferrule is normally inserted into the main by means of an 'under-pressure' tapping machine as shown in Figure 14.2. The machine bores a hole into the main and taps a screw-thread in the hole. By rotating the head of the machine, the ferrule is brought into position over the hole and screwed into place. The ferrule has a plug in it which can be screwed down, thus cutting off the supply.

Service pipe connections for houses are usually to a standard size according to the practice of the water utility, 15 or 20 mm nominal bore being the typical size for a one-family house or flat which is within 30 m of the distribution main. Although required to meet level of service standards for pressure (Section 14.2), in practice and subject to the physical characteristic and operating condition of the network, utilities typically try to maintain pressure in the mains between about 25 and 35 m. Higher operating pressures are maintained in the USA, where there are buildings greater than three storeys and where a water utility is committed to provide higher mains pressures for contractual or fire-fighting reasons.

Service Pipe Materials

Materials used for service pipes include polyethylene, PVC-U, steel, copper, lead and lead alloys.

Polyethylene pipes. Polyethylene (PE), as described in Section 15.17, is now the principal material used in the UK for service pipes and is being installed increasingly internationally, where locally manufactured or where material costs are comparable with other pipe materials. BS EN 12201:2003 specifies the requirements for the five most commonly used sizes for cold potable water services at pressures of up to 120 m and temperatures of up to 20 °C. The standard refers to blue pigmented pipes typically used for below ground systems or where the pipes are protected from sunlight by enclosure in ducts or buildings, and to black pigmented pipes generally laid above ground. Dimensions of polyethylene service pipes are given in Table 14.3.

PVC-U pipes. Unplasticized polyvinyl chloride (previously known as uPVC) piping is described in Section 15.18. It is used for service piping in temperate climates and is increasingly being installed for cold water domestic plumbing pipework. It is not an approved material for hot water plumbing

Forcing screw

Heads for ratchet
arm for boring

Drill spindle

Ferrule spindle

Spindle head rotates
to bring ferrule spindle
beneath forcing screw

Ferrule

Drill

Note:
The drill bit bores and
taps a hole in the pipe
wall. It is withdrawn,
the spindle head
rotated, and then the
ferrule screwed in.

Waterway to
service pipe

Plug screwed up to
open waterway
through ferrule

Screwed into main

FIGURE 14.2

Under-pressure tapping in a service main showing the type of ferrule inserted.

and is not wholly suitable for hot climates because the maximum working pressure reduces by about 2% per 1 °C above 20 °C. However, it does represent a low-cost pipe where locally made. The pipes should be stored under cover to protect them from the ultraviolet rays of sunlight. The pipes are corrosion-resistant, light to handle and easy to joint using either solvent cement or compression joints. Solvent cement joints are not now favoured; if a solvent joint is broken it cannot be remade. PVC-U pipes are particularly susceptible to damage due to poor workmanship during laying and backfilling. Dimensions of PVC-U service pipes to BS 1452:2000 are given in Table 14.3.

Steel pipes are no longer used for service connections in the UK. However they are still used in many countries because they are one of the cheapest forms of service pipe and can sustain high pressures. They may be supplied 'black' (i.e. untreated) or galvanized and with a range of internal and external protection systems. When laid unprotected in aggressive ground conditions, the life expectancy of steel can be as little as 2–5 years. Poor installation, particularly associated with the rigid screw-thread joints, can result in high leakage from an early stage in the life of the pipe. Many older properties have steel internal plumbing pipes. If steel pipes must be used for underground

Table 14.3 Polyethylene and PVC-U for cold potable water up to 20°C and steel pipes

Nominal Bore (mm)	Polyethylene Pipes to BS EN 12201:2003 Part 2: Tables 1 & 2				PVC-U Pipes to BS EN 1452-2:2000 Part 2: Tables B1 & B2				Steel Pipes to BS EN 10255:2004		
	Mean OD		Wall thickness		Mean OD		Wall thickness		Outside diameter		Wall Thickness (mm)
	Min. (mm)	Max. (mm)	Min. (mm)	Max. (mm)	Min. (mm)	Max. (mm)	Class PN12[a] (mm)	Class PN15[a] (mm)	Min. (mm)	Max. (mm)	
15	20.0	20.3	2.0	2.3	21.2	21.5	—	1.7	21.0	21.8	3.2
20	25.0	25.3	2.3	2.7	26.6	26.9	—	1.9	26.5	27.3	3.2
25	32.0	32.3	3.0	3.4	33.4	33.7	—	2.2	33.3	34.2	4.0
32	40.0	40.4	3.7	4.2	42.1	42.4	2.2	2.7	42.0	42.9	4.0
40	50.0	50.4	4.6	5.2	48.1	48.4	2.5	3.1	47.9	48.8	4.0
50	63.0	63.4	5.8	6.5	60.2	60.5	3.1	3.9	59.7	60.8	4.5
75	75.0	75.5	6.8	7.6	88.7	89.1	4.6	5.7			

[a]For working pressures Class PN12 (122.4 m), Class PN15 (153.0 m)

service connections, heavy grade steel tubes to BS EN 10255:2004 should be used, galvanized and, if laid in corrosive ground, wrapped with protective adhesive tape. Permissible working pressures are ample for the highest waterworks distribution pressures likely to be met in practice. Dimensions of steel service pipes are given in Table 14.3.

Copper pipes are widely used for plumbing and sometimes for service pipes where ground conditions are corrosive to iron. They are strong, durable, resistant to corrosion, easily jointed and capable of withstanding high internal pressures, but are expensive. Pipes and fittings are jointed either by compression or solder/capillary joints. However, as a consequence of the changes to the standards for acceptable concentrations of lead in drinking water in Europe and the USA, there is concern about the contribution of soldered joints to the lead concentrations in water, particularly in the first draw after water has been standing in the service pipe overnight (Section 6.33).

Lead pipes. Lead and lead alloy pipes and fittings are no longer installed but many older service pipes in the UK are of lead and still continue in service. An EC Directive requires the maximum value for lead in drinking water to be reduced to10 μg/l by year 2013 (Section 6.33). UK water utilities have been relying mainly on treatment solutions to achieve the standard. Where treatment solutions alone will not achieve the EC standard, utilities will need to implement significant lead service pipe replacement programmes. In addition if a consumer replaces his lead supply pipe with another pipe material, the utility is required to replace any lead piping used in the communication pipe. The utilities have been replacing lead communication pipes systematically in parts of their networks when carrying out mains rehabilitation and service improvement works.

In the USA US EPA has set a maximum level 15 μg/l for lead. If this is exceeded the water has to be treated to reduce its plumbosolvency and, where this treatment is not effective, lead pipe replacement must be undertaken.

14.6 DOMESTIC FLOW REQUIREMENTS AND DESIGN OF SERVICE PIPES

Flow Requirements

The maximum demand rate from a house depends on the amount of storage, if any, provided on the premises. In most modern UK plumbing systems the only storage provided is that on the hot water system and in WC flushing cisterns. In such cases all cold water supplies to taps, showers, WC ball valves and the cold water feed tank to the hot water system are fed directly from the mains. In other cases, generally older properties, only the cold water taps in the kitchen and to a bathroom washbasin are direct off the mains; the WC cisterns and showers being fed via the cold water storage tank. The maximum rate of flow required to a property therefore depends upon the type of plumbing arrangement adopted.

BS 6700:2006 recommends that systems are designed to deliver a total supply not exceeding 0.3 l/s (18 l/min) from any group of outlets and that the maximum velocity does not exceed 3.0 m/s. Simultaneous discharges are likely to cause reduced flows and could cause appliances to malfunction. If the pressure in the mains is high this effect may be less noticeable but when it is below about 30 m the impact is observable. A householder does not normally expect to get the maximum flow at all his water consuming facilities simultaneously: he knows that fully opening the kitchen tap tends to reduce the cold flow to washbasin and bath taps on the floor above.

Conversely, the recent trend to install pump assisted (power) showers has significantly increased the potential peak demand rate from individual dwellings. However in many cases the capacity of

Table 14.4 Design flow rates and headlosses through fittings

	Rate of flow (l/m)		Headloss (m) at flow rates		
	Design	Minimum	7.5 l/min	15 l/min	25 l/min
Kitchen tap 1/2 inch (DN 15)	12	6	0.7	1.7	4.9
Kitchen tap 3/4 inch (DN 20)	18	12	0.3	0.7	2.4
Washing machine	12	9			
Dish washing machine	9	6			
Handbasin (pillar/mixer taps)	6	4.2			
Handbasin (spray/mixer spray taps)	3	1.8			
Bath 3/4 inch (DN 20)	18	12			
Shower head	12	6			
WC cistern to fill in 2 minutes	7.8	6			
Ferrule 13 mm			0.3	0.7	2.4
Ferrule 19 mm			0.15	0.4	1.1
Float valve to tank—6 mm orifice			1.7	4.7	—
Float valve to tank—9 mm orifice			0.7	1.8	4.0

Other data:

Service pipe fittings headloss	Values of K in $V^2/2g$
Joints in 13 mm pipes	1.2
Joints in 19 mm pipes	0.8
Elbows	1.0
Short radius bends	0.8
Equal tees—flow to branch	1.5
Screwdown stopcock—fully open	8.2
Gate type stopcock	0.5

Note: Losses for service pipe fittings are for average condition

the existing service pipework will restrict the maximum flow rate to the unit. Table 14.4 gives the British Standard recommended design and minimum flow rates for a range of fittings.

For 'high-rise' flats the cold water supply from the mains is usually boosted by a small pump to a roof tank which then feeds all the supplies to the flats below. The peak demand from the group of flats determines the amount of storage needed and the pump characteristics and controls.

Service Pipe Design

For the design of a service pipe the peak demand rate must be estimated and the minimum mains pressure must be known. The flow and headloss through the piping can be calculated using one of the formulae described in Sections 12.4 and 12.5. Section 12.4 suggests some friction coefficients that can be used in the Colebrook–White formula for new small diameter service pipes. For service pipes that have been in use for some (or many) years, friction coefficients may be much higher but are too variable to be able to quote reliable values, for example the headloss in an old service pipe can range between 0.5 to 2.5 m/m of pipe at a flow of 9 l/min. Therefore, a 'realistic' friction value should be used bearing in mind that the k_s values in the Colebrook–White formula are meant to represent 'equivalent sand grain size' of the pipe's internal roughness and that allowance must also be made for the relatively high number of joints and fittings occurring on service pipes and internal plumbing.

Table 14.4 also gives headlosses expected through typical fittings. Pressure losses in the tapping to the main, in stopcocks, and at ball valves and taps may represent the major pressure losses on the delivery line. Losses at fittings which have been in use for a number of years may be very high. It must be appreciated that no formula gives consistently identical results to those obtained even in laboratory tests and, in practice, discrepancies will be not less than ±10%. Hence, for design purposes actual flows should be taken as something less than the calculated value and headlosses as something more.

BS EN 14154-1:2005 requires that the maximum headloss across water metes within "Rated Operating Conditions" should not exceed 6.4 m; ranging from 1 m upwards depending on the Pressure Loss Class of the meter (Class ΔP 10 to Class ΔP 63); the class being chosen by the meter manufacturer. It is therefore best to consult manufacturers' literature for actual headlosses.

All supply meters need to be tested for accuracy every few years. A typical small meter-testing bench is shown in Fig. 16.16. Large industrial supply meters may be tested in situ using a turbine or electromagnetic flow probe as described in Section 16.30.

14.7 WATERWORKS BYELAWS

Water utilities may have regulations or byelaws setting out a variety of requirements with respect to consumers' use of water and the materials and design of plumbing systems. In England and Wales all the water companies' individual byelaws were repealed and replaced by the Water Supply (Water Fittings) Regulations 1999. These Regulations, (issued by the Secretary of State for the Environment, Food and Rural Affairs), follow requirements laid down by the EC and, although not as prescriptive in detail as some of the previous UK byelaws, they maintain many of the previous byelaw provisions, such as the requirement that "*no water fitting shall be installed, connected, arranged or used in such a manner which causes or is likely to cause waste, misuse, undue consumption or contamination of water supplied by a water undertaker*". Additional provisions of the 1999 Regulations (in Section 5) include a requirement to notify the water undertaker of any proposal to install certain features likely to increase water demand including, inter alia: a bath of capacity over 230 litres; a pump or booster drawing more than 12 l/min; 'construction of a pond or swimming pool of capacity greater than 10 000 litres…designed to be replenished by automatic means and is to be filled with water supplied by the undertaker'.

Schedule 2 to the Regulations gives details of requirements for plumbing fittings for cold and hot water supply systems. Among the principal new provisions is that WC flushing cisterns installed after 1 January 2001 must not give a single flush exceeding 6 litres. Until then cisterns of 7.5 litres (the current requirement) could be installed; but there is no time limit to the replacement of cisterns installed before 1 July 1999 by ones of smaller capacity. Both flushing cisterns and *pressure* flushing cisterns are permitted; but pressure flushing *valves* may only be installed in a building which is not a house. As siphonic flushing cisterns are no longer mandatory, flushing cisterns equipped with flap valves are permitted (Table 1.4).

An overall requirement is that fittings and their installation must conform to "an appropriate British Standard or some other national specification of an EEA [European Economic Area] State which provides an equivalent level of protection and performance". Further guidance is given in the Water Industry Act, 1999 and the Guidance Document to the Water Supply Regulations 1999 (HMSO, 1999; WRSA, 1999).

DISTRIBUTION SYSTEM MAINTENANCE

14.8 NETWORK PERFORMANCE

Day-to-day management of a distribution system is an ongoing process. As described in Section 14.3 the key people responsible for the ongoing management and daily network maintenance are the zone engineer and district inspector. The zone engineer manages the network as a whole while the district inspector is the utility's daily interface with its customers. Between them they risk assess and authorize physical interventions on the network for planned and emergency maintenance activities and deal with consumer complaints. All their actions and decisions result from analysing and assessing data and following procedures that form part of the utility's overall asset and business management process.

A consumer's complaint is generally investigated by the district inspector. Pressure data would be retrieved from instrumentation at the nearest critical monitoring point; typically the highest supply point within the DMA or supply zone at which a permanent level of service pressure monitor is installed and measurements would also be taken at the consumer's tap. Flow data would be retrieved from the district and revenue meters. A water sample might also be taken from the tap for testing. Further investigations may then include analysing the system's performance using a hydraulic model. The objective is to determine whether the problem is an operational one, for which immediate remedial measures can be initiated, or whether it is due to a gradual deterioration of the main requiring asset renewal that will need longer term planned and financing to resolve. Interruptions to supply generally result either from equipment failure, for example pump and electrical fault or pipe burst or from planned maintenance of short duration. The impact of interruptions can be minimized either by repairing the equipment or rerouting supplies to restore the service. Flow, pressure and water quality complaints may result from interruptions to supply or they may be caused by longer term deterioration of the network in which case a more comprehensive analysis will be required to understand the problem and develop the optimum rehabilitation or renewal solution. Dirty water complaints are often related to the disturbance of corrosion deposits from old cast iron mains caused by some change in the hydraulic conditions in the network which could be rectified by mains cleaning or, if persistent, by mains renewal or relining. Some complaints may concern the presence of animals or insects due to larvae having entered the system at the source works.

14.9 MAINS REHABILITATION AND CLEANING

Rehabilitation techniques can be divided into short and longer term measures. Short term actions include re-zoning supply areas to improve pressures and prevent interruption to supply; repairing visible leaks; active leakage control; and mains flushing to resolve dirty water problems. Valve maintenance should also be included since, for relatively small cost, air and gate valves can be repaired and leaking glands re-packed. Longer term techniques include mains rehabilitation or replacement. Rehabilitation techniques can comprise non-aggressive and aggressive cleaning methods or use of non-structural and structural linings; while pipe replacement may utilize trenchless technology techniques.

Mains Cleaning Methods

Mains cleaning techniques include flushing, air scouring, swabbing, pressure jetting and scraping, of which the first three are non-aggressive. Flushing, air scouring and jetting are only effective for pipes up to 300 mm diameter.

Flushing is a well-established technique for small diameter mains where there is adequate mains water pressure (Plate 25(b)). It is an effective technique for removing loose sediment and for flushing polluted water from a pipe. It is not effective or efficient where there is an underlying persistent problem with sediments and discoloration which can only be resolved by eliminating the source of the contaminants. For effective flushing, the system hydraulics has to be managed so that the velocity at the pipe invert is sufficient to pick up material and hold it in suspension. This depends on the specific gravity of the material, water velocity and pipe diameter. In practice pipes up to about 100 mm diameter can be flushed through one or two hydrants supplied from both ends provided there is adequate pressure in the main. For larger diameter pipes, flushing generally needs to be through three or more adjacent hydrants, supplied from one direction only at a pressure of at least 4 bar. Achievement of this may be difficult due to the system hydraulic conditions. Flushing is unlikely to be effective on mains over 300 mm diameter. When planning a flushing programme the following issues need to be taken into account:

- the real source of the problem;
- the characteristics of the material to be removed;
- the maximum achievable pipe velocity under realistic hydraulic conditions;
- the discharge capacity and locations of the hydrants or washouts in relation to the size of pipe to be flushed;
- acceptable methods of disposing of the flushing water.

Flushing a system must be carried out systematically from upstream to downstream, taking into account likely secondary water quality problems created by isolating lengths of pipe. However, flushing is unlikely to be effective for a heavily turberculated main.

Air scouring can be used to generate higher velocities during the flushing process without using as much water. The injection of filtered compressed air forces slugs of water along the pipe which cause the disturbance of loose deposits on the pipe walls as the slugs form and collapse. The procedure is suitable for pipes up to 200 mm diameter and for lengths of up to 1000 m; its effectiveness relies on the skill of the operator in forming suitable air/water mixtures.

Foam swabs can be used to remove soft or loose material such as organic debris, iron and manganese deposits, sand and stones. The process usually involves using a series of increasingly hard plastic swabs. The swab should have a diameter 25–75 mm larger than the bore of the pipe and a length of 1.5–2.0 times the bore. The softer swabs can pass through butterfly valves. They can be inserted into the main via a hydrant branch, but for harder or larger swabs the main must be opened. The optimum velocity of the swab is generally about 1 m/s. Its speed can be controlled by regulating the outflow at the discharge end, the speed of the swab being usually one-half to three-quarters of the water velocity because of flow past the swab. Swabbing can greatly improve the flow characteristics of a slimed main and is frequently used to clean out a newly laid main before being put into service.

Scraping and relining is applicable to old tuberculated cast iron mains, which are structurally sound (Plate 25(a)). Pipe scraping methods include drag scraping, power boring and pressure scraping. Pigs or aggressive swabs comprise plastic swabs incorporating grades of wire brush or studs. For harder encrusted materials, several passes may be necessary and it is essential to have a good flow of water past the device to prevent debris accumulating in front of it. An electrical transmitter incorporated in the swab or pig assists in tracking its progress along the pipe. Scraping with pigs is effective in removing hard deposits but can also damage and remove linings. The latter may increase PAH levels in the water by exposing old coal tar linings (Section 6.42). Aggressive cleaning is usually followed by applying a secondary lining to the pipe.

Long lengths of main can be aggressively cleaned in 'one go' if the run is fairly straight and no obstacles to the passage of the device exist. However, because of the past practice of using tapers either side of reduced diameter valves, the length is frequently limited to the distance between valve positions. Service connections off the main must be isolated and, after cleaning and relining, must be 'blown back' with fresh water to clean. The cleaning and relining processes is also likely to dislodge, damage or block ferrule connections, requiring them to be excavated and repaired thereby increasing the cost.

14.10 PIPE LINING METHODS

Cement mortar and epoxy resin linings applied to the cleaned internal surfaces of cast iron mains can improve hydraulic capacity and reduce discoloration caused by corrosion. In situ linings can be applied to pipes from 75 mm diameter upwards. Cement mortar linings are typically 4–6 mm thick. They tend to increase the pH of the water. For pipes up to 150 mm diameter, the reduction in pipe bore and relative roughness of the mortar surface may not provide adequate hydraulic capacity.

It is technically feasible to reline mains with epoxy resins up to 2000 mm diameter. The finished lining is 1–2 mm thick, has a projected life of up to 75 years and is very smooth providing a good hydraulic performance. The lining can be applied to most pipe materials, not only iron and steel mains. However, mixing the two part components must be carefully controlled and care has to be taken to ensure the epoxy lining adheres properly to the wall of the pipe.

Structural lining methods are adopted where the structural integrity of the pipe has deteriorated. These lining methods involve inserting a flexible lining into the main by methods such as 'sliplining', '*Rolldown*', or '*Swagelining*'; the latter two being patented proprietary techniques. Rolldown involves inserting a polyethylene pipe, which has previously been reduced in diameter by up to 10% by squeezing in a 'rolldown' machine (Plate 26(a)). When in position, the pipe's 'elastic memory' is activated by internal pressure and it reverts to its former larger diameter to make a close-fit with the existing pipe wall (Plate 26(b)). The process is suitable for re-lining pipes of 75–500 mm diameter. Sliplining and Swagelining can be used for pipes of up to 1200 mm or more. All these techniques reduce the internal bore of the host pipe but can significantly improve its hydraulic performance because of the smoothness of the lining. They are quicker to install and less disruptive than relaying the pipe in open cut, but require each service pipe to be reconnected. In the *'Cured-in-place'* lining method a new felt liner 'sock', impregnated with resin and reinforced with glass fibre for pressure pipe rehabilitation, is usually inverted into the pipe to be rehabilitated using water, steam, or air pressure. The resin is then cured by heating the water in the 'sock', continuing to pass the steam through, or allowing ambient temperature cure. The inversion pressure is maintained until the resin has cured. The method is more suited to rehabilitating sewer pipes, although it is suitable for pipes with difficult access and complex shapes. However, the finished hydraulic surface is not as smooth as that of other methods (Plate 25(c)).

14.11 PIPE REPLACEMENT

The main reason for replacing a pipe is structural failure. Causes of failure include fracturing due to severe cold weather periods, prolonged dry spells causing ground movement, mining settlement, poor original workmanship when laid, age of pipes or their faulty manufacture, internal and external corrosion, inadequate cover for increased traffic loading, original class strength too low for current pressures, or because supply operating conditions have caused cyclic stressing of pipes. Analysing the record of bursts on mains may reveal those which are particularly prone to failure. Where evidence indicates that a main has failed frequently during the last 5 years, typically more than two or three times per annum per kilometre, it will probably need to be scheduled for immediate or early replacement.

The traditional replacement method is to lay a new main adjacent to the pipe to be replaced. The service connections are transferred to the new main as a separate operation to minimize interruption of supply to consumers. As well as the traditional methods of trenching, newer techniques include narrow bucket conventional excavation, rockwheels, chain trenching machines and mole ploughing. Narrow trenching for laying pipes up to 500 mm diameter avoids the need for an operative to enter the excavated trench and results in less handling of excavated material and reinstatement; but it can only be used for piping which is sufficiently flexible for pipe lengths to be jointed before lowering into the trench. It is quicker and cheaper than traditional open cut excavation. However, it does not reduce the risk of disrupting other services and is therefore only suitable for uncongested locations, or to meet restricted site possession or when rapid laying and reinstatement is required. Mole ploughing is only used for pulling small diameter pipes through soft ground.

The high cost of installing, replacing or renovating small diameter underground pipes and services by traditional methods including surface reinstatement, has resulted in the increased use of trenchless construction methods. In addition to the relining methods discussed above, there is a range of 'no dig' or 'low dig' pipe replacement and tunnelling techniques, which include pipe bursting, directional drilling and microtunnelling (Plate 25(d)). These techniques are typically used in urban areas, where the presence of many other underground services or much surface traffic makes trenching difficult and expensive. In pipe bursting a tapered bursting tool is winched through an existing pipe to break it up and displace the fractured material into the surrounding ground. A new PE liner pipe is pulled in behind the bursting tool. Pipes up to 300 mm diameter can be installed. Extensive prior site preparation may be necessary to excavate or remove steel repair collars, pipe bends and fittings and concrete surrounds, and to disconnect mains and service connections to prevent their damage. Ground disturbance can occur and may be a problem.

Micro-tunnelling machines have been developed for tunnels up to 900 mm diameter. Articulated shields, adjustable by means of remotely controlled jacks in the launch chamber, allow the

(a) Heavily tuberculated cast iron pipe. What is the value of pipe roughness?

(b) Flushing a 100 mm diameter cast iron main. Debris collects in net for grading and analysis. Flow meter records the quantity of water used. (Black & Veatch)

(c) Cure-in-place lining in 600 mm diameter steel pipe. CCTV survey identified defect. (Black & Veatch)

(d) Directional drilling for a 350 mm OD pumping main. (Black & Veatch)

PLATE 25

(a) 757 mm OD PE 'swagelining' entering 800 mm diameter host pipe. (Black & Veatch)

(b) Exit pit—757 mm OD PE still in tension and in its 'swaged' reduced diameter (Black & Veatch)

(c) Pipe jack under railway line for sleeve for 300 mm diameter trunk main. (Black & Veatch)

(d) Globe pressure reducing valve in chamber. (Black & Veatch)

PLATE 26

machines to be accurately steered on course. Muck is removed by an auger or by fluid transport using bentonite or water. A permanent pipe liner is jacked into position behind the shield. When complete the water pipe is installed in the lined tunnel and the annular space filled by grout. Pipe jacking is a similar technique, usually used for pipes over 900 mm diameter. A tunnelling shield and string of tunnel lining is jacked into the ground from the drive shaft (Plate 26(c) to a reception shaft. Lengths of several hundred metres can be achieved up to about 2500 mm diameter. The water pipe is installed inside the jacked 'tunnel'. Special ductile iron pipes without sockets but with in-wall joints have been manufactured for pipe jacking to avoid the use of a primary lining. Auger boring installs a sleeve, up to 1200 mm diameter, usually steel, between two pits. The pipe is installed as for microtunnelling. Directional drilling can be used to install pipes under an obstruction, such as a road, railway, or river. The technique involves drilling a bore between two points either side of the obstruction, using a steerable drill string. When complete the hole is reamed out to the required size and the pipe is pulled through. Depending on the capacity of the pulling equipment, length of pull and size of pipe, pulls of up to 1500 m and 800 mm diameter of continuously welded steel pipes have been achieved. Care needs to be taken when pulling PE pipes through so that its permissible tensile stress is not exceeded. The radius of curvature of PE and steel pipe when installed should not exceed 50 times the pipe diameter.

14.12 CONTROLLING WATER LOSSES

Water losses, a component of non revenue water (NRW) (Fig. 1.1), are made up of apparent and real losses and unbilled authorized consumption. The two components of unbilled authorized consumption, unbilled metered and unmetered consumption, can be managed effectively either by installing permanent meters to measure consumption or through regular monitoring exercise to assess demand. Customer metering inaccuracies, usually the more significant component of apparent losses, can be minimized by maintaining the meters (by inspecting and recalibrating or replacing) and managing the billing procedures to minimize recording errors. Unauthorized consumption, also a component of apparent losses, is more difficult to manage because of the difficulties in quantifying the unauthorized consumption and locating illegal connections. Consequently these losses are hidden in the global water balance and per capita figures. Leakage from the trunk and distribution mains and service connections, the real losses, generally represent the majority of non-revenue water.

Reducing water losses and NRW involves both the utility and the consumer. The utility needs to provide adequate numbers of properly trained staff and resources to detect and repair leaks and identify and quantify the apparent losses and to educate the consumers in how to avoid leakage and wastage within their premises.

Network leakage is managed either by a passive or active control strategy. Passive control involves reacting to a reported burst or unexpected change in monitored flow or pressure; leaks are detected, located and repaired only when the utility has been made or becomes aware that there is a leak or a supply problem in the network. Where this approach is practiced, as in many developing countries or where water resources are plentiful, the undetected leaks and hence underlying system losses will gradually rise and the network assets will deteriorate. This is called the '*passive policy*' or '*detectable natural rate of rise*'. However, increasingly financial and environmental pressures are forcing water utilities to abandon passive leakage control and to adopt a more proactive approach.

Leaks continuously break out and therefore the quantity of water leaking from a system is the aggregate sum of the flow rate of each leak multiplied by the time it runs before repair. Some level of leakage is unavoidable. For active leakage control (ALC), the '*total natural rate of rise*' is defined as the growth in leakage which would occur if neither detected nor reported leaks were repaired. For each system, there is a '*baseline*' or '*policy minimum*' level: the minimum leakage that can be achieved economically, with reasonable resources and conventional ALC methods and technology. The baseline is used to calculate the Economic Level of Leakage (ELL). The ELL is the level at which the cost per m^3 of constructing, treating and delivering additional water resources equals the operational and capital cost per m^3 of leak detection and repair to reduce loss. There are various methodologies for calculating the ELL, but there is increasing concern that the calculation does not sufficiently recognize the real impact of environmental and social issues (*the triple bottom line*), some components of which are difficult to quantify in economic terms but are fundamental to the principles of sustainable management of water resources. When new methodologies for evaluating the environmental and social factors are agreed, the value of water is likely to increase, thereby justifying more effort to reduce leakage before new resources are developed. Hence the calculation defines the level of effort and thereby resources necessary for minimizing leakage for a defined set of conditions.

Where water losses are to be managed within acceptable limits in terms of economics, sustainability and customer levels of service, leak detection and repair need to be continuous operational activities, based on targets set by calculations such as the ELL. ALC is a proactive approach for monitoring for leakage, detecting, locating and repairing unreported leaks as they develop and for repairing visible leaks efficiently, in order to minimize leakage and meet targets. Where a utility is regulated, targets may be set by or agreed with the regulator. Utilities not subject to external regulation may set themselves internal performance targets for a variety of reasons including managing their supply and demand water balance, recognising the need for environmental sustainability or managing their customers' and shareholders' perceptions of their performance. Targets can be set either at the company level or, where the data are available and leakage levels justify a detailed calculation, for individual supply zones and District Meter Areas (DMAs). It is more common to calculate the ELL at company level and use the value to determine individual targets for DMAs.

The scale of the problem of detecting and locating leaks in the thousands of kilometres of pipework that make up a typical water distribution system should not be under-estimated. Modern leak detection equipment can be used to assist in locating leaks and flow monitoring can assist in identifying components of system losses, but they are only tools which aid leak detection. Reducing leakage is therefore inevitably a labour intensive process which has to be pursued continuously.

Table 1.13 in Chapter 1 presents some typical 'background' leakage levels from UK water distribution networks; Table 1.12 shows the likely burst frequency rates for a range of materials in different countries worldwide.

14.13 LEAKAGE STRATEGY

A strategy for reducing water losses should comprise:

- asset renewal to maintain and replace mains and service connections;
- pressure management to minimize supply pressures and leakage;
- district metering to monitor system flows and for water losses;
- Active Leakage Control (ALC) for targeted detection and location of unreported leaks;
- minimizing repair times for visible and detected leaks.

Asset renewal and pressure management address the asset and operational conditions to reduce leakage. Asset renewal is intended to target those pipes that are more vulnerable to bursts (visible leaks) or leaks (unreported leaks) and for which it would be more cost effective to replace the asset rather than renew it. Research has shown that the quantity lost through a leak or burst is proportional to the operational pressure of the system. Therefore, reducing pressure, particularly during periods of low demand, will have a significant impact on leakage. This is usually achieved by use of a pressure reducing valve (PRV) on the main suppling a DMA or PMA (Plate 26(d)). Pressure management can also ensure that network assets are subjected to reduced continuous and transient pressures resulting in reduced stress on pipes and fittings thereby extend the life of the asset. Reduced distribution pressures at times of low demand will also provide the consumer with a more consistent supply pressure and hence flow rate from fittings. Pressure management can also impose water conservation on the consumer without him necessarily noticing. The flow rate from pressure dependent fittings supplied directly off the mains is reduced for example showers, taps and toilets that are equipped with flush valves rather than flushed through a cistern.

The objective of ALC is to shorten the duration of leaks and so reduce the quantity of water lost by minimizing:

- *awareness time*: the time between the start of the leak and the utility becoming aware of it;
- *location time*: the time to detect the leak and locate its position;
- *repair time*: including scheduling the work, obtaining permissions and carrying out enabling works where necessary.

District metering, ALC and minimizing repair times all contribute to reducing the time between when a developing leak can be detected and then repaired. The process involves the sequence of:

Monitor—Target—Detect—Locate—Repair—Monitor

District metering provides the infrastructure for monitoring network performance and enabling the leakage technician to analyse and understand system and minimum night flows and consumption, and to identify those areas of the network where leakage is greatest. This process is also used to monitor flows at the end of the sequence to ensure that leakage has been reduced and assess the effectiveness of the overall process. Once located a leak must be repaired speedily within the constraints of obtaining necessary permissions from third party organisations responsible for the public highways and open areas or private land owners through which the pipes are laid. Generally such permissions take longer to obtain than it would to organize the repair gang to execute the repair unless the burst or leak is sufficiently large to justify invoking emergency powers allowing the utility to commence the repair immediately. Visible leaks (bursts) tend to be repaired as they are reported; the same day, or within a few days if the size of the leak is assessed to be small and there are more urgent repairs to be carried out. Detected leaks on mains and leaks and bursts on service pipes can take longer typically 10 to 15 days for mains and 30 to 45 days for minor leaks on service pipes.

Once leakage has been reduced to a given target quantity—termed the *exit level*—ALC and repair activities can be suspended. However, the area should continue to be monitored to detect as and when the supply into the area increases. The flow is analysed to assess whether an increase is a result of increased demand or leakage and, if the latter, whether it has risen to a level at which it is necessary to re-enter the area and to repeat the detection and repair process. This threshold level, termed the '*intervention level*', is determined by a combination of factors including; the ALC resources available within the organisation; the unit cost of water in an area compared with the unit cost of reducing non-revenue water in that area; other savings that can be made by reducing losses;

and whether other areas should take priority. The aim of setting targets and thresholds is to maximize water loss savings by targeting the area(s) which will deliver the higher value or immediate returns. As the characteristics of each area are better understood, targets and thresholds can be reviewed and revised, taking into account the latest cost information on water production and manpower and the resources required to achieve the targets. Where a target has been exceeded or significant rehabilitation has taken place, it may be appropriate to reduce both the target and threshold to reflect the improved physical condition of the asset. Political and environmental considerations also influence intervention levels; a utility may need to show it has reduced leakage down to a target set by a regulating authority or to justify a proposal to develop a new source. Targets therefore need to be re-assessed periodically to ensure that they remain relevant to the prevailing network and business conditions and strategies.

14.14 DISTRICT METERING

Targeting ALC activities relies on monitoring the system for evidence of leakage. The most effective method of monitoring is to divide the system into three levels of control:

- zones of 10 000 to 25 000 connections or perhaps more;
- DMAs of typically 500–3000 connections;
- waste districts of a few hundred connections.

The data derived from monitoring a zone is generally too coarse for detailed targeting and efficient deployment of ALC resources. However, zone monitoring can be considered as first stage monitoring for utilities that are starting to address leakage. Typically a zone would comprise a number of DMAs. Monitoring at the DMA level of control is the most common method. The data provides sufficient detail to be able to target ALC activities efficiently at individual DMAs. Each DMA can include one or more waste districts. Waste districts are generally too small for efficient monitoring but are effective subdivisions for detecting and locating leaks within a DMA.

Ideally the design of these control areas should incorporate any existing system divisions and natural boundaries; physical or hydraulic. Permanent flow meters are installed on the feed mains to zones and district meter areas. Preferably there should be a single feed main for each zone or DMA, although this is not always possible. However, DMAs fed by more than three meters are generally difficult to monitor and the flow analyses can be unreliable. For the waste districts, a by-pass in a chamber is constructed around a valve on the main feed to the district, so that a waste meter can be temporarily inserted in the by-pass when undertaking waste metering by step-testing (Section 14.15). The boundary valves dividing zones and DMAs are normally kept closed unless, for operational reasons, some DMAs need to be combined. Pressure monitoring equipment is also installed at selected 'critical monitoring points' in the system.

Historically a venturi meter or a shorter form, the 'Dall tube' (Section 12.16), was installed in a large pipe feeding a zone; but these are no longer used because of their high cost. Instead the electromagnetic flowmeter is now generally used. Inferential meters may continue to be used in smaller mains, equipped with electrical pulse recording equipment that can log either the total flow or rates of flow at stated intervals, but are progressively being replaced by electromagnetic meters (Section 16.28).

Zone and DMA flow and pressure measurement data can be transferred to an Operational Control Centre to be processed, stored and monitored using an integrated telemetry system, hard wired, by radio or GPS technology. The data is used to monitor daily demand and minimum night flows and is compared with historic data or calculated theoretical flows to identify differences and anomalous data. The data processing would typically be completed overnight, the output being a summary report which identifies areas where flows, pressures and processed data are outside pre-defined parameters, '*exclusion reporting*', and indicating where further data analysis or investigation may be necessary. Alternatively, the data can be stored in on-site data loggers that are interrogated periodically, and their data transferred to the flow monitoring database for analysis. The data from on-site loggers is typically downloaded on a 28-day cycle. Hence there is inevitably a delay before the information can be processed.

For many large water utilities in under-developed countries, setting up district metering or waste districts could be very expensive, because of the need to purchase and install new meters and boundary valves and to replace or repair many existing valves needed to create the discrete hydraulic boundaries. Implementing district metering and flow monitoring has therefore to be a long term plan starting with the zone metering of supplies, and progressively developing DMAs where zonal flows indicate losses are probably highest. For utilities only able to provide intermittent supplies, quite different methods for checking levels of consumption and losses have to be adopted, as described in Section 14.17.

14.15 LOCATING LEAKS

Many visible leaks are reported first by the public. However more can be found by district inspectors looking for signs of leaks, particularly in areas with a history of pipe failures and leakage. Damp patches, trickles of running water and extra vegetation growth close to pipe alignments, valves or fire hydrants, or above ferrules or stop taps, may be indicative of a leak. The inspector is also able to detect signs of consumer wastage, such as overflows discharging outside properties. However, beneath metalled roads, or where mains are laid in freely draining ground, or adjacent to or below a water course, even large leaks may not show on the surface. Trunk main routes through open country should also be inspected periodically for signs of leakage.

Unreported leaks can be detected and located using ALC. The monitoring equipment and analysis processes will initiate detection and location activities as the leakage flow rate increases beyond the trigger threshold. The leak can then be located using one or a combination of methods and technologies available including waste metering/step-testing, 'sounding', leak noise correlation, acoustic logging, ground penetrating radar, in-pipe inspection tools, tracers and gas injection.

Waste Metering and Step-Testing

Waste metering and step-testing is a method of localising a leak by metering the night flow to a small part of the distribution system. Sections of network are systematically shutting down within the area in 'steps' in a predefined sequence, while monitoring the drop in flow as each main is isolated. A larger flow reduction than expected from the properties that are connected to the main may indicate that the main that has just been isolated has a leak. The test results are interpreted and passed to the district inspector who would arrange for further investigations to locate possible leaks. After all suspected leaks have been located and repaired the night test is repeated to confirm that the exercise has been effective.

The cost of installing and maintaining meters and valves for waste metering and step testing tends to be high. A by-pass including a 'waste meter' and isolating valve is installed on the feed main into the waste area (or district). During the test, all flow into the area is passed through the meter with all other feeds shut off. A DMA meter can also be used for step testing provided it is sufficiently sensitive to measure the small changes of flow as each main is shut down.

Step testing is time consuming and generally has to be carried out at night when legitimate network flows are low. Computing and GPS technology developments have however enabled the inspector to monitor and analyse flows during the test remotely and thereby assess there and then whether the drop in flow relates reasonably to the number of domestic connections on the main and any legitimate non-domestic night-time demand or if there is a suspected leak.

Step-testing is only practicable on 24-hour supply systems, is time consuming, generally requires working at night and increases the risk of creating a water quality incident by disturbing sediments when closing and opening valves. For these reasons acoustic logger surveys are replacing step testing for locating leaks.

Locating Leaks by Sound

Monitoring the sound made by water leaking from a pipe is the basis of the majority of leak detection techniques and electronic equipment currently available. The traditional 'listening stick' is a light solid metal bar about 1.5 m long or the bar of a valve key. Purpose-made listening sticks can be equipped with an earpiece. One end of the stick is placed on an exposed part of the main or service pipe, such as a valve or hydrant spindle or stopcock, and the other is placed against the ear. The sound emitted by a leak, if audible, is 'a low drumming noise' or 'a continuous buzzing sound' and tends to be continuous without any change of audibility or quality. It stops abruptly when, and if, the water can be turned off. An experienced waste inspector using a listening stick can detect even a small leak at a distance of 10–15 m, if it is making a sound. By listening at another point of contact on the main, the inspector can judge by the difference in sound volume the probable location of the leak. Sounding is frequently carried out at night when background noise should be low. However, an experienced operator can still be successful during the day away from main roads and when there are lulls in traffic noise. Daytime sounding may also be preferred because, at night, parked cars can prevent access to valves and stopcocks.

Sounding all pipes in a system is not usually adopted because it is very labour intensive, relatively ineffective on non-metallic pipes and not efficient at identifying new or increased leakage as it occurs. In the UK where many service connections do not leak an inspector can sound up to 200 connections per shift. Overseas where pressures are lower and background conditions are less favourable, an experienced inspector should be able to sound 80–120 connections per day on average.

Ground microphones, first introduced in the 1960s, amplify the sound of a leak and are effective for leaks from non-ferrous pipes. It must be borne in mind, however, that not all leaks emit a sound, and the volume of sound emitted is mostly not related to the size of the leak.

Leak noise correlators have been available for about 30 years. They are electronic devices used to analyse the sound of a leak picked up by sensors in contact with the pipe. The software analyses the sounds from the two points of contact to estimate the positions of the leaks, using the sound time-delay between the sensors, using pipe material, diameter and distance between sensors as input data. Within the last few years the technology has improved such that the new digital correlators are, readily portable, effective for all pipe materials and easy to use by less experienced operators.

Acoustic loggers are used both to detect and locate leaks, either as permanently installed instruments at defined locations or temporarily deployed for a survey of an area. When permanently deployed, they are monitoring continuously for changes in noise characteristics, an indication of changes in network flows and hence possible leakage. The loggers are downloaded periodically either remotely or manually or specifically to investigate an identified anomaly. When temporarily deployed, sets of loggers, typically eight or ten loggers per set, are installed in an area at predefined points. They are deployed for 24 hours and programmed to log during the minimum night flow period. A whole supply zone may be surveyed over two or more consecutive nights. Deployment may be for longer periods where the leakage team is attempting to reduce the baseline leakage level in the zone.

The data retrieved from both types of installation are analysed to filter out the background and 'normal' system noises and then combined with pipe diameter and material data and the length of pipe between logger locations, to correlate the location of possible leaks by triangulation. The same monitoring points are always used for regular repeat surveys within a system, System noise profiles can be developed and used to assess system changes and predict the need for intervention. A group of acoustic loggers can successfully detect and locate a number of leaks at different locations within the triangulation of the deployed loggers. The equipment suppliers suggest that the location accuracy provides sufficient detail for the utility to be able to instruct the repair contractor without the potential financial risk of digging a *dry hole*; excavating down to the pipe where there is no leak. However, where the utility is constrained by third parties in opening excavations in the public highway, located leaks should be confirmed on site using a leak noise correlator before instructing the repair contractor to proceed.

Some utilities are proposing to use acoustic loggers instead of setting up district metering, relying on the measurements from zonal metering to analyse flows. Although this does represent a short term expedient where resources are constrained, the level of detail available from zone metering is too coarse to understand variations in subsystem flows and demands and hence prioritising the critical leaks for location and repair. Therefore, in the longer term acoustic logging represents a key tool in leakage management but will not replace the detailed understanding of network operations and performance derived from monitoring flows in the smaller DMAs.

Other Leak Location Technologies

Pipeline integrity management systems use continuously monitored flow and pressure data linked to hydraulic modelling software to provide on-line active leak detection on a pipeline. The technology, developed from the oil and gas industry, is suitable for application to lengths of trunk main where pipe failure could be damaging and would need immediate, automatic shutdown of a section of the main. The technology analyses flow and pressure measurements continuously to assess the performance of the main between the sensing points and compares the measured values against normal pipeline performance characteristics. Where the software detects abnormal measurements, the software calculates the location of a leak and initiates the closure of appropriate valves in order to isolate the fractured length of pipe. The equipment is expensive to install and maintain but cost effective in terms of the consequential damage that might otherwise occur if a large pipeline were to develop a serious leak.

Other techniques that have been developed include:

- intelligent pigs inserted into live mains to inspect the internal condition of the pipe using closed circuit television and to pinpoint leaks;
- tracers and gas injection; gas injection is suitable for low pressure systems and non-metallic pipes for supply and small diameter pipes;
- ground penetrating radar and thermal imaging to identify changes in soil moisture.

These more sophisticated tools tend to be expensive to deploy and usually require a specialist contractor and are therefore more appropriate for more complex surveys, such as for trunk mains and large pipes where the leaks are difficult to detect and locate.

14.16 REPAIRING LEAKS

Confirmation of the existence of a leak is only obtained when a pipe is exposed and the leak located. However, pipe repairs can be expensive, especially if the leak is under a heavily trafficked road or road junction, and special planning of the repair operation may be necessary to minimize traffic disruption. The relevant road authorities have to be informed and all their requirements met. If the leak is known to be small and unlikely to increase in size rapidly, if the repair cost estimate is high and if the work would cause serious traffic disruption, the water utility may choose to delay repairing the pipe and accept the water loss until the leak can be repaired in a more cost effective and efficient way. One of the 'no-dig' rehabilitation techniques discussed in Sections 14.10 and 14.11 may offer a solution if internal access is feasible.

Where leak detection is let out to a detection contractor, the contract conditions should require the repair contractor to confirm the existence of detected leaks by excavation before the detection contractor is credited with finding the leak. Payment for detection services should be made against measured reduction in leakage flows, not the number of 'detected' leaks and theoretical leak rates. Repair contractors should be paid for all pipes excavated to repair a located leak, including 'dry holes'. The cost of excavating and refilling dry holes should be charged back to the detection contractor.

14.17 REHABILITATION, LEAK DETECTION AND DEVELOPMENT OF DISTRIBUTION SYSTEMS IN DISREPAIR

Worldwide numerous large public water utilities operate distribution systems in need of rehabilitation and repair. Total non-revenue water of 50–60% of the supply can be reported, made up of unknown proportions of distribution leakage, consumer wastage, and failure to meter or bill all consumers taking a supply. It is not easy to improve the performance of a large system in a short time; resources invariably represent the major constraint and inevitably the utility has to adopt a strategy of progressive stages of improvement as finance becomes available. Ideally rehabilitation becomes self financing as water losses are converted into revenue. In practice there is always the danger that rehabilitation can not keep up with the rate of deterioration and therefore the asset life continues to reduce. An indication of the work involved in starting off a rehabilitation project is given in Figure 14.3.

The difficulties of rehabilitation are increased where supplies are intermittent because of the scarcity of resources. Resource shortages are often exacerbated by large losses of water and revenue through leaks, wastage, illegal connections and unpaid supplies. Traditional methods of leak detection suitable for 24-hour supplies, such as waste metering, step-testing and acoustic logging, become impracticable because many consumers leave taps open, ready to discharge water into storage receptacles to cover periods of non-supply. Even the procedure of valving off lengths of main to put them under a pressure test for leakage after shutting down all service pipes, may be vitiated by lack of stop-valves on service pipes, by leaking ferrule connections or tappings to the main or by many illegal or unknown connections. Often a first attempt at a pressure test results in a burst main

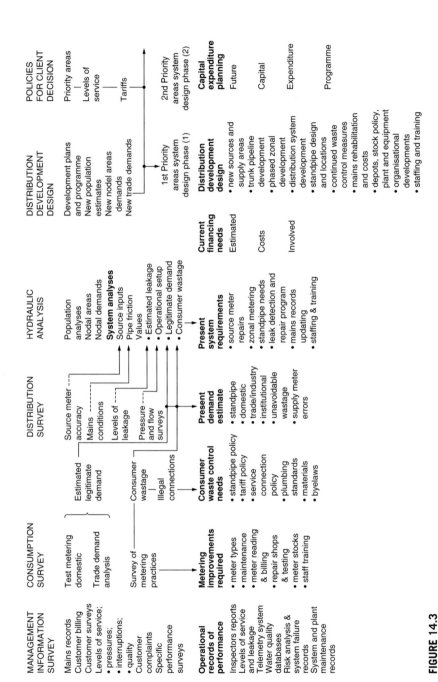

FIGURE 14.3

Schematic for distribution system rehabilitation and development project.

because the system has long been on intermittent supply at low pressures, or it may be difficult to gain any pressure because of leaking boundary valves or the existence of some unknown connection which has not been marked on the mains records.

A more productive approach can be to expose the soffit of lengths of pipe in selected areas to find the principal causes of loss and to investigate supply pipes and meters. Often it will be found that high losses are due to one or more of the following: badly made service pipe tappings on the main; illegal and unknown connections; illegal by-passes to customer meters; leaking service pipes not fitted with a stopcock; service pipes continuously taking water because of waste on consumers' premises.

Exposing the soffit of pipes is time consuming and relatively expensive. However, it enables location and reduction of leakage directly and helps to find leaking joints. In many countries it is not as expensive as relaying a main since it is primarily labour-intensive work involving the use of less expensive repair materials and plant. Attempts to find all leaks on a pipe network known to be in poor condition using various methods of surface detection equipment and flow monitoring can be very frustrating because of the repeated failure to achieve an acceptable result each time the system is re-tested. Repeated non-success after so much work reduces motivation to continue leak detection activities. The direct exposure approach produces positive results and, if pursued, can increase efficiency in finding leaks through experience and increases the amount of the system that is rehabilitated.

REFERENCE STANDARDS

British Standards (BSI)

BS 750:2006. *Underground fire hydrants and surface box frames and covers.*

BS EN 806:2005. *Specification for installations inside buildings conveying water for human consumption; 2:2005 Part 2 Design and 3:2006 Part 3 Pipe sizing—Simplified method.*

BS EN 1452:2000. *Plastics piping systems for water supply—Unplasticized poly (vinyl chloride) (PVC-U), Part 2 Pipes.* (Note: the earlier standard remains current.)

BS 6700:2006. *Design installation, testing and maintenance of services supplying water for domestic use within buildings and their curtilages—Specification.*

BS EN 10255:2004. Non-alloy *steel tubes suitable for welding and threading—Technical delivery conditions.*

BS EN 12201:2003. *Plastic piping systems for water supply—Polyethylene (PE), Part 2:Pipes.* (Note: the earlier standards (BS 6572 and BS 6730) also remain current.)

BS EN 14154:2005. *Water Meters; Part 1—General requirements, Part 2—Installation and conditions of use, Part 3—Test methods and equipment.*

REFERENCES

Bernis, J. and Galan, F. (1976). Special Subject 7, Water Requirements for Firefighting. *Proc. IWSA Congress,* J11–J17.

FSA (2003). *Fire Services Act.* HMSO.

NFPA (2007). *NFPA1, Uniform Fire Code.* 2006 Edn. (*including: NFPA14 2007: Standard for the Installation of Standpipe and Hose System, 2007 Edition and NFPA1141 2003: Standard for Fire Protection in Planned Building Groups,* 2003 Edn.). NFPA.

Ofwat (2007). *June Return 2007 Reporting Requirements*. Chapter 2. Ofwat.

Water Industry Act (1999). HMSO.

Water Regulations Advisory Scheme (1999). *Guidance Document for Water Supply (Water Fittings) Regulations 1999 and Water Byelaws 2000*. Scotland. HMSO.

Water UK (2007). *National Guidance Document on the Provision of Water for Fire Fighting*, 3rd Edn. Local Government Association and Water UK.

Pipeline Design and Construction

15.1 PIPE DEVELOPMENT

Low pressure earthenware pipes have been found at Knossos and in Mesopotamia. Copper is known to have been used for unburied water pipes from the same early period. Lead was used by the Greeks to seal joints in earthenware pipes and the Romans used it for pipe. They also made pipe from clay and hollowed out tree trunks and such pipe persisted in use (being cheaper then lead) in the UK until the 17th century. Wood had a brief renaissance in North America at the beginning of the 20th century as a material for large diameter pressure pipelines for industrial use, but not for potable water. The method used preservative treated Douglas fir staves bound together by threaded hoops. The method became defunct on the ready availability of large diameter steel pipe with reliable welding and coatings.

Iron pipe has been in use in Europe for over 500 years, cast initially in horizontal moulds and later in vertical moulds. Centrifugal casting was introduced in about 1920. All these iron pipes were joined by caulking yarn into the annular space at a spigot and socket joint and followed up by molten (run) lead but, after about 1950, flexible rubber joint rings were the norm. Ductile iron began to replace cast iron pipe in the UK a few years later.

Reinforced concrete pressure pipe was developed after the material's invention in the early 20th century while asbestos cement pipes were developed in Italy at about the same time. Prestressed concrete pipe was developed in the 1950s and has allowed concrete to be used for higher pressures in very large diameters. Plastic pipe materials were introduced in the latter half of the 20th century and have since widened considerably in applicability with improved knowledge and quality of these materials.

15.2 MATERIALS AND POTABLE WATER

Pipes, lining materials and joints must not cause a water quality hazard. In the UK, materials for customers' installations are covered by the Water Supply (Water Fittings) Regulations 1999—and their equivalents in Scotland and Northern Ireland. Materials used in public supply systems are covered by Regulation 25 of the Water Supply (Water Quality) Regulations 1989 and amendments—and their equivalents in Scotland and Northern Ireland. Materials considered to comply on the water supply side are those (a) approved by the UK Government (DETR) Committee on Chemicals and Materials of Construction for use in Public Water Supply and Swimming Pools and (b) products verified and listed under the UK Water Regulations Advisory Scheme (formerly the Water Byelaws Scheme), administered by WRc.

15.3 TYPES OF PIPES AND ORGANISATIONS SETTING STANDARDS

Pipes found in water supply systems are generally of the following materials: cast or 'grey' iron; ductile iron (DI); steel; polyethylene (PE); PVC (polyvinyl chloride); GRP (glass reinforced plastic); prestressed concrete, cylinder or non-cylinder (PSC); reinforced concrete cylinder (RC) and asbestos cement (AC, no longer produced in the UK).

Other materials include galvanized iron, copper and lead which tend to be found in service pipes, plumbing, common connections and other small diameter pipes. Lead, extensively used in the past, is no longer installed because of the risk of plumbo-solvency (Section 6.33), but many lead pipes remain in service for house connections and internal plumbing. Copper pipe can also give rise to plumbo-solvency problems from lead solder. In the UK these materials now tend to be superseded by polyethylene and, where lead in water is a problem, lead pipes are generally being replaced by polyethylene.

A number of organisations have set standards for pipes and their fittings. Table 15.1 lists the terms most widely used to define pipe characteristics or pressures applied to pipes or pipe systems (as in this chapter).

PIPELINE DESIGN

15.4 INTRODUCTION

Hydraulic design, including loss calculation and diameter selection, surge analysis, longitudinal profile and air transport are discussed in Chapter 13. Management of air is discussed in Chapter 16. This chapter deals with pipe material selection, structural design and construction.

15.5 STRUCTURAL DESIGN OF PIPES

Pipelines are required to withstand internal hydrostatic pressure, external loads from soil, surcharge and traffic and are required to be safe against buckling (Young, 1983 and 1986; Clarke, 1968). The load imposed by the soil increases with depth and surcharge; traffic loads reduce with depth. Main road traffic load depends on the wheel load and configuration considered but typically accounts for less than 10% of the total load at 6 m depth. The minimum total external pipe load under main roads typically occurs at about 1.8 m depth. Loading from traffic differs between standards; for example, pressures determined at the crown of the pipe from traffic under normal roads according to BS EN 1295-1 and from HS-20 loading according to AWWA M45 differ

Table 15.1 Abbreviations for pipe and pipe system standards	
Abbreviations used in standards and in this book:	
Pipes	
DN	nominal diameter in millimetres, e.g. DN 400
PN	nominal pressure in bar. For flanges it is the same as PFA—see below.
Pipe pressures (BS EN 805)	
PFA	allowable operating pressure, excluding surge
PMA	allowable maximum operating pressure, including surge
PEA	maximum allowable hydrostatic test pressure after installation;
Pipe system pressures[a] (BS EN 805)	
DP	design pressure—maximum operating internal pressure of a system or zone fixed by the designer but excluding surge
MDP	as DP but including surge—designated MDPa when there is a fixed allowance for surge or MDPc where surge is calculated
STP	system test pressure as applied after installation

[a]The pressure rating for a pipeline system may well be limited by fittings and particularly flange pressure ratings.

significantly for pipe cover less than 5 m. At 1.0 m cover the pressure using BS EN 1295-1 is about 3 times higher and at 1.8 m cover the ratio is about 2.5 times reducing to about 1.2 times at 5 m cover (Little, 2004).

Pipes can be grouped as rigid, flexible and semi rigid. As a general rule pipes classed as flexible are sized by outside diameter (OD) while rigid and semi-rigid pipes are sized by internal diameter (ID); GRP pipes are sized by ID or OD depending on the manufacturing method (Section 15.19). Rigid pipes deflect little; their load carrying capacity is derived from ring bending strength (as determined from crushing tests) and can be increased by bedding factors for various standardized bedding and surrounds. Rigid pipes tend to attract more load than more flexible pipes, particularly in wide trench situations. Flexible and semi rigid pipes deflect under load in inverse relation to pipe stiffness and overall soil modulus. Flexible pipes derive their support primarily from passive soil resistance which develops as the pipe ovalizes under vertical load and deflects horizontally into the side fill. The contribution of pipe stiffness is small. Ring bending stress is traditionally not taken into account for steel pipes but, under the current UK standards is addressed as combined hoop and bending stress for thermoplastic (PE and PVC) pipes. Bending strain is calculated for GRP pipelines.

Overall soil modulus is a function of the pipe surround (embedment) type, compaction and depth and depends to some extent on the modulus of the native (trench wall) soil. If the trench wall material is weak, the trench width may need to be increased or the backfill material and compaction improved or both. Safety against buckling of flexible pipes also depends on overall soil modulus. Flexible pipes offer economical solutions and are therefore widely used. However, the more flexible the pipe the greater the care needed in selection and control of embedment construction. Other pipe types also require attention to backfill materials and workmanship. Semi rigid pipes derive their support partly from the soil and partly from the pipe stiffness. Their design takes into account ring bending stress. Permissible deflection limits for ductile iron pipes are set to limit ring bending stress in the pipe wall and for joint and lining performance.

Rigid pipes include concrete and asbestos cement. Flexible pipes include thin walled steel (diameter to thickness ratio more than about 120) and plastics. Ductile iron pipes are classed as semi-rigid. Flexibility is determined by ring bending stiffness (specific stiffness), S (kN/m^2) = EI/D^3 = $1000.E$ $(t/D)^3/12$, where E is the modulus of elasticity (N/mm^2), t is the mean pipe wall thickness and D is the mean diameter of the pipe (outside diameter less thickness). Ring bending stiffness may also be expressed in units of N/m^2. This diametrical value is one eighth of the stiffness based on radius, used in Germany.

Typical stiffness values for semi rigid and flexible pipes in the range DN 100 to 2000 are 1350 to 16 kN/m^2 for ductile iron; 580 to 4 kN/m^2 for steel; and, irrespective of diameter, (except for small diameters) 80 to 16 kN/m^2 for PVC-U, 17 to 8 kN/m^2 for MOPVC (Section 15.18), 10 to 5 kN/m^2 for GRP and 80 to 4 kN/m^2 for PE. In comparison, reinforced concrete pipe stiffness is typically about 500 to 600 kN/m^2 for DN 800 and above and stiffer at smaller diameters. Very flexible pipes (stiffness less than about 4 kN/m^2) are at risk of deflection to a "squared" shape (as the pipe is deflected first inwards and upwards and then downward and outward as backfill is built up in layers). Pipes, particularly of plastic, of stiffness less than about 4 kN/m^2 are not recommended without great care in construction (Janson, 1995). This stiffness value is also adopted under the European standards developed for plastics pipes. It is equivalent to a diameter/thickness (D/t) ratio of about 160 for steel pipes. This is similar to the D/t range of 160 to 180 quoted as the maximum suitable for centrifugal cement mortar or spun concrete lining. Above this range, pipes need flexible lining or should be lined with cement mortar lining in situ. Despite case histories of much greater values, D/t for steel pipes should in no case be more than 200 (stiffness 2 kN/m^2) to retain sensible control during installation. Preferences and economics vary widely but some design and build contractors prefer stiffer pipe and pay less attention to backfill. The D/t ratio selected then depends on overall installed cost. Wall thickness can be determined by design for internal pressure and buckling and choice of material strength.

Characteristics of plastics pipes change with time; in particular, the modulus of elasticity (E) reduces very considerably with time. 'Short term' is defined as the one hour modulus. Long term values are obtained by extrapolation of tests over various periods to obtain 50 year values. Long term values must be taken into account in design. For PVC-U and PE respectively, E values may typically be 50% and 25% of the short term value, but depend on material formulation. 'Ultra short' term (10 second) values are significantly greater than the one hour values and are applied when calculating pressure surges.

Structural design in general is based on the pipe cross section. Longitudinal bending arising from uneven trench support is not usually specifically considered except where pipes are designed to be supported at discrete intervals (Section 15.7) or at connections to structures and where differential settlement is expected. For rigid pipes with flexible joints (spigot and socket or flexible coupling) differential movement can be accommodated by a short length of pipe between couplings (e.g. the rocker pipe at exit from a structure). In addition ring bending for buried pipe tends not to be included specifically in steel pipe design. Such factors can be taken as being covered in the factors of safety but, if necessary, can be analysed separately according to the appropriate design codes.

Work by the European standards organisation (CEN) to develop a common calculation approach found that the several European methods (and American methods) were similar; principal differences were in coefficients and in values used for soil stiffness and assumptions for soil support. There are differences in design strategy, for example use of low factors of safety with upper and lower bound loads and strength, or higher factors of safety to mean loads and strength. BS EN 1295-1 summarizes the various European design methods and its UK National annex A sets out procedures for the three pipe categories. PD CEN TR1295-2 gives details of six European methods. PD 8010-1 and -2 set out design considerations for pipelines on land and those under water.

15.6 FLEXIBLE PIPE DESIGN

Flexible pipe design principles apply to steel and plastics but can also be applied to large diameter ductile iron pipe under large loads. For steel pipes greater than about DN 750 (or D/t about 120 to 140, depending on conditions) the theoretical pipe thickness to meet usual internal pressures is frequently less than the thickness required to limit deflection under backfill load. To save adding extra steel thickness for stiffening the pipe against backfill load, the pipe may be laid under 'controlled backfill' conditions with the pipe being laid on a thin layer of uncompacted sand or on a preformed circular invert (60° width) in sand or fine gravel and selecting and carefully compacting the sidefill in shallow equal layers either side of the pipe to achieve a required soil stiffness.

At large diameters, temporary jacks may be inserted (in steel pipes) to maintain circularity or to predeflect the pipe upwards but care must be taken to spread jack loads to avoid damage to the lining. As the sidefill progresses the jacks may be removed depending on the degree of predeflection. Deflection is measured during embedment construction and, if it exceeds the permissible value, the backfill should be removed and replaced to obtain the necessary circularity. Measurements should be continued periodically after installation and checked at critical sections before the pipe is filled for testing.

Deflection of flexible pipes is usually calculated using the Spangler formula:

$$\frac{\Delta}{D} = \frac{k(P_e D_1 + P_s)}{8S + 0.061E'}$$

where:

Δ = pipe deflection (assuming horizontal and vertical deflections are equal);

D = mean diameter of the pipe;

k = a constant, dependent on the angle, between contact points, over which the trench bed supports the pipe (typically 0.1 for 65°, 0.083 for 180°);

P_e = soil load per unit area (kN/m²);

P_s = surcharge or traffic load per unit area (kN/m²);

D_1 = deflection lag factor, dependent on soil type and compaction

S = diametrical ring bending stiffness = 1000. E $(t/D)^3/12$ (kN/m²);

E = modulus of elasticity of the pipe wall (N/mm²);

E' = soil stiffness (kN/m²); and

0.061 is derived from assumed (parabolic) loading over a 100° lateral support angle.

For granular soils D_1 is unity. Long term, in clay, the value may be 3 or more. Where the cover is less than 2.5 m and where the pipeline will be under sustained pressure within a year of installation, long term deflection may be reduced by a re-rounding factor $D_R = 1 - (P_i/40)$, where P_i is the internal pressure in bar. Diurnal variations in pressure in distribution mains should be taken into account in deciding the re-rounding factor.

Embedment soil stiffness E'_2 depends on the nature of the soil, the degree of compaction, the amount of overburden and degree of saturation. Values for granular soils are given in BS EN 1295-1 for different degrees of compaction. These values may be considered to be suitable for the least favourable situations—soil fully saturated and with little cover. Stiffness values quoted by AWWA and AASHTO tend to be rather higher and some sources show variation with soil cover and ground water level (Little, 2004). Compaction is quoted either as per cent Proctor (modified Proctor density, M_p—which corresponds to the heavy compaction test to BS 1377) or, for granular materials, to relative density. Soil stiffness can also be derived from laboratory tests. The overall soil modulus is modified to take account of the trench side soil if the native soil modulus is less than 5 MN/m^2 and if the trench width is less than 4.3 times the pipe diameter. Native soil stiffness (Little, 2004) may be estimated from SPT or other test results with correction for depth or to undrained shear strength. AWWA M11 quotes accuracy of predicted deflections for different soil compaction; it does not cover native soil modulus.

Ring bending stress is traditionally not addressed in steel pipe design but is included in design of PVC and PE (thermoplastics) pipe; bending strain is considered for design of GRP (thermosetting) pipe. Pipe thickness for plastic pipe is usually expressed as a minimum but thickness tolerance is not covered. Bending stress is given by $\sigma_{bs} = E\, D_f\, (\Delta/D)\, (t/D)$, where E is the flexural modulus of elasticity of the pipe material and D_f is a strain factor, dependent on pipe and soil stiffness and is given in BS EN 1295-1 and other references. For PVC and PE pipe the sum of bending stress and hoop stress is required to be less than the design value. The approach is similar for GRP but uses a criterion of strain. Designs for plastics pipes need to take into account both the initial short term and the long term characteristics.

Deflection limits vary according to pipe material. Deflection limits quoted in AWWA M11 for steel pipe are as follows:

- mortar lined and coated   2% of diameter
- mortar lined and flexible coated   3% of diameter
- flexible lined and coated   5% of diameter

Deflections can be allowed to exceed the limit of 2% frequently quoted where pipes are lined with cement mortar after installation or for flexible linings. However, if large deflections are to be permitted then for steel pipes it would seem consistent to carry out a more detailed analysis, including ring bending stress, but adopting a design factor greater than the traditional value of 0.5 and taking into account the stress strain characteristics which differ from those applying to plastics.

15.7 ABOVE GROUND PIPELINES

Loadings on above ground pipelines are usually limited to self weight, pipe contents, snow, wind, internal pressure (including end restraint and Poisson's ratio effects), support loads and thermal stresses. Superimposed loads need not otherwise be taken into account unless the pipe is to carry an access walkway. Pipe bridges can be designed as arch or suspension structures using bridge design methods. However, the pipe arrangement most likely to be encountered by the pipeline engineer is the support of the pipe on a series of saddles.

Without the constraint of embedment in a trench pipelines should be treated as mechanisms. This applies particularly to pipes with flexible joints but also to pipes with flanged joints since such joints are not absolutely rigid. Suitable lateral restraints must be included in the design since small thrusts at slight misalignments can develop into large forces through the multiplying effect of increased deflections and consequent increased misalignment. Pipes with flexible joints such as spigot and socket DI pipe and GRP should be supported at least at every joint. For large diameter pipes with flexible joints additional supports may be needed to avoid excessive distortion and failure of the joint ring.

The ability of a pipe to support itself between saddle supports is determined by bending theory. However, buckling of the pipe at supports may occur. DI pipe is best supported just behind the socket which provides additional stiffness and anchors the flexible joint. For large diameter steel pipes with large spans additional stiffness is necessary. AWWA M11 provides guidance for the design of ring girder stiffeners and of saddle supports for steel pipes as well as design guidance for the pipe bending condition.

BS EN 545 requires that flanged joints of DI pipe (and therefore the pipe also) should be able to resist a bending moment as well as internal pressure without visible leakage for a given duration. For pipe equal to or larger than DN 300 with screwed or welded flanges the values of pressure and moment are PN $\times$ 2 bar and approximately $884 \times (DN/1000)^{1.8}$ kN-m respectively with a test duration of 15 minutes. This moment is equivalent to four times that arising in a full pipe simply supported over a length of 12 m. For the equivalent service condition test (duration 2 hours) the applied moment is 25 per cent of the above value at a pressure of PN $\times$ 1.5 + 5 bar.

Above ground pipes are exposed to the elements. Greater provision for thermal movements and movement due to pressure (and cyclic loads due to pressure transients) has to be made. Use of special expansion bellows joints may be necessary. To protect against freezing insulation may have to be provided in cold climates; smaller pipes might need to be trace heated as well; AWWA M11 provides guidance for pipes in freezing conditions. Above ground, pipes are also exposed to vandalism and terrorism and may, in poor countries, be considered a source of free water.

IRON PIPES

15.8 CAST OR 'GREY' IRON PIPES

Many cast iron pipes made towards the end of the nineteenth century are still in use; their walls were relatively thick and not always of uniform, 'Spun' grey iron pipes were formed by spinning in a mould and produced a denser iron with pipes of more uniform wall thickness; they comprise a large proportion of the distribution mains in many countries. Three classes of such pipes were available: B, C, and D for working pressures of 60, 90, and 120 m respectively; classes B and C were more widespread. Carbon is present in the iron matrix substantially in lamellar or flaky form; therefore, the pipes are brittle and relatively weak in tension and liable to fracture. The manufacture of grey iron pipes has been discontinued in most countries, except for the production of non-pressure drainage pipes.

15.9 DUCTILE IRON PIPES

Ductile iron pipes are normally cast by centrifugally spinning molten iron in high quality steel moulds; fittings are cast in static moulds. The iron contains small quantities of magnesium to transform the lamellar form of carbon into a spheroidal form, thereby increasing tensile strength

Table 15.2 PFA, PMA and PEA values for K9 ductile iron pipes

Nominal diameter (DN) mm	Table 14 works leak test pressure bar	PFA bar	PMA bar	PEA bar
<200	50	64	77	96
200	50	62	74	79
300	50	49	59	64
400	40	42	51	56
600	40	36	43	48
900	32	31	37	42
1200	25	28	34	39
1600	25	27	32	37

and ductility. Ductile iron pipes are available in sizes up to DN 1600 with socket and spigot ends suitable for forming push fit type joints, plain ends suitable for jointing with flexible couplings, or with flanged ends formed by welding on ductile iron flanges complying with BS EN 1092-2. Pipes larger than DN 2000 are manufactured but with joints other than the push fit type. Standard lengths vary depending on the diameter of the pipe, the type of ends required, and where the pipe is manufactured. In the UK standard lengths for socket and spigot ended pipes are 5.5 m up to DN 800 and 8 m for DN 900 and larger. Plate 27(a) shows DI pipe being laid.

BS EN 545 requires wall thickness, e in mm, for pipes and fittings to comply with the formula:

$$e = K(0.5 + 0.001 \text{ DN})$$

where DN is in mm; $K = 9$ for socket/spigot, plain ended and welded flange straight pipe; $K = 12$ minimum for fittings without branches, e.g. bends and tapers; $K = 14$ minimum for fittings with branches e.g. tees. BS EN 545 also sets out tolerances on thickness.

The allowable operating pressures excluding surge (PFA) and including surge (PMA), and the maximum allowable test pressure after installation (PEA), vary with the diameter of pipes as shown in Table 15.2 for selected sizes, together with the corresponding works leak test pressures for centrifugally cast pipe.

The PFA values allow for a safety factor of three on the ultimate tensile strength of ductile iron of 420 N/mm². PMA, the allowable maximum operating pressure including surge is approximately 20% more than the PFA. PEA, the maximum allowable hydrostatic test pressure after installation, is in general PMA plus 5 bar, except for pipes smaller than DN 200 where PEA is 1.5 times PFA. When ordering pipes and fittings, specification of the operating and test pressures is vital, so that pipes and fittings, particularly tees, can be made to suit. For very high pressures (PFA typically 25 bar or more, but depending on the diameter and types of fittings) availability should be checked. Joints are required to be watertight at PEA and to durably withstand without leakage the PMA pressure under service conditions, including angular, radial and axial movement. BS EN 545 sets out a design methodology and gives a table of allowable external loads for selected backfill conditions.

15.10 EXTERNAL COATINGS AND INTERNAL LININGS

Current UK practice for external coating of ductile iron pipe comprises spray coating with zinc onto the exterior of the pipes, followed by a coating of bitumen paint. For aggressive soils polyethylene sleeving, either factory or site applied, is often adopted, plus imported backfill where appropriate. Minor damage and puncturing of the sleeving does not impair the efficiency of the protection, although any such tear should be patched with adhesive tape; the sleeving holds a virtually static body of water against the pipe. The effectiveness of the wrapping, even though the sheeting is not watertight, is ascribed to insulation of the pipe from uneven soil contact (which can produce galvanic cells) due to uneven pipe bed and surround, particularly in the case of clay soils which are hard to compact near the bottom of the pipe. For highly aggressive soils the pipes can be wrapped with a heavy duty PVC backed bitumen adhesive tape, overlapped 25 mm or 55%. Cathodic protection is not normally recommended but, if used, requires bonding across the joints to provide electrical continuity and tape wrap to reduce current and anode consumption. Details of corrosion protection provided, especially for joints, can vary—manufacturer's advice should be sought for each application. A points system, depending on soil resistivity, ground water level and other characteristics, can be used to judge the potential severity of aggression in a given location.

Earlier ductile iron pipes were coated with cold or hot applied bitumen or hot applied bitumen-based material sprayed or brushed onto the pipe metal. Use of coal-tar products for pipe coatings and linings was discontinued because coal-tar can give rise to PAHs in potable water (Section 6.42). Internal lining of ductile iron pipes in use in UK comprises spun mortar lining, using sulphate resisting cement plus, for pipes DN 800 and smaller, an epoxy seal coat on top of the cement mortar. Details of cement mortar lining are given in Section 15.14.

15.11 JOINTS FOR IRON PIPES

Types of joints in use are shown in Figure 15.1.

The run lead joint for spigot and socket pipes is now superseded by the various proprietary mechanical joints, but many mains still in use have been laid with this joint. Skilled workmanship is required to make the joint properly. The lead is heated to 400 °C (at this point strong rainbow colours are revealed when the surface scum is drawn aside). A clip is placed around the pipe against the annular space of the socket and the lead is poured in one continuous pour through an opening left at the top of the clip. The operation is potentially dangerous and the socket space must be completely dry to avoid blowback of the lead by steam. The lead solidifies almost immediately after pouring and is then caulked up using a series of chisels. The joint is rigid and even slight movement tends to cause such joints to weep; however, countless mains have been satisfactorily laid with this type of joint.

Flanged joints are covered by BS EN 1092-2. Flanges are machined from heavy castings so that the face is perfectly flat. Bolt hole spacings, position and diameter are set by appropriate standards according to pressure. Flange gaskets should be manufactured from high grade, non-biodegradable natural or synthetic rubber to a thickness of 3.2 to 4.8 mm.

FIGURE 15.1

Joints for iron pipes.

Bolted or screwed gland joints work by forcing a rubber ring into an annular space with a cast iron gland ring which is drawn by bolts or screwed into the socket but have not been used for water pipes for several years.

Flexible couplings such as the Viking Johnson (or Dresser) couplings are used for connecting together lengths of plain ended pipe. This joint also uses rubber rings compressed into an annulus and can be used with ductile iron pipes and is very widely used with steel pipes. This type of joint can accept slight angular joint deflection (reducing with increasing pipe diameter). If specified without a central register the coupling can be moved along the pipe, allowing removal of a section.

The Victaulic coupling is used in conjunction with shouldered ends of pipes, thus holding them together longitudinally. The joint can be unmade and remade without difficulty and is most often used for temporary pipelines laid above ground.

The push fit joint includes a specially shaped dual hardness synthetic rubber ring gasket fitted into the socket of a pipe before the spigot of the next pipe is pushed in. A little lubricant must be used on the inside of the gasket and on the outside of the pipe spigot before forcing it into the socket. Complete cleanliness of the socket, gasket, and spigot is essential. These joints are flexible: BS EN 545 requires a minimum possible angular deflection of 3.5° up to DN 300, 2.5° for DN 350 to 600 and 1.5° for larger sizes. Allowable deflections for pipes produced in the UK are 5° up to DN 300 and 4° for larger sizes. These joints are the norm for ductile iron pipes due to their low cost, simplicity and the ease of installation.

STEEL PIPES

15.12 STEEL PIPE MANUFACTURE AND MATERIALS

BS 534 covers carbon steel pipes, joints and specials (bends and other fittings) but is partly replaced by BS EN 10224 (pipe ranging from 26.9 to 2743 mm outside diameter using steel of yield strengths 235, 275 and 355 N/mm^2) and by BS EN 10311 for joints. BS EN 10312 covers stainless steel pipe. EN standards have been published and others are under development for polyethylene, galvanized, liquid epoxy and polyurethane coatings, mortar linings and external concrete and insulating coatings, but reference can be made to BS 534 for coatings and linings. BS EN 10224 uses pipe consistent with BS EN 10216-1, 10217-1 and 10220, but pipe to ISO 3183 (API 5L) and other standards can be used. CP 2010 Part 2 for design and construction of steel pipes on land remains current. BS EN 1295-1 covers structural design of buried pipelines for the water industry. Eurocode 3: BS EN 1993-4-3, issued in 2007, applies to design of steel pipelines which are not treated by other European standards covering particular applications; it can be used as soon as its national annex is published. This code requires consideration of 5 ultimate limit states, including fatigue, and three serviceability limit states including vibration. PD 8010-1 and -2 are intended primarily for oil and gas pipelines but apply to and provide useful design information for the water industry.

BS EN 10224 covers four principal welding methods for manufacture: butt (BW)—outside diameter up to 114.3 mm; electric (resistance) welded (EW)—outside diameter up to 610 mm; seamless (S)—outside diameter up to 711 mm and submerged arc welded (SAW)—outside diameter 168.3 to 2743 mm. In ISO 3183 the designations EW and SAW are recognized but seamless pipe is designated SMLS and LW means laser weld.

Steel pipes are fabricated from steel plate bent to a circular form or they may be continuously produced from a coil of steel strip bent to a spiral and butt welded along the spiral seam. Joints between coil ends of spiral welded pipes are known as skelp end welds. Butt welded pipes are made from rolled strip with a longitudinal seam furnace butt welded by a continuous process. Lengths of pipe are usually in the range of 9 to 12 m dependent on manufacture, transport and project requirements. Weld beads must be machined flush with the pipe surface at pipe ends to make them suitable for joint couplings. Spigot and socket ends, where shaped, are formed by die. Weld bead height needs to be limited for coating and lining. Electric (resistance) welding is done by passing electric current (by induction or direct contact) across the edges which are joined under pressure, without filler metal. Heat treatment at least of the weld zone is usual in sizes larger than DN 200. EW pipes now tend to be known as HFI (high frequency induction) pipes. Inspection typically includes chemical and mechanical material tests, ultrasonic inspection of plate and welds, radiography of welds and hydraulic pressure tests.

There are no standard classes for steel pipes: wall thickness above about DN 750 is designed for handling; internal pressure; buckling under external pressure and internal sub-atmospheric pressure; and to limit deflection when buried. External load carrying capacity in trunk mains is mostly a function of the backfill and compaction design. BS 534 sets out nominal wall thicknesses considered to be the minimum for handling and typical buried installations.

Steel grades as designated in ISO 3183 and, as from 2008, the American Petroleum Institute standard API 5L are designated by grade and by yield stress in thousands of psi, as Table 15.3. Grades less than grade B would not normally be used. Grades up to about X60 can normally be welded without special heat treatment. Their price is only marginally above that for grade B and

Table 15.3 Steel grades to API 5L / ISO 3183

Grade		A25	A	B	X42	X46	X52	X56	X60	X65	X70	X80
Yield strength	psi	25 400	30 500	35 500	42 100	46 400	52 200	56 600	60 200	65 300	70 300	80 500
	N/mm²	175	210	245	290	320	360	390	415	450	485	555

provide good economy where high pressure or (typically for pipes above ground or installed underwater) significant longitudinal bending resistance is required.

AWWA M11 gives a range of thicknesses and pressures and steels for diameters up to 4000 mm. Sizes in M11 are designated by outside diameter below 30 inches (762 mm), otherwise by inside diameter.

Pipe wall thickness, t (mm) for internal pressure is determined by hoop stress, as follows:

$$t = \frac{PD}{2a\sigma e}$$

where P is the internal pressure (N/mm²); D is the external diameter (mm); a is the design or safety factor; σ is the minimum yield stress (N/mm²); and e is the joint factor. The design factor, joint factor and definition of wall thickness depend on the design code. Design factors for hoop stress typically range from 0.4 to 0.8; the joint factor is 1.0 for SAW pipes and certain codes require the negative tolerance to be deducted from wall thickness. ASME codes B31.4 and B31.8 quote a basic design factor of 0.72 and state that this includes for thickness tolerance. For water supply under normal conditions, it is suggested here that the design factor of 0.5 (as given in AWWA M11 and the WRc pipes selection manual) is overly conservative and that, for high pressure long distance pipelines, a factor of 0.72 is realistic (after deducting thickness tolerance and any corrosion allowance) and up to 0.83 may be considered in some circumstances (PD 8010, BS EN 14183). For many water supply pipelines wall thickness is determined by handling and installation and the need to control deflection.

Further consideration can be given where particular conditions warrant: for example the American Society of Mechanical Engineers (ASME) code B31.8 quotes design factors for a variety of laying conditions. Where necessary the analysis can be elaborated to include ring bending, longitudinal bending, longitudinal stress from temperature changes, Poisson's ratio effects on buried (and thus restrained) pipe under hoop tension, combined (equivalent) stresses and where appropriate, for example for underwater pipes, can include strain based design.

BS EN 10224 and BS 534 give dimensions for common fittings, for example bends and branches. However, fittings can be made to any dimensions required, bends being made by cutting and welding together sections of pipe. For outside diameters up to 1016 mm, bends can be made by forming. Design of fittings and of any reinforcement needed is described in AWWA M11.

15.13 EXTERNAL AND INTERNAL PROTECTION OF STEEL PIPE

External and internal surfaces of pipes and specials are protected against corrosion by a coating and lining. It is not normally economic in design of water pipelines to provide a corrosion allowance (although, for comparison, corrosion allowance and corrosion inhibitors may be adopted in petrochemical practice). Corrosion protection may be carried out by the manufacturer at the place of manufacture or elsewhere or by a specialist. Pipe out of roundness and straightness can affect application of coatings and linings; both parameters must be specified to suit the process of application of the corrosion protection.

Principal options for *external coating* to water pipes are bitumen sheathing; fusion bonded epoxy (FBE); three layer polyethylene (3LPE) (Plate 27(b)) and liquid coatings (paints). In all cases cleaning and preparation by grit blasting to BE EN ISO 8501 (BS 7079) second quality (or Swedish Standard Sa 2.5) is a prerequisite.

Bitumen sheathing is included in BS 534 and consists of a hot applied bitumen with an inert filler, reinforced if required with a woven glass cloth, to a thickness of 3 mm for small diameter pipes rising to 6 mm for diameters exceeding 350 mm. Bitumen enamel wrapping consists of a hot applied bitumen containing a mineral filler with an inner wrapping of glass tissue and an outer wrapping of bitumen impregnated reinforced glass tissue or composite glass fibre fabric to the same protection thickness as for bitumen sheathing. All reinforcement materials are spirally wound onto the pipe with adequate overlaps. The pipes are painted with emulsion to act as a reflective and non stick surface; alternatively, sacrificial galvanized steel sheet may be used for both solar reflection and protection against damage during transit. Bitumen can biodegrade and as an alternative, coal tar has been used, particularly for underwater pipelines, but has lost favour in view of health and safety issues. Both bitumen and coal tar lose volatiles on exposure and then crack. Bitumen still presents a viable option for buried pipelines but FBE and PE coatings are now preferred. Alternatives should be used for exposed pipelines.

FBE has been in use since the early 1970s for *external* and *internal coatings*. The pipe surface is prepared by grit blasting and by phosphate or chromate pre-treatment. The FBE coating is applied as a powder which is fused onto the preheated pipe surface and cured chemically to form a layer thickness typically about 450 ± 100 μm thick for pipelines onshore and 650 ± 100 μm for pipelines offshore. FBE coatings generally have excellent resistance to cathodic disbonding when used on cathodically protected pipelines. Protection at joints is completed by heat shrink sleeves, tape wrap, by paint or by fusion bonded epoxy.

Three layer polyethylene systems have been in use since the early 1980s, more in Europe than in the USA, and supersede the two layer system. The three layer system is more expensive than FBE but is considered cost effective in view of potential damage during shipping over long distances. The system comprises the following typical thicknesses:

fusion bonded epoxy: 150 to 200 μm;
copolymer adhesive: 200 to 350 μm, and finally
polyethylene: 2.5 mm up to DN 750 and 3.0 mm for larger sizes.

Total thickness for the three layer system is thus between about 3 mm up to DN 750 and 3.5 mm for larger sizes. After surface preparation by grit blasting and phosphate or chromate pre-treatment, the pipe is preheated and application of all three components takes place while the pipe is rotated about its axis and moved forward through a specially designed booth. Timing is vital and curing takes place before the pipe reaches supporting rollers. Completion at joints is by heat shrink sleeves (favoured for larger pipe diameters) or by tape wrap.

Paint options include epoxy and polyurethane and may include a zinc rich base layer. Thickness typically is similar to that for FBE; application is by airless spray. Polyeurethane lining would typically be about 2 mm thick. Coal tar blends with epoxy or eurethane are not suitable for lining water supply pipes. Good preparation and protection is needed for fittings, which tend to be anodic and more prone to corrosion due to the additional welding and working. Small diameter steel pipes are galvanized. Polypropylene is not generally used for water supply. Other developments continue including blends of FBE, adhesive and PE and some manufacturers may also provide a thin additional coating of concrete for mechanical protection. Cement mortar external coating is covered by American standards.

The continuity of the applied protection (other than cement-mortar) is checked with a 'holiday' detector. A scanning electrode in the form of a brush containing a high voltage electrical charge is passed over the coating and lining; the voltage is set so as to produce a spark length of 10 mm or double the specified minimum thickness of the protective material whichever is the greater. Pinholes or breakages are disclosed by an electrical discharge to the steel of the pipe and this can be arranged to cause a buzzer to sound.

Internal surfaces of steel pipes can be protected with concrete or cement-mortar (Section 15.14), epoxy or polyeurethane. Market forces and perceived quality issues in the UK currently favour epoxy. Epoxy coatings may be hot applied FBE or airless spray liquids and high solids systems. Thickness is typically about 0.5 mm and surface preparation is by grit blasting to a high standard. Bitumen was a popular and successful option but is no longer used in the UK. A hot applied bitumen with or without an inert filler is sprayed or brushed onto the pipe, after first priming it with a compatible priming coat; usual practice was to apply a minimum thickness of 1.5 mm for diameters up to 300 mm and 6 mm for diameter exceeding 1000 mm.

15.14 MORTAR AND CONCRETE LININGS

Steel and ductile iron pipes can be lined with cement-sand mortar or spun concrete and these form some of the best and most cost effective types of interior protection. The cement mortar or concrete lining generates a high pH environment which passivates the metal and prevents corrosion. Cement mortar lining (CML) may be either factory applied by centrifugal spinning or may be applied in situ by spraying (orange peel finish) followed, at large diameters and if required at small diameters, by trowelling to a smooth (washboard) finish. BS 534 covers both mortar and spun concrete and AWWA C205 and C602 cover shop applied and in situ mortar respectively. BS EN 10298 covers mortar linings. Concrete lining has been chosen in preference to CML on some projects because it is thicker (typically 25 mm instead of 12 mm at DN 1400 and larger) and because the aggregate is larger, therefore reducing shrinkage and presenting a more robust and durable surface.

Aggressive waters (as indicated by the Langelier index, Section 10.41) dissolve the cement and increase pH, causing water quality problems particularly for small diameter pipelines. However, the cement lining may be sealed with an approved thin (150 to 250 μm) layer of epoxy paint. The seal coat provides a barrier to solution of the cement mortar by aggressive low pH, low calcium waters but it is not particularly robust. Where pH cannot be raised, for example by dosing with lime or, where dosing cannot be guaranteed due to materials supply or operational uncertainties, epoxy or other lining should be used.

Cement mortar linings and concrete linings are thicker than epoxy. They therefore result in a slight reduction in pipe diameter (or require a slightly larger external diameter to obtain the same finished internal diameter). Although cement mortar or concrete linings may, after a while in service and with biofilm formation, attain a hydraulic roughness similar to epoxy, they in general tend to have a greater initial hydraulic roughness than epoxy. These factors may lead to epoxy being a more cost effective solution if energy costs are taken into account, depending on the circumstances. However, for typical civil engineering applications no distinction is made between lining types when considering long-term roughness values. The current trend away from concrete and mortar lining and towards epoxy lining for new steel pipe may be linked to market forces, and also to the need for material in contact with water used for the public supply (including raw water mains) to meet the required approval process. The approval process is time consuming and entails continued control of materials sources, including cement, as for concrete for water retaining structures and linings for ductile iron pipes. Epoxy lining on its own would be an expensive solution for ductile iron due to the surface roughness after grit blasting; the pipes are cement mortar lined at all sizes and in the UK have a seal coat on pipes only of DN 800 and smaller.

Bare metal pipe is spun at high speed on rollers and the mortar is poured as a slurry into the interior. The lining builds up by centrifugal force and when the required quantity has been poured the speed of rotation of the pipe is increased. The mortar compacts further and surplus water runs off as the pipe is tilted very slightly. The mortar mix comprises sand or fine crushed rock aggregate mixed with Portland or sulphate-resisting cement in the ratio of 2.5:1 or 3:1 by weight. The lining is usually so well compacted after spinning that the pipe can be immediately taken off the spinning bed and is then cured in a damp warm atmosphere for 21 days. The mortar lining thickness for steel is generally thicker than for ductile iron pipes of the same diameter because of the flexibility of steel pipes compared with ductile iron pipes. All fittings associated with mortar lined pipes are also mortar lined but this has to be trowelled on. The discontinuity of the lining at joints in steel pipes should be filled if the pipe is of large enough diameter to give inside access. Steel pipes which are not large enough for this should either be jointed using flexible joints or be jointed by externally welded collar joints. With the latter the lining is brought flush with the pipe ends which are butted together; the heat input is not sufficient to damage the lining. The collar is coated before assembly; any water between pipe and collar is largely stagnant giving low risk of corrosion.

15.15 JOINTS FOR STEEL PIPES

The following types of joints are in common use: welded sleeve joints (Figs 15.2(a) and (b)); butt welded joints; flexible couplings (e.g. Viking Johnson type) (Fig. 15.1); flanged joints (Fig. 15.2(c)) and push fit joints.

The BS EN 10311 Type 1 taper sleeve joint (Fig. 15.2(a)) is common in the UK. It permits a small angular deviation (up to 1°) as long as the joint is arranged so that there is a penetration of at least four times the pipe wall thickness after deflection. The increased gap around the outside of the pipe requires the weld to be buttered: filler rod can be used. The internal joint gap should be fairly constant as the mating surfaces are brought together. Pipes of DN 700 and larger should be welded internally and externally. The internal weld is the strength weld and the outer is a seal weld. The joint can then be tested by pressurising the annular space between the two welds (usually with nitrogen at a pressure of 200 kPa), which permits pipe joints to be tested before backfilling the trench. On pipes smaller than DN 700 the weld should be made on the outside only. Consideration needs to be given to access and the completion and inspection of the internal protection. On pipes DN 600 and smaller, the alternative of using mechanical or other joints can be considered.

BS EN 10311 Type 2 collar joints comprise a short sleeve usually in two parts which is slipped around the pipe end and joined by fillet welding. It is used for jointing small diameter pipes where internal access is not possible and for jointing closing lengths on larger diameters. Where access is possible welds should be made on the inside as well and each side should be air tested.

Sheathing and lining made complete after test

(a) Welded short sleeve joint for steel pipes

(b) Welded spherical sleeve joint for steel pipes

(c) Flanged joint for steel pipes

FIGURE 15.2

Joints for steel pipes.

The spherical joint (Fig. 15.2(b)) was developed to permit angular deflection without unduly increasing the size of the external weld and has been found practicable and more satisfactory than the short sleeve joint because of the smaller welds required. It is not recognized in BS EN 10311 but has been frequently and successfully used on large diameter pipes outside UK. It allows larger deflections than the Type 1 joint, thus allowing gentle curves to be followed without use of manufactured bends. There must be an overlap of at least four times the pipe wall thickness.

For butt welded joints the ends of pipes are prepared by forming a 30° bevel on the full thickness of the pipe wall except for the inner 1.6 mm; the resulting V-groove between pipes is then filled with weld metal which is finished to stand proud of the external pipe surface. Butt welding is usually only adopted when the pipe wall thickness is substantial or when full longitudinal continuity of strength is required (such as in high pressure applications), in underwater pipes or if pipes are to be snaked into the trench.

Flexible couplings permit modest angular deviation of pipes at joints. Longitudinal movement is also possible but is restricted to the amount allowed by shear movement within the rubber. Further movement will drag the rubber over the pipe surface eventually damaging the joint. Once filled and conveying water, temperature movements are usually small even on an exposed pipeline in tropical climates; therefore it is possible to use these couplings, spaced at say two pipe joints (24 m centres) with an anchored joint between, to avoid the need for special contraction or expansion (bellows or slip) joints.

The continuity of the internal lining needs to be completed at the joint. This can be done by entry into the pipe when its diameter exceeds 600 mm or from outside when its diameter is smaller. In the latter case a former is pushed into the pipe on a rod and expanded against the pipe wall at the joint; protection material is injected through a hole in the coupling to fill the void at the collar. Alternatively a lining train can be used to apply a measured quantity of paint by roller or spray. The outside coating is completed by heat shrink sleeves or by tape wrap with an overlap of at least 50%. For bitumen sheathed pipes the external protection is completed by 'flood coating'—putting a mould over the coupling, filling it with bitumen and allowing it to cool before removing the mould.

Pipes for jointing with couplings or but welds must be truly circular at the joint. Pipes that are intended to be cut for jointing must be ordered as 'true diameter throughout'. Out of roundness can cause difficulty in jointing when inserting a branch into an existing pipeline at a point where its diameter may not be true.

15.16 CATHODIC PROTECTION

Clays, sulphate-bearing soils, moorland acid waters and saline ground-waters are the principal causes of aggressive ground conditions. Cathodic protection provides an additional measure of security to cover breakdown and deficiencies in the external coating.

Cathodic protection uses electrolytic chemical action to protect pipes by providing the pipeline at intervals with sacrificial anodes or, alternatively, by impressing a (negative) voltage on the pipeline so that it becomes cathodic with respect to buried anodes in the medium of the wet soil which acts as a weak electrolyte. Corrosion takes place at the anode and not at the cathode. Sacrificial anodes are usually made of zinc, magnesium or aluminium. They are connected electrically to the pipe and cause a weak current to flow in the right direction since they are 'anodic' with respect to iron. They are buried a few metres away from the pipeline at 50 to 100 m intervals and are connected to it by insulated cables. If impressed current is used a d.c. negative voltage is applied to the pipeline and the anodes wired to it can be made of iron as the impressed voltage is sufficient to drive current in the right direction. Impressed current anodes are silicon iron, titanium and more currently, mixed metal oxide. Graphite is no longer used. This type of cathodic protection is used where the soil resistivity is high or where long stretches (typically more than 5 km) of pipe need protection since it can provide much higher potential driving forces than the sacrificial system. The lengths of pipe to be protected by a given anode must be electrically connected together. Where joints use rubber rings the pipes either side must be electrically connected by an insulated conductor welded to each.

Cathodic protection should not be used without pipe coating to which it provides an additional defence. Where other services exist it cannot be used without care since effectiveness of the system may be reduced by draining of current to the other services and since corrosion of those services might be induced.

Where a steel pipeline passes parallel to and up to about 200 m away from overhead high voltage lines, stray electric potential may be induced in the pipeline. This requires the pipeline to be earthed, using either zinc anodes or through diodes which limit the current flow to one direction and only above a predetermined limit. A steel pipeline should be earthed during construction to prevent welding current from damaging coatings and to earth any induced currents. The length of pipe strung above ground should also be limited typically to about 400 m—and a strict safety regime implemented. Exact precautions need to be determined according to circumstances.

PLASTIC PIPES

15.17 POLYETHYLENE PIPES

Polyethylene (PE) is a thermoplastic and has been widely used for the production of pipes by an extrusion process for over 40 years. Pipe is produced in the UK up to DN 1000 and is available from Europe up to DN 1600. PE pipe is specified by nominal (minimum) outside diameter (OD), by standard dimensional ratio (SDR—the ratio of nominal outside diameter/minimum wall thickness, typically 11, 17.6 and 26) and by material. Material classifications in current use are PE 80 and PE 100, where the number refers to the minimum required strength (MRS) in bar (80 bar = 8 N/mm^2). The MRS is the pipe wall burst stress at 20 °C at 50 years, extrapolated from shorter duration tests.

Materials terminology is confusing: current trend is to refer to grade (PE 80 or PE 100). PE 80 includes both medium density polyethylene (MDPE) and high density polyethylene (HDPE). MDPE was brought into service in the UK in the 1980s to replace low density polyethylene (LDPE) and high density polyethylene (HDPE). To take account of the risk of crack propagation (RCP) reported for gas pipes, PE 80 pipes DN 250 and larger were derated by about 20% and, above DN 315, were further derated for pipes containing more than 10% free air. This derating has been reviewed in the light of recent research; derating for fatigue is no longer considered necessary for modern PE 80 pipes that meet the stress crack requirements of relevant Water Industry Standards (WISs) but is required for pipes that do not meet these high toughness requirements. PE 100 has been produced since 1990 and is also known as high performance polyethylene (HPPE). Compared with PE 80 it has greater resistance to RCP for which it needs no derating. PE 100 costs more than PE 80 but its greater strength allows higher pressures or thinner walls and hence provides economy under certain conditions.

Maximum operating pressures are controlled by the sum of hoop stress and ring bending stress, which is limited to 0.8 times MRS (BS EN 12201-1). Maximum operating pressure (PMA) includes surge. Recommendations were that the range of surge pressure should be limited to half the maximum operating pressure. However, since the transient pressure wave speed is lower in PE pipe surge pressures are typically about 25% of those in steel for the same event. Research shows that PE pipes can sustain surge pressures at least twice their static rating (the actual amount depending on the material and increasing with rate of strain). Attention must be given to derating for operating temperatures above 20 °C; BS EN 12201-1 gives derating coefficients up to 40 °C. Long term deflection is allowed up to 6% of the diameter. However, this must be calculated taking into account creep in the soil and using long term pipe stiffness—which may be one sixth of the initial stiffness.

PE pipe in Europe is to be specified to BS EN 12201. Whilst the relevant UK WISs are likely to be withdrawn eventually, UK practice with respect to certain aspects of some of these is recognized in BS EN 12201.

Advantages of PE pipes are that they are light and easy to handle, flexible but strong and resistant to cracking, do not corrode, are chemically resistant, have a low frictional resistance to water and can easily be cut to length. SDR 11 and 17.6 pipes can be bent safely to radii 25 times OD, increasing to 35 times in cold weather and reducing to lower values in warm weather. For SDR 26 and 33 pipes these radii need to be 50% greater. Small diameter pipe can be supplied in coils and straight lengths can be joined above ground and snaked into narrow non-man-entry trenches. Disadvantages are that the strength of pipe, defined as its ability to withstand hoop stress, decreases with time and reduces with increasing temperature. PE pipes are also liable to UV degradation if exposed overlong to sunlight. In the UK blue PE (for water supply) can be stored above ground for up to a year; for longer periods they should be covered. Under certain conditions they may be degraded or may be slightly permeable (and therefore give rise to taste problems) due to oils, organic solvents and by strong concentrations of halogens or acids. CP 312 Part 1 gives a list of chemicals and suitability with plastics. Where soil is polluted by hydrocarbons, typically in former gas works sites, fuel stations, scrap yards and other industrial areas, it may be enough to use imported backfill and impermeable membranes: alternatively other measures or other pipe materials may be needed. The coefficient of linear expansion of PE 80 is about 1.5×10^{-4}, which is more than ten times greater than for steel and is equivalent to about 9 mm for 10 °C in a 6 m pipe length. PE pipes are not traceable underground with metal detectors and do not transmit leak noises like metal pipes. For distribution mains PE pipes are supplied in 6 m or 12 m straight lengths; their flexibility makes the use of small angle bends largely unnecessary.

PE pipes can be jointed mechanically with push fit joints (with longitudinal strength) but are more usually jointed by butt fusion or electrofusion thus forming a continuous string and do not need thrust blocks. Butt fusion requires a purpose-designed powered machine in which the prepared and cleaned pipe ends are clamped and held under longitudinal pressure against a thermostatically controlled heated plate (Plate 27(d)). After a given heating time the ends are pulled back, the plate is removed and the two melted ends of the pipe are brought together under longitudinal pressure to fuse and cool. Growth of the bead of melted material is observed during heating. After jointing the external bead is removed, numbered with the joint and twisted at several positions. Any split in the bead requires the joint to be cut out and made again; any further defect requires investigation and cleaning of the equipment and new trial joints before production is allowed to continue. Internal weld beads can be removed if required. Good quality assurance and quality control procedures are essential and should include tracking to monitor performance of individual and gang workmanship. Experienced and disciplined inspection is essential. Electrofusion joints are used at fittings and comprise a spigot and socket with in-built electrical heating coils which are energized to locally melt and fuse the fitting and pipe together. Equipment can be provided to read a bar code fixed to the fitting which contains information to control the weld. The equipment can then store weld data for downloading onto a computer database for traceability and quality assurance. Flow of melt material from indicator holes in the fitting gives visual evidence of weld completion.

Pressure testing must take into account the viscoelastic properties of the pipe material. Traditional pressure test techniques are not suitable due to creep and stress relaxation; pressure in a closed PE pipe falls with time even when the pipe is leak free. The accepted method is to fill the pipe with water, raise the pressure, measure the pressure decay with time, derive decay parameters using a 'three point analysis' (more points are strongly suggested) and compare with a range of standard values according to pipe restraint. The pipe must be sensibly free of air and not above 30 °C. Pressure should be released slowly. If a retest is required, a rest time, typically of about 5 times the previous test duration, must be allowed beforehand for the pipeline to recover from the previous conditions. Recommended test pressures are:

- for systems rated up to 10 bar, 1.5 times *Rated Pressure*
- for 12 and 16 bar systems, 1.5 times *Working Pressure*.

The test pressure should not exceed 1.5 times the maximum rated pressure of the lowest rated component.

15.18 POLYVINYL CHLORIDE (PVC) PIPES

Three types of PVC pipes available in the UK are:

- PVC-U: unplasticized PVC;
- MOPVC: molecular orientated PVC and
- PVC-A: PVC with added impact modifiers—also known as PVC-M (modified PVC).

MOPVC is first formed at about half the diameter and double the final thickness and is then heated and expanded to the final diameter. The result increases the strength particularly in the hoop direction and allows use of reduced wall thicknesses (slightly more than half of those for PVC-U). It also increases significantly the resistance to failure from cyclic loading. MOPVC pipes are about 30% of the stiffness of PVC-U pipes due to their reduced thickness but taking into account their greater modulus. PVC-A is an alloy of PVC-U, polyethylene (PE) and acrylics: wall thicknesses are about 80% of those for PVC-U.

A number of failures of early PVC-U pipes led to lack of confidence in their use. The problems are now understood to result from local bending stress from point loads, typically uneven bedding. Material improvements have restored confidence but:

- pipe should be derated for temperatures above 20 °C, derating coefficients are given in BS EN 1452-2 for up to 45 °C;
- pipe with significant internal or external scratches should be rejected due to notch sensitivity;
- care is required with installation;
- care is required during tapping and not to over tighten saddles, and
- derating is required for repeated cyclic (fatigue) loads (distinct from diurnal variations), particularly for pumping mains.

PVC pipes are classified by nominal outside diameter and by pipe series (S) where: $S = (SDR-1)/2$. PVC-U pipes are available in the UK up to DN 630 and are supplied in 6 m lengths, with spigot and socket rubber ring push fit joints. Solvent joints have not been used since the mid 1980s and are no longer permitted for underground use (primarily due to problems arising from movement during bedding and backfilling) but are permitted for above ground use. BS EN 1452 replaces BS 3505 and the relevant Water Industry Standards (WISs) which will be withdrawn, but UK practice with respect to certain aspects of these is recognized in BS EN 1452. Typical PVC-U pipe material has a relative density of 1.4 and a softening point of about 80 °C. The coefficient of linear expansion is about 6×10^{-5} per °C which is more than five times the value for steel.

Wall thickness for PVC-U is based on an allowable design stress of 10 N/mm^2 for pipe OD up to 90 mm and 12.5 N/mm^2 for OD 110 mm and higher. These compare with the minimum required strength (MRS) specified in BS EN 1452-2 of 25 N/mm^2 and the required minimum bursting stress of 35 N/mm^2 at 100 hours. The design stresses for MOPVC and typically for PVC-M are 22.5

and 18 N/mm^2 respectively. Full structural design would not normally be needed for DN 315 and smaller.

Maximum operating pressures are controlled by the sum of hoop stress and ring bending stress which should not exceed the design stress. Maximum operating pressure includes surge. Recommendations were that the range of surge pressure should be limited to half the maximum operating pressure, but because the wave speed in the pipe wall is lower, surge pressures in PVC pipes are less (typically 30% less) than those in steel for the same event. Research shows that PVC pipes can sustain surge pressures about twice their static rating (about 1.8 times for PVC-A and 2.2 times for MOPVC and PVC-U, the actual amount increasing with rate of strain). Care should be taken for pumping mains larger than DN 315 and fracture toughness tests should be specified as set out in BS EN 1452.

The main advantage offered by PVC is its resistance to corrosion; hence its use for chemical transfer lines in water treatment works. It is not suitable for use in ground contaminated or likely to be contaminated by detergents or solvents or from oil storage areas. It is light in weight, flexible and has easily made joints. Small size pipe can be joined above ground and 'snaked' into narrow non-man-entry trenches. Care should be taken when handling pipe at temperatures below freezing due to reduced impact resistance.

Fittings can be made in PVC-U or metal, usually ductile iron. PVC-U pipes are degraded by ultraviolet light, the effect increasing with temperature so that the pipes must not be exposed to sunlight in hot climates for more than a day or two. In the UK, MOPVC, PVC-U and PVC-A can be stored above ground for up to a year; for longer periods they should be covered.

15.19 GLASS REINFORCED PLASTICS (GRP)

GRP pipes are lightweight and corrosion resistant and are made from a polyester resin, glass fibre reinforcement and usually a filler of silica sand. GRP pipes have been in use, primarily for trunk mains, since 1970. Current standards are AWWA C950, BS EN 1796 and BS 8010 Section 2.5. Filler inclusion allows increase in wall thickness at modest cost and thus provides additional stiffness, an advantage for laying. By adjusting the amount of glass fibre and the wall thickness, the pressure rating and stiffness can be varied independently, unlike other flexible pipes. GRP is thermosetting (sets irreversibly under heat) as opposed to thermoplastic (PE and PVC, which melt on heating).

Pipes may comprise layers of different resins and different quantities and qualities of glass through the pipe wall: typically using resin rich and corrosion resistant qualities for the surface layers. Where different resins are used, the inner surface layer comprises a more flexible and chemically-resistant resin, e.g. bisphenol, and the structural and outer surface layers use a cheaper, stronger resin, e.g. isophthalic or orthophthalic. Several formulations are used, different for each resin supplier and may be peculiar to each pipe manufacturer. The outer surface layer should be formulated to provide handling robustness and to resist environmental conditions. As an alternative to thermosetting resin BS EN 1796 allows use of a thermoplastic liner and allows the internal diameter of such pipe to be up to 3.5% smaller than the nominal size (DN). Care is required to specify and ensure that the resins are suitable for encasement in concrete, if required; alternatively fittings in contact with concrete may be made of ductile iron or steel. Materials and the process must be suitable for potable water.

Like paints, thermosetting resins are slightly permeable. Consequently the whole pipe wall thickness is exposed to some extent to the liquid under pressure from one surface. Therefore, the types of resin and the glass (Types 'E' and 'C' are covered in BS EN 1796) should be selected to suit. Early problems with delamination have largely been overcome by improved materials and manufacturing processes but care has to be taken in selection of resins and in surface preparation of the glass fibres to ensure full wetting by the resin and resistance to corrosion.

Pipes can be made by the Hobas centrifugal casting process or may be filament wound on a mandrel. In the Hobas process metered weights of liquid resin plus filler and randomly orientated chopped glass strand are fed to the inside of a rotating mould. This produces a dense pipe wall with similar characteristics in the hoop and longitudinal direction, although fibre orientation can be adjusted if required to increase tensile capacity in a particular direction. The constant outside diameter allows pipes to be cut and joined at any position.

Filament wound pipes in the UK are made in discrete lengths on a rotating mandrel. For reasons of production economy, this arguably allows more choice in design of the pipe wall but may require more glass fibre to retain (fluid) material as the mandrel rotates. Modern filament wound processes allow filler to be to be incorporated to produce economic pipe of higher stiffness. The Drostholm process comprises a cantilevered mandrel wrapped by a continuous steel strip which advances helically around and along the mandrel, providing a continuous rotating surface before returning up the inside of the mandrel and commencing the helix again. Pipe materials are built up on this surface, being added sequentially along the pipe as the helix advances. Material is heated so that the pipe cures on the mandrel and can be cut into discrete lengths shortly after it leaves the mandrel. The Drostholm process is not used in the UK but pipe made by this process has been used on some major projects.

There is less experience with long, large diameter high pressure trunk pipelines in GRP than with steel or ductile iron. Manufacture is a complicated process which must come under skilled control and should preferably be automated. With improved materials and manufacturing control previous problems of delamination are now rare but deflections need to be controlled to avoid strain corrosion. Where such problems are suspected inspection using ultrasonic techniques should be adopted. Pipes must be laid under strictly controlled backfill conditions to prevent unacceptable distortion of the pipe wall or bending and must be designed for both short and long term conditions, typically using 50 year pipe characteristics extrapolated from shorter duration tests.

Pressure rating and stiffness can be tailored independently to suit project specific requirements; this benefits economy. The pipes are flexible and typically are designed to BS EN 1796 and installed to BS 8010 Section 2.5. The standard states that pipes of stiffness less than 1000 N/m^2 are not intended for laying directly in the ground. However, pipes of stiffness less than 5000 N/m^2 require use of expensive embedment materials and very close control of placing and compacting the embedment (Plate 27(c)). Unless it is clear that such measures will be effective stiffness should not be less than 5000 N/m^2.

Where GRP pipes are subject to cyclic loads the manufacturer must be informed of the detailed performance requirements. Thrust blocks at bends and fittings, on large diameter pipes particularly, need to be fully reinforced around the pipe because of the high strain in GRP under pressure. GRP pipes also need to be wrapped locally with elastic material at the exit from rigid structures; manufacturers can provide suitable details.

Lengths up to 18 m are produced (Plate 27(c)). BS EN 1796 sets out the initial minimum longitudinal tensile strengths for two cases: where the pipe (a) is, or (b) is not, required to resist internal pressure acting on a closed end. Where such resistance is required (a) the minimum tensile strengths imply a short term safety factor of about 2.8. It should be noted that the values set out in the BS for case (b) are higher in most cases than those obtained by application of case (a). Therefore, if case (a) applies the specifications should require the pipe to meet both requirements.

Joints generally are the flexible push fit spigot and socket or collar type. Alternatives are resin adhesive or flanges and screw threads. Angular joint deflection limits for push fit joints are at least

3° for DN 500 and less, 2° for DN 500 to 800, 1° for DN 900 to 1700 and 0.5° for DN 1800 and greater. Actual values at installation should not exceed half these values. Where required, rope or other locking strip can be threaded circumferentially into a preformed groove in push fit joints to lock and provide tensile strength across the joint, for anchorage and for installation under water.

GRP fittings such as bends and branches may be fabricated or moulded; the latter being made by building up filament, resin and filler in a mould. Fabricated fittings are made from GRP pipe which is cut, glued together and laminated with glass fibre across the joint. Any bend angle can be supplied as well as multiple fittings as a one piece unit, for example a duckfoot bend and integral bellmouth. Alternatively ductile iron and PVC-U fittings can be used for DN up to 600 with Series (Section 15.18) selected accordingly. Flanges can be fabricated in GRP but any axial or bending loads must be specified.

GRP has a key advantage for carrying aggressive waters and or for laying in exceptionally aggressive ground conditions where both iron and concrete pipes would be severely attacked. A balance has to be drawn between resins with the best chemical resistance, such as vinyl ester but which may be brittle and crack due to deflection in service—and those with less chemical resistance but better physical flexibility. Modern practice is to specify GRP pipe by performance and leave resin choice to the manufacturer, who has links to resin suppliers and can tailor resin according to the pipe manufacturing process.

CONCRETE PIPES

15.20 PRESTRESSED CONCRETE PRESSURE PIPES

Prestressed concrete pressure pipes are covered by BS EN 639 and BS EN 642. The nominal size range is from DN 200 to 4000 and is the design internal diameter. American Standards AWWA C301-92 and C304-92 also cover large diameters. Examples of such pipe are typically DN 4000 up to 20 bar and DN 3600 up to 26 bar for North Africa. Design is restricted in principle only by material properties and capability of the structural section. Pipes are made by tensioning high tensile wire wound spirally around a cylindrical core. This core may consist either of concrete, which is prestressed longitudinally, or of a thin steel cylinder, which has a thick spun concrete lining to the interior as shown in Figure 15.3. For large diameter prestressed concrete pipes the core is cast vertically around the inside and outside of the steel cylinder. Pipes with steel cylinders in the core are called 'prestressed concrete cylinder pipes', the others with longitudinal prestressing are termed 'non-cylinder'. When the wires have been wound on to the core, stressed and anchored, a relatively thin but dense cement rich mortar coating (20 mm minimum) is applied pneumatically or by machine at high velocity as the pipe is rotated externally. The resulting hard dense mortar coating provides mechanical and corrosion protection.

The chief advantage of prestressed concrete pressure pipes is that they can offer a cost advantage over other pipes, particularly for large diameter and higher pressures. Prestressed pipes can be made to withstand higher pressures simply by increasing the number of turns of prestressing wire per unit length or by doubling up the layer of prestressing wire. A second advantage is that such pipes are proof against certain corrosive conditions that would attack iron and steel, although they may need special protection if the groundwater is high in sulphates and chlorides or otherwise aggressive to concrete. In very aggressive cases the pipes may need to be cathodically protected. This would involve the provision of electrical continuity to the prestressing wires, the steel cylinder and

FIGURE 15.3

Joint in prestressed concrete cylinder pipe.

joint rings at the ends of each pipe so that these can be bonded together electrically after the pipe is laid. Alternatively, or in addition, the pipes can be coated with urethane or epoxy. Coal tar enamel and or coal tar epoxy are not suitable as internal protection for pipes carrying water intended for human consumption. The pipes can also be made using sulphate-resisting cement. The pipes are used in the marine environment where buried and permanently submerged but attention is needed to prestressing steel quality and to combating aggressive conditions in the tidal and surf zone where the water level varies and where the oxygen supply is high.

The prestressed pipe steel joint rings provide a high performance O-ring seal allowing joint rotations of up to 0.9° in any direction at each joint while maintaining water tightness. Deep joints are also available which can accommodate up to 1.4° of rotation at any joint. Full bevel and half bevel joints which allow several degrees of deflection can be built into standard sections of pipe, enabling prestressed concrete pipes to deal with severe changes of grade and alignment (up to about 5°) before having to use preformed bends. Joints can be welded if required and can then carry longitudinal force.

The pipes are rigid thus providing a reasonable degree of resistance against rough handling and poor backfilling techniques. However, in the larger diameters the pipes are very heavy and difficult to align in very soft ground unless they are placed on a prepared bed of granular material of adequate thickness to prevent uneven settlement. Connections can be made after the pipeline has been laid and while the pipe is in operation although to avoid shutdown, tees have to be incorporated in the line as it is laid.

Cracking of the mortar coating does not occur under normal operating conditions (up to the design operating pressures and external loads), but it may occur if the design operating conditions are accidentally exceeded through pipeline operational errors. However, the elastic behaviour of prestressed concrete pipe is such that the cracks close up after the abnormal conditions have been corrected. The mortar coating cannot be prestressed directly but does receive some compressive strain as the concrete core continues to shrink after completion of prestressing since the coating is applied immediately after the prestressing operation and is mechanically bonded to the prestressing wires. This feature together with a feature known as tensile softening allows the mortar coating to take much more strain than the standard calculated plain strain before cracking. If prestressed concrete pipes are to be tested for long durations (say 24 hours) at a pressure greater than the design operating pressure, then the pipe must be designed accordingly: that is, the sustained test pressure should be considered as the design pressure.

The works hydrostatic proof pressure is the design working pressure plus surge (MDP) or 1.5 times the design working pressure (1.5 times PFA), whichever is the greater. Works hydrostatic proof pressure is required to be applied for at least one minute.

Joints for prestressed concrete pressure pipes are usually of the socket and spigot O-ring push fit type. The socket is normally mortared up afterwards and the joint then becomes rigid. Where ground conditions are known to cause differential settlement the joints may alternatively be filled with a bitumen mastic which provides the necessary protection while allowing joint rotation to occur.

15.21 REINFORCED CONCRETE CYLINDER PIPES

Reinforced concrete cylinder pipes are similar to prestressed concrete pipes, except that there is no prestress and instead of using high tensile steel wire for circumferential reinforcement, carbon steel rod is used. Size range is from 250 mm to 4000 mm internal diameter with wall thicknesses from 78 mm to 320 mm respectively, giving external diameters from 406 to 4640 mm. Effective lengths in the UK are about 6 m up to DN 2100, reducing to 2 m at DN 4000 and are available up to 10 m from DN 400 to 1000. Up to DN 1000 in the UK the reinforcing rod is wound spirally under low tension on to a rotating steel cylinder, being welded to the cylinder at both ends. The reinforced cylinder is then covered internally and externally with concrete, the internal lining being spun on and the external coating being applied by impact and smoothed in layers. At larger sizes the pipe is cast vertically (wet mix) using a preformed spot welded reinforcement cage with either one or two layers of reinforcement according to design requirements. In the design of the pipe both the steel cylinder and the reinforcement are assumed to resist hoop tension stresses.

The principal advantages of reinforced concrete cylinder pipes are:

- they can be designed for relatively high heads;
- they require less sophisticated manufacturing techniques than do prestressed concrete pipes;
- the concrete acts as a good protection to the steel and thus permits some saving over the amount of steel that would be used in equivalent steel pipes where a corrosion allowance is applied, and
- the pipes are rigid thus providing a reasonable degree of resistance against rough handling and poor backfilling techniques.

They can be used in the marine environment, albeit with care in the tidal and surf zone. Disadvantages are as mentioned for prestressed concrete pipes but, in addition, cracks in reinforced concrete cylinder pipes caused by operational overload may not close up in the same way as for prestressed concrete. Some degree of shrinkage cracking is normal after manufacture and can be accepted—typically up to 1 mm in surfaces which will be continually wet after installation (internal surfaces and surfaces constantly below water level) as they will close after wetting and seal by the process of autogenous healing. Reinforcement size and spacing are designed to limit structural crack widths under load as in water retaining, maritime and other codes. For carrying axial forces joints generally are welded and therefore are rigid requiring careful pipe bedding to limit differential movement. Rubber ring push fit joints and other joints can be made if required. Fittings are coated internally and externally with reinforced concrete.

BS EN 639 covers reinforced concrete pipes and BS EN 641 covers reinforced concrete cylinder pipes. Reinforced concrete non cylinder pipes are produced in the UK but are used for gravity flow and drainage.

AWWA C300 addresses reinforced concrete cylinder pipes that are not prestressed or pretensioned. It covers pipe from DN 760 to DN 3600, typically with wall thicknesses between 89 and 305 mm (3.5 inches to 12 inches). This type of pipe has a relatively thick welded cylinder towards the inside and a layer of bar reinforcement towards the outside separated by concrete which renders the pipe rigid. AWWA C303 covers reinforced concrete cylinder pretensioned pipes. The pipe is reinforced with a steel cylinder that is helically wrapped with mild steel bar reinforcement under moderate tension, in sizes from DN 250 to DN 1520 (10 inches to 60 inches) inclusive and for working pressures up to 27 bar. The lining thickness is 13 mm minimum for DN 250 to DN 410 and 19 mm minimum for DN 460 to DN 1520. The cement mortar coating thickness is 19 mm over the reinforcing bar or 25 mm over the cylinder, whichever results in the greater thickness of coating. Compared with reinforced concrete cylinder pipes, such pipes have thicker steel shells and thinner concrete and are semi rigid. They are not used in the UK.

15.22 ASBESTOS CEMENT PIPES

Asbestos cement (AC) pipes are made of Portland cement and asbestos fibre mixed into a slurry and deposited in layers on a cylindrical mandrel. When the required thickness has been built up, pipes are steam or water cured, cleaned, the ends turned down to an accurate diameter for some 150 mm, and then usually dipped into cold bitumen. Although they have been in use for over 50 years and continue to be produced in some countries, the danger to health posed by the handling of asbestos in pipe manufacture and on site, particularly due to the risks of inhaling asbestos dust during cutting and handling pipes, has resulted in cessation of production in many countries. There is however no evidence that their use for conveyance of drinking water presents a direct danger to health. UK standards warn against the hazards (asbestosis, lung cancer and mesothelioma) of breathing asbestos dust and refer to prohibitions on the supply of products containing amosite (brown asbestos) and crocidolite (blue asbestos). AC pipes made in the UK since 1982 are said to contain only chrysolite (white asbestos). Production of AC pipes in the UK ceased in 1986. BS EN 512 applies to two types: AT (Asbestos Technology), containing chrysolite asbestos and NT (Non-asbestos Technology) containing fibres other than asbestos.

Asbestos cement pipes can often be produced from local resources more cheaply than other types of pipe, especially in countries which have to import steel or iron. They are also resistant to internal and external corrosion except in sulphated soils which attack the cement unless it is protected with bitumen. The principal disadvantage of asbestos cement pipes is that they are brittle; hence, they need careful handling in transit, their bedding and surround must contain no large stones and they should preferably not be laid beneath roads subject to vibration from heavy traffic unless surrounded by concrete. Ferrules tapped directly into a main without the use of a saddle are frequent causes of leakage. Service pipe connections are made using a metal saddle clamped onto the pipe.

BS EN 512 gives standards for fibre cement pipes up to DN 2500. Pipes up to DN 1000 are classified according to nominal pressure up to 20 bar. Preferred classes are 4, 6, 10 and 16 bar but pipe classes greater than 20 bar can be supplied if required. The required nominal pressure rating is decided in relation to hydraulic and operating conditions and external load. Pipes exceeding DN 1000 are designed to suit particular requirements of the pipeline. The works test pressure (PT) is specified as double the nominal pressure up to DN 500 and 1.67 times the nominal pressure for larger sizes and is held for 30 seconds, or if the test pressure is increased by 10%, 10 seconds. Sizes for water supply are commonly DN 100 to 900 for working pressures of 75, 100 and 125 m. Nominal lengths are typically up to 5 m for DN 300 and smaller and up to 6 m for larger sizes.

Although some fittings are asbestos cement, common practice is to use steel or standard ductile iron fittings with spigots to match the class of pipe used. Asbestos cement pipes and fittings are all plain ended and hence are jointed using collars of asbestos cement or by use of flexible couplings. Great care is needed to ensure that O-ring joints have no twist in them when placed, otherwise the joint will leak. A leak under pressure at a joint may in time cut right through the pipe wall. A gap must be left between the ends of pipes to allow for deflection and, where appropriate, thermal movement.

15.23 ANCHORAGE AND THRUST BLOCKS

Thrusts are developed on pipelines at changes in direction, tapers, junctions and closed ends (including valves) due to out of balance forces on surfaces exposed to the pipe fluid. Two types of thrust operate, that due to internal pressure and that due to velocity. The latter force is due to change in momentum and can usually be ignored in water supply pipe systems as velocities are low. The velocity force T on a bend is given by:

$$T = \frac{\pi}{2} D^2 \rho V^2 \, \mathrm{Sin}\frac{\theta}{2}$$

where D is pipe internal diameter, ρ is fluid density and θ is the angle of the bend.

The pressure thrust is proportional to pressure; values for different fittings are given in Figure 15.4 which shows an assembly of the usual fittings. For pipework between flexible joints the relevant diameter is that at the centre of the joint ring. The loadings from internal pressure should be the thrust developed under operating and surge pressure or pipeline testing conditions, whichever gives the more critical result. The designer must ensure that all thrusts on pipework can be taken safely particularly, but not only, where pipe joints are not able to transmit axial load. This is the case for flexibly jointed ductile iron, GRP, PVC or concrete pipe. Inside reinforced concrete structures the various thrusts should be carried to the structure by blocks but the stability of the whole structure under all load combinations must be checked. For buried pipelines thrust blocks (or walls) must be designed for the soil resistance available and must bear against undisturbed ground. Differential movement at thrust and anchor blocks needs attention: rocker pipes may be required.

Blocks or walls to resist horizontal thrust transfer it to the ground. The soil resistance that can be mobilized to resist thrust must be carefully analysed; specialist advice based on site investigation is needed to obtain appropriate geotechnical design parameters. Thrust block design in UK is carried out in accordance with CIRIA Report 128 (CIRIA, 1994). Vertical thrusts have also to be allowed

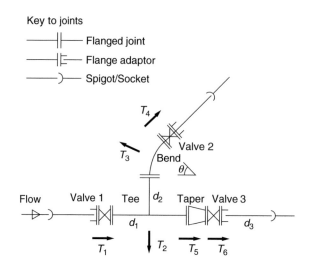

Key to joints

—||— Flanged joint

—|E— Flange adaptor

—)— Spigot/Socket

Case	A	B	C	D*	E	F**	Thrust value
Valve shut	None	1	2	3	2 & 3	All	
Thrust T_1		Yes				Yes (negative)	$\pi/4\ p\ d_1^2$
Thrust T_2	Yes		Yes	Yes	Yes	Yes	$\pi/4\ p\ d_2^2$
Thrust T_3	Yes		Yes	Yes	Yes	Yes	$\pi/2\ p\ d_2^2\ \sin\theta/2$
Thrust T_4			Yes		Yes	Yes	$\pi/4\ p\ d_2^2$
Thrust T_5	Yes		Yes	Yes	Yes	Yes	$\pi/4\ p\ (d_1^2-d_3^2)$
Thrust T_6				Yes	Yes	Yes	$\pi/4\ p\ d_3^2$

d_1, d_2 and d_3 are internal diameters of joint socket or sleeve.
*Net thrust is that from taper only (T_5) as T_2, T_3 and T_4 cancel each other out.
**Hypothetical situation equivalent to a pressure vessel—resultant of all thrusts is zero.

FIGURE 15.4

Thrusts at pipe fittings.

for, upthrust being resisted by tying the fitting down to a reinforced concrete block of sufficient weight. Blocks or walls for horizontal thrusts are usually suitable for isolated fittings where there is room. However, thrust blocks designed following the CIRIA procedure may be large in weak soils, particularly where ground water is high. If space is restricted, for example due to the presence of other services, or where thrusts have to be accommodated from several fittings, at buried valved branches or service reservoir inlet and outlet with several branches for example, alternatives should be found. One alternative is to construct a reinforced concrete thrust slab beneath the pipework concerned and beneath adjacent services where appropriate. Thrusts are transferred by reinforced concrete upstands to the slab which should be designed as a thrust block. Where this solution is impracticable, a joint type that can reliably transmit axial load should be used or that section of pipe should be installed as a coated steel fabrication.

A tied coupling is a flexible coupling across which threaded anchor bars are fixed between lugs or flanges welded to the pipe wall. Some DI pipe manufacturers offer locking flexible push fit joints for their ductile iron pipe. These joints are similar to the standard push fit joint but the joint ring is armed with steel teeth designed to bite into the outside of the pipe wall on tightening, thus providing capacity to carry tension. Allowance must be made for the longitudinal movement that is initially required for the joint teeth to bite. It is maintained that some such joints retain flexibility but users should be aware that any movement due to cycles of expansion and contraction or angular displacement is liable to cause the teeth to drag and consequently reduce locking capacity over time. If corrosion takes place it would further reduce anchor capacity. It is suggested that such joints be used only with caution, and preferably not for larger pipes and higher pressures and where risks are higher and repairs would be awkward or expensive.

Thrust block design (and any concrete surround) must consider hoop strain in the pipe under pressure and the possible need for reinforcement in the concrete and for elastic sleeving at entry and exits of the pipe from the concrete. For low stiffness materials such as GRP, hoop strain can be considerable and enough to split the concrete, rendering the block unserviceable, or burst thin sections locally leaving sharp edges which can damage the pipe over time, particularly under cyclic pressure loading.

PIPELINE CONSTRUCTION

15.24 CHOICE OF PIPES

Principal factors affecting choice of pipe material are technical considerations, price, local experience and skills, ground conditions, preference and standardisation. Only an indication can be given here of general patterns in the choice of pipes.

For new service pipes (DN 50 and less) in the UK, over 99 per cent of material used is MDPE. The remainder tend to be copper (for ground conditions). For the smaller pipe sizes in water distribution systems (DN 51 to 300), in the UK over 80 per cent of new pipes are plastic (about 10% PE100, 50% PE80 and over 20% PVC-U and MOPVC) and over 15% ductile iron. Plastic pipes offer cost advantages at small diameters and, as they can be joined above ground to form continuous lengths which can be snaked into narrow trenches, they also offer advantages of speed and consequently less social and environmental impact during construction. This advantage disappears with increasing diameter and pressure class and in hot climates where temperature derating is necessary.

For the middle range of pipe sizes used in water distribution systems (DN 350 to 800) ductile iron is most widely used because of stiffness, strength, toughness and durability in many kinds of ground. In addition a complete range of compatible, standard dimensioned ductile iron fittings and valves are available to make up a homogeneous pipeline, which simplifies both the design and construction processes. Service pipe connections can also be tapped directly on to the main, under pressure if necessary. Almost any other pipe type can be used as an alternative. Asbestos cement pipes in some countries form the main alternative to ductile iron pipes, their advantages being that they are usually cheaper (especially if ductile iron must be imported). However, asbestos cement pipes are more liable to breakage than are ductile iron pipes, both in transport and when laid, fittings must be mainly of ductile iron, and gunmetal or cast iron saddles have to be strapped to the pipes before service pipe connections can be made. Deterioration of these saddles can bring about severe leakage problems. Asbestos cement pipes have now been phased out of production in many countries. PE can also be highly competitive with DI.

For trunk mains and large diameter pipes no general rules of choice can be laid down since steel, ductile iron, prestressed concrete, concrete cylinder, polyurethane and GRP pipes may each be used in any particular case according to circumstances applying. Steel is predominantly used for trunk mains or high pressure mains, the welded joints offering a distinct advantage in the latter case. Steel pipes are also particularly useful in congested and urban areas where welded joints provide longitudinal strength and avoid the need for large (or any) thrust blocks (Little, 1986). Steel pipes do not normally require rocker pipes at junctions with chambers and minor structures but may need them to accommodate major settlement where temporary excavation has to be backfilled at large structures. Anchorage points are nevertheless required at terminations and at connections to chambers where flexible couplings are to be installed at valves. Alternatively, the steel pipeline may be tied across flexible joints to provide longitudinal continuity of strength; the pipeline can then in principle be allowed to move longitudinally within the chamber and need not be anchored into the chamber walls. Leakage of groundwater into the chamber would have to be addressed but would not normally be an issue in dry conditions. To appreciate the positioning and design of joints and anchorages the pipeline can be considered as a potential mechanism limited only by the pipe structure, friction and soil stiffness. Ductile iron pipe in the largest diameters tends only to be used instead of steel if its price is competitive or where supply of skilled welders may be limited. The alternatives of prestressed concrete, concrete cylinder or GRP pipes tend to be used because of circumstances such as local preference and practice, price competitiveness, tied funding, in-country manufacture as opposed to importing, aggressive ground conditions or aggressive water to be conveyed, or (in the case of concrete pipes only) where a greater margin of safety is required against rough handling and backfilling. GRP pipes have the principal advantage that they are not attacked by ground conditions or by waters, such as desalinated water, which are severely aggressive to both iron and concrete.

15.25 PIPE LAYING AND INSTALLATION

Strict control needs to be exercised over the laying of pipes because of the high capital cost of pipelines and their swift deterioration if not properly laid. Factors key to successful installation of a pipeline are discussed below.

Typically a 'working width' for construction is defined in the contract documents to allow space for access, pipe stringing and separate stockpiles for topsoil and excavated spoil. In addition space is required for pipe storage and office and support facilities. The pipe 'easement' as finally granted relates to the pipe as laid and sets out rights for access and limits subsequent construction and other works. A 'wayleave' is an agreement similar in nature but less specific than an easement, granted by the landowner, permitting the pipeline owner to carry out the works.

Care should be taken to ensure that pipe supports on vehicles transporting pipes are adequate to prevent damage to the external coating (especially in hot climates where the coating may soften) and that flexible pipes, particularly those with concrete or cement mortar linings, are supported to maintain circularity and avoid damage to the lining. The stock dumps for pipes should be properly planned and pipes should only be stacked one above the other if they are properly provided with timber supports and packers. Care should be taken not to damage pipe ends: bevel protectors should be fitted to pipes to be joined by butt welding and flanges should be protected with plywood blanks. Pipes should be stacked in accordance with the manufacturer's recommendations and should not be stacked in areas where long grass may grow: in a dry period this grass can catch fire, thus ruining the exterior protection of the pipes. All pipes should be handled using purpose made lifting slings of a wide fabric material not chains so that the external coating is not damaged and so that the pipe does not slip. When pipes are delivered, and again just before they are lowered into the trench, they should be inspected for flaws. The Holiday detector should be passed over steel pipes. Any coating or lining flaws detected should be made good. The interior of each pipe should be inspected as the pipe is lifted and any debris must be brushed out.

At small diameters, flexible PE and PVC pipes can be joined on the surface and snaked into narrow non-man-entry trenches. Pipes are then usually installed on 100 mm granular bedding and backfilled to 100 mm above the pipe with free flowing granular material, above which the main backfill is placed. This allows trenches to be dug by chain excavator, rock-wheel or narrow bucket. If the pipe is laid directly on the soil then the trench bottom must give sensibly uniform support to the pipe: it must be stable, fine grained and free from flints and large stones or other material which may cause point loads. Additional excavation is required at pipe sockets. If the trench bottom is unsuitable a bedding should be used. A bedding must be provided for support in soft ground. A nominal minimum of 150 mm granular bedding must be provided in rock. The trench width for non-man-entry trenches should be at least 300 mm greater than the outside of the pipe. Trench side-fill must be brought up evenly either side of the pipe with soil selected and compacted to give the required support.

Steel, plastics and to a large extent ductile iron, are flexible conduits which when buried rely on the pipe/soil-structure interaction for their load carrying capacity. Deflection under vertical load is limited by support obtained from the trench sidefill, which in turn transfers load to the trench sides. It is therefore essential that the pipes are bedded evenly and are surrounded in material which is well compacted and can transmit the lateral thrusts from the pipe to the trench sides and that the undisturbed ground does not become overstressed. Design and constructability considerations should determine whether the pipe can be laid on the trench bottom after trimming or whether a bedding must be used.

Successful construction requires care in preparation of the trench bottom, handling the pipe and in placing and compaction of backfill, which in turn requires competent labour and dedicated supervision and quality control. It is vital to make an even bed for pipes, with joint holes previously excavated in the positions required. After trimming the trench bottom or compacting the bedding a depth of 50 mm of uncompacted soft sand may be left loose to form a uniform support for the bottom arc of the pipe. Alternatively the bedding must be shaped and compacted to match the curvature of the bottom of the pipe: the method adopted depends on in situ soils and workmanship. Either way the work is required such that the specified deflections are not exceeded. During backfilling, pipes must not be supported on timber or hard blocks and pegs should not be used in the trench bottom for setting out bedding levels. Similarly, all large stones must be removed from the bed. A large stone can be taken as one sufficient to risk damage to the pipe or coating. Such hard points may damage the coating or overstress the pipe locally, both in ring and longitudinal bending on settling under load. Hard bands of rock across the trench should be covered by sufficient granular material (at least 150 mm) so that the pipe does not suffer uneven support. The pipe must be bedded evenly and all voids beneath the pipe must be filled with compacted fill. The bedding, sidefill and initial backfill 300 mm above the pipe must be free from large stones (larger than say 15 mm) which could damage the pipe coating. In order for the embedment and natural ground to support the pipe as required by the design, it is particularly important to achieve compaction in the "upper bedding", in the narrow section of fill above the pipe invert: 'pogo stick' type rammers may be needed for this purpose. Similarly, voids left by removal of trench supports must be filled by withdrawing trench supports as side fill is raised. Compaction equipment must not come into contact with the pipe or be used within about 150 mm of the top of the pipe.

The provision of bedding or fully surrounding pipes can solve many pipe-laying problems in a sound manner. The bedding not only gives full support to the pipe, protecting it from settlement on hard points and from excessive overburden pressure or traffic loading, but it can also give protection against corrosion in aggressive soils.

The use of boning rods and sight rails for every pipe is essential and work should be stopped until these are provided. Lengths of pipeline should be laid to even grades: good practice is to limit grades to not less than 1:500. This gradient can be readily achieved and monitored during construction. Flatter gradients can be properly carried out only with much greater quality control to level and to backfilling (Plate 27(c)). Backfalls caused by pipe level variations will produce minor ponding—which may not be important—and air traps, which can affect pressure tests. For distribution pipes which have service connections a flat grade does not particularly matter as air will be drawn off via the service connections. On trunk pipelines, however, it is important to arrange even rises to air valves.

Trenches should be backfilled as soon as possible after laying and jointing and long lengths should not be left uncovered. This helps prevent flotation in the event of flood and helps limit thermal movement, which can be particularly important for plastics pipes.

Considerable trouble is experienced when laying pipelines in urban areas where many other services, e.g. gas, electricity, etc., have to be negotiated. Considerable efforts have been made in the last 10 years to record the locations of buried services and to store the information using GIS systems; however, records of these other services are seldom perfect or to the accuracy necessary to avoid all the problems that may be encountered. Close liaison with the utility operators is required and services should be located, before pipe trench excavation, by remote detection equipment from the ground surface, followed by opening of trial pits where appropriate. The trench should preferably be excavated well ahead of pipe-laying, if the road authority permits this, so that the line and level can be adjusted in good time. Small angle deviations can be accommodated at pipe joints but preformed bends should be available to get around obstacles. Gusseted bends (otherwise termed mitred, or lobster back bends) in steel pipes can be fabricated from straight pipe to suit any combination of vertical and horizontal angles.

If two pipelines are to be accommodated in one trench, a minimum spacing of 300 mm should be kept between the two lines. If a parallel pipeline is to be constructed at a future time enough space must be allowed in the pipeline reserve to construct a second trench without disturbing the original works and consideration must be given to unbalanced thrusts as it is unlikely that the original pipeline can be shut down for an extended period.

Sidefilling to pipes should be placed in even layers either side of the pipe up to soffit level and, in addition, for non-rigid pipes the backfill must be carefully compacted to keep the pipe in a true cylindrical form. The backfill material adjacent to the pipe and for 150 mm above its crown must be free of large sharp stones that could puncture the sheathing of the pipe. When the material from the trench is being excavated it should be inspected and instructions should be given for setting aside material that should not be used against the pipe. This material can be used in refilling the trench once the pipe has been properly covered with soft material.

A principal requirement for satisfactory pipelaying is care in making the joints. Achieving cleanliness in a muddy trench is not easy; pipelayers should be provided with the facilities required, such as clean water and buckets, plenty of wiping rags, enough room to work and time to make the joint properly. The reward for taking care with each joint is a pipeline which passes the test requirements at the first test. This can save weeks of extra work.

Cover to pipes should normally be not less than 1.0 m to provide some protection against physical (third party) damage, for frost protection and to limit the effect of seasonal ground movement. Depths of cover need to be increased in climates with very low and prolonged low temperatures. Cover can if necessary be reduced (for example to limit temporary environmental impact or to reduce excavation in rock) where flow in the pipe is continuous, even in areas liable to frost, but requires calculation of heat transfer. Cover may be increased to reduce loading under heavily trafficked roads (Section 15.5). Cover may also be increased for flexible pipes which rely on support of soils that may be removed by excavation for other services—alternatively an easement must be created (and enforced) to prevent encroachment of other excavations on the pipeline soil support (which typically requires no soil movement within two pipe diameters each side of the pipe, total five diameters width). Where shallow burial is necessary below roads, a reinforced concrete slab across and bearing on undisturbed ground each side of the trench can be used to transfer imposed load away from the pipe.

Care is needed in backfill compaction and in selection of backfill material and grading, particularly on steep gradients, to ensure that the pipe trench does not become a drain, washing out trench fill and collecting at low points, softening the trench fill and weakening the combined pipe and soil structure. Support blocks are required where pipes are laid at gradients of 1 on 6 or steeper and may be needed on slopes between 1 on 6 to 1 on 12, depending on ground conditions. In weak soils particularly, backfill grading should be selected using normal filter rules (Section 5.9); in certain circumstances, it may be necessary to contain the pipe backfill within a membrane of filter fabric. Care should be taken during installation to ensure that joints are not put under undue strain. Pipelines with flexible joints at close spacings provide inherently more capacity for natural minor movement after installation: flexible joints at wide spacings are fewer but can increase local movement and shear forces. To allow for movement after laying, angular deflections at flexible joints on installation should be limited to 50% of the maxima allowable.

Flanges must be carefully aligned before the bolts are inserted and the flanges pulled together. The alignment must be almost as precise as that adopted for aligning motor couplings. To pull up misaligned flanges is likely to cause fracture of the pipe or flange. A rubber gasket is inserted between the flanges, of such diameter that it lies inside the bolt circle but does not intrude into the pipe bore. The faces of the flanges and the gasket must be perfectly clean before assembly and the bolts must be tightened up little by little in the sequence recommended by the pipe manufacturer so that an even pressure is maintained all round. No grease, paint, oil, dirt, grit, or water should be permitted on the flange or rubber ring faces. The contact should be between clean dry metal and clean dry rubber. When making joint faces which are vertical, some difficulty may be experienced in keeping the rubber ring flat against the vertical flange face and, to counteract this, a little clear rubber solution may be used to tack the rubber ring on to the metal face. This is the only material whose use can be permitted in connection with rubber rings. If a greasy material such as a bitumen-based adhesive is used, tightening the flanges may cause the rubber ring to extrude into the pipeline and lead to leakage.

15.26 TESTING OF PIPELINES

The test usually consists first of filling the pipeline with water and allowing it to stand and stabilize under working pressure. The test pressure is then applied slowly by pumping. BS EN 805 covers water supply systems, including raw and treated water pipelines, outside buildings and specifies the (site) system test pressure (STP). BS EN 805 requirements for STP may be interpreted as:

Surge pressures determined by analysis and included in maximum design pressure (MDPc):

■ MDPc + 1 bar.

Surge pressure not determined by analysis but for which an allowance of not less than 2 bar is included in maximum design pressure (MDPa):

lower of:

- MDPa × 1.5, or
- MDPa + 5 bar.

The BS EN 805 test pressures differ somewhat from those set out in older but current standards such as BS 8010, Section 2.1 for DI pipe. PD 8010 also covers pipe testing and sets out procedures for allowing for air. Procedures for testing plastic pipes, in particular polyethylene, need to allow for creep and reference should be made to the manufacturer. In the UK PE pipe is tested to 1.5 times Rated Pressure for 6 and 10 bar systems and 1.5 times Working Pressure for 12 and 16 bar systems. For steel pipelines AWWA M11 says that the test pressure must not cause stress exceeding 75% of yield.

BS EN 805 gives two options for test method, both with test duration of 1 hour:

a. water loss method with measurement, either by (i) volume drawn off to replicate observed pressure drop in test period or (ii) volume pumped in to restore pressure, or
b. pressure loss method.

In both cases the acceptability criterion is independent of test pressure. The allowable pressure loss for method (b) is 0.2 bar for normal pipelines. For method (a) the acceptable water loss (or volume change) is calculated using a formula in which a change of pressure of 0.2 bar is inserted. The allowance factor used (1.2) results in the acceptable loss being 20% higher than the elastic volume change due to the pressure change. The formula includes terms for pipe wall thickness and mechanical properties. These should be taken from data provided by the manufacturer. If the pressure loss method is used a preliminary test (pressure drop test) should be carried out to check that the amount of air left in the pipeline is within acceptable limits. If the criterion is not met measures should be taken to eliminate the excess air and the drop test repeated. If the criterion cannot be met the water loss method should be used. The preparation for the main test is:

- after flushing and venting pipeline allow to relax for at least 1 hour;
- raise pressure continuously to STP in less than 10 minutes;
- maintain STP (by pumping) for 30 minutes whilst observing for leaks;
- allow pipe to relax under pressure for 1 hour and measure remaining pressure;
- if pressure drop is less than 30% of STP proceed to main test.

Pipe testing should proceed quickly after installation and results should be linked to payment for the pipeline. Initial tests should be made starting with short lengths for each mainlaying gang and pipe size and material. Test lengths can gradually be increased typically to 5 or 10 km or between section valves as a successful track record is developed. Test lengths depend on topography and availability of water.

Fluctuating test pressure results are likely to be caused by air locks in the pipe. To avoid air locks there must be suitable air valves on the pipeline (Section 16.24). Filling must proceed slowly, particularly on falling gradients, to vent air and avoid hydraulic jumps and entraining air. An equivalent velocity between about 0.2 to 0.5 m/s is sensible but can be varied to suit circumstances. Source of water and pumping arrangements should be designed to allow any entrained air to escape and to prevent air entrainment into the pipe.

Air must not be used for testing water mains. The test must be hydrostatic and take place between blank flanges, bolted or welded to pipe ends, or caps may be used if fully supported by anchor blocks. Where pipes have flexible joints the end pipe must be fully anchored. Testing should not take place between closed valves because if the valve is already inserted in the line it is not possible to detect any leakage past the valve and, if the valve is exposed at the end of a section of the line, it would be in the 'open end condition' and may leak unless it is designed for the 'closed end condition' (Section 16.6).

When a pipeline fails its test and it has been backfilled, searching for leaks can be troublesome. It is best to leave the pipeline under pressure for a day or two so that, possibly, a wet patch on the surface of the ground may indicate the whereabouts of failure. The pipe may also be sounded for leakage using leak detection methods (Section 14.15). It is possible to use some kind of tracer element in the water but this is a skilled matter requiring a specialist. In practice, therefore, it is usual to expose the joints where leakage is suspected.

15.27 MAKING CONNECTIONS

If a socket and spigot tee has to be inserted into an existing pipeline the length cut out of the latter must be slightly greater than the overall length of the tee. The socket of the tee is pushed up to fit one end of the cut pipe and the resulting gap between the two spigots at the other end is joined by using a collar. If a double-socketed tee is used this must be inserted using a plain piece of pipe on one side, again joined by a collar to the pipeline.

An alternative is to use an under-pressure connection as shown in Figure 14.2. A split collar is clamped on to the main, the collar having a flanged branch on it to which is bolted a gate valve. A cutting machine is attached to the valve, the latter is then opened, and the cutter is moved forward through the valve and trepans a hole in the side of the pipe. The cutter is withdrawn with the trepanned piece of pipe wall and the valve is closed. The cutting machine is then removed and the branch connection can be made. Steel and iron pipes can be cut in situ using a rotating cutting tool which is clamped onto the main. A manually operated wheel cutter can be used on small diameter cast iron mains of 80 or 100 mm size. Oxyacetylene cutting of steel and iron pipes can be used but the cut is ragged and difficult to make exactly at right-angles to the axis of the pipe. Cast and ductile iron pipes can be cut above ground using a hammer and chisel. The pipe is placed on a timber baulk below the line of cut and is rolled back and forth as the chiselling proceeds: first to 'mark' the cutting line and then to deepen the chiselled groove. At a certain stage the pipe will come apart at the chiselling line.

15.28 UNDERWATER PIPELINES

Steel and PE are materials commonly used for pipelines laid below water for the crossings of rivers or estuaries. Both materials can be welded, thus providing longitudinal continuity of strength. Other materials can also be used including concrete, GRP, PVC and ductile iron but require particular attention to tolerance on bed preparation, practicalities of level and position measurement underwater, plus the strength and deflection limits of flexible joints. Pipelaying is normally carried out by laybarge, reel barge, bottom pull or float and sink. The laybarge is a 'factory' for progressively adding pipes to a string whilst winching the barge along the pipeline route so that the string hanging in a catenary from the back of the barge is gradually lowered onto the river or sea bed or into a pre-dredged trench. The reel barge method is similar but is used for unreeling lengths of plastic tubing as the barge moves along the pipeline route. In the bottom pull method lengths of pipe, prefabricated onshore, are joined to form a pipe string which is progressively pulled into the water by a winch, mounted on a pontoon or on the far shore, until the crossing is complete. In the float and sink method lengths of pipe are made up into strings at a remote fabrication yard, the string is towed at or below the water surface to the crossing location where it is aligned into position and sunk by removing supporting buoyancy tanks or by filling with water.

Internal operating conditions are similar to those described for land pipelines but special attention needs to be given to external conditions as, once laid, access to a pipeline for remedial work is unlikely to be available. Attention has to be paid to stresses during laying, soil conditions, the effects of currents and waves (including soil liquefaction), sea bed topography and morphology, protection against corrosion and protection against damage by ships' anchors, dropped objects, fisheries (trawl boards) and other activities. Physical protection in shallow water typically requires burial with some cover below the lowest likely level of bed movement and depth of penetration of any ship's anchor. In deep water, where wave and current action is small, the pipeline may be laid without burial, subject to limitations of profile and hazards.

Because of their weight, pipelines are in general installed empty. Positive submerged weight ('negative buoyancy') is required to prevent the pipes from floating and for stability under the action of waves and currents. Pipes installed by float and sink would be towed in strings and manoeuvred into position using temporary buoyancy. Weight is added typically, for steel pipes, as a continuous concrete cladding; for polyethylene pipes concrete collars are used, spaced, shaped and sized to give the required underwater specific gravity and stability. Design considerations include buckling under external hydrostatic pressure, pull forces, longitudinal bending as the pipe is towed or as it conforms to the underwater profile, and ring bending under backfill loads. The cladding thickness is typically about 10 per cent of the pipe diameter. It is sometimes considered simply as temporary weight coat but it does serve to protect the external corrosion coat and therefore can act as part of the permanent structure. The cladding can contribute significantly to bending stiffness particularly in ring bending; therefore differences at joints can raise stresses locally. Above about DN 1500 the cladding is stiff and thick enough to be designed as a reinforced concrete structural ring. This can be used to support and reduce the thickness required to withstand buckling of the steel shell, typically by 40 per cent, and carry external backfill loads. The combined structure can also be designed to withstand ships' anchors, thus saving considerably on the costs and environmental impacts of deep burial (Little, 1989). Backfill must be designed not to liquefy and to be stable under wave action. Rock armour can also be chosen and designed to save on burial costs. With care, considerable economies can therefore be achieved.

REFERENCE STANDARDS

British Standards (BSI).

BS 534 *Steel pipes, joints and specials for water and sewage.*

BS 1377 *Methods of test for soils for civil engineering purposes.*

BS 3505 *Unplasticized polyvinyl chloride (PVC-U) pressure pipes for cold potable water.*

BS 7079 *Preparation of steel substrates before application of paints and related products.*

BS 8010 *Code of Practice for Pipelines. Pipelines on land: design, construction and installation:*

    Section 2.1 Ductile Iron.

    Section 2.5 Glass reinforced thermosetting plastics.

(a) Twin 1000 mm diameter ductile iron pipe laying in Middle East. (Black & Veatch)

(b) 1524 mm diameter steel pipeline awaiting backfilling in Middle East. (Binnie Black & Veatch)

(c) 900 mm diameter GRP pipeline installation in Middle East—note attention to sidefill compaction. (Binnie Black & Veatch)

(d) Automatic butt fusion jointing of PE pipe. (Courtesy of The Fusion Group, Chesterfiled, UK)

PLATE 27

(a) Metal seated gate valve.
(Glenfield Valves Ltd.,
Kilmarnock, Scotland)

(b) Resilient seated DN 150 butterfly valve.
(By courtesy of ERHARD GmbH & Co. KG)

(c) Diaphragm sealed globe control valve—flow
left to right. (Singer Valve Inc., British
Columbia, Canada)

(d) Resilient seated (Multamed)
gate valve. (By courtesy of
ERHARD GmbH & Co. KG)

PLATE 28

BS EN 512 *Asbestos cement pipes.*

BS EN 545 *Ductile iron pipes, fittings, accessories and their joints for water pipelines. Requirements and test methods.*

BS EN 639 *Common requirements for concrete pressure pipes including joints and fittings.*

BS EN 641 *Reinforced concrete pressure pipes, cylinder type, including joints and fittings.*

BS EN 642 *Prestressed concrete pressure pipes, cylinder type, including joints and fittings.*

BS EN 805 *Water supply—Requirements for systems and components outside buildings.*

BS EN 1092 *Flanges.*

BS EN 1295-1 *Structural design of buried pipelines under various conditions of loading.*

BS EN 1452 *Plastics piping systems for water supply. Unplasticized poly vinyl chloride (PVC-U).*

BS EN 1796 *Plastics piping systems for water supply with or without pressure. Glass reinforced thermosetting plastics (GRP) based on unsaturated polyester resin.*

BS EN 10216 *Seamless steel tubes for pressure purposes. Technical delivery conditions.*

BS EN 10217 *Welded steel tubes for pressure purposes. Technical delivery conditions.*

BS EN 10220 *Seamless and welded steel tubes. Dimensions and masses per unit length.*

BS EN 10224 *Non-alloy steel tubes and fittings for the conveyance of water and other aqueous liquids. Technical delivery conditions.*

BS EN 10298 *Steel tubes and fittings for on and offshore pipelines. Internal lining with cement mortar.*

BS EN 10311 *Joints for the connection of steel tubes and fittings for water and other aqueous liquids.*

BS EN 10312 *Welded stainless steel tubes for the conveyance of aqueous liquids including water for human consumption. Technical delivery conditions.*

BS EN 12201 *Plastics piping systems for water supply. Polyethylene (PE).*

CP 312 *Code of practice for plastics pipework (thermoplastics material).*

CP 2010 *Code of Practice for Pipelines, Part 2 Design and construction of steel pipelines in land.*

Eurocodes.

BS EN 1993-4-3 *Design of steel structures. Pipelines.*

Draft European standards Published Documents.

PD CEN/TR 1295-2 *Structural design of buried pipelines under various conditions of loading. Summary of nationally established methods of design.*

PD 8010-1, -2 *Code of practice for pipelines. Part 1—Steel pipelines on land; Part 2—Subsea pipelines.*

American Waterworks Association (AWWA):

AWWA M11 *Manual: Steel Pipe—A Guide for Design and Installation.*

AWWA M45 *Fiberglass pipe design.*

AWWA C205 *Cement-mortar protective lining and coating for steel water pipe-4 in (100 mm) and larger—shop applied.*

AWWA C300 *Reinforced Concrete Pressure Pipe, Steel Cylinder Type.*

AWWA C301 *Prestressed Concrete Pressure Pipe, Steel Cylinder Type.*

AWWA C303 *Concrete Pressure Pipe, Bar Wrapped, Steel Cylinder Type.*

AWWA C304 *Design of Prestressed Concrete Cylinder Pipe.*

AWWA C602 *Cement-mortar lining of water pipelines in place-4in (100 mm) and larger.*

AWWA C950 *Fiberglass pressure pipe.*

American Society of Mechanical Engineers (ASME):

ASME B31.4 *Pipeline Transportation Systems for Liquid Hydrocarbons and Other Liquids.*

ASME B31.8 *Gas Transmission and Distribution Piping Systems.*

American Petroleum Institute (API):

API 5L *Specification for line pipe. (Identical to ISO 3183).*

REFERENCES

Clarke, N. W. B. (1968). *Buried pipelines—a manual for structural design and installation.* Maclaren.

CIRIA (1994). *Thrust blocks for buried pressure pipelines.* Report 128. CIRIA.

Janson, L. E. (1995). *Plastics pipes for water supply and sewage disposal.* Borealis, Stockholm.

Little, M. J. (1986). New pipelines on land and across Hong Kong Harbour. *JIWES,* June. pp. 271–286.

Little, M. J. and Duxbury, J. A. (1989). Tolo channel submarine pipelines, Hong Kong, *Proc. ICE,* Part 1, 1989, Apr., 395–412.

Little, M. J. (2004). Pipeline design—a UK perspective, *AWWA conference, Florida.* USA. June.

Young, O. C. and O'Reilly, M. P. (1983). *A guide to design loadings for buried rigid pipes.* HMSO.

Young, O. C. and O'Reilly, M. P. (1986). *Simplified tables of external loads on buried pipelines.* HMSO.

Valves and Meters

16

PART I VALVES

16.1 VALVE DEVELOPMENT

Valves have been in use on water pipe and lifting systems for over 2000 years. Bronze and brass valves dating from the Roman period have been found; these included plug cocks and flap valves. Leather flap valves may have been used in ancient Egypt in association with pumps and simple flaps have been used for several centuries on hand well pumps.

Valve technology took a leap forward at the end of the 18th century with the advent of steam and with improved metal working techniques. Plug cock, butterfly, flap, gate and pressure relief valves became widely used. By 1950 valves had become increasingly sophisticated and several new types of valve had been developed—diaphragm, ball and lubricated plug. Needle, sleeve, air release and hollow jet valves were also invented in the 20th century and are now available in numerous variants.

16.2 VALVE FUNCTIONS

Valve type should be selected according to the required function. Valves for industrial process are classified in Europe as: isolating; regulating and control valves. The same classifications may be adopted for water supply valves. There are also other common functions such as air release, non-return, pressure relief and hydrant:

Isolating	Set either closed or fully open and normally not operated in flow conditions
Regulating	Set with any degree of opening to regulate flow and capable of periodic adjustment to opening
Control	Used with autonomous or external systems to respond to changes in flow or pressure conditions so as to achieve a set result which itself is capable of being reset
Non-return	Prevents reverse flow when downstream pressure is higher
Venting	To exhaust or admit air

It should be noted that the 'regulator' according to ANSI/ISA S75.05 is equivalent to the autonomous control valve—using the pressure of the fluid to actuate movement.

Operation to achieve the required function imposes conditions that the valve should be able to withstand. Such conditions may include load on components, vibration, wear, erosion and cavitation. Some of these conditions may, in certain circumstances, exacerbate corrosion.

16.3 ISOLATION

Isolating valves are usually required to be drop tight so that work can be carried out on parts of the system in the dry. Therefore, seals should be resilient. For safety reasons an isolating valve should be lockable so that it cannot be inadvertently opened. An isolating valve should not be left part closed unless designed for this condition, in which case it would be classed as a regulating valve. It is good practice to provide a means of checking valve status. Foam injected into the surface box shows if a valve might have been moved. More sophisticated is to use a 'WIZKEY' which is placed of the valve cap during operation and records, for later downloading to a database, the action taken.

16.4 REGULATION

Where necessary to reduce downstream head, a valve may be left part closed (throttled). The degree of closure would not be adjusted, except occasionally, therefore head loss is a function of flow. When flow stops the downstream head equals the upstream head. This decreases the utility of a regulating valve in a distribution system where flow varies considerably during the day. However, when used for regulation the valve must be capable of withstanding the expected conditions for long periods whilst throttling flow. Unless designed to be drop-tight on closure such valves cannot be relied on to be drop tight and therefore cannot protect the system downstream from high static pressures.

16.5 CONTROL

When the setting of the valve has to be adjusted frequently to suit varying conditions this is usually arranged by some automatic means. This allows the valve to maintain a pre-set flow, pressure or water level. Control valves fulfil this function and must be capable of frequent movement under the range of conditions expected. The suitability of a valve type for control depends on its flow characteristic, its 'rangeability' (percentage of opening over which flow can be adequately controlled), mechanical play intrinsic to the design and, in some cases, resistance to cavitation. Information on the 'rangeability' of control valves is available from manufacturers but guidance is available in *Inherent Flow Characteristics and Rangeability of Control Valves* (ANSI/ISA S75.11, 1991). Control valves are not usually drop-tight on closure and therefore cannot be relied on protect the system downstream from high static pressures.

Defined 'intrinsic' (i.e. due to valve only) flow characteristics include 'quick opening', 'linear' and 'equal percentage'. The first type produces a high flow rate at a relatively small spindle movement from closed and is not suitable for control valves, but is used for some on-off duties. A linear intrinsic flow characteristic is beneficial where system losses are small. The 'equal percentage' intrinsic characteristic is so called because any equal increment in opening produces an equal percentage increase in flow and shows as a straight line when plotted on semi-log paper. When combined with appreciable system losses it produces a near linear combined characteristic. Flow characteristics vary with the opening and are affected by body shape, any porting or special trim and by seating arrangements. For a given head loss, flow through a valve varies approximately with the square of the opening area. Figure 16.1 shows intrinsic flow characteristics through different types of valve ignoring the effect of the system in which they may be used. However, actual flow through a valve in a real situation depends on the overall system losses.

The sum of the head loss through the valve and the head lost in the system equals the head available. This may be fixed (for example between reservoirs) or variable as in a pumped system. Figure 16.2 illustrates the combined characteristics of DN 300 butterfly, ball and globe valves in pipelines of the same diameter having lengths zero, 10 m, 100 m, 1 km and 10 km and roughness k_s of 0.6 mm and with available head of 10 m. For good control, the combined characteristic should be as near linear as possible, particularly over the range in which the valve is expected to operate. The type of valve and any porting arrangement should be selected to achieve this in the system in which it is to

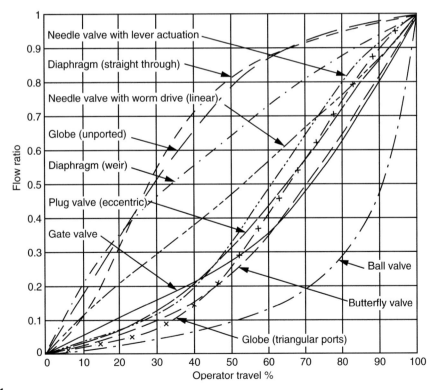

FIGURE 16.1

Valve intrinsic flow characteristics.

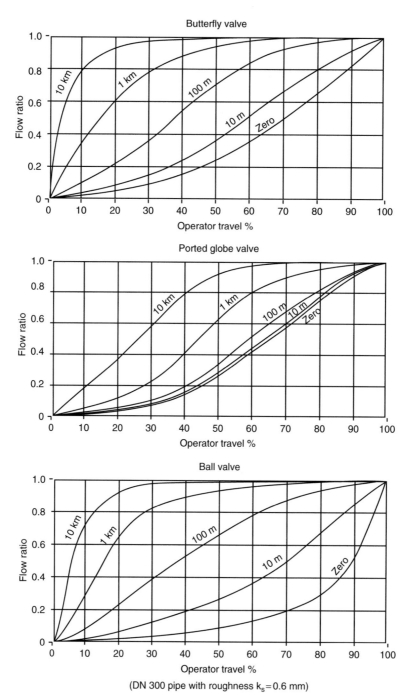

FIGURE 16.2

Effect of system length on valve flow characteristics.

be installed. The shape of ports (triangular) in the ported globe valve assumed for Figures 16.1 and 16.2 influences the shape of the characteristic. It is possible to produce almost any desired characteristic by means of shaped ports, holes of varying spacing or other special trim. Globe, sleeve and, to a lesser degree, needle valves are designed to accommodate such trim. Even more complex (and costly) porting arrangements can be applied to control valves of any type to reduce noise and vibration. Where the arrangements dissipate head in stages, the valve is less susceptible to cavitation.

Automatic control requires the valve to respond to deviation from a set point (for example water level at a downstream discharge point). The signal initiating this response may be from a mechanical linkage (float valve), a hydraulic relay system (altitude valve) or a computer controller (PLC) to which a measured value is transmitted from a level, pressure or flow transducer. Whatever method is employed the system designer needs to ensure that there is no instability sufficient to cause 'hunting'. Hunting is an oscillation of valve opening which causes increased valve and actuator wear and could, in extreme cases, lead to system failure. In a distribution system hunting is undesirable for the consumer. There should be sufficient delay and damping in the control arrangement to deal with slowness of response of the system.

Baumann (1998) discusses control characteristics and maintains that controllers need change of flow per unit change of valve opening to be constant (linear characteristic) within certain limits. However, he suggests that controllers can cope (without need for resetting) with changes of flow between 5 and 15% for each 10% change in valve opening. Just as important to successful holding of a set point is tightness in the actuator and operator arrangements. In some types of valve such as the gate valve there is a degree of play in the operating mechanism. This, together with unfavourable flow characteristics of this type, renders the gate valve unsuitable for control. Any gear mechanism is a potential source of play (backlash) after wear and of flow induced vibration.

16.6 VALVE SELECTION AND SPECIFICATION

Valves should be selected to suit the function and duty and the materials used should be appropriate for the fluid being conveyed. For larger valves and those operating in difficult conditions selection is a specialist task. Helpful guidance is provided in the *Valve Selection Handbook* (Smith, 2004) and in *An Introduction to Valve Selection* (Smith, 1995). For control valves the reader should refer to *Control Valves* (Borden, 1998) or *Control Valve Primer* (Baumann, 1998). The authors' suggestions for the use of the main types of valve employed in the water industry are given in Table 16.1.

Metal seated valves are designed to seal at a specified unbalanced pressure and may seal less well at higher or lower pressures. This is why the leakage test is carried out at a specified pressure. Valves with large bodies (particularly those of large diameter) should be specified as designed for the 'open end test' or the 'closed end test'. (Note that BS EN 12266 does not distinguish between these test conditions and implies open end.) A valve designed for drop tight closure under open end conditions may not close drop tight for closed end conditions due to body strain. Closed end siting is defined as one which prevents valve expansion under load, as in the case of a double-flanged valve connected into rigidly held pipework. Open end siting allows the valve to expand, for instance where there is flange adaptor or other flexible coupling on or near one side, and is the usual situation.

Some valve types such as gate, plug and ball are heavy, bulky and therefore relatively expensive. It may be advantageous to use a valve 5/8 to 3/4 the diameter of the pipeline with carefully selected tapers each side to minimize headloss. This can reduce cost, weight, height and operating torque but prevents transit of swabs for cleaning (Section 14.9).

Table 16.1 Suggested valve utilisation for water supply[a]

Valve Type	Largest size (mm)	Typical valve use[b]	Waterway[c]	Cavitation resistance[d]	Head loss (fully open)	Manual closing speed[e]	Comments
Gate (sluice)	2400	Isolation (to DN 600), regulation at low head in smaller sizes	Clear	Poor	Very low	Very slow	Heavy, tall, simple and reliable
Knife gate		Isolation at low heads, cannot be buried	Clear	Very poor	Very low	Slow	Slim, tall—good for slurries
Butterfly	>4 m	Isolation (above about DN 300), regulation and control	Obstructed	Poor	Low	Moderate	Light and compact
Globe[f]	900	Control in onerous conditions—PRV	Obstructed	High	High	Fast	Piston seal type
Globe[g]	900	Control under moderate head—PRV	Obstructed	Moderate	High	Fast	Diaphragm seal type
Ball	300	Isolation (friction significant under higher heads)	Clear[h]	Poor	Very low	Very fast	Bulky
Plug (or cone)	1800	Isolation, eccentric plug for larger sizes	Obstructed[i]	Poor	Low	Slow	Heavy
Diaphragm	350	Isolation, regulation and control	Obstructed[j]	Not applicable	Moderate	Slow	Slurry and chemicals
Pinch	300	Isolation and regulation	Obstructed	Not applicable	Moderate	Slow	Slurry and chemicals
Needle[k]	1800	Regulation and control in onerous conditions	Obstructed	Moderate	Low—moderate	Moderate	Bulky, expensive
Sleeve[l]	1600	Regulation and control in very onerous conditions	Obstructed	High	High	Moderate	Bulky, expensive

[a]Valve selection is for specialists. This table is intended as an approximate guide to the non-expert. Cost is an important factor but relative costs depend on valve size; [b]Porting (by means of a toothed or perforated sleeve) can assist control characteristics for some valves where a sleeve can be fitted; [c]Many valves do not present a clear opening for the transit of swabs. To keep costs low, a valve of smaller diameter than the pipeline may be used, but it obstructs swab passage; [d]Cavitation resistance takes into account valve and flow geometry. Performance can be improved by arranging staged head dissipation or by using ported or other valve trim. Generally, good cavitation performance is obtained at the expense of high fully open head loss; [e]Valve closing speed varies with valve size, operating pressure and is affected by any gearing provided. Un-motored closure times would vary from 1 to 90 minutes for 'very slow' valves such as gate valves (but these are normally actuated in larger sizes) to 1 to 5 seconds for 'very fast' quarter turn ball valves; [f]The characteristics are for a ported globe valve with flow inwards through the orifice; [g]The characteristics are for an un-ported valve with flow outwards through the orifice; [h]Ball valve openings are usually sized as the inlet pipe diameter; [i]Plug valves usually have rectangular or trapezoidal orifices. Valves with circular water ways of the same diameter as the pipe are available up to about 200 mm diameter; [j]In the weir type of diaphragm valve the waterway is obstructed; [k]The needle valve indicated is one with no sleeve although they are designed to accommodate sleeves which can improve cavitation performance; [l]The sleeve valve is of the angled or in-line type where flow is via circular orifices from the outside to the inside.

Some valves such as gate and plug valves require high operating torque to overcome friction. A bypass around the valve may be fitted to allow pressure to be equalized before opening the main valve; it is also useful for pipeline filling. Bypasses need not be very large in diameter; a DN 100 bypass would normally be suitable for a DN 800 valve. However, a bypass is of little use where the pressure downstream cannot build up quickly, for example where the pipe network downstream is very extensive or where there are appreciable outflows that cannot be stopped. A permanently installed power driven actuator may be used to operate a valve, mounted either on the valve directly or on a headstock and coupled to the valve through an extension spindle. Alternatively, a portable actuator may be used if a suitable stub and gearbox are provided on the headstock.

16.7 GATE VALVES

Metal seated wedge-gate valves have not altered substantially over 100 years except that toroidal (O ring) stem seals were introduced for smaller valves. However, gate valves with a stuffing box at the top of the valve are still available and often preferred for larger sizes. Figure 16.3 shows a 300 mm diameter metal seated valve with a gland packing stuffing box; a large metal seated valve is illustrated in Plate 28(a). The machined sealing faces (seats) are usually made of gunmetal or phosphor-bronze and the seats are forced together by wedge action to produce a seal. Resilient seat gate valves (available in sizes up to DN 600) have a gate which is encapsulated in rubber (Plate 28(d)) and seals

FIGURE 16.3

Metal seated gate valve (DN 300).

against a clear full bore typically without grooves in which dirt can collect and prevent full closure. A further development is the boltless design for the upper valve body: this facilitates mass production and, for valves up to and including DN 300, is cheaper to replace than refurbish. Parallel slide or 'knife gate' valves are used above ground for infrequent, low head isolating duties where space is restricted and are widely used with slurries and as block valves in power stations. Gate valves may be buried but are tall; pipelines of large diameter need to have more than the minimum cover where a gate valve is to be sited.

Gate valves have a low fully open head loss coefficient and good shut off; therefore, they are useful for isolation duties, particularly in pumped systems. However, they are not suitable for flow regulation or control due to poor flow characteristics, vibration and play in the stem. Gearing or special torque reduction devices such as ball-bearing thrust collars (which can reduce torques by 50% may be needed to overcome unseating forces and friction in large gate valves depending on the head. The disadvantage is that manual full opening or closing of a large gate valve requires considerable work and time, over an hour for a valve larger than DN 600. Regular operation of gate valves is advisable to keep the grooves clear and the stem and nut threads clean.

16.8 BUTTERFLY VALVES

Butterfly valves tend to be cheaper than gate valves because they require less material and less civil works. They are also easier to operate against unbalanced water pressures as the disc pivots about an axis on or near the pipe axis. Consequently butterfly valves are now commonly used in water distribution systems. Butterfly valves can be metal seated or resilient seated: in the latter case the seat is usually made of natural or synthetic rubber and is commonly fixed to the body of valves of smaller sizes or to the disc. Plate 28(b) shows a resilient seated butterfly valve.

Resilient seated valves can remain virtually watertight, even after prolonged use in silty water. Therefore, resilient seats are usually specified for isolating valves in distribution systems. Resilient seated valves may also be used for control purposes but, if operated at small openings, the seal may be damaged. Solid rubber is the material usually used for resilient seatings. Inflatable seals have been used on very large valves but not always with success. Metal seated butterfly valves do not have tight shut-off characteristics and are mainly intended for flow control purposes where they need to be held in the partially open position.

Distribution network pipe systems are now designed to produce self cleaning velocities at least once every 24 hours and should not need swabbing as part of normal operation. A transfer pipeline may need to be swabbed periodically. Butterfly valves on the line prevent the passage of foam swabs (except for very soft ones) but this does not usually pose a problem if the valves are spaced sufficiently far apart to allow the pipe to be cleaned in sections. Short lengths of pipe either side of the valve are made removable so that the cleaning apparatus can be inserted and removed.

Butterfly valves should normally be mounted with the spindle horizontal since this allows debris in the pipe invert to be swept clear as the valve is closed. Where the spindle is vertical solids can lodge under the disc at the spindle and cause damage to the seal. Disc position indicators are useful and strong disc stops integral with the body should be specified, so that the operator can feel with certainty when the disc is fully closed or fully open.

Butterfly valves have been made to very large diameters (10 m or more) operating under very high heads and at high water velocities (20 m/s or more) and have proved successful in use. However when a butterfly valve is to be used for flow control purposes the maximum velocity of approach to the valve should be limited to 5 m/s. Resilient seated valves can be specified to have no visible leakage on seat test but the range of acceptable seat leakage rates for metal seated valves varies from about 0.004 to 0.04 l/hr per 100 mm of nominal diameter (DN), at the specifier's choice. However, a low rate for a high pressure differential would be expensive to achieve and difficult to maintain with metal seats. For some control applications, an acceptable seat leakage rate of about 0.4 l/hr per 100 mm DN may be appropriate.

If a valve may be required to remain in place closed on removal of the pipe on one side for a temporary operation, it must be flanged for bolting to a pipe flange on the other side. 'Wafer' butterfly valves whose bodies are sandwiched between pipe flanges do not achieve this. Use of such valves for isolation of air valves allows maintenance to be carried out on the air valve in situ with the pipeline in service but does not allow removal and replacement of the air valve under pressure. Since replacement of air valves is likely to be cheaper than in situ refurbishment, flanged isolating valves are preferred in such situations.

16.9 GLOBE VALVES

A globe valve consists of a circular orifice, usually with its axis at right angles to the pipe axis, against which a piston or disk obturator makes a seal. Movement of the obturator reveals a cylindrical opening which can be ported (provided with a serrated or perforated sleeve). The obturator is driven by a shaft, which can be operated by a device such as a spring. Spring actuation allows the globe valve to serve as a non-return valve. When configured for autonomous (self-acting) control, valve opening can be regulated automatically via a secondary hydraulic circuit acting on the shaft, using pilot valves and differential pressure across a diaphragm or piston. Globe valves are available in sizes up to DN 600 and are widely used for control.

Flow direction depends on the type of globe valve. Usually flow is outwards through the circular opening of valves with diaphragms but inwards through the opening of globe valves with secondary pistons. This difference is important where cavitating conditions are likely, since cavitating vapour pockets generated in a constriction should be prevented from collapsing against the valve body or other components where erosion damage could take place. This disadvantage of the outflow configuration in cavitating conditions can be overcome by dissipating some of the head just upstream of the circular orifice or by use of a sleeve with circular ports, but only at the expense of increased headloss when fully open. Plate 28(c) shows a cut-away illustration of a diaphragm sealed globe valve. Figure 16.4 shows an inflow type of globe valve with piston seal arranged as a pressure reducing valve (PRV).

16.10 SCREWDOWN VALVES

Screwdown valves are normally made only in small sizes but their operation is similar to that of the globe valve. The bib tap is a typical example. The body of the valve is so cast that the water must pass through an orifice which is normally arranged in the horizontal plane. A plug or diaphragm or, in the case of a bib tap or stopcock, a 'jumper' can then be forced down on to this orifice by a screwed handle, as shown in Figure 16.5. In small sizes, high pressures can be controlled, as in the case of the ordinary domestic tap. However, screwdown valves are not

1 Valve element
2 Upper cylinder
3 Isolating cocks
4 Strainer
5 Orifice
6 Needle cock
7 Indicator
8 Relay valve
9 Diaphragm
10 Spring
11 Adjusting screw
12 Pressure gauge

Operation. If P_2 increases, valve 8 tends to close, increasing P_3, causing main valve 1 to close (and *vice versa*).

FIGURE 16.4

Globe valve with piston seal—with auxiliary circuit for PRV duty.

Stuffing box

Washer

Seating

FIGURE 16.5

Screwdown valve (for service pipes).

suitable for in-line flow regulation or isolation since they cause high headloss and their seatings need periodic renewal.

16.11 BALL VALVES

Ball valves consist of a spherical obturator with a cylindrical hole, usually of the same diameter as the pipe, although it can be smaller. Operation is by rotation (1/4 turn) of a shaft mounted (often horizontally) with its axis at right angles to the cylindrical hole. Seals are usually resilient and can

FIGURE 16.6

Ball valve (2-part body).

provide drop-tight shut off. Ball valves are commonly used in small diameters (up to DN 300) although at least one manufacturer can make ball valves up to DN 1200. Ball valves are manufactured in one-piece, top entry, two piece (Figure 16.6) and three piece bodies. A top entry body allows access to the ball and seats for maintenance without the need to remove the valve and is preferred for larger sizes.

16.12 PLUG VALVES

The principle of the plug valve is similar to that of the ball valve in that closure is effected by a 1/4 turn of a spindle which, in this case, is usually mounted vertically to allow the weight of the plug to be taken on a bottom bearing. The opening is often rectangular rather than circular; it requires transitions upstream and downstream and prevents the passage of harder swabs. Plugs with circular openings are available but, except for the smallest sizes, the opening is smaller than the pipe. The plug may be a complete frustum of a cone or a sector only, in which case the obturator is called an eccentric plug or cam. The plug can be removed from the top of the valve body. Some designs allow the plug to be partly supported by upstream hydraulic pressure, thereby reducing wear. Eccentric plug valves are available up to DN 1800 and full cone plug valves can be made in similar sizes if required. Plate 29(a) shows a cut-away of an eccentric plug valve.

16.13 DIAPHRAGM VALVES

In a diaphragm valve the diaphragm is forced down onto a weir in the valve body (weir type—Figure 16.7) or onto the invert (straight-through type—Figure 16.8). The action is created by rotation of a threaded spindle of the rising type and is transmitted to the diaphragm though a shaped platen. Diaphragm movement for the weir type is less; consequently this type is preferred for higher pressure or partial vacuum conditions. As the fluid is separated from the moving parts the diaphragm valve is particularly useful for chemical liquids (if the body is lined) and those carrying solids. Diaphragm valves are made in sizes from about 6 mm to about DN 350.

FIGURE 16.7

Diaphragm valve (weir type).

FIGURE 16.8

Diaphragm valve (straight type).

16.14 PINCH VALVES

A pinch valve consists of a diaphragm which is deflected by a cam or roller to reduce the gap between the diaphragm and the base of the valve opening. In some designs, an upper and a lower diaphragm are employed, both squeezed at the same time. Valves of this sort are particularly useful with very aggressive fluids if the liner/diaphragm is made of suitably resistant material.

16.15 NEEDLE VALVES

In a typical needle valve flow passes around the housing for a piston located centrally within the valve body. Flow can be gradually reduced by advancing the piston towards a circumferential seat in the downstream end of the valve body, often through a sleeve. The piston and central bulb may be streamlined to reduce losses with the valve fully open. Plate 29(b) shows an actuated needle valve.

The full open head loss of needle valves can be low whilst affording good control and cavitation performance—particularly when a perforated or slotted sleeve is employed.

For 'free discharge' high pressure duties such as occur during draw-off at the base of a dam, the needle valve should be adapted with a jet disperser. This consists of a vaned insert located downstream of the valve in order to break up the jet and impart a twist to the flow. Provided the jet has sufficient height and space to break up into falling water droplets, little protection of 'soft' river beds is needed.

16.16 SLEEVE VALVES

A sleeve valve consists of a fixed perforated sleeve where the openings are revealed by movement of a piston. In-line sleeve valves may be mounted with their axes on the pipe axis or at an angle. Flow in in-line sleeve valves is from the outside to the inside which provides a good basis for cavitation resistance. Sleeve valves are particularly suitable for control duties where high heads have to be dissipated. However, facilities should be provided for access to the sleeve for cleaning unless a self-cleaning mode is built into the design.

Where discharge of high head water into a tank is required a submerged discharge valve is often used. This is usually fitted with a sleeve and discharges radially outwards through the sleeve at the base of a concrete pit where excess energy is destroyed in turbulence.

16.17 HOLLOW JET DISCHARGE VALVES

The hollow-jet discharge valve consists of a cone mounted in the water way with its apex upstream on a shaft with a gearbox and drive spindle through the side of the body. The cone is moved downstream revealing an annular orifice. The jet of water issuing from the valve is diverging and is able to dissipate energy in a shorter distance than the needle valve. The original design was by Messrs Howell and Bunger in 1935 but similar valves are now made by several companies. The Howell-Bunger valve may be fitted with a hood to limit lateral spread of the jet and to cut down spray, although this partly negates the advantage of this design which can otherwise deal with heads up to 425 m and can be made in diameters up to 4.25 m.

16.18 NON-RETURN (OR CHECK) VALVES

Non-return or check valves are of six basic designs:

- a disc with a single hinge or off-centre pivot, usually closing against an inclined orifice;
- a disc similar to that of a butterfly valve but with an off-centre pivot and a counter weight (Plate 29(c));
- a spring loaded disc closing linearly against a circular orifice either mounted in the horizontal plane as in a globe valve or in-line as in the needle or nozzle type;
- a split disc hinged in the middle like butterfly wings;
- falling ball type;
- conical diaphragm type.

With all these designs forward flow tends to move the obturator out of the way while the valves are designed to close as soon as possible after forward flow ceases. It is desirable for closure to be achieved without slamming, which generates surge pressures and may damage the valve or other

Brass air plug

Flow

FIGURE 16.9

Non-return check valve.

parts of the system. Non-slam characteristics are achieved by either arranging for closure to be complete before reverse flow commences or by slowing closure down so that any reverse flow is reduced gradually. The latter method usually relies on air or hydraulic damping while the former may be assisted by external aids such as springs or weights. However, any assisted closure tends to force the obturator into the flow and therefore increases headloss. All non-return valves are thus a compromise between low head loss and speed of closure. It should be noted that damped closure can allow reverse flow and thus cause a pump upstream to turbine.

The most common type of non-return valve consists of a flat disc within the pipeline pivoted so that it is forced open when the flow of water is in one direction and forced shut against a seating when the flow tries to reverse (Fig. 16.9) and is often termed a 'swing check valve'. The seating is arranged slightly out of perpendicular when the valve is to be inserted into a horizontal pipe so that the flap closes by gravity when there is no flow. A more inclined seating reduces flap travel and speeds closure. Another way to reduce travel is to provide two or more smaller and lighter flaps in a single bulkhead but at the penalty of some increase in headloss. The globe type of non-return valve has a very short travel and with low spring pressures can close very quickly with effectively no reverse flow. However, it has high losses and is not usually employed on pumped systems due to wastage of energy. Spring loaded valves of the needle or nozzle type have lower losses (k of 0.75 for DN 300) than other types of non-return valve.

16.19 FLAP VALVES

Flap valves are used at drain or washout outfalls where a pipe discharges to a body of water with varying level. This is to prevent back flow at high water levels and consequent contamination. Flap valves can be circular or rectangular and are usually mounted against the outer face of a concrete wall, but can be flange mounted. The materials used for gate and frame are usually cast iron but stainless steel is used for some large fabricated gates.

Factors to be considered in the selection (and design) of flap gates include:

- robustness and resistance to distortion under high (tidal) back pressure, particularly with debris trapped under the gate;
- ability to seal adequately after long use in difficult conditions; 'double-hung' (two hinges in series—see Figure 16.10) flaps overcome some of the effects of wear and temperature changes;

FIGURE 16.10

Flap valve.

- avoidance of water and silt traps in the gate construction that can increase resistance to opening and aggravate corrosion;
- effects of waves and damage to seals and hinges; a stilling chamber should be considered.

16.20 **CAVITATION AT VALVES**

Cavitation is the generation of pockets (cavities) of water vapour and their subsequent collapse (Section 13.12). A part open valve presents an orifice which produces a high velocity low-pressure jet. The onset of cavitation is marked by a fall-off in discharge coefficient as flow is increased, as downstream pressure is decreased or as the orifice is made smaller since any of these factors tends to depress the pressure in the orifice. The fall-off in discharge coefficient occurs because vapour pockets start to occupy the orifice opening. This effect can be used to determine when incipient cavitation occurs and is identified as the point where the curve of headloss coefficient against cavitation number begins to rise markedly. At this level there is minimal noise, vibration or risk of damage. Another method of assessing the onset and severity of cavitation is by taking noise and vibration measurements in the valve. Three stages of increasing cavitation are described in *Control Valves* (Borden, 1988): incipient, full and super cavitation. In this context incipient cavitation is characterized by the irregular occurrence of cavitation instances at vortices. As cavitation increases, there is constant production and collapse of cavities and noise is steady and reaches a maximum level, after which (in super-cavitation) noise decreases somewhat due to the dampening effect of the volume of vapour present and as the collapse area is pushed downstream.

At small openings the internal shape of the valve (according to type) has little effect on the head loss coefficient. Therefore, in such circumstances critical cavitation conditions are likely to present themselves at very similar loss coefficients irrespective of valve type. The shape of the opening has a minor effect on the onset of cavitation. However, both the nature of the opening and the shape of valve internals significantly affect the risk of damage once cavitation is occurring. Valves in which collapse of vapour pockets occurs away from the valve body and other components should not be damaged by cavitation. The distance a jet travels before head recovers sufficiently to collapse vapour pockets depends on jet size. Therefore, porting the opening to produce numerous jets can help keep the area of collapse away from critical components. An added advantage of numerous small jets is that noise and vibration are reduced.

614 **CHAPTER 16** Valves and Meters

Some materials commonly used in valve construction (cast iron and brass) are not particularly resistant to cavitation damage but some bronzes are more resistant. Various stainless steels, when hardened, particularly Duplex types, provide superior resistance (Borden, 1988). In very extreme cases titanium may have to be used.

16.21 VALVE OPERATING EQUIPMENT

Manual closing of a valve is usually by clockwise rotation of a hand wheel or tee key. However, for historic reasons some utilities may use valves with anticlockwise closing in the whole or part of their systems. Clear labelling of the operating direction is essential. Where the closing direction is in doubt the valve can be 'sounded' to detect at which end of its travel there is noise of water rushing through a narrow opening. Manual operation of valves is by hand wheel or tee key. Hand wheels may be mounted directly on the valve. However, if this would put the operator below ground, where confined spaces procedures would be needed (to avoid danger of suffocation or noxious gases), a means of operation from above ground should be provided. This may require a headstock and, in many cases, an extension spindle running in brackets rigidly attached to the chamber walls. Spindles which are to be immersed in water, such as those for operating valves inside a reservoir, should be of manganese bronze or stainless steel. Although many old manually-operated large gate valves without gearing are still in use, they have be operated by large bar and ring key, sometimes by a team of operators. Valves for manual operation should now be specified to allow operation by one person with a force on each side of a handwheel or tee-key of not more than 200 N or, for unseating, 400N.

If manual operation is not feasible (due to very long operating times) for a large geared valve for example, or where remote or automatic control is required, an actuator should be fitted. Actuators for pipeline valves are usually electric but pneumatic and hydraulic systems are used where a large number of actuators can be powered from a central compressed air supply or hydraulic power pack. Hydraulic systems are mainly used for very large installations—hydropower for example. Pneumatic actuators are more suited to on-off operation as positioning is not precise enough for control. They are very useful where fail-safe (open or closed) operation is needed as there is sufficient stored energy in the system to close or open a valve after power failure.

Electric actuators are usually provided with a hand wheel to allow manual operation in case of power failure and should be provided with local operation buttons. Where remote or automatic operation is needed signal cables are required. The sizing of actuators needs to be done with care since there should be enough torque to overcome resistance but not so much as to overstress the valve spindle or the obturator stops. For most types of valve, electric actuators can be position-limited whilst for gate valves they need to be torque-limited. It is usual to arrange pinned couplings which should shear at a torque above that required to operate the valve but below that which could cause damage.

16.22 VALVE CLOSURE SPEED

Valve closure is one of the causes of pressure transients (surge) in a pipe system. Closure decelerates the flow and causes a pressure build up according to the Joukowsky equation:

$$\Delta H = - c \Delta V / g$$

where: ΔH is the change in head, c is the speed of a pressure wave, g is the acceleration of gravity and ΔV is the change in velocity (negative decrease).

Surge is discussed in more detail in Section 13.11. The magnitude of the pressure rise is roughly proportional to $(1/T_c)^{1.5}$ where T_c is the time of closure of the valve. Graphs published (by Thorley, 2004) indicate that, in a pipeline with initial flow velocity of 2 m/s, in order to limit the rise in head on valve closure to about 50 m, T_c needs to be about 200 times the length (L) in metres of the pipeline upstream for gate valves. The equivalent figure for butterfly and ball valves is about 100L. For an initial velocity of 1 m/s these closure times would be about half of the above values. Pipe diameter has little effect on these requirements; therefore it can be seen that there is greater risk of high pressures on closure of small valves since they can be closed more quickly. On the other hand, for ductile iron pipelines there is considerable spare capacity for higher pressures in smaller diameters. For actuated valves the closure speed should be specified taking into account the acceptable pressure rise. Where necessary surge analyses should be carried out to check that the selected closure speeds are appropriate, using the characteristics of the valve to be supplied (Section 13.11).

16.23 WASHOUTS

Despite the name, a washout is seldom used for scouring or 'washing out' a pipeline because its diameter is usually too small to create sufficient flow velocity in the main to wash out debris; its principal use is for emptying the pipeline or for the removal of stagnant or dirty water. Discharges may be subject to consent, as in UK, but the relevant authority should always be consulted whether the discharge is to be to sewer, watercourse or other water body. If chlorinated water is involved its dechlorination may be required. In a major pipeline primary washouts may be installed to drain the majority of the length between section valves; secondary washouts of smaller diameter can then be used to empty undrained low points. Sizes, particularly of primary washouts, should be calculated according to the required drain down time, which should typically not be longer than one working shift. Factors are the number of washouts, head available and limits on discharge, access and resources. Initial flows could be very high unless the washout valve is throttled but such throttling could cause cavitation unless the type of valve is selected accordingly (Table 16.1). The drain down time is dominated by the low head available during the later stages. The washout diameters given below should allow the last 200 m length of a pipeline to be emptied in about one hour in typical situations:

Main pipeline diameter	Washout branch diameter
Up to 300 mm	80 mm
400 to 600 mm	100 mm
700 to 1000 mm	150 mm
1100 to 1400 mm	200 mm
1500 to 1800 mm	250 mm

In open country it is usual to install washouts at every low point with additional washouts being provided in each section where a main is sub-divided into sections by stop valves. Each washout should discharge, wherever possible, by gravity to the nearest watercourse. The discharge should be to a concrete pit with overflow to the watercourse in order to prevent scour from the high velocity discharge. The washout branch on the main should be a 'level invert tee'. In flat country washouts should be spaced 2 to 5 km apart, depending on pipeline gradients and valving. Where it is not possible to get a free discharge to a watercourse the washout will have to discharge to a chamber from which the water can be pumped out to some other discharge point. In this case a means of prevention of backflow should be provided in addition to the washout valve. This could be a flap valve on the outlet to the chamber.

In distribution systems principal feeder mains are usually provided with washouts wherever this is convenient, regard being paid to the position of valves on the main and any branch connections and to the need to be able to empty any leg of the main in a reasonable time of one or two hours. On small mains, washouts are not normally provided because fire hydrants can be used to help empty the system. However, it would be usual to lay a specific washout to empty a part of the system where a convenient watercourse exists. Although pipe supply systems should be designed to avoid dead ends, spur mains may be necessary in some situations. Washouts should be placed at the end of every spur main; these usually comprise fire hydrants even though they may not be officially paid for by the fire authority and designated as such. They should be operated regularly to sweeten the water at the end of the main.

Care is essential in the design of washouts since, under high heads, the velocity of discharge can be very high and the consequent jet discharge destructive and dangerous. Manholes receiving the jet should be of substantial construction and valves should be lockable and slow opening for high heads. Two valves, one guard and one operating may be installed where prolonged throttling at high head is required.

16.24 AIR VALVES

To fill a pipeline with water, there must be means for releasing air from it. Where hydrants are installed, for example in a distribution system, these can be opened to help release air (and to exhaust the first flush of water if required). However, unattended air release points should be provided with automatic valves which must close as soon as there is no more air present so that no water is lost. An air release valve contains a ball or other shape float which rises to close an orifice as the last air is excluded. The design must prevent the float from being sucked onto the orifice by high velocity air and must prevent oscillation of the float. Two sorts of air release valve are used, one being a 'large orifice' air valve, designed to release or admit large quantities of air at low differential pressure when a pipeline is being filled or emptied; this type does not open to release air under any appreciable pressure. The other is a 'small orifice' type (Fig. 16.11) which is designed for continuous operation, releasing small quantities of air under operating pressure as it collects at high points. Double orifice air valves (Fig. 16.12 and Plate 29(d)) combine one of each type in the same unit and are used at locations where both duties are required.

For raw water pipelines (as for sewage rising mains) it is important that air valve action is not obstructed by debris in the water. Air valves made for this purpose reduce the risk of float jamming or orifice obstruction. It is seldom necessary to put small orifice air valves on distribution mains since air is removed via the service pipe connections which should be soffit connected. Exceptions

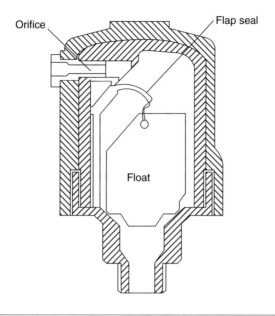

FIGURE 16.11

Small orifice flap type air valve. *Based on information from:* Aqua-Gas AVK.

FIGURE 16.12

Double orifice air valve.

(a) Eccentric plug valve (Cam-centric)
(Val-Matic Valve & Mfg. Corp.,
Illinois, USA)

(b) Larner Johnson (needle) valve – inlet
450 mm, throat 229 mm. (Blackhall
Engineering Ltd., UK)

(c) Tilting disc check valve. (By courtesy
of ERHARD GmbH & Co. KG)

(d) Double orifice 80 mm air valve
(HiVent). (GA Valves Sales Ltd.)

PLATE 29

(a) Electromagnetic district flowmeter installation. (Black & Veatch)

(b) District metering roadside cabinet. Control unit, data logger and pressure gauge. Telemetry and modem can also be installed. (Black & Veatch)

(c) Electromagnetic flowmeter insertion probe with pressure transducer and data logger. (Black & Veatch)

(d) Semi-positive rotary piston meter (V100) for 15 mm to 40 mm supply pipes. (Manufacturer: Elster Metering Ltd., Luton, UK)

PLATE 30

are where there is a sudden hump in the main, such as when it is laid over a bridge. For other pipelines, Lescovitch (1972) may be referred to but the authors advise the following:

1. Large orifice air valves are required for filling and emptying:
   a. at high points;
   b. at a steepening in gradient in a falling pipeline;
   c. (possibly) at a flattening in gradient in a rising pipeline;
   d. at about 2 km intervals on long lengths of pipeline (with no fire hydrants), and
   e. between any intermediate line valves.

2. Small orifice valves are required at similar locations and additionally:
   f. at high points relative to the slope of the hydraulic gradient;
   g. on access manhole covers and other local 'humps' where air may collect;
   h. at other locations on a pipeline supplied by pump if likely to introduce air:
      - downstream of the pumps,
      - downstream of any pressure reducing valve if there is a substantial reduction of pressure, and
      - (possibly) at 0.5 km intervals on downward legs.

Large orifice air valves are usually sized to suit the maximum outflow expected during use of washouts, to cover the fastest likely rate of filling and in some cases to limit sub-atmospheric pressures in the event of a burst. However, fast filling is inadvisable and a planned filling rate equivalent to a pipeline flow velocity of about 0.3 m/s would be acceptable (AWWA M51). Where two or more large orifice air valves are available to exhaust the air the rate may be increased. The limitation is aimed at avoiding damage on closure of the air valve once all the air has been removed. It is suggested that air valves be sized for a differential pressure of 0.25 bar. The flow capacity of air valves should be taken from manufacturers' catalogues. With some designs the orifice size is smaller than the connection size. Table 16.2 shows quoted air valve capacities at 0.25 bar pressure differential; they vary considerably, therefore it is important to specify performance:

An isolating valve (sluice valve, butterfly valve or stopcock) should be sited below each air valve, thus making it possible to remove the air valve for repair or replacement without shutting down the main. The restricting effect of valves and fittings between the pipeline and the air valve must be taken into account in assessing air valve capacity as installed. Although each pipeline should be treated specially, Table 16.3 provides a guide to the minimum air valve capacities for pipelines of selected diameters.

Small orifice air valves have an orifice size which should be related to the operating pressure in the pipeline: the higher the pressure the smaller the orifice. If it is assumed that 5% of the dissolved air in a pipeline may come out of solution (actual amounts should be much less in a well designed system), the total small orifice air valve capacity provided should equate to about 0.1% by volume of the water flow. A greater amount of small orifice air valve capacity at the upstream end of a pipeline than downstream should ensure early release of air. Capacities of small orifice air valves should be taken from manufacturers' catalogues, using the operating pressure at the air valve location.

On large pipes, air valves may be fitted to a blank flange on a DN 600 or larger tee to allow access to the pipeline. All air valves should be sited above the highest possible groundwater level that can occur in any pit; the pit should be free-draining with minimal maintenance; if this is not

Table 16.2 Large orifice air valve capacities

Branch size (mm)	Range of quoted air flows at standard temperature and pressure for 0.25 bar differential (m³/min)	
	Out	In
80	12 to 51	8 to 45
100	21 to 70	16 to 60
150	50 to 150	50 to 135
200	50 to 300	50 to 200

Table 16.3 Suggested minimum large orifice air valve capacities

Pipeline diameter (mm)	Outflow capacity (m³/min)	Inflow capacity (m³/min)
300	8	12
600	20	25
900	40	50
1200	70	80
1800	150	180

done, polluted water may enter via the air valve when the pipeline is emptied. Where the location of an air valve is not possible in the road (as cover at high points is usually a minimum), it may be necessary to connect the branch to an air valve in the verge, where air valve chambers can be raised above ground level if necessary.

On thin walled pipelines of steel or other flexible material anti-vacuum valves may be essential in order to admit air to limit sub-atmospheric pressures and prevent pipeline collapse on emptying or on a burst. If below atmospheric pressures are expected during a surge transient, this type of protection may be needed for large diameter or flexible pipes. Large orifice air valves can be used for this duty but special anti-vacuum valves may be necessary for very large air flows.

16.25 VALVE CHAMBERS

Large valves on trunk pipelines should be located in accessible chambers to facilitate maintenance. Smaller gate valves, for example on distribution systems, are usually buried. It is preferable to site butterfly valves in chambers so that the gearbox can be maintained, but for large butterfly valves some water companies put a chamber around the gearbox only.

A simple valve chamber for a gate valve is shown in Figure 16.13. The valve is anchored on the upstream side by an anchor flange in the chamber wall. Downstream of the valve a flange

FIGURE 16.13

Valve chamber.

adaptor allows removal of the valve. The valve body is therefore free to expand downstream when the valve is closed under pressure and it must therefore be specified for the open end test conditions (Section 16.6). The pipe through the downstream wall is provided with a puddle flange to reduce ingress of ground water. In order that the chamber can resist the thrust from the anchor flange, it should be designed in the same way as a thrust block, transferring the load to the soil through friction or cohesion and passive earth resistance (Section 15.23). Two flexible joints, with an intervening 'rocker pipe' two pipe diameters in length, should be provided on either side of the chamber to avoid damaging pipework if any differential settlement of the pipeline, relative to the chamber, should occur. The chamber cover can be made of precast reinforced concrete slabs, provided with sockets or other arrangements to receive lifting devices. Valve chambers should not be sited in roads if they can be located in the verge or other open ground instead but it is wise to ensure that the cover is strong enough to take the loading from heavy vehicles which may run off roads.

Small valves, i.e. those DN 300 and smaller, are usually buried and a pipe (DN 80 or 100) is fitted as a sleeve for the spindle which is accessed through a surface box for tee key operation. Valves up to and including DN 600 may also be buried, depending on location and preference: the choice being on maintenance and the civil works required for giving access to the valve.

PART II MEASUREMENT OF FLOW AND CONSUMPTION

16.26 PURPOSES OF FLOW MEASUREMENT

There are three reasons for measuring flow:

- control (either manual or automatic) of a process;
- data acquisition for legal, record or operational purposes;
- billing to customers.

Control of a process using flow is based on instantaneous measurement by a flowmeter instrument whose output is converted to an analogue (4 to 20 mA) or digital signal (Section 17.34). The signal is then transmitted to a PLC for automatic control of a valve, pump or other device or to a monitoring screen for manual intervention.

The data acquisition application is similar, but the signal is transmitted to a monitoring and control centre for recording or it may be recorded locally on a data logger. The use of chart recorders for this purpose has almost ceased. Typical uses include metering of pump output and of water supply zones or District Metering Areas (DMAs) which is essential for checking leakage and for monitoring consumption.

Billing of customers requires measurement of consumption at a water meter. For domestic and small commercial and industrial consumers the meter is usually of the mechanical volumetric type. These meters are increasingly provided with short range transmitters to enable remote reading without having to enter premises. Water meters for larger consumers may use the same principle as flowmeters.

16.27 TYPES OF FLOWMETER

Flowmeters are classed as volumetric or inferential, the latter term referring to meters that determine velocity from other variables such as pressure differences across a device such as an orifice plate. There is a large variety of flow measurement device, using numerous physical principles. Full discussion of the whole range of flow measurement device is out of the scope of this book but the reader will find a comprehensive reference in the *Flow Measurement Handbook* (Baker, 2000). Table 16.4 gives typical information on some of the flow meters usually encountered in the water industry.

The principles of orifice and venturi meters are discussed in Section 12.16. Two other kinds of inferential (or momentum) meter are the Dall tube and the V cone venturi. In both, flow accelerates through a constriction and leads to a pressure drop. The pressure difference is measured in the Dall tube and the V cone venturi as an indicator of velocity (and so flow) in the same way as for an orifice. The V cone venturi design is claimed to have a turn-down ratio of 25:1 and to be less affected by conditions upstream and downstream and can be fitted into shorter lengths of straight pipe than is recommended for other meter types. Further types of momentum meter are indicated in Table 16.5.

16.28 VOLUMETRIC FLOWMETERS

Electromagnetic flowmeters. This type of meter (Plate 30(a)) is based on Faraday's Law of electromagnetic induction. The voltage induced in a conducting fluid as it moves through a magnetic field is measured between electrodes in the pipe wall and is proportional to the mean velocity of the fluid, the strength of the magnetic flux and the size of the conduit. Calibration produces a relationship between induced voltage and total flow through the conduit. The output is integrated across the pipe section and is not affected by small differences in velocity profile across the section. The magnetic field is generated by coils usually arranged as saddles around the pipe, powered by battery or, for large meters, from the mains electricity supply.

This type of meter is increasingly being used in the water industry owing to zero head loss and good accuracy and reliability. It does require fluid conductivity to be more than 5 µS/cm and air entrainment to be low but it can be used for a large variety of transported materials other than water. Deposits on the electrodes should be cleaned off periodically and any deposits on the pipe wall near the electrodes need to be rendered non-conducting. Both can be achieved by application of appropriate dc and ac voltages.

Table 16.4 Characteristics of meter types (data from manufacturers' catalogues)

Type	Typical pipe diameter (D) range (mm)	Typical velocity range (m/s)	Typical accuracy (% of full scale)	Typical repeatability (% of full scale)	Typical head loss coefficient (K)	Comment
Volumetric						
Electro-magnetic	25–2500	0.5–10	0.25	0.1	0	
Ultrasonic	Up to 7000	0.05–4	1	0.5	0	Sensitive to vibration and head alignment
Coriolis	15–80	0.5–5	1 (of mass)		15	May include density measurement with accuracy about 0.4%
Vortex	15–300	0.4–10	1	0.2	2	Not suitable for pulsating flow
Turbine	15–400	0.6–10	0.5	0.1	0.8	
Rotating vane	15–150	0.02–6	1		5.5	
Rotating piston	15–20	0.03–4.4	2		4	
Paddle wheel	15–1000	0.3–6	1	0.5	5.5	
Inferential					*Head loss as % of pressure difference*	
Venturi tube	50–1200		1.25		10–20	Length = 5D
Dall tube	150–2000		1.25		30	Length = 1.75D
V cone venturi	15–3000		0.5	0.1	Up to 75	
Orifice plate	25–200		3	1	Up to 90	

Table 16.5 Further types of momentum meter

Type	Principle
Vortex meter	Detection by various means of the frequency of vortices shed by a "bluff" body into the flow. Flow being proportional to frequency. Accuracy is claimed to be almost as good as for the ultrasonic type but there is some head loss due to the obstruction.
Swirl meter	Generation of a spiral flow stream in a throat by a vane array upstream and its detection via the pressure fluctuations caused at a tapping in the throat.
Fluidic meter	A jet of water is flipped from side to side of diverging conduits under the action of feedback from each side. The frequency of oscillations is proportional to velocity and is detected by pressure sensors.
Variable area meter	A "float" sits in upward flow in a tube with increasing diameter and whose position is detected by optical transducer (principle used for the chlorinator).
Viscous meter	Measurement of the pressure difference across an array or labyrinth of fine tubes in the conduit.
Drag plate meter	The force or moment on a plate inserted into the flow is detected at its support in the wall.

Calibration is best done 'wet' against a master meter since this is more reliable, although dry calibration (by measuring the magnetic field across the whole section) can be done (see also 'digital fingerprinting'—Section 16.29).

Ultrasonic flowmeters make use of sound waves of frequency over 20 kHz. The transit time (time-of-flight) meter measures time differences due to the difference in velocity of sound waves transmitted upstream and downstream through the flow. The time difference is a function of average flow velocity along the beam. Doppler meters measure the change in frequency in sound waves reflected back from particles in the flow and, therefore, cannot be used for pure or treated water.

Ultrasonic flowmeters of the transit time type are reasonably accurate if installed and operated correctly. The pipe wall must be of a hard material that transmits sound well. Deteriorated and porous pipe linings and deposits are likely to affect the accuracy of ultrasonic flow measurement and may cause drift. Accuracy is adversely affected by transducer misalignment and vibrations in the system. Doppler meters are generally less accurate. Ultrasonic meters do not intrude into the flow and cause no head loss but their accuracy is not as good as that of the magnetic meter. Clamp-on ultrasonic meters require no intrusion into a pipeline and, although their accuracy (about 3%) is not as good as for complete meters, arc becoming increasingly used particularly for large conduits (over DN 500) and for temporary installations and check measurements.

The accuracy of single path ultrasonic flowmeters is affected by uneven flow distribution across the section. In situations where this may be a problem, meters using multiple paths are recommended.

Propeller or turbine meters use a freely-rotating, bladed rotor positioned in the flow. The speed of rotation is measured, usually by a pick-up coil mounted in the housing to sense the passage of the rotor blades. Pulses are thus generated which can be counted over a known time. There is a small amount of resistance to rotation due to friction and the sensor. Therefore, the turbine does not rotate until velocity exceeds a threshold value. Above this level, the speed of rotation is proportional to the fluid velocity so that it is necessary to know the flow area to calculate flow. For large conduits it may be necessary to carry out velocity traverses across the section to establish the relationship between velocity at a particular location and the total flow rate. Provided the velocity profile is well known, the accuracy of the turbine meter for flow measurement is very good.

16.29 PERMANENT FLOWMETER INSTALLATIONS

Source meters and meters on large mains have historically been of the Venturi or Dall tube type and a number of these may still be in use. Venturi and Dall tubes can have an accuracy of ±1.25%, but their performance is affected by upstream and downstream features such as bends and valves. If properly maintained they can continue to give satisfactory readings but the throats of such meters should be inspected from time to time to ensure that they are clean and free of slime or deposits, since they affect meter accuracy. For this purpose a hatch is usually provided over the throat of a venturi. The recording equipment, usually mechanical, also needs keeping in good order. To be certain that source outputs are known as accurately as possible, *in situ* volumetric testing of such meters is always advisable.

Most Venturi and Dall tube meters have been replaced with more modern types such as the electromagnetic meter (Plate 30(a)) which is simpler to install, more accurate and which can be equipped with a full telemetry interface capability for district metering, customer billing, leakage control and treatment works applications. Electromagnetic meters have an accuracy of ±1% but recent developments have enabled some manufacturers to quote ±0.25% accuracy for their meters. However, accuracy in the field is affected by pipe features. Plate 30(b) shows a pillar for housing district metering electronic equipment.

The flowmeter manufacturer provides a specification and a flow calibration certificate from a recognized test laboratory. However, installation conditions usually differ markedly from the bench tests and invariably lead to differences in performance. Meters are frequently sited too close to fittings, valves or tees, or are affected by protruding gaskets and other factors such as vibration, flooding or large ambient temperature swings. To overcome the worst of installation effects manufacturers recommend that the length of straight pipe upstream and downstream of meters should be a certain multiple of pipe diameter. A commonly used figure for electromagnetic meters is 5D upstream and 3D downstream. For other types of meter and for precision installations a larger multiple and possibly flow straighteners may be needed. Extensive test data from various sources on the effects of nearby features on the accuracy of various types of meter is set out in the *Flow Measurement Handbook* (Baker, 2000). The data show that the effect of a pipe feature on meter accuracy depends on the nature of the feature (i.e. bend or part open valve), its distance from the meter, its orientation with respect to meter sensor orientation and on the type of meter.

In situ testing is often impractical and installed meters are likely to produce errors. Therefore, estimating leakage from the difference in flows measured at different locations can produce quite large errors. 'Digital fingerprinting' has been introduced to check for changes in calibration with time. This tests the electrical characteristics of a magnetic meter. Provided the electrical field remains constant and other key electrical parameters within the circuitry are also stable, it is possible to relate the electrical 'fingerprint' of a meter and transmitter back to a change in meter calibration. The meter is self checking and the results can be sent back to the manufacturer annually for audit. In situ testing should be traceable in situ and should be conducted by an experienced engineer.

16.30 TEMPORARY FLOW MEASUREMENT DEVICES

Insertion probe flow meters are installed for temporary measurement of flow for consumption surveys or for distribution networks analyses. These instruments are either the turbine or electromagnetic (EM) type, the latter becoming more common. Both are inserted into the pipe where flow measurement is required. The turbine type uses a small rotating vane at the end of a probe to record flow velocity. The vane is susceptible to damage, in which case the instrument has to be returned to the manufacturer for repair and recalibration. The turbine meter is inserted through a 40 mm diameter tapping in the pipe which has to be of at least 200 mm diameter. The EM probe (Plate 30(c)) uses an electromagnet at its end to apply a magnetic field to the water. Electrodes either side of the probe pick up the induced EMF in the water which is proportional to the velocity past the electrodes. The tapping for an EM insertion probe is 20 mm diameter and can usually be installed in pipes of diameter 150 mm and greater. EM probes are made up to 1 m long; therefore, they cannot be used for pipes of diameter greater than 900 mm and are restricted to flow with velocity less than about 1.75 to 2.0 m/s due to the flexibility of the probe.

Insertion probes measure the velocity at the position of the measuring device. This can be at the pipe centre line or at defined points along the diameter. The measured velocity has to be converted to mean pipe velocity of flow by relating the measured value to the average velocity across the whole pipe. For this a velocity profile for the pipe is used, determined by using the same instrument to record velocities at set points across the diameter from crown to invert. The recorded measurements are corrected to take account of the disturbance caused by the instrument itself (increased local velocity). The disturbance coefficients are unique to each instrument and are provided by its manufacturer. For the conditions usually encountered the ratio of the mean velocity to the centerline velocity is 0.83 but can range from 0.7 to 1.0. Values differing widely from 0.83 should be viewed with caution and the cause investigated. However, satisfactory results should be obtained if the internal diameter is measured accurately and if the number of flow profile readings is sufficient—five for pipes of DN 150, nine for DN 300 and 13 for larger pipes. The measurements should be repeated at least three times to ensure the ratio is consistent and repeatable; the flow must be relatively consistent during each profile run. In practice poor field conditions often make precise measurement difficult so that several attempts may be necessary. Once the profile is established satisfactorily the instrument is set at the pipe centre line and the data logger is attached. In pipes in poor condition the velocity profile changes with flow and can render very inaccurate measurements of flows significantly different from that at which the velocity profile was established.

Making tappings and installing insertion probes pose a risk to water quality. Although such risks can be managed, ultrasonic strap-on flow meters are being used increasingly as an alternative and avoid the tedious exercise of velocity profiling. Versions of these meters can be installed on all sizes and materials of pipe used in distribution systems.

16.31 SUPPLY (REVENUE) METERS

Water supply meters in common use on consumer's service pipes are of two types: semi-positive and inferential. The semi-positive meter, typically sized in the range 15–40 mm, is almost universally used in the UK for metering domestic and small trade supplies. The most usual of the semi-positive meters in UK is the rotary piston (or rotary cylinder) meter which has an eccentrically pivoted, light weight, freely moving cylinder which is pushed around by the water inside the cylindrical body, opening and closing inlet and outlet ports as it turns (Plate 30(d)). This movement operates a counter mechanism which summates the total flow. Another type uses the nutating disc which wobbles around in a circle as water is drawn through the meter and endeavours to pass above or below the disc. This meter does not measure low flows to the same accuracy as the rotary piston meter but is widely used in the USA where domestic flows are generally higher. Other types of semi-positive meter include the single or multiple orifice vane meters and the paddle wheel type (Fig. 16.14). All semi-positive meters should incorporate a strainer upstream as the meter is only suitable for water free from grit or other suspended matter.

The inferential meter has a bladed turbine which is turned by the flow. The quantity of water is 'inferred' by counting the revolutions of the turbine (Fig. 16.15) and must be calibrated at the maker's factory. It is primarily used on industrial supplies, being suitable for large flows. It can summate the total flow either through counter gearing or each revolution of the propeller may initiate an electrical pulse which an electrical logger can summate for the total flow and also record flow rates over short time intervals. This data can be locally recorded or transmitted elsewhere by telemetry. For the measurement of widely fluctuating flows, beyond the range of any single meter, past practice has been to use two meters of different sizes in parallel; an automatic device ensures

Single jet meter, 15 or 20 mm   Multi jet meter, 15 to 50 mm

FIGURE 16.14

Vane meters.

FIGURE 16.15

Turbine meter.

FIGURE 16.16

Small meter testing bench.

that the smaller flows pass through the small meter only. However, such combination meters were expensive to install and have now been superseded by the electromagnetic meter (Section 16.28) which gives good accuracy over a wide flow range and is cheaper to install than the mechanical volumetric type.

All supply meters need to be regularly tested for accuracy every few years. A typical small meter-testing bench is shown in Figure 16.16. Large industrial supply meters may be tested in-situ using a turbine or electromagnetic flow probe as described in Section 16.30.

16.32 THE ACCURACY OF WATER METERS

BS EN 14154-1 sets out the criteria for accuracy of water meters in UK. The accuracy classes defined in the superseded BS 5712-1 have no equivalent in the new standard. Instead manufacturers are expected to offer meters which suit typical applications, stating the values of Q_1 (minimum flow), Q_2 (transitional flow), Q_3 (permanent flow) and Q_4 (overload flow). BS EN 14154-1 requires the measurement error not to exceed 5% in the lower flow range Q_1 to Q_2 and 2% in the upper flow range Q_2 to Q_4. There is no requirement for the minimum flow at which the meter must start to register but some manufacturers state approximate values. The smaller meters used for domestic supply are usually 'semi-positive meters' in which an eccentrically pivoted plastic cylinder is caused to rotate by the through-flow of water. Larger meters, predominantly used on trade supplies, are usually 'inferential' meters of the rotating vane type. Domestic supply meters would normally have a maximum capacity of between 2 and 5 m^3/h. These values may be compared to the design flow rates shown in Table 14.4 for different domestic fittings.

Small water meters typically over measure in the upper flow range and under measure in the lower flow range; indeed they cease to turn at flows below about 3 l/h, even when new. This cut off should be compared with common situations in domestic premises: a relatively fast dripping tap (4 drips/s) would waste 3–4 l/h, and the thinnest continuous stream about 6 l/h. In a direct system all cold taps and WC cisterns are fed direct from the mains. In indirect systems only the cold water drinking taps in the kitchen and bathroom are fed direct from mains, the rest are fed from a float-valved roof storage tank. As a ball float valve approaches closure flow decreases and for a finite period before full closure flow is less than that which the meter can measure. The WRc found that low flows caused by near-closed float valves to WC cisterns and storage tanks were seriously under-recorded, resulting in a mean under-registration of 2.5% for 'direct' supply systems and 6% for 'indirect' systems (Welton, 1984). The accuracy of all meters also deteriorates with age. Most tests show the great majority of domestic meters under-record total consumption.

In the National Metering Trials in England 1989–1992 (Section 1.10) it was found that, of 200 meters withdrawn annually for testing, approximately 20% had failed, most by under-recording; about one-sixth of them due to blockages from particles in the flow (Hall, 1992). Estimates of the under-recording of supply meters reported by undertakings in England and Wales for 1998–99 are mostly 3–4%. The two water companies having the largest number of domestic meters installed, namely Anglian and Severn Trent, estimated their household meters under-registered by 2.9 and 4.1%, respectively and their trade meters by 2.6 and 6.6%, respectively (Ofwat, 1998–99). Generally 3% under-recording would be considered an average for semi-positive meters and 5% under-recording for inferential meters; but if meters are over 10 years old and have not been regularly removed for testing and refurbishment, substantially greater under-recording must be suspected.

Accuracies considerably better than 2% may be claimed by some manufacturers for the upper flow range, particularly for larger inferential (vane) meters. However, such meters have to be carefully sited because an adjacent upstream bend or tee can seriously affect their accuracy. Because of this WRc thinks that their under-recording in practice is greater than that of domestic supply meters. The 'multi-jet' meter (dividing the flow into several streams) was developed to improve the accuracy of such meters.

16.33 FUTURE TRENDS IN METERING

Automatic meter reading (AMR) technology is being trialled by a number of utilities worldwide. AMR comprises automatic remote reading of consumer meters and transmitting the data to the utility's billing database. Automatic reading can also be achieved by the meter reader when walking or driving past the premises. Reading devices, handheld or attached to the vehicle, automatically read the meter. The data can be downloaded from the reader using a wire connection, GPS radio telemetry or power line carrier transmission. Provided the meter location does not require access to the property, AMR offers potentially significant cost and manpower savings. However, application is currently constrained by the significant investment required.

Smart metering is an extension of AMR whereby more sophisticated meters can be read automatically to help the utility to operate and manage the distribution network. The systems are able to measure, analyse and store consumptive data throughout the day or at preset times and thus enable the utility to understand patterns of demand and set variable tariffs according to diurnal and seasonal consumption. Development is at present constrained by issues of cost and the accuracy of meters at different flow rates.

REFERENCES

ANSI/ISA S75.05 (1983). *Control Valve Terminology*. Research Triangle Park, NC: American National Standards Institution (ANSI), Washington; Instrument Society of America (ISA), North Carolina, USA.

ANSI/ISA S75.11 (1991). *Inherent Flow Characteristics and Rangeability of Control Valves*. Research Triangle Park, NC: ISA, USA.

AWWA M51 (2001). *Air-release, Air/vacuum and Combination air valves*. AWWA. USA.

BS EN 12266. Industrial valves. Testing of valves. BSI.

BS EN 14154. Water meters. BSI.

Baker, R. C. (2000). *Flow Measurement Handbook*. Cambridge University Press.

Baumann, H. (1998). *Control Valve Primer*. Instrument Society of America.

Borden, G. and Friedmann, P. G. (1998). *Control Valves*. 3rd Edn. Instrument Society of America.

Hall, M. (1992). Technological developments in metering. IWEM Symposium 'Paying for Water', Jan 1992. *J IWEM*, Aug. p. 517.

Lescovitch, J. E. (1972). Locating and sizing air release valves, *J AWWA*, Water Technology/Distribution, July.

Ofwat (1998–99). *Report on Leakage and Water Efficiency*. Ofwat.

Smith, E. and Vivian, B. E. (1995). *An Introductory Guide to Valve Selection*. Mechanical Engineering Publications Limited, London.

Smith, P. and Zappe, R. W. (2004). *Valve Selection Handbook*, 5th Edn. Elsevier.

Thorley, A. R. D. (2004). *Fluid Transients in Pipeline Systems*. 2nd Edn. Professional Engineering Publishing Ltd.

Welton, R. J. and Goodwin, S. J. (1984). *The Accuracy of Small Revenue Meters*. Report TR 221, WRc.

Pumping; Electrical Plant; Control and Instrumentation

17

PART I   PUMPS

17.1 PUMPING PLANT

Using machines to lift water is a very ancient art, developed to satisfy the most basic of human needs, water for domestic use and irrigation of crops. Pumping machinery has developed a long way from the hand, foot, or animal-driven shadufs and water-wheel pumps of the ancient world but most modern pumps strongly resemble those of 100 years or so ago, although with many improvements. Examples are better materials, improved bearings, protective coatings, better designed and finished hydraulic passages and better methods of drive and control. These have contributed to great improvements in performance and reliability.

The laws of physics dictate the minimum power required to lift a given mass of water through a given height in a given time (1 m^3/s lifted 1 m requires 9.81 kW at 100% efficiency). High pump efficiency is almost always of great importance in water supply since most water supply pumps operate for long continuous periods. Pumping costs generally dominate the running costs of water supply systems and the lifetime cost of owning pumps in nearly continuous use is heavily dependent on the cost of the power they use.

17.2 CENTRIFUGAL PUMPS

Pumps which operate by rotary action are called rotodynamic pumps and the centrifugal pump is the first type to be considered. Other types of pump still have their uses, particularly for pumping chemicals, slurries and sludge (Sections 7.20 and 9.4), but the centrifugal pump is the most commonly used for pumping water because of the wide range of duties possible and the comparatively high efficiency and low cost.

Centrifugal pumps are available in many different arrangements, as single or multistage units, and mounted vertically (Plate 31(a)) or horizontally (Plate 31(c)) to suit particular needs. The centrifugal pump comprises an impeller which is rotated at high speed in a casing. The impeller usually consists of two discs with a number of shaped blades between them; it is manufactured as a single casting and different materials are used for different applications. For fresh water bronze is a satisfactory material for resistance to corrosion, abrasion and cavitation damage, combined with ease of casting, good machining properties and moderate cost. One of the discs is fixed to the shaft of the pump and the other has a central hole in it, making an annular space around the shaft—the 'eye'. When the impeller rotates, water is drawn in through the eye, passes through the impeller and is thrown radially off the tips of the vanes, which adds high kinetic energy. In the diffuser chamber around the impeller, part of the energy is converted to pressure, part to forward movement of the water through the connected system and part is lost in turbulence and friction. Plate 31(b) shows a cut-away view of a double suction centrifugal pump. The efficient conversion to useful pressure rise and forward movement of the water is done by careful design of the impeller and diffuser chamber.

With good design the maximum efficiency of the centrifugal pump can exceed 80%, including the energy lost in bearing friction as well as the hydraulic losses within the pump. Efficiency depends on several factors including the size of the pump; larger sizes are generally more efficient than small. The efficiency of a centrifugal pump must vary at different flow rates. Maximum efficiencies of 80–90% are possible with special designs and large machines but these may be obtained, amongst other things, by fine clearances between the moving and static parts of the pump so that efficiency may fall sharply with wear. In practice, all centrifugal pump efficiencies are likely to fall with time; a figure of 1% per annum is often used as a basis for planning but the actual rate depends on hours run and operating conditions, solids in suspension and cavitation. The operating efficiency is a useful indicator of need for refurbishment or replacement, since the efficiency inversely affects the running cost.

17.3 TYPES OF CENTRIFUGAL PUMP

All centrifugal pumps work on the principle set out above but their construction varies greatly according to the duty required. For example, multistage pumps (Plate 31(d)), used to generate high heads, consist of several impellers and diffuser chambers arranged in series, the impellers being fixed to one shaft. The water from one diffuser chamber is led to the impeller of the next stage so that the pressure developed increases stage by stage. For general waterworks duties the maximum pressure normally developed by one impeller may be between 80 and 100 m. Higher head (H) can be produced by higher speeds of rotation and larger impellers (H is proportional to D^2 for impellers of different diameter (D) in the same body), although this increases the cost and is limited by available suction conditions.

If it is known that increased head will be required in the future a multi-stage pump with a dummy first stage can be specified. This is simply a diffuser chamber without an impeller, allowing the later addition of an impeller to develop more pressure. The efficiency of a pump is not much altered by the 'dummy stage' but making provision for it can prolong its useful life. The driving motor must of course have enough power to drive the pump when the impeller is added, or a new motor will be needed.

A 'split casing' (Plate 32(a)) is the preferred centrifugal pump arrangement in which the upper half of the casing, which is easy to remove, gives access to the impellers and diffuser chambers for inspection and any needed maintenance without disconnecting the pipework or the driver.

Axial thrust arises in a centrifugal pump due to pressure imbalance on the two plates. Several ways can be used to balance this thrust, which would otherwise quickly cause wear on the pump and shorten its life. Small pumps can absorb the end thrust by the use of thrust bearings. For larger pumps a double-entry, back-to-back impeller design may be used, with the water entering the impeller from both sides; the end thrusts are then effectively balanced. A multistage pump does not normally have double entry for each impeller, although special designs have been successful for very high duties in which there is more than one double-entry impeller on the same shaft, with sometimes a balanced arrangement of single-entry impellers as well. Another common device for overcoming end thrust on smaller multistage centrifugal pumps is to incorporate a balancing disc on the shaft, high pressure water being led to one side of it so that most of the end thrust is taken by the disc.

Vertical spindle pumps (Fig. 17.1 and Plate 32(b)) are often used for pumping water from a well and for intakes. The driving motor is at the surface, mounted above flooding level, but the pump is immersed in the water. With enclosed shafting, the spindle rotates within a protective tube or sleeve, perhaps 75 to 125 mm diameter, and held centrally in the riser pipe by 'spider' bearings. The pumped water is delivered to the surface through the annular space between the sleeving and the riser. A typical arrangement would be a 250 mm diameter riser pipe in 3 m lengths bolted together with flanged joints, the sleeve tube being perhaps 100 mm diameter, with bearings for the spindle at every joint in the riser pipe. These bearings are nearly always water lubricated (oil or grease lubrication risks contamination of the pumped flow), the water being taken from the delivery pipe via filters and fed through the sleeving to the bearings. The alternative arrangement of open 'line shafting' exposes the shaft and its bearings to the pumped fluid and is not suitable for water containing solids. The whole weight of spindle and pump impellers and the hydraulic thrust generated, is taken by a Michell thrust bearing at the top of the shafting, just below the coupling to the motor. Pumps of this type are very reliable, being robust and suitable for continuous heavy duty. However, they are expensive and take time and skill to dismantle or erect when repairs are necessary. Their capital cost may be double that for a horizontal spindle pump and they are now much less common with the increased use of cheaper submersible pumps.

Submersible Pumps

Submersible pumps should strictly be termed 'submersible-motor' pumps or 'submersible pump-sets'. The motor design (Plate 32(c)) is the main difference from more conventional designs. The pump, driven by a submersible motor, is very similar to a pump driven by a vertical spindle 'dry' motor, although some differences are given below. Submersible pumps gained in popularity because they usually result in a cheaper installation than one using dry motors. The disadvantages of having a submerged motor (out of the sight and hearing of any attendant and less reliable than a dry motor when the submersible machine was first introduced) have been largely overcome by improvements in the motor design, particularly in the insulation and in the instrumentation used for monitoring pump performance. Properly chosen submersible pumps have proved reliable in service over many years; submersible designs are now available from specialist manufacturers for a very wide range of duties.

FIGURE 17.1

Vertical spindle (lineshaft) pump and motor for borehole.

Many submersible pumps in water supply are installed in drilled boreholes. The high cost of drilling is affected by the borehole diameter; therefore the diameter of the submersible pump is of great importance. Designers must produce pumps and motors of small diameter. Mixed flow pumps produce more flow at a given casing diameter than radial flow pumps and are suitable for borehole pumps. However, they produce less head and more pump stages are needed. This results in pumps longer and narrower than more conventional designs. For the same reasons, submersible motors are longer than equivalent dry motors. They are nearly always 2-pole designs (Section 17.19) to develop more power from a given size motor and to run at the highest available speed to maximize pump output, hence reducing overall cost. Naturally the mechanical design of the pump, especially its bearings, must be appropriate for the chosen speed. The disadvantages of a higher speed are increased wear, particularly if the pumped water contains abrasive solids, and reduced suction capability, so that deeper submergence may be required. As usual, each installation needs careful consideration before the type and speed of the pumps is finally decided.

In a typical borehole installation, the pump is directly coupled to the submersible motor, which is underneath, and power is supplied to the motor through waterproof cables clipped to the outside of the riser pipe. The water inlet is between pump and motor with the outlet from the final pump stage leaving axially. The motor is normally a water-cooled fixed-speed caged induction motor, specially designed for underwater running. However, where varying output is needed, a variable frequency power supply may be used (Section 17.22), although at significant extra cost. If there is any risk that inflow to the borehole could be predominantly from a higher level than the pump inlet, or if the pump is installed in a large body of water so that the pumped flow does not pass over the motor, a motor shroud should be used to ensure a cooling flow passes over the motor.

Submersible pumps are relatively quick and easy to install. The rising main is free of the spindle and sleeving needed with the vertical spindle pump; a large thrust bearing to support the heavy rotating parts is not required. Submersible pumps need not be installed truly vertically, which may be a big advantage in very deep wells. They are even sometimes used horizontally as booster pumps in distribution mains (Plate 32(d)). Submersible-pump reliability in non-corrosive waters has been proven over the years; even in corrosive waters they can be withdrawn for attention or replacement more easily than pumps of the vertical spindle design. Some modern borehole installations are now designed without any surface housings, although provision still needs to be made for access for a mobile crane or sheerlegs for withdrawal of the pumpset and its riser pipe. This simplification can make substantial cost savings.

Submersible pumpsets may be less efficient than the vertical spindle design, partly because of the special design of the motor but also because of the higher number of stages needed to achieve a given duty. This can be important if the pumping duty is wrongly estimated, because of the pronounced peak in the efficiency curve with the multistage unit. However, submersibles gain by avoiding the transmission shaft losses of the vertical spindle design.

17.4 CHARACTERISTICS OF CENTRIFUGAL PUMPS

Typical characteristic curves for a typical true radial flow centrifugal pump are shown in Figure 17.2. The head/flow curve is relatively flat up to the design duty point and the power at zero flow is only about 40% of that required at the design duty. This allows the centrifugal pump to be started against a closed valve, the power required to start it being much less than the power required at the duty

(a) Vertically mounted close coupled centrifugal pump—20 500 m³/h @ 21 m lift. (Sulzer Pumps (UK) Ltd.)

(b) Cut-away diagram of single stage double suction split case centrifugal pump. (Courtesy of Clyde Pumps Ltd. incorporating Weir Pumps, Glasgow)

(c) Horizontally mounted double suction centrifugal pumps—1650 m³/h @ 53 m lift. (Sulzer Pumps (UK) Ltd.)

(d) Cut-away diagram of horizontally mounted five stage centrifugal pump. (Courtesy of Clyde Pumps incorporating Weir Pumps, Glasgow)

PLATE 31

(a) Vertically mounted split casing centrifugal pumps—900 m³/h @ 119 m lift. (Sulzer Pumps (UK) Ltd.)

(b) Two stage line shaft wet well mixed flow pumps—up to 40 000 m³/h @ up to 100 m lift. (Courtesy of Clyde Pumps Ltd. incorporating Weir Pumps, Glasgow)

(c) Cut-away diagram of mixed flow submersible motor intake pump. (Bedford Pumps Ltd., UK)

(d) Submersible motor in-line booster pump — 1500 m³/h @ 10 m lift. (Bedford Pumps Ltd., UK)

PLATE 32

Note: The peak in the flow characteristic that could cause instability.

FIGURE 17.2

Characteristic curves for a radial flow centrifugal pump at constant speed.

point. Centrifugal pumps are commonly started against a closed delivery valve and shut down with the valve closed first. The pump is unaffected provided the valve is opened (or the pump stopped) before the pump becomes overheated. The reduced power required for starting reduces the starting current when an electric motor is used to drive the pump; the pressure rise in the delivery system on starting can be managed by controlling the rate of opening of the valve.

The maximum head generated by the pump, for a given speed, is usually not greatly in excess of the design duty head. However, in the example shown, the head-output curve is unsatisfactory since for heads higher than the duty head there are two possible outputs. The pump is therefore unstable which could cause trouble if operated in parallel with another pump. The head-output curve should preferably fall continuously from the 'shut-valve' head with increasing output. If the curve is steep, which is often specified, the pump output will not vary much if the head alters somewhat during operation. However, the maximum head at zero flow (the 'shut valve' head) should not be excessive since this governs the design head of the connected system. The efficiency curve should be reasonably flat about the design duty point so that there is no great reduction in efficiency if the actual head is slightly different from the expected duty, which is often difficult to estimate exactly at the design stage and may alter in operation.

Affinity Laws

When the rotational speed (N) of a centrifugal pump is changed, there may be little change in efficiency, depending on the amount of the change, but the output (Q), head developed (H) and power required (P) are altered according to the following relationships, known as the 'Affinity Laws':

$$Q_1/Q_2 = N_1/N_2$$
$$H_1/H_2 = (N_1)^2/(N_2)^2$$
$$P_1/P_2 = (N_1)^3/(N_2)^3$$

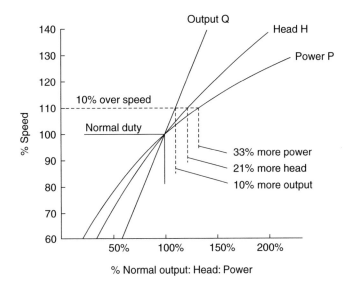

FIGURE 17.3

Effect of changing speed of a centrifugal pump.

The theoretical effect of changing the speed of the pump is shown in Figure 17.3.

Gas (air or water vapour), even 1%, in a water pump reduces its efficiency; 5% can cause choking (Rayner, 1995). Dissolved air will come out of solution if the pressure is lowered below that at which the water is saturated with air. This is normally at some pressure below atmospheric depending on previous exposure to air (Section 13.10). Therefore, there should normally be no loss of efficiency due to air coming out of solution if the pump inlet pressure is above atmospheric. However, obtaining such conditions is not always feasible. Whether air is present or not water vaporizes at a pressure which depends on temperature. This pressure is lower than that at which air may come out of solution so that, as pressure drops in the pump, air bubbles form first and then the water starts to vaporize. On subsequent rise in pressure water vapour pockets collapse instantaneously causing noise and erosion if adjacent to a solid surface. However, air does not go back into solution as quickly and bubbles remain, albeit compressed, as pressure rises. Remaining bubbles form a cushion which can reduce cavitation damage, depending on flow patterns in the pump, but air does cause a hissing noise. A hard crackling noise heard from a pump, almost as if there is gravel inside it, is an indication of cavitation. If such a sound does not disappear shortly after starting, the cause should be investigated and eradicated before damage results.

NPSH

Each pump has a minimum 'net positive suction head requirement', sometimes abbreviated to NPSHR, which varies with flow. This is the head that causes water to flow into the eye of the impeller and is the minimum suction pressure required to prevent cavitation. Its value varies with the speed and capacity of the pump and will normally be given by the manufacturer, based on the results of tests. By convention NPSHR is the NPSH at which the pump head reduction (due to cavitation) is 3%. NPSHR curves rise as pump flow increases with lower head. These are the critical conditions for pump cavitation since the NPSH available (NPSHA) is reduced due to increased suction losses at higher flows. NPSHA is given by:

$$NPSHA = Z + \frac{(P_a + P_{vp})}{\gamma} - h_f - V_s^2/2g$$

where Z is the difference between the pump impeller eye level and the suction water level, P_a is the absolute atmospheric pressure, P_{vp} is the absolute vapour pressure of the liquid at the pumping temperature, γ is the specific weight of liquid at pumping temperature, h_f is the head lost in the suction pipework and $V_s^2/2g$ is the suction velocity head.

Although water supply pumps normally pump cold water at modest altitudes, terms are included in the above expression to take account of both atmospheric pressure and water temperature. The value of NPSHA must always be greater than NPSHR and a margin of perhaps 1 m is usually specified to cover any tendency for cavitation at slightly higher NPSH values, although some sources recommend a margin of 1.5 m (Sanks, 1998). This allows for any minor differences between calculated and actual figures, as well as changes with time.

Specific Speed

To classify geometrically similar pumps, the numerical quantity 'specific speed' has been adopted. Specific speed is the speed required for delivery of unit flow against unit head; it varies in accordance with the system of units used.

$$N_s = \frac{NQ^{1/2}}{H^{3/4}}$$

where: N_s is specific speed; N is pump impeller speed in rpm; Q is output at maximum efficiency (m³/s); H is delivery head at maximum efficiency (m). Specific speeds (in metric units) fall approximately into the following categories:

Type of pump	Specific speed.
Radial flow	10–90.
Mixed flow	40–160.
Axial flow	150–420.

17.5 AXIAL FLOW AND MIXED FLOW PUMPS

Axial flow pumps are of the propeller type, in which the rotation of the impeller forces the water forward axially, and do not strictly qualify as centrifugal pumps. Mixed flow pumps act partly by centrifugal action and partly by propeller action, the blades of the impeller being given some degree of 'twist'. However, in practical terms there are no precise dividing lines between radial flow (centrifugal), mixed flow and axial flow pumps. In general, axial and mixed flow pumps are primarily suited for pumping large quantities of water against low heads, while centrifugal pumps are best for pumping moderate outputs against high heads. Axial flow pumps have poor suction capability

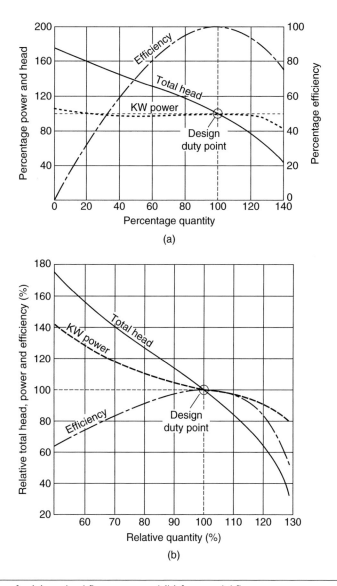

FIGURE 17.4

Characteristic curves for (a) a mixed flow pump and (b) for an axial flow pump.

and must be submerged for starting. They are most often used for land drainage or irrigation or for transferring large quantities of water from a river to some nearby ground-level storage. Mixed flow pumps are shown in Plates 32(b) and (c).

Characteristic curves for typical mixed flow and axial flow pumps are given in Figure 17.4. The starting power required by the mixed flow pump shown is about the same as the duty power, but for the axial flow pump the starting power is substantially greater than the duty power. Axial flow pumps are therefore not started against a closed valve, which would overload a motor correctly sized for the expected duty. They are either started against an open valve to minimize the starting power and current required, or installed in systems specially designed to ensure no delivery valve is needed, for example with a siphonic delivery.

17.6 RECIPROCATING PUMPS

Most pumps used for water supply nowadays use a rotating impeller, but the reciprocating pump still has its uses. The ram pump is the most common form of reciprocating pump; it consists of a piston reciprocating within a cylinder provided with water inlet and outlet valves. Water is drawn in by one stroke of the pump and forced out by the next. With all ram pumps the output fluctuates cyclically, but if three or more cylinders are used, a reasonably constant flow is maintained. The type of pump with three cylinders is termed a 'triple throw ram pump', the pistons being connected to the operating crank 120° apart.

Ram pumps were common for high-head duties in water supply before the development and widespread introduction of centrifugal pumps. Their slow speed suited the reciprocating steam engines which usually drove them. Driven nowadays by an electric motor, the triple-throw ram pump may sometimes be used where exceptionally high heads must be produced, since no special design is necessary other than making the working parts strong enough to sustain the pressure. Efficiencies of more than 90% can be obtained, falling off with piston and cylinder wear. However, since very high head centrifugal pump designs are now available, the use of new ram pumps for high-head duties in water supply has almost completely ended.

Small ram pumps, usually driven by variable-speed electric motors are, however, still in widespread use for the injection of the chemical solutions needed for dosing in water treatment. Their property of delivering a constant volume of liquid per stroke makes them well suited to the metering duty required. With appropriate design of the pump and its driving system, both the pump speed and its stroke are varied to maintain any desired rate of chemical injection per unit volume of water.

The bucket pump is another form of the ram pump, arranged to operate vertically for drawing water from a well or borehole. Instead of a piston a 'bucket' is given a vertical reciprocating motion in the rising main. The bucket incorporates a valve, which opens as the bucket descends below water level. When the bucket rises, the valve closes and water is lifted. An alternative arrangement makes the descent of the bucket force water up the rising main through a series of valves. This type of pump is known as a 'lift and force' pump. The village pump is a single-throw bucket pump, operated manually; it illustrates one of the chief advantages of the bucket pump apart from its low cost: it is always ready for use and able to function even if erratically operated. Modern versions of these machines are still being installed for village water supplies in parts of the world where an electro-submersible borehole installation would be too costly. Small single-throw bucket pumps are also still used, for example to provide water supplies for remote farmland, sometimes with wind turbines driving them, again illustrating the reliability of these pumps under intermittent operating conditions.

The hydrostat or *hydraulic ram* pump is another example of reciprocating pump still in occasional use. A large volume of water flowing in one pipe is used to drive a ram which is connected to a smaller ram pump, which pumps part of the water to a higher elevation through a branch pipe. The flow which continues down the main pipeline suffers a loss of head. The proportion of flow pumped depends on the relative pressures. Hydraulic rams are not convenient in public water supply systems because any increase of flow on either the high or low pressure side quickly invalidates the set-up. However, they do find use in special situations, such as when a small volume of water is to be lifted in a remote location without an accessible power supply.

17.7 CHOICE OF PUMPS FOR WATER SUPPLY

The horizontal centrifugal pump is suitable for nearly all waterworks duties, except handling very large volumes of water against low heads and pumping from wells and boreholes. The main advantages of the horizontal pump are that it is relatively low in first cost and it can be arranged to provide easy access for maintenance. A great variety of designs are available to meet required pumping conditions but the most common arrangement is to use the horizontal, split-casing double-suction design (Plate 31(b)), which has been developed over many years of water supply duties. For a single unit the output can range from about 50 Ml/d at 60 m head to 10 Ml/d at 200 m head. The most common waterworks duty is from 10 to 25 Ml/d per unit at 30 to 120 m head, and in this range the horizontal pump is cheapest.

The pump should have stable characteristics and should be 'non-overloading', i.e. the power absorbed should not increase much if the delivery head drops. This is not always possible to arrange and protection against overload must be provided for the motor. Low head could result if the delivery main were to burst near to the pumping station or when pumping into an empty rising main. The efficiency curve should also indicate no severe fall in efficiency for moderate variations of flow and head about the duty point. When specifying a pump, the manufacturer must be told of the complete range of duties the pump is intended to meet, including whether series or parallel operation is required (Section 17.8). To specify the duty point only, i.e. the theoretical normal conditions under which the pump will operate, could lead to large efficiency loss if the actual running conditions vary from the theoretically calculated duty.

For wells and boreholes, the choice lies between vertical spindle pumps (Fig. 17.1 and Plate 32(b)) and submersible motor pumps (Plate 32(c)). The vertical spindle pump may be regarded as a 'heavy duty' pump and can be driven by a synchronous electric motor (Section 17.18) which enables the power factor to be brought to unity. The consequent cost of power is reduced, although the motor is more expensive initially and costs more to maintain. Variation in speed is sometimes necessary when pumping from a well or borehole because the output from the well may have to be kept in step with the demand, irrespective of seasonal fluctuation in water levels in the well. A thorough appraisal of all the possible operating conditions must be made before choosing the right pump and drive for the duty.

The vertical spindle motor driving a mixed flow pump (Fig. 17.1) is suitable for pumping large quantities of water against low heads; the pump is immersed in the water and the motor sited above the highest flooding level. An alternative arrangement for pumping water from a tank or main is to site the cheaper horizontal centrifugal pump in a dry well, with its centreline below the bottom water level in the tank so that the pump is always primed. Centrifugal pumps should preferably not be sited so that they are subject to any suction lift, because their efficiency drops and cavitation may result (Section 17.4). However, if essential and depending on the design and NPSHR curve, up to 4 m lift can be managed, with 5.5 m the maximum possible.

For high lift pumping the most common choice is the fixed-speed, horizontal multistage centrifugal pump (Plate 31(d)). Submersible pumps are also used for this duty, either for pumping direct from a well to a high level tank or for inserting as boosters in a pipeline in a pit below ground level, thus avoiding construction of a building to house the normal arrangement of horizontal pump and motor.

Standby Pumping Plant

In water supply pumping stations, a single pump is seldom relied on for the full output; adequate standby is essential to ensure continuity of supply. If the full duty can be handled by a single pump then a duplicate of equal capacity should be installed; this is a common arrangement when the unit sizes are small. However, if the pumps are each sized for 50% of total required output, the installation of three similar pumps ensures 100% output on breakdown of any one machine, 50% in the much more unlikely event of two pumps failing at the same time. This reduces the cost of the standby plant. When pumps of different sizes are needed, for example when pumping into a system with limited storage or when large fluctuations in demand must be satisfied and where variable speed drives are not appropriate it is normal to provide at least one pump of each size as standby.

17.8 BOOSTING

Booster pumping augments the pressure or quantity of water delivered through an existing system. The use and control of boosting is discussed in Section 13.2. The three most important boosting arrangements are:

- addition of a fixed extra flow to an existing supply;
- addition of a fixed extra pressure to an existing supply;
- maintenance of a given pressure, irrespective of the flow.

Addition of Fixed Extra Flow or Pressure

To increase the flowrate, two similar pumps can be connected in parallel; to increase the pressure two similar pumps can be connected in series. For lesser increases the pump added in parallel or series is of a smaller rating. In both cases the characteristics of the system into which the pumps are to deliver must be considered. Figure 17.5 shows the characteristic flow-pressure curve for a pump A.

FIGURE 17.5

Output of two pumps connected in series or parallel.

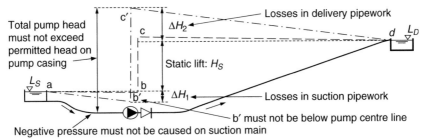

FIGURE 17.6

Effect of boosting in a rising main to a reservoir.

Combined characteristic curve (A + A) is for series working of two such pumps and combined characteristic curve (A‖A) is for parallel operation. The head-flow relationship for the system into which the pumps are delivering is shown as system curve S. The points of intersection of the curves (A + A) and (A‖A) with the system curve S indicate the resulting output of the two pumps A operating in series and parallel.

A hydraulic gradient must be drawn for the system in which the pumps are to work to confirm that the siting of the added pump is correct. Referring to Figure 17.6, if pump A initially draws water from reservoir R_1 and pumps it to reservoir R_2 the hydraulic gradient will be the line *abcd*. If the pumped flow is increased the hydraulic gradient must change to some line *ab'c'd* where *b'* is lower in elevation than *b*, and *c'* is higher in elevation than *c*. The difference *cb* is the pump lift from pump A alone and the difference *c'b'* is the new lift developed by pumps (A + A). The following conditions must be satisfied:

1. The level of point *b'* must satisfy NPSH requirements (Section 17.4) and any requirements for self priming. When pumping treated water there should be no negative pressure in the suction pipeline to avoid risk of infiltration of groundwater. If necessary the suction pipeline could be duplicated to reduce losses and raise point *b'*.
2. Point *c'* must not imply a pressure beyond the safe rated working pressure of the pump bodies, valves, fittings and pipeline to reservoir R_2. If necessary the second pump could be inserted sufficiently far along the delivery pipeline so that no components are overpressurized. This would avoid the need to replace the delivery pipeline.
3. If the pumps are to be operated in series the joint duty must not cause pump A to be overloaded (if this is so then the difficulty might be overcome by changing its impeller).
4. The efficiencies of the pumps at the new duty point must be checked. Any significant reduction in efficiency might mean it would be better to replace pump A with a single new pump, capable of managing the whole of the enhanced duty.
5. If the pumps are to be run in parallel they must have stable running characteristics.

Maintenance of a Given Pressure

The operation of this kind of booster is discussed in Section 13.2. The booster pump is usually arranged to start automatically when the pressure downstream of the pump reaches a certain low value. Safeguards should be included in the pump controls to prevent hunting and ensure correct interpretation of events such as a burst main near the pumping station. The duties required from a booster pump of this kind are usually too wide for a fixed speed machine and variable speed drives may be needed. Alternatively a range of pumps could be provided so that these can be brought into use one by one. In both cases, pressure surges must be controlled (Section 13.11).

17.9 INCREASING PUMPING STATION OUTPUT

If an increase in pumping station output is needed there are several ways of doing this without the substantial cost of complete replacement of the pumps and making large alterations to the building:

Running some of the standby plant. This method is simplest, since no modifications are needed. However, it may not provide all the required extra flow and the loss of standby may be unacceptable except for very short periods.

Replacing the pumpsets with units of larger capacity. Usually building design allows installing somewhat larger machines with relatively minor modifications, but the extent of equipment replacement needs careful consideration. The pump inlet and outlet pipework and valves may also need replacement and suction conditions may need to be enhanced to ensure adequate NPSH is provided for the increased flow. The new pumps will use more power and may need new motors and starters and the complete electrical system will need careful review. New higher power pumps and motors may weigh more; therefore, the capacity of any station crane and its supports will need checking. Similarly, structural support for the new machines when installed will need to be checked.

Replacing only the pump impellers. Replacing only the pump impellers to increase the pump output should be possible if the maximum size impeller for the casing has not already been installed. Impeller replacement is cheaper than pump replacement but power and NPSH requirements still need consideration. The manufacturer must be consulted to ensure that larger impellers can be fitted and that other pump components are capable of the new duty.

Replacing the motors to drive the existing pumps at a higher speed. Very occasionally this may be possible but careful checks in collaboration with the manufacturer are essential. As well as considerations of power, weight and NPSH, other components of the system should be checked, including the driving arrangements, which may need replacement for higher speed operation, particularly if long vertical shafts are involved.

17.10 STATION ARRANGEMENT AND PLANT LAYOUT

Pumping stations may be arranged in many different ways, depending on such things as land availability, station duty, required standards and the preferences of the engineers responsible. Cost is always a consideration but compromising by saving initial cost at the expense of future benefits from reduced running costs or inaccessibility for maintenance should be avoided. The station layout naturally depends on the type of pumps to be used. Wet well pumps require a sump providing appropriate hydraulic conditions (Section 17.11). Some other principles to be considered at the design stage are as follows.

1. Accommodating the plant needed to meet the required duty comes first; this takes account of the site, the type and number of pumpsets, their drives, valves and pipework, power supplies, methods of starting and control, access for maintenance and any needed surge protection. Once the plant detailed requirements are decided, but never before, design of the pumping station buildings can proceed.

2. Facilities for lifting and loading must be provided. A travelling crane is essential for all but the smallest pumping units; power operation can usually be justified. The crane should be arranged to serve all the heavy plant and to cover the loading area, which should be big enough for access by a vehicle suitable for carrying the largest plant. For borehole installations a mobile crane may sometimes be judged adequate. The main pump hall must leave adequate working space around the machinery and there should be a clear space big enough for complete stripping down of one of the units, which can be the same space as a loading area.

3. High voltage switchgear and transformers are usually needed for all but the smallest pumping stations. HV switchgear must be sited in a locked room separate from the main rooms, with access restricted to authorized personnel. Transformers are usually mounted outside the building although for resin insulated units in cold climates, use can be made of the heat they generate to provide some background heating to the station if they are inside. Main switchgear, both high and low voltage, and control panels, must have adequate space both in front and behind them. About 1 m should be regarded as the minimum at the back, and 2 m in the front. All switchgear should as far as possible be in line so that the main bus bars are kept short; cable routes from switchgear to motors should also be kept as short as practicable. Instrumentation and control panelling is nowadays often combined with the low-voltage switchgear panels to achieve a neat installation; this needs similar accessibility.

4. All cabling and pipework should be in conduits of ample size and runs must be carefully laid out allowing for minimum bending radii for cables. Pipework T-junctions or $90°^-$ bends in the line of flow should be avoided. Radiused junctions and large radius bends should be used to reduce station hydraulic losses. All moving machinery must be securely bolted to adequate foundation blocks. If pumps are sited over a well or tank they should preferably be fixed on properly designed steel joists, rather than bedded on concrete slabs.

5. If air vessels for surge protection are needed, these can be placed outside the pumping station if the climate permits, or inside in an unheated part of the station. Outdoor vessels may affect the appearance of the station and in severe climates need frost protection. This can be done with insulation (which may only delay the onset of freezing) or by arranging a trickle flow of warmer water from the supply main to keep the temperature above freezing. In extreme cases electric heaters may be needed. If the vessels are inside, condensation may occur on their surfaces and cause staining of floors and rusting of the vessels.

6. Facilities for chlorination may be needed, but dosing points and chlorination equipment must be independent of the main parts of the building, preferably in a separate building with restricted access (Section 11.12).

7. Some facilities are usually provided for the station attendants even if the site is not permanently manned; a small workshop may be justifiable and a messroom containing a sink and means for cooking light meals. Sanitation of the highest standard is needed, together with hot water and washing facilities.

17.11 PUMP SUCTION DESIGN

Pumps should be sited so that below atmospheric (negative) pressure does not develop on the suction side. Negative pressure can cause a reduction in the performance of the pump and may prevent the pump from being automatically primed. If negative pressure is really unavoidable a 'self-priming' pump must be specified but this should be avoidable in most cases by suitable design. Figure 17.7

shows a bad design and a good design for a wet well. The aim is to minimize pressure losses in the suction system, prevent air being released and avoid vortexing, which can be reduced by fitting flow-straightening vanes at the inlet. *The Hydraulic Design of Pump Sumps and Intakes* (Prossor, 1977) provides guidance on wet well pump arrangements but it may be necessary to check flow patterns by modelling (Chapter 12, Appendix). The reader may also like to refer to *Pump Intake Design* (HI 9.8, 1998).

The diameter of pump suction pipes is usually larger than the delivery pipe diameter to reduce head losses. Suction pipes for raw water may need to be equipped with a strainer which should have an effective area of opening at least double that of the suction pipe. Strainers should be kept in good condition. Foot valves, i.e. non-return valves, fitted at the inlet to vertical suction pipes, are a potential source of trouble and should be avoided where possible. If installed they must be of high quality and be well maintained or they may tend to stick open. They are sometimes installed for keeping a pump primed when it is idle but often they are not effective if the pump is stopped over a long period; other priming methods are better. Foot valves are also sometimes used to prevent reverse rotation on stoppage of a vertical spindle pump, which has a long rising

FIGURE 17.7

Pump suction arrangements.

main; a high speed of reverse-rotation can be caused by the falling column of water. Restarting a pump when it is rotating in reverse may cause shaft breakage, which must be prevented. Instead of a foot valve, a pump may be specified as suitable for being 'turbined' under reverse flow; a time delay switch is then incorporated in the switchgear to prevent restarting before the reverse rotation has stopped.

17.12 THERMODYNAMIC PUMP PERFORMANCE MONITORING SYSTEM

As a pump gets older, its performance inevitably declines, until its upgrading by servicing or replacement is needed to maintain high efficiency and minimize operating costs. The regular monitoring of individual pump performance is needed to do this and a record must be kept of pump output compared with energy consumed. The record needs to be regularly updated, perhaps annually, but most water supply pumping stations are not equipped to monitor individual pumpsets although the total station power consumption and flow delivered are usually logged. The difficulty lies in measuring the individual pump flow.

Thermodynamic pump performance monitoring provides one possible solution, which has been developed and improved so that good accuracy of measurement is now possible. Most of the pump energy losses, which lead to reduced efficiency, result in heating the pumped water by a small amount. Thermodynamic monitoring relies on accurate sensors to detect the resulting small temperature rise across a pump, which is proportional to the wasted energy. The great advantage of this method is that there is no need for an individual flowmeter to measure the pumped flow directly to determine the pump efficiency.

These devices are not yet suitable for every installation. They work best with higher head machines because the resulting temperature rise is then higher for the same pump efficiency and power. They are not always very effective when used with borehole pumps because of difficulty in measuring the inlet water temperature. Development work is continuing however, and the performance of these devices continues to improve.

In a typical installation, temperature probes and pressure transducers are mounted on the pump pipework (Yates, 1989) to detect the inlet and outlet conditions (Fig. 17.8). A micro-processor,

FIGURE 17.8

Pump performance monitoring.

usually mounted in portable equipment, is used to analyse the data and to display efficiency, flow, head and power consumption. Thermodynamic pump testing is accepted in *Pump Test Standard* IS0 5198 (1987) as a precision-class test.

17.13 CAVITATION DAMAGE

Cavitation occurs when the absolute pressure in the pumped water falls below its vapour pressure (Section 13.12). Pockets of water vapour are then released in the impeller. On subsequently entering a higher pressure zone, against the blade of the impeller or elsewhere, they collapse. A stream of vapour pockets continuously collapsing on the same area quickly erodes the blade material. Cavitation damage can be avoided by good design of the pump and suction pipework and by maintaining a positive pressure on the suction side, wherever possible, by siting the pump below or as near as possible to the suction-side water level.

17.14 CORROSION PROTECTION

Corrosion protection can be provided for most pumping plant by the correct choice of pump materials to suit the water. Care should be taken to avoid use of metals with markedly different electropotentials. A more recent development has been the application of specially developed coatings to pump internal parts. One commonly applied material is the glass-filled resin-based coating. Such coatings have been successfully used for application to the internal surfaces of pipework, valves, pump bowls, and pump impellers to restore performance following corrosion or erosion attacks. They are also sometimes applied to new pumps to prevent damage and improve efficiency by providing smoother surfaces. Metal surfaces require careful surface preparation to obtain the full benefits of the coating material. Since the coatings have significant thickness, care is needed to ensure that pump passages, particularly at the pump inlet, are not reduced in area so much that performance is impaired. Clearly this is more significant with smaller pumps and, for this reason, combined with the cost, coatings are not justified on pump sizes of less than about 300 mm branch diameter.

17.15 TRANSIENT PRESSURES: WATER HAMMER AND SURGE

The key causes, features and consequences of surge are discussed in Section 13.11. The most critical case is usually a power failure causing simultaneous stopping of the pumps. Most modern low-inertia pumps stop producing forward flow of water in a few seconds when the power fails. Measures to control the magnitude of pressure transients are set out in Section 13.11 and are described in Thorley (2004). Those that would need to be accommodated at the pumping station are listed below:

1. Reduce delivery valve closure speed.
2. Increase pump inertia by fitting fly-wheels to the pumps.
3. Have surge vessels, accumulators or surge shafts connected to the delivery pipeline just downstream of the pumping station.
4. Incorporate by-pass pipework around the pumps.

With the emphasis of pump manufacturers on producing lighter pumps in order to reduce motor starting torques and currents, fly-wheels and added inertia are out of fashion; but in the right circumstances they are the most reliable and effective form of protection. More commonly on pumping systems a surge vessel or accumulator is used.

Pump delivery non-return valves need particular attention to ensure they are suitable for the system and its transient response, especially if a surge vessel is also provided because the flow in the connecting pipe to the air vessel may reverse very quickly. Ideally, the non-return valve should shut at the moment of flow reversal but if it reacts more slowly the reversed flow may slam the valve shut with the generation of a high shock pressure. The dynamic response of the non-return valves should thus be matched to the transient characteristics of the pipeline system.

17.16 EFFICIENCIES AND FUEL CONSUMPTIONS

Quoting efficiencies and fuel consumptions is difficult because they vary so widely and every individual case has something special about it. The figures in Tables 17.1 and 17.2 are intended to give a guide to the efficiencies and fuel consumptions normally to be expected. There are wide variations according to the power rating and type of pump and motor.

Table 17.1 Efficiencies

Pumps:	
Horizontal centrifugal	Medium size 80% to 82%, perhaps 85% large size. Even higher with special construction but at higher price.
Vertical spindle shaft driven	Tending towards about 3% less than the horizontal centrifugal.
Submersible	75% to 81% and can be lower, to about 70% for small sizes. Generally about 3% less again than the vertical spindle pump; the reason being that the pump is restricted in diameter.
Electric motors:	
For horizontal pumps	93% to 95%, for fixed speed a.c. induction
For vertical pumps	90% to 94%, for fixed speed a.c. induction
For submersibles	85% to 89%. Less than the above because of the restrictions imposed on the design
Variable speed	About 3% to 5% less than with a caged a.c. motor.

Table 17.2 Overall Fuel Consumptions

Electrically driven pumps	About 1.0 kW for every 0.75 kW of water power output, this implies an overall efficiency of about 75% which would be usual. Up to 1.3 kW per 0.75 kW water power output or higher for small pumps or variable speed pumps.
Diesel engines	0.21 kg of diesel fuel oil consumed per kWh of engine power exerted would be considered good, 0.28 kg per kWh being not unusually high. For lubricating oil add 5% to fuel oil cost.

17.17 PUMP DRIVES

Pumps may be driven by almost any prime mover of suitable power and speed. Most water supply pumps are nowadays driven by electric motors but there are still some driven by other means, usually a diesel engine. Where large powers are needed, there are some water supply pumps powered by steam turbine. However, the widespread availability of reliable and economically priced electricity supplies and the comparative cheapness, wide choice of designs and sizes, ease of operation and control of electric motors make them clear favourites in most circumstances.

Two disadvantages of electric motor drives nevertheless need mentioning. Firstly, the electricity supply must be secure to ensure the constant availability of pumping, which is nearly always important in water supply systems. Secondly, the cheaper motor designs are inflexible since they are normally able to run only at a fixed speed.

PART II ELECTRICAL PLANT

17.18 ELECTRIC MOTORS FOR PUMP DRIVES

The three phase alternating current (a.c.) motor is the most common type of electric motor used for driving pumps. Alternating current motors can be classified as follows:

- Induction motors, either (i) 'cage' or 'caged' (formerly known as 'squirrel cage' motors) or (ii) 'wound rotor' motors (sometimes called 'slip ring' motors)—see below;
- synchronous motors;
- commutator motors.

Induction motors of the caged type are the most widely used because of their simplicity, robustness, reliability and relatively low cost. They are inherently fixed-speed machines when connected to conventional fixed-frequency supplies. This is a handicap for centrifugal pumps, as the pumps themselves are capable of a wide range of duties without modification, if the speed can be varied. Since many water supply systems require pumping installations of output varying at different times, fixed-speed pumps are often a disadvantage. However, variable speed drives have disadvantages including reduced overall efficiency, limited speed range, limited power capability and high capital cost but the continuing development of variable frequency drives has overcome most of these disadvantages. Such methods of varying pump output are now much more common although cost is still an important consideration as are space and cooling requirements. Motors designed for operating at two different fixed-speeds are also available and sometimes can provide an economical method of altering pump performance, although with obvious limitations.

The *synchronous motor* costs more because of the need for more complex control equipment. A d.c. supply is also required (supplied by an 'exciter' driven by the motor itself) for the rotor. This results in the rotor turning at the same speed as the rotating magnetic field created by the stator. The synchronous motor (which has to be started up by short circuiting the rotor windings in such a way that it acts as an induction motor) is normally used for applications requiring constant speed operation under varying load conditions, or where its ability to provide power factor correction or improve the voltage regulation of the supply system is justified by the extra cost. If used, it is usually only for pumping large steady outputs.

The *commutator motor*, which provides variable speed, has largely been replaced by the electronic variable frequency drive using the cage type motor, or the slip energy-recovery type using the wound rotor induction motor.

Because by far the most commonly used motors for driving pumps are induction motors of the caged or wound rotor type; only these are described in more detail below.

17.19 **THE INDUCTION MOTOR**

The induction motor comprises a stator which incorporates a distributed winding, connected to the three phase electrical supply, and a laminated steel rotor which in its simplest form has embedded large section bars which are short-circuited at each end. The motor thus consists of two electrically separate windings which are linked by a magnetic field forming a transformer, with an air gap magnetic circuit. The three phase current in the stator winding produces a smoothly rotating magnetic field whose rotational speed N (synchronous speed in rpm) is given by the equation:

$$N = supply\ frequency\ (Hz) \times 120/number\ of\ poles$$

Thus for a supply frequency of 50 Hz a 2-pole motor has a field speed of 3000 rpm, a 4-pole motor 1500 rpm, and a 6-pole motor 1000 rpm and so on.

The rotating magnetic field cuts the rotor bars and induces a current in them. The rotor current produces a magnetic field which interacts with the stator field, producing an accelerating torque. The motor will run up to a speed at which the developed torque is equal to the torque required by the driven load, including that required to overcome friction and windage losses.

In practice, the rotor cannot reach synchronous speed with the magnetic field when loaded, because under such a condition no current would be induced in the rotor conductors and hence no magnetic flux and torque would be produced. This explains why, for example, a loaded 4-pole motor may actually run at say 1440 rpm. The difference between the actual speed of rotation and synchronous speed with the magnetic field is termed the slip. The slip increases with the load and is normally expressed as a percentage of the synchronous speed. The slip at full load typically varies from about 6% for small motors to 2% for large motors. The starting torque and speed characteristics of the cage induction motor of basic standard design are shown in Figure 17.9.

The *cage induction motor* has a rotor core which is made up of laminations and conductors of aluminium, copper or copper-alloy non-insulated bars in semi-enclosed slots, the bars being short-circuited at each end by rings. For the smaller motors, the rotor bars and end rings are often cast. The advantages are simple construction, low cost and low maintenance. The limitations of the cage induction motor of basic standard design are low breakaway torque and high starting current, the former typically ranging from 0.5 to 2.0 times and the latter 3 to 7 times rated values, depending on motor rating and number of poles. These limitations can be improved by the use of motors with multi-cage rotors. The multi-cage rotor, in its simplest form, comprises a low reactance and high resistance outer or starting cage, whose influence predominates during the starting period and results in increased torque and reduced current, and an inner cage of lower resistance which is dominant under running conditions.

The *wound rotor induction motor* design incorporates a three phase, star configuration rotor winding connected to slip rings, to which external resistance is connected for starting. This type of motor is used when high starting torque, reduced starting current and controlled acceleration

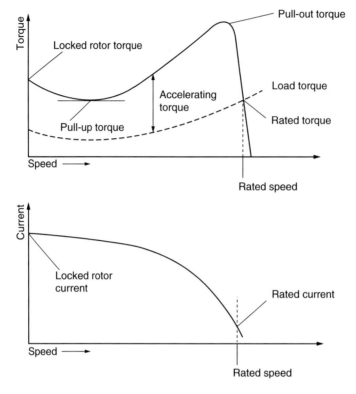

FIGURE 17.9

Torque and speed characteristics of the standard cage induction motor.

characteristics are required. The magnitude of starting torque and current, and acceleration period are determined by the value of the external resistance. Under running conditions the slip rings are shorted-out. Disadvantages of the wound rotor motor are higher cost and additional slip ring and brush gear maintenance.

Rated Output, Starting Torque and Start Frequency

The *rated output* of an induction motor is for the designated duty when the ambient temperature of the coolant air does not exceed 40 °C or, for a water cooled motor, when the temperature of the water entering the heat exchanger does not exceed 25 °C, both at a height above sea level not exceeding 1000 m. The supply voltage is allowed to deviate between 95% and 105% of the rated voltage of the motor without affecting the rated output.

When the motor is operated in conditions different from the reference values, motor rated output will be affected. This must be considered at the time of motor selection (EN 60034, BS 4999). The normal requirement is for the motor to operate continuously at voltages not differing from the rated value by more than plus and minus 5%. When the motor is required to operate under varying and cyclic load conditions, which may include periods of either no-load or standstill, reference must be made to the manufacturer giving details of the load inertia and the load/time duty sequence.

Starting torque. The standard cage induction motor is designed to produce a starting torque over the range between standstill and that at which pull-out (minimum) torque occurs, which is not less than 1.3 times a torque characteristic which varies as the product of the square of the unit speed and the rated torque. This is representative of the run-up characteristic of a typical centrifugal pump. The factor of 1.3 is chosen to allow for a voltage of 90% rated value at the motor terminals during the starting period. The load and motor torque characteristic during the starting period must be considered at the time of motor selection. This is especially important where the motor is required to start with a voltage drop greater than 10%, typically for direct-on-line starting control.

Frequency of starting. The standard cage induction motor is designed to allow two starts in succession (running to rest between starts) from cold or one start from hot after running at rated conditions. Further starting is permissible only if the motor temperature does not exceed the steady state temperature at rated load. Because the number of starts directly affects motor service life, they should be kept to a minimum. More onerous starting requirements must be considered at the time of motor selection. For the wound-rotor induction motor, the starting frequency is generally dictated by the short-time rating of the starting resistor.

17.20 INDUCTION MOTOR STARTING METHODS

The starting current drawn by an a.c. induction motor depends on its type, rating, voltage and starting method. When starting a motor the resulting current causes a voltage dip which has to be kept within defined limits, stipulated by the electricity supply company, to avoid affecting other consumers and the maloperation of connected equipment. The limits will depend on the degree of voltage variation and frequency of occurrence. A detailed electrical system analysis is required in special cases to assess whether unwanted effects are likely. However, experience shows that general guidelines can be applied based on permitted transient voltage dip at the point of common coupling (that is the busbar from which other consumers are supplied) and the frequency of starting. Typical permitted voltage dip limits are 3% for motors started infrequently, say at intervals longer than 2 hours, and 1% for motors started frequently. In special cases, for example motors which are either infrequently started or are located in remote areas, the electricity supply company may permit a transient voltage dip greater than 3%. Where large motors are to be installed it is important to establish the transient voltage dip and starting frequency criteria at the design stage to decide the method of starting. In addition to complying with voltage dip criteria at the point of common coupling, the ability of the motor to start and accelerate if the voltage drop at the motor terminals is greater than 10% and the effect of motor starting on other parts of the consumer's installation must be assessed. These aspects can be significant if the electricity supply is taken from a low fault level rural distribution system or from small capacity transformers.

The starting method depends on motor starting torque and current requirements and, in some applications, the need to control acceleration. Starting methods for the cage induction motor can be classified as:

- full voltage, or direct-on-line;
- reduced voltage (star-delta, auto-transformer and electronic soft start);
- rotor resistance starting (for wound rotor motors only).

Direct-on-line starting is the cheapest, simplest, and most reliable method, and is therefore the most widely used. However, direct-on-line starting causes a high starting current and may not be suitable if the electricity supply company requires a reduced starting current (or limited voltage dip) if shock-free controlled starting is required. With direct-on-line starting, the applied voltage, starting torque and current are 100% rated values and rapid run-up at maximum available torque is achieved. An advantage is that direct-on-line starting simplifies motor construction, as only one end of each phase winding need be brought out to terminals. For low voltage motors, the size that can be started by the direct-on-line method is often arbitrarily limited by the station designer to the smaller ratings, typically up to 15 kW. However, unless the maximum size of motor is stipulated by the electricity supply company, or is limited by supply transformer transient loading considerations, there is no restriction on the size of motor which can be started direct-on-line, provided transient voltage dip and the mechanical impact loading criteria are met. These design limitations apply equally to high voltage motors.

Star-delta starting is the most usual method used when reduced voltage starting is needed and involves connecting the stator winding of the motor initially in 'star' until an optimum speed is achieved, when it is switched to 'delta'. For a 3-phase a.c. supply in star connection the stator windings are connected phase to neutral and in delta connection the windings are connected phase to phase. When connected in star, the voltage across each phase winding is 58% ($1/\sqrt{3}$) of the supply voltage and the starting current and torque are reduced to 33.3% of the full voltage values.

This starting method is relatively simple and inexpensive but its use is limited to low inertia drives because of the reduced starting torque available from the motor. The disadvantage of this starter type is the possible high transient torque and current which could occur when switching from star to delta. The number of starts per hour is not normally restricted by the starter, although consideration has to be given to the characteristics of the main circuit, short circuit and thermal over-load protection. To eliminate or reduce the high transient torque and current when switching from star to delta it is appropriate to use the closed transient star-delta (Wauchope) type starter in which resistance is inserted when changing over from star to delta to provide a no-break transition. This closed-transition type starter provides three steps of acceleration against the two steps provided by the standard star-delta starter. Because of the additional resistors and control equipment, the closed transition starter is more expensive than the standard starter. The number of starts per hour is limited by the short-time thermal rating of the transition resistors.

Auto-transformer starting provides more flexibility than star-delta, because the applied voltage and hence the starting torque and current can be varied by changing the transformer voltage tappings. The motor starting torque and current are a function of the square of the transformer tapping with respect to rated voltage. For example, at 50% voltage tapping, the torque and current are 25% of the full voltage values, and at 80% voltage tapping, the torque and current are 64% of the full voltage values. The ability to adjust the voltage of the auto-transformer provides a convenient means for closely matching the starting torque to the driven load and for reducing the starting current. For this reason, the auto-transformer method is suitable for starting larger motors than the star-delta method.

The disadvantages of the auto-transformer starter are cost (it is considerably more complex than the star-delta starter) and the possibility of high transient torque and current when switching from reduced to full supply voltage. To eliminate or reduce the high transient torque and current when switching from reduced to full voltage, a closed transition configuration can be used. Its operating principle is similar to that of the closed transition star-delta starter. It involves the use of a more costly transformer and additional control equipment. The number of starts per hour is limited by the short-time thermal rating of the auto-transformer.

Electronic soft starting. The disadvantages of the above electro-mechanical starting methods can be partly mitigated by electronic soft starting. With this method, the motor supply voltage is gradually increased linearly up to rated value, providing smooth acceleration, controlled and reduced starting current (typically between 200% and 300% of full load value) and controlled motor torque. At full speed, the electronic controls would normally be by-passed and the motor connected directly to the mains supply. Electronic soft starters are more complex and expensive than the electro-mechanical methods discussed above but are being continuously improved, both in ability to provide reduced starting current and for reliability and durability.

Rotor resistance starting of wound rotor induction motors is used where the applied load has large inertia and requires a high starting torque, where the starting current needs to be limited by supply system voltage drop limits or if controlled acceleration is required. With the connection of external resistance into the rotor circuit via the slip rings, high starting torque and low starting current can be obtained. Typically, for full load torque at standstill, the starting current is in the order of 125% of the full load value. By selection of resistance, starting torques of 200% to 250% can be attained with corresponding currents of 250% to 300%. The resistance is either the multi-stage metallic non-inductive grid or the liquid type, the latter providing smooth control of acceleration. The number of starts per hour is limited by the short-time thermal rating of the rotor resistance.

17.21 INDUCTION MOTOR PROTECTION

The function of the protection is to initiate disconnection of the motor from the supply to prevent or limit damage caused by overheating due to abnormal load, failure of winding insulation or some other fault. For low voltage motors, the protection provided needs to be chosen by considering factors such as motor cost and the characteristics and importance of the drive. The degree of protection provided can range from the thermal overload relay to a motor protection relay providing high-set over current, overload, earth fault, negative phase sequence (unbalance) and stall protection. A comprehensive motor protection relay would not normally be used for motors rated below 50 kW. High voltage motors, irrespective of rating, are provided with high-set over current, overload, earth fault, negative phase sequence (unbalance) and stall protection. For the larger motors, typically 1000 kW and above, high speed differential protection is often provided to minimize damage to the stator core in the event of a stator winding fault. For motors where a reversal in the direction of rotation could cause damage to the driven load, phase reversal protection should be provided.

Additional motor winding and bearing protection can be provided by thermistors, thermocouples and resistance elements. These methods give protection against faulty conditions which are not reflected in the line current of the motor. For applications requiring temperature indication, in addition to alarm and motor tripping initiation, the thermocouple or resistance element is used. Thermocouples can be located in stator slots, stator end windings, cooling air circuits and bearings. The location of resistance elements is the same as for the thermocouple, except for the stator end winding. The degree of protection provided by the three devices is good and response to temperature change is fast.

17.22 SPEED CONTROL OF INDUCTION MOTORS

The standard induction motor is a constant speed machine, the speed being determined by the number of poles and the power supply frequency (Section 17.19). When speed control is needed, no simple solution is of universal application; costs and benefits must be assessed for each case. The two most commonly used methods of speed control are based on changing the number of poles or changing the supply frequency, although limited speed variation can be obtained by using the wound rotor type induction motor and varying the rotor resistance. However, the use of variable rotor resistance has now largely been abandoned because of its inefficiency and the development of slip energy recovery systems. The most commonly used methods of speed control are summarized below.

Pole changing motor. The simplest form of pole changing motor has a single-tapped winding, also known as a Delanger winding, which provides two speeds in a ratio of 2:1, for example 3000:1500 rpm for a motor connected to a 50 Hz supply. A variant on the single-tapped winding is a motor with two separate windings, which can provide any combination of two speeds. By combining the tapped and two winding arrangements, up to four speeds can be obtained, for example 3000, 1500, 1000, 500 rpm at a supply frequency of 50 Hz.

Pole amplitude modulated motors. The pole amplitude modulated (PAM) motor is a development of the single-tapped 2:1 speed ratio winding; speed change is achieved by reversing one half of each phase which changes the magnetic flux distribution and produces a resultant field of different polarity. Various combinations of speed ratios can be obtained with this motor design. Other than speed ratio, the advantages of PAM motor design over the pole changing type are better utilisation of active materials resulting in smaller physical size for a specific speed ratio and rated power and improved efficiency and power factor.

Variable frequency drive. This form of drive uses a standard cage induction motor with variation of the supply frequency. The variable frequency drive controller is of the static design, the basic components of which are a rectifier connected to the a.c. supply, an inverter to provide a variable frequency supply to the motor and a d.c. link between the rectifier and the inverter. The motor speed can be regulated typically over the range 10% to 200% motor rated speed, although mechanical design considerations make speeds of greater than 100% very unusual. The rectifier produces harmonic currents, which reflect into the electricity supply system and can cause interference with other consumers or be detrimental to connected plant such as capacitors, generators and motors. Electricity supply companies lay down guidelines for the permitted magnitude of individual harmonic currents and/or harmonic voltage distortion. Methods available for reducing harmonic currents are to: (a) increase the rectifier from 6 pulse to 12 or 24 pulse, depending on the power rating of the drive; (b) install filters or (c) use phase shifting supply transformers. Increasing rectifier pulse number or installing filters considerably increases the cost of the drive. When selecting an induction motor for use with a variable frequency drive controller, the following factors need to be considered:

- increased losses and hence heating in the magnetic circuit caused by the harmonics in the inverter output waveform;
- power and torque requirements throughout the speed operating range, and
- possibly impaired cooling at low speed operation (an important consideration for constant torque loads) since motor cooling fans are normally driven by the motor shaft and so produce lass cooling at lower motor speeds; separately powered fans can be used to avoid this difficulty.

Slip energy recovery. The slip energy recovery variable speed drive, also known as a Kramer drive, uses a wound rotor type induction motor and operates by recovering rotor energy and feeding it back into the supply. The mechanism for doing this is to convert the slip ring frequency power to d.c. and, for the static type drive, return the power to the mains supply via an inverter. An alternative to the inverter would be a d.c. motor driven asynchronous generator. A direct connection between the rotor and the mains is not possible because both the rotor voltage and frequency vary with motor speed. Harmonic currents are produced by the inverter and the considerations above for the variable frequency drive apply.

17.23 EFFECT OF ELECTRICITY TARIFFS

The charges made by electricity supply companies are designed to favour the consumer who takes a steady supply of electricity at high power factor. The consumer who wants large power intermittently is penalized. Tariffs often comprise two, three, or even four separate charges as listed below. If the supply is required for a new installation in a remote area, the supply company will usually also ask for a connection charge to recover the cost of providing the supply.

1. A 'maximum demand charge' per kW or kVA of maximum demand in that month above a given figure (and sometimes, more onerously, per kW or kVA of maximum demand for the last twelve months). The charge may differentiate between summer and winter maximum demand or may be varied month by month.
2. A 'unit charge', which is a monthly charge for the number of units of electricity consumed, the charge per unit being substantially higher for daytime consumption. It is usually on a sliding scale allowing reductions when large numbers of units are consumed.
3. A 'fuel clause', which increases the unit price of electricity according to the increase of basic fuel cost.
4. A 'power factor clause', which increases the maximum demand charge if the power factor drops below a certain figure (usually 0.90).

The effect of the maximum demand charge (especially when based on twelve months) can be onerous. The larger the gap between the maximum demand and the average demand, the more penal the maximum demand charge becomes and the more costly the charges per unit. Hence, it is very important for pumping stations to be run at a high 'load factor', i.e. the average power consumed should be as close as possible to the maximum power required at any time during a month (or year). This means that pumps should be run for long periods rather than short; pumping for short intermittent periods at high outputs is expensive. Similarly, the occasional running of pumps at full load when they normally run only at part load, or bringing in extra pumps for short periods can also be expensive. Even the testing of pumps when first installed—perhaps running several together on test—can bring a heavy extra cost for so brief a run as 30 minutes.

Choosing the right duty for the pumps, the right hours of working, and the right amount of standby to be provided for any pumping station are matters that must be thoroughly investigated. The benefits due to adequate storage in the supply system can be very significant. These include safeguarding continuity of the supply, levelling out pumping rates thus reducing friction losses and the maximum power demanded of the motors, and permitting long steady running at a high load factor to minimize electricity charges. With modern competitive electricity supply arrangements, the water supply company may be in a strong position to obtain especially favourable rates from the electricity supplier. Unlike many users, water companies are usually large electricity consumers with a steady load for most of the year.

17.24 ELECTRICAL POWER SUPPLIES

The preferred source of electricity supply for a pumping station or other water supply facility is from the local electricity supply company. If the maximum demand is not high (typically below 250 kW), the supply would be arranged from the nearest existing substation and the supply voltage would normally be at low voltage. However, for larger maximum demands, the supply may require the construction of new transmission lines and substations dedicated for the facility. For sites with large maximum demand higher voltage supplies will be required. Early negotiation between the user and the electricity provider is essential for any capital project. Information needed for these discussions includes the load and the nature of the load, types of motor starting proposed and intended use of any variable frequency devices which could induce harmonics in the electrical distribution system.

Public electricity supplies are to the national standard with respect to frequency and voltages available. Voltages and frequencies of electricity supplies vary considerably around the world and even from city to city in some countries. *Technical Manual 5-688* provides a comprehensive listing of supply voltages and frequencies (USACE, 1999). The requirements for the supply include the load and the required voltage which depends on the voltage suitable for large units such as motors. It also depends on the distance from the facility to the point where the new electricity supply is connected to the existing electricity supply company distribution or transmission system. Use of higher supply voltages may be necessary in order to limit energy losses and voltage drop in the transmission lines. Except for the very smallest installations supplies will normally be three phase. For a very large load of strategic importance, supplies from two independently fed substations may be sought in locations where power supply failure is an unacceptable risk.

For installations where all electrical components may operate at the same time the connected load may be used for sizing the power supply. However, this rarely applies to water supply facilities where standby units such as pumps are on line in case of failure or need for maintenance. In addition, other loads are not usually expected to operate simultaneously and diversity factors should be applied to allow for this. The electrical load should be determined using a comprehensive list of all drives and other loads. Pump motor loads will tend to dominate but on-site ozone and hypochlorite generation will also add significant load. All other plant should be covered and allowances should be made for building services, including heating, ventilation and air conditioning as well as lighting and small power.

The electricity supply company will require the supply be metered and a disconnection switch provided. The location of these facilities is often at the site boundary but, in any case, has to be secure while being accessible to the supplier who will notify such requirements.

Electrical loads such as motors which comprise windings have inductance. The consequence is a power factor (kW/kVA) which is less than unity. Electricity suppliers often impose a penalty on low power factor loads. Therefore, it is common to correct low power factors by introducing capacitance of a value sufficient to raise the power factor enough to avoid the penalty. The capacitors may be installed for individual motors or as a group connected to the user's main switchboard busbars.

17.25 STANDBY AND SITE POWER GENERATION

Although in many places the reliability of electrical power supplies has greatly improved, so that often no consideration need be given to outages of any long duration, there are still installations where power failure must be taken into account. A typical example is where severe weather can affect transmission lines to remote pumping stations. Some form of standby generating plant will then be needed. Except for very large unit sizes such as 5 MW and greater, for which gas-turbine generators are usually used, diesel-driven generators are preferred. Where a single large water source works provides a major part of the water supply, there may be no escaping the costly provision of a fixed, dedicated standby generating station, possibly serving high and low-lift pumps and a water treatment works as well. The generating capacity to be installed depends on such fundamentals as the duration of the outage to be covered (which affects the fuel storage required), how much of the plant must be kept operational, what size is the largest motor to be started and so on. If providing a large generating plant is essential, substantial capital investment in plant will be needed. It could stand idle for most of its life but, alternatively, it could be operated continuously to provide the base load required; unused generating installations are prone to deterioration unless regularly operated and the operations staff need to have experience of running the plant.

The cost-saving practice, called peak lopping, of switching to standby generators when the amount of electrical energy being used approaches a higher tariff band is an option to be considered to allow the standby generation plant to be usefully operated and thereby avoid financial penalties associated with certain tariffs. When considering peak lopping as an option the capacities of standby generator and bulk fuel storage should be carefully considered.

An alternative, which may sometimes be possible when a number of source works contribute to the total water supply, is to set up a pool of mobile generating sets, which can be quickly transported to any site where a power outage has occurred, if the same outage is not expected to affect all works simultaneously. This reduces the amount of standby generating capacity needed and makes for more cost effective use of the generators but transporting large generators can be awkward and expensive. The solution may not always be effective since severe storm or other conditions causing the original electrical outage might affect more than one installation and could also make roads impassable. Simple and speedy arrangements for receiving and connecting the generators, and for fuelling them, must be made at each point of use and more than one size of generator may be needed. Arrangements must also be made for safe storage and maintenance of the generators to ensure they are always ready for emergencies.

If pumping stations and treatment works have to be built where there is no public electricity supply available within an economic distance, power may have to be provided by a diesel or gas turbine engine on site. If more than one pump has to be operated, the diesel engine would drive a generator to make electrical power available to all pumps and provide power for other uses such as lighting and instrumentation. In rare cases of small isolated pumping stations with only one pump, a small diesel engine can be direct coupled to the pump. Normally, however, the choice is between building a power station to supply the works, which will have to include some standby capacity, or negotiating with the electricity supply authority to make new supply arrangements. The economics of these situations can be complex, being dependent on the charge the electricity supplier makes for bringing a power line to site, the terms of the supply, and the need to take into account likely future developments in both power and water needs. The capital, operating and maintenance costs of diesel-generating power plant and the need to provide adequate fuel storage tend to be high, so that such a set-up is rarely economic in developed countries where high-voltage electrical networks make supplies widely available.

17.26 **TRANSFORMERS**

Transformers are alternating current devices which, in water supply installations, are normally used for two purposes: voltage change and isolation of instrumentation. Voltage or power transformers have primary and secondary windings around an iron core in which a magnetic field is set up when alternating voltage is applied to the primary winding.

Winding insulation is either 'liquid' or 'dry'. Liquid insulation consists of oil of low flammability; dry insulation is a resin. Oil used in 'liquid' transformers is a potential source of fire so such transformers are always located in enclosures separated from other building spaces (Plate 33(a)). In case of leaks of oil, oil insulated transformers are installed on plinths surrounded by open gravel in a concrete bund so that any spillage is contained. Resin insulated transformers may be located within buildings subject to any noise and cooling constraints. Smaller transformers below 200 kVA on incoming supplies are often pole mounted at the supply connection.

Power transformers are normally provided with off-load tap changers to cope with long term voltage changes. Such taps are usually 2.5% to 5% above and below normal supply voltage. Alternatively, for certain supplies, which have fluctulating voltages, a more expensive on-load automatic tap changer could be considered to maintain a more stable secondary voltage supply.

Transformer impedance for transformers above 500 kVA is usually in the range 5 to 8%. Impedance has differing effects, for example cost, short circuit currents, switchgear ratings and motor starting voltage drop, and needs careful consideration when designing and selecting components for the electrical distribution system.

17.27 HV AND LV SWITCHBOARDS

Switchboards are enclosed panels (usually of coated metal) arranged in sections with an interconnecting bus bar which usually runs at the top of the switchboard in a segregated compartment. Panels may be arranged for top or bottom entry for site cabling. For the latter a cable trench is necessary and for the former high level cable trays/ladder would run to the panel. The switchboards are normally provided with segregated sections which comprise circuit breakers, fuses, switches, indicators and meters. High voltage (HV) switchboards should be located in a room separate from low voltage and other panels, as is usually required by regulations, since different safety procedures and permits usually apply to HV switchgear. It is sensible for main switchboards to be arranged with a bus-coupler in order for each for each section to be isolated individually. This permits maintenance on one side of the switchboard without de-energising the complete panel. This principle would normally be adopted for improving the security of supply and the provision of two incoming transformers.

17.28 MOTOR CONTROL CENTRES

A motor control centre (MCC) (Plate 33(b)) is an electrical panel similar to a switchboard but including motor starters and controls. It may also contain PLCs and HMIs (Section 17.37) and a number of analogue or digital indicators and lights. The MCC is usually divided into vertical sections which have separate compartments for different functional devices. The designer needs to decide the Form of the MCC which is normally a Form 4 construction in the UK water industry. An MCC designed for Form 4 construction will have all sections and devices which are accessible without having to interfere with other sections. MCC panel layout should be logical and should be designed for either front or rear entry.

Traditional MCCs with individual starter door furniture such as indicating lamps, pushbuttons and ammeters can take up a large amount of space in the switchroom. An alternative is the intelligent MCC which minimizes the door furniture and therefore reduces the space requirement by providing a HMI to allow the operator to start/stop the pumps/motors from a touch screen rather than pushbuttons on individual starters.

17.29 ELECTRICAL CABLING

Power cables comprise a number of conducting cores of copper or aluminium, insulation and, for some situations which require mechanical protection, armouring. Insulation is usually XLPE (cross linked polyethylene), LSF (low smoke and fume—for occupied buildings) or EPR (ethylene-propylene). Armouring is usually steel wire which itself is then sheathed to prevent corrosion. Cables serving the motors of large pumps or other large plant have a large cross-sectional diameter and require a large bend radius to be considered for the installation. This affects the space needed in cable trenches under panels and at other changes in direction such as draw-pits.

17.30 HEATING AND VENTILATION

Electric motors have losses that generate heat and noise. Very large motors are usually water cooled. Other electrical plant such as transformers and variable frequency drives also give off heat and are usually air cooled. Such heat needs to be removed and this is normally done by ventilating the room housing the plant. The number of air changes per hour needed should be calculated from the heat output and permissible temperature rise above ambient summer conditions, taking into account solar gain and other effects.

Condensation is potentially harmful to electrical plant. This can arise in any season when the temperature of equipment such as motors and panels falls below the dew point of the adjacent air. Electric motors are usually fitted with built-in anti-condensation heaters, which are energised automatically when the motor is stopped. These are very effective and take little power: perhaps 0.25 kW for motors under 50 kW and up to 1.5 kW for very large motors. Switchboards and MCCs are also equipped with anticondensation heaters for the same reason.

PART III   CONTROL AND INSTRUMENTATION (C&I)

17.31 INTRODUCTION

The control and instrumentation of pumping plant, distribution network flows and water treatment plant has changed rapidly in the water industry with the introduction of stricter requirements for service levels, maintenance of water quality and the development of many specialized and advanced water treatment processes. The last alone requires the processing of thousands of signals and large amounts of data; a typical modern water treatment works control system includes ten to fifteen thousand signals to be processed. The automation of such plant is becoming essential due to this complexity and the amount of data to be handled. Automatic control systems are also seen as a way to reduce labour operating and supervisory costs, whilst maintaining product quality.

The increasing use of computers in control applications has also been beneficial, particularly in the control of water treatment processes and pumping systems. Regimes of use of high power plant such as pumps have been developed to minimize power costs. This can be done by making the best use of electricity supply tariffs, concentrating pumping at times of low-cost electricity and avoiding incurring high maximum demand charges. The calculations involved in such optimisations may be complex and may involve negotiation with the electricity supply authority to achieve the best results. Such computerized control systems can also be readily designed to take account of a wide variety of conditions. Examples include preventing the over-frequent starting of electric motors; sharing the hours run equally between a number of pumpsets; coordinating the operation of different pumping stations serving the same supply area; and regulating the operation of pumping stations arranged in series along a long transfer main.

17.32 CONTROL

Current C and I development has resulted from changes in the electronics industry which have reduced the capital cost of providing complex monitoring and control systems. The water industry has moved away from traditional 'hard wired' control systems to the use of more adaptable programmable electronic systems. Three levels of control are practised as described below.

Manual control is the most basic form of controlling plant and relies primarily on the operator to oversee and to react to any condition that requires alteration of plant operation. Although manual control may be impracticable for more complex plant, it must still be available as a 'fall-back' method for emergencies and when other systems are unsuitable or fail. Manual control facilities are commonly located adjacent to the equipment being controlled. For motorized plant with starting equipment in a motor control centre (MCC), the manual control facilities are typically located on the starter section of the MCC.

Semi-automatic control aids the operator to cope with treatment processes which are too complicated for control without assistance. Examples are ozone production and the overall start-up of plant processes applying numerous chemicals. Some operations, such as filter washing, which were traditionally manual, are now frequently specified as semi-automatic. Semi-automatic facilities tend to be located centrally, but not necessarily remote, to the equipment being controlled.

Automatic. Full automation is used to control processes with the minimum of operator intervention. This type of control must ensure a high quality, safe product and that corrective action is taken on equipment failure. Operator intervention is usually limited to entering key set point parameters related to output quality and quantity. Plant operators are usually provided with facilities for automatic control from a central monitoring point such as a remote control room. Although the control logic performing the automatic functionality may be located within controllers distributed around the plant, the control room facilities are linked with the plant by means of data communication.

17.33 AUTOMATION

Automation of water treatment plant involves the control system opening and closing valves and starting and stopping equipment in predefined sequences to complete specific tasks or to provide the desired process plant output. To achieve these results the automation system relies on signals from correctly selected and placed instruments, on devices such as actuators and motor control circuits and a reliable control logic. The degree of automation to be used is fundamental to developing an automation system.

A *functional design specification* (FDS) must be prepared in detail setting out all the functional requirements for each item of equipment and the controls to be applied. The FDS is critical to achievement of a satisfactory monitoring and control system that meets the requirements. In the information technology industry the following terms are in common use:

Hardware (Section 17.37) means the device (sometimes programmable) which controls and monitors the operation of an item of plant.

Software is the programming logic or set of instructions which designates the tasks to be carried out by a group of hardware. These tasks can range from starting a pump to calculating, from monitored process variables, the appropriate hardware actions to be taken.

Firmware means a permanent form of software (embedded in hardware) which is normally not available for alteration by the operator but may have features permitting selection from a standard menu.

'Application' or 'bespoke' software is software which is custom designed for a specific application. Most control systems involve the use of some bespoke software.

Where two items of instrumentation equipment are required to operate as a system the interface between them must ensure proper data communication. When selecting components for a system, the designer must ensure that they will operate satisfactorily together.

17.34 INSTRUMENTATION

Only a few process parameters make up the majority of those measured in the water industry. Of these parameters, flow, level, pressure and water quality are the most common. Common types of flow, level and pressure instruments are described in more detail below. Parameters such as chlorine residual, colour, turbidity and pH are among the most frequently used water quality measurements. Permanently installed on-line instruments can be selected to measure these parameters and provide electrical signals for use within a control and monitoring system.

Measurement of process parameters can be digital or analogue. Digital is used for reporting the status of a plant e.g. whether running or stopped. Analogue measurement is used to represent variable values, such as chlorine residual or pump output. There are several standard analogue interface standards. The water industry uses the 4 to 20 mA loop as the standard. The 4 mA value usually represents the lowest end of the value range and the 20 mA the highest. The 4 mA value is referred to as 'live zero', so called that an actual zero milliamp value can be attributed to a problem such as a broken connection. The instrumentation industry is working towards standardising inter-instrument communication with the development of standard protocols (the communication terms of engagement between two components). Two such protocol standards are HART (highway addressable remote transducer) and 'Field Bus'. Also, intelligent instruments have been developed to operate on their own standard data communication network: 'Profibus' and 'ControlNet' are two such standards. Although instruments with these capabilities have been available for some time, only recently has the water industry begun to use them extensively.

The operation and characteristics of flow meters are described in Chapters 12 and 16. Common methods used to measure other key parameters in the water industry are described below.

Level

Ultrasonic level measurement is a non-contact method used for measuring the distance between a surface and the instrument sensing element. Ultrasonic waves are transmitted from and received at the sensing element; the time taken is used to calculate the distance to the surface. Once the sensing element is installed over the surface the instrument is programmed with datum information to allow the process surface level, in a tank or channel, to be measured in useful terms. When programmed with tank or container dimensional information, the ultrasonic level instrument can provide volumetric readings. Precautions must be taken when locating the sensing element to avoid reflections off fixed surfaces that may confuse the instrument and make the reading unreliable. A drawback is that floating debris and foam can adversely affect accuracy.

Hydrostatic pressure instruments use the static head of the process fluid to deflect a diaphragm in a sensing element which is placed at the bottom of the fluid being measured. The deflection is monitored and, based on the specific gravity of the fluid, a signal is produced in proportion to the depth of the fluid above the element. This type of instrument is unaffected by floating debris and surface foam but may require frequent cleaning in some applications.

Differential pressure level instruments are frequently used to measure fluid surface level in pressurized vessels where hydrostatic measurement would be useless. The differential pressure instrument has a high-pressure port and a low-pressure port. For this application the high-pressure port is connected to the bottom of the vessel and the low-pressure port is connected to the top of the vessel. The instrument calculates the level of the interface between the fluid in the vessel and the space above the fluid by finding the difference between the two pressures. This type of instrument can also operate in open-air tanks by simply leaving the low-pressure port open to atmosphere.

Conductivity level probes are commonly used to detect discrete surface level points. Conductivity is established between a reference (earth) probe and the probe set to the desired switching level when the process fluid wets both probes. The conductivity is lost when the fluid level drops below the switching level probe. The signal produced can be used for control purposes.

Float operated level switches are simple devices that operate by tilting a floating bulb with an internal gravity operated switch. The float switch is connected to a flexible cable that acts as the tether for the float and for transmiting the signal to the control system. These switches are suitable when precise level switching is not required but are reliable and easy to maintain.

Pressure

Pressure measurement using diaphragm deflection involves the use of variable electronic resistors fixed to a flexible surface (diaphragm). As the pressure deflects the diaphragm an electronic circuit detects minute changes in a group of resistors, known as a 'Wheatstone Bridge', and produces a signal proportional to the pressure causing the deflection.

Most pressure gauges in the water industry are Bourdon tube type. The principle utilizes a curved tube that straightens as pressure is applied internally. The pressure gauge display pointer is fastened mechanically to the end of the Bourdon tube.

Water Quality

Water quality instruments commonly used in the water industry are:

- Chlorine residual—(e.g. potentiostatic)
- Colour—(e.g. light diffusion)
- Turbidity—(e.g. light scatter)
- pH—(e.g. oxidation reduction chemistry).

The technologies associated with these water quality instruments are not discussed in this text. Selecting the correct type of instrument for a given application is critical to the reliability and cost effectiveness of the application and should be done by an instrumentation expert taking into account the operating conditions, environment, maintenance, consumable components and overall cost, drawing on experience and research of products.

17.35 SYSTEMS

A 'system' may involve only one measurement, a controlled device and a controller to maintain a set point of a single parameter; more commonly it will handle data from many monitoring points and propagate many plant control commands. For any given treatment process, the combination of these items defines the control system. Usually the control logic currently used for automation is installed in programmable devices evolved from the electronics and information technology industry. However, modern control systems still use traditional hard wired monitoring methods, primarily as a back up, fail safe, system for a critical process or piece of plant. *Shut down* systems have to prevent damage and danger to plant or personnel and ensure there is no failure to maintain output quality. *Alarms* notify the need for operator attention when plant, process, personnel or output quality and quantity may be at risk. *Alerts* draw operator attention to any state that may develop into an alarm if left unattended (e.g. tank water level high or process flow low etc). *Events* inform the operator of a 'normal' operating occurrence that may or may not require attention (e.g. a pumping sequence completed or a valve opening).

Hard wired. Traditional hard wired systems utilize direct wiring between electromechanical relays (Plate 33(c)) to perform control logic operations and are directly wired to the controlled equipment. These systems do not rely on programming software and are therefore rigid in terms of functionality and often very difficult to modify. However, well designed hard wired systems have a reputation for being very reliable. For these reasons independent hard wired systems often supplement programmable systems, primarily for simple back up controls and safety critical applications such as alarm and emergency shut down systems.

Programmable control. There are many programmable devices used in control systems. Most electronic equipment now has some form of programmable functionality. Any device that can be programmed to carry out predefined sequences to provide control outputs could be called a programmable controller. A program that performs specific logic or tasks is referred to as a software 'algorithm'. The industry work horse is the programmable logic controller (PLC) (Section 17.37). When networked with others, a system of PLCs can be created that is capable of controlling even the largest and most complex water treatment works.

Distributed. A 'distributed control system' (DCS) is now the type most commonly used (Fig. 17.10). In it the control logic is distributed in outstations around the plant being controlled. Until it was practicable to locate processors at outstations the entire computing power was held centrally. Such centrally based systems are now mostly obsolete. An up-to-date DCS comprises a number of discrete process control units, e.g. PLCs, outstations or remote terminal units (RTUs), linked by a 'state-of-the-art' communication network (Fig. 17.10). The automatic control logic for each process area is contained in the control units and can therefore function independently of the overall monitoring and control system. A DCS links discrete process blocks to achieve:

- transfer of data round the system, including changes to control;
- changing parameters at RTUs, such as adjustment of set points or duty plant selection;
- monitoring of individual RTUs;
- display and recording of data, events and alarms;
- safe reaction to changes or failure of individual RTUs which could affect product criteria, process performance or plant safety.

SCADA (supervisory control and data acquisition) refers to virtually any data acquisition system, but usually one which exercises monitoring and supervisory control of a number of sites from a

(a) Oil insulated transformer with gravel filled sump.

(b) MCC—sections from left to right: incomer and spare; 4 × plant motor starters; cable way; 2 × pump starters; 2 × pump starters.

(c) ICA section of MCC with conventional hard-wired relay logic and PLC.

(d) ICA section of MCC with small PLCs and telemetry equipment.

PLATE 33

(a) Rectangular reinforced concrete monolithic service reservoir under construction in UK. (Engineer: Binnie & Partners)

(b) 45 000 m^3 circular steel plate ground storage tanks under construction in the Middle East. (Engineer: Binnie & Partners).

(c) 2270 m^3 reinforced concrete water tower, Malaysia. (Engineer: Binnie & Partners in association with SMHB)

(d) 700 m^3 welded steel water tank on reinforced concrete tower, Malaysia. (Engineer: Binnie & Partners in association with SMHB)

PLATE 34

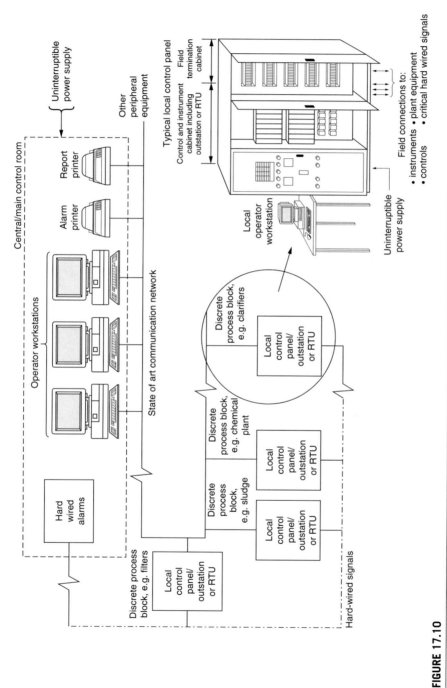

FIGURE 17.10

Typical distributed control system (DCS) linking a number of discrete process block control units.

control centre. Such systems are widely used in the water industry so that a 24-hour manned control centre can react to any problems arising at sources or throughout the distribution system.

Telemetry is the system for transmitting data from one location to another. The distance between the two locations can vary from a few metres to thousands of kilometres; with only the limitations of the transmission medium applying (e.g. hardwired data communication networks, telephone networks, microwave systems or radio systems etc.). Telemetry systems used by most monitoring and control systems include DCS and SCADA systems (Plate 33(d)).

Web-based technology has changed the way in which control system data is transmitted around the world. The introduction of web-based global internet technology is likely to change the way in which control systems are used in the water industry. Already, there are web-based applications that provide remote monitoring and control facilities and allow plant to be interrogated, controlled and modified from any computer that has web access and the appropriate software. For monitoring purposes this technology offers limitless benefits but, for control of critical plant, use of the internet raises concerns about security.

17.36 COMMUNICATIONS

Networks generally. Communications networks have been used to transmit instructions and data for process monitoring and control for many years. Process control networks have used various technologies and topologies, as have business and computer networks in less critical situations. Process control networks require robustness, determinacy and compatibility.

Robustness is a measure of the reliability of the network to perform its function throughout the life of the installation. The designer of a network must evaluate the need for redundancy of network components and cabling, as well as error testing and error correction facilities, to provide a network that meets the needs of the system it serves. *Determinacy* is a specific guarantee that messages enter the network and reach their destination within known times. Non-deterministic networks cannot guarantee message delivery in a specific time but recent technologies have allowed some non-deterministic networks (such as Ethernet) to function satisfactorily in the process control industry. *Compatibility* describes the ability of the network to communicate between equipment from various manufacturers/suppliers without conflicts.

The obvious advantage of using a network rather than traditional cabling is that all the data is passed along a single cable therefore reducing significantly the number of cables to be installed.

Ethernet. CSMA/CD (carrier sense multiple access/collision detection) performance has evolved to a very high standard since its introduction in 1973. Although it is non-deterministic, CSMA/CD allows every device on a network to check if any other device is transmitting before attempting to transmit and, if multiple devices transmit simultaneously, the collision detection discards corrupted messages (packets) and instructs the devices to resend their messages after random wait times. The network is non-deterministic because there is no guarantee that a message will reach its destination is a specific time. However, industrialized Ethernet networks have become so reliable that they are the standard in the industry for linking site-wide process control systems. The Ethernet design specification must be followed regarding connectivity and cable lengths, for example: using a maximum of 90 m of copper cable from a network switch.

Ethernet networks can form part of local area networks (LAN) or wide area networks (WAN). Although there are no specific definitions, a site-wide SCADA system for a process plant would generally be referred to as a LAN and a regional network linking several sites would be considered to be a WAN.

A *bus network* is an arrangement in a LAN where a single cable links multiple devices. The cable is the 'bus' to which 'nodes' are connected. Each node generally corresponds to a specific item of equipment. Bus networks are simple and reliable and reasonable fail-safe. If one node device fails the bus continues to operate with the remaining functional devices. Only if the bus cable itself is broken would there be serious communication problems in the network. Bus networks provide a simple means of expansion because generally they allow nodes to be added fairly easily.

The limitations of bus networks are primarily the physical properties of the bus cable itself. As cable length increases, the losses affect the reliability of the data being transmitted. Therefore the topology of the bus network needs good design. Other network topologies such as 'ring' or 'star' may sometimes provide better flexibility and may be cheaper.

Profibus is one bus technology that the industry has adopted. Profibus is a high-speed digital communication system that utilizes a single cable (bus) to link devices. Many manufacturers of electrical, electromechanical and instrumentation equipment now provide Profibus compatible products. It is common to link a number of related plant items by a Profibus network while also linking other areas of plant by a larger network such as Ethernet. As with Ethernet networks it is important for the design of a Profibus network to comply with Profibus specifications. There are two types of Profibus namely Distributed Protocol (DP) and Process Automation (PA); both types communicate over two-wire cables but there are several differences. Profibus DP is a voltage-based communication using the RS-485 standard; whereas Profibus PA uses a current loop technology to communicate. Profibus DP is available with copper and fibre optic cable capabilities but Profibus PA is only available for use with copper cables. Profibus DP is not able to offer intrinsically safe solutions whilst Profibus PA can. The choice of type of Profibus network depends on review of all functional requirements for the network.

Wireless data communication networks have become very popular for business and domestic use and for some process monitoring applications but there remain many safety and security concerns with its use for control of plant. For these reasons, wireless networks should be used for control of plant only after careful consideration and when directly connected systems are impractical or impossible to install.

Wireless technologies such as low-power radio and microwave have been used for control of some equipment but should always be used with caution as risks associated with plant protection and personnel safety must always be considered. Typically, these technologies are used over relatively small areas and primarily for transmitting status monitoring signals.

17.37 HARDWARE

The essential components of ICA system hardware are described briefly below. *MCCs* are described in Section 17.28. These, particularly intelligent MCCs, may house complex ICA equipment or it may be located in dedicated panels.

PLCs (Programmable Logic Controllers) are computers with components similar to those of the desktop PC, but configured for multiple inputs and outputs and capable of withstanding a wider range of environmental conditions such as temperature, dust and vibration. *Programmable Logic Controllers* (Bolton, 2006) provides a useful guide to PLCs. Plates 33(c) and (d) show small PLCs in ICA sections of an MCC.

HMIs (Human Machine Interfaces) provide an interface between an operator and the plant and allow the operator to monitor, control, diagnose and manage the plant. In the traditional MCC, key plant parameters are displayed on analogue or digital instruments mounted on the front, with plant control by push button and switch. The modern HMI makes use of VDUs and keyboards or touch screens and allows MCC size to be reduced.

Control rooms are dedicated areas where functions are gathered to allow plant to be monitored and controled by the minimum number of operators. In the traditional control room there may be a mimic diagram consisting of a large fixed display of the system, with live plant data indicated at appropriate points, and a control console with indicators and push button controls. The modern arrangement consists of a more simple control desk arranged with VDUs and other HMI elements. The VDUs are likely to be large flat screen plasma or LCD monitors, arranged to display schematics of sections of the plant, which can be selected by the operator. The operator will also be able to call up information and trends from memory and will be able to produce a variety of reports of plant operation and failures.

The HSE publishes guidance on certain industrial plant and includes on its web site a Technical Measures Document on *Control room design*—www.hse.gov.uk. Relevant reference standards include ISO/DIS 11064 (1997), BS EN 894 (1997), BS EN 60073 (1997) and NUREG-0700 (1996).

17.38 ANCILLARY EQUIPMENT

Uninterruptible power supplies (UPS) are often used in support of water industry monitoring and control systems to maintain system operation for a predefined period during a power supply failure. Plant behaviour under power failure conditions must always be reviewed; each piece of equipment must be controlled into a fail-safe condition when power supply fails. Ancillary plant such as mechanical or pneumatic back up may be necessary to fulfill these requirements. In addition, plant start up following a power failure is another key consideration; and automatic start up or manual start up must be reviewed accordingly and specified clearly.

17.39 OPERATION AND MAINTENANCE

Operators of automated treatment facilities need an understanding of electronic control and software programs, as well as a full knowledge of the treatment processes. Maintenance of automatic control systems is a specialist field: this aspect of modern system installation is often overlooked. The system buyer must fully appreciate the maintenance commitment that is involved. The shift from traditional plant, with a high level of operator input, to plant which has complex software and electronics control has not significantly reduced overall operational and maintenance costs of water supply, because maintenance costs associated with C and I systems for a treatment works tend to offset the savings due to reduction in plant operational personnel needed. On the other hand the wide variety of specialist processes now involved in water treatment require more complex systems of monitoring and control than would be possible manually.

REFERENCES

Bolton, W. (2006). *Programmable Logic Controllers*. Elsevier Science.

BS 4999 (1997). *General Requirements for Rotating Electrical Machines*. BSI.

BS EN 894 (1997). *Safety of machinery. Ergonomics for the design of displays and control actuators*. BSI.

BS EN 60034. *Rotating electrical machines*. BSI.

BS EN 60073 (1997). *Basic and safety principles for man-machine interface, marking and identification*. BSI.

HI 9.8 (1998). *American National Standard for Pump Intake Design*, ANSI/HI 9.8.

IS0 5198 (1987). *Specification for Acceptance Tests for Centrifugal, Axial and Mixed Flow Pumps: Precision Class Tests* (BS 5316-3, 1988). ISO.

ISO/DIS 11064 (1997). *Ergonomic design of control centres*. Draft. ISO.

NUREG-0700 (1996). *Human-system interface design review guidelines*. Rev. 1. US Nuclear Regulatory Commission. Washington.

Prossor, M. J. (1977). *The Hydraulic Design of Pump Sumps and Intakes*. CIRIA and BHRA Report. July.

Rayner, R. (1995). *Pump Users Handbook*. 4th Edn. Elsevier Science.

Sanks, R. L. (1998). *Pumping Station Design*. 2nd Edn. Butterworth-Heinemann. USA.

Thorley, A. R. D. (2004). *Fluid Transients in Pipeline Systems*. 2nd Edn. Professional Engineering Publications Ltd. London.

Yates, M. A. (1989). *A Meter for Pump Efficiency Measurement*. World Pumps, January.

USACE (1999). *Technical manual 5-688. Foreign Voltage and Frequencies Guide*. Publication of the Headquarters. USACE.

Treated Water Storage

18.1 FUNCTIONS OF TREATED WATER STORAGE

Treated water storage became important with the expansion of piped water supply systems in the 19th century. Its subsequent development has been driven by the need to protect water supplies from contamination and deterioration and by increasingly sophisticated operation of supply systems. Treated water storage may be provided at the treatment works or further downstream; it can be located at ground level or in elevated tanks.

Treated water storage is used to balance relatively consistent or stepped changes in source output with the variable and less predictable demand of consumers. The storage covers diurnal demand variations and may additionally, during peak seasonal demand periods, provide a balance over a week. Treated water storage is also used to maintain supplies during failure of a source works or critical pipeline or to meet fire or other emergency demands. Such storage is located strategically to ensure resilience within the trunk main network and to support local demand centres.

The need for treated water storage depends on the facilities supplying water to a distribution zone and on variation in demand in the zone. It is seldom possible or economic for a water treatment works to provide a fluctuating output in step with demand; treatment processes need to be run 24 hours a day with only infrequent, carefully controlled changes of output. For maximum efficiency and to avoid risk of cavitation, pumps should be operated near their design duty point; electricity tariffs may influence pump running times (Section 17.23). It is not usually economic for a long supply pipeline to have a capacity large enough to meet the peak demand of a few hours duration. Introducing water storage to the system can reduce whole life costs and overcome such technical difficulties. The various storage, resilience and siting options should be evaluated technically and in terms of whole life costs: financial, environmental and social (Sections 2.15 and 2.20).

The minimum water level feeding distribution by gravity should generally be just high enough to maintain the required minimum pressures in the distribution system at peak flows during the planned life of the scheme. A balance has to be struck between having an elevation just high enough to maintain the required pressures and having some reserve elevation to meet future forecast needs at the expense of increasing pumping costs.

18.2 STORAGE CAPACITY REQUIRED

Minimum Storage to Even Out Hourly Demand

A typical graph of the hourly variation of demand for a UK town of about 75 000 population is shown in Figure 18.1(a). The demand reaches a peak typically between 0700 and 0900 and remains high until after midday. It slackens off in the afternoon but rises to a second peak in the evening when people wash and prepare an evening meal. In summer the evening peak may be higher and more prolonged due to garden watering. The demand pattern is usually different at week-ends and on holidays with peaks occurring later and lasting longer. Daily demand profiles in tropical countries or where there are different industrial, work or social regimes should be assessed. Garden watering can form a large part of peak demand and may require special attention.

The average demand for the day from Figure 18.1(a) is assumed to represent the input to the reservoir; Figure 18.1(b) shows the consequent net outflow starting at 0700 when demand rises above input. A total of 3750 m³ of stored water is used for an average demand of 960 m³/h. This is almost 4 hours' storage and is representative for towns of this size in temperate climates. The number of hours of storage needed for balancing demand through the day is more for smaller systems and less for larger systems, particularly those with significant 24 hour industrial demand.

In practice a greater storage volume will be required because:

a. daily peak demands and diurnal profile will vary from day to day and with the season;
b. a full reservoir cannot be guaranteed before the start of the peak demand period; and
c. provision must be made for contingencies (see below) and some storage volume must be retained in the bottom of the reservoir for settling sediments.

As a guide, for a town with a population of about 75 000, adequate flexibility is provided if the capacity of the reservoir is increased by about 50%, i.e. to about 6 hours' supply or 25%

FIGURE 18.1

Typical variation in demand and use of storage for 23 Ml/d supply.

of the average daily demand. This is not, however, sufficient to safeguard the continuance of the supply against all contingencies and applies to locations where flow from source works is continuous.

Contingency Storage

Some contingency storage should be provided to cover intermittent source operation, breakdowns at sources, loss of supplies after major bursts, and consumption for major fires. Pollution incidents may cause a source to be shut down until the hazard has passed (Section 7.1). The storage required for these contingencies depends on the ability to change source output, the availability of alternative sources, the layout of the pipe supply network; and what fire regulations or safety precautions may require.

Risk assessments should be carried out for critical sources and pipelines; the likelihood of a failure event, its consequences and any possible mitigation should be determined. The results should be covered in emergency action plans. The loss of water following a major burst should also be considered and how such loss would be regained. A major fire can use 5000–15 000 m^3 of water, but it should be possible to increase the output from sources accordingly. Allowance should be made for the event to occur when the reservoir is already low.

Shut downs for planned maintenance and due to power supply cuts must be allowed for, particularly for sources such as borehole pumps or booster stations with a single source of electricity supply and no standby generator. Such outages may last a few hours.

Pipeline bursts cause supply to stop for rather longer—the following figures are indicative only of the least time that may be needed; if things go wrong, double the time may elapse before supply can be restored.

Burst Reported	1/2 hour (say)
Mobilising repair gang and closing valves	2 hours
Repairing pipeline:	
– up to 600 mm diameter	6–8 hours
– over 600 mm	8–12 hours
Refilling and disinfecting	2–4 hours

Taking into account the need for diurnal, seasonal and strategic contingency storage for routine and emergency conditions, network resilience and source works operational constraints, the overall desirable storage within a system is 18 to 24 hours—subject to risk assessment, levels of service commitments and availability of land. This is a useful target but larger amounts of strategic storage increase the risk of water quality deterioration unless the storage is carefully managed to ensure regular turnover. The location of the storage should be selected so that supplies can be re-routed to areas cut-off by any burst pipe or by maintenance work. It may be necessary to employ a water balance/strategic transfer model, especially where a reservoir forms one component of a complex integrated system. Each water authority should decide on suitable peak hour and peak week demand factors; on seasonal demand variations, for example in holiday areas where visitors can cause local demands to increase by as much as 50%; on allowance for high seasonal garden watering; and on policy for use of cheap-rate off-peak electricity supplies for pumping to minimize pumping costs (Section 17.23).

18.3 GROUND OR ELEVATED STORAGE

If storage is required there is usually some flexibility in selection of its location. It may be possible to site it a little way from the ideal point of connection to the system. This could allow it to be located on a hill and thus allow it to maintain the pressure required in the distribution system. However, where there is no convenient high ground, some other solution is necessary. An alternative is to provide elevated storage at a water tower, but local objections may prevent granting of necessary consents. If a water tower is unacceptable in such circumstances, the only alternative is to provide ground level storage and boost the water into local supply. Except for environmental acceptability, the choice of ground or elevated storage is a matter of economics and operational factors. Elevated storage is more expensive to construct and maintain but might allow shorter connection pipelines. There is a practical limit to the size of elevated storage so that, in theory, the choice is between a large ground level service reservoir serving a large area and several elevated water towers each serving smaller areas. The choice is then influenced by the configuration of the distribution system (Section 13.2).

18.4 STATUTORY CONSENTS AND REQUIREMENTS

Reservoirs and associated structures require the consent of the national and local Planning Authority. Regulations vary from country to country but in the UK reference has to be made, under the Town and Country Planning Act 1990, for consent. This action normally attracts comment from various statutory bodies, local organizations, environmental groups and other stakeholders depending on the sensitivity of the location. Early consultation with these bodies is necessary to avoid delay or refusal of the planning application. Section 5.27 sets out the relevant legislation in UK and the types of body that may need to be consulted.

Where the proposed works may affect the apparatus of other statutory undertakers, they must be consulted, and Notices may need to be served (in the UK—under the New Roads and Street Works Act 1991) for any street works. Where public roads are affected, the Highway Authority should be consulted at an early stage. Discharge consent from the Environment Agency will be required for overflow and drain-down discharge into a sewer, soakaway or watercourse. A range of non-statutory consultation may also be required (e.g. The Countryside Agency for landscape, and English Heritage for archaeology).

In the UK, if a service reservoir will store more than 25 000 m^3 above natural ground level of any part of the land adjoining the reservoir, its design and construction have to be supervised by a Construction Engineer appointed by the water company and who is one of the qualified civil engineers on a panel of engineers appointed under the Reservoirs Act 1975 and associated Regulations of 1985 and 1986. The Construction Engineer is required to issue certificates during construction and after completion of the reservoir, specifying the water level(s) to which the reservoir may be filled. During the life of the reservoir, it has to be inspected at intervals not exceeding 10 years by an Inspecting Engineer appointed also from the panel (Section 5.23).

18.5 WATER QUALITY CONSIDERATIONS

There is increasing awareness of the influence of service reservoirs on water quality. High storage times can be detrimental to water quality by allowing decay of disinfectant residuals and growth of disinfectant by-products such as trihalomethanes (THMs) (Section 6.25). Of particular concern is the potential for stagnant regions to form. Reservoirs with common inlet and outlet pipework are particularly susceptible to water quality deterioration since demand tends to be met direct from the reservoir inlet; this leads to low turnover of reservoir contents and high stored water age. To maintain good water quality, it is important to ensure that there is good turnover of water throughout the reservoir. This can be achieved by:

a. installing baffles to promote plug flow; or
b. positioning inlets and outlets to create good mixing and prevent stratification; and
c. using the operational regime to generate diurnal fluctuations in water level.

Perfect mixing or perfect plug flow in a tank is not achievable and even reasonable mixing or plug flow can be surprisingly difficult to engineer. The tank characteristics required for plug flow are very different to the characteristics required for good mixing. It is therefore best to decide which type of flow the basic tank design is most likely to promote and then modify the design to optimize those flow characteristics, using suitable modelling. This is now best achieved with Computational Fluid Dynamics (CFD) (Chapter 12, Appendix). Useful guidance is given in the AwwaRF report *Water Quality Modeling of Distribution System Storage Facilities* (Grayman, 2000).

18.6 SAMPLING AND WATER TESTING

Monitoring of water quality in service reservoirs and water towers is required in UK under the Water Supply (Water Quality) Regulations 2000 but, irrespective of any statutory requirement, routine monitoring of water quality is good practice to ensure that water in a service reservoir has not become polluted. At sites with more than one service reservoir or compartment, each unit should be sampled separately unless they are sufficiently inter-connected that a sample at the combined outlet is representative of both storage units. The sampling arrangements should always ensure that the water sampled is from the body of the reservoir or that leaving the reservoir. If there is no suitable location for a sample tap at the reservoir site, then a tap should be provided on the outlet main at the nearest possible point to the reservoir. Tanks with combined inlet and outlet pipework should be sampled at different depths. Dip sampling is not recommended and should only be used as an emergency measure. Some water companies have particular sampling requirements and may make it necessary to provide several sample points from each compartment.

18.7 INSTRUMENTATION

The following instruments and equipment are normally required (see Section 18.12 for terminology).

1. A stilling well, either within or external to the reservoir, in which the water level measuring equipment is installed. If the reservoir is divided into two compartments, an externally-mounted stilling well with valved connections to each compartment or separate stilling wells

for each compartment should be provided. The well and valving are normally sited in the valve house.

2. Equipment for displaying the level measurement in either digital or analogue form in the valve house or in a weatherproof and vandal-proof enclosure installed adjacent to the stilling well.

3. Level electrodes or float switches at TWL (high) and BWL (low) for normal on/off pump control and alarm purposes. Additional switches at intermediate levels may be required for some pump control schemes.

4. Level electrodes or float switches at OWL (extra high) and just below BWL (extra low) for alarm purposes and emergency control of pumps feeding to or from the reservoir.

5. A display panel (wall- or floor-mounted as appropriate) for level and alarm indications, power supplies and the like. Provision should be made for interfacing with a telemetry outstation which, if required, may be mounted inside the panel.

6. A telemetry link to the pumping station or control centre to transmit water levels, flow rates through the inlet and outlet pipes, valve positions, power supply failure, telemetry failure and to carry signals for an intruder alarm and for remote control of valves, where appropriate.

If required by the reservoir's Construction Engineer or where foundation or structural movements are thought to be a risk, facilities for monitoring should be provided. These may include physical reference points at selected external locations on the structure and at nearby datum points well clear of the zone of influence of the reservoir. If necessary, strain gauges and inclinometers can be installed.

18.8 OVERFLOW AND DRAIN DOWN CAPACITY

Inflow controls are normally provided to prevent water level rising above a set normal maximum or top water level (TWL). The control may be via telemetry on source pumps or inlet valves or may use mechanical or hydraulic devices such as float or altitude valves. However, failure of such controls is possible so that some fail safe means of restricting water level has to be provided by means of an overflow. Reservoir roofs are not normally designed for uplift and flow out of vents or access openings is undesirable. Both could lead to structural damage or even reservoir failure.

The capacity of the overflow should be not less than the maximum likely inflow. For small reservoirs with small inlet pipework in a simple system it may be sufficient to set design overflow rate at the maximum inflow that the source could produce. However, provision of such overflow capacity and disposal in an urban area may be very onerous for a large reservoir. If the system supplying the reservoir is complex a risk analysis should be carried out with the aim of making reasonable provision for overflow without necessarily covering combined risks of very low probability, particularly where overwhelming of the overflow would not lead to loss of life or other serious consequences.

When it is necessary to drain a service reservoir down as much of the contents as possible should be released to supply via the usual outlet arrangements. However, separate facilities should be provided to allow the rest of the contents to be drained. These may have to be used for emptying the complete contents if contamination prevents use of the water in supply. Drawdown times of 8–12 hours for reservoir compartments under 2000 m³, and between 24 hours and 3 days for larger compartments, may be appropriate.

18.9 VENTILATION

Ventilation of reservoir compartments is needed to maintain a fresh supply of air above the water surface, for temperature control of that air and to admit or release air displaced by varying water levels in the compartment. The capacity of the ventilation system should be subject to risk analysis. It should be sufficient for the fastest rate of fall (or rise) of water level that is likely. Drain down must be allowed for but pipe burst may also need to be taken into account to avoid roof collapse.

The cross-sectional area of ventilation ducts or openings should be based on a suitable air speed (say 15 m/s). The air vents need to be insect proof and allowance should be made for the reduction in effective air passage area caused by insect screens. In many cases, traditional mushroom-type roof ventilators have been found unsatisfactory in long-term service and to be a potential source of pollution. 'Vented' access covers have sometimes been used for small reservoirs but are also vulnerable to pollution. Alternatives include piped systems above the reservoir roof, leading to one or more ventilation chambers or ventilation ducts.

At unmanned sites or at very sensitive installations and where serious malicious intrusion is a risk ventilation (and access) facilities should be designed against introduction of chemicals and explosives. The way this is done should be agreed with the reservoir owner but should not be made public (Section 18.13).

18.10 WATER RETAINING CONCRETE DESIGN

British practice since 1987 has been to follow the procedures set out in BS 8007 for the design of liquid retaining structures. However, by 2010 in UK, when BS 8110 and BS 8007 are due to be withdrawn, design of concrete structures will have to follow BS EN 1992-3 Eurocode 2. As with BS 8007 in relation to BS 8110, BS EN 1992-3 is based on limit state philosophy as used for the design of reinforced concrete structures to BS EN 1992-1-1. BS EN 1992-3 defines four classes of water tightness (0, 1, 2 and 3) for checking cracking serviceability. The first applies where some degree of leakage is acceptable or irrelevant. Tightness Class 1 is the case usually applying to water retaining structures where small leakage leading to damp patches or staining is acceptable. Tightness classes 2 and 3 require appearance not to be impaired by leakage or, in the latter case, there to be no leakage. For both classes 2 and 3 measures are needed to ensure that part of the concrete section remains in compression at all times or that a supplementary barrier such as a liner is applied. For Tightness Class 1 the width of any crack that is expected to pass right through the section is to be limited to:

$$w_{k1} = 0.875/(h_D/h) + 0.025$$

(with limiting values of 0.2 and 0.05 mm), where h_D is the hydrostatic pressure and h is the concrete section thickness.

Such cracks can be expected to heal in time subject to certain provisos. Where Tightness Class 2 applies (perhaps for exposed surfaces of tank walls) the minimum thickness of the concrete to be always in compression is the lesser of 50 mm and $0.2h$. The procedure involves the determination of crack widths and the reinforcement needed. Checks are made for other serviceability limit states as well as ultimate limit states. BS EN 1992-3 limits steel tensile stresses and bar spacing for different design crack widths for sections under axial tension.

Under BS 8007 water retaining structural concrete should have at least 325 kg/m³ of cement with a maximum water/cement ratio of 0.55, reduced to 0.50 when pulverized fuel-ash (PFA) or GGBS forms part or all of the cementitious material. It also requires cover to reinforcement to be not less than 40 mm. There are no equivalent requirements in BS EN 1992-3 but the durability requirements of BS EN 206-1 and BS 8500, and any general requirements of BS EN 1992-1-1, should be followed in respect of exposure to water as well as aggressive ground, liquids or environments. Additional protective measures (APMs) may be required for protection of concrete; examples are: (a) application of an external waterproof membrane—usually a bitumastic material on a carrier film, protected either by a fibre board at least 10 mm thick or by concrete blockwork or mass concrete; and (b) increased concrete cover. It should be stressed that good quality control is essential for achieving satisfactory water retaining concrete.

Wherever service reservoirs are built outside UK any relevant local codes of practice should be followed. It is prudent to take account of local constructional abilities and the quality of materials available and amend factors of safety if necessary. Design practices may need to be modified to take account of local prices so that an economic construction results.

18.11 WELDED STEEL PLATE DESIGN

Large steel plate tanks are usually cylindrical with their axes vertical. By 2010 steel tank design in UK must follow BS EN 1993-4-2, Eurocode 3, subject to issue of the UK National Annex. Under this Eurocode steel tanks for water come under Consequence Class 1 for which membrane theory may be used for determining principle stresses and simplified expressions may be used to determine local bending effects. Loadings are to be as defined in BS EN 1990, Eurocode 0. The shell is to be checked for plastic limit, cyclic plasticity, buckling and fatigue. Serviceability checks are to be made for deformations, deflections and vibrations. Minimum carbon steel plate thickness for tank bottoms excluding corrosion allowance is to be 5 mm for butt welded plates and 6 mm for lap welded plates. Reinforcement at openings is calculated by the area replacement method.

Otherwise design of tanks for storing water can be to AWWA D100 which, for certain details makes reference to API standard 650 (oil tanks). Both these codes include basic and refined design procedures. The basic procedures use conservative allowable stresses (the same in both codes) and are based on simplified design rules. With these, the steel plate selected is usually the cheapest that satisfies the rules for the intended service although a wide range of steel grades are permitted. The refined design procedures recognize the benefits of higher grade steels, an advantage for higher loaded members such as walls. The steel grade must be weldable and suitable for the stress and temperature ranges expected. Toughness needs to be taken into account for higher strength steels; it reduces with increased thickness but is improved for fine-grained steels and for those with higher manganese content.

One difference between the AWWA and API codes is that the latter allows the excess thickness at the top of one wall plate to be taken into account in determining the thickness of the plate above it. D100 requires the plate thickness to be based on the stress in the highest loaded extremity. Loads to be allowed for in the tank walls include those from the contained liquid and either wind or seismic effects. Shape factors are given in the standards and provide a convenient means of determining wind pressure. Such lateral loads may induce buckling in tall shell cylinders if the roof provides insufficient bracing. The factor of safety against buckling should be calculated and wind girder stiffeners included if necessary (Rajagopalan, 1990). AWWA D100 contains factors for determining increases in plate thickness for different tank heights and seismic accelerations. Sloshing of tank contents may need to be considered separately since it can occur in earthquakes, depending on the tank size and the frequency of seismic oscillations. Distortion may result but failure is rare. When a tall tank is empty wind loads can cause floor lift; anchors should be provided to prevent this.

Under AWWA D100 roof loads should include dead weight, snow or other live loads, wind and vacuum or internal pressure commensurate with the capacity of the ventilation system. Buckling safety factors should be determined for spherical or ellipsoidal roofs. Minimum plate thicknesses given in the codes are 6.35 mm (1/4") for floor and 4.76 mm (3/16") for the roof. Care should be taken with penetrations of the shell, floor or roof where additional stiffening or reinforcement may be needed. Guidance on these and on other details is given in both AWWA D100 and API 650. Care should also be taken at the joints between wall shell and both roof and floor. Five designs of the latter are covered:

1. Tank founded on 75 mm of oiled sand with the shell located on and bedded in grout on a reinforced concrete ring beam.
2. Tank founded on a reinforced concrete slab covered by 25 mm of oiled sand or 13 mm of joint filler.
3. Tank founded on 150 mm of oiled sand with the shell located inside a reinforced concrete ring beam leaving a gap of at least 19 mm.
4. Tank founded on a platform of graded stone or gravel with side slopes of 1 in 1.5 and surrounded by a level berm at least one metre width.
5. As for Type 4 but with a steel retaining ring extending into the gravel platform.

Rigorous corrosion protection is required including: meticulous surface preparation; use of galvanized plates; zinc rich coating after welding; and paint systems such as epoxy resin, glass flake resin and elastomeric coatings. Access has to be provided for the inspection and maintenance of coatings in all areas.

SERVICE RESERVOIRS

18.12 RESERVOIR SHAPE AND DEPTH

The most cost effective shape of a reservoir is circular in plan but the area of land required is greater. Except where a storage facility comprises several storage tanks, service reservoirs are generally built with at least two compartments so that one can be drained for maintenance. Reservoirs which are circular in plan are less suitable for subdivision. Nevertheless circular tanks permit the use of pre-stressed concrete or steel, which may offer cost advantages. For a two-compartment rectangular reservoir the most economic plan shape is usually obtained when its length (measured perpendicular to the division wall) is 1.5 times its breadth. These proportions may need alteration in the light of the shape and slope of the site, the cut and fill balance, pipework configuration for circulation, and any future extension likely or amenity requirements. If significant or abnormal soil settlements are expected, there may be advantages in providing two adjacent structurally independent single-compartment reservoirs instead of one two-compartment reservoir.

There is an economic depth for any service reservoir of a given storage capacity. The greater the depth the less length of wall and area of roof and floor is needed, though the unit cost of the wall increases with increased water depth. There can, however, be other constraints on the depth such as foundations, the character of the available site or the desirable range of distribution pressures. Depths most usually used for rectangular concrete reservoirs are:

Size (m³)	Depth of water (m)
Up to 3500	2.5–3.5
3500–15 000	3.5–5.0
Over 15 000	5.0–7.0

The following parameters are key to the design:

- top water level (TWL); usually the level at which the supply into the reservoir is to be shut off;
- the overflow weir level (OWL); giving a small margin above (a);
- the maximum water level (MWL) needed to discharge the maximum possible inflow over the overflow weir;
- bottom water level (BWL), being the lowest level to which the water should be allowed to fall for the purposes of supply;
- lowest roof soffit level—allowing for roof slope.

A freeboard between MWL and the roof soffit is required for ventilation and should not be less than 150 mm above MWL or less than 300 mm above TWL. Settlement of precipitated or suspended solids may occur in reservoir compartments. To prevent turbid water being drawn into supply, BWL should be not less than 150 mm above the highest level of the floor. It may need to be higher, depending on the outlet arrangements (Section 18.23).

18.13 COVERING AND PROTECTING RESERVOIRS

In temperate climates flat-roofed concrete reservoirs are usually covered over with earth and grass for appearance and temperature insulation. This involves maintenance and grass cutting, but an uncovered reservoir may result in amenity objections. The earth cover to the roof should comprise grassed top soil 150 mm thick, over a fabric filter membrane laid over 100 mm of single size 20 mm round gravel forming a drainage layer. The drainage layer is laid over a waterproof membrane (Section 18.21). A 150 mm thick gravel layer with no topsoil can be used in arid countries if there are no amenity objections. This provides thermal insulation. The earth banks against the external reservoir walls must be designed to stable slopes and, for ease of grass cutting, should not be steeper than 1 on 2.5. Topsoil cover to banks should be not less than 150 mm vertical thickness.

Special attention has now to be paid to ensuring service reservoirs are secure against vandalism, acts of terrorism and theft. Guidance is given in the UK water industry *Code of practice for the security of service reservoirs, 1997* (this is a restricted document, accessible through managers responsible for security in each of the water companies). An impact and vulnerability assessment should be undertaken to determine the level of risk and hence the security measures necessary. Secure perimeter fencing can minimize ordinary vandalism but is not sufficient protection by itself. Access manholes to the reservoir can be screwed and locked down; but, if possible, concealment is better, their location being known. Roof air vents are a problem because they are a potential source of pollution and access (Section 18.9). Valve and instrument houses and their doors should be of strong construction and should have no windows. Sampling points are necessary on both inlet and outlet mains and they too should be protected. All reservoirs should be visited frequently to ensure that none of the protective measures have been tampered with.

18.14 SERVICE RESERVOIR STRUCTURES

The earliest service reservoirs were built in masonry (usually brick) on a concrete or masonry base. Roofs were commonly vaulted and supported on masonry columns. Many such reservoirs are still in use in the UK—some, such as Honor Oak in South London, being very large. Smaller brick reservoirs were often mortar lined to assist water tightness. Masonry is a flexible material that can accommodate movement but the cracking that may result and the gradual erosion of mortar lead to ongoing maintenance problems so that old masonry reservoirs may need to be replaced (in reinforced concrete).

With the advent of reinforced concrete in the first half of the 20th century, this material became very widely used for service reservoirs. Until about 1980 reservoirs were jointed but after that monolithic construction became common because:

1. Joints require considerable care in construction and frequently are the source of poor concrete and leaks.
2. Monolithic construction usually requires less concrete and reinforcement.
3. There is a better understanding (with modern codes) of the shrinkage and stress cracking of concrete.
4. There is a better understanding of soil-structure interaction and a better ability to model it to establish whether monolithic construction can take the movements.
5. Piling has become cheaper and has enabled monolithic construction on poor ground to be feasible.
6. Site practice can now achieve satisfactory concrete with high wall pours and large distances between construction joints.

The materials adopted for reservoirs today depend on their availability and unit cost, local skills, client preferences and on the topography and geology of the site. The materials that should be considered are the following (starting with those suitable for large reservoirs and ending with those for small tanks): concrete (reinforced or prestressed), steel, GRP and polypropylene.

18.15 RECTANGULAR JOINTED CONCRETE RESERVOIRS

Jointed concrete reservoirs normally have joints: between lengths of wall; between the top of walls and the roof; between floor panels and their junction with wall bases and columns; and in roofs dependent on their area. The floor and roof are usually parallel so that walls and columns are of constant height.

Walls are usually reinforced concrete free-standing and cantilevered from a substantial base and stable against sliding and overturning (Figure 18.2) under soil or water loadings. An unreinforced mass concrete wall may be used if reinforcement is locally difficult to obtain for some reason. A sliding joint is normally provided between the top of the wall and the roof to prevent transfer of load due to roof thermal movements. Vertical joints with waterstops and sealing grooves are provided in the wall at spacings of about 12.5 m for contraction joints and not exceeding 30 m for expansion joints.

Columns of reinforced concrete are normally arranged on a rectangular (usually square) grid pattern. A column spacing of 5 m results in a flat-slab roof of economic thickness without the need for dropped panels. The side dimension or diameter should be not less than 300 or 350 mm, respectively and not less than one-twentieth of the height from reservoir floor to bottom of column head.

The *floor* is cast as a single or two layer slab in square panels having a side length equal to the column spacing. The single layer slab, typically 175 mm thick, is suitable for founding on a firm, non-compressible material. It is laid on a membrane of low frictional resistance; it may be unnecessary to provide reinforcement if the subsoil is firm and of uniform bearing capacity. The two layer slab has an upper layer, typically 175 mm thick, over a lower layer, typically 100–125 mm thick. A membrane between the layers permits sliding of the upper layer. This design is suitable for a clay subsoil. Usually only the top layer is reinforced, the reinforcement being discontinuous through the contraction joints. With these types of jointed floors, uplift pressures must be prevented by provision of an underdrainage system which has a free discharge to a lower level.

Joints separating floor slabs should be of the 'complete contraction' type, incorporating a joint sealant at the water face (Fig. 18.3). Externally placed waterstops are generally used on the underside of the base slab since these allow better compaction of the concrete at the waterstop than with the centrally placed type. In a two-layer floor, the joints in each layer should be staggered to avoid vertical alignment. Where possible, the upper floor slab should not be cast until the reservoir, including

FIGURE 18.2

Section of jointed reinforced concrete service reservoir with 2-layer floor slab.

EXPANSION JOINTS

Flexible expansion joint membrane
Polysulphide sealant compound
HDPE waterproof membrane
Compressible non-absorbent, closed cell, polythene joint filler
Centre bulb waterstop
PVC capping strip

Roof

Compressible non-absorbent, closed cell, polythene joint filler
Internal face
No steel continuity
External face
Polysulphide sealant compound on both sides
Centre bulb waterstop

Wall

Polysulphide sealant compound
Compressible non-absorbent, closed cell, polythene joint filler
No steel continuity
Polythene separating layer
75 mm blinding
External expansion type waterstop

Floor

Sealant compound
No steel continuity
Separating layer of polythene
75 mm blinding
External waterstop

Floor

ROOF-WALL JOINTS

2 layers sliding/bearing strip
Mortar bed, steel float finished
25 dia. filler
Compressible filler
32 dia. expansion strip
20 × 10 sealer
Waterproof membrane and polythene sheet secured with mastic in 20 × 20 rebate

Jointed construction

HD gauge polythene above waterproof membrane
Mortar bed, steel float finished
Second stage concrete
20 × 10 sealer
Waterproof membrane and polythene sheet secured with mastic in 20 × 20 rebate
2 layers sliding/bearing strip

Monolithic construction

CONTRACTION JOINTS

Sealant compound on both faces
No steel continuity
Waterstop

Wall & roof

FIGURE 18.3

Typical joints for reinforced concrete reservoirs (and other water retaining structures).

the roof, is substantially complete. This helps to avoid excessive shrinkage, temperature movements and joint damage and fouling before the joint sealant is applied.

The *roof* is a reinforced concrete slab of uniform thickness, minimum 200 mm, and is monolithic with the column heads. This is acceptable because the columns are flexible enough to permit roof expansion/contraction. If expansion joints are needed (depending on exposure and insulation) waterstops must be of the centre bulb type; such waterstops must be provided at any other joints such as construction joints. The roof design must allow for the impact loading of construction plant placing gravel and soil on the roof and for any other live loading that may occur.

18.16 MONOLITHIC CONCRETE RESERVOIRS

A monolithic concrete reservoir has reinforced concrete walls, floors, columns and roof, in which there are few (if any) permanent movement joints. In some cases the walls and floor are monolithic but there are sliding joints between roof and top of walls. This type of design has been found to be structurally economical in most situations where the underlying ground (after improvement if necessary) can support the load without risk of appreciable differential settlement. The reservoirs are normally rectangular in plan (Fig. 18.4) but circular and other shapes are feasible.

External walls are usually vertical or near vertical on the inner face but battered on the outer face to give the tapered section appropriate to the form of loading (Fig. 18.5). Depending on the height of the walls and the length of the roof slab, monolithic connections with floor and roof slabs can result in lower bending moments and shear forces (especially in the vertical plane) than is the case with jointed structures. The roof is normally constructed in two stages; the second stage at the wall interface being cast after the initial thermal shrinkage has taken place in the roof slab. Within the walls, joints are usually restricted to partial contraction joints (discontinuities in the concrete with 50% of the main horizontal reinforcement passing through) with a sealing groove on each face. The maximum spacing of partial contraction joints should be 7.5 m to avoid unacceptable cracking. For operational reasons, the division wall is usually full-height and can therefore assist in supporting the roof. The columns are arranged on a square grid, the span to external walls being typically reduced to three-quarters of the normal spacing.

FIGURE 18.4

Plan of monolithic reinforced concrete service reservoir.

FIGURE 18.5

Sections of monolithic reinforced concrete service reservoir.

An economical form of floor is a reinforced concrete slab of uniform thickness except at the perimeter, where it should be thickened to cater for moments transferred from the walls or resulting from differential vertical movements as between perimeter and centre (Fig. 18.5). Local thickening of the floor below columns should be avoided as it can be awkward and costly to construct; instead additional reinforcement under the columns can be used to increase the shear strength. Local thickening is usually required at drainage channels and sumps where these are included. Joints are normally restricted to construction joints. Plate 34(a) shows an internal view of a monolithic RC reservoir nearing completion.

18.17 CIRCULAR REINFORCED CONCRETE RESERVOIRS

The circular reservoir makes the most efficient use of materials since it needs a minimum wall length for a given plan area. Part of the water load on the walls is taken in tension by hoop reinforcement. As the reservoir size increases crack control becomes more difficult but this design has been used extensively for smaller reservoirs, both buried and unburied. However, the curved formwork required for the walls and the double curvature formwork usually needed for the roof are expensive and tend to outweigh the savings in concrete materials by adopting the circular shape.

Inlet and outlet pipework is usually arranged through the base with access and ventilation through the roof. If a division wall is required it can be in the form of a concentric inner wall with an internal radius about 70% of that of the outer wall. If a diagonal division wall is used thickening is needed at the joints with the circular wall to cope with the horizontal moments; this reduces the simplicity of the design. Circular reinforced concrete reservoirs are of monolithic construction with stiffening beams at the top, and bottom if necessary, of the wall. Roofs may be of the self-supporting thin shell type or flat with columns. A full hemispherical roof imposes no radial load on the wall but this shape is undesirable for aesthetic and planning reasons and is difficult to construct. A satisfactory span to height ratio is about 8:1 but requires thrust to be taken at the top of the wall by a ring beam.

18.18 PRE-STRESSED CONCRETE RESERVOIRS

For larger circular reservoirs hoop tension in the walls needs to be resisted by stressed tendons to eliminate cracks in the concrete. The compression in the concrete is usually arranged to match the tension caused by the maximum internal water load plus a margin to ensure that the concrete is always in compression. Two methods of wall circumferential pre-stressing have been used, both strictly speaking being post-tensioning. With the first, tendons inserted in ducts cast into the concrete are later stressed. The second is where tendons or cables are wrapped around the outside of the wall and later covered with sprayed concrete. A particular case of this involves winding the cables under tension. This allows the cable spacing to be varied up the wall to provide exactly the required load profile. Otherwise, the tendons are arranged in groups and provide a stepped load profile which sets up secondary vertical bending moments in the wall.

It should be remembered that, as with any post tensioning, tendons stressed earlier in the process lose tension as other cables are stressed later. This has to be allowed for in the design, along with concrete creep, temperature and moisture content effects, either by providing excess tension at the outset or by returning to apply additional load. Friction is also a serious issue and leads to load at the jack exceeding load in the middle of the tendon by a significant margin. Distortion of the cylindrical shell can arise unless the tendon arrangement and order of stressing are carefully planned.

Were it not for the need to ensure a good seal at the base of the wall, a frictionless sliding joint would be ideal since it induces no shear or bending in the wall base. However, a completely frictionless joint cannot be achieved so that some horizontal load transfer takes place, even with a sliding joint. To deal with this, additional pre-stress has to be applied at the base of the wall to compensate and thereby achieve the correct wall profile. A more usual joint is the pinned joint. This may take various forms, some involving 'pinning' after cable stressing. Sealing is achieved by use of a waterstop and sealants. Where the joint is monolithic the resulting shears and moments have to be allowed for. For pinned or fixed walls over about eight metres high the walls are usually pre-stressed in the vertical direction. Otherwise vertical reinforcement is sufficient. For large reservoirs, where radial loads on the base slab would otherwise be difficult to resist, the slab may be radially pre-stressed, with the cables anchored through the base of the wall in some cases. Precast panel walls have been used, an example is the 4.5 m high 3880 Ml tank in Truro, Cornwall. This used plane concrete 'staves' with exterior tensioned cables. Joints were cement-mortar packed and grouted after tensioning.

Roof construction may be either self-supporting spherical or ellipsoidal shells or flat slabs supported on columns as described in Section 18.16. Precast concrete profiled planks have been used as a permanent shutter for an in situ concrete roof. This reduces the amount of internal falsework needed. In theory, a shell roof need only be lightly reinforced to carry the load. However, thermal effects (usually not uniform) result in movement and cracking. The working of such cracks and their penetration by foreign matter over time can cause gradual increase in load on the wall ring beam and may result in settlement of the roof profile. Such settlement can further increase the load on the top of the wall and lead to failure.

While much used in the second half of the 20th century, pre-stressing of water tanks is now less common. One reason is that the thinner concrete sections and the use of pre-stressing require a high degree of skill and control in construction. Another reason is the ongoing inspection requirement, particularly with the 'wire wound and gunited' form of construction, as this is subject to concrete spalling and wire corrosion. A significant factor is the increased attention to health and safety in the built environment. High locked-in stresses in pre-stressed concrete represent a considerable hazard during demolition since the energy released can be damaging and unpredictable.

Although published in 1961, Creasy's book *Prestressed concrete cylindrical tanks* (Creasy, 1961) remains a very valuable reference to those coming across this form of construction.

18.19 **STEEL PLATE RESERVOIRS**

Circular steel above ground reservoirs have been used for water storage since before World War II. Steel reservoirs with capacity up to 100 000 m³ are now in use in many countries, particularly in North America and the Middle East (Plate 34(b)) where that form of construction is also used for petro-chemical storage. The design is of all welded steel plate (Fig. 18.6) and is very similar

FIGURE 18.6

Welded steel ground tank.

to that used for oil storage but greater attention is paid to coatings. The floor of a circular steel reservoir is made up of rectangular plates of thickness sufficient to take any radial tension and provide some contingency for corrosion, where required by design standards. The plates are either butt-welded together or lapped and fillet welded on one side only. Walls are made of butt welded rectangular plates but are thicker. Thickness is matched to circumferential tension according to the position in the wall. The wall base is fillet welded to a ring plate at the edge of the floor. This is thick and wide enough to prevent rotation and lifting of the floor under water load. Roofs are either conical or of low rise spherical shape, sometimes with a tighter curvature at the perimeter (the torispherical shape). Purlins and light truss supports on columns may be used or the roof may be self supporting.

The tank base is usually founded on a concrete slab for small tanks or on a bed of sand—oiled sand has been used in arid locations—or fine gravel. Where foundation loadings require it the granular bed should be retained by a reinforced concrete ring beam placed centrally under the wall shell. This prevents local shear failure of the foundation. In common with all ground tanks, foundation settlements should be evaluated. These may be more in the centre than at the perimeter. Differential settlements around the perimeter tend to cause the shell to cant or twist and the walls to get out of vertical. Jacking distorted tanks back into shape has been successful where movement was excessive but is best avoided.

Overflow and inlet and outlet pipework is usually arranged to penetrate the wall 'shell' via circular nozzles around which the shell plate is reinforced. Washout and drain down pipework usually exit via the ring plate into a sump below the wall. However, thermal movements of the tank perimeter may be large and may need to be accommodated at pipework by special expansion and movement joints. Access is via flanged manholes in the shell. Access to the roof is often provided by external ladders or stairs. In both hot and temperate climates the water temperature is increased by action of the sun on the steel surfaces and can lead to bacteriological issues.

The light construction cannot resist uplift forces when the tank is empty; therefore the design is not suitable for sites subject to flooding or high ground water. The tank walls should be left exposed, for inspection of coatings, with a level space 1.2 m wide all round. The finished ground should be battered back from the level strip if the tank base is below original ground level and suitable drainage should be provided.

The glass coated steel tank has become more popular in recent years. The plates are pre cut, drilled and then coated with a bonded glass coating. The plates are bolted together on site to a pattern with a sealer applied at the contact. The tanks appear to perform well: any leaks at joints are visible and can be resealed. However, the coating is delicate and is difficult to make good if damaged.

18.20 OTHER TYPES OF GROUND LEVEL TANK

Panel or sectional tanks were developed for military purposes to allow mass production, adaptation to many capacities and configurations and for easy transport and assembly. Originally, sections were of pressed steel with flanges drilled for bolting together through a gasket. Panel sizes were then 4 foot square and this size is still available (1.22 m) but 1.00 m square and rectangular panels are now produced. Coatings of steel panels are now usually epoxy or borosilicate glass. GRP panels are increasingly used but their flanges are not as robust as those of steel and need to be treated with care. Other plastics have been tried but are not favoured due to temperature distortion.

The use of internal stiffening braces theoretically allows any capacity to be achieved, albeit limited to a depth of about 6 m for steel (or 4 m for GRP). Sectional ground tanks are raised off the ground on dwarf walls at spacings to suit the panel size. These allow access for assembly and maintenance. Panels can be made with flanged nozzles for pipe entries and instrumentation.

18.21 DRAINAGE AND WATERPROOFING CONCRETE SERVICE RESERVOIRS

Where the reservoir external wall is designed as a free-standing cantilever or is of mass concrete, the backfill against the wall should comprise a vertical drainage layer of gravel about 300 mm, extending the full height of the wall and continuing down over the wall heel to link with a drainage system. Where the wall is monolithic with the reservoir floor it is still advantageous to retain the vertical gravel layer to control the ground water level and also to transmit water draining from the roof (Fig. 18.5). The wall and roof drainage systems must be kept separate from any underfloor drainage system. Wall drainage pipework should discharge into an observation chamber to help locate any leaks.

Ingress of water and pollution through the roof must be prevented by a positive waterproof membrane. For new reservoirs an adhesive membrane such as Bituthene DW is recommended, protected by heavy duty polyethylene sheet under the gravel drainage layer. The roof gradient should be no flatter than 1:250 for drainage; the floor slope should be made parallel to it so as to maintain constant wall and column heights. The simplest way to achieve this is to provide the slope in one direction only. Where there is no vertical wall drainage layer and where support from the embankment is essential for wall stability, a low peripheral kerb provided along the lowest edge of the roof can act as a collector for piping the water away. Inside the reservoir, the floor should have a shallow collecting channel leading to a drainage sump to aid cleaning of the floor of the reservoir.

The underfloor drainage system is usually laid to a rectangular pattern, normally comprising porous pipes surrounded in gravel in a trench below the floor. The layout should make it possible to observe drainage or leakage flow from separate areas of the floor. The porous pipes from each area are continued in ductile iron piping laid in concrete below the wall and the embankment and discharge individually to collector manholes, from which there is a free outfall pipe to some lower point. With jointed reservoir construction there must be no possibility of the drain outfall being submerged. If, on the other hand, conditions are such that uplift below the floor is unavoidable, then monolithic construction must be adopted, with the unit area weight of the floor, columns and roof being made greater than the design uplift pressure.

18.22 ACCESS TO SERVICE RESERVOIRS

Access to each reservoir compartment is needed for personnel, plant and materials. Access openings are usually sized to allow entry by a person wearing breathing apparatus. Access openings for plant and materials should be larger. Upstands should be provided around each opening to prevent surface water entering the reservoir. Covers to all openings must be robust but they do not normally need to be designed to support heavy loadings. They must be secure to prevent unauthorized access and must not allow rainwater to enter the reservoir. Lift-off covers risk introduction of mud and debris into the reservoir; therefore hinged covers are preferred but they must have an effective system for holding them in the open position when the access is in use.

For personnel entry into the reservoir the preferred arrangement is an inclined ladder leading to a platform about 2.5 m below the roof and a stairway leading from the platform to the floor. Where a stairway height exceeds 3 m, an intermediate landing is required. Reinforced concrete construction is recommended for platforms and stairways as this needs less long-term maintenance. The platforms can either be supported on columns or, in some cases, cantilevered from the walls. Alternatively the platforms and stairways can be fabricated in galvanized steel or anodized aluminium alloy. The same material should be used for the ladder. Typically two separate human accesses should be provided into each compartment, near opposite corners to assist ventilation of the compartment when work is in progress and to provide an escape route in an emergency.

Access for plant and materials has to be unobstructed to allow items to be lowered vertically to the compartment floor. The clear opening needed for small plant and materials for normal maintenance should be not less than 1.5 m × 1.0 m to allow a wheelbarrow to be lowered. Consideration should be given to the provision of removable handrailing around such openings, or of sockets into which it could be fitted. For reservoir compartments exceeding about 10 000 m³ a second and larger access for plant and materials should be considered if larger mechanical equipment might be needed for cleaning or major repairs. It is important to ensure that unauthorized vehicles cannot reach the roof or be used outside any specially strengthened areas of the roof.

18.23 SERVICE RESERVOIR PIPEWORK

Reservoir pipework normally comprises: inlet(s), outlet(s), overflow, drawdown, reservoir bypass and drainage pipes. The outlet may comprise a suction main to a site pumping station. An 80 mm diameter valved pipe through the division wall of a two-compartment reservoir should be provided so that water is available for hosing down a compartment when taken out of service for cleaning. Unless separate connecting pipes are used for flow in each direction, the control valve on the connection must be operable from outside the reservoir. Flexible joints should be incorporated between embedded or rigid pipes and external pipelines to accommodate differential settlement (Section 16.25).

Inlet and outlet pipes should bifurcate to serve each reservoir compartment equally. The inlet pipe can discharge at top water level (TWL) or near bottom water level (BWL). One of the disadvantages of the latter is that, in the event of a burst on the incoming main, the reservoir contents will be lost unless a suitable non-return valve is provided. On the other hand, if the incoming supply is pumped, a high-level entry will forfeit the energy savings potentially available when the reservoir is operating below TWL.

Inlet Pipework

The inlet piping arrangement needs to either achieve complete mixing of the inflow with the stored water or produce plug flow and thus avoid build-up of stagnant water areas. This involves suitable siting of the inlet and outlet pipes and, if necessary, the use of baffle walls. For the design of large reservoirs, and where there are water quality problems, it is becoming more common to use 3-D (CFD) modelling techniques (Chapter 12—Appendix) to optimize inlet and outlet arrangements and baffle wall (if any) placement. Options for encouraging circulation comprise: placing the inlet and outlet at opposite ends of the compartment (Fig. 18.4); distributing the incoming flow as evenly as possible along an end wall by the use of a long inlet weir; using a tapered diffuser pipe with several openings, or delivering to a semi-circular terminal box with slotted outlets.

Outlet Pipework

The most common (and simplest) outlet system uses only one draw-off point per compartment, but this is likely to leave some potentially stagnant areas in one (or both) corners at the outlet end of the compartment. To avoid this, the outlet may draw water from a number of points along an end wall if flow distribution is used as the sole means of avoiding stagnation.

  If the inlet and outlet pipelines are to terminate at opposite sides of the reservoir compartment, it may be appropriate to have separate inlet and outlet valve houses. However, for economy and ease of operation, a single valve house containing controls for both inlet and outlet is usually preferable. With this arrangement, one pipeline (usually the inlet) is normally laid within the reservoir compartment to feed water to the far end of it. This pipeline should be placed alongside the wall and encased in concrete to avoid 'dead' spaces and to inhibit external corrosion of the pipes. If the reservoir is of the jointed design, the internal pipework (and its surround) must have flexible joints corresponding with the joints in the structure.

  The outlet pipe can be laid horizontally, either through the reservoir compartment wall, or under the floor with a 90° vertical bend. It is usual to provide an entry bellmouth, to reduce hydraulic losses. The outlet bellmouth must be sufficiently submerged at BWL to prevent the entrainment of air into the flow, particularly where the flow will be pumped. For a bellmouth in the horizontal plane (i.e. vertical axis), a safe rule for minimum submergence of the bellmouth lip is:

$$S/D = 1.0 + 2.3F \tag{18.1}$$

where S = submergence below BWL; D = bellmouth diameter at lip; and F is the Froude number $V_D/(gD)^{0.5}$, where V_D = the average flow velocity through the bellmouth opening. With a bellmouth in the vertical plane (horizontal axis) the same equation may be used, but S is measured from the bellmouth axis.

  For a gravity supply outlet, the submergence requirement can be somewhat relaxed depending on the acceptability of air entry into the pipeline, but should not be less than D. The required submergence may create an uneconomic depth of 'dead' water unless the reservoir outlet is lowered by means of a sump in the floor. The sump should be generously sized to avoid undesirable hydraulic turbulence. The bottom of the sump collects floor deposits and should be not less than 300 mm below the bellmouth lip. Safety features for maintenance personnel need to be considered in the detailed design of a sump. The outlet sump can also serve as the drain sump.

  Where a service reservoir has a common inlet/outlet main, circulation inside the reservoir can be achieved by dividing the common main into inlet and outlet pipes before these pipes bifurcate to each compartment. If a low-level entry design is used, both inlet and outlet must be fitted with non-return valves. With the high-level type entry, a non-return valve is required on the outlet pipes only. Common inlet/outlet arrangements are not preferred due to the risk to water quality from stratification and stagnation (Section 18.5).

Overflow and Draindown Arrangements

Adequate overflow arrangements must be provided in case of an inflow control malfunction. Each compartment should be provided with an overflow capacity equal to the maximum likely inflow possible into that compartment with the other in or out of service (Section 18.8). The simple provision of a vertical pipe with bellmouth attached as an overflow has limited capacity and a horizontal weir is usually required. A convenient arrangement in a two-compartment reservoir is a weir box in the central division wall with weir entries from each compartment. The weir box often discharges to a pipe laid through the valve house, which can also receive the washout pipework connections.

The combined overflow/washout system should preferably discharge into an open watercourse or, failing that, to a sewer; both of which must be of adequate capacity. It may be necessary to consult with the land drainage authority or sewerage agency concerned for any permissions needed. A break pit must be provided before final discharge to allow levels to be monitored and dechlorination to be carried out if necessary. Drainage pipes should be connected to a drainage sump in each compartment and sized to allow emptying in an acceptable time (Section 18.8).

A reservoir bypass (between inlet and outlet pipes) is necessary in the case of a single-compartment reservoir, or where the whole reservoir may need to be taken out of service. It is normally possible to accommodate the valve on this within the valve house.

Valves

Stop valves (gate or butterfly) must be provided on inlets, outlets, scour pipes and the reservoir bypass but must not be provided on the overflow or on any wall or underfloor drainage systems. Gate valves become impracticable for normal reservoir use above about 600 mm diameter, when resilient-seated butterfly valves should be provided (Section 16.8). The valve size can be less than that of the pipeline, though the saving in cost of the valve is at least partly offset by the need for tapers and the increased space occupied by the pipework in a valve house. If a smaller size of valve is selected, a check should be made that the maximum velocity through the valve does not exceed that recommended by the valve manufacturer.

Autonomous over-velocity valves, designed to close automatically when the water velocity in the pipeline exceeds a predetermined rate, have fallen out of general favour because of their high cost and infrequent use. They may still be appropriate in special circumstances, for example where a large reservoir provides the major supply to a distribution area, or where the loss of water from a failed outlet main would be severe because of high head. The possible need for such valves should therefore be reviewed in reservoir planning and electrically operated butterfly valves should be considered as an alternative.

Wherever they are located, all butterfly valves and special control valves should be installed in chambers or houses so that they are accessible for maintenance. Important gate valves (such as the isolating valves on any pipes connecting into the reservoir) should also be placed in chambers but others can be buried.

Isolating valves on pipes leading into or out of the reservoir should be bolted to flanged pipes cast into the reservoir wall. Otherwise any differential movement between reservoir and valve could cause a joint to fail and release of the entire reservoir contents. The same principles apply to outlet or drain pipework built into the reservoir floor.

18.24 VALVE HOUSES FOR SERVICE RESERVOIRS

It is often convenient for all reservoir control valves to be concentrated in one chamber or valve house. For security of supply, the valve house should be as close as possible to the reservoir and is usually part of the reservoir structure. Access to the valve house may be by top entry through the roof or side entry through a wall. Top entry may result in the whole of the interior of the house being classified as a 'confined space', with consequent safety constraints on entry. These could give rise to unacceptable delays in gaining access if, for example, it becomes necessary to isolate the reservoir in an emergency. Side-entry valve houses are therefore preferred but may give rise to unacceptable visual impact in environmentally sensitive areas.

The pipework within the valve house must be arranged so that it is possible to install, maintain or remove any valve without great difficulty. If a valve is too heavy to be manhandled, it is important that there is clearance for a straight vertical lift out of the building (if top entry is provided) or to a position where it can be transferred to a trolley or road vehicle (if side entry is provided). Fixed monorail hoists, lifting eyes, davits or portable hoists may need to be provided.

In addition to pipework, valve houses are often used to accommodate sampling pumps and pipework, level recording and indicating equipment, telemetry and site monitoring equipment for flowmeters on inlet and outlet pipework, ventilation and dewatering equipment. Provision must also be made for dealing with any water resulting from spillages during maintenance work or leakage from pipework components. As a minimum, this should comprise a sump into which the suction hose of a portable pump can be inserted.

18.25 BAFFLES IN SERVICE RESERVOIRS

Baffle walls or curtains aim to achieve plug flow through the reservoir by directing flow from inlet to outlet by a circuitous route. The optimum arrangement of baffles depends on the shape of the compartment and the most convenient positions for the inlet and outlet. For construction convenience, baffles are normally installed between the columns supporting the reservoir roof.

Solid baffle walls are normally made of reinforced concrete, brickwork or blockwork. Lightweight, hollow blockwork should be avoided because the free chlorine normally present in a reservoir air-space has been known to cause deterioration of some blocks of this type. Plastic or rubber curtains are sometimes used. Where appropriate, openings must be provided along the bottom of the baffle walls to allow all areas of the reservoir floor to drain and to facilitate cleaning and maintenance. Openings should also be provided at the top of the walls to assist with ventilation. Joints in solid baffle walls must be provided at all points where they bridge movement joints in the floor or roof. Additional joints may be needed to allow for thermal movements or for the flexing of the reservoir walls as water levels rise and fall.

Baffle curtains may be less robust but easier to install. The material should be fibre reinforced to reduce elasticity and should be resistant to chlorinated water and approved for use with potable water. The edges of each curtain are formed by a seam in which holes are cut out at intervals for attachment points and through which a stainless steel rod is inserted. The recommended method for supporting the top edge of the curtain is to build a dove-tail galvanized or stainless steel channel into the roof soffit for the subsequent attachment of hangers although, for columns, lashing or strapping may be acceptable. The bottom edge must also be firmly anchored, preferably by fixing it to steel or

GRP angles bolted to the floor. The use of pre-cast concrete blocks which are not fixed to the floor is not recommended. Curtain alignment should be chosen so that they are not subject to high velocity or turbulent flow such as may occur near an inlet. At the inlet a dwarf concrete wall, bonded to the floor, should be provided below the curtain to protect it when the reservoir is filling. This wall should be about 500 mm high and the bottom of the curtain should be anchored to the top of the wall.

18.26 TESTING SERVICE RESERVOIRS

Service reservoirs should be tested for water tightness before being put into service. The test should be carried out before placing any backfill or banks against the outside walls unless the wall design relies on the embankment to resist hydraulic forces. The roof should be complete including any second-stage concrete.

Each reservoir compartment should be tested separately, with the other compartment empty. The compartment should be filled with treated water, to a test level about 75 mm below the overflow sill, at a uniform rate not exceeding 2 m vertical rise in water level per 24 hours. It should then be left to stand for at least 7 days to allow for absorption into the concrete. Longer periods (up to 21 days) may be required by some specifications. The water level should then be measured and recorded using a hook gauge with vernier control, or by other approved means of no less accuracy, and the water allowed to stand under test for 7 days. At least once each day during this period, the water level should be measured and recorded. During the 7-day test period, the effects of evaporation from the water surface can be reduced by closing all air vents and access openings (except for one vent left open for pressure balance).

Any flows in the underdrain and wall drain systems should be measured and recorded throughout the test, from a time at least 24 hours before beginning to fill, until 24 hours after emptying or on completion of a final water level measurement. Taking such measurements in chambers on the drain systems normally requires safety precautions appropriate to confined spaces. The outfalls of all pipes connected to the reservoir should be inspected during the test to ensure that all isolating valves are shut tight. Any significant leakage through them should be measured. In some circumstances it may also be necessary to keep records of evaporation losses from the water surface.

The test may be deemed successful if the drop of water level over the 7-day test period does not exceed the lesser of $1/500 \times$ average water depth or 10 mm, after deducting any measured leakage through valves and making allowance for any evaporation or condensation. If the test fails, any increase in underdrain or wall drain flow during the test period should be investigated to identify, if possible, the part of the reservoir that leaked. The test compartment should then be emptied and closely inspected for faults likely to cause the leakage. Investigating reservoir leakage can be troublesome and time consuming. The interior of the reservoir, especially any joints, should be closely inspected before filling with water.

18.27 SEARCHING FOR LEAKS

It is not easy to make a large reservoir fully watertight and the following notes may help track down the point of leakage.

1. Flows from underdrains should be examined. If they serve known areas of floor and are not joined to one common outlet point they may help narrow the search.

2. The inlet and outlet valves must be observed to ensure that they are not passing water into or out of the reservoir. The only secure way of knowing this is by withdrawing flange adaptors from the valve or by removing or temporarily leaving out a section of pipe next to the valve. Any outflow should be measured.

3. The rate of leakage at full depth, half depth, and with about 0.6 m of water in the reservoir should be measured. It is not likely that any revealing mathematical relationship between rate of leakage and depth of water in the reservoir will be established but leakage is usually less when the depth of water is less. No leakage at all below a certain level points to a leak at higher sections of wall.

4. After attempts to narrow the search for leaks the reservoir must be emptied and subjected to the most careful internal inspection with good lights, adequate ladders, plenty of time and a consistent pattern of examination. It is very easy to miss a faint crack in the wall or floor. Walls (particularly the joints next to the corners) should receive special attention for it is here that there is most likelihood of movement having occurred. After emptying, reservoir walls should be kept under observation when drying off since temporarily visible damp patches may be evidence of water held in underlying cracks or poor concrete. If first inspection does not reveal the causes of the leaks, it is worth repeating the inspection with even more care.

5. The floor joints should be inspected. Jointing material should be examined to see if it has sunk, has holes in it, or has come away from or failed to bond to the concrete of the sealing grooves. The majority of leakages arise from defects of this kind. Wall joints should similarly be examined.

6. If failure still results, about 0.6 m of water should be put into the reservoir and be left to stand until the water is quite still. Then crystals of potassium permanganate may be dropped into the reservoir, widely spaced, and left for a considerable time. Observed from a pre-arranged walkway (so as not to disturb the water) with a good light, streaks of colour may be noticed from the permanganate crystals showing some definite flow towards a point of leakage.

7. As an alternative to method (6), about 150 mm of water can be put into the reservoir, a hole or several holes bored through the floor, and compressed air can be introduced under the floor. In certain conditions of floor foundation air bubbling upwards through the water may indicate where faulty floor joints occur.

8. If, despite all these attempts, the cause of leakage is still unaccounted for then more drastic measures may have to be undertaken, such as digging pits in the bank to inspect the rear of the wall joints, placing further sealing strips (such as glass fibre embedded in bitumen) over joints, or even rendering wall face areas. Sealing of leakage through a reservoir floor has been achieved by gravity grouting. About 450 mm of thin grout mix is put into the reservoir and the cement is kept in suspension by continually sweeping the floor and disturbing the water with squeegees for two successive days. Thus grout passes into the unknown paths of leakage and the cement sets. It should not be necessary, however, to adopt these measures unless poor construction has taken place.

WATER TOWERS

18.28 USE OF WATER TOWERS

Water towers are used as a local source of water at times of peak demand where it would not be economical to increase the size of the supply pipeline and add a booster pump installation. In undulating terrain ground level storage could provide the pressure needed but in areas of flat topography the storage must be elevated. Many shapes and design features are possible but the designer should aim to produce a structure that meets the requirements of the water supply and planning authorities, bearing in mind that it will become a landmark in the community which it serves. Ancillary equipment including pipework, valves, ladders, instrumentation and booster pumps, if required, can all be hidden in the cylindrical shaft.

The optimum depth/diameter ratios should be determined taking into account the most efficient shape and the needs of the distribution system. It is usually advisable to avoid large pressure fluctuations in distribution that may be caused by drawdown or filling in excessively deep tanks.

18.29 CONCRETE WATER TOWERS

Concrete water towers are built with capacities up to about 5000 m³. They are usually circular in plan although rectangular concrete towers have been built (Plate 34(c)). The diameter of circular water towers is not usually sufficient to warrant the use of pre-stressing since cracks can be controlled by applying normal water retaining concrete criteria. Concrete water towers allow some scope for architectural statement so that the result can be regarded as a visual asset. Typical dimensions adopted for the reinforced concrete design shown in Figure 18.7 are:

Size (m³)	Depth of water (m)	Internal diameter (m)
1200	7.5	17.0
2000	9.1	19.4
3000	10.2	22.6

FIGURE 18.7

Reinforced concrete water tower.

FIGURE 18.8

Reinforced concrete water tower (Intze type).

Rectangular water towers are designed as small monolithic service reservoirs with the floor slab supported on some form of open column and beam framework or a hollow vertical shaft, itself founded on a base slab, piled if necessary. Wind and seismic loads should be taken into account in the design of tank, supports and piles. Circular concrete water towers allow more scope for different styling from a simple cylinder with a flat base to a sophisticated form such as the hyperbolic-paraboloid of the 39 m high Sillogue tower near Dublin airport built in 2006. In this case the vase shape resembles an inverted version of the nearby control tower. The Intze type water tower (Rajagopalan, 1990) is designed so that bending moments are as near zero as possible at all sections (Fig. 18.8). The radial thrusts from the outer conical section of base on the supporting ring balance those from the spherical centre section. Roofs may be flat for small tanks, conical or, for larger tanks, spherical as described in Section 18.17.

18.30 WELDED STEEL WATER TOWERS

Relatively small welded tanks have been used for over 100 years for industry and rail transport. These were usually small radius cylinders supported on a framework of steel columns with braces or ties. Welded steel water towers of capacities up to 15 000 m³ are now available and have been widely used all over the world—particularly in North America, the Middle East and the Far East. These are now constructed of butt welded steel plate in several configurations:

- spheroids or ellipsoids on tubular columns belled out at the base;
- cylinder or spherical shapes with conical bases and supported on wide steel columns which help resist seismic loads and provide space for plant rooms or offices or on a reinforced concrete frame (Plate 34(d)).

Whilst the forms available for welded steel water towers do not offer much scope for architectural treatment, the coatings provide an opportunity for decoration and can be attractive.

18.31 SEGMENTAL PLATE TANKS

The type of steel or GRP panel construction described in Section 18.20 can also be used for elevated storage. However, it is unlikely that segmental plate tanks would be used for anything other than industrial or emergency water storage since their poor visual appearance is exaggerated by height. Where they are used, the bases are placed on a series of beams which are supported on a framework of braced columns.

18.32 PIPEWORK AND ACCESS FOR WATER TOWERS

Pipework and access facilities below water towers are usually concealed within or obscured by the tank supports. A dry access shaft in the centre of the tank allows access to equipment above water level such as water level instruments and inlet float valves and may permit access out onto the roof. Old designs used to provide facilities for external access including circular walkways and revolving ladders. However, such facilities themselves need maintenance and cannot easily meet modern safety standards. This means that external coatings for steel tanks must be of the highest quality to minimize time to first maintenance.

REFERENCE STANDARDS

AWWA D100 *Steel tanks for water storage*. AWWA.

ANSI/API STD650 *Welded steel tanks for oil storage*. ANSI.

BS 8007 *Code of Practice for Design of Structures for Retaining Aqueous Liquids*. BSI.

BS 8110-1 *Structural Use of Concrete. Code of Practice for Design and Construction*. BSI.

BS 8500-1 *Concrete. Method of specifying and guidance for the specifier*. BSI.

BS EN 206-1 *Concrete. Specification, performance, production and conformity*. BSI.

BS EN 1990 Eurocode 0. *Basis of structural design*. BSI.

BS EN 1992-1-1 Eurocode 2. *Design of concrete structures. General rules and rules for buildings*. BSI.

BS EN 1992-3 Eurocode 2. *Design of concrete structures. Liquid retaining and containing structures*. BSI.

BS EN 1993-4-2 Eurocode 3. *Design of steel structures. Tanks*. BSI.

REFERENCES

Creasy, L. R. (1961). *Prestressed concrete cylindrical tanks*. John Wiley & Sons.

Grayman, W. M. et al (2000). *Water Quality Modeling of Distribution System Storage Facilities*. Denver, Colo. AwwaRF and AWWA.

Rajagopalan, Dr. K. (1990). *Storage structures*. A. A. Balkema.

Conversion Factors

	Metric to US	US to Metric
Length:	1 mm = 39.37 thou (in/1000) 1 m = 39.37 inches (in) 1 m = 3.2808 ft = 1.0936 yd 1 km = 0.61237 miles	1 in = 25.40 mm 1 foot (ft) = 304.80 mm 1 yard (yd) = 0.9144 m 1 mile = 1.60934 km
Area:	$1 \text{ m}^2 = 1.196 \text{ yd}^2 = 10.764 \text{ ft}^2$ 1 ha = 2.471 acres $1 \text{ km}^2 = 0.3861$ sq. miles *$1 \text{ km}^2 = 100$ ha* *$1 \text{ ha} = 10\,000 \text{ m}^2$*	$1 \text{ ft}^2 = 0.0929 \text{ m}^2$ $1 \text{ yd}^2 = 0.8361 \text{ m}^2$ 1 acre = 0.4047 ha $1 \text{ sq. mile} = 2.590 \text{ km}^2$ *$1 \text{ acre} = 4840 \text{ yd}^2$*
Volume:	$1 \text{ ml} = 0.06102 \text{ in}^3$ $1 \text{ litre (l)} = 61.02 \text{ in}^3$ 1 litre = 0.2642 gal (US) $1 \text{ m}^3 = 35.315 \text{ ft}^3$ $1 \text{ m}^3 = 1.3080 \text{ yd}^3$ 1 Ml = 0.2642 Mgal (US) *1 litre = .0220 gal (UK)*	$1 \text{ in}^3 = 16.387$ ml $1 \text{ ft}^3 = 28.32$ l 1 quart (US) = 0.9463 l $1 \text{ yd}^3 = 0.76456 \text{ m}^3$ 1 gal (US) = 3.785 l *1 gal (US) = 0.8326 gallon (UK)* *1 gal (UK) = 4.546 l*
Mass:	1 g = 0.03527 ounces (oz) 1 kg = 2.2046 pounds (lb) 1 tonne = 19.684 hundredweight (cwt) 1 t = 0.9842 tons	1 lb = 0.4536 kg 1 cwt = 50.802 kg 1 ton = 1.01605 tonne (t)
Mass unit area:	$1 \text{ kg/m}^2 = 0.2048 \text{ lb/ft}^2$ $1 \text{ g/mm}^2 = 22.76 \text{ oz/in}^2$	$1 \text{ lb/ft}^2 = 4.882 \text{ kg/m}^2$ $1 \text{ oz/ft}^2 = 305.2 \text{ g/m}^2$
Force:	1 Newton (N) = 0.2248 lb(f) 1 kN = 0.10036 ton(f) *9.81 N = 1 kg(f) at sea level*	1 lb(f) = 4.4482 N 1 ton(f) = 9.964 kN
Pressure:	$1 \text{ N/mm}^2 = 145.038 \text{ lb(f)/in}^2$ (psi) $1 \text{ N/mm}^2 = 9.324 \text{ ton(f)/ft}^2$ $1 \text{ kN/m}^2 = 20.885 \text{ lb(f)/ft}^2$ 1 m head of water = 1.422 lb(f)/in² *$1 \text{ N/m}^2 = 1$ pascal (pa)* *$1 \text{ N/mm}^2 = 1$ bar*	$1 \text{ lb(f)/in}^2 = 6895 \text{ N/m}^2$ $1 \text{ ton(f)/ft}^2 = 10.937 \text{ t(f)/m}^2$ $1 \text{ ton(f)/ft}^2 = 107.25 \text{ kN/m}^2$ $1 \text{ lb(f)/ft}^2 = 47.88 \text{ N/m}^2$
Density:	$1 \text{ kg/m}^3 = 0.06243 \text{ lb/ft}^3$ $1 \text{ t/m}^3 = 0.7525 \text{ ton/yd}^3$	$1 \text{ lb/ft}^3 = 16.018 \text{ kg/m}^3$ $1 \text{ ton/yd}^3 = 1.329 \text{ t/m}^3$

Notes:
1. US masses (oz, lb) are avdp (avoirdupois)
2. US ton and cwt are 'long' (2240 lb and 112 lb respectively). 1 'short' US ton = 2000 lb

Conversion Factors *(continued)*

	Metric to US	US to Metric
Unit weight:	$1 \text{ kN/m}^3 = 6.3658 \text{ lb(f)/ft}^3$	$1 \text{ lb(f)/ft}^3 = 0.15709 \text{ kN/m}^3$
Velocity:	$1 \text{ m/s} = 3.2808 \text{ ft/s} = 2.237 \text{ mph}$ $1 \text{ km/h} = 0.6214 \text{ mph}$	$1 \text{ ft/s} = 0.3048 \text{ m/s}$ $1 \text{ mph} = 0.447 \text{ m/s} = 1.609 \text{ km/h}$
Acceleration:	$1 \text{ m/s}^2 = 3.2808 \text{ ft/s}^2$	$1 \text{ ft/s}^2 = 0.3048 \text{ m/s}^2$
Flow:	$1 \text{ m}^3/\text{s (cumec)} = 35.315 \text{ ft}^3/\text{s (cfs)}$ $1 \text{ m}^3/\text{h} = 4.403 \text{ gal (US)/min (gpm)}$ $1 \text{ l/s} = 15.851 \text{ gpm (US)}$ $\text{Ml/d} = 0.2642 \text{ Mgal (US)/d (mgd)}$	$1 \text{ ft}^3/\text{s} = 0.0283 \text{ m}^3/\text{s}$ $1 \text{ gpm (US)} = 0.2271 \text{ m}^3/\text{h}$ $1 \text{ gpm (US)} = 0.0631 \text{ l/s}$ $1 \text{ mgd (US)} = 3.785 \text{ Ml/d}$ *$1 \text{ mgd (US)} = 0.83264 \text{ Mgd(UK)}$*
Power:	$1 \text{ kW} = 1.3410 \text{ hp}$ $1 \text{ W} = 0.73756 \text{ ft.lb(f)/s}$ *$1 \text{ J/s} = 1 \text{ watt (W)}$*	$1 \text{ hp} = 0.7457 \text{ kW}$ $1 \text{ ft.lb(f)/s} = 1.3558 \text{ W}$
Energy:	$1 \text{ Joule (J)} = 0.7376 \text{ ft.lb(f)}$ *$1 \text{ kJ} = 0.9478 \text{ Btu}$* *$1 \text{ Wh} = 3.412 \text{ Btu}$*	$1 \text{ ft.lb(f)} = 1.3558 \text{ J}$ *$1 \text{ Btu} = 1055 \text{ J}$*
Filtration rate:	$1 \text{ m}^3/\text{m}^2.\text{h} = 0.4091 \text{ gal (US)/ft}^2.\text{min}$	$1 \text{ gal (US)/ft}^2.\text{min} = 2.444 \text{ m}^3/\text{m}^2.\text{h}$

Membrane flux:	$1 \text{ l/m}^2.\text{h} = 0.5891 \text{ gal (US)/ft}^2.\text{d}$
	   (lmh) (gfd)
		$1 \text{ gal (US)/ft}^2.\text{d} = 1.697 \text{ l/m}^2.\text{h}$
		   (gfd) (lmh)

Hydrological:	$1 \text{ mm on } 1 \text{ km}^2 = 0.8107 \text{ acre-ft}$ $1 \text{ l/s per km}^2 = 0.09146 \text{ ft}^3/\text{s per 1000 acres}$	$1 \text{ acre-foot} = 1.2335 \text{ mm.km}^2$ $1 \text{ acre-foot} = 1233.5 \text{ m}^3$ $1 \text{ ft}^3/\text{s per 1000 acres} = 6.997 \text{ l/s per km}^2$
Permeability:	$10^{-6} \text{ m/s} = 103.46 \text{ ft/yr}$ $1 \text{ lugeon} = 1 \text{ l/m of test length/min at 10 bar injection pressure} = 0.01076 \text{ ft}^3/\text{ft/min at 142 psi}$ *$1 \text{ lugeon is approximately equivalent to permeability of } 1 \times 10^{-7} \text{ m/s}$*	$1 \text{ ft/yr} = 9.665 \times 10^{-9} \text{ m/s}$
Dynamic viscosity:	$1 \text{ N.s/m}^2 = 0.02089 \text{ lb(f).s/ft}^2$	$1 \text{ lb(f) s/ft}^2 = 47.88 \text{ N.s/m}^2$
Kinematic viscosity:	$1 \text{ m}^2/\text{s} = 10.764 \text{ ft}^2/\text{s}$ *$1 \text{ m}^2/\text{s} = 1 \times 10^6 \text{ centiStokes (cSt)}$*	$1 \text{ ft}^2/\text{s} = 0.0929 \text{ m}^2/\text{s}$
Temperature:	$t \text{ °C} = 5/9(t\text{°F} - 32)$	$t\text{°F} = 9/5 \, t\text{°C} + 32$

Index

701

Printed in the United States
By Bookmasters